Lecture Notes in Computer Science 13982

Founding Editors

Gerhard Goos
Juris Hartmanis

Editorial Board Members

The series Lecture Notes in Computer Science (LNCS), including its subseries Lecture Notes in Artificial Intelligence (LNAI) and Lecture Notes in Bioinformatics (LNBI), has established itself as a medium for the publication of new developments in computer science and information technology research, teaching, and education.

LNCS enjoys close cooperation with the computer science R & D community, the series counts many renowned academics among its volume editors and paper authors, and collaborates with prestigious societies. Its mission is to serve this international community by providing an invaluable service, mainly focused on the publication of conference and workshop proceedings and postproceedings. LNCS commenced publication in 1973.

Jaap Kamps · Lorraine Goeuriot · Fabio Crestani ·
Maria Maistro · Hideo Joho · Brian Davis ·
Cathal Gurrin · Udo Kruschwitz ·
Annalina Caputo
Editors

Advances in Information Retrieval

45th European Conference on Information Retrieval, ECIR 2023
Dublin, Ireland, April 2–6, 2023
Proceedings, Part III

 Springer

Editors
Jaap Kamps 🆔
University of Amsterdam
Amsterdam, Netherlands

Lorraine Goeuriot 🆔
Université Grenoble-Alpes
Saint-Martin-d'Hères, France

Fabio Crestani 🆔
Università della Svizzera Italiana
Lugano, Switzerland

Maria Maistro 🆔
University of Copenhagen
Copenhagen, Denmark

Hideo Joho 🆔
University of Tsukuba
Ibaraki, Japan

Brian Davis 🆔
Dublin City University
Dublin, Ireland

Cathal Gurrin 🆔
Dublin City University
Dublin, Ireland

Udo Kruschwitz 🆔
Universität Regensburg
Regensburg, Germany

Annalina Caputo 🆔
Dublin City University
Dublin, Ireland

ISSN 0302-9743   ISSN 1611-3349 (electronic)
Lecture Notes in Computer Science
ISBN 978-3-031-28240-9   ISBN 978-3-031-28241-6 (eBook)
https://doi.org/10.1007/978-3-031-28241-6

This Springer imprint is published by the registered company Springer Nature Switzerland AG
The registered company address is: Gewerbestrasse 11, 6330 Cham, Switzerland

Preface

The 45th European Conference on Information Retrieval (ECIR 2023) was held in Dublin, Ireland, during April 2–6, 2023, and brought together hundreds of researchers from Europe and abroad. The conference was organized by Dublin City University, in cooperation with the British Computer Society's Information Retrieval Specialist Group (BCS IRSG).

These proceedings contain the papers related to the presentations, workshops, and tutorials given during the conference. This year's ECIR program boasted a variety of novel work from contributors from all around the world. In total, 489 papers from authors in 52 countries were submitted to the different tracks. The final program included 65 full papers (29% acceptance rate), 41 short papers (27% acceptance rate), 19 demonstration papers (66% acceptance rate), 12 reproducibility papers (63% acceptance rate), 10 doctoral consortium papers (56% acceptance rate), and 13 invited CLEF papers. All submissions were peer-reviewed by at least three international Program Committee members to ensure that only submissions of the highest relevance and quality were included in the final program. The acceptance decisions were further informed by discussions among the reviewers for each submitted paper, led by a senior Program Committee member. In a final PC meeting all the final recommendations were discussed, trying to reach a fair and equal outcome for all submissions.

The accepted papers cover the state of the art in information retrieval: user aspects, system and foundational aspects, machine learning, applications, evaluation, new social and technical challenges, and other topics of direct or indirect relevance to search. As in previous years, the ECIR 2023 program contained a high proportion of papers with students as first authors, as well as papers from a variety of universities, research institutes, and commercial organizations.

In addition to the papers, the program also included 3 keynotes, 7 tutorials, 8 workshops, a doctoral consortium, the presentation of selected papers from the 2022 issues of the Information Retrieval Journal, and an industry day. Keynote talks were given by Mounia Lalmas (Spotify), Tetsuya Sakai (Waseda University), and this year's BCS IRSG Karen Spärck Jones Award winner, Yang Wang (UC Santa Barbara). The tutorials covered a range of topics including conversational agents in health; crowdsourcing; gender bias; legal IR and NLP; neuro-symbolic representations; query auto completion; and text classification. The workshops brought together participants to discuss algorithmic bias (BIAS); bibliometrics (BIR); e-discovery (ALTARS); geographic information extraction (GeoExT); legal IR (Legal IR); narrative extraction (Text2story); online misinformation (ROMCIR); and query performance prediction (QPP).

The success of ECIR 2023 would not have been possible without all the help from the team of volunteers and reviewers. We wish to thank all the reviewers and meta-reviewers who helped to ensure the high quality of the program. We also wish to thank: the short paper track chairs: Maria Maistro and Hideo Joho; the demo track chairs: Liting Zhou and Frank Hopfgartner; the reproducibility track chair: Leif Azzopardi; the workshop track chairs: Ricardo Campos and Gianmaria Silvello; the tutorial track chairs: Bhaskar Mitra and Debasis Ganguly; the industry track chairs: Nicolas Fiorini and Isabelle Moulinier; the doctoral consortium chair: Gareth Jones; and the awards chair: Suzan Verberne. We thank the students Praveen Acharya, Chinonso Osuji and Kanishk Verma for help with preparing the proceedings. We would like to thank all the student volunteers who helped to create an excellent experience for participants and attendees. ECIR 2023 was sponsored by a range of research institutes and companies. We thank them all for their support.

Finally, we wish to thank all the authors and contributors to the conference.

April 2023   Lorraine Goeuriot
                                                      Fabio Crestani
                                                       Jaap Kamps
                                                      Maria Maistro
                                                        Hideo Joho
                                                    Annalina Caputo
                                                     Udo Kruschwitz
                                                      Cathal Gurrin

Organization

General Chairs

Annalina Caputo   Dublin City University, Ireland
Udo Kruschwitz   Universität Regensburg, Germany
Cathal Gurrin   Dublin City University, Ireland

Program Committee Chairs

Jaap Kamps   University of Amsterdam, Netherlands
Lorraine Goeuriot   Université Grenoble Alpes, France
Fabio Crestani   Università della Svizzera Italiana, Switzerland

Short Papers Chairs

Maria Maistro   University of Copenhagen, Denmark
Hideo Joho   University of Tsukuba, Japan

Demo Chairs

Liting Zhou   Dublin City University, Ireland
Frank Hopfgartner   University of Koblenz-Landau, Germany

Reproducibility Track Chair

Leif Azzopardi   University of Strathclyde, UK

Workshop Chairs

Ricardo Campos   Instituto Politécnico de Tomar/INESC TEC, Portugal
Gianmaria Silvello   University of Padua, Italy

Tutorial Chairs

Bhaskar Mitra   Microsoft, Canada
Debasis Ganguly   University of Glasgow, UK

Industry Day Chairs

Nicolas Fiorini   Algolia, France
Isabelle Moulinier   Thomson Reuters, USA

Doctoral Consortium Chair

Gareth Jones   Dublin City University, Ireland

Awards Chair

Suzan Verberne   Leiden University, Netherlands

Publication Chairs

Brian Davis   Dublin City University, Ireland
Joachim Wagner   Dublin City University, Ireland

Local Chairs

Brian Davis   Dublin City University, Ireland
Ly Duyen Tran   Dublin City University, Ireland

Senior Program Committee

Omar Alonso   Amazon, USA
Giambattista Amati   Fondazione Ugo Bordoni, Italy
Ioannis Arapakis   Telefonica Research, Spain
Jaime Arguello   University of North Carolina at Chapel Hill, USA
Javed Aslam   Northeastern University, USA

Krisztian Balog	University of Stavanger & Google Research, Norway
Patrice Bellot	Aix-Marseille Université - CNRS (LSIS), France
Michael Bendersky	Google, USA
Mohand Boughanem	IRIT University Paul Sabatier Toulouse, France
Jamie Callan	Carnegie Mellon University, USA
Ben Carterette	Spotify, USA
Charles Clarke	University of Waterloo, Canada
Bruce Croft	University of Massachusetts Amherst, USA
Maarten de Rijke	University of Amsterdam, Netherlands
Arjen de Vries	Radboud University, Netherlands
Giorgio Maria Di Nunzio	University of Padua, Italy
Laura Dietz	University of New Hampshire, USA
Shiri Dori-Hacohen	University of Connecticut, USA
Carsten Eickhoff	Brown University, USA
Tamer Elsayed	Qatar University, Qatar
Liana Ermakova	HCTI, Université de Bretagne Occidentale, France
Hui Fang	University of Delaware, USA
Nicola Ferro	University of Padova, Italy
Ingo Frommholz	University of Wolverhampton, UK
Norbert Fuhr	University of Duisburg-Essen, Germany
Debasis Ganguly	University of Glasgow, UK
Nazli Goharian	Georgetown University, USA
Marcos Goncalves	Federal University of Minas Gerais, Brazil
Julio Gonzalo	UNED, Spain
Jiafeng Guo	Institute of Computing Technology, China
Matthias Hagen	Friedrich-Schiller-Universität Jena, Germany
Martin Halvey	University of Strathclyde, UK
Allan Hanbury	TU Wien, Austria
Donna Harman	NIST, USA
Faegheh Hasibi	Radboud University, Netherlands
Claudia Hauff	Spotify, Netherlands
Ben He	University of Chinese Academy of Sciences, China
Jiyin He	Signal AI, UK
Dietmar Jannach	University of Klagenfurt, Austria
Adam Jatowt	University of Innsbruck, Austria
Hideo Joho	University of Tsukuba, Japan
Gareth Jones	Dublin City University, Ireland
Joemon Jose	University of Glasgow, UK
Jaap Kamps	University of Amsterdam, Netherlands

Noriko Kando	National Institute of Informatics, Japan
Jussi Karlgren	Spotify, Sweden
Liadh Kelly	Maynooth University, Ireland
Dominik Kowald	Know-Center, Austria
Udo Kruschwitz	University of Regensburg, Germany
Oren Kurland	Technion, Israel Institute of Technology, Israel
Jochen L. Leidner	Coburg University of Applied Sciences/University of Sheffield/KnowledgeSpaces, Germany
Yiqun Liu	Tsinghua University, China
Craig Macdonald	University of Glasgow, UK
Joao Magalhaes	Universidade NOVA de Lisboa, Portugal
Philipp Mayr	GESIS, Germany
Donald Metzler	Google, USA
Alistair Moffat	University of Melbourne, Australia
Yashar Moshfeghi	University of Strathclyde, UK
Ahmed Mourad	University of Queensland, Australia
Henning Müller	HES-SO, Switzerland
Marc Najork	Google, USA
Jian-Yun Nie	Université de Montreal, Canada
Kjetil Nørvåg	Norwegian University of Science and Technology, Norway
Michał Olek	Wrocław University of Science and Technology, Poland
Harrie Oosterhuis	Radboud University, Netherlands
Iadh Ounis	University of Glasgow, UK
Javier Parapar	IRLab, University of A Coruña, Spain
Gabriella Pasi	Università di Milano-Bicocca, Italy
Raffaele Perego	ISTI-CNR, Italy
Benjamin Piwowarski	CNRS/Université Pierre et Marie Curie, France
Fiana Raiber	Yahoo Research, Israel
Paolo Rosso	Universitat Politècnica de València, Spain
Mark Sanderson	RMIT University, Australia
Philipp Schaer	TH Köln (University of Applied Sciences), Germany
Ralf Schenkel	Trier University, Germany
Christin Seifert	University of Duisburg-Essen, Germany
Chirag Shah	University of Washington, USA
Gianmaria Silvello	University of Padua, Italy
Fabrizio Silvestri	University of Rome, Italy
Mark Smucker	University of Waterloo, Canada
Laure Soulier	Sorbonne Université-ISIR, France
Hussein Suleman	University of Cape Town, South Africa

Paul Thomas	Microsoft, Australia
Nicola Tonellotto	University of Pisa, Italy
Theodora Tsikrika	Information Technologies Institute, CERTH, Greece
Julián Urbano	Delft University of Technology, Netherlands
Suzan Verberne	LIACS, Leiden University, Netherlands
Gerhard Weikum	Max Planck Institute for Informatics, Germany
Marcel Worring	University of Amsterdam, Netherlands
Andrew Yates	University of Amsterdam, Netherlands
Jakub Zavrel	Zeta Alpha, Netherlands
Min Zhang	Tsinghua University, China
Shuo Zhang	Bloomberg, Norway
Justin Zobel	University of Melbourne, Australia
Guido Zuccon	University of Queensland, Australia

Program Committee

Shilpi Agrawal	Linkedin, India
Qingyao Ai	Tsinghua University, China
Dyaa Albakour	Signal AI, UK
Mohammad Aliannejadi	University of Amsterdam, Netherlands
Satya Almasian	Heidelberg University, Germany
Omar Alonso	Amazon, USA
Ismail Sengor Altingovde	Middle East Technical University, Turkey
Giuseppe Amato	ISTI-CNR, Italy
Enrique Amigó	UNED, Spain
Sophia Ananiadou	University of Manchester, UK
Linda Andersson	Artificial Researcher IT GmbH, TU Wien, Austria
Vito Walter Anelli	Politecnico di Bari, Italy
Negar Arabzadeh	University of Waterloo, Canada
Arian Askari	Leiden Institute of Advanced Computer Science, Leiden University, Netherlands
Giuseppe Attardi	Università di Pisa, Italy
Maurizio Atzori	University of Cagliari, Italy
Sandeep Avula	Amazon, USA
Mossaab Bagdouri	Walmart Global Tech, USA
Ebrahim Bagheri	Ryerson University, Canada
Georgios Balikas	Salesforce Inc, France
Krisztian Balog	University of Stavanger & Google Research, Norway
Alvaro Barreiro	University of A Coruña, Spain

Alberto Barrón-Cedeño	Università di Bologna, Italy
Roberto Basili	University of Roma Tor Vergata, Italy
Alejandro Bellogin	Universidad Autonoma de Madrid, Spain
Patrice Bellot	Aix-Marseille Université - CNRS (LSIS), France
Alessandro Benedetti	Sease, UK
Idir Benouaret	CNRS, Université Grenoble Alpes, France
Klaus Berberich	Saarbruecken University of Applied Sciences (htw saar), Germany
Sumit Bhatia	Adobe Inc., India
Paheli Bhattacharya	Indian Institute of Technology Kharagpur, India
Valeriia Bolotova	RMIT University, Australia
Alexander Bondarenko	Friedrich-Schiller-Universität Jena, Germany
Ludovico Boratto	University of Cagliari, Italy
Gloria Bordogna	National Research Council of Italy - CNR, Italy
Emanuela Boros	University of La Rochelle, France
Florian Boudin	Université de Nantes, France
Leonid Boytsov	Amazon, USA
Martin Braschler	ZHAW Zurich University of Applied Sciences, Switzerland
Pavel Braslavski	Ural Federal University, Russia
Timo Breuer	TH Köln (University of Applied Sciences), Germany
Sebastian Bruch	Pinecone, USA
George Buchanan	University of Melbourne, Australia
Fidel Cacheda	Universidade da Coruña, Spain
Ricardo Campos	Ci2 - Polytechnic Institute of Tomar; INESC TEC, Portugal
Christian Cancedda	Politecnico di Torino, Italy
Iván Cantador	Universidad Autónoma de Madrid, Spain
Shubham Chatterjee	University of Glasgow, UK
Catherine Chavula	University of Texas at Austin, USA
Tao Chen	Google Research, USA
Xuanang Chen	University of Chinese Academy of Sciences, China
Hao-Fei Cheng	Amazon, USA
Adrian-Gabriel Chifu	Aix Marseille Univ, Université de Toulon, CNRS, LIS, France
Stephane Clinchant	Naver Labs Europe, France
Paul Clough	University of Sheffield, UK
Fabio Crestani	Università della Svizzera Italiana (USI), Switzerland
Shane Culpepper	RMIT University, Australia
Arthur Câmara	Delft University of Technology, Netherlands

Célia da Costa Pereira   Université Côte d'Azur, France
Duc Tien Dang Nguyen   University of Bergen, Norway
Maarten de Rijke   University of Amsterdam, Netherlands
Arjen de Vries   Radboud University, Netherlands
Yashar Deldjoo   Polytechnic University of Bari, Italy
Gianluca Demartini   University of Queensland, Australia
Amey Dharwadker   Meta, USA
Emanuele Di Buccio   University of Padua, Italy
Giorgio Maria Di Nunzio   University of Padua, Italy
Gaël Dias   Normandie University, France
Laura Dietz   University of New Hampshire, USA
Vlastislav Dohnal   Faculty of Informatics, Masaryk University, Czechia
Zhicheng Dou   Renmin University of China, China
Antoine Doucet   University of La Rochelle, France
Pan Du   Thomson Reuters Labs, Canada
Tomislav Duricic   Graz University of Technology, Austria
Liana Ermakova   HCTI, Université de Bretagne Occidentale, France
Ralph Ewerth   L3S Research Center, Leibniz Universität Hannover, Germany
Guglielmo Faggioli   University of Padova, Italy
Anjie Fang   Amazon.com, USA
Hossein Fani   University of Windsor, Canada
Yue Feng   UCL, UK
Marcos Fernández Pichel   Universidade de Santiago de Compostela, Spain
Juan M. Fernández-Luna   University of Granada, Spain
Nicola Ferro   University of Padova, Italy
Komal Florio   Università di Torino, Italy
Thibault Formal   Naver Labs Europe, France
Ophir Frieder   Georgetown University, USA
Ingo Frommholz   University of Wolverhampton, UK
Maik Fröbe   Friedrich-Schiller-Universität Jena, Germany
Norbert Fuhr   University of Duisburg-Essen, Germany
Michael Färber   Karlsruhe Institute of Technology, Germany
Petra Galuščáková   Université Grenoble Alpes, France
Debasis Ganguly   University of Glasgow, UK
Dario Garigliotti   No affiliation, Norway
Eric Gaussier   LIG-UJF, France
Kripabandhu Ghosh   Indian Institute of Science Education and Research (IISER) Kolkata, India
Anastasia Giachanou   Utrecht University, Netherlands

Lorraine Goeuriot	Univ. Grenoble Alpes, CNRS, Grenoble INP, LIG, France
Nazli Goharian	Georgetown University, USA
Marcos Goncalves	Federal University of Minas Gerais, Brazil
Julio Gonzalo	UNED, Spain
Michael Granitzer	University of Passau, Germany
Adrien Guille	ERIC Lyon 2, EA 3083, Université de Lyon, France
Nuno Guimaraes	CRACS - INESC TEC, Portugal
Chun Guo	Pandora Media LLC., USA
Dhruv Gupta	Norwegian University of Science and Technology, Norway
Christian Gütl	Graz University of Technology, Austria
Matthias Hagen	Friedrich-Schiller-Universität Jena, Germany
Lei Han	University of Queensland, Australia
Preben Hansen	Stockholm University, Sweden
Donna Harman	NIST, USA
Morgan Harvey	University of Sheffield, UK
Maram Hasanain	Qatar University, Qatar
Claudia Hauff	Spotify, Netherlands
Mariya Hendriksen	University of Amsterdam, Netherlands
Daniel Hienert	GESIS - Leibniz Institute for the Social Sciences, Germany
Orland Hoeber	University of Regina, Canada
Frank Hopfgartner	Universität Koblenz, Germany
Gilles Hubert	IRIT, France
Juan F. Huete	University of Granada, Spain
Bogdan Ionescu	University Politehnica of Bucharest, Romania
Radu Tudor Ionescu	University of Bucharest, Faculty of Mathematics and Computer Science, Romania
Adam Jatowt	University of Innsbruck, Austria
Faizan Javed	Kaiser Permanente, USA
Renders Jean-Michel	Naver Labs Europe, France
Tianbo Ji	Dublin City University, Ireland
Noriko Kando	National Institute of Informatics, Japan
Nattiya Kanhabua	SCG CBM, Thailand
Sarvnaz Karimi	CSIRO, Australia
Sumanta Kashyapi	NIT Hamirpur, USA
Makoto P. Kato	University of Tsukuba, Japan
Abhishek Kaushik	Dundalk Institute of Technology, Ireland
Mesut Kaya	Aalborg University Copenhagen, Denmark
Roman Kern	Graz University of Technology, Austria

Manoj Kesavulu	Dublin City University, Ireland
Atsushi Keyaki	Hitotsubashi University, Japan
Johannes Kiesel	Bauhaus-Universität Weimar, Germany
Benjamin Kille	Norwegian University of Science and Technology, Norway
Tracy Holloway King	Adobe, USA
Udo Kruschwitz	University of Regensburg, Germany
Lakhotia Kushal	Outreach, USA
Mucahid Kutlu	TOBB University of Economics and Technology, Turkey
Saar Kuzi	Amazon, USA
Wai Lam	The Chinese University of Hong Kong, China
Birger Larsen	Aalborg University, Denmark
Dawn Lawrie	Johns Hopkins University, USA
Jochen L. Leidner	Coburg University of Applied Sciences/University of Sheffield/KnowledgeSpaces, Germany
Mark Levene	Birkbeck, University of London, UK
Qiuchi Li	University of Padua, Italy
Wei Li	University of Roehampton, UK
Xiangsheng Li	Tsinghua University, China
Shangsong Liang	Sun Yat-sen University, China
Siyu Liao	Amazon, USA
Trond Linjordet	University of Stavanger, Norway
Matteo Lissandrini	Aalborg University, Denmark
Suzanne Little	Dublin City University, Ireland
Haiming Liu	University of Southampton, UK
Benedikt Loepp	University of Duisburg-Essen, Germany
Andreas Lommatzsch	TU Berlin, Germany
Chenyang Lyu	Dublin City University, Ireland
Weizhi Ma	Tsinghua University, China
Sean MacAvaney	University of Glasgow, UK
Andrew Macfarlane	City, University of London, UK
Joel Mackenzie	University of Queensland, Australia
Khushhall Chandra Mahajan	Meta Inc., USA
Maria Maistro	University of Copenhagen, Denmark
Antonio Mallia	New York University, Italy
Thomas Mandl	University of Hildesheim, Germany
Behrooz Mansouri	University of Southern Maine, USA
Jiaxin Mao	Renmin University of China, China
Stefano Marchesin	University of Padova, Italy
Mirko Marras	University of Cagliari, Italy
Monica Marrero	Europeana Foundation, Netherlands

Miguel Martinez	Signal AI, UK
Bruno Martins	IST and INESC-ID - Instituto Superior Técnico, University of Lisbon, Portugal
Flavio Martins	INESC-ID, Instituto Superior Técnico, Universidade de Lisboa, Portugal
Maarten Marx	University of Amsterdam, Netherlands
Yosi Mass	IBM Haifa Research Lab, Israel
David Maxwell	Delft University of Technology, UK
Richard McCreadie	University of Glasgow, UK
Graham McDonald	University of Glasgow, UK
Dana McKay	RMIT University, Australia
Paul McNamee	Johns Hopkins University, USA
Parth Mehta	IRSI, India
Florian Meier	Aalborg University, Copenhagen, Denmark
Ida Mele	IASI-CNR, Italy
Zaiqiao Meng	University of Glasgow, UK
Donald Metzler	Google, USA
Tomasz Miksa	TU Wien, Austria
Ashlee Milton	University of Minnesota, USA
Alistair Moffat	University of Melbourne, Australia
Ali Montazeralghaem	University of Massachusetts Amherst, USA
Jose Moreno	IRIT/UPS, France
Alejandro Moreo Fernández	Istituto di Scienza e Tecnologie dell'Informazione "A. Faedo", Italy
Philippe Mulhem	LIG-CNRS, France
Cristina Ioana Muntean	ISTI CNR, Italy
Henning Müller	HES-SO, Switzerland
Suraj Nair	University of Maryland, USA
Franco Maria Nardini	ISTI-CNR, Italy
Fedelucio Narducci	Politecnico di Bari, Italy
Wolfgang Nejdl	L3S and University of Hannover, Germany
Manh Duy Nguyen	DCU, Ireland
Thong Nguyen	University of Amsterdam, Netherlands
Jian-Yun Nie	Université de Montreal, Canada
Sérgio Nunes	University of Porto, Portugal
Diana Nurbakova	National Institute of Applied Sciences of Lyon (INSA Lyon), France
Kjetil Nørvåg	Norwegian University of Science and Technology, Norway
Michael Oakes	University of Wolverhampton, UK
Hiroaki Ohshima	Graduate School of Applied Informatics, University of Hyogo, Japan

Salvatore Orlando	Università Ca' Foscari Venezia, Italy
Iadh Ounis	University of Glasgow, UK
Pooja Oza	University of New Hampshire, USA
Özlem Özgöbek	Norwegian University of Science and Technology, Norway
Deepak P.	Queen's University Belfast, UK
Panagiotis Papadakos	Information Systems Laboratory - FORTH-ICS, Greece
Javier Parapar	IRLab, University of A Coruña, Spain
Pavel Pecina	Charles University, Czechia
Gustavo Penha	Delft University of Technology, Brazil
Maria Soledad Pera	TU Delft, Netherlands
Vivien Petras	Humboldt-Universität zu Berlin, Germany
Giulio Ermanno Pibiri	Ca' Foscari University of Venice, Italy
Francesco Piccialli	University of Naples Federico II, Italy
Karen Pinel-Sauvagnat	IRIT, France
Florina Piroi	TU Wien, Institue of Information Systems Engineering, Austria
Marco Polignano	Università degli Studi di Bari Aldo Moro, Italy
Martin Potthast	Leipzig University, Germany
Ronak Pradeep	University of Waterloo, Canada
Xin Qian	University of Maryland, USA
Fiana Raiber	Yahoo Research, Israel
David Rau	University of Amsterdam, Netherlands
Andreas Rauber	Vienna University of Technology, Austria
Gábor Recski	TU Wien, Austria
Weilong Ren	Shenzhen Institute of Compuiing Sciences, China
Zhaochun Ren	Shandong University, China
Chiara Renso	ISTI-CNR, Pisa, Italy, Italy
Thomas Roelleke	Queen Mary University of London, UK
Kevin Roitero	University of Udine, Italy
Haggai Roitman	eBay Research, Israel
Paolo Rosso	Universitat Politècnica de València, Spain
Stevan Rudinac	University of Amsterdam, Netherlands
Anna Ruggero	Sease Ltd., Italy
Tony Russell-Rose	Goldsmiths, University of London, UK
Ian Ruthven	University of Strathclyde, UK
Sriparna Saha	IIT Patna, India
Tetsuya Sakai	Waseda University, Japan
Eric Sanjuan	Laboratoire Informatique d'Avignon—Université d'Avignon, France
Maya Sappelli	HAN University of Applied Sciences, Netherlands

Jacques Savoy	University of Neuchatel, Switzerland
Harrisen Scells	Leipzig University, Germany
Philipp Schaer	TH Köln (University of Applied Sciences), Germany
Ferdinand Schlatt	Martin-Luther Universität Halle-Wittenberg, Germany
Jörg Schlötterer	University of Duisburg-Essen, Germany
Falk Scholer	RMIT University, Australia
Fabrizio Sebastiani	Italian National Council of Research, Italy
Christin Seifert	University of Duisburg-Essen, Germany
Ivan Sekulic	Università della Svizzera italiana, Switzerland
Giovanni Semeraro	University of Bari, Italy
Procheta Sen	University of Liverpool, UK
Mahsa S. Shahshahani	Accenture, Netherlands
Eilon Sheetrit	Technion - Israel Institute of Technology, Israel
Fabrizio Silvestri	University of Rome, Italy
Jaspreet Singh	Amazon, Germany
Manel Slokom	Delft University of Technology, Netherlands
Mark Smucker	University of Waterloo, Canada
Michael Soprano	University of Udine, Italy
Laure Soulier	Sorbonne Université-ISIR, France
Marc Spaniol	Université de Caen Normandie, France
Damiano Spina	RMIT University, Australia
Andreas Spitz	University of Konstanz, Germany
Torsten Suel	New York University, USA
Kazunari Sugiyama	Kyoto University, Japan
Dhanasekar Sundararaman	Duke University, USA
Irina Tal	Dublin City University, Ireland
Lynda Tamine	IRIT, France
Carla Teixeira Lopes	University of Porto, Portugal
Joseph Telemala	University of Cape Town, South Africa
Paul Thomas	Microsoft, Australia
Thibaut Thonet	Naver Labs Europe, France
Nicola Tonellotto	University of Pisa, Italy
Salvatore Trani	ISTI-CNR, Italy
Jan Trienes	University of Duisburg-Essen, Germany
Johanne R. Trippas	RMIT University, Australia
Andrew Trotman	University of Otago, New Zealand
Theodora Tsikrika	Information Technologies Institute, CERTH, Greece
Kosetsu Tsukuda	National Institute of Advanced Industrial Science and Technology (AIST), Japan

Yannis Tzitzikas	University of Crete and FORTH-ICS, Greece
Md Zia Ullah	Edinburgh Napier University, UK
Kazutoshi Umemoto	University of Tokyo, Japan
Julián Urbano	Delft University of Technology, Netherlands
Ruben van Heusden	University of Amsterdam, Netherlands
Aparna Varde	Montclair State University, USA
Suzan Verberne	LIACS, Leiden University, Netherlands
Manisha Verma	Amazon, USA
Vishwa Vinay	Adobe Research, India
Marco Viviani	Università degli Studi di Milano-Bicocca - DISCo, Italy
Ellen Voorhees	NIST, USA
Xi Wang	University College London, UK
Zhihong Wang	Tsinghua University, China
Wouter Weerkamp	TomTom, Netherlands
Gerhard Weikum	Max Planck Institute for Informatics, Germany
Xiaohui Xie	Tsinghua University, China
Takehiro Yamamoto	University of Hyogo, Japan
Eugene Yang	Human Language Technology Center of Excellence, Johns Hopkins University, USA
Andrew Yates	University of Amsterdam, Netherlands
Elad Yom-Tov	Microsoft, Israel
Ran Yu	University of Bonn, Germany
Hamed Zamani	University of Massachusetts Amherst, USA
Eva Zangerle	University of Innsbruck, Austria
Richard Zanibbi	Rochester Institute of Technology, USA
Fattane Zarrinkalam	University of Guelph, Canada
Sergej Zerr	Rhenish Friedrich Wilhelm University of Bonn, Germany
Fan Zhang	Wuhan University, China
Haixian Zhang	Sichuan University, China
Min Zhang	Tsinghua University, China
Rongting Zhang	Amazon, USA
Ruqing Zhang	Institute of Computing Technology, Chinese Academy of Sciences, China
Mengyisong Zhao	University of Sheffield, UK
Wayne Xin Zhao	Renmin University of China, China
Jiang Zhou	Dublin City University, Ireland
Liting Zhou	Dublin City University, Ireland
Steven Zimmerman	University of Essex, UK
Justin Zobel	University of Melbourne, Australia
Lixin Zou	Tsinghua University, China

Guido Zuccon   University of Queensland, Australia

Additional Reviewers

Ashkan Alinejad
Evelin Amorim
Negar Arabzadeh
Dennis Aumiller
Mohammad Bahrani
Mehdi Ben Amor
Giovanni Maria Biancofiore
Ramraj Chandradevan
Qianli Chen
Dhivya Chinnappa
Isabel Coutinho
Washington Cunha
Xiang Dai
Marco de Gemmis
Alaa El-Ebshihy
Gloria Feher
Yasin Ghafourian
Wolfgang Gritz
Abul Hasan
Phuong Hoang
Eszter Iklodi
Andrea Iovine
Tania Jimenez
Pierre Jourlin

Anoop K.
Tuomas Ketola
Adam Kovacs
Zhao Liu
Daniele Malitesta
Cataldo Musto
Evelyn Navarrete
Zhan Qu
Saed Rezayi
Ratan Sebastian
Dawn Sepehr
Simra Shahid
Chen Shao
Mohammad Sharif
Stanley Simoes
Matthias Springstein
Ting Su
Wenyi Tay
Alberto Veneri
Chenyang Wang
Lorenz Wendlinger
Mengnong Xu
Shuzhou Yuan

Contents – Part III

Demonstration Papers

Workshops

Doctoral Consoritum

CLEF Lab Descriptions

Reproducibility Papers

Knowledge is Power, Understanding is Impact: Utility and Beyond Goals, Explanation Quality, and Fairness in Path Reasoning Recommendation

Giacomo Balloccu[1], Ludovico Boratto[1], Christian Cancedda[2],
Gianni Fenu[1], and Mirko Marras[1](✉)

[1] University of Cagliari, Cagliari, Italy
{giacomo.balloccu,ludovico.boratto,mirko.marras}@acm.org, fenu@unica.it
[2] Polytechnic University of Turin, Turin, Italy
christian.cancedda@studenti.polito.it

Abstract. Path reasoning is a notable recommendation approach that models high-order user-product relations, based on a Knowledge Graph (KG). This approach can extract reasoning paths between recommended products and already experienced products and, then, turn such paths into textual explanations for the user. Unfortunately, evaluation protocols in this field appear heterogeneous and limited, making it hard to contextualize the impact of the existing methods. In this paper, we replicated three state-of-the-art relevant path reasoning recommendation methods proposed in top-tier conferences. Under a common evaluation protocol, based on two public data sets and in comparison with other knowledge-aware methods, we then studied the extent to which they meet recommendation utility and beyond objectives, explanation quality, and consumer and provider fairness. Our study provides a picture of the progress in this field, highlighting open issues and future directions. Source code: https://github.com/giacoballoccu/rep-path-reasoning-recsys.

Keywords: Recommender systems · Knowledge graphs · Replicability

1 Introduction

Recommender systems (RS) are a popular strategy to enable personalized users' experience [29]. Historical data (e.g., browsing activity and ratings) and product characteristics (e.g., title and description) are well-recognized data sources to train RSs. Product information is often augmented with *Knowledge Graphs* (KGs) [8,27]. These KGs include *entities* (e.g., users, movies, actors) and *relations* between entities (e.g., an actor starred a movie). Integrating KGs within RSs has led to a gain in recommendation utility [34,36], especially under sparse data and cold-start scenarios [18]. Their inclusion is essential to make RS explainable and turn recommendation into a more transparent social process [33,43].

© The Author(s), under exclusive license to Springer Nature Switzerland AG 2023
J. Kamps et al. (Eds.): ECIR 2023, LNCS 13982, pp. 3–19, 2023.
https://doi.org/10.1007/978-3-031-28241-6_1

Notable recommendation methods based on KGs include *path reasoning methods* [1,23,25,26,31,37,37,40,44]. To guide RS training, they rely on paths that model high-order relations between users and products in the KG, and identify those deemed as relevant between already experienced products and products to recommend. Such paths are also used to create explanations, through explanation templates or text generation. In the movie domain, the path "user$_1$ watched movie$_1$ directed director$_1$ directed^{-1} movie$_2$" might lead to the template-based explanation "movie$_2$ is recommended to you because you watched movie$_1$ also directed by director$_1$". Path reasoning methods are in contrast to regularization methods, which weight product characteristics based on their importance for a given recommendation but do not provide any explanation [16,19,34–36,42].

An abundance of KGs were proposed for recommendation, along with path reasoning methods, to produce both recommendations and explanations [5]. However, evaluation protocols were heterogeneous (e.g., different train-test splits) and limited to a narrowed set of evaluation data sets and metrics. Prior works often showed that a novel method led to a higher recommendation utility, compared to (non) knowledge-aware baselines. None of the them deeply analyzed beyond utility goals (e.g., coverage, serendipity) nor monitored consumer (i.e., end users) and provider fairness. Hence, it remains unclear whether path reasoning methods emphasize any trade-off between goals unexplored so far. Being the landscape convoluted and polarized to utility, there is a need for a common evaluation ground to understand how and when each method can be adopted.

In this paper, we conduct a replicability study (different team and experimental setup) on unexplored evaluation perspectives relevant to path reasoning methods. In a first step, we scanned the proceedings of top-tier conferences and journals, identifying seven relevant papers. We tried to replicate the original methods based on the released source code, but only three of them were replicable. In a second step, we defined a common evaluation protocol, including two public data sets (movies; music), two sensitive attributes (gender; age), and sixteen metrics pertaining to four perspectives (recommendation utility; beyond utility goals; explanation quality; fairness). We evaluated path reasoning methods under this protocol and compared them against other knowledge-aware methods. Results reveal that, despite of an often similar utility, path reasoning methods differ in the way they meet other recommendation goals. Our study calls for a broader evaluation of these methods and a more responsible adoption.

2   Research Methodology

In this section, we describe the collection process for path reasoning methods, the steps for their replication, and the common evaluation protocol.

2.1   Papers Collection

To collect existing path reasoning methods, we systematically scanned the recent proceedings of top-tier information retrieval events (CIKM, ECIR, ECML-PKDD, FAccT, KDD, RecSys, SIGIR, WSDM, WWW, UMAP) and journals

Table 1. Path reasoning methods deemed as relevant in our study.

Method	Year	Status[a]	Experimental setting				
			Data sets[a]	Split size[c]	Split method[d]	Recommendation[e]	Explanation[e]
PGPR [40]	2019	*RE*	*AZ*	70-00-30	*Rand*	NDCG, R, HR, P	–
EKAR [31]	2019	*RE̅*	*ML, LFM, DB*	60-20-20	*Rand*	NDCG, HR	–
CAFE [41]	2020	*RE*	*AZ*	70-00-30	*Rand*	NDCG, R, HR, P	–
UCPR [32]	2021	*RE*	*ML, AZ*	60-20-20	*Rand*	NDCG, R, HR, P	PPC
MLR [38]	2022	*RE̅*	*AZ*	70-00-30	*Rand*	NDCG, R, HR, P	–
PLM-Rec [13]	2022	*R̅E̅*	*AZ*	60-20-20	*Time*	NDCG, R, HR, P	–
TAPR [46]	2022	*R̅E̅*	*AZ*	60-10-30	*Rand*	NDCG, R, HR, P	–

[a] **Status** *RE* : Replicable and Extensible; *RE̅* : Replicable but not Extensible; *R̅E̅* : Not Replicable nor Extensible.

[b] **Data Set** *AZ* : Amazon [22]; *ML* : MovieLens 1M [15]; *LFM* : LastFM [30]; *DB* : DBbook2014 [9].

[c] **Split Size** reports the percentage of data for training, validation, and test, respectively.

[d] **Split Method.** *Rand* : Random based; *Time* : Time based.

[e] **Metrics** *R* : Recall; *HR* : Hit Ratio *P* : Precision; *PPC* : Path Pattern Concentration

edited by top-tier publishers (ACM, Elsevier, IEEE, Springer). The adopted keywords combined a technical term between "*path reasoning recommender systems*" and "*explainable recommender system*" and a non-technical term between "*explainable AI*" and "*knowledge enabled AI*". We marked a paper as relevant if (a) it addressed recommendation, (b) it proposed a KG-based method, and (c) the method could produce reasoning paths. Papers on other domains or tasks, e.g., non-personalized rankings or mere entity prediction tasks (w/o any recommendation) were excluded. We also excluded knowledge-aware methods unable to yield reasoning paths, although we will use some representatives of this class for comparison. Seven relevant papers were selected for our study (Table 1).

We attempted to replicate the method of each relevant paper, relying as much as possible on the original source code. To obtain it, we first tried to search for the source code repository into the original paper and on the Web. As a last resort, we sent an e-mail to the original authors. We considered a method to be replicable in case a fully working version of the source code was obtained and needed minor changes to accept another data set and extract recommendations (and reasoning paths). Three out of the seven relevant papers were replicable with a reasonable effort. As per the non-replicable ones, three did not provide any source code[1]. The other one included unavailable external dependencies [20].

2.2   Methods Replication

For each relevant paper, we analyzed the rationale of the proposed method and the characteristics of the experimental setting, as summarized in Table 1.

---

[1] Note that the source code of these papers might appear soon online as an effect of our e-mails to the original authors. We leave their replication as a future work.

PGPR [40] (original source code: https://github.com/orcax/PGPR) was based on the idea of training a reinforcement learning (RL) agent for finding paths. During training, the agent starts from a user and learns to reach the correct products, with high rewards. During inference, the agent directly walks to correct products for recommendation, without enumerating all the paths between users and products. The original experiments were done on four AZ data sets [22] and on a KG built from product metadata and reviews.

EKAR [31] (original source code not available) modeled the task as a Markov decision process on the user-item-entity graph and used deep RL to solve it. The user-item-entity graph is treated as the environment, from which the agent gets a sequence of visited nodes and edges. Based on the encoded state, a policy network outputs the probability distribution over the action space. Finally, a positive reward is given if the agent successfully finds those products consumed by the target users in the training set. The novelty lays in using an LSTM for the policy network and a reward function that makes training stable and encourages agent exploration. Only this study included data sets from three diverse domains: movies (ML1M), music (LFM), and books (DB).

CAFE [41] (original source code: https://github.com/orcax/CAFE) follows the coarse-to-fine paradigm. Given the KG, a user profile is created to capture user-centric patterns in the coarse stage. To conduct multi-hop path reasoning guided by the user profile, the reasoner is decomposed into an inventory of neural reasoning modules. Then, these modules are combined based on the user profile, to efficiently perform path reasoning. Original experiments followed the PGPR experimental setting (same data sets, data split, and evaluation metrics).

UCPR [32] (original source code: https://github.com/johnnyjana730/UCPR/) introduces a multi-view structure leveraging not only local sequence reasoning information, but also a view of the user's demand portfolio. The user demand portfolio, built in a pre-processing phase and updated via a multi-step refocusing, makes the path selection process adaptive and effective. The original experimental setting covered the movie (ML1M) and e-commerce domains (AZ). This study was the only one assessing an explanation quality property, i.e., to what extent the KG relation type differs among the selected paths.

MLR [38] (source code shared by e-mail, but external dependencies missing) is another RL framework that leverages both ontology-view and instance-view KGs to model multi-level user interests. Through the Microsoft Concept Graph (MCG) [20], the method creates various conceptual levels (e.g., Prada is an Italian luxury fashion brand). The reasoning is then performed by navigating through these multiple levels with an RL agent. The authors provided the source code, but the MCG was no longer online and the provided KG dump referred only to the originally used AZ data sets.

PLM-Rec [13] (original source code not available), given a KG, extracts training path sequences under different hop constraints. By leveraging augmentations of language features with semantics, the method obtains a series of training data sequences. A transformer-based decoder is then used to train an auto-regressive

Table 2. Interaction and knowledge information for the two considered data sets.

Interaction	ML1M	LFM1M	Knowledge	ML1M	LFM1M
Users	6,040	4,817	Entities (Types)	13,804 (12)	17,492 (5)
Products	2,984	12,492	Relations (Types)	193,089 (11)	219,084 (4)
Interactions	932,295	1,091,275	Sparsity	0.0060	0.0035
Density	0.05	0.01	Avg. degree overall	28.07	25.05
Gender (Age) groups	2 (7)	2 (7)	Avg. degree products	64.86	17.53

path language model. This method could limit previous methods' recall bias in terms of KG connectivity. Only AZ data sets were used in the experiments.

TAPR [46] (original source code not available) proposed another path reasoning approach based on RL, characterized by the incorporation of a temporal term in the reward function. This temporal term guides a temporal-informed search for the agent, to capture recent trends of user's interests. Original results (on AZ data sets) showed a gain in utility compared to PGPR, although the model was evaluated using a random split, which is not ideal for time-aware models.

2.3 Evaluation Protocol

To ensure evaluation consistency and uniformity across methods, given the heterogeneous original experimental settings, we mixed replication and reproduction [7], but use only the term "replicability" for convenience throughout this paper. Specifically, we relied on the source code provided by the original authors to run their methods, and our own data and source code to (a) pre-process the input data sets as per their requirements and (b) compute evaluation metrics based on the recommendations and reasoning paths they returned.

Data Collection. We conducted experiments on two data sets: MovieLens (ML1M) [15] and LastFM (LFM1B) [30]. Given our interest in the fairness perspective, we selected data sets that provide (or make it possible to collect) users and providers' demographic attributes. We therefore discarded other data sets, such as the Amazon ones [22], where this was not reasonably possible. The selected data sets are all public and vary in domain, extensiveness, and sparsity, providing novel insights on the generalizability of the replicated path reasoning methods under a common ground, with respect to their original settings (see Table 1). For ML1M, we used the KG generated in [9] from DBpedia, while we generated the KG from the Freebase dump extracted by [45] for LFM1B.

Data Preprocessing. Concerning the ML1M data set, both gender and age sensitive attributes for the consumers, but not for the providers, were originally provided in [15]. Being directors considered as movie providers in prior work [6], we relied on their sensitive attribute labels collected in that study. In LFM1B, gender and age labels were attached only to a small subset of end users. We therefore discarded all those users whose sensitive attributes were not available.

Given that the original papers included only data sets far smaller than LFM1B and that our preliminary experiments uncovered a low scalability for those methods[2], we then uniformly sampled a subset of the filtered LFM1B, ensuring that users (products) had at least 20 (10) interactions. We will refer to this data subset as LFM1M throughout the paper and results. This sampling allowed us to obtain a data set size comparable to ML1M and avoid cold-start scenarios, which are not our focus. Since providers' sensitive attributes were not attached to the original data set (in music RS, artists are commonly considered as providers), we crawled them from Freebase and released them with our study.

Both KGs were pre-processed as performed in [2] to make triplets uniformly formatted. More specifically, we only consider triplets composed of a product as the entity head and an external entity as the entity tail, to obtain a common ground data set for the analysis of both knowledge-aware and path-based methods. Considering triplets having external entities or products as the head and tail entities would have required to craft additional meta-paths (needed for path-based methods) compared to the reproduced studies, going beyond the scope of our work. In addition, to control sparsity, we removed relations having a type represented in less than 3% of the total number of triplets. Concerning the user-product interactions, we discarded products (and their interactions) which are not present in the KG. Pre-processed data set statistics are collected in Table 2.

Data Preparation and Split. For each data set, we first sorted the interactions of each user chronologically. We then performed a training-validation-test split, following a time-based hold-out strategy, with the 60% oldest interactions in the training set, the following 20% for validation, and the 20% most recent ones as the test set. The aforementioned pre-processed data sets were used to train, optimize, and test each benchmarked model. This allowed us to carry out the evaluation procedure in a realistic setting, in which the trends that might determine interaction patterns are non-stationary and evolve over time.

Comparative Knowledge-Aware Models. Path reasoning methods belong to a subclass of the knowledge-aware recommendation class. To better contextualize our study, we therefore decided to provide comparisons (when interesting) against two knowledge-aware models based on knowledge embeddings, namely CKE [42] and CFKG [1], and a knowledge-aware model based on propagation, namely KGAT [36]. These three models, unable to provide reasoning paths to users, were replicated and evaluated under the same protocol[3]. For conciseness, we do not explain their replication in detail and refer the reader to our repository.

---

[2] Solving substantial scalability issues goes beyond the scope of our replicability study.

[3] For conciseness, we did not include non-knowledge-aware methods (e.g., BPR), which were compared against path reasoning methods under some metrics (e.g., NDCG) in studies like [4]. Nevertheless, this is an important aspect for future work.

Table 3. Evaluation metrics covered in our replicability study.

Perspective	Metric	Acronym	Range	Description
Consumers utility	Normalized Discounted Cumulative Gain	NDCG	[0, 1]	The extent to which the recommended products are useful for the user (1 means more useful)
	Mean Reciprocal Rank	MRR	[0, 1]	The extent to which the first recommended product is useful for the user (1 means more useful)
Consumers beyond utility	Coverage	COV	(0, 1)	The percentage of products overall recommended at least once (1 means high coverage)
	Diversity	DIV	(0, 1)	The percentage of product categories covered in the recommended list (1 means high diversity)
	Novelty	NOV	(0, 1)	Inverse of the popularity of products recommended to a user (1 means low popularity, so high novelty)
	Serendipity	SER	[0, 1]	The percentage of the recommended products not suggested also by a baseline (1 means more unexpected)
Consumers explanation quality	Fidelity	FID	[0, 1]	The percentage of the recommended products that can be explained (1 means all products can be explained)
	Linking Interaction Recency	LIR	[0, 1]	The recency of the past interaction in the paths accompanying recommended products (1 means recent)
	Linking Interaction Diversity	LID	[0, 1]	The number of distinct past interactions in the paths accompanying recommended products (1 means different)
	Shared Entity Popularity	SEP	[0, 1]	The popularity of the shared entity in the paths accompanying recommended products (1 means popular)
	Shared Entity Diversity	SED	[0, 1]	The number of distinct shared entities in the paths accompanying recommended products (1 means different)
	Path Type Diversity	PTD	[0, 1]	The percentage of distinct path types within paths accompanying recommended products (1 means different)
	Path Type Concentration	PTC	[0, 1]	The extent to which the distinct path types representation is equally balanced (1 means balanced)
Providers utility	Exposure	EXP	[0, 1]	Exposure of the items of a given provider in the recommended list (1 means high exposure)

Hyper-parameter Fine-Tuning. Given a data set and a model, we selected the best hyper-parameters setting via a grid search that involved those hyper-parameters (and their values) found to be sensitive in the original papers. In certain cases, given our findings from preliminary experiments, we extended the grid of values to better adhere to the characteristics of the data set at hand. Full details on the hyper-parameters and their values in our grid search are reported in our repository. Models obtained via different hyper-parameter settings were evaluated on the validation set, selecting the one achieving the highest NDCG.

Evaluation Metrics Computation. Given a model and a data set, we monitored recommendation utility, beyond utility objectives, explanation quality, and both consumer and provider fairness, on recommended lists with the well-known size of $k = 10$ (e.g., [7]), based on the corresponding test set. We describe each metric in Table 3 and refer to the repository for implementation details.

Concerning recommendation utility for consumers, we monitored the Normalized Discounted Cumulative Gain (NDCG) [39], using binary relevance scores and a base-2 logarithm decay, and the Mean Reciprocal Rank (MRR) [11]. Differently from recall and accuracy, NDCG takes into account the position of the relevant products in the recommended list. MMR instead considers the position of the first relevant product only, giving us a perspective different than NDCG.

In our work, we focused also on four well-known beyond utility goals [21]. We monitored the extent to which the generated recommendations cover the catalog of available products (coverage). High coverage may increase users' satisfaction and the sales. Another goal, diversity, was found to be relevant for human understanding [12] and content acceptance [10]. We computed it as the percentage of distinct product categories in the recommended list. Further, serendipity measures recommendation surprise [17]. Given our offline setting, we compared the recommendations with those of a baseline model, i.e., a most popular recommender [24]. The more the recommendations differ between the benchmarked and the baseline model, the higher the serendipity. Finally, we estimated novelty as the inverse of product popularity (as per the received ratings), assuming that products with low popularity are more likely to be surprising [47].

With regard to explanation quality, we considered the proportion of explainable products in a recommended list (fidelity) [28]. In addition, we monitored reasoning paths properties concerning recency, popularity, and diversity [3]. Recent linking interactions can help the user to better catch the explanation. The linked interaction recency measures the recency of the past interaction presented in the explanation path, whereas the linked interaction diversity monitors how many distinct past interactions are present. The second perspective is related to shared entities, assuming that more popular shared entities have a higher chance of being familiar to the user. The shared entity popularity measures the popularity (node degree in the KG) of the shared entity in an explanation path. Conversely, the shared entity diversity monitors the distinct shared entities. Finally, path type diversity focuses on how many distinct path types are included. Path type concentration monitors whether path types are equally balanced.

With the increasing importance received by fairness, we also assessed fairness with respect to a notion of demographic parity [7,14]. For consumer fairness, given a metric, we computed the average value of that metric for each demographic group and monitored the absolute pairwise difference between groups[4]. Concerning provider fairness, we computed the average exposure given to products of providers in a given demographic group [6]. Again, we finally computed the average absolute pairwise difference between provider groups.

3   Experimental Results

Our study aimed to investigate multiple evaluation perspectives of path reasoning methods, by answering to the following research questions:

RQ1. Do path reasoning methods trade recommendation utility and/or beyond utility objectives for explanation power?

RQ2. To what extent can path reasoning methods produce explanations for all the recommended products, depending on the recommended list size?

RQ3. How does the quality of the selected paths vary among path reasoning methods, based on the path type and characteristics?

3.1   Trading Recommendation Goals for Explanation Power (RQ1)

In a first analysis, we investigated whether there exists any substantial difference in recommendation utility and beyond utility objectives between the considered path reasoning methods (PGPR, CAFE, UCPR) and relevant knowledge-aware but not explainable methods (KGAT, CKE, CFKG). To assess statistical significance, t-tests were carried out for each metric, considering the two categories of methods as the two separate groups, under each data set. This allowed us to discern behavior also with respect to the sparsity of the KG. Although the total sample size (six methods) is rather small, this setup made it possible to notice some preliminary characteristics of the two method classes. Still, further studies on a broader set of methods should be run to assess results generalizability more. P-values obtained for each test are reported in Table 5.

---

[4]   Our data includes sensitive attributes pertaining to the gender (Male, Female) and age (Under 18, 18–24, 25–34, 35–44, 45–49, 50–55, 56+), as per the data set labels.

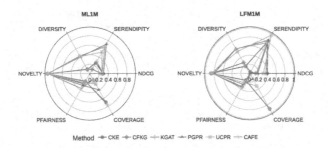

Fig. 1. Comparison on recommendation utility and beyond utility goals [RQ1].

Table 4. Metric scores for recommendation utility and beyond utility goals [RQ1].

Method	ML1M							LFM1M						
	NDCG ↑	MMR ↑	SER ↑	DIV ↑	NOV ↑	PF[a] ↓	COV ↑	NDCG ↑	MMR ↑	SER ↑	DIV ↑	NOV ↑	PF[a] ↓	COV ↑
CKE	**0.29**	**0.23**	0.26	0.10	**0.93**	0.19	<u>0.70</u>	**0.40**	**0.34**	<u>0.82</u>	0.18	0.88	0.18	**0.91**
CFKG	<u>0.26</u>	0.21	0.11	0.11	0.92	0.25	0.16	0.13	0.10	0.04	0.27	0.86	<u>0.34</u>	0.02
KGAT	**0.29**	**0.23**	0.29	0.10	**0.93**	0.19	**0.75**	<u>0.37</u>	<u>0.31</u>	0.79	0.19	**0.88**	0.18	<u>0.89</u>
PGPR	0.28	0.21	0.78	0.42	0.93	0.27	0.42	0.31	0.25	0.81	0.54	0.82	0.32	0.20
UCPR	<u>0.26</u>	0.20	0.53	<u>0.42</u>	**0.93**	0.22	0.25	0.34	0.27	**0.94**	<u>0.57</u>	<u>0.87</u>	0.22	0.41
CAFE	<u>0.26</u>	0.18	<u>0.63</u>	**0.44**	**0.93**	0.36	0.21	0.15	0.09	0.75	**0.58**	0.84	**0.36**	0.11

For each dataset: best result in **bold**, second-best result <u>underlined</u>.
[a] **Metrics** *PF*: Provider Fairness.

Table 5. T-test p-values to assess statistically significant differences between path-based and knowledge-aware methods across evaluation metrics [RQ1].

Data set	NDCG	MMR	SER	DIV	NOV	PF	COV
ML1M	0.33	0.08	**0.01**	0	0.422	0.21	0.33
LFM1M	0.77	0.65	0.38	0	0.16	0.38	0.34

P-values below 0.05 are reported in **bold**.

Figure 1 depicts our evaluation results in terms of utility (NDCG, MMR) and beyond utility goals (serendipity, diversity, novelty, and coverage). We also report provider fairness estimates, whereas we will discuss consumer fairness later on. Even though Fig. 1 makes the comparison easier, we refer to Table 4 for specific values. Concerning recommendation utility, path reasoning methods achieved comparable scores (0.26 to 0.28 NDCG; 0.18 to 0.21 MRR) to knowledge-aware non-explainable methods (0.26 to 0.29 NDCG; 0.21 to 0.23 MRR) in ML1M. These observations did not hold in LFM1M, where path reasoning methods led to recommendations of lower utility (0.15 to 0.34 NDCG; 0.09 to 0.27 MMR) than knowledge-aware baselines (0.13 to 0.40 NDCG; 0.10 to 0.34 MMR). However, in both cases and under both data sets, no statistical differences were found in terms of recommendation utility between method classes (all p-values were greater than 0.05).

With regard to beyond utility objectives, path reasoning methods achieved substantially higher serendipity (0.53 to 0.78 ML1M; 0.75 to 0.94 LFM1M) than knowledge-aware non-explainable baselines (0.11 to 0.29 ML1M; 0.04 to 0.82

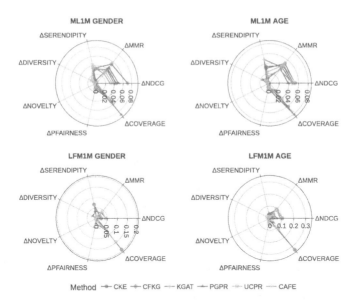

Fig. 2. Disparate impacts between groups (gender and age) on recommendation utility, beyond utility objectives, and provider fairness. The lower it is, the fairer [RQ1].

LFM1M). Interestingly, it was confirmed that path reasoning methods tend to perform worse in LFM1M than ML1M. Furthermore, our statistical tests show that, on ML1M, the two method classes performed differently in terms of serendipity (p-value equal to 0.011), with path reasoning methods showing higher serendipity than the knowledge aware methods, on average. Similar patterns were found for diversity. Path reasoning methods led to a higher diversity on average (respectively for path-reasoning and knowledge-aware methods, 0.56 and 0.21 on LFM1M; 0.43 and 0.10 on ML1M), on both ML1M and LFM1M (p-value ≈ 0). Conversely, on coverage, the best path reasoning method showed a decrease of 44% on ML1M and 54.9% on LFM1M than the best knowledge-aware method. This might be due to the low number of paths in the KG available to the path-reasoning methods. Except for CFKG, there is evidence that knowledge-aware methods should be preferred in case someone aims to optimize for coverage. Finally, novelty scores were similar between the two classes of methods.

From a provider fairness perspective, path reasoning methods led to a fairer exposure of provider groups (0.22 to 0.36 ML1M; 0.22 to 0.36 LFM1M), compared to the other family (0.19 to 0.25 ML1M; 0.18 to 0.34 LFM1M). Surprisingly, CFKG reported the second best provider fairness score, despite of its low recommendation utility. On the other hand, consumer fairness estimates according to the considered evaluation metrics[5] are collected in Fig. 2. Being the patterns comparable across data sets and demographic groups, we describe only the

---

[5] Differences between the demographic groups achieving the best and worst score on avg. were all statistically significant under t-test (if applicable) or h-test otherwise.

results obtained in ML1M and the gender groups. For the latter, all the models presented some yet low levels of unfairness in term of utility (both NDCG and MRR). Path reasoning methods (PGPR, UCPR, CAFE) achieved higher levels of unfairness on coverage, respectively 0.084, 0.045, 0.032, compared to baseline methods (KGAT reached the highest coverage unfairness, with 0.06). For the other metrics, we uncovered small yet comparable differences.

Findings RQ1. Path reasoning methods trade recommendation utility and coverage for explanation power, especially in LFM1M. Conversely, they resulted in higher estimates on other beyond utility objectives and provider fairness than knowledge-aware non-explainable baselines.

3.2 Producing Explanations for All Recommended Products (RQ2)

In a second analysis, we were interested in understanding the extent to which path reasoning methods can produce explanations for all the recommended products across recommended lists of different sizes. This property, fidelity, is essential for a method which yields reasoning paths (and produces explanations). Table 6 shows fidelity scores for the two data sets on lists of size 10, 20, 50 and 100.

Concerning PGPR, paths were attached to almost every product of the recommended list, until the size of 50. Under a size of 100, fidelity remarkably decreased to 78%. This decay in fidelity was exacerbated in LFM1M. In the latter data set, already with a size of 20, only 74% of recommended products were explained. With regard to UCPR, we observed a similar but reversed pattern, compared to PGPR, on the two considered data sets. UCPR was challenged to produce explanations even under a size of 10 on ML1M (only 61% of products were explained). Conversely, in LASTFM, the same model obtained a higher fidelity than PGPR, with 99% of explained products under a size of 10. Surprisingly, CAFE was able to provide a reasoning path for each recommended product until a list of size 100. It should however be noted that, to make this happen, the list size must be specified in advance during training. CAFE indeed automatically adapts the size of the neighbourhood to search around, according to the list size. Hence, this method would be the best choice when the list size is known in advance, constant, or up to a certain limit. Although fidelity could be controlled, doing this led to a smaller NDCG for CAFE (see Sect. 3.1).

Table 6. Explanation fidelity analysis across cut-offs $k = \{10, 20, 50, 100\}$ [RQ2].

Method	ML1M				LFM1M			
	10	20	50	100	10	20	50	100
PGPR	1.00	0.99	0.99	0.78	0.98	0.74	0.31	0.15
CAFE	1.00	1.00	1.00	1.00	1.00	1.00	1.00	1.00
UCPR	0.61	0.34	0.14	0.07	0.99	0.98	0.68	0.35

Findings RQ2. Path reasoning methods show very different patterns in terms of fidelity. CAFE's fidelity is high and stable across data sets and recommended list sizes. On the other hand, PGPR provides higher but rapidly decaying fidelity in ML1M than LFM1M, viceversa for UCPR.

3.3   Differences on Explanation Quality (RQ3)

In a final analysis, we investigated how the quality of the selected paths (and so of the resulting explanations) varied based on the path characteristics. To this end, Table 7 collects seven explanation path quality perspectives (LIR, LID, SEP, SED, PTD, PTC, PPC) for each reasoning path method and data set.

Concerning the recency dimension, we did not observe any substantial difference in linked interaction recency among the three methods, with the maximum (minimum) value 0.44 (0.34) achieved by PGPR (CAFE) on ML1M (similarly on LFM1M). Whereas, in terms of linked interaction diversity, it can be interestingly noted that PGPR (0.84 ML1M; 0.77 LFM1M) and UCPR (0.82 ML1M; 0.84 LFM1M) led to higher diversity than CAFE.

Moving to the popularity perspective and, the shared entity popularity in particular, CAFE was able to obtain the highest SEP in both data sets (0.75 ML1M; 0.77 LFM1M), meaning that it had the tendency to yield paths with more popular shared entities. Compared to CAFE, PGPR and UCPR showed instead substantially lower values. Estimates on SED were very high (0.92 to 1 ML1M; 0.78 to 0.98 LFM1M) for all the methods. These methods had hence the ability to include a good variety of shared entities in their reasoning paths.

Patterns regarding path types were particularly interesting. Specifically, both path type diversity and path type concentration were higher in CAFE and UCPR than PGPR. This highlights that the explanations of the former were, on average, richer in terms of path types (e.g., starred by, directed by), while PGPR's explanations were limited, on average, to a narrow set of different path types.

Figure 3 depicts pairwise differences in the average score of a given metric between demographic groups (consumer fairness). Again, under the same conditions of RQ1, all the differences were statistically valid. For conciseness, we discuss only the results for gender groups. PGPR and CAFE showed very low unfairness estimates across all metrics, with all scores lower than 0.01 in UCPR and 0.02 in PGPR on both data sets. Differently, UCPR emphasized unfairness on PTD, SED and PPC. Remarkably, the strongest disparate impact was

Table 7. Explanation quality analysis [RQ3].

Model	ML1M							LFM1M						
	LIR ↑	LID ↑	SEP ↑	SED ↑	PTD ↑	PTC ↑	PPC ↑	LIR ↑	LID ↑	SEP ↑	SED ↑	PTD ↑	PTC ↑	PPC ↑
PGPR	**0.44**	**0.84**	<u>0.43</u>	<u>0.99</u>	0.12	<u>0.03</u>	0.12	0.49	<u>0.77</u>	0.61	<u>0.94</u>	<u>0.30</u>	0.05	0.24
CAFE	0.34	0.16	**0.75**	**1.00**	**0.33**	**0.73**	**0.37**	0.49	0.25	**0.77**	**0.98**	0.25	**0.62**	**0.50**
UCPR	<u>0.40</u>	<u>0.82</u>	0.35	0.92	<u>0.24</u>	0.01	<u>0.24</u>	0.49	**0.84**	<u>0.67</u>	0.78	**0.42**	<u>0.24</u>	<u>0.34</u>

For each dataset: best result in **bold**, second-best result <u>underlined</u>.

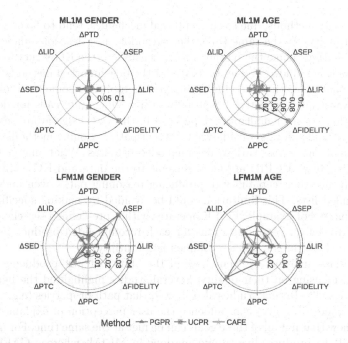

Fig. 3. Disparate impacts between groups (gender and age) on explanation quality pertaining to recency, popularity, and diversity. The lower it is, the fairer [RQ3].

reported on FID (0.12), PTD and PPC (0.05) for both data sets, and SED (0.03). We conjecture that these estimates of unfairness could be driven by data imbalance across the different demographic groups.

Findings RQ3. Path reasoning methods often yield substantially different paths in terms of recency, popularity, and diversity. Although they exist, no remarkable disparate impacts on explanation quality were found.

4   Discussion and Conclusion

In this section, we connect the main findings coming from the individual experiments and present the implications and limitations of our replicability study.

In the first analysis, we analyzed how methods able to produce explanations (path reasoning) compared against knowledge-aware non-explainable baselines in terms of utility, beyond utility objectives, and provider fairness. Results show that these two classes of methods diverge slightly in terms of utility, although explanations may have a persuasive effect which could not be captured offline. Further studies should investigate how explanations impact user decisions and consequently utility. Considering beyond accuracy objectives, we observed that path reasoning methods, due to their internal mechanics, tend to favor serendipity and diversity. At the same time, the methodological decisions made to produce explanations make the methods more sensible to the KG structure, consequently resulting in low product coverage for the benchmarked methods. Results also show that all methods (including baselines) emphasize some levels of unfairness in almost all perspectives, especially utility. Our study calls for debiasing methods that consider multiple perspectives in the knowledge-aware setup.

In the second analysis, we analyzed whether path reasoning methods can produce explanations across various recommended list sizes. What emerged is that some models (e.g., UCPR) are more sensible to the data and KG composition, which influence their capability of producing reasoning paths even under short recommended lists. This limitation could be avoided by making specific model design choices. For example, CAFE operates with a pre-defined search space for each user to deliver reasoning paths for each recommended product (although this design choice might affect recommendation utility).

In the last analysis, we went beyond the ability of just producing reasoning paths, focusing on their quality. Several studies highlighted the benefits of explanations [33]. Recent studies also showed that path properties (e.g., recency, popularity, and diversity) can influence the user perception of explanations [3]. Results show that not all of these goals can be met at the same time. For instance, PGPR fails to produce diverse explanations in ML1M, whereas CAFE yields explanations based on a tiny set of past user interactions. Future studies should address this aspect through in- and post-processing methods and look at other explanation perspectives (e.g., persuasiveness, trust, and efficiency).

Overall, our analyses showed that replicating research in this area is still a challenging task. In future work, we plan to explore in detail the impact of KG characteristics on the considered perspectives, as well as devise novel path reasoning methods robust to the KG structure and effective on multiple objectives.

References

1. Ai, Q., Azizi, V., Chen, X., Zhang, Y.: Learning heterogeneous knowledge base embeddings for explainable recommendation. Algorithms **11**(9), 137 (2018)
2. Balloccu, G., Boratto, L., Fenu, G., Marras, M.: Hands on explainable recommender systems with knowledge graphs. In: Proceedings of the 16th ACM Conference on Recommender Systems, RecSys 2022, pp. 710–713. Association for Computing Machinery, New York, NY, USA (2022)
3. Balloccu, G., Boratto, L., Fenu, G., Marras, M.: Post processing recommender systems with knowledge graphs for recency, popularity, and diversity of explanations. In: Proceedings of the 45th International ACM SIGIR Conference on Research and Development in Information Retrieval, SIGIR 2022, pp. 646–656. Association for Computing Machinery, New York, NY, USA (2022)
4. Balloccu, G., Boratto, L., Fenu, G., Marras, M.: Reinforcement recommendation reasoning through knowledge graphs for explanation path quality. Knowl.-Based Syst. **260**, 110098 (2023)

5. Barredo Arrieta, A., et al.: Explainable artificial intelligence (XAI): concepts, taxonomies, opportunities and challenges toward responsible AI. Inf. Fusion **58**, 82–115 (2020)
6. Boratto, L., Fenu, G., Marras, M.: Interplay between upsampling and regularization for provider fairness in recommender systems. User Model. User Adapt. Interact. **31**(3), 421–455 (2021)
7. Boratto, L., Fenu, G., Marras, M., Medda, G.: Consumer fairness in recommender systems: contextualizing definitions and mitigations. In: Hagen, M., et al. (eds.) Advances in Information Retrieval, pp. 552–566. Springer, Cham (2022). https://doi.org/10.1007/978-3-030-99736-6_37
8. Cao, Y., Hou, L., Li, J., Liu, Z.: Neural collective entity linking. In: Proceedings of the 27th International Conference on Computational Linguistics, pp. 675–686. Association for Computational Linguistics, Santa Fe, New Mexico, USA, August 2018
9. Cao, Y., Wang, X., He, X., Hu, Z., Tat-seng, C.: Unifying knowledge graph learning and recommendation: towards a better understanding of user preference. In: WWW (2019)
10. Carbonell, J., Goldstein, J.: The use of MMR, diversity-based reranking for reordering documents and producing summaries. In: Proceedings of the 21st Annual International ACM SIGIR Conference on Research and Development in Information Retrieval, SIGIR 1998, pp. 335–336. Association for Computing Machinery, New York, NY, USA (1998)
11. Craswell, N.: Mean Reciprocal Rank, pp. 1703–1703. Springer, Cham (2009). https://doi.org/10.1007/978-0-387-39940-9_488
12. Gedikli, F., Jannach, D., Ge, M.: How should i explain? A comparison of different explanation types for recommender systems. Int. J. Hum Comput Stud. **72**(4), 367–382 (2014)
13. Geng, S., Fu, Z., Tan, J., Ge, Y., de Melo, G., Zhang, Y.: Path language modeling over knowledge graphs for explainable recommendation. In: Proceedings of the ACM Web Conference 2022, WWW 2022, pp. 946–955. Association for Computing Machinery, New York, NY, USA (2022)
14. Gómez, E., Zhang, C.S., Boratto, L., Salamó, M., Marras, M.: The winner takes it all: geographic imbalance and provider (un)fairness in educational recommender systems. In: SIGIR 2021: The 44th International ACM SIGIR Conference on Research and Development in Information Retrieval, Virtual Event, Canada, 11–15 July 2021, pp. 1808–1812. ACM (2021)
15. Harper, F.M., Konstan, J.A.: The MovieLens datasets: history and context. ACM Trans. Interact. Intell. Syst. **5**(4), 1–19 (2015)
16. He, G., Li, J., Zhao, W.X., Liu, P., Wen, J.R.: Mining implicit entity preference from user-item interaction data for knowledge graph completion via adversarial learning. In: Proceedings of the Web Conference 2020, pp. 740–751 (2020)
17. Herlocker, J.L., Konstan, J.A., Terveen, L.G., Riedl, J.T.: Evaluating collaborative filtering recommender systems. ACM Trans. Inf. Syst. **22**(1), 5–53 (2004)
18. Huang, C., Gan, Z., Ye, F., Wang, P., Zhang, M.: KNCR: knowledge-aware neural collaborative ranking for recommender systems. In: IEEE International Conference on Dependable, Autonomic and Secure Computing, International Conference on Pervasive Intelligence and Computing, International Conference on Cloud and Big Data Computing, International Conference on Cyber Science and Technology Congress, DASC/PiCom/CBDCom/CyberSciTech 2020, Calgary, AB, Canada, 17–22 August 2020, pp. 339–344. IEEE (2020)

19. Huang, J., Zhao, W.X., Dou, H., Wen, J.R., Chang, E.Y.: Improving sequential recommendation with knowledge-enhanced memory networks. In: The 41st International ACM SIGIR Conference on Research & Development in Information Retrieval, SIGIR 2018, pp. 505–514. Association for Computing Machinery, New York, NY, USA (2018)

20. Ji, L., Wang, Y., Shi, B., Zhang, D., Wang, Z., Yan, J.: Microsoft concept graph: mining semantic concepts for short text understanding. Data Intell. **1**(3), 238–270 (2019)

21. Kaminskas, M., Bridge, D.: Diversity, serendipity, novelty, and coverage: a survey and empirical analysis of beyond-accuracy objectives in recommender systems. ACM Trans. Interact. Intell. Syst. **7**(1), 1–42 (2016)

22. Linden, G., Smith, B., York, J.: Amazon.com recommendations: item-to-item collaborative filtering. IEEE Internet Comput. **7**(1), 76–80 (2003)

23. Ma, W., et al.: Jointly learning explainable rules for recommendation with knowledge graph. In: The World Wide Web Conference, pp. 1210–1221 (2019)

24. Murakami, T., Mori, K., Orihara, R.: Metrics for evaluating the serendipity of recommendation lists. In: Satoh, K., Inokuchi, A., Nagao, K., Kawamura, T. (eds.) JSAI 2007. LNCS (LNAI), vol. 4914, pp. 40–46. Springer, Heidelberg (2008). https://doi.org/10.1007/978-3-540-78197-4_5

25. Musto, C., de Gemmis, M., Lops, P., Semeraro, G.: Generating post hoc review-based natural language justifications for recommender systems. User Model. User-Adap. Inter. **31**(3), 629–673 (2021)

26. Ni, J., Li, J., McAuley, J.J.: Justifying recommendations using distantly-labeled reviews and fine-grained aspects. In: Inui, K., Jiang, J., Ng, V., Wan, X. (eds.) Proceedings of the 2019 Conference on Empirical Methods in Natural Language Processing and the 9th International Joint Conference on Natural Language Processing, EMNLP-IJCNLP 2019, Hong Kong, China, 3–7 November 2019, pp. 188–197. Association for Computational Linguistics (2019)

27. Oramas, S., Ostuni, V.C., Di Noia, T., Serra, X., Di Sciascio, E.: Sound and music recommendation with knowledge graphs. ACM Trans. Intell. Syst. Technol. **8**(2), 21:1–21:21 (2017)

28. Peake, G., Wang, J.: Explanation mining: post hoc interpretability of latent factor models for recommendation systems. In: Proceedings of the 24th ACM SIGKDD International Conference on Knowledge Discovery & Data Mining, KDD 2018, pp. 2060–2069. Association for Computing Machinery, New York, NY, USA (2018)

29. Ricci, F., Rokach, L., Shapira, B.: Recommender Systems Handbook, vol. 1–35, pp. 1–35. Springer, Cham (2010). https://doi.org/10.1007/978-0-387-85820-3

30. Schedl, M.: The LFM-1B dataset for music retrieval and recommendation. In: Proceedings of the 2016 ACM on International Conference on Multimedia Retrieval, ICMR 2016, pp. 103–110. ACM, New York, NY, USA (2016)

31. Song, W., Duan, Z., Yang, Z., Zhu, H., Zhang, M., Tang, J.: Ekar: an explainable method for knowledge aware recommendation. CoRR abs/1906.09506 (2022)

32. Tai, C.Y., Huang, L.Y., Huang, C.K., Ku, L.W.: User-centric path reasoning towards explainable recommendation, pp. 879–889. Association for Computing Machinery, New York, NY, USA (2021)

33. Tintarev, N., Masthoff, J.: A survey of explanations in recommender systems. In: Proceedings of the 23rd International Conference on Data Engineering Workshops, ICDE 2007, pp. 801–810. IEEE Computer Society (2007)

34. Wang, H., et al.: RippleNet: propagating user preferences on the knowledge graph for recommender systems. In: Proceedings of the 27th ACM International Conference on Information and Knowledge Management, CIKM 2018, pp. 417–426. Association for Computing Machinery, New York, NY, USA (2018)
35. Wang, H., Zhang, F., Xie, X., Guo, M.: DKN: deep knowledge-aware network for news recommendation (2018)
36. Wang, X., He, X., Cao, Y., Liu, M., Chua, T.S.: KGAT: knowledge graph attention network for recommendation. In: Proceedings of the 25th ACM SIGKDD International Conference on Knowledge Discovery & Data Mining, KDD 2019, pp. 950–958. Association for Computing Machinery, New York, NY, USA (2019)
37. Wang, X., Wang, D., Xu, C., He, X., Cao, Y., Chua, T.S.: Explainable reasoning over knowledge graphs for recommendation. In: Proceedings of the Thirty-Third AAAI Conference on Artificial Intelligence and Thirty-First Innovative Applications of Artificial Intelligence Conference and Ninth AAAI Symposium on Educational Advances in Artificial Intelligence. AAAI Press (2019)
38. Wang, X., Liu, K., Wang, D., Wu, L., Fu, Y., Xie, X.: Multi-level recommendation reasoning over knowledge graphs with reinforcement learning. In: Proceedings of the ACM Web Conference 2022, WWW 2022, pp. 2098–2108. Association for Computing Machinery, New York, NY, USA (2022)
39. Wang, Y., Wang, L., Li, Y., He, D., Liu, T.: A theoretical analysis of NDCG type ranking measures. In: COLT 2013 - The 26th Annual Conference on Learning Theory, 12–14 June 2013, Princeton University, NJ, USA. JMLR Workshop and Conference Proceedings, vol. 30, pp. 25–54. JMLR.org (2013)
40. Xian, Y., Fu, Z., Muthukrishnan, S., de Melo, G., Zhang, Y.: Reinforcement knowledge graph reasoning for explainable recommendation. In: Proceedings of the 42nd International ACM SIGIR Conference on Research and Development in Information Retrieval, SIGIR 2019, pp. 285–294. Association for Computing Machinery, New York, NY, USA (2019)
41. Xian, Y., et al.: CAFE: coarse-to-fine neural symbolic reasoning for explainable recommendation. In: Proceedings of the 29th ACM International Conference on Information & Knowledge Management, CIKM 2020, pp. 1645–1654. Association for Computing Machinery, New York, NY, USA (2020)
42. Zhang, F., Yuan, N.J., Lian, D., Xie, X., Ma, W.Y.: Collaborative knowledge base embedding for recommender systems. In: Proceedings of the 22nd ACM SIGKDD International Conference on Knowledge Discovery and Data Mining, KDD 2016, pp. 353–362. Association for Computing Machinery, New York, NY, USA (2016)
43. Zhang, Y., Chen, X.: Explainable recommendation: a survey and new perspectives. Found. Trends® Inf. Retrieval 14(1), 1–101 (2020)
44. Zhao, K., et al.: Leveraging demonstrations for reinforcement recommendation reasoning over knowledge graphs, pp. 239–248. ACM, New York, NY, USA (2020)
45. Zhao, W.X., et al.: KB4Rec: a data set for linking knowledge bases with recommender systems. Data Intell. 1(2), 121–136 (2019)
46. Zhao, Y., et al.: Time-aware path reasoning on knowledge graph for recommendation. ACM Trans. Inf. Syst. 41, 1–26 (2022)
47. Zhou, T., Kuscsik, Z., Liu, J., Medo, M., Wakeling, J.R., Zhang, Y.C.: Solving the apparent diversity-accuracy dilemma of recommender systems. Proc. Natl. Acad. Sci. 107, 4511–4515 (2010)

Stat-Weight: Improving the Estimator of Interleaved Methods Outcomes with Statistical Hypothesis Testing

Alessandro Benedetti[(✉)] and Anna Ruggero

Sease Ltd., London, UK
{a.benedetti,a.ruggero}@sease.io

Abstract. Interleaving is an online evaluation approach for information retrieval systems that compares the effectiveness of ranking functions in interpreting the users' implicit feedback. Previous work such as Hofmann et al. (2011) [11] has evaluated the most promising interleaved methods at the time, on uniform distributions of queries. In the real world, usually, there is an unbalanced distribution of repeated queries that follows a long-tailed users' search demand curve. This paper first aims to reproduce the Team Draft Interleaving accuracy evaluation on uniform query distributions [11] and then focuses on assessing how this method generalises to long-tailed real-world scenarios. The replicability work raised interesting considerations on how the winning ranking function for each query should impact the overall winner for the entire evaluation. Based on what was observed, we propose that not all the queries should contribute to the final decision in equal proportion. As a result of these insights, we designed two variations of the Δ_{AB} score winner estimator that assign to each query a credit based on statistical hypothesis testing. To reproduce, replicate and extend the original work, we have developed from scratch a system that simulates a search engine and users' interactions from datasets from the industry. Our experiments confirm our intuition and show that our methods are promising in terms of accuracy, sensitivity, and robustness to noise.

Keywords: Interleaved comparison · Interleaving · Implicit feedback · Online evaluation · Academia-Industry collaborations

1 Introduction

In information retrieval, online evaluation estimates the best ranking function for a system, targeting a live instance with real users and data. In previous works interleaved methods have been evaluated on a uniform distribution of queries [11]. In real-world applications, the same query is executed multiple times by different users and in different sessions, leading to a distribution of collected implicit feedback that is not uniform across the query set.

This paper aims to reproduce and then replicate the Team Draft Interleaving experiments from Hofmann et al. (2011) [11], investigating the effect that

© The Author(s), under exclusive license to Springer Nature Switzerland AG 2023
J. Kamps et al. (Eds.): ECIR 2023, LNCS 13982, pp. 20–34, 2023.
https://doi.org/10.1007/978-3-031-28241-6_2

different query distributions have on the accuracy of this approach. The reason we chose this work [11] is that it is one of the most prominent surveys on inter-leaved methods and presents an experimental setup that felt perfect to evaluate the long-tailed real-world scenario. Specifically, three research questions arise:

- *RQ1*: Is it possible to reproduce the original paper experiments?
- *RQ2*: How does the original work generalise in the real-world scenario where queries have a long-tailed distribution?
- *RQ3*: Does applying statistical hypothesis testing improve the evaluation accuracy in such a scenario?

Thanks to the insights collected during the replicability work, we designed two novel methods that enrich the interleaving scoring estimator with a pre-liminary statistical analysis: *stat-pruning* and *stat-weight*. The idea is to weigh differently the contribution of each query to the final winner of the evaluation. We present a set of experiments to show interesting perspectives on the original work's reproducibility, confirm that the original work generalises quite accurately to the considered real-world scenario, and validate the intuition that our statis-tical analysis-based methods can improve the accuracy. The concepts 'ranking function', 'ranking model', and 'ranker' are used interchangeably.

The paper is organized as follows: Sect. 2 presents the related work. Section 3 details the experimental setup, datasets, and runs used for reproduction and replication. Section 4 introduces the theory behind the proposed improvements and describe our *stat-pruning* and *stat-weight* implementations. Section 5 dis-cusses the experiments' runs and the obtained results. The paper's conclusions and future directions are listed in Sect. 6.

2   Related Work

Evaluation of Information Retrieval systems follows two approaches: offline and online evaluation.

For the offline, the most commonly used is the Cranfield framework [4]: an evaluation method based on explicit relevance judgments. The relevance judg-ments are provided by a team of trained experts and this is why this process is expensive. Collecting these judgments requires a lot of effort and there is the possibility that they do not reflect the same document relevance perceived by the common users. Users' interactions are easier to obtain and come with a minimal cost. Being performed directly by the end-users, they can be used to represent their intent closely, bypassing the domain experts' indirection. Implicit feedback is a very promising approach but, as a drawback, it could be noisy and therefore requires some further elaboration [2,3,21,24].

Implicit feedback is collected in real-time and it's at the base of the interleav-ing process. Despite the fact that the most common method of online evaluation is still AB testing, interleaving is experiencing a growing interest in research. This type of testing uses a smaller amount of traffic, with respect to the commonly used AB testing, without losing accuracy in the obtained result [2,16,23]. Interleaving was introduced by Joachims [14,15] and from then, many improvements followed [2,20–22,24].

Team-Draft Interleaving is among the most successful and used interleaving approaches [21] because of its simplicity of implementation and good accuracy. It is based on the strategy that captains use to select their players in team matches. TDI produces a fair distribution of ranking functions' elements in the final interleaved list. It has also been shown to overcome issues of a previously implemented approach, Balanced interleaving, in determining the winning model [2]. Other types of interleaved methods are Document Constraint [9], Probabilistic Interleaving [11], and Optimized Interleaving [20].

Document constraint infers preference relations between pairs of documents, estimated from their clicks and ranks. The method compares the inferred constraints to the original result lists and assesses how many constraints are violated by each. The list that violates fewer constraints is deemed better. This method proved more reliable than either balanced interleave or team draft on synthetic data, but it's more computationally expensive.

In probabilistic interleaving, both the choice of the model that contributes to the interleaving list and the document to put in the list, are selected based on probability. This approach is more complex but shows higher reliability, efficiency, and robustness to the noise with respect to the others.

Optimized interleaving proposes to formulate interleaving as an optimization problem that is solved to obtain the interleaved lists that maximize the expected information gained from users' interactions.

It's worth mentioning that a generalized form of the team draft interleaving has been proposed [17] and that additional research has been performed by Hofmann et al. with a new interleaving approach that aims to reduce the bias related to the way results are presented to the users [10] and studies on the fidelity, soundness, and efficiency of interleaved methods [11].

In the studies mentioned so far, ad hoc uniform query distributions are generally used when evaluating ranking models. Balog et al. [1] address the issue of real-world long-tailed query distributions by reducing the evaluation to the most popular queries, the ones that effectively provide a good amount of data to work on, reserving an analysis of the totality of queries for the future.

3   Experiments

This paper aims to reproduce and then replicate under different scenarios experiment 1 from one of the most prominent surveys on interleaved methods [11].

In the first set of experiments, we address $RQ1$ with the same settings and data as the original work (uniform query distribution). Despite the original paper showing the probabilistic interleaving to be superior, as a baseline, we evaluate the Team Draft Interleaving (TDI) method only. The choice has been made because it is much simpler to implement and more popular in the industry. Furthermore, our focus is on the Δ_{AB} score estimator which is in common between the two approaches. We examine if the accuracy of the TDI method matches the published results and discuss the details that we found to hold up, and the ones that could not be confirmed. In addition, we assess how the traditional TDI accuracy compares with two novel methods for Δ_{AB} score calculations (*stat-pruning* and *stat-weight*) under the uniform distribution conditions.

In the second set of experiments, we address *RQ2* and *RQ3* introducing a long-tailed query distribution. The aim is to examine how well the traditional TDI method generalises to real-world query distributions extracted from anonymized query logs and how *stat-pruning* and *stat-weight* perform for comparison.

Finally, the last set of experiments introduces a realistic click model simulator to assess how well TDI, *stat-pruning* and *stat-weight* methods respond to noise.

The datasets used are detailed in Sect. 3.1. The experimental setup is described in Sect. 3.2 and the experiment runs are explained in Sect. 3.3.

3.1 Datasets

All experiments make use of the MSLR-WEB30k Microsoft learning to rank (*MSLR*) dataset[1]. It consists of feature vectors extracted from query-document pairs along with relevance judgment labels. The relevance judgments take 5 values from 0 (irrelevant) to 4 (perfectly relevant). The experiments use the training set of fold 1. This set contains 18,919 queries, with an average of 119.96 judged documents per query. The average number of judged documents differs from what has been written by Hofmann [11] and the authors confirmed it was a mistake in the original paper. For each feature, we define a ranker (identified by the feature id) that sorts the search results descending by the feature value.

The experiments involving the long-tailed distributions use an industrial dataset we call *long-tail-1*. It consists of a list of anonymised query executions extracted from the query log of an e-commerce search engine, over a period of time. Each query is associated with the number of times it was executed. The amount of users collected per query is capped to 1,000. This threshold has been chosen to maintain a realistic long-tailed distribution while keeping a sustainable experimental cost.

From this dataset, we derive the long tail in Fig. 1.

3.2 Experimental Setup

To reproduce the original experiments we designed and developed a system[2] that simulates a search engine with users submitting queries and interacting with the results (clicks). The experiments are designed to evaluate the interleaved methods' ability to establish the better of two ranking functions based on (simulated) user clicks.

---

[1] Download from http://research.microsoft.com/en-us/projects/mslr/default.aspx.

[2] https://github.com/SeaseLtd/statistical-interleaving.

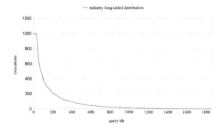

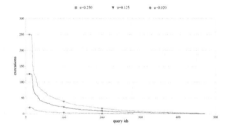

Fig. 1. Total unique queries: 1 861, total executions:156 550

Fig. 2. Long-tailed query distributions used in the experiments

Each experiment run repeats a number of simulations. When an experiment evaluates r ranking functions, it evaluates the first r, ordered by ascending id. Specifically, given a set of ranking functions, the number of simulations s in the run is the number of unique pairs in the set, where the pairs are subject to the commutative property (AB = BA). The system simulates a user submitting a query from the set of available queries in the distribution (in long-tailed distributions each query is submitted multiple times). The search engine responds with an interleaved result list that is presented to the user. The user clicks are randomly generated following the probability distribution that the click model assigns to the relevance judgments provided. Once the simulation completes, the ranking function preference of each click collected is evaluated and the Δ_{AB} score is computed to establish the winner. The ground truth winner is calculated as the ranking function with the best Normalised Discounted Cumulative Gain (NDCG) [12,13] averaged over the query set, using the explicit relevance judgments provided with the dataset $MSLR$. The winning ranker identified by the Δ_{AB} score is compared to the ground truth winner: when they match we have a correct guess. To assess the accuracy of the interleaved evaluation method we count the number of correct guesses over the total number of simulations s in the run showing at least one click.

Below we describe the query distributions, the click models, and the NDCG we used in our experiments.

Query Distributions. The query distribution in input to the simulation establishes the number of queries submitted to the system. We use two types of query distributions in our experiments: uniform and long-tailed.

In the uniform query distribution, each unique query is executed a constant number of times.

In the long-tailed query distribution, each query is executed a variable number of times. Starting from the *long-tail-1* distribution from the industry, we scaled down the number of queries and their executions by a factor $u \leq 1$ (see Fig. 2, Table 1). $u \leq 1$ has been introduced to experiment with different instances of realistic long-tailed distributions and to act within our computational limits.

Table 1. Scaling the query executions

u	Unique queries	Total executions
0.020	283	1 247
0.125	472	7 681
0.250	455	14 449

Click Models. The click model simulates user interactions according to the Dependent Click Model (DCM) [7,8], an extension of the cascade model [5]. It establishes the probability of a search result being clicked given its explicit relevance label (ground truth).

We use two models proposed by the original research [11]: the perfect and the realistic model.

NDCG. The NDCG metrics we use in our experiments are the complete NDCG (the complete search results list for a query) and the NDCG@10 (cut-off at 10). Using NDCG@10 is quite common in the industry as many search engines show 10 documents on their first page. When comparing the complete NDCG with NDCG@10 we noticed that the average difference between the pair of rankers to evaluate is smaller, making it more difficult for the interleaved methods to correctly guess the best ranker.

3.3 Runs

We divided the runs into four groups: reproduction, uniform query distribution, long-tailed query distribution, and realistic click model. When considering q queries in a run, we refer to the first q query ids, in the same order as they appear in the dataset *MSLR* rows. We define a ranker for each of the 136 individual features provided with the *MSLR* dataset. The results report the percentage of pairs for which the methods correctly identified the better ranker.

Reproduction. The scope of this set of runs is to reproduce *experiment 1* from the original research and answer *RQ1*. We exhaustively compare all 9, 180 distinct pairs derived from the 136 rankers. For each ranker pair, the user submits 1, 000 queries. The click model used is the *perfect model.*

- **Run 1**: exactly reproduce the original experiment, users click on the top-10 results for each query, to determine NDCG for the ground truth we use the complete NDCG.

We observed some inconsistencies with the original work results so we added to this group two additional runs:

- **Run 2**: users click on the complete list of search results for each query. To determine NDCG for the ground truth we use the complete NDCG.
- **Run 3**: users click on the top-10 results for each query. To determine NDCG for the ground truth we calculate NDCG@10.

Uniform Query Distribution The scope of this set of runs is to evaluate how the *stat-pruning* and *stat-weight* methods compare with the TDI baseline. We exhaustively compare all $9,180$ distinct pairs derived from the 136 rankers. The query distribution used is uniform, each run uses a different number q of queries. The click model used is the *perfect model*. Users click on the top-10 results for each query. Unless stated otherwise, to determine NDCG for the ground truth we calculate NDCG@10.

- **Run 4**: the query set consists of $1,000$ queries. Each query is executed once.
- **Run 5**: the query set consists of 100 queries. Each query is executed once.
- **Run 6**: the query set consists of 100 queries. Each query is executed 10 times.
- **Run 7**: the query set consists of 100 queries. Each query is executed 10 times. To determine NDCG for the ground truth we use the complete NDCG.

Long-Tailed Query Distribution. The scope of this set of runs is to evaluate how the *stat-pruning* and *stat-weight* methods compare with the TDI baseline over long-tailed query distributions and answer *RQ2* and *RQ3*. Unless stated otherwise, we exhaustively compare all $9,180$ distinct pairs derived from the 136 rankers and calculate NDCG@10 to determine NDCG for the ground truth. The click model used is the *perfect model*. Each run uses a different long-tailed query distribution. Users click on the top-10 results for each query.

- **Run 8**: the query set consists of 283 unique queries repeated following the long-tailed distribution with $u = 0.020$.
- **Run 9**: the query set consists of 472 unique queries repeated following the long-tailed distribution with $u = 0.125$.
- **Run 10**: the query set consists of 455 unique queries repeated following the long-tailed distribution with $u = 0.250$. We exhaustively compare all $2,415$ distinct pairs derived from the first 70 rankers.
- **Run 11**: the query set consists of 283 unique queries repeated following the long-tailed distribution with $u = 0.020$. To determine NDCG for the ground truth we use the complete NDCG.

Realistic Click Model. The scope of this set of runs is to evaluate how the *stat-pruning* and *stat-weight* methods compare with the TDI baseline over long-tailed query distributions with noisier clicks. We exhaustively compare all $9,180$ distinct pairs derived from the 136 rankers. The click model used is the *realistic model*. The query distribution used is long-tailed. The query set consists of 283 unique queries repeated following the long-tailed distribution with $u = 0.020$. Users click on the top-10 results for each query.

- **Run 12**: to determine NDCG for the ground truth we calculate NDCG@10.
- **Run 13**: to determine NDCG for the ground truth we use the complete NDCG.

4   Improving the Overall Winner Decision

In TDI, all the winners (i.e., all the queries) are considered equal when aggregating the results to establish the overall winning ranker (Eq. 1). This may include preferences that are obtained with few clicks or preferences that are not strong enough given the number of clicks collected.

To mitigate this problem, this paper proposes two variations for the Δ_{AB} score: *stat-pruning* and *stat-weight*. They assign to each query a credit inversely proportional to the probability of obtaining by chance at least the same number of clicks, assuming the two rankers are equivalent.

4.1   Statistical Hypothesis Testing

In classic TDI, to assess the overall winner between $ranker A$ and $ranker B$, the Δ_{AB} score is computed as [2]:

$$\Delta_{AB} = \frac{wins(A) + \frac{1}{2}ties(A,B)}{wins(A) + wins(B) + ties(A,B)} - 0.5 \qquad (1)$$

where:

- $wins(A)$ is the number of queries in which $ranker A$ is the winner
- $wins(B)$ is the number of queries in which $ranker B$ is the winner
- $ties(A,B)$ is the number of queries in which the two rankers have a tie

A Δ_{AB} score < 0 means $ranker B$ is the overall winner, a Δ_{AB} score $= 0$ means a tie, a Δ_{AB} score > 0 means $ranker A$ is the overall winner. While performing our replicability research we observed two problems:

- some queries have many interactions, but a very weak preference for the winning ranker
- some queries have a strong preference for the winning ranker but few interactions (the long tail)

The overall winner decision may be polluted by the aforementioned queries. The approach we suggest is to assign a different credit to each query. The idea is to exploit statistical hypothesis testing to estimate if the observations for a query are reliable and to what extent. This happens after the computation of the clicks per ranker (h_a and h_b [2]) and before the computation of the Δ_{AB} score. The theory behind our approach is statistical hypothesis testing [25]. A statistical test verifies or contradicts a null hypothesis based on the collected samples. A result has statistical significance when it is very unlikely to have occurred given the null hypothesis [18].

The *p-value* of an observed result is the probability of obtaining a result at least as extreme, given that the null hypothesis is true. The result is statistically significant, by the standards of the study, when

$$p\text{-}value <= \alpha$$

Such a scenario leads to the rejection of the null hypothesis and acceptance of the alternate hypothesis. The significance level, denoted by α is assigned at the beginning of a study [6].

To run this test we need a null hypothesis, a *p-value*, and a significance level. Our null hypothesis is that the two ranking functions we are comparing are equivalent i.e. the probability of each ranking function winning is 0.5.

For each query:

- n is the total number of clicks collected
- the winning ranker is the ranker that collected more clicks
- k is the clicks collected by the winning ranker.
- p is 0.5 (null hypothesis).

Given we are limiting our evaluation to two ranking functions:

$$k \geq \frac{n}{2}$$

When $k = \frac{n}{2}$, there is a draw, the query doesn't show any preference since each ranker collected the same amount of clicks.

The *p-value* is calculated through a binomial distribution as the probability of obtaining exactly that number of clicks k assuming the null hypothesis is true:

$$P(X = k) = \binom{n}{k} p^k (1 - p)^{n-k} \tag{2}$$

When $k > \frac{n}{2}$, the query shows a preference for a ranking function. We are testing whether the clicks are biased towards the winning ranking function, a single-tailed test is used because the winner is known already.

The *p-value* is calculated through a binomial distribution as the probability, for the winning model, to obtain at least that number of clicks k assuming the null hypothesis is true:

$$P(X \geq k) = 1 - P(X < k)$$
$$= 1 - \sum_{i=0}^{k-1} \binom{n}{i} p^i (1 - p)^{n-i} \tag{3}$$

4.2   Stat-Pruning

The first approach we designed is the simplest and most aggressive: the statistical significance of each query is determined by comparing the *p-value* with a significance level $\alpha = 0.05$. This is the standard threshold used in most statistical tests. If the *p-value* is below the threshold, the result is considered significant. The queries not reaching significance are discarded before the Δ_{AB} score calculation. The downside of this approach is that is strictly coupled to the significance level α hyper-parameter. Other works explore this aspect [19, 26].

4.3   Stat-Weight

The credit associated with each win or tie in the original Δ_{AB} score formula (Eq. 1) is a constant 1. The idea is to assign a different credit to each win and tie. This credit is the estimated probability of the win/tie to have happened not by chance.

$$credit(q_x) = 1 - p\text{-}value(q_x)$$

The *p-value* for a query q_x that presents a tie is calculated with the Eq. 2.

The *p-value* for a query q_x that presents a win is calculated with the Eq. 3 and it is normalised with a min-max normalization ($min = 0$ and $max = 0.5$) to be between 0 and 1.

The proposed updates to the Δ_{AB} score formula are the following:

$$wins(A) \Rightarrow \sum_{a=0}^{wins(A)} credit(q_a)$$

$$wins(B) \Rightarrow \sum_{b=0}^{wins(B)} credit(q_b)$$

$$ties(A, B) \Rightarrow \sum_{t=0}^{ties(A,B)} credit(q_t)$$

q_a is in the query set showing a preference for the *ranker A*.
q_b is in the query set showing a preference for the *ranker B*.
q_t belongs to the query set showing a tie.

5   Results and Analysis

5.1   Reproduction

Table 2. Query distribution: uniform 1,000 queries, click model: *perfect model*, 136 rankers (9,180 pairs)

id	NDCG	Clicks	Accuracy	Original-accuracy
1	complete	top-10	0.852	0.898
2	complete	complete	0.825	0.898
3	top-10	top-10	**0.902**	0.898

run-1 follows the same experimental setup, dataset, and parameters from the original research, but it fails to reproduce the originally recorded accuracy of TDI (see Table 2). We think that the difference can be caused by a dis alignment between the published paper and the NDCG and clicks generation parameters

Table 3. Complete NDCG and NDCG@10 averaged over the 1,000 queries vs original-paper NDCG

ranker	NDCG	NDCG@10	Original-paper NDCG
1	0.550	0.195	0.231
14	0.536	0.179	0.201
64	0.600	0.294	0.301
84	0.574	0.239	0.256
97	0.564	0.234	0.303
106	0.606	0.295	0.253
134	0.614	0.333	0.341

used at run-time. For these reasons we executed two additional runs, trying to explain the possible causes of this failed reproduction. The closer we got to the originally recorded accuracy is with *run-3*, but not exactly the same (see Table 2).

The average ground truth NDCGs calculated for the rankers, do not align with the ones reported by the original paper (Table 3).

Also, the average NDCG@10 over the 1,000 queries are closer but not exactly matching the original work ones (Table 3).

After long discussions with the original authors, we could ascertain that the NDCG formula used is the same as ours. However, we weren't able to check if the query set or NDCG cut-off correspond since it has not been possible to access the original paper code. Our best guesses are therefore the following:

- **NDCG**: the published paper clearly specifies it is the complete NDCG, but the original experiments used a different cut-off. This could explain why NDCG@10 scores are much closer to the reported ones.
- **Queries**: the query set used is not the same as our runs i.e., not the first distinct 1,000 queries as occurring in the *MSLR* dataset rows. This could explain why NDCG@10 scores are closer but not exactly the same as the reported ones.

5.2 Uniform Query Distribution

Table 4. Clicks: top-10, click model: *perfect model*, 136 rankers (9,180 pairs)

id	NDCG	Queries	Users	Accuracy		
				TDI	stat-pruning	stat-weight
4	top-10	1 000	1	**0.902**	N/A	0.886
5	top-10	100	1	**0.812**	N/A	0.790
6	top-10	100	10	0.857	0.853	**0.883**
7	complete	100	10	0.828	0.839	**0.857**

From Table 4, *run-4* and *run-5* show a better accuracy for the original TDI method. In these scenarios there are very few clicks per query, the accuracy for *stat-pruning* is not available as it removes aggressively all the queries as deemed not significant. *stat-weight* does not shine as well: it has too few clicks per query to work on. So *run-6* explores what happens if the distribution is uniform and each query is executed 10 times.

In this scenario, *stat-weight* does better than the baseline, with a 3% increase in accuracy (it correctly guessed 239 additional pairs). Comparing *run-5* and *run-6* we notice that by increasing the number of users running the queries uniformly, all the methods improve their accuracy and converge more quickly. This is expected as we get more clicks per query and it's interesting to notice that *stat-weight* is able to better handle the additional interactions discerning where they are reliable or not to identify the best ranker.

run-7 makes the task more difficult as the complete NDCG presents less difference between the rankers, so it's more challenging for the interleaving methods to guess correctly. *stat-weight* demonstrated to be more sensitive in this challenging scenario identifying correctly 267 additional pairs.

5.3   Long-Tailed Query Distribution

Table 5. Clicks: top-10, click model: *perfect model*

				Accuracy		
id	*NDCG*	*u*	*rankers*	*TDI*	*stat-pruning*	*stat-weight*
8	top-10	0.020	136	0.880	0.860	**0.897**
9	top-10	0.125	136	0.892	0.900	**0.904**
10	top-10	0.250	70	0.904	0.910	**0.911**
11	complete	0.020	136	0.827	0.817	**0.837**

From Table 5, *run-8*, *run-9* and *run-10* explore different long-tailed distributions. Due to computational limits we had to limit the amount of rankers, the closer we were getting to the original *long-tailed-1* distribution.

The steepest the long-tail, the better *stat-pruning* performs. This is expected as we assume the long part of the tail to add uncertainty for TDI, an uncertainty that is cut by the *stat-pruning* and mitigated by the *stat-weight* approach. This confirms the intuition that statistical hypothesis testing improves the classic TDI Δ_{AB} score accuracy in the long-tailed scenario.

run-11 explores again the harder problem of closer rankers with the complete NDCG. We can see that *stat-weight* confirms its sensitivity and it is able to identify correctly 91 additional pairs.

Table 6. Query distribution: long-tailed $u = 0.020$, clicks: top-10, click model: *realistic model*, 136 rankers (9, 180 pairs)

		Accuracy		
id	NDCG	TDI	stat-pruning	stat-weight
12	top-10	0.818	0.708	**0.833**
13	complete	0.782	0.693	**0.795**

5.4   Realistic Click Model

From Table 6, *run-12* introduces noisier and fewer clicks. *stat-weight* demonstrated to be robust to noise with a 1.8% increase (137 additional pairs) in comparison to the classic TDI. Finally, *run-13* tests the methods with the complete NDCG. The overall scores across the three methods are smaller, but the *stat-weight* keeps consistently the lead.

6   Conclusions and Future Directions

RQ1 has not been satisfied. Reproducing the original research turned out to be challenging from many angles: it was easy to align with the datasets but it required a substantial amount of work and discussions with the original authors to try to figure out the exact parameters and code used in the original runs. We had to design and develop from scratch the experiment code to cover all the necessary scenarios. Unfortunately, it was not possible to exactly reproduce the reported accuracy for TDI due to missing information and code unavailability.

RQ2 has been satisfied. We verified that it is possible to generalise the original TDI evaluation to long-tailed query distributions with good accuracy.

RQ3 has been satisfied. Applying statistical hypothesis testing has shown to be promising: it adapts quite well to various real-world scenarios and doesn't add any big overhead in terms of performance. *stat-weight* performs consistently well across realistic uniform and long-tailed query distributions, it's sensitive to small differences between the rankers and it is robust to noise. *stat-pruning* performs well in some realistic scenarios, but it felt generally too aggressive and too coupled with the hyper-parameter α that can be tricky to tune.

We validated the intuitions of our analysis and our proposed methods using experiments based on a simulation framework developed from scratch.

Applying *stat-weight* to other interleaved methods in real-world scenarios is an interesting direction for future works. Also calculating the query credit with different statistical approaches and normalizations could be explored. Finally, it would be interesting to run experiments with bigger numbers and many seeds to see how the different evaluation methods perform.

References

1. Balog, K., Kelly, L., Schuth, A.: Head first: living labs for ad-hoc search evaluation. In: Proceedings of the 23rd ACM International Conference on Conference on Information and Knowledge Management, pp. 1815–1818 (2014)
2. Chapelle, O., Joachims, T., Radlinski, F., Yue, Y.: Large-scale validation and analysis of interleaved search evaluation. ACM Trans. Inf. Syst. (TOIS) **30**(1), 1–41 (2012)
3. Chuklin, A., Serdyukov, P., De Rijke, M.: Click model-based information retrieval metrics. In: Proceedings of the 36th International ACM SIGIR Conference on Research and Development in Information Retrieval, pp. 493–502 (2013)
4. Cleverdon, C.W., Mills, J., Keen, E.M.: Factors determining the performance of indexing systems, (Volume 1: Design), p. 28. College of Aeronautics, Cranfield (1966)
5. Craswell, N., Zoeter, O., Taylor, M., Ramsey, B.: An experimental comparison of click position-bias models. In: Proceedings of the 2008 International Conference on Web Search and Data Mining, pp. 87–94 (2008)
6. Dalgaard, P.: Power and the computation of sample size. In: Introductory Statistics with R, pp. 155–162. Springer, Cham (2008). https://doi.org/10.1007/0-387-22632-X_8
7. Guo, F., Li, L., Faloutsos, C.: Tailoring click models to user goals. In: Proceedings of the 2009 workshop on Web Search Click Data, pp. 88–92 (2009)
8. Guo, F., Liu, C., Wang, Y.M.: Efficient multiple-click models in web search. In: Proceedings of the Second ACM International Conference on Web Search and Data Mining, pp. 124–131 (2009)
9. He, J., Zhai, C., Li, X.: Evaluation of methods for relative comparison of retrieval systems based on ClickThroughs. In: Proceedings of the 18th ACM Conference on Information and Knowledge Management, pp. 2029–2032 (2009)
10. Hofmann, K., Behr, F., Radlinski, F.: On caption bias in interleaving experiments. In: Proceedings of the 21st ACM International Conference on Information And Knowledge Management, pp. 115–124 (2012)
11. Hofmann, K., Whiteson, S., De Rijke, M.: A probabilistic method for inferring preferences from clicks. In: Proceedings of the 20th ACM International Conference on Information and Knowledge Management, pp. 249–258 (2011)
12. Järvelin, K., Kekäläinen, J.: Cumulated gain-based evaluation of IR techniques. ACM Trans. Inf. Syst. (TOIS) **20**(4), 422–446 (2002)
13. Järvelin, K., Kekäläinen, J.: IR evaluation methods for retrieving highly relevant documents. In: ACM SIGIR Forum, vol. 51, pp. 243–250. ACM New York, NY, USA (2017)
14. Joachims, T.: Optimizing search engines using clickthrough data. In: Proceedings of the Eighth ACM SIGKDD International Conference on Knowledge Discovery and Data Mining, pp. 133–142 (2002)
15. Joachims, T., et al.: Evaluating retrieval performance using clickthrough data (2003)
16. Kharitonov, E., Macdonald, C., Serdyukov, P., Ounis, I.: Using historical click data to increase interleaving sensitivity. In: Proceedings of the 22nd ACM International Conference on Information & Knowledge Management, pp. 679–688 (2013)
17. Kharitonov, E., Macdonald, C., Serdyukov, P., Ounis, I.: Generalized team draft interleaving. In: Proceedings of the 24th ACM International on Conference on Information and Knowledge Management, pp. 773–782 (2015)

18. Myers, J.L., Well, A.D., Lorch, J.: Developing the fundamentals of hypothesis testing using the binomial distribution. Research Design and Statistical Analysis, pp. 65–90 (2010)
19. Queen, J.P., Quinn, G.P., Keough, M.J.: Experimental Design and Data Analysis for Biologists. Cambridge University Press, Cambridge (2002)
20. Radlinski, F., Craswell, N.: Optimized interleaving for online retrieval evaluation. In: Proceedings of the Sixth ACM International Conference on Web Search and Data Mining, pp. 245–254 (2013)
21. Radlinski, F., Kurup, M., Joachims, T.: How does clickthrough data reflect retrieval quality? In: Proceedings of the 17th ACM Conference on Information and Knowledge Management, pp. 43–52 (2008)
22. Schuth, A., et al.: Probabilistic multileave for online retrieval evaluation. In: Proceedings of the 38th International ACM SIGIR Conference on Research and Development in Information Retrieval, pp. 955–958 (2015)
23. Schuth, A., Hofmann, K., Radlinski, F.: Predicting search satisfaction metrics with interleaved comparisons proceedings of the 38th international ACM SIGIR Conference on Research and Development in Information Retrieval, Santiago, Chile, 9–13 August 2015, Ricardo. ACM (2015)
24. Schuth, A., Sietsma, F., Whiteson, S., Lefortier, D., de Rijke, M.: Multileaved comparisons for fast online evaluation. In: Proceedings of the 23rd ACM International Conference on Conference on Information and Knowledge Management, pp. 71–80 (2014)
25. Sirkin, R.M.: Statistics for the Social Sciences. Sage, London (2006)
26. Sproull, N.L.: Handbook of Research Methods: A Guide for Practitioners and Students in the Social Sciences. Scarecrow Press, Metuchen (2002)

A Reproducibility Study of Question Retrieval for Clarifying Questions

Sebastian Cross$^{(\boxtimes)}$, Guido Zuccon, and Ahmed Mourad

The University of Queensland, St Lucia, Australia
`sebastian.cross@uq.net.au`, `g.zuccon@uq.edu.au`

Abstract. The use of clarifying questions within a search system can have a key role in improving retrieval effectiveness. The generation and exploitation of clarifying questions is an emerging area of research in information retrieval, especially in the context of conversational search.

In this paper, we attempt to reproduce and analyse a milestone work in this area. Through close communication with the original authors and data sharing, we were able to identify a key issue that impacted the original experiments and our independent attempts at reproduction; this issue relates to data preparation. In particular, the clarifying questions retrieval task consists of retrieving clarifying questions from a question bank for a given query. In the original data preparation, such question bank was split into separate folds for retrieval – each split contained (approximately) a fifth of the data in the full question bank. This setting does not resemble that of a production system; in addition, it also was only applied to learnt methods, while keyword matching methods used the full question bank. This created inconsistency in the reporting of the results and overestimated findings. We demonstrate this through a set of empirical experiments and analyses.

1 Introduction

Creating a single query that is complex and detailed enough to retrieve the required information accurately is a difficult task. Failure often requires users to recreate and rewrite the query several times to get their desired information. This issue has led to the development of systems designed to assist the user with query formulation [6,11,15,24–26]. These systems implement multiple methods, one being asking for *clarifying questions* [3]. Clarifying questions help to identify a user's information-seeking intent by identifying if their query meets an ambiguity threshold [12]. If this is the case, the system poses a clarifying question to the user, expecting their answer to clarify aspects of their query. Asking clarifying questions has been recognised as an increasingly useful feature for conversational search [1–4,12,13,20,27,29]. In this context, the search agent often can only present a limited set of results (e.g., one, a handful, or even an answer synthesised from some top results) and thus the need for clear intent-driven queries is further exacerbated.

Developing methods for asking clarifying questions has become a recent focus in information retrieval, with Aliannejadi et al.'s work [3] being a key milestone.

J. Kamps et al. (Eds.): ECIR 2023, LNCS 13982, pp. 35–50, 2023.
https://doi.org/10.1007/978-3-031-28241-6_3

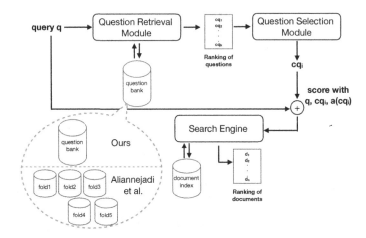

Fig. 1. Overview of a retrieval pipeline that exploits clarifying questions. Our work specifically focuses on the question retrieval module. In red, we highlight the key issue related to data preparation in the context of Aliannejadi et al.'s paper we reproduce [3]: we show the difference between our data preparation (i.e. use the whole question bank to retrieve against) and Aliannejadi et al.'s data preparation (i.e. divide the question bank into 5 folds to retrieve against). (Color figure online)

This work modelled solutions for clarifying questions as a two-step process: question retrieval and question selection. They then contributed methods that adhere to the two-step process along with an evaluation methodology and a rich dataset (Qulac). The work by Aliannejadi et al. [3] has been key for the establishment of the task of clarifying questions and has been relied upon by many others to build upon their research [2,4,10,12,13,20,27,29,31,35,39].

Our paper aims to reproduce that original work [3], and focuses specifically on the key step of *question retrieval*, see Fig. 1. The question retrieval module receives the user query as input and retrieves from a question bank a list of candidate clarifying questions. The original paper explored a number of methods for tackling this task. Methods implementation, data preparation and raw results were not made available with the original work, and we sought to then create reference implementations and data preparations for others to build upon and compare – with the intention of building ourselves our future work on this task based on these implementations. In reproducing this work, and working closely with two of the original authors who shared their data, we were able to identify a key divergence between common data preparation practice and what the authors did. Data preparation was necessary as 5-fold cross validation was used in the empirical experiments. The original authors partitioned into folds both the query set and the target data for retrieval (i.e. the index of clarifying questions), while standard practice in information retrieval is to partition the query set only, and instead maintain the index on which retrieval is performed unchanged across the folds. The data preparation they adopted was not documented explicitly in the original paper. This issue, as we show in this paper, leads to a considerable overestimation of the effectiveness of some of the methods studied in the paper.

In addition to this key issue, we then further identified other issues related to the features used for learning to rank and ties in retrieval scores – which however had less impact on results. Overall, the data preparation issue has important implications because many others have built upon Aliannejadi et al.'s milestone paper, but apparently did so without reproducing that work.

The remainder of the paper is organised as follows. In Sect. 2 we introduce the task of question retrieval for clarifying questions, along with the methods originally proposed by Aliannejadi et al., which adapted well-known information retrieval methods to the retrieval of questions from a question bank [3]. In Sect. 3 we describe the key issue regarding data preparation present in the original work, along with what we argue a more common and realistic[1] data preparation for this task should have been. Section 4 lists the experimental settings used to reproduce the original work and to study the differences in results caused by the differences in data preparation. We then report and analyse the experimental results in Sect. 5. For this, we develop the analysis along 3 main directions: (1) the effectiveness of methods on the data preparation of Aliannejadi et al., (2) their effect on what we repute as the correct data preparation for this task, and (3) an analysis of issues related to the topics and question-bank, their representation and the difference that these generate across keyword matching and learnt models.

2   Question Retrieval for Clarifying Questions

2.1   The Question Retrieval Task

Figure 1 provides an overview of a system that uses clarifying questions to improve the search effectiveness of document retrieval. In the context of this paper, we focus on the first module: the question retrieval module. In this context, the question retrieval module takes as input the original user query q and uses it to retrieve candidate clarifying questions organised as a ranking $\mathcal{R}_{cq} = <cq_1, cq_2, \ldots, cq_k>$ from a question bank $\mathcal{QB}$. We note that the question bank $\mathcal{QB}$ is not tailored a-priori to the query q: that is, $\mathcal{QB}$ contains a large set of clarifying questions, some of which applicable to any of the users' queries (or any of the queries that are deemed to needing a clarifying question). This is an important aspect to stress because, as we shall discuss in Sect. 3, it should be reflected in how the data in the $\mathcal{QB}$ should be prepared for experimentation. We further note that the question retrieval module would not be necessary if questions were generated on the fly given the input (e.g., via a generative language model), rather than retrieved from a question-bank [1,32,35,38]; this is however not the setting considered by the reference paper we aim to reproduce.

2.2   Methods for Question Retrieval

In the original paper, Aliannejadi et al. adapted the question retrieval task methods from three broad families of retrieval models: keyword-matching models [23, 36], learning to rank models [16,19], and transformer-based models [17,30].

---

[1] In that it resembles what a production system may look like.

As keyword matching models, they considered the common Query Likelihood (QL) with Dirichlet smoothing [36], BM25 [23], and the use of RM3 pseudo-relevance feedback method on top of QL [14].

As learning to rank models, they considered LambdaMART and RankNet [16]; both are pairwise models. As features to represent a query-question pair, they used the QL, BM25 and RM3 scores. This representation choice should be kept in mind, as we further analyse the implication of this later when examining the results: while this specific choice is not the focus of our paper, we do argue that this choice is problematic if considering the broader generalisation of these learning to rank methods for question retrieval.

As transformer model, the original authors introduced BERT-LeaQuR, which constituted one of the key original contributions of that work. From the description of BERT-LeaQuR, we understand that the model structure is similar to that of a typical cross-encoder ranker like monoBERT [22], but where the pre-trained language model is directly fine-tuned on the target dataset for question retrieval (see Sect. 4 for a description of the dataset). To clarify the implementation of BERT-LeaQuR we contacted the original authors to also acquire the corresponding source code. We were however told that one of their follow-up work yielded a stronger transformer model [1,2], and we were advised to use that for reproduction instead of BERT-LeaQuR.

3   Issues with Data Preparation

The experiments in the original paper used the Qulac question bank. According to the original paper, this question bank contained 2,649 clarifying questions, but the question bank made available by the authors in the repository associated with the paper contained 2,593 questions; we used this available question bank. These questions were assembled through crowdsourcing tasks and with respect to 198 target queries from the TREC 2009, 2010, and 2011 Web Track collection [9,28]. For each topic, only a small subset of the clarifying questions in the question bank is relevant to the specific topic.

3.1   Early Investigation of Reproducibility

In our early attempts to reproduce the question retrieval methods from the original paper, we were failing to obtain similar results as in the original experiments for the learnt models (LabdaMART, RankNet, BERT). On the other hand, we were able to obtain similar results for the keyword-matching models. In particular, our results on some of the learnt models had lower effectiveness compared to the keyword matching models: an unexpected result, especially in light of the results reported in the original paper. This triggered an in-depth analysis of the dataset and runs. Yet, we could not identify specific faults in our implementations or use of toolkits such as RankLib[2].

---

[2] https://sourceforge.net/p/lemur/wiki/RankLib.

We then decided to contact the authors for advice. While they also could not identify why we were unable to reproduce the results, with a genuine collaborative and supportive spirit, they were able to retrieve from back-ups and share the feature files and the saved models they created for learning to rank. We then turned to examine these files. We started by running their saved models on their test files, which returned similar results to those originally reported. We then retrained the learning to rank models using our settings and their dataset files (train, validation and test files) – obtaining the same models and results of the original experiment. Yet, we were unable to reproduce the results if we changed to our dataset files (train, validation and test files).

While we expected minor differences in effectiveness due to different random splitting[3] of topics into train, validation and test files (the dataset was split into 5 folds to allow for 5-fold cross-validation), we could not reconcile this being the reason for the remarkable drop in effectiveness, rendering the trends we observed being widely different from those in the original work. This then triggered a review of the train, validation and test files for the learning to rank models.

3.2 Analysis of Data Preparation and Differences Identified

The train, validation and test files for learning to rank contained a list of query-question pairs for several topics. Each pair was represented by three features: the BM25, QL and RM3 scores. A binary label was associated with each pair: 1 if the clarifying question was for that query-question pair, 0 otherwise. For this data, pairs originated from the Qulac question bank.

When we examined and compared the original and our train, validation and test files we identified two differences.

The first was a minor difference. Our features were the BM25, QL and RM3 scores as computed by the models (or, more precisely, by the Anserini toolkit[4]). This meant for example that if a candidate clarifying question did not contain any query term, the BM25 score we assigned to the question and therefore we used for the corresponding feature was 0. This was not the case however in the files given to us by Aliannejadi et al. In these files, retrieval scores appeared to have been smoothed – we believe by adding an ad-hoc $\epsilon \neq 0$ value (akin to Laplace smoothing in language modelling [37]). While this smoothing has, in practice, no effect on the learning to rank models that were created, this highlighted as most of the query-question pairs for a topic had a feature representation that was zero-valued (in our file) – so many pairs had the same exact representation. We comment on this aspect in Sect. 5.3.

The second instead was a major difference. The experimental setup used by Aliannejadi et al. required to train and test learnable models (learning to rank, transformers) using a 5-fold cross-validation setup (60% train, validation 20%,

---

[3] Note this was true in early experiments, but in the experiments reported in this paper, we were able to reproduce the exact split of topics into folds as they had.

[4] We note that different information retrieval toolkits follow different reference implementation of some of the keyword matching methods, e.g. of BM25.

test 20%). Therefore, when we prepared the data, we created 5 folds by dividing the topics. However, we did not divide the question bank into folds. This meant that, for example, the test file for a fold contained 40 topics (queries). For each topic, the file contained 2,593 candidate query-question pairs, i.e. all the possible clarifying questions in the question bank. This is akin to the common practice in information retrieval when performing n-fold cross-validation: topics are divided into n folds, but retrieval occurs over the whole candidate set (the entire index[5]).

The setup we produced in our data preparation mimics that of a production system. In this case, a question bank would not be limited to a set of queries. Attempts would be made instead to source questions that can cover a large portion of queries that users would issue. Thus, when experimenting with methods for question retrieval, the entire set of candidate clarifying questions should be considered, i.e. retrieval should take place from the whole question bank.

However, when examining the train, validation and test files from the original work, we noticed that: (1) each of the train, test, and validation not only contained a subset of topics, as expected, but it also contained just a subset of all candidate clarifying questions, (2) these subsets always contained all the relevant questions for a given topic, (3) the folds did not contain 2,593 clarifying questions as in their released question bank, but 2,609, and we could not exactly map one to the other because of different identifiers been used. In other words, while our data files contained, for each topic, the same 2,593 clarifying questions, their data files contained on average 1,558.8, 521.8, and 521.8 clarifying questions for train, validation and test files, respectively. These statistics are clearly compared side by side in Table 1. This difference, also visualised in Fig. 1, had two implications, which we discuss in Sect. 3.3.

We already highlighted the difference between the question bank statistics reported in the original paper (size: 2,639 questions), the learning to rank files provided to us (2,609), and the question bank made available publicly in their repository[6] (2,593). When asked, the authors recalled that they modified the data after publication of the original paper. Our experiments, including when examining both our and their data partition, are based on the question bank with 2,593 questions. We, therefore, expect *minor* differences in evaluation metrics' absolute values when compared to the results reported in the original paper.

3.3 Implications of the Differences in Data Preparations

The first implication is that at *training time*, for any given topic, we provide to the model an average of 13.1 relevant clarifying questions and 2,579.9 non-relevant clarifying questions. In Aliannejadi et al.'s experiments, instead, because

---

[5] Note that commonly in learning to rank, feature files are created for the top-k candidate documents. This however is not because retrieval only considers k documents. Learning to rank is unfeasible for large collections, and is therefore part of a cascade pipeline where full index retrieval occurs first with a cheaper model, and then learning to rank is applied to the top-k. Yet, retrieval considers the full index, not an arbitrary subset that – what the chances – contains all relevant documents.

[6] https://github.com/aliannejadi/qulac.

Table 1. Statistics of the original data preparation (folds) by Aliannejadi et al. [3] compared to Ours. The number of topics for train/validation/split are the same across the two preparations. Differences are found with regards to the number of clarifying questions per topic. Note that Aliannejadi et al.'s data preparation statistics computed on the learning to rank data they provided differ from those reported in their paper: they reported having 2,639 questions in total in their question-bank, while we could count only 2,609; yet the question bank released in their repository contained 2,593.

	Average number of clarifying questions per topic		Average number of topics per fold	
	Aliannejadi et al.	Ours	Aliannejadi et al.	Ours
Train	1,558.8	2,593	118.8	118.8
Validation	521.8	2,593	39.6	39.6
Test	521.8	2,593	39.6	39.6

of the different way of preparing the data, they provided the same number of *relevant* clarifying questions as we do, but *far less non-relevant* clarifying questions (on average, 523.9). This may make our learnt models weaker than theirs because our training data is much more imbalanced. However, we believed this not to be the case, because we observed most of the non-relevant clarifying questions have a feature representation that consists of zero-valued weights (i.e. there are no matching keywords in the questions).

The second and most important implication is that at *retrieval time (test)*, our learnt model has to rank 2,593 candidate clarifying questions per topic, of which, on average, only 13.1 are relevant. The learnt model in Aliannejadi et al.'s experiments instead had to rank only 537 candidate clarifying questions per topic, despite the number of relevant candidate clarifying questions per topic being the same. This means that our ranker is more likely to obtain lower effectiveness than their ranker. But this is not necessarily because it is a worse ranker. In fact, given a ranker $\mathcal{R}$ and an equal number of relevant candidate questions across two candidate sets S_1 and S_2, with $S_2 \supseteq S_1$ (and thus also $|S_1| < |S_2|$, where $|S_1|$ is the size of S_1), $\mathcal{R}$ is more likely to produce a ranking with higher effectiveness when applied to S_1 than when applied to S_2, a (significantly, in the case of these experiments) larger superset of candidate questions than S_1. This is obvious for example when considering a relevant question with a zero-valued feature representation. In this case, $\mathcal{R}$ ranks the question at the bottom of the ranking[7]. In our data preparation, this means this relevant question is ranked at rank 2,593. However, in Aliannejadi et al.'s data preparation this same question, when ranked at the bottom of the ranking, would be at rank $\approx$521.8. This difference would translate into a sizable difference in effectiveness as measured for example by MAP. In fact, in the case of our data preparation, the MAP's gain [7] contributed by this question is 3.8565×10^{-4}, while in the case of Aliannejadi et al. is 1.9164×10^{-3} – one order of magnitude larger contribution to MAP.

---

[7] Possibly tied with other questions that also have a zero-valued feature representation, which, in the dataset considered, are the majority of them.

Given these differences in data preparations, we asked ourselves if this could be the reason why we could not reproduce the original results, or at least trends. To investigate these aspects and the empirical differences induced by the two data preparations, we set up the experiments described and analysed in Sect. 5.

4   Experimental Settings

Datasets. The Qulac dataset used for the question bank and the topics have already been described in Sect. 3. Topics were split into folds for 5-fold cross-validation following those supplied by Aliannejadi et al.[8]. Additionally, we created two question bank preparations: one containing all clarifying questions from the question bank (referred to as Ours), and one following the division of clarifying questions into separate folds (referred to as Aliannejadi et al.).

Evaluation Metrics. We chose to use the same evaluation metrics as the original work. For question retrieval, these are: mean average precision (MAP) and recall for the top 10, 20, and 30 retrieved questions (Recall@10, Recall@20, Recall@30). In addition, we also report Success@1 and Precision@5: this is to understand the suitability of the rankings produced by question retrieval if questions were issued to users (without further refinement from the question selection model). In such a case, it is likely that 1 to 5 questions are asked to a user in a conversational or web setting.

We used the widespread `trec_eval` tool for computing metrics. However, `trec_eval` has an odd treatment of items with a tied score: the rank position information in the result file is discarded, and tied items are ordered alphanumerically [5,18,21,34]. This is often ignored in information retrieval experiments, however, this also arises as a (minor) issue in the experiments in this paper. We explain why this is the case in Sect. 5.3. To avoid ties, we post-process all results, including those from learning to rank, to assign to each question a unique score such that decreasing ranks correspond to decreasing scores.

For statistical significance analysis, we use a paired two-tails t-test with Bonferroni correction and regard a difference as significant if $p < 0.05$; however no significant differences were found in our experiments.

Models Implementation. For the keyword matching models, the original authors used the implementations from the Galago search engine toolkit [8]. In our reproduction, we instead use the implementation of these methods from the Anserini/Pyserini toolkit [33]. The use of these different libraries is likely to have caused minor differences in the runs produced, e.g., due to implementations, parameter settings[9], stemming and stop-listing[10]. Because of this difference in

---

[8] Once we obtained the feature files for learning to rank, we knew which topics were grouped together in which fold, and thus could recreate the same topic-wise division.

[9] Ours: (BM25) $k_1 = 0.9$, $b = 0.4$, (QL) $\mu = 1000$, (RM3) $fb_{terms} = 10$, $fb_{docs} = 10$ $original\_query\_weigh = 0.5$. They do not report parameter values.

[10] We used Porter Stemmer and Anserini's default stop-list. They do not report their settings.

toolkits, it is important to not directly compare the absolute numbers obtained by the methods in the original work vs. in our reproduction: comparison should instead take place with respect to the trends that are observed when comparing across models.

For the learning to rank models, we used the RankLib toolkit, as did Alian-nejadi et al.[11]. They do not indicate if feature normalisation was performed. We experimented with the three normalisations made available in the toolkit (zscore, sum, and linear) and no normalisation. We used those that gave us the best effectiveness and were closer to the original values: no normalisation for LambdaMART and zscore for RankNet. As per features, we directly used the scores of QL, BM25 and RM3, with no further processing (e.g., smoothing) as no further processing was reported by Aliannejadi et al. Regardless, we found in testing that smoothing did not affect results.

For the BERT model, we used the `bert-base-uncased`[12] checkpoint made available by the Huggingface library. The architecture of the model resembled that of a monoBERT cross-encoder ranker [22], with the difference that inputs were pairs of query-question rather than query-document. The checkpoint was then fine-tuned on the Qulac dataset; fine-tuning occurred on the training por-tion of a fold. The implementation of this method was made publicly available[13] by Aliannejadi et al. in a separate publication [1,2].

5 Experiments and Results Analysis

Next, we report and analyse the experimental results obtained when attempting to reproduce the question retrieval component from Aliannejadi et al. For this, we develop the analysis along 3 main directions, as indicated in Sect. 1.

5.1 Experiment 1: Aliannejadi et al.'s Data Preparation

We start by attempting to replicate the results reported by Aliannejadi et al., using their data preparation based on the splitting of the question bank into subsets across train, validation and test. Our results should be compared with Table 2 of the original paper. We do not expect to obtain the same exact values of evaluation measures: (i) we know minor differences would be present because of Galago vs. Anserini – this may influence absolute values, but not the trends, (ii) their BERTleaQuR and the BERT cross-encoder we implemented may have minor differences, (iii) they reported their question-bank having 2,639 but the one we have access to has 2,593. However, we expect to observe the same trends, i.e. that learning to rank methods are superior to keyword-matching methods, and that the BERT-based method largely outperforms all others. We believe they have a mismatch in the data in the paper and associated repository, but we confirmed they ran their experiments on the data they gave us. Also, if there was

---

[11] We used version 2.17; Aliannejadi et al. did not report the version.

[12] https://huggingface.co/bert-base-uncased.

[13] https://github.com/aliannejadi/ClariQ.

Table 2. Question retrieval results for Aliannejadi et al.'s data preparation, which splits the question bank into 5 folds. These results are the replication of the results reported by Aliannejadi et al.'s original work in their Table 2 [3].

Aliannejadi et al. data preparation						
Method	MAP	Recall@10	Recall@20	Recall@30	Success@1	Precision@5
QL	0.7183	0.6426	0.7376	0.7394	0.9795	0.9329
BM25	0.7198	0.6426	0.7376	0.7393	0.9795	0.9380
RM3	0.7198	0.6426	0.7376	0.7393	0.9795	0.9380
LambdaMART	0.7274	0.6299	0.7253	0.7323	0.9697	0.9364
RankNet	0.7406	0.6352	0.7372	0.7498	0.9697	0.9354
BERT	0.8352	0.6868	0.8345	0.8673	0.9848	0.9657

a mismatch, it would likely affect only a handful of queries – regardless, it would be expected to impact absolute values but not trends. Results are reported in Table 2; we make the following observations:

1. QL, BM25, RM3: we were able to obtain consistently higher effectiveness than that reported in the original paper, across all metrics (e.g., for MAP they reported QL: 0.6714, BM25: 0.6715, RM3: 0.6858 [3]). While explanations for this could be because of points (i) and (iii) above, we do not believe these are the core reasons. Instead, we believe Aliannejadi et al. did not execute the keyword matching retrieval against the same data preparations (and thus, subdivisions of the question bank) they use for the learnt models (i.e. the ones used in these experiments). Instead, we believe the results they reported for keyword matching methods were obtained against the whole question bank: this is the setup we argue should have been used to evaluate *all* methods. We investigate this setup in Sect. 5.2. We show that in that setup we obtain effectiveness values for keyword-matching methods that are much closer to the ones they originally reported.
2. LambdaMART and RankNet: we were not able to obtain the same effectiveness reported in the original paper, but values are close (e.g., the reported MAP for LambdaMART is 0.7218, for RankNet is 0.7304 [3]). Differences may likely be due to the feature files they used for the paper containing more questions than the ones they gave us. The absence in our question bank of these additional questions made that effectiveness higher: intuitively this is because most of them are not relevant for most topics, and thus removing them improves rankings if they appeared before a relevant question.
3. BERT: we obtained values that are close to the ones they reported in the original paper (e.g., MAP 0.8349 [3]). While there were minor differences, we ascribe these differences mainly to points (ii) and (iii) above.

Overall, with minor discrepancies, we were able to obtain similar results to the ones reported in the original paper for the learned models (LambdaMART, RankNet, BERT), but not for the keyword-matching models (QL, BM25, RM3). Trends across models were as they reported: BERT is the best model, followed by

Table 3. Question retrieval results for our data preparation. These results strongly differ from those reported by the original work of Aliannejadi et al. in their Table 2 [3] for the learned methods, i.e. LambdaMART, RankNet, BERT.

Our data preparation						
Method	MAP	Recall@10	Recall@20	Recall@30	Success@1	Precision@5
QL	0.6975	0.6152	0.7218	0.7238	0.9442	0.9177
BM25	0.6979	0.6167	0.7201	0.7321	0.9492	0.9187
RM3	0.6979	0.6167	0.7201	00.7321	0.9492	0.9187
LambdaMART	0.6728	0.5882	0.6947	0.7068	0.9394	0.8889
RankNet	0.6851	0.6028	0.7051	0.7171	0.9293	0.9020
BERT	0.7512	0.6349	0.7686	0.7979	0.9596	0.9131

the learning to rank methods, with the keyword matching models being the worst – though gains over keyword matching models were not as large as those they reported. We believe they however incorrectly reported the values for keyword-matching models. Specifically, we believe values for keyword matching models were obtained when retrieving on the whole question bank, rather than the smaller splits they created for the learnt methods, see next.

5.2 Experiment 2: Our Data Preparation

We now consider our data preparation, where question retrieval occurs against a unique question bank, which contains all possible clarifying questions for all topics. Results are reported in Table 3; we make the following observations:

1. QL, BM25, RM3: the results we obtained when searching on the whole question bank appear to be more akin to those Aliannejadi et al. reported for their experiments (e.g., for MAP they reported QL: 0.6714, BM25: 0.6715, RM3: 0.6858 [3]) than in the data preparation setup of Sect. 5.1. Differences could be ascribed to tools (Anserini vs. Galago), model parameters, and question bank size.
2. LambdaMART and RankNet: our data preparation setup led to learning to rank methods achieving lower effectiveness than keyword matching models. This is the opposite trend of that reported in the original work, and also is opposite to what we found for Aliannejadi et al.'s data preparation in Sect. 5.1.
3. BERT: we found that on our data preparation, BERT performed worst than on theirs. While BERT was still the best method across all those considered, the gains over keyword matching were sensibly lower, e.g., for MAP a 7.64% gain in ours vs. 24.33% in theirs compared to BM25.

Overall, the results of these experiments (Table 3) differ greatly from those reported for Aliannejadi et al. 's data preparation (Table 2). Specifically, we found that learning to rank models cannot outperform keyword matching when retrieval occurs on the whole question bank, and in this setting, BERT does provide improvements over keyword matching, but not at the rate reported. Importantly,

Table 4. Statistics of the data preparations by Aliannejadi et al. and ours. While both data preparations have the same number of relevant documents and relevant documents with keyword matching score equal to zero, they consistently differ in terms of the number of non-relevant documents with keyword matching score above zero.

Statistic	Aliannejadi et al	Ours
avg. # relevant questions per query	13.1	13.1
avg. # relevant questions per query with zero score	3.5	3.5
avg. # non-relevant questions per query with score > 0	1.9	13.3

these improvements are not statistically significant. This result empirically demonstrates that the two data preparations lead to different estimations of effectiveness and different overall findings.

5.3 Further Analysis: Zero Scores, Ties

Next, we analyse two aspects of the data and experiments of this paper: the use of keyword matching scores as only features in learning to rank, and the presence of tied scores in the rankings.

Zero Score. The scores of QL, BM25 and RM3 are used in the learning to rank methods as only features for representing query-question pairs. This resulted in two characteristics arising: (1) there were a number of pairs with the same non-zero representation, (2) many pairs had a representation that was zero-valued for all three features. Characteristic 2 occurred often for non-relevant questions, but sporadically it occurred also for relevant questions: in fact on average each query had 3.5 relevant questions that had their features being all zeros (see Table 4). This fact, combined with the fact that items that had a non-zero feature representation often had their representation been the same as another item, meant that the learning to rank methods often ended up assigning to pairs at test time one of two scores: 0 or 1. This caused many ties in the ranking (see below). The analysis of the features files also revealed another problem when comparing Aliannejadi et al.'s data preparation and ours: in theirs on average there was only 1.9 non-relevant questions that had features that were not all zeros, while in ours there were 13.3 – and this would have made ranking harder in our data preparation.

Ties. As mentioned above, the keyword matching methods and the learning to rank methods produced rankings with a large number of ties. The trec_eval tool behaves oddly when ties are present (see Sect. 4), while RankLib considers the actual rank position recorded in the ranking file. We are unsure which tool Aliannejadi et al. used to report their results, and if mixing trec_eval for keyword matching and BERT methods and RankLib for learning to rank, the evaluation would have been inconsistent. We show this in Table 5 for Lamba-MART. In our evaluations we transformed scores as a function of their rank and used trec_eval, so that no ties were present.

Table 5. Differences in MAP between evaluation tools when analysing the LambdaMART run on our data preparation: RankLib evaluation, `trec_eval`, and `trec_eval` with ties removed by converting scores to a function of rank.

RankLib eval	0.6728
`trec_eval`	0.7233
`trec_eval` no ties	0.6728

6 Conclusions

The use of clarifying questions within a search system can have a key role in improving retrieval effectiveness and user interaction with the system, especially if this is a conversational search system. In this paper, we attempted to reproduce the work by Aliannejadi et al., which is a key milestone in the area of clarifying questions for search. Working closely with the original authors and thanks to their sharing of data, we identified a fundamental issue related to data preparation. In particular, their practice of dividing the question bank containing clarifying questions into folds is, we believe, unrealistic for a production system, and is also different from standard information retrieval experimentation practice. Throughout our experiments and analyses, we have shown how this issue affected the results reported in the original work.

We found that learning to rank models cannot outperform keyword matching when retrieval occurs on the whole question bank. We also found that while BERT does outperform keyword matching methods in this setting, it does so with much smaller gains than what was originally reported and, importantly, these differences are not statistically significant. We do not believe this is a generalisable result. Specifically, we believe this result is due to: (i) the amount of training data being too little for those models, especially for BERT, and (ii) the feature representation being particularly poor for learning to rank models, where most questions had identical representation. We would expect that if these two points were addressed, learnt models would provide consistently better results than keyword matching, as it often occurs in other information retrieval tasks.

This work demonstrates that it is critical to be able to communicate and share resources among researchers to facilitate the reproduction of methods and results and the identification of possible factors that may have influenced results beyond the intentions of the original researchers.

We make code, data preparations, run files and evaluation files publicly available at www.github.com/ielab/QR4CQ-question-retrieval-for-clarifying-questions.

Acknowledgments. This work was partially supported by Australian Research Council DECRA Research Fellowship (DE180101579).

References

1. Aliannejadi, M., Kiseleva, J., Chuklin, A., Dalton, J., Burtsev, M.: ConvAI3: Generating Clarifying Questions for Open-Domain Dialogue Systems (ClariQ). arXiv:2009.11352 (2020)
2. Aliannejadi, M., Kiseleva, J., Chuklin, A., Dalton, J., Burtsev, M.: Building and evaluating open-domain dialogue corpora with clarifying questions. In: Proceedings of the 2021 Conference on Empirical Methods in Natural Language Processing, pp. 4473–4484 (2021)
3. Aliannejadi, M., Zamani, H., Crestani, F., Croft, W.B.: Asking clarifying questions in open-domain information-seeking conversations. In: Proceedings of the 42nd International ACM SIGIR Conference on Research and Development in Information Retrieval, pp. 475–484 (2019)
4. Bi, K., Ai, Q., Croft, W.B.: Asking clarifying questions based on negative feedback in conversational search. In: Proceedings of the 2021 ACM SIGIR International Conference on Theory of Information Retrieval, pp. 157–166 (2021)
5. Cabanac, G., Hubert, G., Boughanem, M., Chrisment, C.: Tie-breaking bias: effect of an uncontrolled parameter on information retrieval evaluation. In: Agosti, M., Ferro, N., Peters, C., de Rijke, M., Smeaton, A. (eds.) CLEF 2010. LNCS, vol. 6360, pp. 112–123. Springer, Heidelberg (2010). https://doi.org/10.1007/978-3-642-15998-5_13
6. Cai, F., De Rijke, M., et al.: A survey of query auto completion in information retrieval. Found. Trends® Inf. Retrieval 10(4), 273–363 (2016)
7. Carterette, B.: System effectiveness, user models, and user utility: a conceptual framework for investigation. In: Proceedings of the 34th International ACM SIGIR Conference on Research and Development in Information Retrieval, pp. 903–912 (2011)
8. Cartright, M.A., Huston, S.J., Feild, H.: Galago: a modular distributed processing and retrieval system. In: Proceedings of the SIGIR 2012 Workshop on Open Source Information Retrieval, pp. 25–31 (2012)
9. Clarke, C.L., Craswell, N., Soboroff, I.: Overview of the TREC 2009 web track. In: Proceedings of TREC (2009)
10. Dubiel, M., Halvey, M., Azzopardi, L., Anderson, D., Daronnat, S.: Conversational strategies: impact on search performance in a goal-oriented task. In: The Third International Workshop on Conversational Approaches to Information Retrieval (2020)
11. Fails, J.A., Pera, M.S., Anuyah, O., Kennington, C., Wright, K.L., Bigirimana, W.: Query formulation assistance for kids: what is available, when to help & what kids want. In: Proceedings of the 18th ACM International Conference on Interaction Design and Children, pp. 109–120 (2019)
12. Kim, J.K., Wang, G., Lee, S., Kim, Y.B.: Deciding whether to ask clarifying questions in large-scale spoken language understanding. In: 2021 IEEE Automatic Speech Recognition and Understanding Workshop (ASRU), pp. 869–876. IEEE (2021)
13. Krasakis, A.M., Aliannejadi, M., Voskarides, N., Kanoulas, E.: Analysing the effect of clarifying questions on document ranking in conversational search. In: Proceedings of the 2020 ACM SIGIR on International Conference on Theory of Information Retrieval, pp. 129–132 (2020)
14. Lavrenko, V., Croft, W.B.: Relevance-based language models. In: ACM SIGIR Forum, vol. 51, pp. 260–267. ACM, New York (2017)

15. Lee, C.-J., Lin, Y.-C., Chen, R.-C., Cheng, P.-J.: Selecting effective terms for query formulation. In: Lee, G.G., et al. (eds.) AIRS 2009. LNCS, vol. 5839, pp. 168–180. Springer, Heidelberg (2009). https://doi.org/10.1007/978-3-642-04769-5_15
16. Li, H.: Learning to rank for information retrieval and natural language processing. Synth. Lect. Hum. Lang. Technol. **7**(3), 1–121 (2014)
17. Lin, J., Nogueira, R., Yates, A.: Pretrained transformers for text ranking: BERT and beyond. Synth. Lect. Hum. Lang. Technol. **14**(4), 1–325 (2021)
18. Lin, J., Yang, P.: The impact of score ties on repeatability in document ranking. In: Proceedings of the 42nd International ACM SIGIR Conference on Research and Development in Information Retrieval, pp. 1125–1128 (2019)
19. Liu, T.Y., et al.: Learning to rank for information retrieval. Found. Trends® Inf. Retrieval **3**(3), 225–331 (2009)
20. Lotze, T., Klut, S., Aliannejadi, M., Kanoulas, E.: Ranking clarifying questions based on predicted user engagement. In: MICROS Workshop at ECIR 2021 (2021)
21. McSherry, F., Najork, M.: Computing information retrieval performance measures efficiently in the presence of tied scores. In: Macdonald, C., Ounis, I., Plachouras, V., Ruthven, I., White, R.W. (eds.) ECIR 2008. LNCS, vol. 4956, pp. 414–421. Springer, Heidelberg (2008). https://doi.org/10.1007/978-3-540-78646-7_38
22. Nogueira, R., Cho, K.: Passage re-ranking with bert. arXiv preprint arXiv:1901.04085 (2019)
23. Robertson, S., Zaragoza, H., et al.: The probabilistic relevance framework: BM25 and beyond. Found. Trends® Inf. Retrieval **3**(4), 333–389 (2009)
24. Russell-Rose, T., Chamberlain, J., Shokraneh, F.: A visual approach to query formulation for systematic search. In: Proceedings of the 2019 Conference on Human Information Interaction and Retrieval, pp. 379–383 (2019)
25. Scells, H., Zuccon, G., Koopman, B.: A comparison of automatic boolean query formulation for systematic reviews. Inf. Retrieval J. **24**(1), 3–28 (2021)
26. Scells, H., Zuccon, G., Koopman, B., Clark, J.: Automatic boolean query formulation for systematic review literature search. In: Proceedings of the Web Conference 2020, pp. 1071–1081 (2020)
27. Sekulić, I., Aliannejadi, M., Crestani, F.: Towards facet-driven generation of clarifying questions for conversational search. In: Proceedings of the 2021 ACM SIGIR International Conference on Theory of Information Retrieval, pp. 167–175 (2021)
28. Soboroff, I.M., Craswell, N., Clarke, C.L., Cormack, G., et al.: Overview of the TREC 2011 web track. In: Proceedings of TREC (2011)
29. Tavakoli, L.: Generating clarifying questions in conversational search systems. In: Proceedings of the 29th ACM International Conference on Information & Knowledge Management, pp. 3253–3256 (2020)
30. Tonellotto, N.: Lecture notes on neural information retrieval. arXiv preprint arXiv:2207.13443 (2022)
31. Vakulenko, S., Kanoulas, E., De Rijke, M.: A large-scale analysis of mixed initiative in information-seeking dialogues for conversational search. ACM Trans. Inf. Syst. (TOIS) **39**(4), 1–32 (2021)
32. Wang, J., Li, W.: Template-guided clarifying question generation for web search clarification. In: Proceedings of the 30th ACM International Conference on Information & Knowledge Management, pp. 3468–3472 (2021)
33. Yang, P., Fang, H., Lin, J.: Anserini: reproducible ranking baselines using lucene. J. Data Inf. Qual. (JDIQ) **10**(4), 1–20 (2018)
34. Yang, Z., Moffat, A., Turpin, A.: How precise does document scoring need to be? In: Ma, S., et al. (eds.) AIRS 2016. LNCS, vol. 9994, pp. 279–291. Springer, Cham (2016). https://doi.org/10.1007/978-3-319-48051-0_21

35. Zamani, H., Dumais, S., Craswell, N., Bennett, P., Lueck, G.: Generating clarifying questions for information retrieval. In: Proceedings of the Web Conference 2020, pp. 418–428 (2020)
36. Zhai, C.: Statistical language models for information retrieval. Synth. Lect. Hum. Lang. Technol. **1**(1), 1–141 (2008)
37. Zhai, C., Lafferty, J.: A study of smoothing methods for language models applied to information retrieval. ACM Trans. Inf. Syst. (TOIS) **22**(2), 179–214 (2004)
38. Zhao, Z., Dou, Z., Mao, J., Wen, J.R.: Generating clarifying questions with web search results. In: Proceedings of the 45th International ACM SIGIR Conference on Research and Development in Information Retrieval, pp. 234–244 (2022)
39. Zou, J., Kanoulas, E., Liu, Y.: An empirical study on clarifying question-based systems. In: Proceedings of the 29th ACM International Conference on Information & Knowledge Management, pp. 2361–2364 (2020)

The Impact of Cross-Lingual Adjustment of Contextual Word Representations on Zero-Shot Transfer

Pavel Efimov[1]([✉]), Leonid Boytsov[2], Elena Arslanova[3],
and Pavel Braslavski[3,4][iD]

[1] ITMO University, Saint Petersburg, Russia
pavel.vl.efimov@gmail.com
[2] Bosch Center for Artificial Intelligence, Pittsburgh, USA
leo@boytsov.info
[3] Ural Federal University, Yekaterinburg, Russia
contilen@gmail.com, pbras@yandex.ru
[4] HSE University, Moscow, Russia

Abstract. Large multilingual language models such as mBERT or XLM-R enable zero-shot *cross-lingual* transfer in various IR and NLP tasks. Cao et al. [8] proposed a *data-* and *compute-efficient* method for cross-lingual adjustment of mBERT that uses a *small* parallel corpus to make embeddings of related words across languages similar to each other. They showed it to be effective in NLI for five European languages. In contrast we experiment with a topologically diverse set of languages (Spanish, Russian, Vietnamese, and Hindi) and extend their original implementations to new tasks (XSR, NER, and QA) and an additional training regime (continual learning). Our study reproduced gains in NLI for four languages, showed improved NER, XSR, and *cross-lingual* QA results in three languages (though some cross-lingual QA gains were not statistically significant), while *mono-lingual* QA performance never improved and sometimes degraded. Analysis of distances between contextualized embeddings of related and unrelated words (across languages) showed that fine-tuning leads to "forgetting" some of the cross-lingual alignment information. Based on this observation, we further improved NLI performance using continual learning. Our software is publicly available https://github.com/pefimov/cross-lingual-adjustment.

Keywords: Cross-lingual transfer · Multilingual embeddings

1 Introduction

Large *multi-lingual* language models such as mBERT or XLM-R pre-trained on large *unpaired* multilingual corpora enable zero-shot *cross-lingual* transfer

---

P. Efimov and L. Boytsov—Contributed equally to the paper.
L. Boytsov—Work done before joining Amazon.

J. Kamps et al. (Eds.): ECIR 2023, LNCS 13982, pp. 51–67, 2023.
https://doi.org/10.1007/978-3-031-28241-6_4

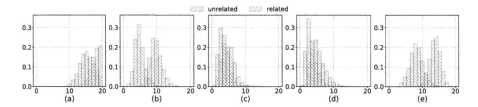

Fig. 1. Histograms of L_2 distances between pairs of mBERT *last-layer* representations for randomly sampled related (i.e., aligned) and unrelated word pairs from WikiMatrix (Hi-En): (a) original, (b) after cross-lingual adjustment, (c) after fine-tuning on English NLI data, (d) after cross-lingual adjustment and subsequent fine-tuning on English NLI data, (e) after cross-lingual adjustment and subsequent *continual* fine-tuning on English NLI data.

[32,44], which is sometimes effective even for languages not seen at the pre-training stage [19,39]. Contextualized word representations produced by the models can be further aligned (across languages) with a modest amount of parallel data, e.g., by finding a rotation matrix using a bilingual dictionary [38]. Such a post hoc alignment can improve zero-shot transfer for text parsing and understanding [28,62,63]. This approach is a more data- and computationally-efficient alternative to training a machine translation system on a large parallel corpus.

Cao et al. [8] used a small parallel corpus for direct cross-lingual adjustment of the mBERT model and found this adjustment procedure to be more effective than the post hoc rotation. However, they experimented only with a single task: cross-lingual NLI, and restricted their study to European languages. In that, we are not aware of any systematic study that tested this procedure on a set of diverse languages and tasks.

To fill this gap, we rigorously evaluated the method using four typologically diverse languages (Spanish, Russian, Vietnamese, and Hindi) and four different tasks all of which are crucial to information retrieval and text mining: natural-language inference (NLI), question-answering (QA), named-entity recognition (NER), and cross-lingual sentence retrieval (XSR). QA and XSR are core information mining tasks directly targeting a user's information need [5,36] while NLI and NER are used in document [59] and query [10,64] analysis. Named entities are the core of entity-oriented [6] and expert search [7]. Entity-oriented retrieval models can also be used to boost recall of traditional keyword-based retrievers [21,57].

Furthermore, existing studies typically avoided statistical testing and presented results for a single seed. In contrast, we performed each experiment with five seeds and assessed statistical significance of the difference from a baseline. In addition to the original method of Cao et al. [8] we evaluated its *continual-learning* extension [42,48,50] where we combined the target task loss with the cross-lingual adjustment loss [9].

In our study, we ask the following research questions:

R1 How does cross-lingually adjusted mBERT fine-tuned on English data and zero-shot transferred to a target language perform on various IR/NLP tasks and typologically different languages?

R2 How does adjustment of mBERT on parallel data and fine-tuning for a specific task affect similarity of contextualized embeddings of semantically related and unrelated words across languages?

R3 Inspired by our observation (see Fig. 1c-1d) that fine-tuning draws embeddings of *both* related and unrelated words closer to each other, which may negatively affect the cross-lingual transfer, we wonder if continual learning— with an auxiliary cross-lingual adjustment loss—can improve effectiveness of the zero-shot transfer.

Our experiments demonstrated the following:

– The cross-lingual adjustment of mBERT improves NLI in four target languages; NER, XSR, and *cross-lingual* QA in three languages (though some cross-lingual QA gains are not statistically significant). Yet, there is no statistically significant improvement for *mono-lingual* QA and a statistically significant deterioration on three out of eight QA datasets.

– When comparing L_2 distances between contextualized-embeddings of words across languages (Fig. 1b), we see that the cross-lingual adjustment of mBERT decreases the L_2 distance between related words while keeping unrelated words apart, which is in line with prior work [68].

– However, we have found no prior work that inspected histograms obtained after fine-tuning. Quite surprisingly, we observe that fine-tuning of mBERT for a specific task draws embeddings of *both* related and unrelated words much closer to each other (Fig. 1c and Fig. 1d). Thus, fine-tuning causes the model to "forget" some of the cross-lingual information learned during adjustment.

– In that, continual learning allows the model to learn a target task while maintaining the separation between related and unrelated words (Fig. 1e). Continual learning consistently improves performance on NLI data. Aside from NLI, it improves XSR and NER *only* for Russian and there are no improvements on other QA or NER datasets.

In summary, our study contributes to a better understanding of (1) cross-lingual transfer capabilities of large multilingual language models and of (2) effectiveness of their cross-lingual adjustment across various tasks and languages. We believe our results support a conjecture that the cross-lingual adjustment of Cao et al. [8] is more beneficial for cross-lingual tasks. Our software is publicly available https://github.com/pefimov/cross-lingual-adjustment.

2 Related Work

2.1 Cross-Lingual Zero-Shot Transfer with Multilingual Models

The success of mBERT in cross-language zero-shot regime on various tasks has inspired many papers that attempted to explain its cross-lingual abilities and limitations [3,11,11,14,17,25,32,44,66]. These studies showed that the multilingual models learn high-level abstractions common to all languages, which make transfer possible even when languages share no vocabulary. However, the gap between performance on English and a target language is smaller if the languages are cognate, i.e. share a substantial portion of model's vocabulary, have similar syntactic structures, and are from the same language family [30,66]. Moreover, the size of target language data used for pre-training and the size of the model vocabulary allocated to the language also positively impacts cross-lingual learning performance [3,30].

Zero-shot transfer of mBERT or other multilingual transformer-based models from English to a different language was applied to many tasks including but not limited to cross-lingual information retrieval, POS tagging, dependency parsing, NER, NLI, and QA [23,34,44,63,66]. XTREME data suite [24] and its successor XTREME-R [52] are dedicated collections of tasks and corresponding datasets for evaluation of zero-shot transfer capabilities of large multilingual models from English to tens of languages. XTREME includes XSR, NLI, NER, and QA datsets used in the current study. Although transfer from English is not always an optimal choice [33,60], English still remains the most popular source language. Furthermore, despite there have been developed quite a few new models that differ in architectures, supported languages, and training data [16], mBERT remains the most popular cross-lingual model.

2.2   Cross-lingual Alignment of Embeddings

Mikolov et al. [38] demonstrated that vector spaces can encode semantic relationships between words and that there are similarities in the geometry of these vectors spaces across languages. A variety of approaches have been proposed for aligning monolingual representations based on bilingual dictionaries and parallel sentences. The most widely used approach—which requires only a bilingual dictionary—consists in finding a rotation matrix that aligns vectors of two monolingual models [38]. Lample et al. [29] proposed an alignment method based on adversarial training, which does not require parallel data. Ruder et al. [53] provide a comprehensive overview of alignment methods for pre-Transformer models.

Schuster et al. [55] applied rotation to align contextualized ELMo embeddings [43] using "anchors" (averaged vectors of tokens in different contexts) and bilingual dictionaries. They showed improved results of cross-lingual dependency parsing using English as source and several European languages as target languages. Wang et al. [62] aligned English BERT and mBERT representations using rotation method and Europarl parallel data [26]. They employed the resulting embeddings in a cross-lingual dependency parsing model. The parser with aligned embeddings consistently outperformed zero-shot mBERT on 15 out of 17 target languages. Instead of aligning on a word level, Aldarmaki and Diab [2] performed a sentence-level alignment of ELMo embeddings and evaluated this approach on the parallel sentence retrieval task.

Cao et al. [8] proposed to directly modify the mBERT model by bringing the vectors of semantically related words in different languages closer to each other. This was motivated by the observation that embedding spaces of different languages are not always isometric [58] and, hence, are not always amenable to alignment via rotation. The authors showed that mBERT simultaneously adjusted on five European languages consistently outperformed other alignment approaches on XNLI data. In the current study, we implement the approach with some modifications.

Liu et al. [35] showed that combining continual learning with fine-tuning improved zero-shot transfer performance for NER and POS tagging. In that, they used cross-lingual sentence retrieval (XSR) and/or masked-language model (MLM) task as additional tasks. Although XSR can be seen as an alternative to the cross-lingual adjustment of Cao et al. [8], the authors did not evaluate the effectiveness of zero-shot transfer after adjusting the model with XSR. In contrast, we evaluate the marginal effectiveness of continual learning with respect to already cross-lingually adjusted mBERT.

Kulshreshtha et al. [28] compared different alignment methods (rotation vs. adjustment) on NER and slot filling tasks. According to their results, rotation-based alignment performs better on the NER task, while model adjustment performs better on slot filling. Zhao et al. [68] continued this line of research and proposed several improvements of the model adjustment method: 1) z-normalization of vectors and 2) text normalization to make the input more structurally 'similar' to English training data. Experiments on XNLI dataset and translated sentence retrieval showed that vector normalization leads to more consistent improvements over zero-shot baseline compared to text normalization. Faisal and Anastasopoulos [22] applied cross-lingually adjusted mBERT and XLM-R to cross-lingual open-domain QA and obtained improvements both on paragraph and span selection subtasks. However, they trained their models on machine-translated data, which is different from our zero-shot settings.

3   Methods

In this study, we use a multilingual BERT (mBERT) as the main model [15]. mBERT is a case-sensitive "base" 12-layer Transformer model [61] with 178M parameters.[1] It was trained with a masked language model objective on 104 languages with a shared WordPiece [67] vocabulary (using 104 Wikipedias). To balance the distribution of languages, high-resource languages were under-sampled and low-resource languages were over-sampled.[2] For a number of NLP tasks, cross-lingual transfer of mBERT can be competitive with training a monolingual model using the training data in the target language.

We align cross-lingual embeddings by directly modifying/adjusting the language model itself, following the approach by Cao et al. [8]. The approach—which differs from finding a rotation matrix—proved to be effective in the NLI task.

---

[1] https://huggingface.co/bert-base-multilingual-cased.

[2] https://github.com/google-research/bert/blob/master/multilingual.md.

However, there are some differences in our implementation. In all cases, we work with one pair of languages at a time while Cao et al. [8] adjusted mBERT for five languages at once.

BERT uses WordPiece tokenization [67], which splits sufficiently long words into *subword* tokens. We first word-align parallel data with *fast_ align* [18] and then average all subword tokens' vectors.[3]

Based on alignments in parallel data, we obtain a collection of word pairs (s_i, t_i): s_i from the source language, t_i from the target one. From these alignments we can obtain their mBERT vector representations $\mathbf{f}(s_i)$ and $\mathbf{f}(t_i)$. Then, we *adjust* the mBERT model on aligned pairs' vectors using the following loss function:

$$L = \sum_{(s_i, t_i)} \|\mathbf{f}(s_i) - \mathbf{f}(t_i)\|_2^2 + \sum_{s_j} \|\mathbf{f}(s_j) - \mathbf{f}^0(s_j)\|_2^2, \tag{1}$$

where the first term "pulls" the embeddings in the source and target language together, while the second (regularization) term prevents source (English) representations from deviating far from their initial values in the 'original' mBERT $\mathbf{f}^0$. Finally, the cross-lingually adjusted mBERT model is fine-tuned for a specific task.

Training neural networks via empirical loss minimization is known to suffer from the "catastrophic forgetting" [37]. The histograms of L_2 distances between embeddings of related and unrelated words in pairs of languages (see Fig. 1 and the discussion in § 5.3), confirm that this is, indeed, the case. Specifically, fine-tuning on a target task—in contrast to the cross-lingual adjustment objective—reduces the separation between related and unrelated words. To counter this effect, we ran an additional experiments in a continual-learning mode [49], which relies on experience replay [48,50].

Technically, this entailed a multi-task training [9] with a combined loss function:

$$L = L_{target} + \alpha L_{align}, \tag{2}$$

where L_{target} was the loss-function for the target task, e.g., NLI, L_{align} was a cross-lingual loss function given by Eq. 1, and $\alpha > 0$ was a small weight. During training, we iterated over the complete (reshuffled) dataset for the target task: After computing L_{target} for a current batch we randomly sampled a small batch of aligned pairs of words $\{(s_i, t_i)\}$ from the parallel corpus and computed L_{align}.

4   Tasks and Data

4.1   Languages and Parallel Data

In our experiments we transfer models trained on English to four languages: Spanish, Russian, Vietnamese, and Hindi. This set represents four different families (including one non-Indo-European language), three scripts, and two different

---

[3] We also experimented with other options reported in the literature – first/last tokens' vectors, as well as aligning subword tokens produced by BERT. Although these choices induced some variations in results, there is no single pattern across all tasks and languages.

Table 1. Language information.

Lang	Family	Script	Word order	Number of Wiki pages
en	IE/Germanic	Latin	SVO	6.3 M
es	IE/Romance	Latin	SVO	1.7 M
ru	IE/Slavic	Cyrillic	SVO	1.7 M
vi	Austroasiatic	Latin	SVO	1.3 M
hi	IE/Indo-Aryan	Devanagari	SOV	150 K

IE : Indo-European; Prevalent word order: SVO – subject-verb-object, SOV – subject-object-verb;

prevalent word orders (see Table 1). All the languages are among languages that were used to train mBERT.[4]

To align embeddings, we use a sample from the parallel corpus WikiMatrix [56] that contains sentences in 1,620 different language pairs mined from Wikipedia.

4.2 Natural Language Inference

Natural language inference (NLI) is a task of determining the relation between two ordered sentences (hypothesis and premise) and classifying them into: entailment, contradiction, or "no relation". English MultiNLI collection consists of 433K multi-genre sentence pairs [65]. The XNLI dataset complements the MultiNLI training set with newly collected 2.5K development and 5K test English examples [13]. They were professionally translated into 15 languages, including all four target languages of the current study. Additionally, for each target language test set, we created a new mixed-language XNLI set by randomly picking either a hypothesis or a premise and replacing it with the original English sentence. Performance on XNLI datasets is evaluated using classification *accuracy*.

4.3 Named Entity Recognition

Named entity recognition (NER) is a task of locating named entities in unstructured text and classifying them into predefined categories such as persons, organizations, locations, etc. In our experiments, we employ the Wikiann NER corpus [46] that is derived from a larger "silver-standard" collection that was created fully automatically [41]. Wikiann NER has data for 41 languages, including all languages in the current study. The named entity types include location (LOC), person (PER), and organization (ORG). The English training set contains 20K sentences. Test sets for Spanish, Vietnamese, and Russian have 10K sentences each; for Hindi – 1K sentences. Performance is evaluated using the *token-level micro-averaged F1*.

---

[4] However, Hindi Wikipedia is an order of magnitude smaller compared to other Wikipedias, which may have led to somewhat inferior contextualized embeddings.

4.4   Question Answering

Machine reading comprehension (MRC) is a variant of a QA task. Given a question and a text paragraph, the system needs to return a continuous span of paragraph tokens as an answer. The first large-scale MRC dataset is the English Wikipedia-based dataset SQuAD [47], which contains about 100K paragraph-question-answer triples. SQuAD has become a *de facto* standard and inspired creation of analogous resources in other languages [51]. We use SQuAD as the source dataset to train MRC models. To test the models, we use XQuAD, MLQA, and TyDi QA datasets. XQuAD [3] is a professional translation of 240 SQuAD paragraphs and 1,190 questions-answer pairs into 10 languages (including four languages of our study). MLQA [31] data is available for six languages including Spanish, Vietnamese, and Hindi (but it does not have Russian). There are about 5K questions for each of our languages. TyDi QA [12] includes 11 typologically diverse languages of which we use only Russian (812 test items).

In addition to monolingual test data, we experimented with parallel/cross-lingual MLQA and XQuAD datasets and explored two directions: (1) question is in a target language, but paragraph is in English; (2) a question is in English, but a paragraph is in a target language.

QA performance is evaluated using a *token-level F1-score*.

4.5   Cross-Lingual Sentence Retrieval

The task of cross-lingual sentence retrieval (XSR) consists in retrieving sentences that are translations of each other. A query is a sentence in one language and a corpus is a set of sentences (translations) in another language (in our case English). For the XSR task, we use a subset of the Tatoeba collection [4] covering 93 languages (including four languages in our study). This subset has one thousand English sentences each of which has a translation in each of the 93 languages. Following Ruder et al. [52], we fine-tune the model on a QA task. A sentence representation is obtained by averaging token embeddings in one of the layers: Representations are compared using the cosine similarity. Because different layers perform differently in different scenarios (original mBERT vs. adjusted vs. adjusted with continual learning), we select the scenario-specific best-performing layer. In that, retrieval performance is measured using *the mean reciprocal rank (MRR)*.

5   Experimental Results and Analysis

5.1   Setup

All experiments were conducted on a single Tesla V100 16 GB. For cross-lingual model adjustment we use the Adam optimizer and hyper-parameters provided by Cao et al. [8]. To obtain reliable results we run five iterations (using different seeds) of model adjustment (for each configuration) followed by fine-tuning on

Table 2. Performance on original and mixed-language NLI datasets (accuracy).

mBERT	es	ru	vi	hi
Original XNLI				
Original	74.20	67.95	69.58	59.03
Adjusted	74.82*	69.45*	70.88*	61.54*
Adj+cont	**75.89****	**71.26****	**72.79****	**63.90****
Mixed-language NLI				
Original	70.93	64.24	62.72	53.53
Adjusted	72.06*	66.56*	66.50*	57.31*
Adj+cont	**73.50****	**69.09****	**69.14****	**61.09****

Statistically significant differences from an original and adjusted mBERT are marked with * and **, respectively (p-value threshold 0.05).

downstream tasks. For each run we sample 30 K sentences from a set of 250 K parallel (WikiMatrix) sentences word-aligned with *fast_align*.[5]

The code to fine-tune mBERT on XNLI, SQuAD, and Wikiann is based on HuggingFace sample scripts,[6] which were modified to support continual learning. We use a standard architecture consisting of a BERT model with a task-specific linear layer [15]. We also reuse parameters provided by HuggingFace, except for the weight $\alpha = 0.01$ in the multi-task loss (Eq. 2), which was tuned on XNLI validation sets. Also note that batch sizes are 32 (for the main target loss) and 16 (for the auxiliary cross-lingual adjustment loss in the case of continual learning).

All reported results are averages over five runs with different seeds. We compute statistical significance of differences between the original and adjusted mBERT using a paired t-test. For XSR, QA, and NLI we first average metric values for each example over different runs and then carry out a paired t-test using averaged values. For NER we concatenate example-specific predictions for all seeds and run 1,000 iterations of a permutation test for concatenated sequences [20, 45].

5.2 Main Results

Results for NLI, NER, QA, and XSR tasks are summarized in Tables 2, 3, 4, and 5, respectively. We can observe consistent and statistically significant improvements (up to 2.5 accuracy point) of aligned models over zero-shot transfer on XNLI for all languages. This is in line with Cao et al. [8] even though

---

[5] We ran the main experiments with 30K parallel sentences. In addition, we conducted experiments with 5 K/10 K/30 k/100 K/250 K Ru-En sentence pairs. Increasing amount of parallel data benefits both NLI and NER, whereas QA performance peaks at roughly 5K parallel sentences and further decreases as the number of parallel sentences increases. Due to limited space we don't report detailed results and analysis here.

[6] https://github.com/huggingface/transformers/tree/master/examples/pytorch.

Table 3. Performance on NER task (token-level F1).

mBERT	es	ru	vi	hi
Original	**73.40**	63.43	71.02	65.24
Adjusted	73.28	65.49*	**71.99***	**68.22***
Adj+cont	72.71**	**66.27****	71.35**	66.07**

Statistically significant differences from an original and adjusted mBERT are marked with * and **, respectively (p-value threshold 0.05).

Table 4. Effectiveness of QA systems (F1-score).

mBERT	Spanish		Russian		Vietnamese		Hindi	
	MLQA	XQuAD	TyDi QA	XQuAD	MLQA	XQuAD	MLQA	XQuAD
Original	**64.96**	**75.59**	67.05	**70.72**	**59.95**	**69.18**	**48.73**	57.56
Adjusted	63.11*	73.99*	67.03	70.58	58.46*	68.63	48.47	57.81
Adj+cont	62.76**	73.44	**67.63**	70.51	57.71**	68.64	48.02**	**57.83**
	Question in target language, paragraph in English							
Original	**67.34**	**75.74**	–	71.54	56.08	65.00	42.48	47.83
Adjusted	66.93*	75.65	–	**71.68**	**56.74***	**66.75***	**44.91***	**50.45***
Adj+cont	66.31**	74.88**	–	70.99	54.51**	64.63**	43.88**	50.13
	Question in English, paragraph in target language							
Original	**67.36**	**76.71**	–	67.31	64.43	68.12	55.32	58.62
Adjusted	66.96*	76.42	–	**68.25***	**65.01***	**68.99**	**55.63**	**58.93**
Adj+cont	66.68**	76.21	–	68.06˙	64.36**	68.54	54.74**	58.22

Statistically significant differences from an original and adjusted mBERT are marked with * and **, respectively (p-value threshold 0.05).

we used a set of more diverse languages, parallel data of lower quality, and a slightly different learning scheme (we adjusted models individually for each pair of languages). In that, employing continual learning lead to additional substantial gains (up to 2.4 accuracy points).

We also evaluated models on the (bilingual) mixed-language XNLI test data (see § 4.2). According to the bottom part of Table 2, compared to the original XNLI, we observe bigger gains for all four languages, especially when we employ continual learning. For Hindi, we obtain a 7.5 point gain by using both the adjustment and continual learning.

NER results are somewhat mixed: We observe statistically significant gains (up to 3 points for Hindi) on all languages except Spanish. In that, continual learning is beneficial only for Russian.

When we fine-tuned a cross-lingually adjusted mBERT on *mono-lingual* QA tasks, there were no statistically significant gains. In that, there was a statistically significant decrease for all Spanish datasets and Vietnamese MLQA. Use of continual learning lead to further degradation in nearly all cases. Note that models were noticeably less accurate on MLQA compared to XQuAD, which is a translation of our training set SQuAD.

Muttenthaler et al. [40] and van Aken et al. [1] showed that QA models essentially clustered answer token vectors and separated them from the rest of the paragraph token vectors using a vector representation of the question. Thus, to solve the QA task, the model learns to rely on *mutual similarities* among question and answer tokens (on English QA data) rather than on their actual vector representations. As a consequence, there may be no need to make representations in the target language to be similar to English-language representations. This may *partially* explain why the cross-lingual adjustment was unsuccessful for *mono-lingual* QA.

This hypothesis—together with the observation of stronger performance on XQuAD compared to MLQA—prompted us to explore whether the cross-lingual adjustment could be more useful for *cross-lingual* QA. We explored two directions: (1) question is in a target language, but paragraph is in English; (2) a question is in English, but a paragraph is in a target language. According to results in the lower part of Table 4, we observe improvements in three languages (except Spanish) and most of these improvements are statistically significant. This is in line with cross-lingual QA results by Faisal and Anastasopoulos [22].

For XSR—another cross-lingual task—using the cross-lingual adjustment lead to substantial and statistically significant improvements in three languages: Spanish, Vietnamese, and Hindi. However, continual learning only marginally helped Russian. Finally, we note that cross-lingual adjustment resulted in bigger gains for the mixed, i.e., cross-lingual XNLI data, than for original one (Table 2). In summary, we believe our results support a conjecture that the cross-lingual adjustment of Cao et al. [8] is more beneficial for cross-lingual tasks.

Table 5. Performance on XSR task (MRR for best-performing layers).

mBERT	es	ru	vi	hi
Original	0.80	0.73	0.67	0.49
Adjusted	**0.82***	0.72*	**0.72***	**0.56***
Adj+cont	0.78**	**0.74****	0.69**	0.53**

Statistically significant differences from an original and adjusted mBERT are marked with * and **, respectively (p-value threshold 0.05).

5.3 Analysis of the Adjusted mBERT for NLI

This section analysis focuses on the NLI task. We calculate L_2 distances between contextualized embeddings in English and other languages.[7] The embeddings are taken from the last layer output (i.e., no prediction heads are used). To this end we sampled semantically related words from parallel sentences (matched via *fast_align*) and unrelated words from unpaired sentences (nearly always unrelated). For each pair of languages and each task, the sampling process was carried out for: (1) the original mBERT, (2) an adjusted mBERT, (3) the original mBERT fine-tuned for the target task, (4) the adjusted mBERT fine-tuned for

---

[7] Although most prior work uses the cosine similarity instead of L_2 [54], it does not distinguish between vectors with the same direction, but different lengths.

the target task, (5) the adjusted mBERT fine-tuned for the target task using *continual* learning.[8]

From Fig. 1 we can see that the cross-lingual adjustment makes embeddings of semantically similar words from different languages closer to each other while keeping unrelated words apart, which is in line with Zhao et al. [68]. However, prior work did not inspect histograms obtained after fine-tuning. Yet, quite surprisingly, fine-tuning of both the original and adjusted mBERT on the English NLI data (Fig. 1c and 1d) makes distributions of related and unrelated words almost fully overlap, i.e. all embeddings become close to each other. Compared to the original mBERT, fine-tuning of the *adjusted* mBERT (Fig. 1d) does result in a better separation of related and unrelated words, but the effect is *quite modest*. We believe this is an example of "catastrophic forgetting" [37], where fine-tuning the model on a target task causes the model to forget some of the knowledge obtained during cross-lingual adjustment.

Continual learning (Fig. 1e) permits fine-tuning for the target task while maintaining a separation between related and related words, which also consistently improves performance for the NLI task. However, when we compared histograms for additional tasks (not shown due to space limit) there was no direct relationship between the degree of separation and the success of cross-lingual transfer among all tasks and training regimes. In the case of NER, the biggest separation was achieved for Spanish, but fine-tuning of the adjusted mBERT resulted in a lower accuracy. More generally, fine-tuning with continual learning *always* led to a better separation of related and unrelated words, but this extra separation was beneficial only for the NLI task.

6   Conclusion

We evaluate effectiveness of an existing approach to cross-lingual adjustment of mBERT [8] using four typologically different languages (Spanish, Russian, Vietnamese, and Hindi) and four IR/NLP tasks (XSR, QA, NLI, and NER). The original mBERT is being compared to mBERT "adjusted" with a help of a small parallel corpus. The cross-lingual adjustment of mBERT improves NLI in four target languages; NER, XSR, and *cross-lingual* QA in three languages (though some cross-lingual QA gains are not statistically significant). However, in the case of *mono-lingual* QA performance never improves and sometimes degrades. We believe our results support a conjecture that the cross-lingual adjustment of Cao et al. [8] is more beneficial for cross-lingual tasks.

Inspired by the analysis of histograms of distances, we obtain *additional* improvement on NLI using continual learning. Our study contributes to a better understanding of cross-lingual transfer capabilities of large multilingual language models and identifies limitations of the cross-lingual adjustment in various IR and NLP tasks.

---

[8] Due to limited space we demonstrate histograms only for Hindi/NLI task, other languages/tasks exhibit similar trends.

Acknowledgment. This research was supported in part through computational resources of HPC facilities at HSE University [27]. PE is grateful to Yandex Cloud for their grant toward computing resources of Yandex DataSphere. PB acknowledges support by the Russian Science Foundation, grant № 20-11-20166.

References

1. van Aken, B., Winter, B., Löser, A., Gers, F.A.: How does BERT answer questions? A layer-wise analysis of transformer representations. In: Proceedings of the 28th ACM International Conference on Information and Knowledge Management, pp. 1823–1832 (2019)
2. Aldarmaki, H., Diab, M.: Context-aware cross-lingual mapping. In: Proceedings of the 2019 Conference of the North American Chapter of the Association for Computational Linguistics: Human Language Technologies, Volume 1 (Long and Short Papers), pp. 3906–3911 (2019)
3. Artetxe, M., Ruder, S., Yogatama, D.: On the cross-lingual transferability of monolingual representations. In: ACL, pp. 4623–4637 (2020)
4. Artetxe, M., Schwenk, H.: Massively multilingual sentence embeddings for zero-shot cross-lingual transfer and beyond. Trans. Assoc. Comput. Linguist. **7**, 597–610 (2019)
5. Baeza-Yates, R., Ribeiro-Neto, B., et al.: Modern information retrieval, vol. 463. ACM press New York (1999)
6. Balog, K.: Entity-Oriented Search, The Information Retrieval Series, vol. 39. Springer (2018). https://doi.org/10.1007/978-3-319-93935-3
7. Brandsen, A., Verberne, S., Lambers, K., Wansleeben, M.: Can BERT dig it? named entity recognition for information retrieval in the archaeology domain. J. Comput. Cult. Herit. **15**(3) (2022)
8. Cao, S., Kitaev, N., Klein, D.: Multilingual alignment of contextual word representations. In: 8th International Conference on Learning Representations, ICLR 2020 (2020)
9. Caruana, R.: Algorithms and applications for multitask learning. In: ICML, pp. 87–95. Morgan Kaufmann (1996)
10. Cheng, X., Bowden, M., Bhange, B.R., Goyal, P., Packer, T., Javed, F.: An end-to-end solution for named entity recognition in ecommerce search. In: AAAI, pp. 15098–15106. AAAI Press (2021)
11. Chi, E.A., Hewitt, J., Manning, C.D.: Finding universal grammatical relations in multilingual BERT. In: Proceedings of the 58th Annual Meeting of the Association for Computational Linguistics, pp. 5564–5577 (2020)
12. Clark, J.H., et al.: TyDi QA: a benchmark for information-seeking question answering in typologically diverse languages. TACL **8**, 454–470 (2020)
13. Conneau, A., et al.: XNLI: Evaluating cross-lingual sentence representations. In: Proceedings of the 2018 Conference on Empirical Methods in Natural Language Processing, pp. 2475–2485 (2018)
14. Conneau, A., Wu, S., Li, H., Zettlemoyer, L., Stoyanov, V.: Emerging cross-lingual structure in pretrained language models. In: Proceedings of the 58th Annual Meeting of the Association for Computational Linguistics, pp. 6022–6034 (2020)

15. Devlin, J., Chang, M.W., Lee, K., Toutanova, K.: BERT: Pre-training of deep bidirectional transformers for language understanding. In: Proceedings of the 2019 Conference of the North American Chapter of the Association for Computational Linguistics: Human Language Technologies, Volume 1 (Long and Short Papers), pp. 4171–4186 (Jun 2019)

16. Doddapaneni, S., Ramesh, G., Kunchukuttan, A., Kumar, P., Khapra, M.M.: A primer on pretrained multilingual language models. arXiv preprint arXiv:2107.00676 (2021)

17. Dufter, P., Schütze, H.: Identifying elements essential for BERT's multilinguality. In: Proceedings of the 2020 Conference on Empirical Methods in Natural Language Processing (EMNLP), pp. 4423–4437 (2020)

18. Dyer, C., Chahuneau, V., Smith, N.A.: A simple, fast, and effective reparameterization of ibm model 2. In: Proceedings of the 2013 Conference of the North American Chapter of the Association for Computational Linguistics: Human Language Technologies, pp. 644–648 (2013)

19. Ebrahimi, A., et al.: AmericasNLI: Evaluating zero-shot natural language understanding of pretrained multilingual models in truly low-resource languages. arXiv preprint arXiv:2104.08726 (2021)

20. Efron, B., Tibshirani, R.J.: An introduction to the bootstrap. Monogr. Stat. Appl. Probab. **57**(1) (1993)

21. Ensan, F., Bagheri, E.: Document retrieval model through semantic linking. In: WSDM, pp. 181–190. ACM (2017)

22. Faisal, F., Anastasopoulos, A.: Investigating post-pretraining representation alignment for cross-lingual question answering. In: Proceedings of the 3rd Workshop on Machine Reading for Question Answering, pp. 133–148 (2021)

23. Hsu, T.Y., Liu, C.L., Lee, H.y.: Zero-shot reading comprehension by cross-lingual transfer learning with multi-lingual language representation model. In: Proceedings of the 2019 Conference on Empirical Methods in Natural Language Processing and the 9th International Joint Conference on Natural Language Processing (EMNLP-IJCNLP), pp. 5933–5940 (Nov 2019)

24. Hu, J., Ruder, S., Siddhant, A., Neubig, G., Firat, O., Johnson, M.: XTREME: A massively multilingual multi-task benchmark for evaluating cross-lingual generalisation. In: International Conference on Machine Learning, pp. 4411–4421 (2020)

25. K, Karthikeyan., Wang, Z., Mayhew, S., Roth, D.: Cross-lingual ability of multilingual BERT: an empirical study. In: ICLR (2020)

26. Koehn, P.: Europarl: A parallel corpus for statistical machine translation. In: MT summit, pp. 79–86 (2005)

27. Kostenetskiy, P.S., Chulkevich, R.A., Kozyrev, V.I.: Hpc resources of the higher school of economics. J. Phys.: Conf. Series **1740**(1), 012050 (2021)

28. Kulshreshtha, S., Redondo Garcia, J.L., Chang, C.Y.: Cross-lingual alignment methods for multilingual BERT: A comparative study. In: Findings of the Association for Computational Linguistics: EMNLP 2020, pp. 933–942 (Nov 2020)

29. Lample, G., Conneau, A., Ranzato, M., Denoyer, L., Jégou, H.: Word translation without parallel data. In: International Conference on Learning Representations (2018)

30. Lauscher, A., Ravishankar, V., Vulić, I., Glavaš, G.: From zero to hero: On the limitations of zero-shot language transfer with multilingual Transformers. In: Proceedings of the 2020 Conference on Empirical Methods in Natural Language Processing (EMNLP), pp. 4483–4499 (2020)

31. Lewis, P., Oğuz, B., Rinott, R., Riedel, S., Schwenk, H.: MLQA: Evaluating cross-lingual extractive question answering. In: ACL, pp. 7315–7330 (2020)

32. Libovický, J., Rosa, R., Fraser, A.: How language-neutral is multilingual BERT? arXiv preprint arXiv:1911.03310 (2019)
33. Lin, Y.H., et al.: Choosing transfer languages for cross-lingual learning. In: Proceedings of the 57th Annual Meeting of the Association for Computational Linguistics, pp. 3125–3135 (Jul 2019)
34. Litschko, R., Vulić, I., Ponzetto, S.P., Glavaš, G.: Evaluating multilingual text encoders for unsupervised cross-lingual retrieval. arXiv preprint arXiv:2101.08370 (2021)
35. Liu, Z., Winata, G.I., Madotto, A., Fung, P.: Preserving cross-linguality of pre-trained models via continual learning. In: RepL4NLP@ACL-IJCNLP, pp. 64–71. Association for Computational Linguistics (2021)
36. Manning, C.D., Raghavan, P., Schütze, H.: Introduction to information retrieval. Cambridge University Press (2008)
37. McCloskey, M., Cohen, N.J.: Catastrophic interference in connectionist networks: The sequential learning problem. In: Psychology of learning and motivation, vol. 24, pp. 109–165. Elsevier (1989)
38. Mikolov, T., Le, Q.V., Sutskever, I.: Exploiting similarities among languages for machine translation. arXiv preprint arXiv:1309.4168 (2013)
39. Muller, B., Anastasopoulos, A., Sagot, B., Seddah, D.: When being unseen from mBERT is just the beginning: Handling new languages with multilingual language models. In: Proceedings of the 2021 Conference of the North American Chapter of the Association for Computational Linguistics: Human Language Technologies. pp. 448–462 (2021)
40. Muttenthaler, L., Augenstein, I., Bjerva, J.: Unsupervised evaluation for question answering with transformers. In: Proceedings of the Third BlackboxNLP Workshop on Analyzing and Interpreting Neural Networks for NLP, pp. 83–90 (2020)
41. Pan, X., Zhang, B., May, J., Nothman, J., Knight, K., Ji, H.: Cross-lingual name tagging and linking for 282 languages. In: Proceedings of the 55th Annual Meeting of the Association for Computational Linguistics (Volume 1: Long Papers), pp. 1946–1958 (2017)
42. Parisi, G.I., Kemker, R., Part, J.L., Kanan, C., Wermter, S.: Continual lifelong learning with neural networks: A review. Neural Netw. **113**, 54–71 (2019)
43. Peters, M.E., Neumann, M., Iyyer, M., Gardner, M., Clark, C., Lee, K., Zettlemoyer, L.: Deep contextualized word representations. In: Proceedings of the 2018 Conference of the North American Chapter of the Association for Computational Linguistics: Human Language Technologies, Volume 1 (Long Papers), pp. 2227–2237 (2018)
44. Pires, T., Schlinger, E., Garrette, D.: How multilingual is multilingual BERT? In: Proceedings of the 57th Annual Meeting of the Association for Computational Linguistics, pp. 4996–5001 (2019)
45. Pitman, E.J.: Significance tests which may be applied to samples from any populations. Suppl. J. R. Stat. Soc. **4**(1), 119–130 (1937)
46. Rahimi, A., Li, Y., Cohn, T.: Massively multilingual transfer for NER. In: Proceedings of the 57th Annual Meeting of the Association for Computational Linguistics, pp. 151–164 (2019)
47. Rajpurkar, P., Zhang, J., Lopyrev, K., Liang, P.: SQuAD: 100,000+ questions for machine comprehension of text. In: Proceedings of the 2016 Conference on Empirical Methods in Natural Language Processing, pp. 2383–2392 (2016)
48. Ratcliff, R.: Connectionist models of recognition memory: constraints imposed by learning and forgetting functions. Psychol. Rev. **97**(2), 285 (1990)

49. Riabi, A., Scialom, T., Keraron, R., Sagot, B., Seddah, D., Staiano, J.: Synthetic data augmentation for zero-shot cross-lingual question answering. In: Proceedings of the 2021 Conference on Empirical Methods in Natural Language Processing, pp. 7016–7030 (2021)

50. Robins, A.: Catastrophic forgetting, rehearsal and pseudorehearsal. Connect. Sci. **7**(2), 123–146 (1995)

51. Rogers, A., Gardner, M., Augenstein, I.: QA dataset explosion: A taxonomy of nlp resources for question answering and reading comprehension. arXiv preprint arXiv:2107.12708 (2021)

52. Ruder, S., et al.: XTREME-R: Towards more challenging and nuanced multilingual evaluation. In: Proceedings of the 2021 Conference on Empirical Methods in Natural Language Processing, pp. 10215–10245 (2021)

53. Ruder, S., Vulić, I., Søgaard, A.: A survey of cross-lingual word embedding models. J. Artif. Intell. Res. **65**, 569–631 (2019)

54. Rudman, W., Gillman, N., Rayne, T., Eickhoff, C.: Isoscore: Measuring the uniformity of embedding space utilization. In: ACL (Findings), pp. 3325–3339. Association for Computational Linguistics (2022)

55. Schuster, T., Ram, O., Barzilay, R., Globerson, A.: Cross-lingual alignment of contextual word embeddings, with applications to zero-shot dependency parsing. In: Proceedings of the 2019 Conference of the North American Chapter of the Association for Computational Linguistics: Human Language Technologies, Volume 1 (Long and Short Papers), pp. 1599–1613 (2019)

56. Schwenk, H., Chaudhary, V., Sun, S., Gong, H., Guzmán, F.: WikiMatrix: Mining 135M parallel sentences in 1620 language pairs from Wikipedia. In: Proceedings of the 16th Conference of the European Chapter of the Association for Computational Linguistics: Main Volume, pp. 1351–1361 (2021)

57. Shehata, D., Arabzadeh, N., Clarke, C.L.A.: Early stage sparse retrieval with entity linking. In: Proceedings of the 31st ACM International Conference on Information & Knowledge Management, pp. 4464–4469. CIKM '22 (2022)

58. Søgaard, A., Ruder, S., Vulić, I.: On the limitations of unsupervised bilingual dictionary induction. In: Proceedings of the 56th Annual Meeting of the Association for Computational Linguistics (Volume 1: Long Papers), pp. 778–788 (2018)

59. Tawfik, N.S.: Text Mining for Precision Medicine: Natural Language Processing, Machine Learning and Information Extraction for Knowledge Discovery in the Health Domain. Ph.D. thesis, Utrecht University, Netherlands (2020)

60. Turc, I., Lee, K., Eisenstein, J., Chang, M.W., Toutanova, K.: Revisiting the primacy of English in zero-shot cross-lingual transfer. arXiv preprint arXiv:2106.16171 (2021)

61. Vaswani, A., et al.: Attention is all you need. In: NIPS, pp. 5998–6008 (2017)

62. Wang, Y., Che, W., Guo, J., Liu, Y., Liu, T.: Cross-lingual BERT transformation for zero-shot dependency parsing. In: Proceedings of the 2019 Conference on Empirical Methods in Natural Language Processing and the 9th International Joint Conference on Natural Language Processing (EMNLP-IJCNLP), pp. 5721–5727. Hong Kong, China (2019)

63. Wang, Z., Xie, J., Xu, R., Yang, Y., Neubig, G., Carbonell, J.: Cross-lingual alignment vs joint training: A comparative study and a simple unified framework. arXiv preprint arXiv:1910.04708 (2019)

64. Wen, M., Vasthimal, D.K., Lu, A., Wang, T., Guo, A.: Building large-scale deep learning system for entity recognition in e-commerce search. In: BDCAT, pp. 149–154. ACM (2019)

65. Williams, A., Nangia, N., Bowman, S.: A broad-coverage challenge corpus for sentence understanding through inference. In: Proceedings of the 2018 Conference of the North American Chapter of the Association for Computational Linguistics: Human Language Technologies, Volume 1 (Long Papers), pp. 1112–1122 (2018)
66. Wu, S., Dredze, M.: Beto, bentz, becas: The surprising cross-lingual effectiveness of BERT. In: Proceedings of the 2019 Conference on Empirical Methods in Natural Language Processing and the 9th International Joint Conference on Natural Language Processing (EMNLP-IJCNLP), pp. 833–844 (2019)
67. Wu, Y., et al.: Google's neural machine translation system: Bridging the gap between human and machine translation. arXiv preprint arXiv:1609.08144 (2016)
68. Zhao, W., Eger, S., Bjerva, J., Augenstein, I.: Inducing language-agnostic multilingual representations. In: Proceedings of *SEM 2021: The Tenth Joint Conference on Lexical and Computational Semantics, pp. 229–240 (2021)

Scene-Centric vs. Object-Centric Image-Text Cross-Modal Retrieval: A Reproducibility Study

Mariya Hendriksen[1]([✉])(iD), Svitlana Vakulenko[2](iD), Ernst Kuiper[3](iD), and Maarten de Rijke[4](iD)

[1] AIRLab, University of Amsterdam, Amsterdam, The Netherlands
m.hendriksen@uva.nl
[2] Amazon, Madrid, Spain
svvakul@amazon.com
[3] Bol.com, Utrecht, The Netherlands
ekuiper@bol.com
[4] University of Amsterdam, Amsterdam, The Netherlands
m.derijke@uva.nl

Abstract. Most approaches to (CMR) focus either on object-centric datasets, meaning that each document depicts or describes a single object, or on scene-centric datasets, meaning that each image depicts or describes a complex scene that involves multiple objects and relations between them. We posit that a robust CMR model should generalize well across both dataset types. Despite recent advances in CMR, the reproducibility of the results and their generalizability across different dataset types has not been studied before. We address this gap and focus on the reproducibility of the state-of-the-art CMR results when evaluated on object-centric and scene-centric datasets. We select two state-of-the-art CMR models with different architectures: (i) CLIP; and (ii) X-VLM. Additionally, we select two scene-centric datasets, and three object-centric datasets, and determine the relative performance of the selected models on these datasets. We focus on reproducibility, replicability, and generalizability of the outcomes of previously published CMR experiments. We discover that the experiments are not fully reproducible and replicable. Besides, the relative performance results partially generalize across object-centric and scene-centric datasets. On top of that, the scores obtained on object-centric datasets are much lower than the scores obtained on scene-centric datasets. For reproducibility and transparency we make our source code and the trained models publicly available.

1 Introduction

Cross-modal retrieval(CMR) is the task of finding relevant items across different modalities. For example, given an image, find a text or vice versa. The main challenge in CMR is known as *the heterogeneity gap* ([5,22]). Since items from

---

Research conducted while the author was at the University of Amsterdam.

different modalities have different data types, the similarity between them cannot be measured directly. Therefore, the majority of CMR methods published to date attempt to bridge this gap by learning a latent representation space, where the similarity between items from different modalities can be measured [57].

In this work, we specifically focus on *image-text* CMR, which uses textual and visual data. The retrieval task is performed on *image-text pairs*. In each image-text pair, the text (often referred to as *caption*) describes the corresponding image it is aligned with. For image-text CMR we use either an image or a text as a query [57]. Hence, the CMR task that we address in this paper consists of two subtasks: (i) *text-to-image retrieval*: given a text that describes an image, retrieve all the images that match this description; and (ii) *image-to-text retrieval*: given an image, retrieve all texts that can be used to describe this image.

Scene-centric vs. Object-centric vs. Datasets. Existing image datasets can be grouped into *scene-centric* and *object-centric* datasets [48,62]. The two types of dataset are typically used for different tasks, viz. the tasks of scene and object understanding, respectively. They differ in important ways that are of interest to us when evaluating performance and generalization abilities of CMR models.

Scene-centric images depict complex scenes that typically feature multiple objects and relations between them. These datasets contain image-text pairs, where, in each pair, an image depicts a complex scene of objects and the corresponding text describes the whole scene, often focusing on *relations and activities*.

Images in object-centric image datasets are usually focused on a single object of interest that they primarily depict. This object is often positioned close to the center of an image with other objects, optionally, in the background. Object-centric datasets contain image-text pairs, where, in each pair, an image depicts an object of interest and the corresponding text describes the depicted *object and its (fine-grained) attributes*.

To illustrate the differences between the two dataset types in CMR, we consider the examples provided in Fig. 1 with an object-centric image-caption (left) and a scene-centric image-caption (right). Note how the pairs differ considerably in terms of the visual style and the content of the caption. The pair on the left focuses on a single object ("pants") and describes its fine-grained visual attributes ("multicolor," "boho," "batic"). The pair on the right captures a scene describing multiple objects ("seagulls," "pier," "people") and relations between them ("sitting," "watching").

Research Goals. We focus on (traditional) CMR methods that extract features from each modality and learn a common representation space. Recent years have seen extensive experimentation with such CMR methods, mostly organized into two groups: (i) contrastive experiments on object-centric datasets [17], and (ii) contrastive experiments on scene-centric datasets [35]. In this paper, we consider representative state-of-the-art CMR methods from both groups. We select two pre-trained models which demonstrate state-of-the-art performance

Seagulls sitting on the ledge of a pier
Multicolor boho batic pants   with people watching

Fig. 1. An object-centric (left) and a scene-centric (right) image-text pair. Sources: Fashion200k (left); MS COCO (right).

on CMR task and evaluate them in a zero-shot setting. In line with designs used in prior reproducibility work on CMR [3] we select two models for the study. Following the ACM terminology [1], we focus on *reproducibility* (different team, same experimental setup) and *replicability* (different team, different experimental setup) of previously reported results. And following Voorhees [55], we focus on relative (a.k.a. comparative) performance results. In addition, for the reproducibility experiment, we consider the absolute difference between the reported scores and the reproduced scores.

We address the following research questions: (RQ1) Are published relative performance results on CMR reproducible? This question matters because it allows us to confirm the validity of reported results. We show that the relative performance results are not fully reproducible. Specifically, the results are reproducible for one dataset, but not for the other dataset).

We then shift to replicability and examine whether lessons learned on scene-centric datasets transfer to object-centric datasets: (RQ2) To what extent are the published relative performance results replicable? That is, we investigate the validity of the reported results when evaluated in a different setup. We find that relative performance results are partially replicable, using other datasets.

After investigating the reproducibility and replicability of the results, we consider the generalizability of the results. We contrastively evaluate the results on object-centric and scene-centric datasets: (RQ3) Do relative performance results for state-of-the-art CMR methods generalize from scene-centric datasets to object-centric datasets? We discover that the relative performance results only partially generalize across the two dataset types.

Main Contributions. Our main contributions are: (i) We are one of the first to consider reproducibility in the context of CMR and reproduce scene-centric CMR experiments from two papers [44,61] and find that the results are only partially reproducible. (ii) We perform a replicability study and examine whether relative performance differences reported for CMR methods generalize from scene-centric datasets to object-centric datasets. (iii) We investigate the generalizability of obtained results and analyze the effectiveness of pre-training on scene-centric datasets for improving the performance of CMR on object-centric datasets, and vice versa. And, finally, (iv) to facilitate the reproducibility of our work, we provide the code and the pre-trained models used in our experiments.[1]

2 Related Work

Cross-modal Retrieval. CMR methods attempt to construct a multimodal representation space, where the similarity of concepts from different modalities can be measured. Some of the earliest approaches in CMR utilised canonical correlation analysis [15,26]. They were followed by a dual encoder architecture equipped with a recurrent and a convolutional component, a hinge loss [12,58] and hard-negative mining [11]. Later on, several attention-based architectures were introduced such as architectures with dual attention [39], stacked cross-attention [31], bidirectional focal attention [36].

Another line of work proposed to use transformer encoders [54] for CMR task [38], and adapted the BERT model [8] as a backbone [13,67]. Some other researchers worked on improving CMR via modality-specific graphs [56], or image and text generation modules [16].

There is also more domain-specific work that focused on CMR in fashion [14, 28–30], e-commerce [19,20], cultural heritage [49] and cooking [56].

In contrast to the majority of prior work on the topic, we focus on the reproducibility, replicability, and generalizability of CMR methods. In particular, we explore the state-of-the-art models designed for the CMR task by examining their performance on scene-centric and object-centric datasets.

Scene-centric and Object-centric Datasets. The majority of prior work related to object-centric and scene-centric datasets focuses on computer vision tasks such as object recognition, object classification, and scene recognition. Herranz et al. [21] investigated biases in a CNN when trained on scene-centric versus object-centric datasets and evaluated on the task of object classification.

In the context of object detection, prior work focused on combining feature representations learned from object-centric and scene-centric datasets to improve the performance when detecting small objects [48], and using object-centric images to improve the detection of objects that do not appear frequently in complex scenes [62]. Finally, for the task of scene recognition, Zhou et al. [66] explored the quality of feature representations learned from both scene-centric and object-centric datasets and applied to the task of scene recognition.

Unlike prior work on the topic, in this paper, we focus on both scene-centric and object-centric datasets for evaluation on CMR task. In particular, we explore

---

[1] https://github.com/mariyahendriksen/ecir23-object-centric-vs-scene-centric-CMR.

how state-of-the-art(SOTA) CMR models perform on object-centric and scene-centric datasets.

Reproducibility in Cross-modal Retrieval. To the best of our knowledge, despite the popularity of the CMR task, there are very few papers that focus on reproducibility of research in CMR. Some rare (recent) examples include [3], where the authors survey metric learning losses used in computer vision and explore their applicability for CMR. Rao et al. [45] analyze contributing factors that affect the performance of the state-of-the-art CMR models. However, all prior work focuses on exploring model performance only on two popular scene-centric datasets: Microsoft COCO (MS COCO) and Flickr30k.

In contrast, in this work, we take advantage of the diversity of the CMR datasets and specifically focus on examining how the state-of-the-art CMR models perform across different dataset types: scene-centric and object-centric datasets.

3   Task Definition

We follow the same notation as in previous work [4,53,65]. An image-caption cross-modal dataset consists of a set of images $\mathcal{I}$ and texts $\mathcal{T}$ where the images and texts are aligned as image-text pairs: $\mathcal{D} = \{(\mathbf{x}_{\mathcal{I}}^1, \mathbf{x}_{\mathcal{T}}^1), \ldots, (\mathbf{x}_{\mathcal{I}}^n, \mathbf{x}_{\mathcal{T}}^n)\}$.

The *cross-modal retrieval* (CMR) task is defined analogous to the standard information retrieval task: given a query $\mathbf{q}$ and a set of m candidates $\Omega_{\mathbf{q}} = \{\mathbf{x}^1, \ldots, \mathbf{x}^m\}$ we aim to rank all the candidates w.r.t. their relevance to the query $\mathbf{q}$. In CMR, the query can be either a text $\mathbf{q}_{\mathcal{T}}$ or an image $\mathbf{q}_{\mathcal{I}}$: $\mathbf{q} \in \{\mathbf{q}_{\mathcal{T}}, \mathbf{q}_{\mathcal{I}}\}$. Similarly, the set of candidate items can be either visual $\mathcal{I}_{\mathbf{q}} \subset \mathcal{I}$, or textual $\mathcal{T}_{\mathbf{q}} \subset \mathcal{T}$ data: $\Omega \in \{\mathcal{I}_{\mathbf{q}}, \mathcal{T}_{\mathbf{q}}\}$.

The CMR task is performed across modalities, therefore, if the query is a text then the set of candidates are images, and vice versa. Hence, the task comprises effectively two subtasks: (i) *text-to-image retrieval*: given a textual query $\mathbf{q}_{\mathcal{T}}$ and a set of candidate images $\Omega \subset \mathcal{I}$, we aim to rank all instances in the set of candidate items Ω w.r.t. their relevance to the query $\mathbf{q}_{\mathcal{T}}$; (ii) *image-to-text retrieval*: given an image as a query $\mathbf{q}_{\mathcal{I}}$ and a set of candidate texts $\Omega \subset \mathcal{T}$, we aim to rank all instances in the set of candidate items Ω w.r.t. their relevance to the query $\mathbf{q}_{\mathcal{I}}$.

4   Methods

In this section, we give an overview of the models included in the study, of the models which were excluded, and provide justification for it. All the approaches we focus on belong to the traditional CMR framework and comprise two stages. First, we extract textual and visual features. The features are typically extracted with a textual encoder and a visual encoder. Next, we learn a latent representation space where the similarity of items from different modalities can be measured directly.

4.1 Methods Included for Comparison

We focus on CMR in *zero-shot setting*, hence, we only consider pre-trained models. Therefore, we focus on the models that are released for public use. Besides, as explained in Sect. 1, we follow prior reproducibility work to inform our experimental choices regarding the number of models. Given the above-mentioned requirements, we selected two methods that demonstrate state-of-the-art performance on the CMR task: CLIP and X-VLM.

Contrastive Language-Image Pretraining(CLIP) [44]. This model is a dual encoder that comprises an image encoder, and a text encoder. The model was pre-trained in a contrastive manner using a symmetric loss function. It is trained on 400 million image-caption pairs scraped from the internet. The text encoder is a transformer [54] with modification from [43]. For the image encoder, the authors present two architectures. The first one is based on ResNet [18] and it is represented in five variants in total. The first two options are ResNet-50, ResNet-101; the last three options are variants of ResNet scaled up in the style of EfficientNet [51]. The second image encoder architecture is a Vision Transofrmer (ViT) [9]. It is presented in three variants: ViT-B/32, a ViT-B/16, and a ViT-L/14. The CMR results reported in the original paper are obtained with a model configuration where vision transformer ViT-L/14 is used as an image encoder, and the text transformer is a text encoder. Hence, we use this configuration in our experiments.

X-VLM [61]. This model consists of three encoders: an image encoder, a text encoder, and a cross-modal encoder. The image and text encoder take an image and text as inputs and output their visual and textual representations. The cross-modal encoder fuses the output of the image encoder and the output of the text encoder. The fusion is done via a cross-attention mechanism. For CMR task, the model is fine-tuned via a contrastive learning loss and a matching loss. All encoders are transformer-based. The image encoder is a ViT initialised with Swin Transformer$_{base}$ [37]. Both the text encoder and the cross-modal encoder are initialised using different layers of BERT [8]: the text encoder is initialized using the first six layers, whereas the cross-modal encoder is initialised using the last six layers.

4.2 Methods Excluded from Comparison

While selecting the models for the experiments, we considered other architectures with promising performance on the MS COCO and the Flickr30k datasets. Below, we outline the architectures we considered and explain why they were not included.

Several models such as Visual N-Grams [32], Unicoder-VL [33], ViLT-B/32 [25], UNITER [6] were excluded because they were consistently outperformed by CLIP on the MS COCO and Flickr30k datasets by large margins.

Besides, we excluded ImageBERT [42] because it was outperformed by CLIP on the MS COCO dataset. ALIGN [23], ALBEF [34], VinVL [64], METER [10] were not included because X-VLM consistently outperformed them. UNITER [6] was beaten by both CLIP and X-VLM. We did not include other well-performing models such as ALIGN [23], Flamingo [2], CoCa [60] because the pre-trained models were not publicly available.

5 Experimental Setup

In this section, we discuss our experimental design including the choice of datasets, subtasks, metrics, and implementation details.

5.1 Datasets

We run experiments on two scene-centric and three object-centric datasets. Below, we discuss each of the datasets in more detail.

Scene-centric Datasets. We experiment with two scene-centric datasets: (i) Microsoft COCO (MS COCO) [35] contains 123,287 images depicting regular scenes from everyday life with multiple objects placed in their natural contexts. There are 91 different object types such as "person", "bicycle", "apple". (ii) Flickr30k contains 31,783 images of regular scenes from everyday life, activities, and events. For both scene-centric datasets, we use the splits provided in [24]. The MS COCO dataset is split into 113,287 images for training, 5,000 for testing and 5,000 for validation; the Flickr30k dataset has 29,783 images for training, 1,000 for testing and 1,000 for validation. In both datasets, every image was annotated with five captions using Amazon Mechanical Turk. Besides, we select one caption per image randomly, and use the test set for our experiments.

Object-centric Datasets. We consider three object-centric datasets in our experiments: (i) Caltech-UCSD Birds 200 (CUB-2000) [59] contains 11,788 images of 200 birds species. Each image is annotated with a fine-grained caption from [46]. We selected one caption per image randomly. Each caption is at least 10 words long and does not contain any information about the birds' species or actions. (ii) Fashion200k contains 209,544 images that depict various fashion items in five product categories (dress, top, pant, skirt, jacket) and their corresponding descriptions. (iii) Amazon Berkley Objects(ABO) [7] contains 147,702 product listings associated with 398,212 images. This dataset was derived from Amazon.com product listings. We selected one image per listing and used the associated product description as its caption. The majority of images depict a single product on a white background. The product is located in the center of the image and takes at least 85% of the image area. For all object-centric datasets, we use the splits provided by the dataset authors and use the test split for our experiments.

5.2 Subtasks

Our goal is to assess and compare the performance of the CMR methods (described in Sect. 5) across the object-centric and scene-centric datasets described in the previous subsection. We design an experimental setup that takes into account two CMR subtasks and two dataset types. It can be summarized using a tree with branches that correspond to different configurations (see Fig. 2). We explain how we cover the branches of this tree in the next subsection.

The tree starts with a root ("Image-text CMR" with label 0) that has sixteen descendants, in total. The root node has two children corresponding to the two image-text CMR subtasks: a text-to-image retrieval (node 1) and image-to-text retrieval (node 2). Since we want to evaluate each of these subtasks on both object-centric and scene-centric datasets, nodes 1 and 2 also have two children each, i.e., the nodes {3, 4, 5, 6}. Finally, every object-centric node has three children: CUB-200, Fashion200k, and ABO datasets {7, 8, 9, 12, 13, 14}; and every scene-centric node has two children: MS COCO and Flickr30k datasets {10, 11, 15, 16}.

5.3 Experiments

To answer the research questions introduced in Sect. 1, we conduct two experiments. In all the experiments, we use CLIP and X-VLM models in a zero-shot setting. Following [55], we focus on relative performance results. In each experiment, we consider different subtrees from Fig. 2. Following [25,32,33,44,61], we use Recall@K where $K = \{1, 5, 10\}$ to evaluate the model performance in all our experiments. In addition, following [50,52,63], we calculate the sum of recalls (rsum) for text-to-image, and image-to-text retrieval tasks as well as the total sum of recalls for both tasks.

For text-to-image retrieval, we first obtain representations for all the candidate images by passing them through the image encoder of the model. Then we

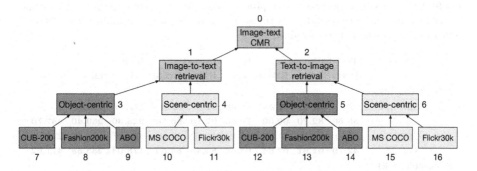

Fig. 2. Our experimental design for evaluating CMR methods across object-centric and scene-centric datasets. The blue colour indicates parts of the tree used in Experiment 1, the green color indicates parts of the tree used in Experiment 2, and the red color indicates parts used in all experiments. (Best viewed in color.) (Color figure online)

pass each textual query through the text encoder of the model and retrieve the top-k candidates ranked by cosine similarity w.r.t. the query.

For image-to-text retrieval, we do the reverse, using the texts as candidates and images as queries. More specifically, we start by obtaining representations of the candidate captions by passing them through the text encoder. Afterwards, for each of the visual queries, we pass the query through the image encoder and retrieve top-k candidates ranked by cosine similarity w.r.t. the query.

In *Experiment 1* we evaluate the reproducibility of the CMR results reported in the original publications (RQ1). Both models we consider (CLIP and X-VLM) were originally evaluated on two scene-centric datasets, viz. MS COCO and Flickr30k. Therefore, for our reproducibility study, we also evaluate these models on these two datasets. We evaluate both text-to-image and image-to-text retrieval. That is, we focus on the two sub-trees $0 \leftarrow 1 \leftarrow 4 \leftarrow \{10, 11\}$ and $0 \leftarrow 2 \leftarrow 6 \leftarrow \{15, 16\}$ (the red and blue parts of the tree) from Fig. 2. In addition to relative performance results, we consider absolute differences between the reported scores and the reproduced scores. Following Petrov and Macdonald [41], we assume that the score is reproduced if we obtain a score value equal to the reported score given a relative tolerance of $\pm 5\%$.

In *Experiment 2* we focus on the replicability of the reported results on object-centric datasets (RQ2). Thus, we evaluate CLIP and X-VLM on the CUB-200, Fashion200k, and ABO datasets. This experiment covers the subtrees $0 \leftarrow 1 \leftarrow 3 \leftarrow \{7, 8, 9\}$ and $0 \leftarrow 2 \leftarrow 5 \leftarrow \{12, 13, 14\}$ (the red and green parts of the tree) in Fig. 2.

After obtaining the results from Experiment 1 and 2, we examine the generalizability of the obtained scores (RQ3). We do so by comparing the relative performance results the models achieve on the object-centric versus scene-centric datasets. More specifically, we compare the relative performance of CLIP and X-VLM on CUB-200, Fashion200k, ABO with their relative performance on MS COCO and Flickr30k. Thus, this experiment captures the complete tree in Fig. 2.

Table 1. Results of experiment 1 (reproducibility study), using the MS COCO and Flickr30k datasets. "Orig." indicates the scores from the original publications. "Repr." indicates the scores that we obtained.

	Model	Text-to-image			Image-to-text			Rsum		
		R@1	R@5	R@10	R@1	R@5	R@10	t2i	i2t	total
MS COCO (5k)										
Orig.	CLIP [44]	37.80	62.40	72.20	58.40	81.50	88.10	172.40	228.00	400.40
	X-VLM [61]	**55.60**	**82.70**	**90.00**	**70.80**	**92.10**	**96.50**	**228.30**	**259.40**	**487.70**
Repr	CLIP	21.59	40.22	49.80	24.36	44.13	53.41	111.61	121.90	233.51
	X-VLM	**42.79**	**67.61**	**67.64**	**64.60**	**84.48**	**84.50**	**178.04**	**233.58**	**411.62**
Flickr30k (1 k)										
Orig.	CLIP [44]	68.70	90.60	95.20	**88.00**	**98.70**	99.40	254.50	**286.10**	540.60
	X-VLM [61]	**71.90**	**93.30**	**96.40**	85.30	97.80	**99.60**	**261.60**	282.70	**544.30**
Repr.	CLIP	**74.95**	**93.09**	**96.15**	**77.02**	**94.18**	**96.84**	**264.19**	**268.04**	**532.23**
	X-VLM	37.82	82.36	82.48	63.30	91.10	91.10	202.66	245.50	448.16

6 Results

We focus on the reproducibility (different team, same setup) and replicability (different team, different setup) of the CMR experiments reported in the original papers devoted to CLIP [44] and X-VLM [61]. To organize our result presentation, we refer to the tree in Fig. 2. We traverse the tree bottom up, from the leaves to the root.

6.1 RQ1: Reproducibility

To address RQ1, we report on the outcomes of Experiment 1. We investigate to what extent the CMR results reported in the original papers devoted to CLIP [44] and X-VLM [61] are reproducible. Given that both methods were originally evaluated on two scene-centric datasets, viz. MS COCO and Flickr30k, we evaluate the models on the text-to-image and image-to-text tasks on these two datasets. Therefore, we focus on the two blue sub-trees $0 \leftarrow 1 \leftarrow 4 \leftarrow \{10, 11\}$ and $0 \leftarrow 2 \leftarrow 6 \leftarrow \{15, 16\}$ from Fig. 2.

Results. The results of Experiment 1 are shown in Table 1. We recall the scores obtained in the original papers [44,61] ("Orig.") and the scores that we obtained ("Repr."), on the MS COCO and Flickr30k datasets. Across the board, the scores that we obtained (the "reproduced scores") tend to be lower than the scores obtained in the original publications (the "original scores").

On the MS COCO dataset, X-VLM consistently outperforms CLIP, both in the original publications and in our setup, for both the text-to-image and the image-to-text tasks. Moreover, this holds for all R@n metrics, and, hence, for the Rsum metrics. Interestingly, the relative gains that we obtain tend to be larger than the ones obtained in the original publications. For example, our biggest relative difference is for the image-to-text task in terms of the R@1 metric: according to the scores reported in [44,61], X-VLM outperforms CLIP by 21%, whereas in our experiments the relative gain is 165%.

On average, the original CLIP scores are as much as ∼70% higher than the reproduced scores; the original scores for X-VLM are ∼20% higher than the reproduced ones. When considering the absolute differences between the original scores and the reproduced scores and assuming a relative tolerance of ±5%, we see that, on the MS COCO dataset, the scores are not reproducible for both models.

On the Flickr30k dataset, we see a different pattern. For the text-to-image task, the original results indicate that X-VLM consistently outperforms CLIP, on all R@n metrics, but according to our results, the relative order is consistently reversed. For the image-to-text task, we obtained mixed outcomes: for R@1 and R@5, the original order (CLIP outperforms X-VLM) is confirmed, but for R@10 the order is swapped. According to our experimental results, however, CLIP consistently outperforms X-VLM on all tasks, and on all R@n metrics (and hence also on the Rsum metrics).

On the Flickr30k dataset, the CLIP scores are reproduced on the text-to-image and image-to-text retrieval tasks when the model is evaluated on R@5 and R@10. On the text-to-image task, the reproduced R@5 score is 2.7% higher than the original score; the reproduced R@10 score is 1% higher than the original score. For the image-to-text retrieval task, the reproduced R@5 score is 4% lower than the original score; the reproduced R@10 score is 2% lower than the original score.

Answer to RQ1. In the case of the CLIP model, the obtained *absolute* scores were reproducible only on the Flickr30k dataset for the text-to-image and the image-to-text tasks when evaluated on R@5 and R@10. For X-VLM, we did not find the absolute scores obtained when evaluating the model on the MS COCO and Flickr20k datasets to be reproducible, neither for the text-to-image nor the image-to-text tasks.

The *relative* outcomes on the MS COCO dataset could be reproduced, for all tasks and metrics, whereas on the Flickr30k dataset they could only partially be reproduced, that is, only for the image-to-text task on the R@1 and R@5 metrics; for the text-to-image task, X-VLM outperforms CLIP according to the original scores, but CLIP outperforms X-VLM according to our reproduced scores.

Upshot. As explained in Sect. 4, in this paper we focus on CMR in a zero-shot setting. This implies that the differences that we observed between the original scores and the reproduced scores must be due to differences in text and image data (pre-)processing and loading. We, therefore, recommend that the future work includes (as much as is practically possible) tools and scripts used in these stages of the experiment with the publication of its implementations.

6.2   RQ2: Replicability

To answer RQ2, we replicate the originally reported text-to-image and image-to-text retrieval experiments in a different setup, i.e., by evaluating CLIP and X-VLM using object-centric datasets instead of scene-centric datasets. Thus, we evaluate CLIP and X-VLM on the CUB-200, Fashion200k, and ABO datasets and focus on the green subtrees $0 \leftarrow 1 \leftarrow 3 \leftarrow \{7, 8, 9\}$ and $0 \leftarrow 2 \leftarrow 5 \leftarrow \{12, 13, 14\}$ from Fig. 2.

Results. The results of Experiment 2 (aimed at answering RQ2) can be found in Table 2. On the CUB-200 dataset, CLIP consistently outperforms X-VLM. The biggest relative increase is 124% for image-to-text in terms of R@10, while the smallest relative increase is 1% for text-to-image in terms of R@1. Overall, on the text-to-image retrieval task, CLIP outperforms X-VLM by 38%, and on the image-to-text retrieval task, the relative gain is 70%.

Table 2. Results of Experiment 2 (replicability study), using the CUB-200, Fashion200k, and ABO datasets.

Model	Text-to-image			Image-to-text			Rsum		
	R@1	R@5	R@10	R@1	R@5	R@10	t2i	i2t	total
CUB-200									
CLIP	**0.71**	**2.38**	**4.42**	**1.23**	**3.40**	**5.48**	**7.51**	**10.11**	**17.62**
X-VLM	0.70	2.28	2.45	1.16	2.35	2.45	5.43	5.96	11.39
Fashion200k									
CLIP	**3.05**	**8.56**	**12.85**	**3.43**	**9.82**	**14.56**	**24.46**	**27.81**	**52.27**
X-VLM	2.80	6.62	6.70	1.84	3.96	4.04	16.12	09.84	25.96
ABO									
CLIP	**6.25**	**13.90**	**18.50**	**7.99**	**18.96**	**25.57**	**38.65**	**52.52**	**91.17**
X-VLM	3.10	6.48	6.56	3.20	7.42	7.50	16.14	18.12	34.26

On Fashion200k, CLIP outperforms X-VLM, too. The smallest relative increase is 9% for text-to-image in terms of R@1, the biggest relative increase is 260% for image-to-text in terms of R@10. In general, on the text-to-image retrieval task, CLIP outperforms X-VLM by 52%; on the image-to-text retrieval task, the relative gain is 83%.

Finally, on the ABO dataset, CLIP outperforms X-VLM again. The smallest relative increase is 101% for text-to-image in terms of R@1, the biggest relative increase is 241% for image-to-text again in terms of R@10. In general, on the text-to-image retrieval task, CLIP outperforms X-VLM by 139%; on the image-to-text retrieval task, the relative gain is 190%. All in all, CLIP outperforms X-VLM on all three scene-centric datasets. The overall relative gain on CUB-200 dataset is 55%, on Fashion200k dataset −101%. The biggest relative gain of 166% is obtained on the ABO dataset.

Answer to RQ2. The outcome of Experiment 2 is clear. The original relative performance results obtained on the MS COCO and Flickr30k (Table 1) are only partially replicable to the CUB-200, Fashion200k, and ABO datasets. On the latter datasets CLIP consistently outperforms X-VLM by a large margin, whereas the original scores obtained on the former datasets indicate that X-VLM mostly outperforms CLIP.

Upshot. We hypothesize that the failure to replicate the relative results originally reported for scene-centric datasets (viz. X-VLM outperforms CLIP) is due to CLIP being pre-trained on more and more diverse image data. We, therefore, recommend that future work aimed at developing large-scale CMR models quantifies and reports the diversity of the training data used.

6.3   RQ3: Generalizability

To answer RQ3, we compare the relative performance of the selected models on object-centric and scene-centric data. Thus, we compare the relative performance of CLIP and X-VLM on CUB-200, Fashion200k, ABO with their relative performance on MS COCO and Flickr30k. We focus on the complete tree from Fig. 2.

Results. The results of our experiments on the scene-centric datasets are in Table 1; the results that we obtained on the object-centric datasets are in Table 2. On object-centric datasets, CLIP consistently outperforms X-VLM. However, the situation with scene-centric results is partially the opposite. There, X-VLM outperforms CLIP on the MS COCO dataset.

Answer to RQ3. Hence, we answer RQ3 by stating that the relative performance results for CLIP and X-VLM that we obtained in our experiments only partially generalize from scene-centric to object-centric datasets. The MS COCO dataset is the odd one out.[2]

Upshot. Given the observed differences in relative performance results for CLIP and X-VLM on scene-centric vs. object-centric datasets, we recommend that CMR be trained in both scene-centric and object-centric datasets to help improve the generalizability of experimental outcomes.

7   Discussion and Conclusions

We have examined two SOTA image-text CMR methods, CLIP and X-VLM, by contrasting their performance on two scene-centric datasets (MS COCO and Flicrk30k) and three object-centric datasets (CUB-200, Fashion200k, ABO) in a zero-shot setting.

We focused on the *reproducibility* of the CMR results reported in the original publications when evaluated on the selected scene-centric datasets. The reported scores were not reproducible for X-VLM when evaluated on the MS COCO and the Flickr30k datasets. For CLIP, we were able to reproduce the scores on the Flickr30k dataset when evaluated using R@5 and R@10. Conversely, the relative results were reproducible on the MS COCO dataset, for all metrics and tasks, and partially reproducible on the Flickr30k dataset only for image-to-text task when evaluated on R@1 and R@5. We also examined the *replicability* of the CMR results using three object-centric datasets. We discovered that the relative results are replicable when we compare the relative performance on the object-centric datasets with the relative scores on the Flickr30k dataset. However, for

---

[2] On the GitHub repository for CLIP, several issues have been posted related to the performance of CLIP on the MS COCO dataset. See, e.g., https://github.com/openai/CLIP/issues/115.

the MS COCO dataset, the relative outcomes were not replicable. And, finally, we explored the generalizability of the obtained results by comparing the models' performance on scene-centric vs. object-centric datasets. We observed that the absolute scores obtained when evaluating models on object-centric datasets are much lower than the scores obtained on scene-centric datasets.

Our findings demonstrate that the reproducibility of CMR methods on scene-centric datasets is an open problem. Besides, we show that while the majority of CMR methods are evaluated on the MS COCO and the Flickr30k datasets, the object-centric datasets represent a challenging and relatively unexplored set of benchmarks.

A limitation of our work is the relatively small number of scene-centric and object-centric datasets used for the evaluation of the models. Another limitation is that we only considered CMR in a zero-shot setting, ignoring, e.g., few-shot scenarios; this limitation did, however, come with the important advantage of reducing the number of experimental design decisions to be made for contrastive experiments.

A promising direction for future work is to include further datasets when contrasting the performance of CMR models, both scene-centric and object-centric. In particular, it would be interesting to investigate the models' performance on datasets, e.g., Conceptual Captions [47], the Flower [40], and the Cars [27] datasets. A natural step after that would be to consider few-shot scenarios.

Acknowledgements. We thank Paul Groth, Andrew Yates, Thong Nguyen, and Maurits Bleeker for helpful discussions and feedback.

This research was supported by Ahold Delhaize, and the Hybrid Intelligence Center, a 10-year program funded by the Dutch Ministry of Education, Culture and Science through the Netherlands Organisation for Scientific Research, https://hybrid-intelligence-centre.nl.

All content represents the opinion of the authors, which is not necessarily shared or endorsed by their respective employers and/or sponsors.

References

1. ACM (2020) Artifact Review and Badging - Current. https://www.acm.org/publications/policies/artifact-review-and-badging-current Accessed Aug 7 2022
2. Alayrac, J.B., et al.: Flamingo: a visual language model for few-shot learning. arXiv preprint arXiv:2204.14198 (2022)
3. Bleeker, M., de Rijke, M.: Do lessons from metric learning generalize to image-caption retrieval? In: Hagen, M., et al. (eds.) ECIR 2022. LNCS, vol. 13185, pp. 535–551. Springer, Cham (2022). https://doi.org/10.1007/978-3-030-99736-6_36
4. Brown, A., Xie, W., Kalogeiton, V., Zisserman, A.: Smooth-AP: smoothing the path towards large-scale image retrieval. In: Vedaldi, A., Bischof, H., Brox, T., Frahm, J.-M. (eds.) ECCV 2020. LNCS, vol. 12354, pp. 677–694. Springer, Cham (2020). https://doi.org/10.1007/978-3-030-58545-7_39
5. Carvalho, M., Cadène, R., Picard, D., Soulier, L., Thome, N., Cord, M.: Cross-modal retrieval in the cooking context: Learning semantic text-image embeddings. In: The 41st International ACM SIGIR Conference on Research & Development in Information Retrieval, pp. 35–44 (2018)

6. Chen, Y.C., et al.: Uniter: Learning universal image-text representations. In: Computer Vision - ECCV 2020, Springer International Publishing, pp. 104–120 (2020)

7. Collins, J., et al.: Abo: Dataset and benchmarks for real-world 3d object understanding. CVPR (2022)

8. Devlin, J., Chang, M.W., Lee, K., Toutanova, K.: Bert: Pre-training of deep bidirectional transformers for language understanding. arXiv preprint (2018). arXiv:1810.04805

9. Dosovitskiy, A., et al. (2020) An image is worth 16x16 words: Transformers for image recognition at scale. arXiv preprint arXiv:2010.11929

10. Dou, Z.Y., et al.: An empirical study of training end-to-end vision-and-language transformers. In: Proceedings of the IEEE/CVF Conference on Computer Vision and Pattern Recognition, pp. 18166–18176 (2022)

11. Faghri, F., Fleet, D.J., Kiros, J.R., Fidler, S.: Vse++: Improving visual-semantic embeddings with hard negatives (2017). arXiv preprint arXiv:1707.05612

12. Frome, A.: Devise: A deep visual-semantic embedding model. In: Proceedings of the 26th International Conference on Neural Information Processing Systems - Volume 2, Curran Associates Inc., NIPS'13, pp 2121–2129 (2013)

13. Gao, D.: Fashionbert: Text and image matching with adaptive loss for cross-modal retrieval. In: Proceedings of the 43rd International ACM SIGIR Conference on Research and Development in Information Retrieval, pp. 2251–2260 (2020)

14. Goei, K., Hendriksen, M., de Rijke, M.: Tackling attribute fine-grainedness in cross-modal fashion search with multi-level features. In: SIGIR 2021 Workshop on eCommerce, ACM (2021)

15. Gong, Y., Wang, L., Hodosh, M., Hockenmaier, J., Lazebnik, S.: Improving image-sentence embeddings using large weakly annotated photo collections. In: Fleet, D., Pajdla, T., Schiele, B., Tuytelaars, T. (eds.) ECCV 2014. LNCS, vol. 8692, pp. 529–545. Springer, Cham (2014). https://doi.org/10.1007/978-3-319-10593-2_35

16. Gu, J., Cai, J., Joty, S.R., Niu, L., Wang, G.: Look, imagine and match: Improving textual-visual cross-modal retrieval with generative models. In: Proceedings of the IEEE conference on computer vision and pattern recognition, pp. 7181–7189 (2018)

17. Han, X., et al.: Automatic spatially-aware fashion concept discovery. In: Proceedings of the IEEE international Conference on Computer Vision, pp. 1463–1471 (2017)

18. He, K., Zhang, X., Ren, S., Sun, J.: Deep residual learning for image recognition. In: Proceedings of the IEEE Conference on Computer Vision and Pattern Recognition, pp. 770–778 (2016)

19. Hendriksen, M.: Multimodal retrieval in e-commerce. In: Hagen, M., et al. (eds.) ECIR 2022. LNCS, vol. 13186, pp. 505–512. Springer, Cham (2022). https://doi.org/10.1007/978-3-030-99739-7_62

20. Hendriksen, M., Bleeker, M., Vakulenko, S., van Noord, N., Kuiper, E., de Rijke, M.: Extending CLIP for category-to-image retrieval in e-commerce. In: Hagen, M., et al. (eds.) ECIR 2022. LNCS, vol. 13185, pp. 289–303. Springer, Cham (2022). https://doi.org/10.1007/978-3-030-99736-6_20

21. Herranz, L., Jiang, S., Li, X.: Scene recognition with cnns: objects, scales and dataset bias. In: Proceedings of the IEEE Conference on Computer Vision and Pattern Recognition, pp. 571–579 (2016)

22. Hu, P., Zhen, L., Peng, D., Liu, P.: Scalable deep multimodal learning for cross-modal retrieval. In: Proceedings of the 42nd international ACM SIGIR conference on research and development in information retrieval, pp. 635–644 (2019)

23. Jia, C., et al.: Scaling up visual and vision-language representation learning with noisy text supervision. In: International Conference on Machine Learning, PMLR, pp. 4904–4916 (2021)
24. Karpathy, A., Fei-Fei, L.: Deep visual-semantic alignments for generating image descriptions. In: Proceedings of the IEEE conference on computer vision and pattern recognition, pp. 3128–3137 (2015)
25. Kim, W., Son, B., Kim, I.: Vilt: Vision-and-language transformer without convolution or region supervision. In: International Conference on Machine Learning, PMLR, pp. 5583–5594 (2021)
26. Klein, B., Lev, G., Sadeh, G., Wolf, L.: Fisher vectors derived from hybrid gaussian-laplacian mixture models for image annotation. (2014) arXiv preprint arXiv:1411.7399
27. Krause, J., Stark, M., Deng, J., Fei-Fei, L.: 3d object representations for fine-grained categorization. In: 4th International IEEE Workshop on 3D Representation and Recognition (3dRR-13), Sydney, Australia (2013)
28. Laenen, K.: Cross-modal representation learning for fashion search and recommendation. PhD thesis, KU Leuven (2022)
29. Laenen, K., Zoghbi, S., Moens, M.F.: Cross-modal search for fashion attributes. In: Proceedings of the KDD 2017 Workshop on Machine Learning Meets Fashion, ACM, vol 2017, pp. 1–10 (2017)
30. Laenen, K., Zoghbi, S., Moens, M.F.: Web search of fashion items with multimodal querying. In: Proceedings of the Eleventh ACM International Conference on Web Search and Data Mining, pp. 342–350 (2018)
31. Lee, K.H., Chen, X., Hua, G., Hu, H., He, X.: Stacked cross attention for image-text matching. In: Proceedings of the European Conference on Computer Vision (ECCV), pp. 201–216 (2018)
32. Li, A., Jabri, A., Joulin, A., Van Der Maaten, L.: Learning visual n-grams from web data. In: Proceedings of the IEEE International Conference on Computer Vision, pp. 4183–4192 (2017)
33. Li, G., Duan, N., Fang, Y., Gong, M., Jiang, D.: Unicoder-vl: a universal encoder for vision and language by cross-modal pre-training. Proc. AAAI Conf. Artif. Intell. **34**, 11336–11344 (2020)
34. Li, J., Selvaraju, R., Gotmare, A., Joty, S., Xiong, C., Hoi, S.C.H.: Align before fuse: vision and language representation learning with momentum distillation. Adv. Neural. Inf. Process. Syst. **34**, 9694–9705 (2021)
35. Lin, T.-Y., Maire, M., Belongie, S., Hays, J., Perona, P., Ramanan, D., Dollár, P., Zitnick, C.L.: Microsoft COCO: common objects in context. In: Fleet, D., Pajdla, T., Schiele, B., Tuytelaars, T. (eds.) ECCV 2014. LNCS, vol. 8693, pp. 740–755. Springer, Cham (2014). https://doi.org/10.1007/978-3-319-10602-1_48
36. Liu, C., Mao, Z., Liu, A.A., Zhang, T., Wang, B., Zhang, Y.: Focus your attention: A bidirectional focal attention network for image-text matching. In: Proceedings of the 27th ACM International Conference on Multimedia, pp. 3–11 (2019)
37. Liu, Z.: Swin transformer: Hierarchical vision transformer using shifted windows. In: Proceedings of the IEEE/CVF International Conference on Computer Vision, pp. 10012–10022 (2021)
38. Messina, N., Amato, G., Esuli, A., Falchi, F., Gennaro, C., Marchand-Maillet, S.: Fine-grained visual textual alignment for cross-modal retrieval using transformer encoders. ACM Transactions on Multimedia Computing, Communications, and Applications (TOMM) **17**(4), 1–23 (2021)

39. Nam, H., Ha, J.W., Kim, J.: Dual attention networks for multimodal reasoning and matching. In: Proceedings of the IEEE Conference on Computer Vision and Pattern Recognition, pp. 299–307 (2017)

40. Nilsback, M.E., Zisserman, A.: Automated flower classification over a large number of classes. In: Indian Conference on Computer Vision, Graphics and Image Processing (2008)

41. Petrov, A., Macdonald, C.: A systematic review and replicability study of bert4rec for sequential recommendation. In: Proceedings of the 16th ACM Conference on Recommender Systems, pp. 436–447 (2022)

42. Qi, D., Su, L., Song, J., Cui, E., Bharti, T., Sacheti, A.: Imagebert: Cross-modal pre-training with large-scale weak-supervised image-text data. (2020) arXiv preprint arXiv:2001.07966

43. Radford, A., Wu, J., Child, R., Luan, D., Amodei, D., Sutskever, I., et al.: Language models are unsupervised multitask learners. OpenAI blog **1**(8), 9 (2019)

44. Radford, A., et al.: Learning transferable visual models from natural language supervision. In: International Conference on Machine Learning, PMLR, pp. 8748–8763 (2021)

45. Rao, J, et al.: Where does the performance improvement come from?: - A reproducibility concern about image-text retrieval. In: SIGIR '22: The 45th International ACM SIGIR Conference on Research and Development in Information Retrieval, Madrid, Spain, July 11–15, 2022, ACM, pp. 2727–2737 (2022)

46. Reed, S., Akata, Z., Lee, H., Schiele, B.: Learning deep representations of fine-grained visual descriptions. In: Proceedings of the IEEE Conference on Computer Vision and Pattern Recognition, pp. 49–58 (2016)

47. Sharma, P., Ding, N., Goodman, S., Soricut, R.: Conceptual captions: A cleaned, hypernymed, image alt-text dataset for automatic image captioning. In: Proceedings of the 56th Annual Meeting of the Association for Computational Linguistics (Volume 1: Long Papers), pp. 2556–2565 (2018)

48. Shen, Z.Y., Han, S.Y., Fu, L.C., Hsiao, P.Y., Lau, Y.C., Chang, S.J.: Deep convolution neural network with scene-centric and object-centric information for object detection. Image Vis. Comput. **85**, 14–25 (2019)

49. Sheng, S., Laenen, K., Van Gool, L., Moens, M.F.: Fine-grained cross-modal retrieval for cultural items with focal attention and hierarchical encodings. Computers **10**(9), 105 (2021)

50. Song, J., Choi, S. Image-text alignment using adaptive cross-attention with transformer encoder for scene graphs (2021)

51. Tan, M., Le, Q.: Efficientnet: Rethinking model scaling for convolutional neural networks. In: International Conference on Machine Learning, PMLR, pp. 6105–6114 (2019)

52. Ueki, K.: Survey of visual-semantic embedding methods for zero-shot image retrieval. In: 2021 20th IEEE International Conference on Machine Learning and Applications (ICMLA), IEEE, pp. 628–634 (2021)

53. Varamesh, A., Diba, A., Tuytelaars, T., Van Gool, L.: Self-supervised ranking for representation learning. (2020) arXiv preprint arXiv:2010.07258

54. Vaswani, A.: Attention is all you need. In: Advances in Neural Information Processing Systems, vol. 30 (2017)

55. Voorhees, E.M.: The philosophy of information retrieval evaluation. In: Evaluation of Cross-Language Information Retrieval Systems, pp. 355–370. Springer, Berlin Heidelberg (2002)

56. Wang, H., et al.: Cross-modal food retrieval: learning a joint embedding of food images and recipes with semantic consistency and attention mechanism. IEEE Trans. Multimedia **24**, 2515–2525 (2021)

57. Wang, K., Yin, Q., Wang, W., Wu, S., Wang, L.: A comprehensive survey on cross-modal retrieval. arXiv preprint (2016). arXiv:1607.06215

58. Wang, L., Li, Y., Lazebnik, S.: Learning deep structure-preserving image-text embeddings. In: Proceedings of the IEEE Conference on Computer Vision and Pattern Recognition, pp. 5005–5013 (2016)

59. Welinder, P.: Caltech-UCSD Birds 200. Tech. Rep. CNS-TR-2010-001, California Institute of Technology (2010)

60. Yu, J., Wang, Z., Vasudevan, V., Yeung, L., Seyedhosseini, M., Wu, Y.: Coca: Contrastive captioners are image-text foundation models. (2022). arXiv preprint arXiv:2205.01917

61. Zeng, Y., Zhang, X., Li, H.: Multi-grained vision language pre-training: Aligning texts with visual concepts. In: Chaudhuri K, Jegelka S, Song L, Szepesvári C, Niu G, Sabato S (eds) International Conference on Machine Learning, ICML 2022, 17–23 July 2022, Baltimore, Maryland, USA, PMLR, Proceedings of Machine Learning Research, vol 162, pp. 25994–26009 (2022)

62. Zhang, C., et al.: Mosaicos: a simple and effective use of object-centric images for long-tailed object detection. In: Proceedings of the IEEE/CVF International Conference on Computer Vision, pp. 417–427 (2021)

63. Zhang, K., Mao, Z., Wang, Q., Zhang, Y.: Negative-aware attention framework for image-text matching. In: Proceedings of the IEEE/CVF Conference on Computer Vision and Pattern Recognition, pp. 15661–15670 (2022)

64. Zhang, P., et al.: Vinvl: Revisiting visual representations in vision-language models. In: Proceedings of the IEEE/CVF Conference on Computer Vision and Pattern Recognition, pp. 5579–5588 (2021)

65. Zhang, Y., Jiang, H., Miura, Y., Manning, C.D., Langlotz, C.P.: Contrastive learning of medical visual representations from paired images and text (2020). arXiv preprint arXiv:2010.00747

66. Zhou, B., Lapedriza, A., Xiao, J., Torralba, A., Oliva, A.: Learning deep features for scene recognition using places database. In: Advances in Neural Information Processing Systems vol. 27 (2014)

67. Zhuge, M., et al.: Kaleido-bert: Vision-language pre-training on fashion domain. In: Proceedings of the IEEE/CVF Conference on Computer Vision and Pattern Recognition, pp. 12647–12657 (2021)

Index-Based Batch Query Processing Revisited

Joel Mackenzie[1]($\boxtimes$) and Alistair Moffat[2]

[1] The University of Queensland, Brisbane, Australia
`joel.mackenzie@uq.edu.au`
[2] The University of Melbourne, Melbourne, Australia
`ammoffat@unimelb.edu.au`

Abstract. Large scale web search engines provide sub-second response times to interactive user queries. However, not all search traffic arises interactively – cache updates, internal testing and prototyping, generation of training data, and web mining tasks all contribute to the workload of a typical search service. If these non-interactive query components are collected together and processed as a batch, the overall execution cost of query processing can be significantly reduced. In this reproducibility study, we revisit query batching in the context of large-scale conjunctive processing over inverted indexes, considering both on-disk and in-memory index arrangements. Our exploration first verifies the results reported in the reference work [Ding et al., WSDM 2011], and then provides novel approaches for batch processing which give rise to better time–space trade-offs than have been previously achieved.

Keywords: Batch query processing · inverted indexes · experimentation

1 Introduction

Batch processing is a general paradigm aimed at reducing computational costs by avoiding repeated or redundant computation when processing a sequence of related tasks. This idea has been explored in a range of contexts, including in relational database management systems [32], spatial-textual databases [10], caching arrangements [1,18], and within information retrieval indexing and querying systems [4,7,16,29]. Given the immense scale of commercial IR and web search systems, revisiting batch processing may be one way to achieve reductions in computational overhead, and consequently, energy consumption and carbon emissions [11,31,33].

In this work, we revisit the problem of batch processing over *inverted indexes*. Given a set of queries $\mathcal{Q}$, the goal is to process each of those queries minimizing the total cost according to some metric such as time taken or volume of data read. That is done by building a *query execution plan* for $\mathcal{Q}$ that involves techniques such as query reordering and partial preliminary computation of shared results into a temporary cache, so as to reduce the overall cost. In this environment individual query latency is unimportant, and it is the aggregate cost that is of interest. Batch processing over inverted indexes is motivated by a range of scenarios in which query traffic is not required to be processed under stringent service-level agreements [19], including refreshing caches [6,17,20,22,24], internal mining and analytics tasks, collecting training data, and queries from external parties.

Despite the many possible applications of batch processing in information retrieval systems, there has been only limited experimental investigation and algorithmic development. The primary resource in this regard is the 2011 work of Ding et al. [16], who took the queries in $\mathcal{Q}$ to be sets of terms handled via Boolean conjunctions. It is reproduction of their work that forms the basis of the first experiments described here. In particular, we confirm that strategic pre-filling of a given volume of cache with postings lists results in overall savings of data transfer volumes; or, if the same amount of cache holds pre-computed list intersections, allows execution time reductions.

We then develop a new technique for the same task. Instead of selecting content with which to fill a static cache, we reorder the queries, paying careful attention to common subexpressions that can be evaluated once, at the time they are required, reused through multiple queries, and then discarded. The signal benefit of this approach is that only one intermediate list is required at any given time. With careful choice of intermediate results and a new cost estimation heuristic, we are able to outperform the Ding et al. [16] approaches using only a fraction of the cache that their methods require.

2   Background and Related Work

We first describe the problem that is considered, and then the techniques for addressing that problem that have been reported in the paper that is the basis for this reprodicibility study [16].

Definitions and Terminology. We suppose that $\mathcal{Q}$ contains n distinct queries, $\mathcal{Q} = \{Q_i \mid 1 \leq i \leq n\}$, which must each be resolved against a document collection $\mathcal{D}$. Each of those queries consists of a set of $q_i = |Q_i|$ distinct terms; lower-case alphabetic letters a, b, and so on from the beginning of the alphabet will be used to represent specific individual terms. The *vocabulary* V of $\mathcal{Q}$ is the complete set of all of $\mathcal{Q}$'s query terms, $V = \cup_{i=1}^{n} Q_i$. Each term $a \in V$ has an associated collection frequency $f_a \geq 0$, the number of documents in $\mathcal{D}$ that contain a, which also represents the length of the *postings list* $L(a)$ for a that records the identifiers in $\mathcal{D}$ of those f_a documents. The subset of queries that contain $a \in V$ is denoted by $T(a) = \{Q_i \in \mathcal{Q} \mid a \in Q_i\}$. That latter definition is also extended to term pairs: if $a, b \in V$, then $T(a, b)$ is the subset of $\mathcal{Q}$ in which both a and b occur as query terms.

If $L(a)$ and $L(b)$ are the postings lists for terms $a, b \in V$ with $f_a \leq f_b$ then their *intersection list* $L(a, b)$ can be computed using $\omega(f_a, f_b)$ steps of computation and in $O(\omega(f_a, f_b))$ time, where $\omega(x, y) = x \log_2(1 + y/x)$. Algorithms for intersecting lists within these bounds are explored elsewhere [14]. Denote the minimum collection frequency of query Q_i by $\mu_i = \min\{f_a \mid a \in Q_i\}$. Then if $Q_i = \{a, b, c, \ldots\}$ is interpreted as a *conjunctive Boolean bag of words query* the output required is the set of at most μ_i postings that result when all of Q_i's terms are intersected, $L(Q_i) = L(a) \cap L(b) \cap L(c) \cdots$. When we give example queries we will always list terms in increasing collection frequency. For example, in $Q_i = \{a, b, c\}$, we assume $\mu_i = f_a \le f_b \le f_c$.

In the *set-versus-set* approach to multi-way intersection the shortest list – of length μ_i – is used as a starting point, and then each other list is in turn intersected against that reducing set of candidates. If $C_0(Q_i)$ is the cost of evaluating the conjunctive query Q_i starting from the terms' individual postings lists, then $C_0(Q_i) \le \sum_{a \in Q_i} \omega(\mu_i, f_a)$. Hence, if $C(\mathcal{Q})$ is the total cost of evaluating all of the queries in $\mathcal{Q}$, with no shared processing between queries and each query evaluated in isolation, then $C(\mathcal{Q}) \le \sum_{1 \le i \le n} C_0(Q_i)$.

In addition to that computational cost there is also the cost of bringing the required lists from secondary storage into memory. A query Q_i of q_i terms requires that q_i transfer operations be initiated (seeks), and that a total of $D_0(\mathcal{Q}) = \sum_{a \in Q_i} f_a$ postings be transferred. Note that these transfer costs might also apply, albeit to a lesser degree, when a fully in-memory index is employed, with typical hardware configurations having multiple levels of memory.

Ding et al. [16] sought to explore the data transfer and computational cost benefits achievable via the complementary techniques of *list caching* and *intersection caching*, focusing on the context that has been described here – a set $\mathcal{Q}$ of queries which can be handled holistically rather than individually. Our goal in this work is to reproduce those results, and then to also extend them.

List Caching. Assume first that the system's inverted index is stored on disk, but that a known amount of main memory is available to temporarily retain selected postings lists so that they are available to future queries without seek or transfer operations being required. In the case of interactive querying, various strategies have been developed for estimating how best to employ the available memory. These are embodied as decision protocols as to whether or not a newly transferred postings should be retained; and if the decision is to retain, what list(s) to eject from the current cache so as to make space for this new entry.

In the case of non-interactive query batches it is possible to greatly assist list caching, because the queries in the batch can be evaluated in any order, and because the decision protocols can be based on *clairvoyant* knowledge [3]. For example, if a term appears only once in $\mathcal{Q}$, it should not be cached.

Query Orderings. One way in which a query set might be reordered is to first sort the terms of each query (for example, alphabetically by query term) and then sort the set of queries lexicographically [16]. This approach brings all queries with the same lexicographically least term into a consecutive group, meaning that an effective caching strategy – including the clairvoyant approach – will recognize that list and thus result in seek and transfer savings.

Ding et al. [16] also consider a second *clustering* approach to query ordering, noting the proposal of Cheng et al. [9], who study the problem of document

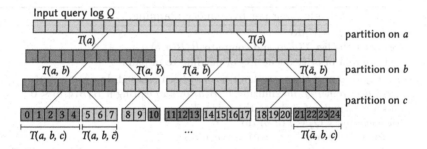

Fig. 1. Recursive Gray code-based query reordering. In this simple example terms a, b, c are ordered $|T(a)| \geq |T(b)| \geq |T(c)| \cdots$, and $T(\bar{a})$ represents the absence of term a, that is, $T(\bar{a}) = \mathcal{Q} \setminus T(a)$.

identifier reordering for inverted indexes. In particular, Cheng et al. propose a partition-based reordering method which recursively clusters an inverted index by considering postings lists in decreasing order of their total size. Translating this to the set $\mathcal{Q}$, we first identify the queries in which each term $a \in V$ occurs, to establish the set $T(a)$. Then, considering those sets in decreasing size order, $\mathcal{Q}$ is recursively partitioned so that all of the queries (within each current partition of $\mathcal{Q}$) containing the next most queried term are contiguous. Each cluster is reversed relative to the previous level of the recursion, so that the cluster labels form a Gray code. Figure 1 illustrates this approach, denoted here as Partitioned.

Intersection Caching. While list caching techniques have the potential to reduce the amount of data transferred from secondary storage, they do not alter the computation required during intersection operations. Suppose now that some pair of terms $a, b \in V$ has had their joint list pre-computed, and that a query $Q_i \in T(a, b)$ is to be resolved, with $f_a \leq f_b$, and hence $|L(a, b)| \leq |L(a)| \leq |L(b)|$. Rather than using $L(a)$ and $L(b)$ in the intersection pipeline, this query should be resolved using the list $L(a, b)$. If $C_{a,b}(Q_i)$ is the computational cost of doing so, then $C_{a,b}(Q_i) \leq C_0(Q_i) - \omega(\mu_i, f_b)$, with the intersection using $L(a, b)$ requiring at most the same time as the original intersection using $L(a)$, but likely less given the avoidance of list decompression operations.

Recall that $T(a, b)$ is the subset of queries in $\mathcal{Q}$ that contain both term a and term b. An estimate of the net saving that might be achieved by precomputing $L(a, b)$ is thus given by $\sum_{Q_i \in T(a,b)} \omega(\mu_i, f_b)$, a benefit which must be debited by $\omega(f_a, f_b)$, the cost of carrying out the pre-computation. Moreover, the list $L(a, b)$ must be stored, and will require as many as $|L(a, b)| \leq f_a$ words of memory, or a fractional equivalent if stored in compressed form.

Term Pair Popularity. This then leads to a first approach to term pair caching: all co-occurring term pairs are tabulated to form the sets $T(a, b)$; the quantity $G(a, b) = -\omega(f_a, f_b) + \sum_{Q_i \in T(a,b)} \omega(\mu_i, f_b)$ is computed for each possible pair a, b, to compute the maximum gain possible by precomputing that pair, and then those values are sorted into decreasing order, to create a static popularity-based ordering of term pairs. Then, assuming that a specified amount of cache is

---

Algorithm 1. Dynamic bang per byte query planning.

---

    for each pair $a, b \in V$ **do**
2:      construct the list $T(a, b)$ and compute $B4B(a, b)$
        add a, b to a priority queue PQ of pending term pairs
4: **while** $|PQ| > 0$ **and** the cache limit has not been reached **do**
        select a, b from PQ, maximizing $B4B(a, b)$, and exiting if $B4B(a, b) \leq 0$
6:      allocate f_a units of cache to the pair a, b
        for each query $Q_j \in T(a, b)$ **do**
8:          **for** each term $c \in Q_j$, with $c \neq a, b$ **do**
            adjust $B4B(a, c)$ (or $B4B(c, a)$, if $f_c < f_a$) to remove Q_j's contribution
10:         adjust $B4B(b, c)$ (or $B4B(c, b)$, if $f_c < f_b$) to remove Q_j's contribution
        remove a, b from PQ

---

available, pairs a, b are taken from that ordered list and given cache allocations, iterating until the given cache limit has been reached. Term pairs for which $G(a, b) \leq 0$ should never be added to the cache.

The query processing plan fetches the postings lists for each of the selected pairs and intersects them to populate the cache; and then processes $\mathcal{Q}$ in any order, checking each Q_i for the appearance of any pairs a, b for which $L(a, b)$ is in the cache, and using pre-computed lists whenever possible. The cache of intersected lists remains constant through that execution sequence.

Static Bang Per Byte. A drawback of the simple popularity-based approach is that high-frequency pairs and low-frequency pairs might have the same value for $G(\cdot, \cdot)$, but a high-frequency pair will consume more cache and thus represent less net value. A second option is thus to normalize the projected gain according to the anticipated storage cost for the intersected list. That is, the list of term pairs is considered instead in decreasing order of $B4B(a, b) = G(a, b)/f_a$.

Dynamic Bang Per Byte. A further refinement is to note that a query Q_i might contain multiple pairs that received high $B4B(a, b)$ values, but that interactions between term pairs means that their independent gains cannot all be achieved. For example, if $Q_i = \{a, b, c\}$ and pairs a, c and b, c are both in the cache, then only one gain of $\omega(\mu_i, f_c)$ can result, making the other illusory. That consideration leads to the mechanism described in Algorithm 1, which is our interpretation of what is described in Sect. 4.2 (page 142, right column) of Ding et al. [16]. A priority queue PQ of pairs not yet placed in the cache is employed, with priorities given by evolving $B4B(a, b)$ values. The weights of pairs are non-increasing, meaning that a "lazy" queue can be employed (rather than an "always up to date" heap), and recomputations of $B4B(c, d)$ deferred until c, d reaches the head of PQ. A flag bit associated with each pair c, d is sufficient to record whether the current $B4B(c, d)$ value is valid or requires recomputation (and possible deferral) should it reach the head of PQ.

This mechanism employs the same query processing plan as the previous two, but should result in a more economical combination of lists $L(a, b)$ in the cache.

On the other hand, the computation described in Algorithm 1 is more complex than is calculation of static $B4B(\cdot, \cdot)$ values.

More Than Just Pairs. The ideas presented above are not limited to term pairs, and intersecting further terms (such as triples or quads) may also lead to improvements. However, Ding et al. [16] found that, due to the power-law distribution of query terms [30], selected pairs were also often overlapping with promising triples and quads, meaning that the pairs provide most of the overall benefit. Furthermore, increasing the number of candidates increases pre-processing costs. In this work, we only consider term pairs, noting that our batch processing framework can be extended to triples and beyond if required.

Other Related Work. Chaudhuri et al. [8] examined how materializing list intersections can reduce querying based on the power-law characteristics of corpus terms. Tolosa et al. [34] have also examined the benefits of caching intersections for in-memory processing scenarios, comparing a number of cache admission strategies. The broad problem of efficiently handling queries over large volumes of data has challenged researchers and practitioners for decades, with ongoing research; see Tonellotto et al. [35] for an overview.

3 Experimental Setup

Hardware and Measurement. Our experiments are conducted on a Linux server with two Intel Xeon Gold 6144 CPUs at 3.5 GHz with 512 GiB of RAM. Only a single processing core is utilized, allowing the use of processing latency as a proxy for total computing cost. The list caching experiments in Sect. 4 measure the volume of compressed postings lists needing to be transferred from secondary storage. Then the intersection caching experiments in Sect. 5 load the whole index into main memory prior to commencement of the query processing plan, and are reported as query batch execution times in elapsed seconds.

Software. One setback in reproducing the reference work is that the previous source code and experimental framework are not available. Thus, we reimplemented the algorithms based on the descriptions provided by Ding et al. [16]. All caching and query planning algorithms were implemented in C++ and compiled with GCC 7.5.0 using -O3 optimization. Our retrieval experiments make use of a version of the efficient PISA search system [28], modified to allow the results of intersections to be stored and used during query processing. Document indexes are built using the Lucene-based Anserini system [36] and are converted to SIMD-BP128 [21] compressed PISA indexes via the common index file format [23]. Indexes are reordered using the recursive partitioning mechanism [15,27].

Document Collections. Two public collections are used: MSMARCO-v1 and MSMARCO-v2 contain 8.8 million and 138.4 million passages respectively, drawn from English web documents [2,13]. The *augmented* version of MSMARCO-v2 is adopted, in which passages are expanded with the document URL, title, and headings [25]. Table 1 provides statistics of the collections after indexing.

Table 1. Index statistics for the two test collections used here, and (last row) the collection employed in the reference work by Ding et al. [16], which is not publicly available.

Collection	Documents	Unique terms	Postings	Size [GiB]
MSMARCO-v1	8,841,823	2,660,824	266,247,718	0.9
MSMARCO-v2	138,364,198	16,579,071	8,629,430,400	22.8
RandomWeb [16]	10,000,000	–	–	4.2

Query Batch. Ding et al. [16] used a total of 1.16 million distinct queries from the Excite query log in their experiments. However, these queries are not likely to be temporally relevant to the MSMARCO passages used in our experimentation. Instead, we use the entire set of 10 million ORCAS queries [12]. These queries were filtered by first applying a normalization process: stemming, case-folding, and stopword removal according to the default Lucene tokenizer. All within-query duplicate terms were then removed, since we only focus on conjunctive matching. Finally, only unique queries which did not contain out-of-vocabulary terms on both MSMARCO-v1 and MSMARCO-v2 were collected into the final batch, resulting in a total of 6,761,892 queries with an average length of 3.2 unique terms.

4   Reducing Data Transfer Volume via List Caching

The first experiment examines the gains possible as a result of in-advance knowledge of the query batch, assuming that the index resides on some form of secondary storage. In this context, data transfer time is an important factor in overall query processing time, and the aim is reduce it via list caching. Ding et al. [16] consider two factors for doing that: the order in which the queries should be processed; and the cache eviction strategy.

Query Ordering. As discussed in Sect. 2, the *order* in which queries are processed can improve term locality. Here, we evaluate three strategies:

– Random: A baseline measurement undertaken using a randomly shuffled log. (Ding et al. [16] started with the *natural* ordering of their query log, but we have no notion of a natural order here.)
– Sorted: Sorts the queries lexicographically, as described in Sect. 2.
– Partitioned: Employs the Gray code-based query reordering shown in Fig. 1. We believe that this corresponds to the "agglomerative clustering" of Ding et al. [16] which is based on the work of Cheng et al. [9, Figure 7].

Ding et al. [16] comment that other clustering methods are possible. To explore that option we also implemented an approximate traveling salesman-based method [9, Figure 4] denoted TSP, which greedily traverses an induced graph of queries, maximizing the similarity between the current query Q_i and all other unvisited neighboring queries Q_j, where neighboring queries are those sharing at least one term with Q_i. Query-to-query similarity is measured as $S(Q_i, Q_j) = \sum_{a \in Q_i \cap Q_j} |L'(a)|$, where $|L'(a)|$ is the length in bytes of the compressed postings list $L'(a)$ for term a, the goal being to maximize the retained posting volume at each query transition. Other similarities were also explored, including ones based on term overlap, but this formulation gave the best results. As an aside, both the TSP and Partitioned algorithms are employed by Cheng et al. [9], and in future work we will also apply these algorithms to document identifier reassignment [5] as a secondary reproducibility effort [26].

Cache Eviction. The second factor explored by Ding et al. was that of clairvoyant cache management. Since the whole query log is known, the cache can always evict the item not required for the longest, denoted here as CV, thereby allowing optimal behavior [3]. The *least recently used* (LRU) policy is used as a baseline mechanism.

Reproducibility Results. In the original evaluation [16] the cache size is measured in *millions* (without units); we assume those to be *millions of uncompressed postings*. Here we assume that cached lists will remain compressed and be decoded on-demand, so as to maximize the number of retained lists, and set the cache size to a sequence of fixed percentages of the original compressed index size. We then count the volume of compressed postings transferred from secondary storage through the whole of Q. Figure 2 shows all eight combinations of log order and cache strategy for the two collections when measured this way.

The patterns of behavior observed in Fig. 2 are consistent with those visible in Fig. 1 of Ding et al. [16] (once the logarithmic vertical scale in Fig. 2 is allowed for). In particular: the Sorted logs outperform the random (baseline) orderings; the recursively partitioned "Gray clustered" logs perform even better than the sorted logs; and in the second aspect of the experimentation, (and completely unsurprisingly, no foresight required) the clairvoyant caching approach results in less data being transferred than does the LRU strategy. Figure 2 also includes the approximate TSP-based query reordering technique, a second form of clustering; it outperforms the Gray code-based reordering process. However, the TSP ordering takes around 55 minutes to compute, whereas the Gray-code partitioning takes only 10 seconds, meaning that the slight reduction in data transferred via TSP is unlikely to be of interest.

5   Reducing Computation via Intersection Caching

We now turn from list caching to intersection caching, seeking to confirm the outcomes reported by Ding et al. [16], again using the two MSMARCO collections.

Methods Tested. Ding et al. add materialized term pairs to a fixed-size cache, seeking to maximize the computational benefit achieved via each fixed volume of pre-computed term intersections, with the computation cost of a query Q_i estimated as $C(Q_i) = \sum_{a \in Q_i} f_a$, which is less precise than the $\omega(f_a, f_b)$ estimator presented in Sect. 2. They then construct an ordering of term pairs (described in their Sect. 4.2) using the mechanism presented in Algorithm 1, which is denoted

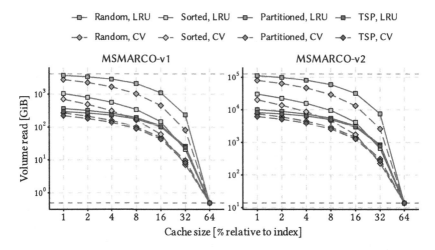

Fig. 2. Index volume transferred by LRU and CV caching for the four different orderings of Q, with cache size set as a percentage of the compressed index size. The orange and blue dashed lines represent the worst-case (read a list from disk each time it is requested) and best-case (read each list just once) I/O performance, respectively.

here as Flexible-B4B. Ding et al. report average query execution speed of approximately 6.8 millisec per query (presumed to be for conjunctive Boolean queries of the kind we are also measuring here, as measured off a highly magnified screenshot of their Fig. 8, with the computed percentage savings corroborated by their Fig. 11); reducing to 5.0 millisec on average using a cache holding 8% of the index terms' original posting, a 26% saving; and further decreasing to approximately 4.6 millisec with a 30% term pair cache, a 33% saving.

We re-implemented the Flexible-B4B scheme, and also tested the static bang-for-byte term-pair selection process (Static-B4B), and an even simpler mechanism based purely on $G(a, b)$, denoted Popularity. Both of these approaches are also described in Sect. 2. In each experiment all of the postings lists were first loaded into memory; then the timing commenced; then all of the planned pre-intersections computed to populate the cache of term pair intersections; then the query batch was executed in query-sorted order (Sorted in the context of Fig. 2); and finally, the timing was ended.

Reproducibility Results. Table 2 shows that the broad relationship between the three approaches can be confirmed. Stepping across each row, increasing the detail in the "merit estimation" process leads to decreased running times; within each column, increasing amounts of cache memory also lead to decreased execution times. Moreover, the gains achieved by the Flexible-B4B approach with a 32% space overhead of 39% for MSMARCO-v1 and 27% for MSMARCO-v2 match the gains achieved by Ding et al. [16], summarized above. Note that we are measuring cache size here as byte-based percentages of the compressed index size, and storing uncompressed intersection lists; whereas Ding et al. measure

Table 2. Reproduced results. Total elapsed time (seconds) to execute the entire batch of queries, as a function of the volume of pre-computed intersections. Query plan generation times were all a few seconds each and are not included.

Space (%)	Popularity		Static-B4B		Flexible-B4B	
	MSM-v1	MSM-v2	MSM-v1	MSM-v2	MSM-v1	MSM-v2
0	735	18,864	735	18,864	735	18,864
1	714	18,686	701	18,523	686	18,333
2	713	18,701	684	18,291	663	18,143
4	702	17,970	661	18,113	628	17,523
8	678	18,417	622	17,606	575	16,750
16	646	18,048	575	16,880	511	15,461
32	597	17,319	516	15,956	447	13,838
64	536	16,291	453	14,598	406	12,547

percentages of posting counts. While these are similar scales, the latter is likely to allow more postings to be stored at each percentage measurement point.

The average query execution times for the Flexible-B4B method with 32% cache are 66.1 *microsec* (MSMARCO-v1) and 2.05 millisec (MSMARCO-v2), with the speed-up relative to Ding et al. likely the result of hardware relativities since 2011. Note also that MSMARCO-v2 is a much larger collection (Table 1).

Strategic Pair-Based Query Reordering. We now introduce a different way in which batches of queries can be expedited. Key to the new proposal is the observation that most queries are relatively short (3.2 terms on average in our test set) and hence having even one cached term pair per query represents a worthwhile target. That insight allows us to reorient the quest from being a search for the best mix of term pairs across the whole set of queries, to a query-by-query search for a good term pair. We call that connection from a query to a single term pair an *association*; and Algorithm 2 describes the process employed to identify them. An initialization phase determines which pairs $a, b \in V$ occur at least two of the queries in Q. Three further processing phases then occur.

In Phase 1, steps 3 to 9, a best pair a', b' is tentatively associated with each query, and at the same time an estimate of the savings that might accrue is made, the latter via the computation at step 10, and credited to the associated pair using accumulator $E(a', b')$. Important new aspects of the estimation are that only one pair is allowed to claim benefit, and that it further hedges the computation by supposing that any of the other *numpairs* available in Q_i might also be used, albeit at a reduced saving. Assuming those other options to be uniformly spread across the range from zero to $\omega(\mu_i, f_{b'})$ means that the marginal benefit of a', b' needs to be discounted by *numpairs*.

Phase 2, steps 11 to 13, then factors in the pre-computation cost, and removes from contention (represented as the set *valid*) any pairs found to not be of benefit. That leaves a reduced set of "known to be definitely useful" pairs available;

Algorithm 2. Associating at most one term pair with each query.

---

    for each pair $a, b \in V$ **do**   ▷ Phase 0: initialization
2:     set $E(a, b) \leftarrow 0$ and set $valid(a, b) \leftarrow |T(a, b)| > 1$

    for each query $Q_i \in \mathcal{Q}$ **do**   ▷ Phase 1: tentative associations
4:     set $numpairs \leftarrow 0$
        for each term pair $a, b \in Q_i$ such that $f_a \leq f_b$ **and** $valid(a, b)$ **do**
6:         set $numpairs \leftarrow numpairs + 1$
           record as a', b' the a, b pair that maximizes f_b / f_a for Q_i
8:     **if** $numpairs > 0$ **then**
        associate the pair a', b' with Q_i
10:       set $E(a', b') \leftarrow E(a', b') + \omega(\mu_i, f_{b'})/numpairs$

    for each pair $a, b \in V$ **do**   ▷ Phase 2: pair pruning
12:     **if** $E(a, b) < \omega(f_a, f_b)$ **then**
        set $valid(a, b) \leftarrow$ **false**
14: repeat steps 3 to 9   ▷ Phase 3: consolidated associations

---

they are used in Phase 3, step 14, which recomputes the associations. Queries previously associated with term pairs that got removed must be re-associated, with any new associations formed certain to be of net benefit. That is, in Phase 3 some queries have the tentative association confirmed, some queries get a revised association, and some queries lose their associations and join a pool of "immune" queries; with every such decision definitely decreasing the estimated execution time. Figure 3 gives an example that illustrates the way in which each query has exactly one associated term pair selected, with immune queries associated with the empty term pair at the top, and benefits gained via the selected edges.

The query processing plan is then simple: each *valid* term pair is computed, all of its associated queries are resolved, and then that intersection is discarded. Only a miniscule amount of cache storage is required – in our MSMARCO-v2 experiments, the peak space needed is just 41.5 million postings.

Table 3 shows the gains resulting from this new approach, with three rows of results: first, using the Phase 1 associations alone, without the pruning step that is part of Phase 2; then without the discounting by *numpairs*, which is too optimistic in its estimations and retains pairs that are of marginal or negative benefit; and then, in the last row, using the mechanism that is described in Algorithm 2, including the discounting. Each variant takes less than 30 s to generate a plan. Not only does this approach allow batch query processing without the need for a large intersection cache, it is also notably faster than the Flexible-B4B method of Ding et al. [16], saving a further 20% and 23% of running time (MSMARCO-v1 and MSMARCO-v2 respectively) compared to the "32% cache space" row in Table 2.

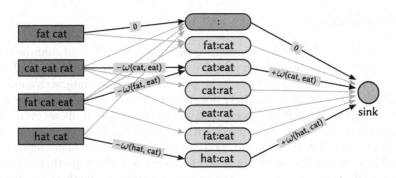

Fig. 3. Associations between queries and term pairs. Each query contains multiple term pairs, of which exactly one is selected (possibly the "empty pair"). Labeled edges represent the net saving of this possible set of associations.

Table 3. Query processing using Algorithm 2. All results are in total seconds, and may be compared with those in Table 2. See the text for details.

Approach	MSM-v1	MSM-v2
Phase 1 only	463	13,138
Phases 1–3, but without discounting by *numpairs*	406	11,825
Phases 1–3, including the discounting at step 10	367	10,684

Finally, Table 4 illustrates why Algorithm 2 is so effective. The important associations derived for MSMARCO-v2 do indeed correspond to term pairs that have natural relationships with each other. Note that all terms were stemmed, and that they are shown with $f_a \leq f_b$ rather than in query appearance order.

Table 4. Popular term pairs in the context of the MSMARCO-v2 collection frequencies: the five with the greatest number of associations (left); and the five with the highest *match percentage* over the $|T(a,b)|$ queries containing a and b (right), where the match percentage in the rightmost column in each of the two groups is the ratio of assigned associations ("Assoc.") as a fraction of $|T(a,b)|$.

a	b	Assoc.	$|T(a,b)|$	%	a	b	Assoc.	$|T(a,b)|$	%
near	me	3,669	30,303	12.1	invent	who	1,042	1,430	72.9
icd	10	3,635	8,084	45.0	forecast	10	1,007	1,411	71.4
between	differ	3,447	7,534	45.8	fargo	well	2,610	3,734	69.9
mean	what	3,149	26,551	11.9	orlean	new	1,321	1,905	69.3
side	effect	2,977	9,632	30.9	depot	home	2,302	3,473	66.3

6 Conclusion

We have reproduced the results presented by Ding et al. [16], and confirmed that
list and intersection caching are effective techniques that can be applied when
batches of conjunctive queries are to be processed in a non-interactive manner.
In addition to having more precisely documented those methods, we have added
a new "term pair association" mechanism to the repertoire. It allows even better
batch execution times to be achieved, without requiring large volumes of cache.
That means that list caching and strategic pair intersection planning can now
be combined, to get the benefit of both enhancements. We hope that our reproducibility effort and public code resource will encourage further investigation
into this area of research.

Software. In the interest of reproducibility, our implementations are available
at https://bitbucket.org/JMMackenzie/batch-conjunctions.

Acknowledgement. This work was supported by the Australian Research Council's
Discovery Projects Scheme (project DP200103136) and a University of Queensland
New Staff Research Grant.

References

1. Albers, S.: New results on web caching with request reordering. In: Proceedings of
SPAA, pp. 84–92 (2004)
2. Bajaj, P., et al.: MS MARCO: a human generated MAchine Reading COmprehension dataset. arXiv:1611.09268v3 (2018)
3. Belady, L.A.: A study of replacement algorithms for a virtual-storage computer.
IBM Syst. J. **5**(2), 78–101 (1966)
4. Benham, R., Mackenzie, J., Moffat, A., Culpepper, J.S.: Boosting search performance using query variations. ACM Trans. Inf. Syst. **37**(4), 41.1–41.25 (2019)
5. Blandford, D., Blelloch, G.: Index compression through document reordering. In:
Proceedings of DCC, pp. 342–352 (2002)
6. Cambazoglu, B.B., et al.: A refreshing perspective of search engine caching. In:
Proceedings of WWW, pp. 181–190 (2010)
7. Catena, M., Tonellotto, N.: Multiple query processing via logic function factoring.
In: Proceedings of SIGIR, pp. 937–940 (2019)
8. Chaudhuri, S., Church, K., König, A.C., Sui, L.: Heavy-tailed distributions and
multi-keyword queries. In: Proceedings of SIGIR, pp. 663–670 (2007)
9. Cheng, C., Chung, C., Shann, J.J.: Fast query evaluation through document identifier assignment for inverted file-based information retrieval systems. Inf. Proc.
Man. **42**(3), 729–750 (2006)
10. Choudhury, F.M., Culpepper, J.S., Bao, Z., Sellis, T.: Batch processing of top-k
spatial-textual queries. ACM Trans. Spat. Alg. Syst. **3**(4), 13.1–13.40 (2018)
11. Chowdhury, G.: An agenda for green information retrieval research. Inf. Proc. Man.
48(6), 1067–1077 (2012)
12. Craswell, N., Campos, D., Mitra, B., Yilmaz, E., Billerbeck, B.: ORCAS: 20 million
clicked query-document pairs for analyzing search. In: Proceedings of CIKM, pp.
2983–2989 (2020)

13. Craswell, N., Mitra, B., Yilmaz, E., Campos, D., Lin, J.: Overview of the TREC 2021 deep learning track. In: Proceedings of TREC (2021)
14. Culpepper, J.S., Moffat, A.: Efficient set intersection for inverted indexing. ACM Trans. Inf. Syst. **29**(1), 1.1–1.25 (2010)
15. Dhulipala, L., Kabiljo, I., Karrer, B., Ottaviano, G., Pupyrev, S., Shalita, A.: Compressing graphs and indexes with recursive graph bisection. In: Proceedings of KDD, pp. 1535–1544 (2016)
16. Ding, S., Attenberg, J., Baeza-Yates, R., Suel, T.: Batch query processing for web search engines. In: Proceedings of WSDM, pp. 137–146 (2011)
17. Fagni, T., Perego, R., Silvestri, F., Orlando, S.: Boosting the performance of web search engines: caching and prefetching query results by exploiting historical usage data. ACM Trans. Inf. Syst. **24**(1), 51–78 (2006)
18. Feder, T., Motwani, R., Panigrahy, R., Zhu, A.: Web caching with request reordering. In: Proceedings of SODA, pp. 104–105 (2002)
19. Hwang, S.-W., Kim, S., He, Y., Elnikety, S., Choi, S.: Prediction and predictability for search query acceleration. ACM Trans. Web **10**(3), 19.1–19.28 (2016)
20. Jonassen, S., Cambazoglu, B.B., Silvestri, F.: Prefetching query results and its impact on search engines. In: Proceedings of SIGIR, pp. 631–640 (2012)
21. Lemire, D., Boytsov, L.: Decoding billions of integers per second through vectorization. Soft. Prac. Exp. **41**(1), 1–29 (2015)
22. Lempel, R., Moran, S.: Predictive caching and prefetching of query results in search engines. In: Proceedings of WWW, pp. 19–28 (2003)
23. Lin, J., et al.: Supporting interoperability between open-source search engines with the common index file format. In: Proceedings of SIGIR, pp. 2149–2152 (2020)
24. Ma, H., Wang, B.: User-aware caching and prefetching query results in web search engines. In: Proceedings of SIGIR, pp. 1163–1164 (2012)
25. Ma, X., Pradeep, R., Nogueira, R., Lin, J.: Document expansions and learned sparse lexical representations for MSMARCO V1 and V2. In: Proceedings of SIGIR, pp. 3187–3197 (2022)
26. Mackenzie, J., Mallia, A., Petri, M., Culpepper, J.S., Suel, T.: Compressing inverted indexes with recursive graph bisection: a reproducibility study. In: Azzopardi, L., Stein, B., Fuhr, N., Mayr, P., Hauff, C., Hiemstra, D. (eds.) ECIR 2019. LNCS, vol. 11437, pp. 339–352. Springer, Cham (2019). https://doi.org/10. 1007/978-3-030-15712-8_22
27. Mackenzie, J., Petri, M., Moffat, A.: Tradeoff options for bipartite graph partitioning. IEEE Trans. Know. Data Eng. (2022, to appear)
28. Mallia, A., Siedlaczek, M., Mackenzie, J., Suel, T.: PISA: performant indexes and search for academia. In: Proceedings of OSIRRC at SIGIR 2019, pp. 50–56 (2019)
29. Marín, M., Navarro, G.: Distributed query processing using suffix arrays. In: Proceedings of SPIRE, pp. 311–325 (2003)
30. Petersen, C., Simonsen, J.G., Lioma, C.: Power law distributions in information retrieval. ACM Trans. Inf. Syst. **34**(2), 8.1–8.37 (2016)
31. Scells, H., Zhuang, S., Zuccon, G.: Reduce, reuse, recycle: green information retrieval research. In: Proceedings of SIGIR, pp. 2825–2837 (2022)
32. Sellis, T.K.: Multiple-query optimization. ACM Trans. Data. Syst. **13**(1), 23–52 (1988)
33. Strubell, E., Ganesh, A., McCallum, A.: Energy and policy considerations for deep learning in NLP. In: Proceedings of ACL, pp. 3645–3650 (2019)
34. Tolosa, G., Becchetti, L., Feuerstein, E., Marchetti-Spaccamela, A.: Performance improvements for search systems using an integrated cache of lists + intersections. Inf. Retr. **20**(3), 172–198 (2017)

35. Tonellotto, N., Macdonald, C., Ounis, I.: Efficient query processing for scalable web search. Found. Trnd. Inf. Retr. **12**(4–5), 319–500 (2018)
36. Yang, P., Fang, H., Lin, J.: Anserini: reproducible ranking baselines using Lucene. J. Data Inf. Qual. **10**(4), 1–20 (2018)

A Unified Framework for Learned Sparse Retrieval

Thong Nguyen[1(✉)], Sean MacAvaney[2], and Andrew Yates[1]

[1] University of Amsterdam, Amsterdam, Netherlands
t.nguyen2@uva.nl
[2] University of Glasgow, Glasgow, Scotland

Abstract. Learned sparse retrieval (LSR) is a family of first-stage retrieval methods that are trained to generate sparse lexical representations of queries and documents for use with an inverted index. Many LSR methods have been recently introduced, with Splade models achieving state-of-the-art performance on MSMarco. Despite similarities in their model architectures, many LSR methods show substantial differences in effectiveness and efficiency. Differences in the experimental setups and configurations used make it difficult to compare the methods and derive insights. In this work, we analyze existing LSR methods and identify key components to establish an LSR framework that unifies all LSR methods under the same perspective. We then reproduce all prominent methods using a common codebase and re-train them in the same environment, which allows us to quantify how components of the framework affect effectiveness and efficiency. We find that (1) including document term weighting is most important for a method's effectiveness, (2) including query weighting has a small positive impact, and (3) document expansion and query expansion have a cancellation effect. As a result, we show how removing query expansion from a state-of-the-art model can reduce latency significantly while maintaining effectiveness on MSMarco and TripClick benchmarks. Our code is publicly available (Code: https://github.com/thongnt99/learned-sparse-retrieval).

Keywords: Neural retrieval · Learned sparse retrieval · Lexical retrieval

1 Introduction

Neural information retrieval has becoming increasingly common and effective with the introduction of transformers-based pre-trained language models [17]. Due to latency constraints, a pipeline is often split into two stages: first-stage retrieval and re-ranking. The former focuses on efficiently retrieving a set of candidates to re-rank, whereas the latter focuses on re-ranking using highly effective but inefficient methods. Neural first-stage retrieval approaches can be grouped into two categories: dense retrieval (e.g., [12, 13, 38]) and learned sparse retrieval (e.g., [7, 40, 44]). Learned sparse retrieval (LSR) methods transform an input text (i.e., a query or document) into sparse lexical vectors, with each dimension

J. Kamps et al. (Eds.): ECIR 2023, LNCS 13982, pp. 101–116, 2023.
https://doi.org/10.1007/978-3-031-28241-6_7

containing a term score analogous to TF. The sparsity of these vectors allows LSR methods to leverage an inverted index. Compared with dense retrieval, LSR has several attractive properties. Each dimension in the learned sparse vectors is usually tied to a term in vocabulary, which facilitates transparency. We can, for example, examine biases encoded by models by looking at the generated terms. Furthermore, LSR methods can re-use the inverted indexing infrastructure built and optimized for traditional lexical methods over decades.

The idea of using neural methods to learn weights for sparse retrieval pre-dates transformers [40,42], but approaches' effectiveness with pre-BERT methods is limited. With the emergence of retrieval powered by transformer-based pre-trained language models [6,17,36], many LSR methods [5,7,8,16,20,23,41] have been introduced that leverage transformer architectures to substantially improve effectiveness. Among them, the Splade [7] family is a recent prominent approach that shows strong performance on the MSMarco [26] and BEIR benchmarks [35].

Despite their architectural similarities, different learned sparse retrieval methods exhibit very different behaviors regarding effectiveness and efficiency. The underlying reasons for these differences are often unclear.

In this work, we conceptually analyze existing LSR methods and identify key components in order to establish a comparative framework that unifies all methods under the same perspective. Under this framework, the key differences between existing LSR methods become apparent. We first reproduce methods' original results, before re-training and re-evaluating them in a common environment that leverages best practices from recent work, like the use of hard negatives. We then leverage this setting to study how key components influence a model's performance in terms of efficiency and effectiveness. We investigate the following research questions:

RQ1: Are the results from LSR papers reproducible?
This RQ aims to reproduce the results of all recent, prominent LSR methods in our codebase, consulting the configuration on the original papers and codes. We find that most of the methods can be reproduced with MRR comparable to the original work (or slightly higher).

RQ2: How do LSR methods perform with recent advanced training techniques?
Splade models [7] show impressive ranking scores on MSMarco. While these improvements could be due to architectural choices like incorporating query expansion, Splade also benefits from an advanced training process with mined hard negatives and distillation from cross-encoders. Our experiments show that with the same training as Splade, many older methods become significantly more effective. Most noticeably, the MRR@10 score of the older EPIC [20] model was boosted by 36% to become competitive with Splade.

RQ3: How does the choice of encoder architecture and regularization affect results?
The common training environment we use to answer RQ2 allows us to quantify the effect of various architectural decisions, such as expansion, weighting, and

regularization. We find that document weighting had the greatest impact on a system's effectiveness, while query weighting had a moderate impact, though query weighting improves latency by eliminating non-useful terms. Notably, we observed a cancellation effect between improvements from document and query expansion, indicating that query expansion is not necessary for a LSR system to perform well.

Our contributions are: (**1**) an conceptual framework that unifies all prominent LSR methods under the same view, (**2**) an analysis of how LSR components affect efficiency and effectiveness, which e.g. leads to a modification that reduces more than 74% retrieval latency while keeping the same SOTA effectiveness, and (**3**) implementations of all studied methods in the same codebase, including simple changes in Anserini [39] that make LSR indexing faster.

2   Learned Sparse Retrieval

Learned sparse retrieval (LSR) uses a query encoder f_Q and a document encoder f_D to project queries and documents to sparse vectors of vocabulary size: $w_q = f_Q(q) = w_q^1, w_q^2, \ldots, w_q^{|V|}$ and $w_d = f_D(d) = w_d^1, w_d^2, \ldots, w_d^{|V|}$. The score between a query a document is the dot product between their corresponding vectors: $sim(q, d) = \sum_{i=1}^{|V|} w_q^i w_d^i$. This formulation is closely connected to traditional sparse retrieval methods like BM25; indeed, BM25 [32,33] can be formulated as:

$$BM25(q,d) = \sum_{i=1}^{|q|} \text{IDF}(q_i) \times \frac{tf(q_i, d) \times (k_1 + 1)}{tf(q_i, d) + k_1 \cdot \left(1 - b + b \cdot \frac{|d|}{\text{avgdl}}\right)}$$

$$= \sum_{j=1}^{|V|} \underbrace{\mathbb{1}_{q(v_j)} \text{IDF}(v_j)}_{\text{query encoder}} \times \underbrace{\mathbb{1}_{d(v_j)} \frac{tf(v_j, d) \times (k_1 + 1)}{tf(v_j, d) + k_1 \cdot \left(1 - b + b \cdot \frac{|d|}{\text{avgdl}}\right)}}_{\text{doc encoder}}$$

$$= \sum_{j=1}^{|V|} f_Q(q)_j \times f_D(d)_j$$

With BM25 the IDF and TF components can be viewed as query/document term weights. LSR differs by using neural models, typically transformers, to predict term weights. LSR is compatible with many techniques from sparse retrieval, such as inverted indexing and accompanying query processing algorithms. However, differences in LSR weights can mean that existing query processing optimizations become much less helpful, motivating new optimizations [21,22,24].

2.1   Unified Learned Sparse Retrieval Framework

In this section, we introduce a conceptual framework consisting of three components (*sparse encoder, sparse regularizer, supervision*) that captures the key differences we observe between existing learned sparse retrieval methods. Later, we describe how LSR methods in the literature can be fit into this framework.

Table 1. Encoder architectures. (Transf: Transformers)

Name	Backbone	Head	Expansion	Weighting
BINARY	Transf. Tokenizer	–	No	No
MLP	Transf. Encoder	Linear(s)	No	Yes
expMLP	Transf. Encoder	Linear(s)	Yes	Yes
MLM	Transf. Encoder	MLM Head + Agg.	Yes	Yes
clsMLM	Transf. Encoder	MLM Head	Yes	Yes

Sparse (Lexical) Encoders. A sparse or lexical encoder encodes queries and passages into weight vectors of equal dimension. This is the main component that determines the effectiveness of a learned sparse retrieval method. There are three distinct characteristics that make sparse encoders different from dense encoders. The first and most straightforward difference is that sparse encoders produce sparse vectors (i.e., most term weights are zero). This sparsity is controlled by sparse regularizers, which we will discuss in the next section.

Second, dimensions in sparse weight vectors are usually tied to terms in a vocabulary that contains tens of thousands of terms. Therefore, the size of the vectors is large, equal to the size of the vocabulary; each dimension represents a term (typically a BERT word piece). On the contrary, (single-vector) dense retrieval methods produce condensed vectors (usually fewer than 1000 dimensions) that encode the semantics of the input text without a clear correspondence between terms and dimensions. Term-level dense retrieval methods like ColBERT [13] do preserve this correspondence.

The third distinction is that encoders in sparse retrieval only produce non-negative weights, whereas dense encoders have no such constraint. This constraint comes from the fact that sparse retrieval relies on software stacks (inverted indexing, query processing algorithms) built for traditional lexical search (e.g., BM25), where weights are always non-negative term frequencies.

Whether these differences lead to systematically different behavior between LSR and dense retrieval methods is an open question. Researchers have observed that LSR models and token-level dense models like ColBERT tend to generalize better than single-vector dense models on the BEIR benchmark [8,35]. There are also recent works proposing hybrid retrieval systems that combine the strength of both dense and sparse representations [3,18,19], which can bring benefits for both in-domain and out-of-domain effectiveness [19].

There are several variants of sparse encoders, which are typically built on a transformer-backbone [36] with additional head layer(s) on top. In Table 1, we summarize a list of common architectures of sparse encoders proposed in the literature. We use the following notation when describing these sparse encoder architectures: v_i denotes the ith term in a vocabulary V; t_j denotes the jth term in an input sequence t (either a query or document) of length L; h_j represents the contextualized embedding of t_j from a transformer encoder; e_i represents the transformer's input embedding of the v_i; $w_i(t)$ represents the weight of v_i in the context of t. The architectures include:

- **BINARY**: The BINARY encoder simply tokenizes the input into terms (word pieces) and considers the presence of terms in the input text. The binary encoder performs neither term expansion nor weighting:

$$w_i(t) = \max_{j=1..L} \mathbb{1}(v_i = t_j) \qquad (1)$$

- **MLP**: This encoder uses a **M**ulti-layer **P**erceptron (usually one layer) on top of each contextualized embedding h_j produced by the transformer-backbone for each input term to generate the term's score. Only terms in the input receive a weight; the other terms are zero.

$$w_i(t) = \sum_{j=1...L} log\left(\mathbb{1}(v_i = t_j)\left(ReLU(h_jW + b) \right) + 1 \right) \qquad (2)$$

where W and b are the weight and bias of the linear head. This MLP architecture focuses on term weighting.
- **expMLP**: This encoder adds a pre-processing step to expand the input with relevant terms before using a MLP encoder. The expansion terms can be selected from an external source/model (e.g., DocT5Query [27]).
- **MLM**: The MLM encoder aggregates term weights over the logits produced by BERT's **M**asked **L**anguage **M**odel head. The weight for each term in the vocabulary is generated as follows:

$$w_i(t) = q(t)log\left(1 + \max_{j=1...L} ReLU\left(h_j^\mathsf{T}e_i + b_i\right)g(t_j) \right). \qquad (3)$$

The ReLU function ensures non-negative weights and can be replaced with e.g. a Softplus, which has similar properties but is differentiable everywhere. The log normalization prevents some weights from getting too large. Term importance and passage quality scores are captured by $g(t_j)$ and $q(t)$, respectively. When present, the $g(t_j)$ and $q(t)$ functions can be modeled by an linear layer on top of contextualized embeddings of input tokens and the [CLS] token. Out of the three approaches using a MLM encoder, only one includes these functions. The choice of max aggregation and ReLU activation makes sparser representations and, at the same time, reduces training time as they disconnect the output from many paths in the computational graph.
- **clsMLM**: This is a simplified version of the MLM encoder that only takes the logits of the [CLS] token, which is at the position 0 of the sequence, as the output vector. Intuitively, this encoder squeezes the information of the whole sequence into a small [CLS] vector, which is then projected into an over-complete set of vocabulary bases:

$$w_i(t) = ReLU(h_0^\mathsf{T}e_i + b_i) \qquad (4)$$

where h_0 is the contextualized embedding of the CLS token.

These encoders are defined independent of input type (i.e., query or document). We can use a single shared encoder to encode both queries and documents or employ two separate encoders mixed-and-matched from the above list.

Sparse Regularizers. Sparse regularizers control the sparsity of weight vectors, which is crucial for query processing efficiency. We describe three common regularization techniques used in learned sparse retrieval methods.

- **FLOPs**: The FLOPs regularizer [29], estimates the average number of floating-point operations needed to compute the dot product between two weight vectors by a smooth function. FLOPs is defined over a batch of N sparse representations as follows:

$$FLOPs = \sum_{i=1}^{|V|} \bar{a}_i^2 = \sum_{i=1}^{|V|} \left(\frac{1}{N} \sum_{j=1}^{N} w_j^i \right)^2 \tag{5}$$

where $\bar{a}_i$ is the estimated activation probability of the ith dimension. Intuitively, the FLOPS regularizer might lead to two side-effects: (1) it forces the weights to be small and (2) it encourages uniform activation probability across all dimension when the square sum is minimized.
- L_p **Norm**: The family of L_p norms has been commonly applied in machine learning to mitigate over-fitting. With LSR, L_p is applied to the output vector rather than to model weights. L_1 and L_2 are two widely used norms.
- **Top-K**: This is a simple pruning technique which only keeps the top-k highest weights and zeroes out the rest. This pruning can be applied at inference time as a post-processing step or at training time with the value of k decreasing over time [20].

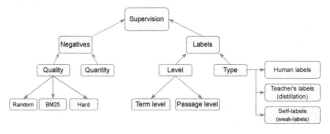

Fig. 1. Aspects of supervision commonly used for learned sparse retrieval.

Supervision. As some published LSR methods have identical sparse encoder(s) and sparse regularizer(s), we consider the supervision component to differentiate them and to consider its effect. As illustrated in Fig. 1, this supervision component is composed of two factors: negative examples and labels.

- **Negatives:** For contrastive learning, the quality and number of negatives used for training have a significant impact on performance [1]. The more and harder the negatives, the better the result. A naive way of selecting negatives is randomly sampling non-positive passages/documents from the corpus [10, 20,26,44], but this tends to create easy, less informative examples. Harder negatives can be selected from the top non-positive documents returned by BM25 or by neural retrieval models [38], which can also be used to filter out false negatives [30].

- **Labels:** Labels for training LSR methods are classified by type and level. Types include human, teacher's, and self-labels. Human labels have good quality but are scarce and costly to collect in large quantities. Teacher's labels are generated by a previously trained model and are referred to as distillation. Self-labels or proxy-labels are generated by the model itself. Label level refers to term-level or passage/document-level labels. Term-level labels provide one score per term, while passage-level labels indicate relevance for query-passage pairs. Most methods use passage-level labels.

2.2 Surveyed Learned Sparse Retrieval Methods

In Table 2, we present a summary of LSR methods fit into our conceptual framework. We cover nearly all transformer-based LSR methods for text ranking in the literature[1], but omit several due to time and space limitations [2,4,11,25]. We group the methods into four groups by their conceptual similarity. We discuss how the methods fit into our framework and point out any small differences that are not described by our three components (e.g., choice of nonlinearity and including term or passage quality functions).

A. Methods without any expansion. DeepCT [5] and **uniCOIL** [16] use an MLP encoder for weighting terms in queries and documents, with a slight modification to Eq. 2 by removing log normalization. Using the MLP means no expansion is applied (to query or document). DeepCT and uniCOIL only differ in supervision. DeepCT is supervised by term-recall, a term-level label defined as the ratio of relevant queries containing a term. On the other hand, uniCOIL uses passage-level labels rather than supervising individual term scores.

B. Methods without query expansion. uniCOIL$_{dT5q}$ [16], **uniCOIL**$_{tilde}$ [16], and **EPIC** [20] replace the MLP document encoder in group **A** with either an $expMLP$ or MLM encoder, which is capable of document expansion. As a pre-processing step, uniCOIL$_{dT5q}$ and uniCOIL$_{tilde}$ expand passages with relevant terms generated by third-party models (docT5query [27], TILDE). Instead of pre-expanding the passages, EPIC is the first to leverage the MLM architecture trained to do document expansion and term scoring end-to-end at once. On the query side, EPIC keeps the log normalization as in Eq. 2. On the document side, the ReLU in Eq. 3 is replaced by a Softplus and both $q(t)$ and $g(t)$ are modeled by a linear layer with a softmax activation.

---

[1] We consider the prominent doc2query document expansion methods [27,28] in the context of pre-processing for document expansion (e.g., combined with uniCOIL), but we do not treat these as standalone *retrieval* methods.

C. Methods without query expansion or weighting. DeepImpact [23], **Sparta** [41], **TILDE** [44], and **TILDEv2** [43] simplify methods in group **B** by removing the (MLP) query encoder, hence have a near-instant query encoding time but no query expansion and weighting capability. DeepImpact and TILDE$_{v2}$ can be viewed as the uniCOIL$_{dT5q}$ and uniCOIL$_{tilde}$ models without a query encoder, respectively. Sparta is simplified from EPIC by (1) removing query encoder and (2) removing $q(t)$ and $g(t_j)$ in Eq. 3. TILDE replaces the MLM head in Sparta with clsMLM.

D. Methods with full expansion and weighting. Splade-max [7] and **distilSplade-max** [7] use a shared MLM architecture on both the query and document side. The MLM enables end-to-end weighting and expansion for both query and document. Instead of selecting top-k terms as in EPIC, this Splade family uses the FLOPs regularizer during training to sparsify the representations. The difference between Splade-max and distilSplade-max is the supervision. While Splade-max is trained with multiple in-batch BM25 negatives, distilSplade-max is trained with a distillation technique using mined hard negatives. Similar to Sparta, $q(t)$ and $g(t_j)$ in Eq. 3 are removed from Splade models.

Table 2. Definition of existing LSR methods. An (s) indicates multiple negatives.

	Method	Query	Passage	Reg.	Supervision		
					Level	Neg	Type
A	DeepCT [5]	MLP	MLP	–	Term	–	–
	uniCOIL [16]	MLP	MLP	–	Passage	BM25(s)	Human
B	uniCOIL$_{dT5q}$ [16]	MLP	expMLP	–	Passage	BM25(s)	Human
	uniCOIL$_{tilde}$ [16]	MLP	expMLP	–	Passage	BM25(s)	Human
	EPIC [20]	MLP	MLM	Top-k	Passage	BM25	Human
C	DeepImpact [23]	BINARY	expMLP	–	Passage	BM25	Human
	TILDE [43]	BINARY	clsMLM	–	Term	–	–
	TILDEv2 [43]	BINARY	expMLP	–	Passage	BM25(s)	Human
	Sparta [41]	BINARY	MLM	–	Passage	BM25	Human
D	SPLADE-max [7]	MLM	MLM	FLOPs	Passage	BM25(s)	Human
	DistilSPLADE-max [7]	MLM	MLM	FLOPs	Passage	Hard	Teacher

3   Experimental Settings

For all experiments, we use Huggingface's BERT implementation with *distilbert-base-cased* [34,37]. We train our models on the MSMarco [26] and TripClick datasets [31]. For models that need hard negative mining and distillation on MSMarco, we use the data provided by SentenceTransformers[2] [30] for training. For TripClick, we use the training triples[3] created by [9]. We evaluate methods

---

[2] huggingface.co/datasets/sentence-transformers/msmarco-hard-negatives.

[3] github.com/sebastian-hofstaetter/tripclick.

Table 3. Reproduced MRR@10 scores on MSMarco dev. (*) Indicates reranking results on BM25 top-1000 passages (following the original work).

	Method	Original MRR	Reproduced MRR	Δ %
A	DeepCT	24.3	24.6	1.234
	uniCOIL	31.5	31.6	0.317
B	uniCOIL$_{dT5q}$	35.2	34.7	−1.420
	uniCOIL$_{tilde}$	34.9	34.8	−0.286
	EPIC$^*_{top1000}$	27.3	28.8	5.495
C	DeepImpact	32.6	31.2	−4.294
	TILDE$^*_{v2}$	33.3	33.7	1.201
	Sparta	−	31.0	−
D	Splade$_{max}$	34.0	34.0	0.000
	distilSplade$_{max}$	36.9	37.9	2.439

with the benchmarks' standard metrics, including MRR@10, NDCG@10, and Recall@1000. In the following sections, we remove the cut-off @K for brevity.

We measure encoding latency on an AMD EPYC 7702 CPU and Tesla V100 GPU. We use a modified version of Anserini [15] for indexing passages and measure retrieval latency on an AMD EPYC 7702 CPU using 60 threads. For **RQ1**, we followed the same hyper-parameters and losses described in the original papers to reproduce LSR methods. For **RQ2** and **RQ3**, we train all methods on a single A100 GPU using the above mined hard negatives, and distillation data for MSMarco or the BM25 triplets for TripClick. Our Github repository contains the full configurations for all experiments.

4   Results and Analysis

In this section we consider our three RQs. We first reproduce LSR methods in their original experimental settings (RQ1), before training them in a common setting (RQ2) and analyzing the impact of architectural differences (RQ3).

4.1   RQ1: Are the Results from LSR Papers Reproducible?

We train the LSR methods using a similar experimental setup described in the original papers and code. The reproduced results are reported in Table 3. For most of the methods, we obtain scores that are slightly higher or comparable to the original work. A slightly higher MRR was observed for DeepCT, uniCOIL, EPIC, TILDE$_{v2}$, and distilSplade$_{max}$, while DeepImpact and uniCOIL$_{dT5q}$ received slightly lower reproduced scores. Sparta was not evaluated on MSMarco in the original paper, so there is no comparison point for our result.

These reproduced results show that DeepCT and uniCOIL (without docT5query expansion) tend to be the least effective approaches, whereas

Table 4. Results with cross-encoder distillation on hard negatives (left) and BM25 negatives (two rightmost columns) on MS MARCO. The **RL** column indicates the latency (ms/q) for query encoding and retrieval.

	Method	MSMarco		DL-2019		DL-2020		Index	RL	BM25 Negs	
		MRR	R	NDCG	R	NDCG	R	GB	ms	MRR	R
A	uniCOIL	$27.3^{†††}_{***}$	$88.0^{†††}_{***}$	59.3	72.9	54.3	77.9	**1.1**	**6.1**	32.1	92.6
B	uniCOIL$_{dT5q}$	$35.0^{†}_{*}$	95.7	65.9	81.0	68.4	84.6	1.8	12.7	34.7	96.4
	uniCOIL$_{tilde}$	$36.1^{†††}_{***}$	$96.8^{††}_{**}$	69.1	82.2	69.4	85.2	2.6	<u>7.1</u>	34.8	<u>96.5</u>
	EPIC$_{top400}$	$37.2^{†††}_{***}$	$97.2^{†††}_{***}$	70.9	<u>87.7</u>	<u>71.8</u>	88.7	9.7	17.7	**35.5**	96.4
C	DeepImpact	32.2_{**}	$94.7^{†††}$	63.1	77.2	63.3	82.1	1.8	16.1	32.2	95.4
	TILDE$_{top400}$	$29.9^{†††}_{**}$	$93.9^{†††}$	65.1	68.5	63.0	69.9	6.4	29.0	21.6	74.5
	TILDE$_{v2}$	$32.9^{††}_{**}$	96.0	66.3	79.7	65.9	83.5	2.6	9.5	33.7	96.1
	Sparta$_{top400}$	$35.3^{†††}$	$96.8^{†††}$	69.1	81.9	68.1	85.8	6.1	26.7	28.3	88.7
D	distilSplade$_{max}$	<u>$37.9^{†††}$</u>	<u>$98.1^{†††}$</u>	**74.8**	**87.9**	**72.5**	**89.5**	6.3	122.5	<u>35.3</u>	**97.0**
	distilSplade$_{sep}$	**38.0**	<u>98.0</u>	<u>74.1</u>	<u>87.7</u>	70.6	<u>89.0</u>	8.0	50.2	–	–

***/† † † $p < 0.01$, **/†† $p < 0.05$, */† $p < 0.1$ with paired two-tailed t-test
Comparing with results in Table 3 (*) and BM25 negatives results (†)

distilSplade$_{max}$ achieves the highest MRR. Interestingly, we observe pairs of methods that have identical architectures, but different training recipes lead to a significant discrepancy in scores. uniCOIL changes the supervision signal of DeepCT from token-level weights to passage-level relevance, making a 28% jump in MRR from 24.6 to 31.6. Apparently, the supervision matters a lot here; using the passage-level labels allows the model to learn the term weights more optimally for passage-level relevance. Similarly, using mined hard negatives and distillation boosts MRR from 34.0 to 37.9 with the Splade model. This change of supervision makes distilSplade$_{max}$ the most effective LSR method considered. Without this advanced training, Splade$_{max}$ performs comparably to uniCOIL$_{dT5q}$ and uniCOIL$_{tilde}$. Looking closely at the group (**B**), EPIC seems to perform under its full capacity because it achieves a MRR substantially below the two uniCOIL variants. This may be due to the fact that EPIC was originally trained on 40000 triples, whereas the other methods were trained on up to millions of samples.

4.2   RQ2: How Do LSR Methods Perform with Recent Advanced Training Techniques?

Variations in environments, as shown in RQ1, make it difficult to fairly compare LSR methods and can lead to inaccurate conclusions. To eliminate these discrepancies, we train all methods in a consistent environment, which we show to be effective in this section. We focus on the most effective supervision setup, which is distilSplade$_{max}$ trained using distillation and hard negatives. Table 4 shows the results of the LSR methods under this setting. Note that several methods (DeepCT and uniCOIL; Splade variants) will have identical scores in this experiment as they collapse into the same model. We only report a representative method in these cases.

Comparing to the results of **RQ1** (Table 3), we find that the least effective methods (DeepCT, now equivalent to uniCOIL) and the most effective method (distilSplade$_{max}$) remain in the same positions. Methods between these two endpoints move around with substantial changes in their effectiveness. Out of 10 methods we reproduced in Table 3, we observe an upward trend on seven methods, while the remaining three methods stay the same or perform worse. The biggest jumps are seen using EPIC and Sparta, with a relative improvement of 8.0 and 4.2 MRR points on MSMarco, respectively. The increase in EPIC's effectiveness, which is due to the combination of longer training time and improved supervision, moves the approach's relative ranking from the second worst to the second best, with metrics competitive with distilSplade$_{max}$ on MSMarco. On TREC DL 2019 and TREC DL 2020, the gap in NDCG@10 between EPIC and distilSplade$_{max}$ is higher. The increased MRR@10 on MSMarco also brings Sparta a nice efficiency-effectiveness trade-off: since there is no query encoder with Sparta, there is no need for a GPU at retrieval time.

In addition to EPIC and Sparta, we also observe positive trends with DeepCT, DeepImpact, uniCOIL$_{dT5q}$ and uniCOIL$_{tilde}$; however, the change is relatively marginal. We observe decreased effectiveness on uniCOIL and TILDE$_{v2}$. While the decline with TILDE$_{v2}$ is small, the drop with uniCOIL ($32.1 \rightarrow 27.3$) is quite large. Indeed, without expansion capability, no soft-matching could be possible, which renders a challenge for uniCOIL to reconstruct the MarginMSE's loss margin produced by a cross-encoder teacher, which is capable of soft-matching.

Regarding architecture types, methods using the MLM architecture, either on the document or query side (EPIC, Sparta, Splade), generally perform better than those using other architectures (clsMLM, MLP, expMLP, BINARY) on all three datasets. However, MLM also increases index size and latency significantly. For instance, EPIC's index is at least 6 GB larger than other methods in the group. Notably, distilSplade$_{max}$ not only creates a large index but also has a notably high retrieval latency, almost 20 times slower than the fastest method.

The latency issue in Splade is related to using the same shared MLM encoder for query and documents, resulting in similar term activation probability between queries and documents. We confirmed this by replacing the shared encoder with two separate ones (distilSplade$_{sep}$), which reduced latency from 122.5 ms to 50.2 ms, a 59% decrease. This benefit of separate encoders was also reported in [14], and our results further support its substantial impact.

In the last two columns of Table 4, we provide additional MSMarco results with training using BM25 in-batch negatives (the same as uniCOIL's original setup). We find that using hard negatives with distillation is generally more effective than using BM25 negatives, though not with uniCOIL or TILDE$_{v2}$.

4.3 RQ3: How Does the Choice of Encoder Architecture and Regularization Affect Results?

In this RQ, we aim to quantify how different factors (*query expansion, document expansion, query weighting, document weighting, regularization*) affect the

Table 5. The effects of architecture and regularizer on MSMarco and TripClick. *We use names that better reflect the architectural differences between methods. Visit our Github repository to see the full configurations and original names.*

Effect		Control	Change	MSMarco		DL 2019	DL 2020	Latency	Index
				MRR	R	NDCG	NDCG	ms	GB
Doc weighting	1_a	Q_{MLM}	$D_{BIN} \rightarrow D_{MLP}$	16.7+18.3	86.0+11.0	44.1+26.8	42.9+24.5	11.4+04.2	0.6+0.7
	1_b	Q_{MLP}	$D_{eBIN} \rightarrow D_{eMLP}$	08.2+27.9	76.2+20.6	30.4+38.7	27.5+41.9	10.8-03.7	1.2+1.4
Query weighting	2_a	D_{cMLP}	$Q_{BIN} \rightarrow Q_{MLP}$	32.9+3.2	96.0+0.8	66.3+2.8	65.9+3.5	09.5-0.9	2.6+0.0
	2_b	D_{MLM}	$Q_{BIN} \rightarrow Q_{MLP}$	35.2+1.9	96.5+0.7	69.4+1.5	69.7+2.1	28.9-7.9	8.6+1.1
Doc expansion	3_a	Q_{MLM}	$D_{MLP} \rightarrow D_{MLM}$	34.9+3.1	97.0+0.9	70.9+3.3	67.4+3.2	15.6+34.6	1.3+6.7
	3_b	Q_{MLP}	$D_{MLP} \rightarrow D_{MLM}$	27.5+10.0	89.7+8.2	59.3+12.0	54.3+17.9	27.5+10.5	1.2+6.9
Query expansion	4_a	D_{MLM}	$Q_{MLP} \rightarrow Q_{MLM}$	38.0+0.0	97.0+0.1	71.3+2.8	72.1-1.3	12.9+37.3	8.0-0.1
	4_b	D_{MLP}	$Q_{MLP} \rightarrow Q_{MLM}$	27.5+7.5	89.7+7.4	59.3+11.6	54.3+13.1	06.1+9.5	1.2+0.1
Regularization	5_a	Q_{MLP} D_{MLM}	$FLOPs \rightarrow Topk$	38.0+0.0	97.9-0.3	71.3+0.8	72.1+0.1	12.8+4.3	8.1-0.7

Effect		Control	Change	TripClick					
				HEAD(dctr)		TORSO(raw)	TAIL(raw)	Latency	Index
				NDCG	R	NDCG	NDCG	ms	GB
Doc weighting	1_a	Q_{MLP}	$D_{BIN} \rightarrow D_{MLP}$	6.5+18.9	69.7+18.4	10.7+17.5	16.2+13.2	2.0-0.1	0.3+0.3
	1_b	Q_{MLP}	$D_{eBIN} \rightarrow D_{eMLP}$	5.7+21.0	67.2+21.1	9.1+20.4	13.9+16.5	2.5-0.3	0.4+0.5
Query weighting	2_a	D_{MLM}	$Q_{BIN} \rightarrow Q_{MLP}$	26.3+3.9	90.0+1.9	31.3+3.3	34.2+3.8	3.2-0.0	1.8-0.1
	2_b	D_{MLP}	$Q_{BIN} \rightarrow Q_{MLP}$	24.2+1.1	87.3+0.8	27.7+0.4	29.4+0.0	2.1-0.2	0.5+0.1
Doc expansion	3_a	Q_{MLM}	$D_{MLP} \rightarrow D_{MLM}$	27.9+2.2	90.9+1.0	32.7+1.5	34.1+3.9	4.6+1.6	0.7+0.7
	3_b	Q_{MLP}	$D_{MLP} \rightarrow D_{MLM}$	25.3+4.7	88.1+3.7	28.2+6.1	29.4+7.9	1.9+1.6	0.6+0.8
Query expansion	4_a	D_{MLM}	$Q_{MLP} \rightarrow Q_{MLM}$	30.0+0.1	91.8+0.1	34.2-0.1	37.4+0.6	3.4+2.8	1.4-0.0
	4_b	D_{MLP}	$Q_{MLP} \rightarrow Q_{MLM}$	25.3+2.6	88.1+2.8	28.2+4.5	29.4+4.6	1.9+2.7	0.6+0.0
Regularization	5_a	Q_{MLP} D_{MLM}	$L1 \rightarrow Topk$	30.0+0.1	91.8+0.1	34.2+0.3	37.4+0.7	3.4-0.2	1.4+0.3

effectiveness and efficiency of LSR systems. To eliminate potential confounding factors due to minor differences between groups (e.g., choice of nonlinearity), we perform a series of controlled experiments in which we make single architectural changes while holding the rest of the architecture constant.

In Table 5, numbers before + or − are the metrics before a change (left side of arrow), while numbers after these symbols show the effect of a change (right). We see that document weighting seems to be the most crucial component since the systems without this component fail on all three datasets. In row $1_{(a,b)}$, the system with a binary document encoder shows very low MRR and NDCG scores regardless of MLM or MLP on the query side. On both MSMarco and TripClick, enabling document weighting (by replacing the binary document encoder with an MLP) improves the effectiveness by a large margin (at least 11 points) with reasonable latency and index size increases. Without document weighting, the models are not able to identify important terms in documents.

Similarly, as shown in rows $2_{(a,b)}$, we control the document side and change the binary query encoder to an MLP query encoder to observe the effect of query weighting. The result suggests that query weighting has a moderate contribution to the ranking metrics overall. Still, interestingly, it causes almost no harm to the index size or even reduces the latency. Note that the latency of the MLP query encoder here is measured on GPU; therefore, the encoding overhead is tiny. The improved overall latency is mostly due to the MLP reducing the weights of some non-useful query terms to zero, making queries shorter. The effect is quite consistent between MSMarco and TripClick collections.

Regarding the expansion factors, we observe the cancellation effect between query expansion and document expansion. Indeed, with the absence of expansion on one side (3_b: Q_{MLP} has no query expansion, 4_b: D_{MLP} has no document expansion), the expansion on the other side largely improves the ranking metrics with at least 7.4(2.6) points and at most 17.9(7.9) points overall on MSMarco (TripClick). The cost of latency, in this case, is rather low. The numbers in rows 3_a and 4_a indicate that query and document expansion have a cancellation effect. That is, query expansion reduces the benefit of performing document expansion and vice versa. Row 4_a shows that when document expansion is in place, query expansion has minimal impact on ranking effectiveness and incurs a relatively high latency overhead (increases of 289% and 82% on MSMarco and TripClick). Row 3_a shows a similar trend, with document expansion making moderate contributions to system effectiveness. On TripClick, the cancellation interaction between the two factors is less strong. Overall, this cancellation effect suggests that including both expansion components may not be necessary.

Lastly, to examine the effect of regularization, we keep the model's architecture constant and change the FLOPs/L_1 regularizer during training to Topk pruning during inference. As shown in rows 5_a, changing the regularization approach does not significantly affect effectiveness or efficiency.

Table 6. Results with only query expansion or only document expansion. * $p < 0.01$ with paired two-tailed t-test

Method	MSMarco-dev				TripClick-HEAD(dctr)			
	MRR	R	Index(GB)	RL(ms)	NDCG	R	Index(GB)	RL(ms)
distilSplade$_{sep}$	38.0	98.0	8.0	50.2	30.1	91.9	1.4	6.3
distilSplade$_{qMLP}$	38.0	97.9	8.1	12.9	30.0	91.8	1.4	3.4
distilSplade$_{dMLP}$	34.9*	97.0*	1.3	15.6	27.9*	90.9*	0.7	4.6

In Table 6, we show the results of systems with expansion only on either the query or the document side. In the table, distilSplade$_{qMLP}$ denotes the distilSplade$_{sep}$ with the MLM query encoder replaced by an MLP query encoder; hence no query expansion is involved. Similar interpretation applies for distilSplade$_{dMLP}$. As can be seen, distilSplade$_{qMLP}$ makes no significant changes on ranking metrics, while reducing the retrieval latency by more than 74% and 46% on MSMarco and TripClick, respectively. distilSplade$_{dMLP}$ exhibits a similar latency improvement, but suffers from a significant drop in effectiveness. In practice, distilSplade$_{qMLP}$ could be viewed as a more efficient drop-in replacement for the full model. This use of $qMLP$ is complementary to other changes (e.g., using a smaller encoder as in [14]) to improve the efficiency of LSR.

5   Conclusion

In this work, we introduced a conceptual framework for learned sparse retrieval that unifies existing LSR methods under the perspective of three components. After reproducing these methods, we carried out a series of experiments to isolate the effect of single changes on a model's performance. This analysis led to several findings about the components, including that we can remove the query expansion from a SOTA system, leading to a significant latency improvement without compromising the system's effectiveness. While this study covered the most prominent transformer-based LSR methods, several others could not be considered due to time and computing constraints (e.g., [2,4,11,25]). We plan to incorporate them into our implementation as future work.

Acknowledgement. We thank Maurits Bleeker from the UvA IRLab for his feedback on the paper.

References

1. Ash, J.T., Goel, S., Krishnamurthy, A., Misra, D.: Investigating the role of negatives in contrastive representation learning. arXiv preprint arXiv:2106.09943 (2021)
2. Bai, Y., et al.: Sparterm: learning term-based sparse representation for fast text retrieval. arXiv preprint arXiv:2010.00768 (2020)
3. Chen, X., et al.: Salient phrase aware dense retrieval: can a dense retriever imitate a sparse one? arXiv preprint arXiv:2110.06918 (2021)
4. Choi, E., Lee, S., Choi, M., Ko, H., Song, Y.I., Lee, J.: Spade: improving sparse representations using a dual document encoder for first-stage retrieval. In: Proceedings of the 31st ACM International Conference on Information & Knowledge Management, pp. 272–282 (2022)
5. Dai, Z., Callan, J.: Context-aware term weighting for first stage passage retrieval. In: Proceedings of the 43rd International ACM SIGIR conference on research and development in Information Retrieval, pp. 1533–1536 (2020)
6. Devlin, J., Chang, M.W., Lee, K., Toutanova, K.: BERT: pre-training of deep bidirectional transformers for language understanding. In: Proceedings of the 2019 Conference of the North American Chapter of the Association for Computational Linguistics: Human Language Technologies, Volume 1 (Long and Short Papers), pp. 4171–4186 (2019)
7. Formal, T., Lassance, C., Piwowarski, B., Clinchant, S.: Splade v2: sparse lexical and expansion model for information retrieval. arXiv preprint arXiv:2109.10086 (2021)
8. Formal, T., Lassance, C., Piwowarski, B., Clinchant, S.: From distillation to hard negative sampling: Making sparse neural ir models more effective. arXiv preprint arXiv:2205.04733 (2022)
9. Hofstätter, S., Althammer, S., Sertkan, M., Hanbury, A.: Establishing strong baselines for tripclick health retrieval. In: Hagen, M., Verberne, S., Macdonald, C., Seifert, C., Balog, K., Nørvåg, K., Setty, V. (eds.) ECIR 2022. LNCS, vol. 13186, pp. 144–152. Springer, Cham (2022). https://doi.org/10.1007/978-3-030-99739-7_17

10. Izacard, G., et al.: Unsupervised dense information retrieval with contrastive learning. arXiv preprint arXiv:2112.09118 (2021)
11. Jang, K.R., et al.: Ultra-high dimensional sparse representations with binarization for efficient text retrieval. In: Proceedings of the 2021 Conference on Empirical Methods in Natural Language Processing, pp. 1016–1029 (2021)
12. Karpukhin, V., et al.: Dense passage retrieval for open-domain question answering. arXiv preprint arXiv:2004.04906 (2020)
13. Khattab, O., Zaharia, M.: Colbert: Efficient and effective passage search via contextualized late interaction over BERT. In: Proceedings of the 43rd International ACM SIGIR Conference on Research and Development in Information Retrieval, pp. 39–48 (2020)
14. Lassance, C., Clinchant, S.: An efficiency study for splade models. In: Proceedings of the 45th International ACM SIGIR Conference on Research and Development in Information Retrieval, pp. 2220–2226 (2022)
15. Lin, J., et al.: Toward reproducible baselines: the open-source IR reproducibility challenge. In: Ferro, N., et al. (eds.) ECIR 2016. LNCS, vol. 9626, pp. 408–420. Springer, Cham (2016). https://doi.org/10.1007/978-3-319-30671-1_30
16. Lin, J., Ma, X.: A few brief notes on deepimpact, coil, and a conceptual framework for information retrieval techniques. arXiv preprint arXiv:2106.14807 (2021)
17. Lin, J., Nogueira, R., Yates, A.: Pretrained transformers for text ranking: bert and beyond. Synth. Lect. Human Lang. Technol. **14**(4), 1–325 (2021)
18. Lin, S.C., Lin, J.: Densifying sparse representations for passage retrieval by representational slicing. arXiv preprint arXiv:2112.04666 (2021)
19. Lin, S.C., Lin, J.: A dense representation framework for lexical and semantic matching. arXiv preprint arXiv:2206.09912 (2022)
20. MacAvaney, S., Nardini, F.M., Perego, R., Tonellotto, N., Goharian, N., Frieder, O.: Expansion via prediction of importance with contextualization. In: Proceedings of the 43rd International ACM SIGIR Conference on Research and Development in Information Retrieval, pp. 1573–1576 (2020)
21. Mackenzie, J., Trotman, A., Lin, J.: Wacky weights in learned sparse representations and the revenge of score-at-a-time query evaluation. arXiv preprint arXiv:2110.11540 (2021)
22. Mackenzie, J., Trotman, A., Lin, J.: Efficient document-at-a-time and score-at-a-time query evaluation for learned sparse representations. ACM Trans. Inf. Syst. (Dec 2022). https://doi.org/10.1145/3576922, https://doi.org/10.1145/3576922, just Accepted
23. Mallia, A., Khattab, O., Suel, T., Tonellotto, N.: Learning passage impacts for inverted indexes. In: Proceedings of the 44th International ACM SIGIR Conference on Research and Development in Information Retrieval, pp. 1723–1727 (2021)
24. Mallia, A., Mackenzie, J., Suel, T., Tonellotto, N.: Faster learned sparse retrieval with guided traversal. In: Proceedings of the 45th International ACM SIGIR Conference on Research and Development in Information Retrieval, pp. 1901–1905. SIGIR '22, Association for Computing Machinery, New York, NY, USA (2022). https://doi.org/10.1145/3477495.3531774, https://doi.org/10.1145/3477495.3531774
25. Nair, S., Yang, E., Lawrie, D., Mayfield, J., Oard, D.W.: Learning a sparse representation model for neural CLIR (2022)
26. Nguyen, T., Rosenberg, M., Song, X., Gao, J., Tiwary, S., Majumder, R., Deng, L.: MS MARCO: a human generated machine reading comprehension dataset. In: CoCo@ NIPs (2016)

27. Nogueira, R., Lin, J.: From doc2query to docTTTTTquery (2019)
28. Nogueira, R., Yang, W., Lin, J., Cho, K.: Document expansion by query prediction. arXiv preprint arXiv:1904.08375 (2019)
29. Paria, B., Yeh, C.K., Yen, I.E., Xu, N., Ravikumar, P., Póczos, B.: Minimizing flops to learn efficient sparse representations. In: International Conference on Learning Representations (2019)
30. Reimers, N., Gurevych, I.: Sentence-bert: sentence embeddings using siamese bert-networks. In: Proceedings of the 2019 Conference on Empirical Methods in Natural Language Processing. Association for Computational Linguistics (11 2019), https://arxiv.org/abs/1908.10084
31. Rekabsaz, N., Lesota, O., Schedl, M., Brassey, J., Eickhoff, C.: Tripclick: the log files of a large health web search engine. In: Proceedings of the 44th International ACM SIGIR Conference on Research and Development in Information Retrieval, pp. 2507–2513 (2021)
32. Robertson, S., Zaragoza, H., et al.: The probabilistic relevance framework: BM25 and beyond. Found. Trends® Inf. Retrieval 3(4), 333–389 (2009)
33. Robertson, S.E., Walker, S., Jones, S., Hancock-Beaulieu, M., Gatford, M.: Okapi at TREC-3. In: Harman, D.K. (ed.) Proceedings of The Third Text REtrieval Conference, TREC 1994, Gaithersburg, Maryland, USA, November 2–4, 1994. NIST Special Publication, vol. 500–225, pp. 109–126. National Institute of Standards and Technology (NIST) (1994). http://trec.nist.gov/pubs/trec3/papers/city.ps.gz
34. Sanh, V., Debut, L., Chaumond, J., Wolf, T.: Distilbert, a distilled version of BERT: smaller, faster, cheaper and lighter. arXiv preprint arXiv:1910.01108 (2019)
35. Thakur, N., Reimers, N., Rücklé, A., Srivastava, A., Gurevych, I.: BEIR: a heterogeneous benchmark for zero-shot evaluation of information retrieval models. In: Thirty-fifth Conference on Neural Information Processing Systems Datasets and Benchmarks Track (Round 2) (2021)
36. Vaswani, A., et al.: Attention is all you need. In: Advances in Neural Information Processing Systems, vol. 30 (2017)
37. Wolf, T., et al.: Transformers: state-of-the-art natural language processing. In: Proceedings of the 2020 Conference on Empirical Methods in Natural Language Processing: System Demonstrations, pp. 38–45 (2020)
38. Xiong, L., et al.: Approximate nearest neighbor negative contrastive learning for dense text retrieval. arXiv preprint arXiv:2007.00808 (2020)
39. Yang, P., Fang, H., Lin, J.: Anserini: reproducible ranking baselines using lucene. J. Data Inf. Qual. (JDIQ) 10(4), 1–20 (2018)
40. Zamani, H., Dehghani, M., Croft, W.B., Learned-Miller, E., Kamps, J.: From neural re-ranking to neural ranking: learning a sparse representation for inverted indexing. In: Proceedings of the 27th ACM International Conference on Information and Knowledge Management, pp. 497–506 (2018)
41. Zhao, T., Lu, X., Lee, K.: Sparta: efficient open-domain question answering via sparse transformer matching retrieval. arXiv preprint arXiv:2009.13013 (2020)
42. Zheng, G., Callan, J.: Learning to reweight terms with distributed representations. In: Proceedings of the 38th International ACM SIGIR Conference on Research and Development in Information Retrieval, pp. 575–584 (2015)
43. Zhuang, S., Zuccon, G.: Fast passage re-ranking with contextualized exact term matching and efficient passage expansion. arXiv preprint arXiv:2108.08513 (2021)
44. Zhuang, S., Zuccon, G.: Tilde: Term independent likelihood model for passage re-ranking. In: Proceedings of the 44th International ACM SIGIR Conference on Research and Development in Information Retrieval, pp. 1483–1492 (2021)

Entity Embeddings for Entity Ranking:
A Replicability Study

Pooja Oza and Laura Dietz[✉]

University of New Hampshire, Durham, NH 03824, USA
pho1003@wildcats.unh.edu, dietz@cs.unh.edu

Abstract. Knowledge Graph embeddings model semantic and structural knowledge of entities in the context of the Knowledge Graph. A nascent research direction has been to study the utilization of such graph embeddings for the IR-centric task of entity ranking. In this work, we replicate the GEEER study of Gerritse et al. [9] which demonstrated improvements of Wiki2Vec embeddings on entity ranking tasks on the DBpediaV2 dataset. We further extend the study by exploring additional state-of-the-art entity embeddings ERNIE [27] and E-BERT [19], and by including another test collection, TREC CAR, with queries not about person, location, and organization entities. We confirm the finding that entity embeddings are beneficial for the entity ranking task. Interestingly, we find that Wiki2Vec is competitive with ERNIE and E-BERT.

Our code and data to aid reproducibility and further research is available at https://github.com/poojahoza/E3R-Replicability.

Keywords: Entity retrieval · Entity embeddings · Knowledge graphs

1 Introduction

We study the problem of entity ranking since users seek entities in response to their queries [11,20], or such entities can be helpful in improving document rankings [6]. The queries can range from short factoid questions (e.g., "Who is the mayor of Berlin?") that seek a particular entity to the queries that request a list of entities (e.g., "Professional sports teams in Philadelphia"). Knowing relevant entities is also helpful when synthesizing relevant information on popular science topics (e.g., "tell me more about horseshoe crabs"). Given a query, the entity ranking task is to return a list of entities ordered by the relevance of each entity to the query. Such entities are taken from a given Knowledge Graph, such as Wikipedia or DBpedia.

Previous work on entity ranking either uses hand-crafted features within a Learning-To-Rank framework [7,22] or leverages information about entities available in a Knowledge Graph such as types [1,2,10,18] and relations [4,5,23]. However, these entity ranking systems consider only lexical matching between the queries and the entity information and disregard any semantic and structural information of the entities. To overcome this, re-ranking models that use

Knowledge Graph embeddings such as TransE [14], and Wiki2Vec [9] have been proposed. Knowledge Graph embeddings capture the structural and semantic information of the entities in the context of the Knowledge Graph. They project the entities and relations in a continuous vector space that preserves information about the structure of the Knowledge Graph. Such Knowledge Graph embeddings have shown to be successful in IR-centric entity ranking tasks, with explicit queries [14].

Additionally, knowledge-enhanced BERT models such as ERNIE [27] and E-BERT [19] have been proposed in recent years which augments the successful BERT model with the entity information through Knowledge Graph embeddings such as TransE [3] and Wiki2Vec [26]. The entity embeddings generated from these models are a fusion of the entity information from the Knowledge Graph and rich contextual information from BERT embeddings. These knowledge-enhanced BERT models have been demonstrated to improve the performance of entity-centric NLP downstream tasks such as relation classification, entity typing, and entity linking.

In this paper, we reproduce and replicate the work of Gerritse et al. [9] (GEEER) which shows that Knowledge Graph embeddings such as Wiki2Vec are beneficial to improve the performance of the entity-oriented search. We choose to reproduce and replicate this work as it is among the first few papers to study the utilization of Knowledge Graph embeddings in an IR-centric entity ranking task. This is a critical work with a high impact in the field of IR and hence reproducibility with further exploration is important. Within the GEEER framework, we explore the efficacy of pretrained entity embeddings in the entity ranking task with different datasets through replicability experiments. We incorporate new entity embeddings, new datasets, and different learning-to-rank methods in the study. In particular, we study the effect of neural fine-tuning of the embeddings for the ranking task.

In the following, we refer to both Knowledge Graph embeddings and entity embeddings of knowledge-enhanced BERT models as entity embeddings.

Experiments: In this work, we perform several sets of experiments to study whether the original findings still hold.

1. **Reproducibility:** Using the same code, entity embeddings, dataset, and entity re-ranking framework as given in the original paper [9] we confirm the findings of the original work.
2. **Replicability:** For these experiments, we re-implement the method of Gerritse et al. [9], using original and additional pretrained entity embeddings, and explore small changes in the setup.
3. **New Dataset:** While the original dataset was asking about people, organizations, and locations, we are adding another dataset, TREC CAR, which asks about other entity types. We confirm that the original findings still hold.
4. **Effect of Fine-tuning:** We further analyze the effect of fine-tuning the embeddings (as opposed to directly using pretrained embeddings). We study fine-tuning with both, point-wise and pair-wise ranking losses and demonstrate that the gains are even more significant.

5. **Study on Missing Entities:** Pretrained entity embeddings often don't contain embeddings for all candidate entities obtained via initial rankings. We quantify performance losses due to missing entity embeddings separately from those due to quality issues with available embeddings.

Findings in the original paper: In the original paper [9], the authors find two important results, of which we focus on the first: (1) Entity embeddings are advantageous to improve the performance of the entity ranking task and (2) Entity embeddings that contain both context and structural information of the Knowledge Graph perform better than the entity embeddings that contain only contextual information.

Findings in our paper: In our paper, we concentrate only on the first finding of the original paper, i.e., entity embeddings are advantageous to improve entity ranking task performance. We are able to reproduce the experiment, and additionally can replicate it under several changes to the setup.

We make the following additional observations: (1) Pretrained and fine-tuned entity embeddings help to improve the performance of entity ranking. While pretrained entity embeddings provide only a slight gain over the baselines, fine-tuned embeddings improve the performance by a significantly large margin. (2) Pretrained and fine-tuned Wiki2Vec embeddings outperform or perform similarly to knowledge-enhanced BERT models ERNIE and E-BERT. (3) In line with prior work [13], we find that in most cases fine-tuning with a pair-wise loss performs better than a point-wise loss for both Wiki2Vec and ERNIE. (4) We find that the pretrained entity embeddings help to improve performance losses of ranking baselines.

2   Related Work

2.1   Knowledge Graph Embeddings

Knowledge Graph embeddings are vector representations of the entities present in the Knowledge Graph. Such embeddings capture the semantic and structural information of the entities. Bordes et al. [3] proposed TransE, a translational-based model, that learns the embeddings of both entities and relations on the modeling assumption that the relation r is a translation between two entities h and t. TransE projects both entities and relations in the same vector space. However, since TransE considers only 1-to-1 relations, it does not work well with 1-to-N, N-to-N, and N-to-1 relations. To overcome this issue, TransH [25] model was proposed that projects each relation r with two vectors. TransR [12] projects each relation r in its own space and projects the entities h and t with respect to the relation r. Recently, Yamada et al. [26] proposed Wiki2Vec that learns entity (and word) embeddings using text and structural information from Wikipedia. We further detail entity embeddings used in our work in Section 3.

2.2 Knowledge-enhanced BERT Models

Recently, knowledge-enhanced BERT models are proposed that infuse knowledge into the BERT model through knowledge graph embeddings such as TransE [3] and Wiki2Vec [26]. ERNIE [27] incorporates entity information in the BERT model through TransE entity embeddings in pretraining, while E-BERT [19] adapts entity embeddings of Wiki2Vec to BERT without any additional pretraining. KEPLER [24] utilizes entity descriptions corresponding to the entities in relation triples and jointly optimizes Knowledge Graph and Language Model representations. KELM [15] injects knowledge in the BERT model via multi-relational subgraphs from the Knowledge Graph and text. ERNIE and E-BERT models are further explained in Sect. 3.

2.3 Entity Retrieval

Retrieval through Pseudo-Relevance Feedback Documents. Prior work of Entity Retrieval uses the unstructured text of pseudo-relevance feedback documents. Dalton et al. [6] uses the entities linked in the feedback documents and the fields of the Knowledge Graph such as entity links and the candidate set of entities for query expansion to retrieve a ranking of documents. Entities and text features such as co-occurrence, and mention features can be combined through a Learning-To-Rank approach [7]. Furthermore, Knowledge Graph links and entity co-occurrence from the feedback runs can be integrated [17].

Retrieval through Knowledge Graph Embeddings. Gerritse et al. [9] use Knowledge Graph embeddings of Wiki2Vec to determine the embedding score between the candidate set of entities and entities linked in the queries. For the final ranking, the embedding score is interpolated with the initial candidate relevance score through a Learning-To-Rank approach. Liu et al. [14] use TransE entity embeddings in the entity retrieval framework. The authors utilize the TransE embeddings to calculate the similarity between the entities in the query and candidate set of entities and further interpolate it through the Learning-To-Rank methods RankSVM and Coordinate Ascent.

Retrieval through Fielded Retrieval Models. A variation of the well-known retrieval method Sequential Dependence Model (SDM) [21,28] uses Knowledge Graph fields such as entity types, names and also documents to determine the relevance of the entities.

3 Approach

We follow the framework of Gerritse et al. [9] to rank entities. To obtain the final ranking of entities, embedding scores are determined using the entity embeddings of Wiki2Vec, ERNIE, and E-BERT which we describe below.

3.1 Entity Embeddings

Wiki2Vec. Wiki2Vec [26] learns a shared embedding space for both, word and entity embeddings, using data from Wikipedia. In particular, the model learns word embeddings using the Word2Vec Skipgram model [16], which uses a fixed-size context to learn the embeddings for each word. The similarity between the embeddings of the two entities is trained to coincide with Wikipedia's link graph. The final element of the model relates both words and entities through anchor text. These three elements are combined linearly to form the final loss function for training.

ERNIE. ERNIE [27] injects TransE entity embeddings in BERT word embeddings to enhance BERT with knowledge. It aligns the entity embeddings of TransE with the BERT word embedding of the first wordpiece token of the corresponding entity mention to generate encoded embeddings in a common embedding space. TransE [3] is a translational model that projects the entities and relations of the Knowledge Graph relation triples in a shared embedding space. The pretraining objective of the model for the knowledge fusion predicts masked entities through aligned tokens. ERNIE is further fine-tuned on NLP tasks of Relation Classification and Entity Typing.

E-BERT. E-BERT [19] infuses Wikipedia knowledge to contextualized BERT wordpiece embeddings by aligning BERT word embeddings with entity embeddings of Wiki2Vec. As word and entity embeddings share the same embedding space in Wiki2Vec, E-BERT uses the word embeddings of Wiki2Vec to learn the weight matrix through the linear transformation of Wiki2Vec word embeddings to BERT-like embeddings. Using the learned weight matrix, it constructs a function to align the entity embeddings of Wiki2Vec with the BERT word embeddings. E-BERT is fine-tuned on the downstream NLP tasks of Relation Classification and Entity Linking.

3.2 Entity Re-Ranking Framework of Gerritse et al.

In this section, we describe the entity re-ranking framework used in the original paper [9]. The authors use a two-stage entity re-ranking framework to identify the relevant entities for the query.

For our reproducibility and replicability experiments, we follow Gerritse et al. [9] and use an existing entity retrieval method to produce a candidate set of entities at the first stage of the framework. In the second stage of the framework, we first get the entity-embedding-based similarity score of each candidate entity with the entities present in the query. Then the candidate set of entities is re-ranked using interpolation.

For a query Q, query entities $E(Q)$ are identified with an entity linker and the link confidence scores $s(e)$ are retained. The entity-embedding-based similarity score for every candidate entity E and query Q is obtained as follows:

$$F(E,Q) = \sum_{e \in E(Q)} s(e) \cdot \cos(\overrightarrow{E}, \overrightarrow{e}) \tag{1}$$

To determine the final score of the entities, we combine the entity-embedding-based similarity score with the relevance score of the first stage entity retrieval method via interpolation as follows (Eq. 6 in the original paper):

$$score_{total}(E, Q) = (1 - \lambda) \cdot score_{other}(E, Q) + \lambda \cdot F(E, Q) \quad \lambda \in [0, 1] \quad (2)$$

Learning-to-rank frameworks can readily learn unnormalized weighted aggregations, through coefficients λ_1 and λ_2 on two features, which is a rank-equivalent reparametrization of the original model.

$$score_{total}(E, Q) = \lambda_1 \cdot score_{other}(E, Q) + \lambda_2 \cdot F(E, Q) \quad \lambda_1, \lambda_2 \in R \quad (3)$$

3.3   Fine-Tuning with Neural Networks

While the original paper determines the re-ranking of the candidate set through interpolation of embedding scores and candidate relevance score retrieved in the first stage, here we describe our fine-tuning approach within end-to-end entity re-ranking.

For a query Q, we identify the entities $E(Q)$ for the query and average their embeddings to obtain a single entity embedding E_Q of the query. We train a similarity metric between query embeddings E_Q and candidate entity embedding E_c as follows: We train a bilinear projection with $E_Q^T W E_c$ to capture correlations across different entries. This is followed by a linear layer to predict the rank score. The model is trained with a point-wise loss (binary cross-entropy loss) and a pair-wise loss (margin ranking loss with **tanh** activation) using the test collection.

4   Experimental Setup

We address the following research questions in our experiments:[1]

- **RQ1:** Can we reproduce the findings of Gerritse et al. [9]?
- **RQ2:** To what extent do the findings of the original paper generalize to other entity embeddings and to another dataset collection that does not focus on frequently used entities such as persons, organizations, or locations?
- **RQ3:** How much improvement can we achieve when we fine-tune the entity embeddings?
- **RQ4:** Missing entities aside, what is the quality of the entity embeddings?

4.1   RQ1: Reproducibility

To reproduce the results, we use the dataset DBpediaV2 which is used in the original paper [9]. The dataset consists of four different types of queries:

---

[1] Our code and data are available at https://github.com/poojahoza/E3R-Replicability.

(1) INEX-LD contains IR-styled keywords. e.g., "electronic music genre"; (2) SemSearchES contains short one entity search type of queries, e.g., "brooklyn bridge" (3); QALD2 consists of natural questions which are answerable by entities, e.g., "who is the mayor of Berlin?"; (4) ListSearch which consists of queries searching for a list of entities, e.g., "Professional sports team in Philadelphia". The dataset consists of 467 queries and has 49280 assessed query-entity pairs.

The existing entity retrieval method used to retrieve the top 1000 candidate set of entities is BM25F-CA, which is the best-performing method for DBpediaV2 and provided by the creators. We use the Wiki2Vec embeddings trained on the 2019-07 dump by the authors of the original paper [9] to calculate the embedding reranking score. We use Wiki2Vec embeddings with 100 dimensions for all reproducibility experiments.

Interpolation: To perform the interpolation, we use the Learning-To-Rank (L2R) approach by utilizing the RankLib library, version 1.12, as used in the original paper. We train the L2R with Co-ordinate Ascent, optimized for NDCG. We perform all experiments on 5-fold cross-validation, on the folds given in the DBpediaV2 collection.

To reproduce the results, we use the code, Wiki2Vec embeddings, and first stage run files provided by the authors of the original paper.[2]

4.2   RQ2 and RQ3: Replicability and Fine-tuning

In addition to DBpediaV2, which focuses on people, organization, and location entities, we include an additional dataset from TREC CAR, that emphasizes other entity types. The TREC Complex Answer Retrieval (CAR) [8] provides test collections for the entity ranking task in *Y2Test*. We use BenchmarkY1-train-automatic for fine-tuning and use *Y2 Test*-automatic for training the interpolation and evaluation. Y2-test consists of 65 topical queries such as "air pollution".

For both datasets, we use binary relevance judgments: 0 (non-relevant) and 1 (relevant) and evaluate with mean-average precision (MAP) and R-precision, i.e., precision at the cutoff of the number of relevant entities. We entity-link the queries using the TAGME entity linker.

Baseline: For a first-stage entity retrieval method and baseline, we use *BM25F-CA* for DBpediaV2 experiments and a high-performing input ranking for ENT-Rank called ExpEcm[3] for the TREC CAR dataset.

Embeddings: We use the Wiki2Vec 100-dimensional embeddings trained by Gerritse et al. on the Wikipedia 2019-07 dump. Additionally, we use the pretrained 100-dimensional ERNIE [27] and 768-dimensional E-BERT [19] embeddings.

---

[2] Source code for GEEER is available at https://github.com/informagi/GEEER.
[3] ExpEcm available at https://www.cs.unh.edu/~dietz/appendix/ent-rank/.

Table 1. (Relevant) candidate entities for which embeddings are not available.

Dataset	Missing Candidate Entities			Missing Relevant Entities		
	Wiki2Vec	ERNIE	E-BERT	Wiki2Vec	ERNIE	E-BERT
TREC CAR	4.06%	16.12%	3.04%	0.24%	0.82%	0.11%
DBpediaV2	16.47%	22.44%	21.89%	5.31%	6.66%	6.74%

Interpolation: We change the learning-to-rank framework to learn linear interpolation. We use the Rank-Lips[4] library optimizing for Mean Average Precision (MAP) with Coordinate Ascent, using five random restarts. Additionally, for DBpediaV2, we use different cross-validation folds. For all the replicability experiments, we re-implement the code of the GEEER entity ranking framework.

Fine-tuning: As an optional step, embeddings are fine-tuned for the entity ranking task with a neural network, we use the same datasets, evaluation, and baselines as we do for the replicability experiments.

We use a batch size of 1000, 10 epochs, 1000 warmup steps, and a 2e-05 learning rate.

We apply two different loss functions, the Margin Ranking loss function for pairwise experiments and BCELogitLoss for pointwise experiments. Since the high dimensionality (768) of E-BERT exceeds the memory of our available hardware, we can not include these experiments.

4.3   RQ4: Entities with Missing Embeddings

Many pretrained entity embeddings are derived from Wikipedia and DBpedia snapshots that differ slightly. As a result, some entities in the candidate set do not have available entity embeddings. This is a practical problem that will be encountered whenever pretrained embeddings are used. In particular, embeddings with many missing entities will obviously obtain lower performance in evaluation results. As it is unclear whether lower performance is due to the missing entities or quality issues of the embeddings, we analyze this in a controlled experiment.

In Table 1, we show the percentage of the candidate entities and relevant entities with unavailable embeddings under each of the three embeddings. We find that up to 7% of relevant entities do not have available embeddings. Furthermore, up to 22% of candidate entities from the baseline retrieval method, are missing in the embedding resource.

We perform an additional experiment where entities whose embeddings are unavailable, are removed from the candidate entities set (and baseline ranking) as well as the qrels. This way we avoid penalizing an embedding for missing entities. As each embedding is missing a different set of entities, we obtain different baseline rankings for each embedding. We only display results that were most affected by this experimental change in Table 4.

---

[4] Rank-Lips is available at https://github.com/TREMA-UNH/rank-lips

Table 2. Overall Reproduction. The reproduced results using BM25F-CA baseline with Entity Ranking Framework of the original paper on DBpediaV2 dataset. $^\triangle$ indicates significant performance improvement compared to * (baseline) using paired t-test with $p < 0.05$. We show the equivalent original results from Gerritse et al. as taken from the paper in the lower half of the table. As seen in the table, we can reproduce the same results as the original paper.

DBpediaV2	INEX_LD		QALD_2		SemSearch		ListSearch		All	
Model	@10	@100	@10	@100	@10	@100	@10	@100	@10	@100
Wiki2Vec	0.217	0.286	0.212	0.282	0.417	0.478	0.211	0.302	0.262	0.335
BM25F-CA	0.439 *	0.530 *	0.369 *	0.461 *	0.628 *	0.720 *	0.425 *	0.511 *	0.461 *	0.551 *
+ Wiki2Vec	0.466	0.552$^\triangle$	0.390 $^\triangle$	0.483 $^\triangle$	0.660$^\triangle$	0.736	0.452$^\triangle$	0.536$^\triangle$	0.487$^\triangle$	0.572$^\triangle$
ESim$_{cg}$	0.217	0.286	0.212	0.282	0.417	0.478	0.211	0.302	0.262	0.335
BM25F-CA	0.439	0.530	0.369	0.461	0.628	0.720	0.425	0.511	0.461	0.551
+ ESim$_{cg}$	0.466	0.552	0.390	0.483	0.660	0.736	0.452	0.535	0.487	0.572

5 Results and Analysis

5.1 RQ1: Reproduction

We reproduce the results of the original paper [9] as shown in Table 2. We are able to generate the same results as given in the original paper. Wiki2Vec method represents the reranking of the candidate entities set based on the embedding score. BM25F-CA+Wiki2Vec model is the linear combination of the candidate entities set retrieved using the BM25F-CA baseline and the entity-embedding-based similarity score method i.e., Wiki2Vec. We observe the same findings: (1) Entity embeddings are beneficial to improve the performance of the entity ranking task. As shown in Table 2, combining entity embeddings with the baseline significantly improves the performance for evaluation metrics of NDCG@10 and NDCG@100, in particular for QALD_2 and ListSearch queries. (2) Entity embeddings do not perform well on their own.

5.2 RQ2: Replicability

We test whether the finding that entity embeddings are beneficial for entity ranking generalizes when re-implemented with slight technical differences as described earlier. We evaluate the performance of Wiki2Vec, ERNIE, and E-BERT through the evaluation metrics of MAP and P@R.

Table 3 shows the results. Methods Baseline+Wiki2Vec, Baseline+ERNIE, and Baseline+E-BERT represent interpolations of the baseline (first stage ranking) with embedding-based similarities. We observe that while untrained embeddings on their own are not performing well, we find several small improvements when interpolated with the baseline. For DBpediaV2, the ERNIE embeddings provide the most consistent gains. For TREC CAR, Wiki2Vec obtains the strongest improvement. Both are significant according to a paired-t-test with 5%.

Table 3. Results on TREC CAR Y2 Test and DBpediaV2 datasets. The best results are marked in bold. Significance results in text. The standard error for fine-tuned embeddings of Wiki2Vec and ERNIE is 1% for both datasets. Fine-tuning E-Bert exceeded the memory available on our GPU.

Dataset	TREC CAR		DBpedia-All		INEX_LD	QALD-2	SemSearch	ListSearch
Model	MAP	P@R	MAP	P@R	MAP	MAP	MAP	MAP
Wiki2Vec	0.084	0.129	0.360	0.382	0.325	0.301	0.428	0.397
ERNIE	0.061	0.101	0.287	0.325	0.243	0.242	0.339	0.328
E-BERT	0.075	0.107	0.346	0.371	0.307	0.289	0.416	0.381
Baseline	0.157	0.223	0.454	0.433	0.420	0.366	**0.606**	0.441
+Wiki2Vec	0.164	0.228	0.450	0.431	0.413	0.371	0.595	0.453
+ERNIE	0.161	0.227	0.459	0.436	0.426	0.371	0.601	0.454
+E-BERT	0.159	0.219	0.455	0.433	0.423	0.367	0.601	0.447
Wiki2Vec-Pair	0.472	0.440	**0.540**	**0.551**	**0.524**	**0.560**	0.521	**0.550**
Wiki2Vec-Point	0.451	0.427	0.504	0.520	0.485	0.528	0.486	0.511
ERNIE-Pair	**0.474**	**0.458**	0.491	0.519	0.454	0.519	0.465	0.512
ERNIE-Point	0.429	0.434	0.485	0.520	0.460	0.528	0.423	0.514

Table 4. Impact on evaluation results when not penalizing for entities for which embeddings are not available (missing removed). The starkest difference for ERNIE and Wiki2Vec is on the weakest method.

Dataset		ERNIE		ERNIE-Pair		Wiki2Vec		Wiki2Vec-Pair	
		MAP	P@R	MAP	P@R	MAP	P@R	MAP	P@R
TREC CAR	Original	0.061	0.101	0.474	0.458	0.084	0.129	0.472	0.440
	Missing removed	0.081	0.129	0.601	0.549	0.104	0.157	0.560	0.508
	% difference	+33%	+28%	+28%	+19%	+24%	+22%	+19%	+15%
DBpediaV2	Original	0.287	0.325	0.491	0.519	0.360	0.382	0.540	0.551
	Missing removed	0.360	0.361	0.597	0.534	0.424	0.405	0.627	0.574
	% difference	+25%	+11%	+21%	+2%	+18%	+6%	+16%	+4%

While results show significant improvements and hence support the replicability of the original findings, without fine-tuning only small gains are obtained over the baseline.

SemSearch. We observe that across all the experiments for SemSearch, the baseline (first stage ranking) performs best—in particular, it is better than or similar to all the three pretrained entity embeddings, including fine-tuned results.

We notice that for several queries in SemSearch, the relevant entities have lexical overlap with query terms, hence being easy to retrieve with keyword search, which might be one of the potential reasons for the baseline to perform the best.

For example, the query "brooklyn bridge" has a total of 14 relevant entities out of which 12 entities contain either one or both query terms. Other such examples are "harry potter" and "nokia e73".

5.3   RQ3: Fine-tuned Embeddings

We examine the performance of the fine-tuned entity embeddings with the baseline and pretrained entity embeddings to study the effect of task-specific fine-tuning. In Table 3 these are listed as Wiki2Vec-Point and ERNIE-Point for results with embeddings that are trained with point-wise loss functions, and equivalently"-Pair" for the pairwise ranking loss. E-BERT exceeded the memory available on our GPU hardware, hence we cannot provide results.

We observe that fine-tuning the existing pretrained entity embeddings significantly improves the performance for both datasets (except for SemSearch, as discussed above). We observe that fine-tuning specifically increases the performance of the TREC CAR dataset, which focuses on entities other than people, organizations, and locations.

Our findings show that the pair-wise ranking loss obtains better results in most cases than the point-wise ranking loss, thus agreeing with the common wisdom.

5.4   RQ 4: Model Performance When Correcting for Missing Entities

We discussed previously that missing entity embeddings of entities from the candidate set can result in lower performance for those embeddings, without providing an insight into the quality of embeddings. To observe the quality of embeddings without the missing entities, we change the experimental setup as described in Sect. 4.3.

We present results for ERNIE and Wiki2Vec in Table 4, and we obtain analogous results for the remaining experiment. We find that while the results change between the two experimental setups, the overarching story is still consistent: Embeddings by themselves are not effective, and interpolation with the baseline yields small gains.

We notice that the difference between the two experimental setups is more pronounced for the weakest and the strongest methods for both ERNIE and Wiki2Vec as shown in Table 4. Compared to Wiki2Vec, we observe a higher increase in the performance of ERNIE which is expected as ERNIE has a higher number of missing entities. This shows that the ERNIE embeddings, when available, are beneficial for the task.

Query-level analysis. We further investigate the performance of the baseline and interpolations with pretrained embeddings at the query-level: We divide the queries into bins based on their difficulty for the baseline measured in MAP. Queries with lower MAP performance are considered to be more difficult queries.

In Fig. 1 and Fig. 2, we observe that interpolating retrieval and the embeddings yield improvements for the difficult queries of (5–75%) for both the TREC CAR dataset and DBpediaV2. This indicates that the embedding scores are a complementary source to the baseline for difficult queries, though they provide

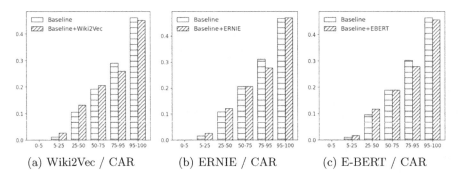

Fig. 1. CAR-Query-level Difficulty Test for MAP Performance, corrected for missing entities. The above figure shows the difficulty test performance of MAP for the TREC CAR dataset, where y-axis is MAP performance and x-axis is the difficulty percentile according to MAP. Here, (a), (b) and (c) compare the entity rankings between the baselines and the linear combination of baselines with embedding scores. Most difficult 5% queries for the baseline are on the left side and the easiest 5% queries are on the right side.

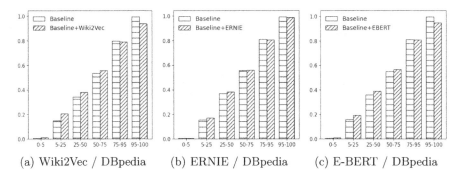

Fig. 2. DBpedia-Query-level Difficulty Test for MAP Performance for all queries. Here, (a), (b) and (c) compare the entity rankings as in Fig 1.

only small gains. For the easy queries, in the 75–100 percentile the retrieval baseline often performs better than the combined methods.

Even after correcting for missing entities, we find that for pretrained entity embeddings Wiki2Vec and E-BERT obtain better performance. Closer inspection shows that they are placing relevant entities above non-relevant entities more often than ERNIE. For fine-tuned embeddings, Wiki2Vec and ERNIE are at par with each other.

6   Conclusion

In this work, we reproduce and replicate the work of Gerritse et al. [9]. Through reproducibility and replication experiments, on the two datasets of TREC CAR and DBpediaV2, we can confirm the findings that the entity embeddings are beneficial for entity ranking. We find that consistent yet small gains are obtained with available pretrained embeddings and, confirming common wisdom, fine-tuning these pretrained embeddings achieves significantly large improvements.

One of the most interesting findings in the reproducibility paper is to use the GEEER framework to evaluate different pretrained entity embeddings. For example, the fact that matrix-factorization based Wiki2Vec embeddings are competitive to transformer-based BERT embedding models, is a sign that none of the currently available pretrained entity embedding models are particularly suitable for an IR task. We speculate that part of the problem is that ERNIE and E-Bert are over-trained on syntactic entity understanding tasks like entity linking, entity typing, and relation extraction for which the entity name fields are informative. In contrast, the entity ranking tasks of DBpediaV2 and (even more so) the TREC CAR datasets require to understand the abstract semantics of entities and their topically related entities. Wiki2Vec was pretrained on lead text, anchor text context, and the general link structure, which is likely to yield entity representations that are more amenable to entity retrieval tasks. A major takeaway from this study is that the IR community needs to train their own entity embedding models that are better suited for topical information retrieval tasks (as opposed to syntactic tasks).

Acknowledgements. This material is based upon work supported by the National Science Foundation under Grant No. 1846017. Any opinions, findings, and conclusions or recommendations expressed in this material are those of the author(s) and do not necessarily reflect the views of the National Science Foundation.

References

1. Balog, K., Bron, M., De Rijke, M.: Query modeling for entity search based on terms, categories, and examples. ACM Trans. Inf. Syst. 29(4) (Dec 2011). https://doi.org/10.1145/2037661.2037667
2. Balog, K., Bron, M., de Rijke, M.: Category-based query modeling for entity search. In: Gurrin, C., et al. (eds.) ECIR 2010. LNCS, vol. 5993, pp. 319–331. Springer, Heidelberg (2010). https://doi.org/10.1007/978-3-642-12275-0_29
3. Bordes, A., Usunier, N., Garcia-Durán, A., Weston, J., Yakhnenko, O.: Translating embeddings for modeling multi-relational data. In: Proceedings of the 26th International Conference on Neural Information Processing Systems - Volume 2, pp. 2787–2795. NIPS'13, Curran Associates Inc., Red Hook, NY, USA (2013)
4. Bron, M., Balog, K., de Rijke, M.: Example based entity search in the web of data. In: Serdyukov, P., et al. (eds.) ECIR 2013. LNCS, vol. 7814, pp. 392–403. Springer, Heidelberg (2013). https://doi.org/10.1007/978-3-642-36973-5_33
5. Ciglan, M., Nørvåg, K., Hluchý, L.: The semsets model for ad-hoc semantic list search. In: Proceedings of the 21st International Conference on World Wide Web, pp. 131–140. WWW 2012, Association for Computing Machinery, New York, NY, USA (2012). https://doi.org/10.1145/2187836.2187855
6. Dalton, J., Dietz, L., Allan, J.: Entity query feature expansion using knowledge base links. In: Proceedings of the 37th International ACM SIGIR Conference on

Research and Development in Information Retrieval, pp. 365–374. SIGIR 2014, Association for Computing Machinery, New York, NY, USA (2014). https://doi.org/10.1145/2600428.2609628

7. Dietz, L.: Ent rank: Retrieving entities for topical information needs through entity-neighbor-text relations. In: Proceedings of the 42nd International ACM SIGIR Conference on Research and Development in Information Retrieval, pp. 215–224. SIGIR 2019, Association for Computing Machinery, New York, NY, USA (2019). https://doi.org/10.1145/3331184.3331257

8. Dietz, L., Gamari, B.: Trec complex answer retrieval. In: Proceedings of Text REtrieval Conference (TREC) (2018)

9. Gerritse, E.J., Hasibi, F., de Vries, A.P.: Graph-embedding empowered entity retrieval. In: Jose, J.M., et al. (eds.) ECIR 2020. LNCS, vol. 12035, pp. 97–110. Springer, Cham (2020). https://doi.org/10.1007/978-3-030-45439-5_7

10. Kaptein, R., Kamps, J.: Exploiting the category structure of wikipedia for entity ranking. Artif. Intell. **194**, 111–129 (2013). https://doi.org/10.1016/j.artint.2012.06.003

11. Lin, T., Pantel, P., Gamon, M., Kannan, A., Fuxman, A.: Active objects: actions for entity-centric search. In: Proceedings of the 21st International Conference on World Wide Web, pp. 589–598. WWW '12, Association for Computing Machinery, New York, NY, USA (2012). https://doi.org/10.1145/2187836.2187916

12. Lin, Y., Liu, Z., Sun, M., Liu, Y., Zhu, X.: Learning entity and relation embeddings for knowledge graph completion. In: AAAI (2015)

13. Liu, T.Y., et al.: Learning to rank for information retrieval. Foundations and Trends® in Information Retrieval 3(3), 225–331 (2009)

14. Liu, Z., Xiong, C., Sun, M., Liu, Z.: Explore entity embedding effectiveness in entity retrieval. In: Sun, M., Huang, X., Ji, H., Liu, Z., Liu, Y. (eds.) CCL 2019. LNCS (LNAI), vol. 11856, pp. 105–116. Springer, Cham (2019). https://doi.org/10.1007/978-3-030-32381-3_9

15. Lu, Y., Lu, H., Fu, G., Liu, Q.: KELM: knowledge enhanced pre-trained language representations with message passing on hierarchical relational graphs. arXiv preprint arXiv:2109.04223 (2021)

16. Mikolov, T., Chen, K., Corrado, G., Dean, J.: Efficient estimation of word representations in vector space. arXiv preprint arXiv:1301.3781 (2013)

17. Oza, P., Dietz, L.: Which entities are relevant for the story? In: CEUR workshop proceedings, vol. 2860 (2021)

18. Pehcevski, J., Thom, J.A., Vercoustre, A.M., Naumovski, V.: Entity ranking in wikipedia: Utilising categories, links and topic difficulty prediction. Inf. Retr. **13**(5), 568–600 (2010). https://doi.org/10.1007/s10791-009-9125-9

19. Poerner, N., Waltinger, U., Schütze, H.: E-BERT: efficient-yet-effective entity embeddings for BERT. In: Findings of the Association for Computational Linguistics: EMNLP 2020, pp. 803–818. Association for Computational Linguistics, Online (Nov 2020). https://doi.org/10.18653/v1/2020.findings-emnlp.71

20. Pound, J., Mika, P., Zaragoza, H.: Ad-hoc object retrieval in the web of data. In: Proceedings of the 19th International Conference on World Wide Web, pp. 771–780. WWW 2010, Association for Computing Machinery, New York, NY, USA (2010). https://doi.org/10.1145/1772690.1772769

21. Raviv, H., Carmel, D., Kurland, O.: A ranking framework for entity oriented search using markov random fields. In: Proceedings of the 1st Joint International Workshop on Entity-Oriented and Semantic Search, pp. 1–6 (2012)

22. Schuhmacher, M., Dietz, L., Paolo Ponzetto, S.: Ranking entities for web queries through text and knowledge. In: Proceedings of the 24th ACM International on Conference on Information and Knowledge Management, pp. 1461–1470. CIKM 2015, Association for Computing Machinery, New York, NY, USA (2015). https://doi.org/10.1145/2806416.2806480

23. Tonon, A., Demartini, G., Cudré-Mauroux, P.: Combining inverted indices and structured search for ad-hoc object retrieval. In: Proceedings of the 35th International ACM SIGIR Conference on Research and Development in Information Retrieval, pp. 125–134. SIGIR '12, Association for Computing Machinery, New York, NY, USA (2012). https://doi.org/10.1145/2348283.2348304

24. Wang, X., et al.: Kepler: a unified model for knowledge embedding and pre-trained language representation. Trans. Assoc. Comput. Linguist. **9**, 176–194 (2021)

25. Wang, Z., Zhang, J., Feng, J., Chen, Z.: Knowledge graph embedding by translating on hyperplanes. In: Proceedings of the AAAI Conference on Artificial Intelligence 28(1) (Jun 2014). https://doi.org/10.1609/aaai.v28i1.8870, https://ojs.aaai.org/index.php/AAAI/article/view/8870

26. Yamada, I., et al.: Wikipedia2Vec: an efficient toolkit for learning and visualizing the embeddings of words and entities from Wikipedia. In: Proceedings of the 2020 Conference on Empirical Methods in Natural Language Processing: System Demonstrations, pp. 23–30. Association for Computational Linguistics (2020)

27. Zhang, Z., Han, X., Liu, Z., Jiang, X., Sun, M., Liu, Q.: ERNIE: enhanced language representation with informative entities. In: Proceedings of the 57th Annual Meeting of the Association for Computational Linguistics, pp. 1441–1451. Association for Computational Linguistics, Florence, Italy (Jul 2019). https://doi.org/10.18653/v1/P19-1139, https://aclanthology.org/P19-1139

28. Zhiltsov, N., Kotov, A., Nikolaev, F.: Fielded sequential dependence model for ad-hoc entity retrieval in the web of data. In: Proceedings of the 38th International ACM SIGIR Conference on Research and Development in Information Retrieval, pp. 253–262 (2015)

Do the Findings of Document and Passage Retrieval Generalize to the Retrieval of Responses for Dialogues?

Gustavo Penha[(✉)] and Claudia Hauff

TU Delft, Delft, The Netherlands
{g.penha-1,c.hauff}@tudelft.nl

Abstract. A number of learned sparse and dense retrieval approaches have recently been proposed and proven effective in tasks such as passage retrieval and document retrieval. In this paper we analyze with a replicability study if the lessons learned generalize to the retrieval of responses for dialogues, an important task for the increasingly popular field of conversational search. Unlike passage and document retrieval where documents are usually longer than queries, in response ranking for dialogues the queries (dialogue contexts) are often longer than the documents (responses). Additionally, dialogues have a particular structure, i.e. multiple utterances by different users. With these differences in mind, we here evaluate how generalizable the following major findings from previous works are: **(F1)** query expansion outperforms a no-expansion baseline; **(F2)** document expansion outperforms a no-expansion baseline; **(F3)** zero-shot dense retrieval underperforms sparse baselines; **(F4)** dense retrieval outperforms sparse baselines; **(F5)** hard negative sampling is better than random sampling for training dense models. Our experiments (https://github.com/Guzpenha/transformer_rankers/tree/full_rank_retrieval_dialogues.)—based on three different information-seeking dialogue datasets—reveal that four out of five findings (**F2–F5**) generalize to our domain.

1 Introduction

Conversational search is concerned with creating agents that satisfy an information need by means of a *mixed-initiative* conversation through natural language interaction, rather than through the traditional search engine results page. A popular approach to conversational search is retrieval-based [3]: given an ongoing conversation and a large corpus of historic conversations, retrieve the response that is best suited from the corpus [11,28,45,47,48]. Due to the effectiveness of heavily pre-trained transformer-based language models such as BERT [4], they have become the predominant approach for conversation response re-ranking [8,28,42,43,53].

J. Kamps et al. (Eds.): ECIR 2023, LNCS 13982, pp. 132–147, 2023.
https://doi.org/10.1007/978-3-031-28241-6_9

The most common evaluation procedure for conversation response re-ranking consists of re-ranking a limited set of n candidate responses (including the ground-truth response(s)), followed by measuring the number of relevant responses found in the first K positions—$Recall_n@K$ [52]. Since the entire collection of available responses is typically way bigger[1] than such a set of candidates, this setup is in fact a selection problem, where we have to choose the correct response out of a few options. This evaluation overlooks the first-stage retrieval step, which retrieves a set of n responses to be re-ranked. If the first-stage model, e.g. BM25, fails to retrieve relevant responses, the entire pipeline fails.

Motivated by a lack of research on the first-stage retrieval step, we are interested in answering in our replicability study whether the considerable knowledge obtained on document and passage retrieval tasks generalizes to the dialogue domain. Unlike document and passage retrieval where the documents are generally longer than the queries, in response retrieval for dialogues the queries (dialogue contexts) tend to be longer than the documents (responses). A second important difference is the structure induced by the dialogue as seen in Table 1.

Table 1. Comparison between passage retrieval and response retrieval for dialogues. In Sect. 3 we define the task of *First-stage Retrieval for Dialogues*. p^+/r^+ are the relevant passage/response.

	Passage retrieval	First-stage retrieval for dialogues
Input	Query q	Dialogue context $\mathcal{U} = \{u^1, u^2, ..., u^\tau\}$
Example	q: *what is theraderm used for*	u^1: *I was in the mood to play Chrono Trigger again [...] Is there a performant SNES emulator that has that feature?* u^2: *{url} allows you to map joypad buttons to keyboard keys and [...]* u^3: *Do the diagonals for the analog stick work correctly for you? [...]*
Output	Ranked list of passages	Ranked list of responses
Example	p^+: *Thera-Derm Lotion is used as a moisturizer to treat [...]*	r^+: *In the"Others" tab, try [...]*

Given the differences between the domains, we verify empirically across three information-seeking datasets and 1.7M queries, the generalizability of five findings (**F1** to **F5**) from the passage and document retrieval literature related to state-of-the-art sparse and dense retrieval models. We are motivated in our selection of these five findings by their impact in prior works (cf. Sect. 2). Our results show that four out of five previous findings do indeed generalize to our domain:

---

[1] While for most benchmarks [52] we have only 10–100 candidates, a working system with the Reddit data from PolyAI https://github.com/PolyAI-LDN/conversational-datasets would need to retrieve from 3.7 billion responses.

F1 $\cancel{\checkmark}^2$ Dialogue context (i.e. query) expansion outperforms a no-expansion baseline [1,18,21,49].

F2 $\checkmark$ Response (i.e. document) expansion outperforms a no-expansion baseline [19,21,25] *if the expansion model is trained to generate the most recent context (last utterance[3] of the dialogue) instead of older context (all utterances).*

F3 $\checkmark$ Dense retrieval in the zero-shot[4] setting underperforms sparse baselines [34,41] *except when it goes through intermediate training on large amounts of out-of-domain data.*

F4 $\checkmark$ Dense retrieval with access to target data[5] outperforms sparse baselines [7,15,34] *if an intermediate training step on out-of-domain data is performed before the fine-tuning on target data.*

F5 $\checkmark$ Harder negative sampling techniques lead to effectiveness gains [46,51] *if a denoising technique is used to reduce the number of false negative samples.*

Our results indicate that most findings translate to the domain of retrieval of responses for dialogues. A promising future direction is thus to start with successful models from other domains—for which there are more datasets and previous research—and study how to adapt and improve them for retrieval-based conversational search.

2   Related Work

In this section we first discuss current research in retrieval-based systems for conversational search, followed by reviewing the major findings of (un)supervised sparse and dense retrieval in the domains of passage and document retrieval.

2.1   Ranking and Retrieval of Responses for Dialogues

Early neural models for response re-ranking were based on matching the representations of the concatenated dialogue context and the representation of a response in a single-turn manner with architectures such as CNN and LSTM [14,23]. More complex neural architectures matching each utterance with the response were also explored [9,22,54]. Heavily pre-trained language models such as BERT were first shown to be effective by Nogueira and Cho [24] for re-ranking. Such models quickly became a predominant approach for re-ranking in IR [21] and were later shown to be effective for re-ranking responses in conversations [28,42].

In contrast, the first-stage retrieval of responses for a dialogue received relatively little attention [29]. Lan et al. [17] and Tao et al. [38] showed that BERT-based dense retrieval models outperform BM25 for first-stage retrieval of

---

[2] $\cancel{\checkmark}$ indicates that the finding does not hold in our domain whereas $\checkmark$ indicates that it holds in our domain followed by the *necessary condition or exception.*

[3] For example in Table 1 the last utterance is u^3.

[4] A zero-shot is a model that does not have access to target data, cf. Table 2.

[5] Target data is data from the same distribution, i.e. dataset, of the evaluation dataset.

responses for dialogues. A limitation of their work is that strong sparse retrieval baselines that have shown to be effective in other retrieval tasks, e.g. BM25 with *dialogue context expansion* [25] or BM25 with *response expansion* [49], were not employed for dense retrieval. We do such comparisons here and test a total of five major findings that have been not been evaluated before by previous literature on the first-stage retrieval of responses for dialogues.

2.2 Dense and Sparse Models for Passage and Document Retrieval

Context for F1. Retrieval models can be categorized into two dimensions: supervised vs. unsupervised and dense vs. sparse representations [19]. An *unsupervised* sparse representation model such as BM25 [35] represents each document and query with a sparse vector with the dimension of the collection's vocabulary, having many zero weights due to non-occurring terms. Since the weights of each term are entirely based on term statistics they are considered unsupervised methods. Such approaches are prone to the vocabulary mismatch problem [6], as semantic matches are not considered. A way to address such a problem is by using query expansion methods. RM3 [1] is a competitive [49] query expansion technique that uses pseudo-relevance feedback to add new terms to the queries followed by another final retrieval step using the modified query.

Context for F2. A *supervised* sparse retrieval model can take advantage of the effectiveness of transformer-based language models by changing the terms' weights from collection statistics to something that is learned. Document expansion with a learned model can be considered a learned sparse retrieval approach [19]. The core idea is to create pseudo documents that have expanded terms and use them instead when doing retrieval. Doc2query [25] is a strong supervised sparse retrieval baseline that uses a language model to predict queries that might be issued to find a document. The predictions of this model are used to create the augmented pseudo documents.

Context for F3 and F4. Supervised dense retrieval models[6], such as ANCE [46] and coCodenser [7], represent query and documents in a small fixed-length space, for example of 768 dimensions. Dense retrieval models without access to target data for training—known as the *zero-shot scenario*—have underperformed sparse methods (**F3**). For example, the BEIR benchmark [41] showed that BM25 was superior to dense retrieval from 9–18 (depending on the model) out of the 18 datasets in the zero-shot scenario. In contrast, when having access to enough supervision from target data, dense retrieval models have shown to consistently outperform strong sparse baselines [7,15,34] (**F4**).

---

[6] A distinction can also be made of cross-encoders and bi-encoders, where the first encode the query and document jointly as opposed to separately [40]. Cross-encoders are applied in a re-ranking step due to their inefficiency and thus are not our focus.

Context for F5. In order to train neural ranking models, a small set of negative (i.e. non-relevant) candidates are necessary as it is prohibitively expensive to use every other document in the collection as negative sample for a query. A limitation of randomly selecting negative samples is that they might be too easy for the ranking model to discriminate from relevant ones, while for negative documents that are harder the model might still struggle. For this reason hard negative sampling has been shown to perform better than random sampling for passage and document retrieval [36,46,51].

3   First-Stage Retrieval for Dialogues

In this section we first describe the problem of first-stage retrieval of responses, followed by the findings we want to replicate from sparse and dense approaches.

Problem Definition. The task of first-stage retrieval of responses for dialogues, concerns retrieving the best response out of the entire collection given the dialogue context. Formally, let $\mathcal{D} = \{(\mathcal{U}_i, \mathcal{R}_i, \mathcal{Y}_i)\}_{i=1}^{M}$ be a data set consisting of M triplets: dialogue context, response candidates and response relevance labels. The dialogue context $\mathcal{U}_i$ is composed of the previous utterances $\{u^1, u^2, ..., u^\tau\}$ at the turn τ of the dialogue. The candidate responses $\mathcal{R}_i = \{r^1, r^2, ..., r^n\}$ are either ground-truth responses r^+ or negative sampled candidates r^-, indicated by the relevance labels $\mathcal{Y}_i = \{y^1, y^2, ..., y^n\}$. In previous work, the number of candidates is limited, typically $n = 10$ [29]. The findings we replicate here come from passage and document retrieval tasks where there is no limit to the number of documents or passages that have to be retrieved. Thus, in all of our first-stage retrieval task experiments n is set to the size of the entire collection of responses in the corpus. The number of ground-truth responses is one, the observed response in the conversational data. The task is then to learn a ranking function $f(.)$ that is able to generate a ranked list from the entire corpus of responses $\mathcal{R}_i$ based on their predicted relevance scores $f(\mathcal{U}, r)$.

F1: Unsupervised Sparse Retrieval. We rely on classic retrieval methods, for which the most commonly used baseline is BM25. One of the limitations of sparse retrieval is the vocabulary mismatch problem. Expansion techniques are able to overcome this problem by appending new words to the dialogue contexts and responses. For this reason, we here translate a query expansion technique to the dialogue domain and perform *dialogue context expansion* with RM3 [1], a competitive unsupervised method that assumes that the top-ranked responses by the sparse retrieval model are relevant. From these pseudo-relevant responses, words are selected and an expanded dialogue context is created and subsequently employed by the sparse retrieval method to rank the final list of responses. **The effectiveness of RM3 in the domain of dialogues is the first finding that we validate.**

F2: Learned Sparse Retrieval. Alternatively, we can expand the responses in the collection with a learned method. To do so we "translate" doc2query [25] into our domain, yielding resp2ctxt. Formally, we fine-tune a generative transformer model G for the task of generating the dialogue context $\mathcal{U}_i$ from the ground-truth response r_i^+. This model is then used to generate expansions for all responses in the collection, $r^i = concat(r^i, G(r^i))$. These expansions are appended to the responses and the collection is indexed again—the sparse retrieval method itself is not modified, i.e. we continue using BM25. This approach (which we coin resp2ctxt) leads to two improvements: term re-weighting (adding terms that already exist in the document) and dealing with the vocabulary mismatch problem (adding new terms). **The effectiveness of doc2query in the domain of dialogues is the second finding that we validate.**

Unlike passage and document retrieval where the queries are smaller than the documents, for the retrieval of responses for dialogues the queries are longer than the documents[7]. This is a challenge for the generative model, since generating larger pieces of text is a more difficult problem than smaller ones as there is more room for errors. Motivated by this, we also explored a modified version of resp2ctxt that aims to generate only the last utterance of the dialogue context: resp2ctxt$_{lu}$. This model is trained to generate u^τ from r_i^+, instead of trying to generate the whole utterance $\mathcal{U}_i = \{u^1, u^2, ..., u^\tau\}$. The underlying premise is that the most important utterance from the dialogue is the last one, and if it is correctly generated by resp2ctxt$_{lu}$, the sparse retrieval method will be able to find the correct response from the collection.

F3: Zero-Shot Dense Retrieval. We rely on methods that learn to represent the dialogue context and the responses separately in a dense embedding space. Responses are then ranked by their similarity to the dialogue context. We rely here on pre-trained language transformer models, such as BERT [4] and MPNet [37], to obtain such representations of the dialogue context and response. This approach is generally referred to as a *bi-encoder* model [21] and is an effective family of models[8]. A zero-shot model is one that is not trained on the target data. Target data is data from the same distribution, i.e. dataset, of the evaluation dataset.

One way of improving the representations of a heavily pre-trained language model for the zero-shot setting is to fine-tune it with intermediate data [33]. Such intermediate data contains triplets of query, relevant document, and negative document and can include multiple datasets. The advantage of adding this step before employing the representations of the language model is to reduce the

---

[7] For example, while the TREC-DL-2020 passage and document retrieval tasks the queries have between 5–6 terms on average and the passages and documents have over 50 and 1000 terms respectively, for the information-seeking dialogue datasets used here the dialogue contexts (queries) have between 70 and 474 terms on average depending on the dataset while the responses (documents) have between 11 and 71.

[8] See for example the top models in terms of effectiveness from the MSMarco benchmark leaderboards https://microsoft.github.io/msmarco/.

gap between the pre-training and the downstream task at hand [26,30,31]. In Table 2 we clarify the relationship between pre-training, intermediate training and fine-tuning.

Table 2. The different training stages and data, their purposes, examples of datasets, and the type of dense model obtained after each stage.

	Pre-training data	Intermediate data	Target data
Purpose	Learn general representations	Learn sentence representations for ranking	Learn representations for target distribution
Model is	Zero-shot	Zero-shot	Fine-tuned
Example	Wikipedia	MSMarco	MANtIS

The intermediate training step learns to represent pieces of text (query and documents) by applying a mean pooling function over the transformer's final layer, which is then used to calculate the dot-product similarity. The loss function employs multiple negative texts from the same batch to learn the representations in a constrastive manner, also known as in-batch negative sampling. Such a procedure learns better text representations than a naive approach that uses the $[CLS]$ token representation of BERT [2,33].

The function $f(\mathcal{U}, r)$ is then $dot(\eta(concat(\mathcal{U})), \eta(r))$, where η is the representation obtained by applying the mean pooling function over the last layer of the transformer model, and $concat(\mathcal{U}) = u^1 \mid [U] \mid u^2 \mid [T] \mid ... \mid u^\tau$, where $\mid$ indicates the concatenation operation. The utterances from the context $\mathcal{U}$ are concatenated with special separator tokens $[U]$ and $[T]$ indicating end of utterances and turns[9]. **The effectiveness of a zero-shot bi-encoder model in the domain of dialogues is the third finding we validate.**

F4: Fine-Tuned Dense Retrieval. The standard procedure is to fine-tune dense models with target data that comes from the same dataset that the model will be evaluated. Since we do not have labeled negative responses, all the remaining responses in the dataset can be thought of as non-relevant to the dialogue context. Computing the probability of the correct response over all other responses in the dataset would give us $P(r \mid \mathcal{U}) = \frac{P(\mathcal{U}, r)}{\sum_k P(\mathcal{U}, r_k)}$. This computation is prohibitively expensive, and the standard procedure is to approximate it using a few negative samples. The *negative sampling* task is then as follows: given the dialogue context $\mathcal{U}$ find challenging responses r^- that are non-relevant for $\mathcal{U}$. Negative sampling can be seen as a retrieval task, where one can use a model to retrieve negatives by applying a retrieval function to the collection of responses using $\mathcal{U}$ as the query.

---

[9] The special tokens $[U]$ and $[T]$ will not have any meaningful representation in the zero-shot setting, but they can be learned on the fine-tuning step.

With such a dataset at hand, we continue the training—after the intermediate step—in the same manner as done by the intermediate training step, with the following cross-entropy loss function[10] for a batch with size B:

$$\mathcal{J}(\mathcal{U}, \mathbf{r}, \theta) = -\frac{1}{B} \sum_{i=1}^{B} \left[f(\mathcal{U}_i, r_i) - \log \sum_{j=1, j!=i}^{B} e^{f(\mathcal{U}_i, r_j)} \right],$$

where $f(\mathcal{U}, r)$ is the dot-product of the mean pooling of the last layer of the transformer model. **The effectiveness of a fine-tuned bi-encoder model in the domain of dialogues is the fourth finding we validate here.**

F5: Hard Negative Sampling. A limitation of random samples is that they might be too easy for the ranking model to discriminate from relevant ones, while for negative documents that are hard the model might still struggle. For this reason, another popular approach is to use a ranking model to retrieve negative documents using the given query with a classic retrieval technique such as BM25. This leads to finding negative documents that are closer to the query in the sparse representation space, and thus they are *harder negatives*. Since dense retrieval models have been outperforming sparse retrieval in a number of cases with available training data, more complex negative sampling techniques making use of dense retrieval have also been proposed [12,46]. **The effectiveness of hard negative sampling for a bi-encoder model in the domain of dialogues is the fifth finding we validate here.**

4   Experimental Setup

In order to compare the different sparse and dense approaches we consider three large-scale information-seeking conversation datasets[11]: MSDialog [32] contains 246K context-response pairs, built from 35.5K information seeking conversations from the Microsoft Answer community, a QA forum for several Microsoft products; MANtIS [27] contains 1.3 million context-response pairs built from conversations of 14 Stack Exchange sites, such as *askubuntu* and *travel*; UDC_{DSTC8} [16] contains 184k context-response pairs of disentangled Ubuntu IRC dialogues.

Implementation Details. For BM25 and BM25+RM3[12] we rely on pyserini implementations [20]. In order to train resp2ctxt expansion methods we rely on the Huggingface transformers library [44], using the *t5-base* model. We fine-tune the T5 model for 2 epochs, with a learning rate of 2e−5, weight decay of

---

[10] We refer to this loss as MultipleNegativesRankingLoss.

[11] MSDialog is available at https://ciir.cs.umass.edu/downloads/msdialog/; MANtIS is available at https://guzpenha.github.io/MANtIS/; UDC_{DSTC8} is available at https://github.com/dstc8-track2/NOESIS-II.

[12] We perform hyperparameter tuning using grid search on the number of expansion terms, number of expansion documents, and weight.

0.01, and batch size of 5. When augmenting the responses with resp2ctxt we follow docT5query [25] and append three different context predictions, using sampling and keeping the top-10 highest probability vocabulary tokens.

For the zero-shot dense models, we rely on the SentenceTransformers [33] model releases. The library uses Hugginface's transformers for the pre-trained models such as BERT [4] and MPNet [37]. For the bi-encoder models, we use the pre-trained *all-mpnet-base-v2* weights which were the most effective in our initial experiments, compared with other pre-trained models[13]. When fine-tuning the dense retrieval models, we rely on the *MultipleNegativesRankingLoss*, which accepts a number of hard negatives, and also uses the remaining in-batch random negatives to train the model. We use a total of 10 negative samples for dialogue context. We fine-tune the dense models for a total of 10k steps, and every 100 steps we evaluate the models on a re-ranking task that selects the relevant response out of 10 responses. We use the re-ranking validation MAP to select the best model from the whole training to use in evaluation. We use a batch size of 5, with 10% of the training steps as warmup steps. The learning rate is 2e−5 and the weight decay is 0.01. We use FAISS [13] to perform the similarity search.

Evaluation. To evaluate the effectiveness of the retrieval systems we use $R@K$. We thus evaluate the models' capacity of finding the correct response out of the whole possible set of responses[14]. We perform Students t-tests at the 0.95 confidence level with Bonferroni correction to compare statistical significance of methods. Comparisons are performed across the results for each dialogue context.

5   Results

In this section, we discuss our empirical results along with the five major findings from previous work (Sect. 1) in turn. Table 3 contains the main results regarding **F1** to **F4**. Table 5 contains the results for **F5**.

F1 ✗ Query expansion via RM3 leads to improvements over not using query expansion [1, 18, 21, 49]. BM25+RM3 (row 1b) does not improve over BM25 (1a) on any of the three conversational datasets analyzed. We performed thorough hyperparameter fine-tuning and no combination of the RM3 hyperparameters outperformed BM25. **This indicates that F1 does not hold for the task of response retrieval for dialogues.**

A manual analysis of the new terms appended to a sample of 60 dialogue contexts by one of the paper's authors revealed that only 18% of them have at least one relevant term added based on our best judgment. Unlike web search

---

[13] The alternative models we considered are those listed in the model overview section at https://www.sbert.net/docs/pretrained_models.html.

[14] The standard evaluation metric in conversation response ranking [8,39,50] is recall at position K with n candidates $R_n@K$. Since we are focused on the first-stage retrieval we set n to be the entire collection of answers.

Table 3. Results for the generalizability of F1–F4. Bold values indicate the highest recall for each type of approach. Superscripts indicate statistically significant improvements using Students t-test with Bonferroni correction. † = *other methods from the same group* 1 = *best from unsupervised sparse retrieval;* 2 = *best from supervised sparse retrieval;* 3 = *best from zero-shot dense retrieval.* For example, in F3 † indicates that row (3d) improves over rows (3a–c), [1] indicates that it improves over row (1a) and [2] indicates it improves over row (2b).

		MANtIS		MSDialog		UDC$_{DSTC8}$	
		R@1	**R@10**	**R@1**	**R@10**	**R@1**	**R@10**
(0)	Random	0.000	0.000	0.000	0.001	0.000	0.001
Unsupervised sparse					**F1**		
(1a)	BM25	**0.133**[†]	**0.299**[†]	**0.064**[†]	**0.177**[†]	**0.027**[†]	**0.070**[†]
(1b)	BM25 + RM3	0.073	0.206	0.035	0.127	0.011	0.049
Supervised sparse					**F2**		
(2a)	BM25 + resp2ctxt	0.135	0.309	0.074	**0.208**	0.028	0.067
(2b)	BM25 + resp2ctxt$_{lu}$	**0.147**[†1]	**0.325**[†1]	**0.075**[1]	0.202[1]	**0.029**	**0.076**
Zero-shot dense (Model$_{IntermediateData}$)					**F3**		
(3a)	ANCE$_{600K-MSMarco}$	0.048	0.111	0.050	0.124	0.010	0.028
(3b)	TAS-B$_{400K-MSMarco}$	0.062	0.143	0.060	0.157	0.019	0.050
(3c)	Bi-encoder$_{215M-mul}$	0.138	0.297	0.108	0.277	0.023	0.076
(3d)	Bi-encoder$_{1.17B-mul}$	**0.155**[†1]	**0.341**[†12]	**0.147**[†12]	**0.339**[†12]	**0.041**[†]	**0.097**[†12]
Fine-tuned dense (Model$_{NegativeSampler}$)					**F4**		
(4a)	Bi-encoder$_{Random(0)}$	**0.130**	**0.307**	**0.168**[123]	**0.387**[123]	**0.050**[12]	**0.128**[123]

where the query is often incomplete, under-specified, and ambiguous, in the information-seeking datasets employed here the dialogue context (query) is quite detailed and has more terms than the responses (documents). We hypothesize that because the dialogue contexts are already quite descriptive, the task of expansion is trickier in this domain and thus we observe many dialogues for which the added terms are noisy.

F2 ✓ Document expansion via resp2ctxt leads to improvements over no expansion [19, 21, 25]. We find that a naive approach to response expansion improves marginally in two of the three datasets with BM25+resp2ctxt (2a) outperforming BM25 (1a). However, the proposed modification of predicting only the last utterance of the dialogue (resp2ctxt$_{lu}$) performs better than predicting the whole utterance, as shown by BM25+resp2ctxt$_{lu}$'s (2b) higher recall values. In the MANtIS dataset the R@10 goes from 0.309 when using the model trained to predict the dialogue context to 0.325 when using the one trained to predict only the last utterance of the dialogue context. **We thus find that F2 generalizes to response retrieval for dialogues, especially when predicting only the last utterance of the context[15].**

---

[15] As future work, more sophisticated techniques can be used to determine which parts of the dialogue context should be predicted.

Table 4. Statistics of the augmentations for resp2ctxt and resp2ctxt$_{lu}$. New words are the ones that did not exist in the document before.

	MANtIS	MSDialog	UDC$_{\text{DSTC8}}$
Context avg length	474.12	426.08	76.95
Response avg length	42.58	71.38	11.06
Aug. avg length - resp2ctxt	494.23	596.99	202.3
Aug. avg length - resp2ctxt$_{lu}$	138.5	135.29	72.57
% new words - resp2ctxt	71%	69%	71%
% new words - resp2ctxt$_{lu}$	59%	37%	63%

In order to understand what the response expansion methods are doing most—term re-weighting or adding novel terms—we present the percentage of novel terms added by both methods in Table 4. The table shows that resp2ctxt$_{lu}$ does more term re-weighting than adding new words when compared to resp2ctxt (53% and 70% on average are new words respectively and thus 47% vs 30% are changing the weights by adding existing words), generating overall smaller augmentations (115.45 vs 431.17 on average respectively).

F3 ✓ Sparse retrieval outperforms zero-shot dense retrieval [34,41]. Sparse retrieval models are more effective than the majority of zero-shot dense models, as shown by the comparison of rows (1a–b), and (2a–b) with rows (3a–c). However, a dense retrieval model that has gone through intermediate training on large and diverse datasets including dialogues is more effective than a strong sparse retrieval model, as we see by comparing row (3d) with row (2b) in Table 3.

For example, while the zero-shot dense retrieval models based only on the MSMarco dataset (3a–b) perform on average 35% worse than the strong sparse baseline (2b) in terms of R@10 for the MSDialog dataset, the zero-shot model trained with 1.17B instances on diverse data (3d) is 68% better than the sparse baseline (2b). When using a bigger amount of intermediate training data[16], we see that the zero-shot dense retrieval model (3d) is able to outperform the sparse retrieval baseline by margins of 33% of R@10 on average across datasets.

We thus show that F3 only generalizes to response retrieval for dialogues if we do not employ a large set of diverse intermediate data. As expected, the closer the intermediate training data distribution is to the evaluation data, the better the dense retrieval model performs. The results indicate that a good zero-shot retrieval model needs to go through intermediate training on a large set of training data coming from multiple datasets to generalize well to different domains and outperform strong sparse retrieval baselines.

F4 ✓ Dense models with access to target training data outperform sparse models [7,15,34]. First, we see that fine-tuning the dense retrieval

---

[16] For the full description of the intermediate data see https://huggingface.co/sentence-transformers/all-mpnet-base-v2.

model, which has gone through intermediate training already, with random sampling—row (4a) in Table 3—achieves the best overall effectiveness in two of the three datasets. **This result shows that F4 generalizes to the task of response retrieval for dialogues when employing intermediate training**[17]. Having access to the target data as opposed to only the intermediate training data means that the representations learned by the model are closer to the true distribution of the data.

We hypothesize that fine-tuning the bi-encoder for MANtIS (4a) is harmful because the intermediate data contains Stack Exchange responses. In this way, the set of dialogues of Stack Exchange that MANtIS encompasses might be serving only to overfit the intermediate representations. As evidence for this hypothesis, we found that (I) the learning curves flatten quickly (as opposed to other datasets) and (II) fine-tuning another language model that does not have Stack Exchange data (MSMarco) in their fine-tuning, bi-encoder$_{bert-base}$ (3c), improves the effectiveness with statistical significance from 0.092 R@10 to 0.205 R@10.

F5 ✓ Hard negative sampling is better than random sampling for training dense retrieval models [46,51]. Surprisingly we find that naively using more effective models to select negative candidates is detrimental to the effectiveness of the dense retrieval model (see Hard negative sampling in Table 5). We observe this phenomenon when using different language models, when switching intermediate training on or off for all datasets, and when using an alternative contrastive loss [10] that does not employ in-batch negative sampling[18].

After testing for a number of hypotheses that might explain why harder negatives do not improve the effectiveness of the dense retrieval model, we found that false negative samples increase significantly when using better negative sampling methods. False negatives are responses that are potentially valid for the context. Such relevant responses lead to unlearning relevant matches between context and responses as they receive negative labels. See below an example of a false negative sample retrieved by the bi-encoder model (row 3d of Table 3):

Dialogue context ($\mathcal{U}$): hey... how long until dapper comes out? [U] 14 days [...] [U] i thought it was coming out tonight
Correct response (r^+): just kidding couple hours
False negative sample (r^-): there is a possibility dapper will be delayed [...] meanwhile, dapper discussions should occur in ubuntu+1

Denoising techniques try to solve this problem by reducing the number of false negatives. We employ a simple approach that instead of using the top-ranked responses as negative responses, we use the bottom responses of the top-ranked responses as negatives[19]. This decreases the chances of obtaining false positives and if $k << |\mathcal{D}|$ we will not obtain random samples. Our experiments in Table 5

---

[17] Our experiments show that when we do not employ the intermediate training step the fine-tuned dense model does not generalize well, with row (3d) performance dropping to 0.172, 0.308 and 0.063 R@10 for MANtIS, MSDialog and UDC$_{DSTC8}$ respectively.

[18] The results are not shown here due to space limitations.

[19] For example, if we retrieve $k = 100$ responses, instead of using responses from top positions 1–10, we use responses 91–100 from the bottom of the list.

reveal that this denoising technique, row (3b), increases the effectiveness for harder negative samples, beating all models from Table 3 for two of the three datasets. **The results indicate that F5 generalizes to the task of response retrieval for dialogues only when employing a denoising technique.**

Table 5. Results for the generalizability of F5—with and without a denoising strategy for hard negative sampling. Superscripts indicate statistically significant improvements using Students t-test with Bonferroni correction . †=*significance against the random sampling baseline,* ‡=*significance against hard negative sampling without denoising.*

	MANtIS	MSDialog	UDC$_{DSTC8}$
	R@10	R@10	R@10
Baseline			
(1) Bi-encoder$_{Random}$	0.307	0.387	**0.128**
Hard negative sampling			
(2a) Bi-encoder$_{BM25}$	0.271	0.316	0.087
(2b) Bi-encoder$_{Bi-encoder}$	0.146	0.306	0.051
Denoised hard negative sampling			
(3a) Bi-encoder$_{BM25}$	0.257	0.358‡	0.121‡
(3b) Bi-encoder$_{Bi-encoder}$	**0.316**†‡	**0.397**†‡	0.107‡

6   Conclusion

In this work, we tested if the knowledge obtained in dense and sparse retrieval from experiments on the tasks of passage and document retrieval generalizes to the first-stage retrieval of responses for dialogues. Our replicability study reveals that while most findings do generalize to our domain, a simple translation of the models is not always successful. A careful analysis of the domain in question might reveal better ways to adapt techniques.

As future work, we believe an important direction is to evaluate learned sparse methods that do weighting and expansion for both the queries and documents [5]—while resp2ctxt is able to both change the weights of the terms in the response (by repeating existing terms) and expand terms (by adding novel terms), it is not able to do weighting and expansion for the dialogue contexts.

Acknowledgements. This research has been supported by NWO projects SearchX (639.022.722) and NWO Aspasia (015.013.027).

References

1. Abdul-Jaleel, N., et al.: Umass at trec 2004: Novelty and hard. Computer Science Department Faculty Publication Series, p. 189 (2004)
2. Aghajanyan, A., Gupta, A., Shrivastava, A., Chen, X., Zettlemoyer, L., Gupta, S.: Muppet: massive multi-task representations with pre-finetuning. arXiv preprint arXiv:2101.11038 (2021)

3. Anand, A., Cavedon, L., Joho, H., Sanderson, M., Stein, B.: Conversational search (dagstuhl seminar 19461). In: Dagstuhl Reports. vol. 9. Schloss Dagstuhl-Leibniz-Zentrum für Informatik (2020)
4. Devlin, J., Chang, M.W., Lee, K., Toutanova, K.: Bert: Pre-training of deep bidirectional transformers for language understanding. arXiv preprint arXiv:1810.04805 (2018)
5. Formal, T., Lassance, C., Piwowarski, B., Clinchant, S.: Splade v2: Sparse lexical and expansion model for information retrieval. arXiv preprint arXiv:2109.10086 (2021)
6. Furnas, G.W., Landauer, T.K., Gomez, L.M., Dumais, S.T.: The vocabulary problem in human-system communication. Commun. ACM **30**(11), 964–971 (1987)
7. Gao, L., Callan, J.: Unsupervised corpus aware language model pre-training for dense passage retrieval. arXiv preprint arXiv:2108.05540 (2021)
8. Gu, J.C., Li, T., Liu, Q., Ling, Z.H., Su, Z., Wei, S., Zhu, X.: Speaker-aware Bert for multi-turn response selection in retrieval-based chatbots. In: Proceedings of the 29th ACM International Conference on Information & Knowledge Management, pp. 2041–2044 (2020)
9. Gu, J.C., Ling, Z.H., Liu, Q.: Interactive matching network for multi-turn response selection in retrieval-based chatbots. In: Proceedings of the 28th ACM International Conference on Information and Knowledge Management, pp. 2321–2324 (2019)
10. Hadsell, R., Chopra, S., LeCun, Y.: Dimensionality reduction by learning an invariant mapping. In: 2006 IEEE Computer Society Conference on Computer Vision and Pattern Recognition (CVPR 2006), vol. 2, pp. 1735–1742. IEEE (2006)
11. Han, J., Hong, T., Kim, B., Ko, Y., Seo, J.: Fine-grained post-training for improving retrieval-based dialogue systems. In: Proceedings of the 2021 Conference of the North American Chapter of the Association for Computational Linguistics: Human Language Technologies, pp. 1549–1558. Association for Computational Linguistics, Online, June 2021. https://doi.org/10.18653/v1/2021.naacl-main.122, https://aclanthology.org/2021.naacl-main.122
12. Hofstätter, S., Lin, S.C., Yang, J.H., Lin, J., Hanbury, A.: Efficiently teaching an effective dense retriever with balanced topic aware sampling. In: Proceedings of the 44th International ACM SIGIR Conference on Research and Development in Information Retrieval, pp. 113–122 (2021)
13. Johnson, J., Douze, M., Jégou, H.: Billion-scale similarity search with GPUs. IEEE Trans. Big Data **7**(3), 535–547 (2019)
14. Kadlec, R., Schmid, M., Kleindienst, J.: Improved deep learning baselines for ubuntu corpus dialogs. arXiv preprint arXiv:1510.03753 (2015)
15. Karpukhin, V., et al.: Dense passage retrieval for open-domain question answering. arXiv preprint arXiv:2004.04906 (2020)
16. Kummerfeld, J.K., et al.: A large-scale corpus for conversation disentanglement. Proceedings of the 57th Annual Meeting of the Association for Computational Linguistics (2019). https://doi.org/10.18653/v1/p19-1374, http://dx.doi.org/10.18653/v1/P19-1374
17. Lan, T., Cai, D., Wang, Y., Su, Y., Mao, X.L., Huang, H.: Exploring dense retrieval for dialogue response selection. arXiv preprint arXiv:2110.06612 (2021)
18. Lin, J.: The simplest thing that can possibly work: pseudo-relevance feedback using text classification. arXiv preprint arXiv:1904.08861 (2019)
19. Lin, J.: A proposed conceptual framework for a representational approach to information retrieval. arXiv preprint arXiv:2110.01529 (2021)

20. Lin, J., Ma, X., Lin, S.C., Yang, J.H., Pradeep, R., Nogueira, R.: Pyserini: a Python toolkit for reproducible information retrieval research with sparse and dense representations. In: Proceedings of the 44th Annual International ACM SIGIR Conference on Research and Development in Information Retrieval (SIGIR 2021), pp. 2356–2362 (2021)

21. Lin, J., Nogueira, R., Yates, A.: Pretrained transformers for text ranking: Bert and beyond. In: Synthesis Lectures on Human Language Technologies, vol. 14(4), 1–325 (2021)

22. Lin, Z., Cai, D., Wang, Y., Liu, X., Zheng, H.T., Shi, S.: The world is not binary: Learning to rank with grayscale data for dialogue response selection. arXiv preprint arXiv:2004.02421 (2020)

23. Lowe, R., Pow, N., Serban, I., Pineau, J.: The ubuntu dialogue corpus: a large dataset for research in unstructured multi-turn dialogue systems. arXiv preprint arXiv:1506.08909 (2015)

24. Nogueira, R., Cho, K.: Passage re-ranking with Bert. arXiv preprint arXiv:1901.04085 (2019)

25. Nogueira, R., Lin, J., Epistemic, A.: From doc2query to doctttttquery. Online preprint 6 (2019)

26. Peeters, R., Bizer, C., Glavaš, G.: Intermediate training of Bert for product matching. Small **745**(722), 2–112 (2020)

27. Penha, G., Balan, A., Hauff, C.: Introducing mantis: a novel multi-domain information seeking dialogues dataset. arXiv preprint arXiv:1912.04639 (2019)

28. Penha, G., Hauff, C.: Curriculum learning strategies for IR: an empirical study on conversation response ranking. arXiv preprint arXiv:1912.08555 (2019)

29. Penha, G., Hauff, C.: Challenges in the evaluation of conversational search systems. In: Converse@ KDD (2020)

30. Poth, C., Pfeiffer, J., Rücklé, A., Gurevych, I.: What to pre-train on? efficient intermediate task selection. arXiv preprint arXiv:2104.08247 (2021)

31. Pruksachatkun, Y., et al.: Intermediate-task transfer learning with pretrained models for natural language understanding: When and why does it work? arXiv preprint arXiv:2005.00628 (2020)

32. Qu, C., Yang, L., Croft, W.B., Trippas, J.R., Zhang, Y., Qiu, M.: Analyzing and characterizing user intent in information-seeking conversations. In: The 41st International ACM SIGIR Conference on Research & Development in Information Retrieval, pp. 989–992 (2018)

33. Reimers, N., Gurevych, I.: Sentence-Bert: Sentence embeddings using SIAMESE Bert-networks. In: Proceedings of the 2019 Conference on Empirical Methods in Natural Language Processing. Association for Computational Linguistics, November 2019. https://arxiv.org/abs/1908.10084

34. Ren, R., et al.: A thorough examination on zero-shot dense retrieval. arXiv preprint arXiv:2204.12755 (2022)

35. Robertson, S.E., Walker, S.: Some simple effective approximations to the 2-poisson model for probabilistic weighted retrieval. In: Croft, B.W., van Rijsbergen, C.J. (eds.) SIGIR 1994, pp. 232–241. Springer, London (1994). doi:https://doi.org/10.1007/978-1-4471-2099-5_24

36. Robinson, J., Chuang, C.Y., Sra, S., Jegelka, S.: Contrastive learning with hard negative samples. arXiv preprint arXiv:2010.04592 (2020)

37. Song, K., Tan, X., Qin, T., Lu, J., Liu, T.Y.: MpNet: masked and permuted pre-training for language understanding. Adv. Neural. Inf. Process. Syst. **33**, 16857–16867 (2020)

38. Tao, C., Feng, J., Liu, C., Li, J., Geng, X., Jiang, D.: Building an efficient and effective retrieval-based dialogue system via mutual learning. arXiv preprint arXiv:2110.00159 (2021)
39. Tao, C., Wu, W., Xu, C., Hu, W., Zhao, D., Yan, R.: Multi-representation fusion network for multi-turn response selection in retrieval-based chatbots. In: WSDM, pp. 267–275 (2019)
40. Thakur, N., Reimers, N., Daxenberger, J., Gurevych, I.: Augmented sbert: data augmentation method for improving bi-encoders for pairwise sentence scoring tasks. arXiv preprint arXiv:2010.08240 (2020)
41. Thakur, N., Reimers, N., Rücklé, A., Srivastava, A., Gurevych, I.: Beir: a heterogenous benchmark for zero-shot evaluation of information retrieval models. arXiv preprint arXiv:2104.08663 (2021)
42. Whang, T., Lee, D., Lee, C., Yang, K., Oh, D., Lim, H.: An effective domain adaptive post-training method for Bert in response selection. arXiv preprint arXiv:1908.04812 (2019)
43. Whang, T., Lee, D., Oh, D., Lee, C., Han, K., Lee, D.H., Lee, S.: Do response selection models really know what's next? utterance manipulation strategies for multi-turn response selection. In: Proceedings of the AAAI Conference on Artificial Intelligence. vol. 35, pp. 14041–14049 (2021)
44. Wolf, T., et al.: Huggingface's transformers: State-of-the-art natural language processing. arXiv preprint arXiv:1910.03771 (2019)
45. Wu, Y., Wu, W., Xing, C., Zhou, M., Li, Z.: Sequential matching network: a new architecture for multi-turn response selection in retrieval-based chatbots. In: ACL, pp. 496–505 (2017)
46. Xiong, L., et al.: Approximate nearest neighbor negative contrastive learning for dense text retrieval. arXiv preprint arXiv:2007.00808 (2020)
47. Yang, L., et al.: IART: intent-aware response ranking with transformers in information-seeking conversation systems. arXiv preprint arXiv:2002.00571 (2020)
48. Yang, L., et al.: Response ranking with deep matching networks and external knowledge in information-seeking conversation systems. In: SIGIR pp. 245–254 (2018)
49. Yang, W., Lu, K., Yang, P., Lin, J.: Critically examining the "neural hype" weak baselines and the additivity of effectiveness gains from neural ranking models. In: Proceedings of the 42nd International ACM SIGIR Conference on Research and Development in Information Retrieval, pp. 1129–1132 (2019)
50. Yuan, C., et al.: Multi-hop selector network for multi-turn response selection in retrieval-based chatbots. In: EMNLP, pp. 111–120 (2019)
51. Zhan, J., Mao, J., Liu, Y., Guo, J., Zhang, M., Ma, S.: Optimizing dense retrieval model training with hard negatives. In: Proceedings of the 44th International ACM SIGIR Conference on Research and Development in Information Retrieval, pp. 1503–1512 (2021)
52. Zhang, Z., Zhao, H.: Advances in multi-turn dialogue comprehension: a survey. arXiv preprint arXiv:2110.04984 (2021)
53. Zhang, Z., Zhao, H.: Structural pre-training for dialogue comprehension. arXiv preprint arXiv:2105.10956 (2021)
54. Zhou, X., et al.: Multi-turn response selection for chatbots with deep attention matching network. In: Proceedings of the 56th Annual Meeting of the Association for Computational Linguistics (Volume 1: Long Papers), pp. 1118–1127 (2018)

PyGaggle: A Gaggle of Resources for Open-Domain Question Answering

Ronak Pradeep[(✉)], Haonan Chen, Lingwei Gu, Manveer Singh Tamber, and Jimmy Lin

David R. Cheriton School of Computer Science, University of Waterloo, Waterloo, ON, Canada
{rpradeep,haonan.chen,l39gu,mtamber,jimmylin}@uwaterloo.ca

Abstract. Text retrieval using dense–sparse hybrids has been gaining popularity because of their effectiveness. Improvements to both sparse and dense models have also been noted, in the context of open-domain question answering. However, the increasing sophistication of proposed techniques places a growing strain on the reproducibility of results. Our work aims to tackle this challenge. In Generation Augmented Retrieval (GAR), a sequence-to-sequence model was used to generate candidate answer strings as well as titles of documents and actual sentences where the answer string might appear; this query expansion was applied before traditional sparse retrieval. Distilling Knowledge from Reader to Retriever (DKRR) used signals from downstream tasks to train a more effective Dense Passage Retrieval (DPR) model. In this work, we first replicate the results of GAR using a different codebase and leveraging a more powerful sequence-to-sequence model, T5. We provide tight integration with Pyserini, a popular IR toolkit, where we also add support for the DKRR-based DPR model: the combination demonstrates state-of-the-art effectiveness for retrieval in open-domain QA. To account for progress in generative readers that leverage evidence fusion for QA, so-called fusion-in-decoder (FiD), we incorporate these models into our PyGaggle toolkit. The result is a reproducible, easy-to-use, and powerful end-to-end question-answering system that forms a starting point for future work. Finally, we provide evaluation tools that better gauge whether models are generalizing or simply memorizing.

Keywords: Open-domain QA · Dense retrieval · Generative readers

1 Introduction

The past few years have seen open-domain QA progress at a breakneck pace. Given the rapid advancement of NLP and IR, both critical aspects of the overall task, this comes as little surprise. Often work needs to come along that can re-examine some of the critical papers, adding veracity to their claims and ensuring that we build on a firm foundation. For instance, Ma et al. [15] confirmed that DPR [9], a foundational work in open-domain QA, is indeed an effective

© The Author(s), under exclusive license to Springer Nature Switzerland AG 2023
J. Kamps et al. (Eds.): ECIR 2023, LNCS 13982, pp. 148–162, 2023.
https://doi.org/10.1007/978-3-031-28241-6_10

dense retrieval approach, but further discovered that the original authors under-reported the effectiveness of BM25 and thus incorrectly arrived at the conclusion that dense–sparse hybrids are worse for popular datasets like Natural Questions (NQ) [10].

As expected, significant progress has been made since the introduction of DPR in the direction of better retrievers, both sparse and dense, and better readers. Our work aims to tackle a broad set of reproducibility and replicability challenges in this context, and all our efforts result in code contributions to our group's Pyserini IR toolkit and PyGaggle open-source library.

For sparse retrieval, Generation Augmented Retrieval (GAR) trained query expansion BART models to generate answers, titles, and sentences relevant to the question. These were then used to enhance the sparse representation of the query. In this work, we train a more powerful GAR model, taking inspiration from the success of T5 [20] in similar information retrieval tasks [18]. We also examine which aspects of a GAR model contribute more to its success, from answer, title, and sentence prediction.

For dense retrieval, Izacard and Grave [4] proposed using signals from down-stream reading comprehension to train a more effective DPR, called DPR_{DKRR}. We perform experiments to replicate this work, integrating this retriever into Pyserini with corresponding prebuilt indexes. Given these strong sparse and dense retrieval models, we can obtain a state-of-the-art hybrid retrieval system with little effort.

On the reader side, we integrate the popular FiD [5] model that aggregates and combines evidence from various documents to generate an answer string in a sequence-to-sequence fashion. We replicate this work in PyGaggle.

Finally, work by Lewis et al. [12] broke down popular open-domain QA bench-marks into subsets to properly examine the source of the gains, be it through training data memorization or true model generalization. To gear our systems and resources to be future-proof, we also integrate scripts that score models based on per-subset effectiveness at the retrieval and the end-to-end QA level; we are the first to consider this on the retrieval side.

To summarize, the main contribution of this work is the integration of a powerful open-domain QA system into an easy-to-use ecosystem that has tightly knit components of query expansion using our proposed, GAR-T5, dense retrieval using DPR_{DKRR}, dense–sparse hybrid retrieval, and generative reading comprehension with FiD.

Code related to our dense–sparse hybrid retrieval experiments and evaluation is packaged in Pyserini, and code related to our generative reading comprehension model as well as end-to-end QA is added to PyGaggle.

2   Data Description and Metrics

We evaluate retrieval and end-to-end QA effectiveness on the open-domain versions of Natural Questions (NQ) [10] and TriviaQA [8]. These two datasets have

Table 1. Number of instances in the different subsets of the NQ and TriviaQA test set.

Method	Total	Question overlap	Answer overlap only	No overlap
(a) NQ	3610	324	315	357
(b) TriviaQA	11313	336	411	254

emerged as the most popular datasets for evaluation, and most open-domain QA papers evaluate the effectiveness of their models on them.

However, specifically for the NQ dataset, there have been some inconsistencies in the pre-processing scripts for the queries and answers, leading to discrepancies in the retrieval effectiveness. Papers before Min et al. [17], building on DPR [9], used their topic sets, which have a few words pre-processed differently. Some that we could discern are words like "don't", "wanna", and "gonna" that took on a new life in "do n't", "wan na", and "gon na", respectively. Another was in the way quotation marks are handled. While these do not seem significant, we find that they bring about a noticeable drop in retrieval effectiveness in Sect. 4 and hence a clear spot where models have invisibly gained or dropped in effectiveness points based on the topic set used.

We integrate both these variants into Anserini and Pyserini, the DPR variants called `dpr-nq-dev` and `dpr-nq-test`, and the NQ variants called `nq-dev` and `nq-test`, the latter of which incorporates the fixes mentioned.

A detailed study by Lewis et al. [12] found that in NQ and TriviaQA, around 33% of the test queries have a near-duplicate paraphrase in their corresponding training set. Similarly, 60–70% of the answers occur somewhere in the training set answers. These observations suggest that many open-domain QA models might have memorized these as facts from training, and a thorough per-category effectiveness check is required to gain deeper insights into a model's generalization capabilities. The main subsets considered by most of the literature are:

- *Question Overlap (QO)*, where for any query, there exists some paraphrase query in the training set.
- *Answer Overlap Only (AOO)*, where there is no *question overlap*, but for the example's ground truth answer set, at least one of the answers is part of some answer set in the training set.
- *No Overlap (NO)*, where there is no train-test overlap. This set probes for open-domain QA generalization.

Table 1 reports how many instances are in each subset of the NQ and TriviaQA test set. Note that the authors only curate these annotations for around 1k examples of each test set. While Lewis et al. [12] only evaluated end-to-end QA effectiveness under these settings, we think it is critical to also look at retrieval effectiveness under this light. To this end, we integrate a tool to measure per-subset retrieval effectiveness into Pyserini and end-to-end QA effectiveness into PyGaggle. The metric used for retrieval effectiveness is the top-k accuracy, i.e., a score of 1 if any top-k document is relevant and 0 otherwise. We leverage the popular exact match (EM) score for end-to-end QA effectiveness.

3 Methods

The retriever–reader pipeline using neural readers was proposed by Chen et al. [1] for open-domain QA tasks. There are two modules, the retriever, which locates relevant passages given the query, and the reader, which processes the retrieved documents to generate an answer.

3.1 Retriever

Given a corpus $\mathcal{C}$ (Wikipedia segments from DPR [9]), the retriever is tasked to return a list of the k most relevant documents ($k << |\mathcal{C}|$). Here, a document is relevant if it contains one of the answer strings as a span up to normalization. In this study, the corpus refers to an English Wikipedia dump from 12/20/2018, and "documents" are non-overlapping 100-word segments of the Wikipedia articles used in DPR [9].

Karpukhin et al. [9] showed that switching the retrieval component out for *dense* representations learned by a BERT encoder (DPR) results in a significant retrieval effectiveness boost. They also investigated hybrid retrieval, combining results from dense retrieval (DPR) and sparse retrieval (BM25) but failed to find any improvements in retrieval effectiveness. However, later work [15] found that DPR under-reported the BM25 effectiveness and hence incorrectly arrived at the conclusion that hybrid retrieval *does* not result in better retrieval and end-to-end QA effectiveness. We explore how to build better sparse and dense baselines.

Sparse Methods. The focus here is Generation-Augmented Retrieval (GAR) by Mao et al. [16], a neural query expansion technique. By no means was this the first work to look into query expansion, RM3 [6] is a traditional technique employed to expand the query with terms from the top retrieved documents after an initial retrieval step. However, GAR is a powerful neural method to enhance retrieval in open-domain QA. They leveraged a pretrained sequence-to-sequence language model, BART [11], to enhance the sparse representation of the query through expansion. More specifically, they first trained three separate models that take as input the query and generate:

1. **The default target answers.** This is similar to recent work in closed-book QA where models generate answers to questions without access to external knowledge [21]. In this case, the model generates *all* possible answers separated by a special delimiter.
2. **A sentence containing the target answer.** They use as the target, the sentence from the highest BM25 ranked document (based on the original query) containing some ground-truth answer. This is an attempt to train the model to generate other words relevant to the question and target answer and help better match results with the sentence-augmented query. While this

method is not infallible to negatives, we leave further exploration to future work.

3. **Titles of passages containing any target answer.** Here, the model is trained to generate *all* titles in the top retrieved documents that contain one of the target answers. The titles are separated by a special delimiter too. Like sentences, these too might contain additional keywords or entities besides the answers that might help the augmented query pull up more relevant results.

During inference, for a particular query q, they generate three separate expanded queries q_{answer}, $q_{sentence}$, and q_{title}. Then, sparse retrieval with BM25 (default parameters) is independently performed on each query using Pyserini [14]. Finally, the top-1k results of the three queries are combined using reciprocal rank fusion (RRF) [2] to return a single ranked list. Since they provide their training code, we first try to reproduce their training and inference results. We dub this GAR-BART.

Given the success of the pretrained sequence-to-sequence transformer model T5 [20] in the similar task of document expansion for passage ranking [18], we also *replicate* their results using the powerful T5-3B model instead. Our training and generation replication uses Google's mesh-tensorflow T5 implementation.[1] We call this model GAR-T5. We train each GAR-T5 model on a Google TPU v3-8 with a constant learning rate of $1 \cdot 10^{-3}$ for 25k iterations with batches of 256. This corresponds to roughly 80 epochs on the training sets. We always use a maximum of 64 input tokens for the query and a maximum of 64, 128, and 128 output tokens for the answer, sentence, and title generation, respectively.

Dense Methods. In this subsection, we build towards the DPR$_{DKRR}$ [4] model, a state-of-the-art dense retriever for open-domain QA retrieval. We begin with the pivotal DPR model that employs a separate query encoder and a passage encoder, both using BERT [3] as the backbone model. However, Izacard and Grave [4] only use a single encoder instead of two and signal a query by prepending the query string with "question:" and signal a document D using the template "title: D^{title} context: D^{text}". Queries and passages are encoded as dense vectors separately, and the relevance score of a query document pair is computed by their inner product. The retriever functionally performs a nearest neighbor search over the representations using Facebook's Faiss library [7].

Izacard and Grave [4] use this DPR model to initialize their retriever and leverage the Fusion-in-Decoder (FiD) reader described in Sect. 3.2. They use an iterative training pipeline where they first train a reader, use it to compute the aggregated cross-attention scores, and leverage them to finetune a retriever. Following this procedure for multiple iterations, they arrive at the more effective DPR$_{DKRR}$ retriever.

We integrate the DPR$_{DKRR}$ retrievers provided by the authors for both NQ and TriviaQA in Pyserini. We also provide prebuilt indexes after the corresponding retriever encodes the entire Wikipedia corpus.

---

[1] https://github.com/google-research/text-to-text-transfer-transformer.

Karpukhin et al. [9] and Ma et al. [15] calculate hybrid scores by linear combinations of dense and sparse scores. Here, we use reciprocal rank fusion (RRF) [2] without any task-specific tuning to fuse the GAR (sparse) and the DPR$_{\text{DKRR}}$ (dense) ranked lists. We find that this allows us to get roughly the same effectiveness without the expensive tuning step on the development set.

3.2 Reader

As is expected in a retriever–reader paradigm, for each query q, the retriever selects k candidate documents. The reader returns the final answer span from the candidate contexts, where each document D_i contains the Wikipedia article title D_i^{title} and its content D_i^{text}.

The reader used in DPR [9] was a BERT-base and takes as input each candidate context C_i concatenated to the question q and is trained to "extract" candidate spans by predicting the start and end tokens of the target answer span. Such readers are generally called "extractive" readers.

In this paper, we use Fusion-in-Decoder (FiD) [5], a state-of-the-art generative reader that uses as backbone a generative model, T5 [20]. This model takes as input the query q and each of the top-100 retrieved documents. More precisely, each query q and document D_i are concatenated and independently run through the encoder to get an encoded representation. These representations across the ranked candidate document list are then concatenated, and the decoder *attends* over the resulting final encoding. Thus evidence fusion happens *only* in the decoder, and hence the name.

The authors experimented with both the T5-base and T5-large variants and trained them in tandem with DPR$_{\text{DKRR}}$ as described in Sect. 3.1. They trained each of these models for 10k iterations with a batch size of 64 using 64 × V100s. This training requires significant compute beyond the scope of our paper. Inference, however, can be run on a single Tesla P40. Thus, we integrate it into PyGaggle, an inference-focused library for knowledge-intensive tasks like open-domain QA.

4   Results

4.1   Retriever

In Table 2, we report retrieval accuracy on both the NQ and TriviaQA datasets. Rows (a)–(e) showcase sparse methods, some of which we augment with query or document expansion, rows (f)–(g) denote dense ones, and rows (h)–(i) denote hybrid ones.

Row (a) represents the BM25 effectiveness using the `dpr-nq-test` topic set, which as mentioned earlier, we move away from, to the better-processed `nq-test` topic set (row (b)). This shift leads to an improvement in retrieval accuracy, highlighting the importance of using the cleaner topic set, and suggesting that previously overlooked gains may merit further examination. Hence, for the rest

Table 2. Retrieval effectiveness comparing different methods on both the NQ and TriviaQA datasets.

Method	NQ					TriviaQA				
	top5	top20	top100	top500	top1000	top5	top20	top100	top500	top1000
Sparse										
(a) BM25' [15]	43.77	62.99	78.23	85.60	88.01	–	–	–	–	–
(b) BM25	44.82	64.02	79.20	86.59	88.95	66.29	76.41	83.14	87.35	88.50
(c) BM25 + RM3	44.16	62.91	77.76	86.01	88.73	62.68	73.12	80.69	85.62	87.17
(d) GAR-BART$_{RRF}$ (repro)	61.16	76.87	86.18	91.08	92.55	71.45	79.47	85.13	88.56	89.51
(e) GAR-T5$_{RRF}$	64.62	77.17	86.90	91.63	92.91	72.82	80.66	85.95	89.07	90.06
Dense										
(f) DPR [9]	68.61	80.58	86.68	90.91	91.83	69.80	78.87	84.79	88.19	89.30
(g) DPR$_{DKRR}$	73.80	84.27	89.34	92.24	93.43	77.23	83.74	87.78	89.87	90.63
Hybrid										
(h) DPR$_{Hybrid}$ [15]	72.52	83.43	89.03	92.16	93.19	76.01	82.64	86.55	89.12	89.90
(i) RRF(GAR-T5$_{RRF}$, DPR$_{DKRR}$)	74.57	84.90	90.86	93.35	94.18	78.63	85.02	88.41	90.29	90.83

of the results, we default to using this "cleaner" topic set, deviating slightly from prior work [9, 16] that used the former topic set.

Row (c) denotes BM25+RM3, a strong traditional IR baseline involving query expansion. We observe that the retrieval accuracy of BM25+RM3 (row (c)) is slightly lower than that of BM25 (row (b).

Rows (d) and (e) represent the BART and T5-based GAR models, respectively. GAR-BART$_{RRF}$ is a reproduction of the entire training and inference pipeline provided by Mao et al. [16]. We find that the retrieval effectiveness of both models is significantly higher than that of the standard BM25 method (row (b)). Comparing the two GAR models, we find that GAR-T5 performs consistently better in both datasets. This improvement is unsurprising, given that T5 is larger and pretrained on a diverse set of tasks, increasing its ability to generalize to new ones. Since our generative reader takes as input the top-100 passages, we aim to maximize the top-100 retrieval accuracy. We find that GAR-T5$_{RRF}$ achieves retrieval effectiveness on par with DPR, a dense model (row (f)).

Row (f) shows the effectiveness of the DPR retriever provided by Karpukhin et al. [9], although we use the `nq-test` topic file instead. A more detailed analysis of the effectiveness of this model is found in Ma et al. [15]. The effectiveness of DPR$_{DKRR}$ is reported in row (g). This model outperforms vanilla DPR (row (f)) across all values of retrieval effectiveness tested for both datasets. It demonstrates a solid ability to retrieve relevant documents, even in early precision settings. The strong top-5 retrieval effectiveness validates this behavior. DPR$_{DKRR}$ also outperforms all the sparse methods considered and performs consistently better than a previous hybrid model described in Ma et al. [15] (row (h)). This DPR$_{Hybrid}$ run is obtained by fusing results from DPR and BM25 using a weighted combination.

Finally, we look at our hybrid model (row (i)) that is obtained by fusing the results of GAR-T5$_{RRF}$ (row (e)) and DPR$_{DKRR}$ (row (g)). This model consistently achieves higher effectiveness than all the other methods considered across

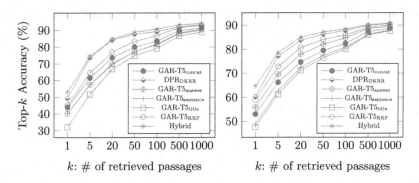

Fig. 1. Top-k accuracy of sparse, dense, and hybrid methods on NQ (left) and TriviaQA (right) as we vary the number of retrieved passages.

Table 3. Comparing the top-100 retrieval accuracy on different QA overlap subsets [12] of the test set for NQ and TriviaQA.

Condition	NQ			TriviaQA		
	QO	AOO	NO	QO	AOO	NO
(a) GAR-T5$_{RRF}$	96.30	88.57	77.59	97.92	93.19	68.50
(b) DPR$_{DKRR}$	95.06	89.84	80.95	97.32	95.13	70.87
(c) RRF(GAR-T5$_{RRF}$, DPR$_{DKRR}$)	96.91	92.38	84.03	98.21	96.10	72.44

all evaluation settings. In fact, a literature survey finds that this model has the highest retrieval accuracy for $k = 100$, the setting we want to maximize for the FiD readers. This demonstrates that hybrid methods continue to play a crucial role in knowledge-intensive tasks. Our results also suggest that sparse methods are less effective than dense methods, and we believe leveraging some learned sparse representations might bridge this gap and lead to better hybrid methods.

Figure 1 primarily shows the effectiveness of various ablations of GAR-T5. Among the individual query expansion models, GAR-T5$_{answer}$ demonstrates effectiveness on par with GAR-T5$_{sentence}$ in NQ and better in TriviaQA, both of which consistently outperform GAR-T5$_{title}$. This drop is perhaps because, in the sentences and titles cases, the model is also more prone to hallucinating additional keywords that might not be relevant to the query. Besides this, the tasks are considerably more challenging, given the lack of context provided to the model. GAR-T5$_{concat}$, which involves just a single query where the answer, sentence, and title are all appended to the original question, displays better effectiveness than each of the individual GAR-T5 models in the NQ case, but it performs worse than GAR-T5$_{answer}$ in TriviaQA. The gap between GAR-T5$_{concat}$ and GAR-T5$_{RRF}$ (where we take the RRF of GAR-T5$_{answer}$, GAR-T5$_{title}$, and GAR-T5$_{sentence}$) in the NQ test set is small, however, in the TriviaQA dataset, the gap is substantial and hence GAR-T5$_{RRF}$ stands as the most effective sparse method. The story with DPR$_{DKRR}$ and the hybrid method is the same as earlier, displaying considerably improved effectiveness compared to the sparse techniques.

Table 4. End-to-end QA effectiveness in terms of the exact match score, comparing different answer span scoring techniques. The "orig" and "repl" columns denote the original and replicated results, respectively.

Condition	NQ		TriviaQA	
	orig	repl	orig	repl
DPR-reader [15]				
(a) DPR	43.5	–	59.5	–
(b) BM25	38.4	–	61.6	–
(c) DPR$_{\text{Hybrid}}$	44.0	–	61.7	–
FiD-base [5]				
(d) DPR$_{\text{DKRR}}$	50.1	49.9	69.3	68.3
(e) GAR-T5	–	49.9	–	66.8
(f) RRF(GAR-T5$_{\text{RRF}}$, DPR$_{\text{DKRR}}$)	–	50.8	–	69.4
FiD-large [5]				
(g) DPR$_{\text{DKRR}}$	54.4	54.7	72.5	72.2
(h) GAR-T5	–	53.3	–	70.4
(i) RRF(GAR-T5$_{\text{RRF}}$, DPR$_{\text{DKRR}}$)	–	55.0	–	72.9

Table 3 reports top-100 retrieval accuracy of our highest scoring sparse, dense, and hybrid models on different QA overlap subsets [12] of the NQ and TriviaQA test set. Here, we gain some more insights into our sparse and dense methods.

Row (a) presents the scores of GAR-T5$_{\text{RRF}}$, and as we can see in terms of *Question Overlap* this model performs better than our effective dense retrieval model (row (b)). As the queries in this setting are paraphrases of some of the questions in the training set, we believe that the higher scores come from the GAR-T5 models remembering the target answers, titles, and sentences and thus generating them verbatim. As a result, it would be relatively easy for a sparse lexical retrieval model which relies on exact matches to pull up relevant documents.

Row (b) presents the scores of DPR$_{\text{DKRR}}$, and as shown, this model demonstrates better effectiveness than GAR-T5$_{\text{RRF}}$ in terms of *Answer Overlap Only* and *No Overlap*. In these settings, since we are looking more for model generalization, it makes sense how more powerful dense methods exhibit higher effectiveness.

Finally, we look at the hybrid model in row (c), and it clearly shows the highest retrieval effectiveness across all settings. The gains are the most in the *No Overlap* subsets, the main testbed of model generalization.

4.2   Reader

Table 4 presents the end-to-end open-domain QA effectiveness in terms of the primary metric, the EM score, for both the NQ and TriviaQA test sets. Here, "orig" refers to the original results in the paper or official repository reported by the authors, and "repl" reports the results we note in our systems.

Rows (a)–(c) report results described in Ma et al. [15] using the extractive reader provided with DPR and varying the retrieval method used. Note that

Table 5. End-to-end QA effectiveness in terms of the exact match score over some of the QA Overlap subsets.

Condition	NQ			TriviaQA		
	QO	AOO	NO	QO	AOO	NO
FiD-large [5]						
(a) DPR_{DKRR}	77.2	49.5	38.1	92.0	74.5	52.4
(b) GAR-T5	76.5	45.1	36.4	92.6	70.6	52.0
(c) RRF(GAR-T5_{RRF}, DPR_{DKRR})	77.8	48.9	39.2	93.8	74.5	54.7

these were the *highest* scores obtained in the paper and involved various components that include post-processing the spans selected by the reader and tuning the weights assigned to each span. Their use of hybrid retrieval methods (row (c)) notes higher end-to-end effectiveness than using the sparse or dense models independently (rows (a), (b)).

Rows (d)–(e) denote the end-to-end effectiveness of the highest-scoring dense, sparse, and hybrid methods described paired with the FiD-base reader model and rows (f)–(h) with the FiD-large reader model. First, we note that the effectiveness of FiD integrated into PyGaggle while in a similar range does not perfectly match the original scores reported on their GitHub page.[2] This difference perhaps has to do with the different implementation choices and modifications to get the reader integrated into PyGaggle that uses HuggingFace Transformers v4.10.0 (vs. the original repository using v3.0.2).

In general, we find that leveraging dense retrieval instead of sparse retrieval results in better end-to-end open-domain QA effectiveness (rows (d) vs. (e) and rows (g) vs. (h)). This boost comes as little surprise as the top-100 retrieval effectiveness is much higher in the dense methods. However, with the FiD-base trained on NQ, this surprisingly does not bring improvements. We see similar phenomena in other knowledge-intensive tasks like multi-stage relevance ranking where improved first-stage retrieval effectiveness does not necessarily translate to better end-to-end effectiveness [19].

Similarly, hybrid retrieval shows even higher end-to-end effectiveness in all cases (rows (f) and (i)). This dominance confirms previous findings in Ma et al. [15] although, in this case, with considerably more effective sparse and dense retrievers.

Finally, moving from the FiD-base (rows (d)–(f)) to the FiD-large variant (rows (g)–(i)) results in an effectiveness boost, as one would expect with bigger language models. Our most effective system involving hybrid retrieval and the FiD-large reader (row (i)) gains 11 points compared to the comparable retriever–reader pipeline in Ma et al. [15] (row (c)) and forms an effective and easy-to-reproduce baseline for the community.

In Table 5, we report the end-to-end open-domain QA accuracy of the QA Overlap [12] subsets when using the FiD-large reader and varying the retrieval

---

[2] https://github.com/facebookresearch/FiD.

method. Dense generally improves over sparse retrieval (rows (a) vs. (b)), and hybrid methods display even higher effectiveness (row (c)). Most of these gains are likely from the improved first-stage retrieval of documents fed into the generative reader. The *No Overlap* subset has huge gains when moving to hybrid retrieval across the datasets. This boost could imply better model generalization capabilities. However, a surprising finding here is that the *Answer Overlap Only* subset sees a drop in effectiveness scores in the NQ case and remains the same in the TriviaQA case. A more thorough analysis of why we note such behavior is needed. Also, as expected, all methods demonstrate the highest effectiveness in the *Question Overlap* subset and the lowest effectiveness in the *No Overlap* subset.

5   Resources

In this section, we discuss the three main types of resources we provide for working with open-domain QA on Wikipedia.

5.1   GAR-T5 Query Expansions

We provide the Generation Augmented Retrieval (GAR) query expansions for NQ and TriviaQA in the widely popular HuggingFace Datasets [13]. This integration comes with various benefits - a unified interface for downloading and wrangling the expansions, single-line data loaders, and easy-to-use integration with HuggingFace Transformers [22]. These expansions for NQ and TriviaQA can be found at `castorini/nq_gar-t5_expansions` and `castorini/triviaqa_gar-t5_expansions`, respectively.

5.2   Reproduction Guides and Commands

Our reproduction guides for both the sparse and dense retrieval methods are linked on the main landing page of Pyserini.

For the main dense retriever evaluated in this paper, the DPR$_{\mathrm{DKRR}}$ model, we provide the retrieval and evaluation instructions for the two test sets.[3] To allow this functionality, we also convert the checkpoints provided by Izacard and Grave [4] to HuggingFace-compatible checkpoints and upload them to the HuggingFace Hub. Given these checkpoints, we encode the entire Wikipedia corpus and provide it as a prebuilt index compatible with Pyserini. Dense retrieval on this index can be performed using a single command:

```
python -m pyserini.search.faiss --topics nq-test \
    --index wikipedia-dpr-dkrr-nq \
    --encoder castorini/dkrr-dpr-nq-retriever \
    --output runs/run.nq-test.dkrr.trec --query-prefix question:
```

---

[3] https://github.com/castorini/pyserini/blob/master/docs/experiments-dkrr.md.

For the sparse retrieval methods considered, i.e., GAR-T5 augmented BM25, the reproduction guide for retrieval and evaluation is provided.[4] We also include instructions for hybrid retrieval using RRF [2]. Given this hybrid run, we also provide a script that can evaluate the per-subset [12] top-k retrieval effectiveness (the evaluation metric we introduced in Sect. 2). We provide a webpage with a two-click (copy and paste) reproduction matrix,[5] that provides commands for reproducing our experimental results in the current or any future version of the Pyserini toolkit. We also include a regression test script which can be easily executed using the following command from Pyserini:

```
python scripts/repro_matrix/run_all_odqa.py --topics <nq or tqa>
```

Our reproduction guides for end-to-end open-domain QA with the generative FiD reader models can be found in the PyGaggle toolkit. This toolkit is the same one where a previous study [15] integrated the extractive reader modules of DPR.

We provide instructions to run reader inference for the NQ and TriviaQA datasets.[6] We can use the following command to run reader inference given the hybrid retrieval file from Pyserini:

```
python -um pygaggle.run.evaluate_fid_reader \
    --model-name nq_reader_large \
    --retrieval-file data/run.nq-test.dkrr.gar.hybrid.json \
    --output-file data/fid_large.nq-test.dkrr.gar.hybrid.out
```

Given this FiD-large reader output file, we integrate scripts that can calculate the per-subset level end-to-end exact match (EM) scores, as described in Sect. 2.

5.3   Interactive End-to-End System

A user can query this system in an end-to-end fashion by issuing the following command:

```
python pygaggle.qa --type openbook --qa-reader fid \
    --reader-model nq_reader_base \
    --retriever-model castorini/dkrr-dpr-nq-retriever \
    --retriever-index wikipedia-dpr-dkrr-nq \
    --retriever-corpus wikipedia-dpr
```

This command pulls up an interactive demo, where a user can ask questions and have the open-domain QA model return answers. The following is an interaction with the system:

```
Enter a question: which concept album by pink floyd was adopted
into a movie
Answer:  The Wall
Enter a question: which year did the movie casablanca come out
Answer:  1942
```

---

[4] https://github.com/castorini/pyserini/blob/master/docs/experiments-gar-t5.md.

[5] https://castorini.github.io/pyserini/2cr/odqa.html.

[6] https://github.com/castorini/pygaggle/blob/master/docs/experiments-fid-reader.md.

6   Conclusion

We address the challenge of reproducibility in increasingly sophisticated open-domain QA systems by developing a state-of-the-art end-to-end pipeline involving a hybrid of sparse and dense retrieval and generative readers.

First, we train a more effective GAR model [16] based on T5, which expands the queries with generated predictions for answers, sentences, and titles and incorporate them into the popular open-source IR toolkit, Pyserini. Second, we replicate the results of the widely-used dense method, DPR_{DKRR} [4] retriever and provide search capabilities and prebuilt indexes through Pyserini. As a result, we can effortlessly put together a state-of-the-art hybrid retrieval system.

We incorporate the Fusion-in-Decoder generative reader into our suite of reading comprehension systems in PyGaggle. This addition allows us to provide a very effective end-to-end open-domain QA system.

The tight integration of Pyserini and PyGaggle allows users to interact directly with our systems and pose factoid questions. Following Lewis et al. [12], we also provide evaluation tools that break down effectiveness across various categories to investigate model generalization and memorization.

We believe that it will be hugely beneficial to the research community to have an effective, easy-to-use, and reproducible open-domain QA baseline and the tools to interact with these systems directly and evaluate model capabilities like the true generalization of the retriever and reader components.

Acknowledgements. This research was supported in part by the Natural Sciences and Engineering Research Council (NSERC) of Canada. Computational resources were provided in part by Compute Ontario and Compute Canada. We thank the Google Cloud and the TPU Research Cloud Program for credits to support some of our experimental runs.

References

1. Chen, D., Fisch, A., Weston, J., Bordes, A.: Reading Wikipedia to answer open-domain questions. In: Proceedings of the 55th Annual Meeting of the Association for Computational Linguistics (ACL 2017), Vancouver, British Columbia, Canada, pp. 1870–1879 (2017)
2. Cormack, G.V., Clarke, C.L.A., Buettcher, S.: Reciprocal rank fusion outperforms condorcet and individual rank learning methods. In: Proceedings of the 32nd International ACM SIGIR Conference on Research and Development in Information Retrieval, New York, NY, USA, pp. 758–759. Association for Computing Machinery (2009)
3. Devlin, J., Chang, M.W., Lee, K., Toutanova, K.: BERT: pre-training of deep bidirectional transformers for language understanding. In: Proceedings of the 2019 Conference of the North American Chapter of the Association for Computational Linguistics: Human Language Technologies, Volume 1 (Long and Short Papers), pp. 4171–4186. Minneapolis, Minnesota (2019)
4. Izacard, G., Grave, E.: Distilling knowledge from reader to retriever for question answering. ArXiv abs/2012.04584 (2021)

5. Izacard, G., Grave, E.: Leveraging passage retrieval with generative models for open domain question answering. In: Proceedings of the 16th Conference of the European Chapter of the Association for Computational Linguistics: Main Volume, pp. 874–880. Online, April 2021
6. Jaleel, N., et al.: UMass at TREC 2004: Novelty and HARD (2004)
7. Johnson, J., Douze, M., Jégou, H.: Billion-scale similarity search with GPUs. IEEE Trans. Big Data **7**(3), 535–547 (2021)
8. Joshi, M., Choi, E., Weld, D., Zettlemoyer, L.: TriviaQA: a large scale distantly supervised challenge dataset for reading comprehension. In: Proceedings of the 55th Annual Meeting of the Association for Computational Linguistics (Volume 1: Long Papers), pp. 1601–1611. Association for Computational Linguistics, Vancouver, Canada (2017)
9. Karpukhin, V., et al.: Dense passage retrieval for open-domain question answering. In: Proceedings of the 2020 Conference on Empirical Methods in Natural Language Processing (EMNLP), pp. 6769–6781 (2020)
10. Kwiatkowski, T., et al.: Natural questions: a benchmark for question answering research. Trans. Assoc. Comput. Linguist. **7**, 452–466 (2019)
11. Lewis, M., et al.: BART: denoising sequence-to-sequence pre-training for natural language generation, translation, and comprehension. In: Proceedings of the 58th Annual Meeting of the Association for Computational Linguistics. pp. 7871–7880. Online, July 2020
12. Lewis, P., Stenetorp, P., Riedel, S.: Question and answer test-train overlap in open-domain question answering datasets. In: Proceedings of the 16th Conference of the European Chapter of the Association for Computational Linguistics: Main Volume, pp. 1000–1008. Online, April 2021
13. Lhoest, Q., et al.: Datasets: a community library for natural language processing. In: Proceedings of the 2021 Conference on Empirical Methods in Natural Language Processing: System Demonstrations, pp. 175–184. Online and Punta Cana, Dominican Republic, November 2021
14. Lin, J., Ma, X., Lin, S.C., Yang, J.H., Pradeep, R., Nogueira, R.: Pyserini: A Python toolkit for reproducible information retrieval research with sparse and dense representations. In: Proceedings of the 44th Annual International ACM SIGIR Conference on Research and Development in Information Retrieval (SIGIR 2021). pp. 2356–2362 (2021)
15. Ma, X., Sun, K., Pradeep, R., Li, M., Lin, J.: Another look at DPR: reproduction of training and replication of retrieval. In: Hagen, M., et al. (eds.) ECIR 2022. LNCS, vol. 13185, pp. 613–626. Springer, Cham (2022). https://doi.org/10.1007/978-3-030-99736-6_41
16. Mao, Y., et al.: Generation-augmented retrieval for open-domain question answering. In: Proceedings of the 59th Annual Meeting of the Association for Computational Linguistics and the 11th International Joint Conference on Natural Language Processing (Volume 1: Long Papers), pp. 4089–4100. Online (2021)
17. Min, S., et al.: NeurIPS 2020 EfficientQA competition: Systems, analyses and lessons learned. In: Escalante, H.J., Hofmann, K. (eds.) Proceedings of the NeurIPS 2020 Competition and Demonstration Track. Proceedings of Machine Learning Research, vol. 133, pp. 86–111. PMLR, 06–12 December 2021
18. Nogueira, R., Lin, J.: From doc2query to docTTTTTquery (2019)
19. Pradeep, R., Nogueira, R., Lin, J.: The expando-mono-duo design pattern for text ranking with pretrained sequence-to-sequence models. arXiv:2101.05667 (2021)
20. Raffel, C., et al.: Exploring the limits of transfer learning with a unified text-to-text transformer. J. Mach. Learn. Res. **21**, 1–67 (2020)

21. Roberts, A., Raffel, C., Shazeer, N.: How much knowledge can you pack into the parameters of a language model? In: Proceedings of the 2020 Conference on Empirical Methods in Natural Language Processing (EMNLP), pp. 5418–5426. Online, November 2020

22. Wolf, T., et al.: Transformers: State-of-the-art natural language processing. In: Proceedings of the 2020 Conference on Empirical Methods in Natural Language Processing: System Demonstrations, pp. 38–45, Online, October 2020

Pre-processing Matters! Improved Wikipedia Corpora for Open-Domain Question Answering

Manveer Singh Tamber[(⊠)], Ronak Pradeep, and Jimmy Lin

David R. Cheriton School of Computer Science,
University of Waterloo, Waterloo, ON, Canada
{mtamber,rpradeep,jimmylin}@uwaterloo.ca

Abstract. One of the contributions of the landmark Dense Passage Retriever (DPR) work is the curation of a corpus of passages generated from Wikipedia articles that have been segmented into non-overlapping passages of 100 words. This corpus has served as the standard source for question answering systems based on a retriever–reader pipeline and provides the basis for nearly all state-of-the-art results on popular open-domain question answering datasets. There are, however, multiple potential drawbacks to this corpus. First, the passages do not include tables, infoboxes, and lists. Second, the choice to split articles into non-overlapping passages results in fragmented sentences and disjoint passages that models might find hard to reason over. In this work, we experimented with multiple corpus variants from the same Wikipedia source, differing in passage size, overlapping passages, and the inclusion of linearized semi-structured data. The main contribution of our work is the replication of Dense Passage Retriever and Fusion-in-Decoder training on our corpus variants, allowing us to validate many of the findings in previous work and giving us new insights into the importance of corpus pre-processing for open-domain question answering. With better data preparation, we see improvements of over one point on both the Natural Questions dataset and the TriviaQA dataset in end-to-end effectiveness over previous work measured using the exact match score. Our results demonstrate the importance of careful corpus curation and provide the basis for future work.

Keywords: Open-domain QA · Dense retrieval · Wikipedia

1 Introduction

Dense Passage Retriever (DPR) [10] has been a pivotal work for open-domain question answering (QA). One of its contributions was the curation of a corpus of passages formed from a Wikipedia XML dump dated December 20, 2018. The authors split articles from this dump into non-overlapping passages, each 100 words long. The corpus, herein referred to as WikiText(100w), served as the

J. Kamps et al. (Eds.): ECIR 2023, LNCS 13982, pp. 163–176, 2023.
https://doi.org/10.1007/978-3-031-28241-6_11

textual knowledge source to allow the DPR retriever–reader pipeline to perform question answering. The corpus continues to be used in subsequent work, and improved retriever and reader models based on it have achieved state-of-the-art results on multiple open-domain QA datasets [7,8,16]. Given such community-wide adoption as the "gold standard" corpus, it is worth questioning if (and how) pre-processing choices affect the trained models.

There indeed seems to be multiple potential drawbacks to how the DPR authors pre-processed the original Wikipedia dump to form the WikiText(100w) corpus. For one, semi-structured data such as tables, infoboxes, and lists were not included in the final passage texts. The corpus contains only unstructured text from the body of the Wikipedia articles. Additionally, splitting the articles into disjoint 100-word passages results in fragmented sentences since the start or end of the passages do not necessarily align with complete sentences. Finally, the choice of disjoint passages (as opposed to overlapping passages) means that some textual information in individual passages may be isolated from associated relevant information that could be useful to provide a more complete context for question answering.

Using Wikipedia as a knowledge source in a retriever–reader pipeline has been explored earlier by Chen et al. [3]. Karpukhin et al. [10] used some of the same pre-processing code, leveraging the WikiExtractor Python library to extract cleaned text from Wikipedia articles. In our efforts, we first tried to closely replicate the corpus preparation steps from the DPR paper. For a fair comparison, we started with the same Wikipedia XML dump (from December 20, 2018) used in Karpukhin et al. [10] and we used some of the same pre-processing code. Specifically, we also used the WikiExtractor library, modifying the code to keep lists in the final cleaned text, and we used the pre-processing code from Chen et al. [3] to filter out disambiguation pages and HTML characters from the Wikipedia dump. We included additional pre-processing steps to add tables, infoboxes, and lists to our corpora and discarded unwanted leftover characters from the XML dump.

In this work, we experimented with Wikipedia corpus variants differing in passage size, overlapping passages, and the inclusion of linearized tables, infoboxes, and lists. One should note that the community has explored the inclusion of tables, infoboxes, and lists from Wikipedia in a text corpus with some success in Oğuz et al. [16], where they extracted Wikipedia tables and infoboxes exclusively from the Natural Questions dataset [11]. Here, we instead extracted the semi-structured data directly from the full Wikipedia XML dump to provide passages that are richer and contain more information. We used existing techniques to train reader and retriever models to answer questions from popular QA datasets. In particular, we started with training DPR models [10] for each corpus variant. We then trained generative Fusion-in-Decoder (FiD) reader models [8] to complete a retriever–reader pipeline.

The main contribution of this work is the replication of DPR and FiD training on our Wikipedia corpus variants. We are able to confirm many of the findings based on the original WikiText(100w) corpus, thereby adding veracity to previous results. Furthermore, we are able to increase the effectiveness of end-to-end QA using our corpus variants, highlighting the importance of corpus curation and providing a foundation for further advances.

We make our corpora and trained retriever and reader models available in HuggingFace.[1] We make our pre-processing code and retrieval code available in the Pyserini toolkit.[2] Our answer extraction code and end-to-end evaluation instructions are provided through the PyGaggle toolkit.[3]

2   Corpus Preparation

2.1   Sliding Window Segmentation

To split Wikipedia articles into passages that we feed to the retriever and reader models, we considered the use of sliding-window segmentation on a sentence level instead of splitting articles into disjoint passages of 100 words, as in the original DPR paper. Sentence-based segmentation has seen success in prior retrieval work that deals with long documents [15,17,18], and has the advantage of generating passages with natural discourse segments.

Given a Wikipedia article with any number of sentences, we employ passage size a and stride b, where $0 < b \leq a$. As a result of this segmentation, the first passage would contain the first a sentences, the second passage would start at sentence $b + 1$ and include the following a sentences, the third passage would start at sentence $2b + 1$ and contain the following a sentences, and so on. In our experiments, we considered two (a, b) configurations, $(6, 3)$ and $(8, 4)$, primarily to keep the passage lengths comparable to the WikiText(100w) corpus. We also prepended each passage with the title of the Wikipedia article to provide some global context.

2.2   Parsing Semi-structured Data

Tables. Our goal here was to convert tables from a semi-structured format into a more linear textual form amenable to downstream readers. We also wanted the linearization to be compatible with the segmentation from Sect. 2.1 since they were likely to be interwoven with "normal" text. Hence, rows from tables were converted into a sentence-like form with the help of table headers. We began with the corresponding column header followed by a colon, then the cell's textual content followed by a comma, and then this process was repeated for each cell in the row. After each row's final cell, we ended with a period. Therefore, each row formed a "sentence". Standard segmentation choices discussed in Sect. 2.1 were applied to these linearized tables also.

It is worth noting that not every table in Wikipedia has a "standard format", where the first row holds column headers and the other rows hold cell values

---

[1] https://huggingface.co/castorini/.

[2] http://pyserini.io/.

[3] http://pygaggle.ai/.

corresponding to those column headers. Some tables, which are not representative of every exception, do not have column headers and instead only have row headers. Nonetheless, our heuristic of the first row being column headers and the remaining rows having values corresponding to the column headers was adequate to linearize most tables successfully.

Name	Relationship	Discipline	Known for	Notes
Gordon Agnew	Professor	Cryptography	Professor	[142]
George Alfred Barnard	Lecturer	Mathematics	Statistics and quality control	[143]
Walter Benz	Professor	Mathematics	Geometer	[144]

```
Name: Gordon | Agnew , Relationship: Professor, Discipline: Cryptography, Known for:
Professor , Notes: , .
Name: George Alfred | Barnard , Relationship: Lecturer, Discipline: Mathematics,
Known for: Statistics and quality control , Notes: , .
Name: Walter | Benz , Relationship: Professor, Discipline: Mathematics, Known for:
Geometer, Notes: , .
```

Fig. 1. Example of a portion of a Wikipedia table (above) and our linearization into sentences (below). Note that the Wikipedia markup does not always align with the visual rendering.

Figure 1 shows an example of how we represent tables as sentences. Given that we include the article title, the table is represented in an easy-to-consume form for both retriever and reader models to reason over, without diverging too much from the standard textual nature of "normal" passages.

Infoboxes. In Wikipedia, infoboxes are table-like content elements that usually appear at the top-right of certain articles summarizing key information about them. The XML dump presents them as labels and their corresponding data. To convert these infoboxes to passages in our corpus, much like tables, we phrased them as labels, followed by a colon, and then the corresponding data. All linearized label–data pairs were then terminated with a period. Standard segmentation choices discussed in Sect. 2.1 were also applied to infoboxes. Figure 2 shows an example; we can see that the textual form captures all the key information from the infobox.

Lists. For Wikipedia lists, we treated each list element as a separate sentence. Standard segmentation choices discussed above were also applied. Figure 3 provides an example. Given that we include the title in every passage, models can easily reason over the provided list information.

2.3   Corpus Variants and Statistics

The culmination of all these pre-processing steps and design choices was a set of Wikipedia corpus variants. While we are aware of the many combinations that can be studied, we limited ourselves to five variants that we believed were particularly worth exploring in detail, as follows:

Johanna Drucker

Johanna Drucker at HyperStudio's Visual
Interpretations Conference @ MIT, 2010

Born	30 May 1952 (age 70) Philadelphia, Pennsylvania, U.S.
Nationality	American
Education	California College of Arts and Crafts, University of California, Berkeley
Known for	artists' books, typography, visual poetry, letterpress, digital humanities

```
birth date: Birth date and age | 1952 | 5 | 30 .
birth place: Philadelphia, Pennsylvania , U.S..
nationality: United States | American .
field: artists' books , typography , visual poetry , letterpress , digital humanities .
training: California College of Arts and Crafts , University of California, Berkeley.
```

Fig. 2. Example of a portion of a Wikipedia infobox (above) and our linearization into sentences (below).

- *cairn*, rhymes with *bairn*, a Northern English and Scottish word meaning child
- *chaos* /ˈ-eɪ.ɒs/, rhymes with *naos*, the inner chamber of a temple
- *circle* /ˈ-ɜːrkəl/, rhymes with *hurkle,* to pull in all one's limbs

```
"cairn", rhymes with "bairn", a Northern English and Scottish word meaning child.
"chaos" , rhymes with "naos", the inner chamber of a temple.
"circle" , rhymes with "hurkle," to pull in all one's limbs.
```

Fig. 3. Example of a portion of a Wikipedia list (above) and our linearization into sentences (below).

Table 1. Statistics of different Wikipedia corpus variants used in this study.

Corpus	# Articles	# Passages	Avg. Len (words)
(1) WikiText(100w)	3.2 M	21.0 M	100
(2) WikiText(100w)*	5.6 M	24.0 M	95
(3) WikiText(6, 3)	5.6 M	34.0 M	109
(4) WikiAll(6, 3)	6.1 M	76.7 M	80
(5) WikiText(8, 4)	5.6 M	25.6 M	141
(6) WikiAll(8, 4)	6.1 M	57.1 M	106

- WikiText(6, 3) is a corpus without tables, infoboxes, and lists with a passage size of 6 sentences and a stride of 3 sentences.
- WikiText(8, 4) is similar to above, but with a passage size of 8 sentences and a stride of 4 sentences.
- WikiAll(6, 3) and WikiAll(8, 4) correspond to the two previous corpora, but with the inclusion of tables, infoboxes, and lists.
- The final corpus, WikiText(100w)*, was our attempt to replicate the WikiText(100w) corpus. Matching WikiText(100w), We used spaCy's `en_core_web_lg` tokenizer to count 100 words ignoring whitespace and punctuation tokens. Additionally, matching WikiText(100w), we augmented the passages at the end of articles that contained fewer than 100 words by looping back to the beginning of the article.

We performed a statistical analysis of our five different corpus variants in comparison to the original corpus; results are shown in Table 1. Comparing rows (1) and (2), we see that our replication of the WikiText(100w) corpus, referred to as WikiText(100w)*, has many more articles, a larger passage count, but the passages are (slightly) shorter than the original. This discrepancy appears to be due to articles with fewer than 100 words not being included in WikiText(100w). We argue that this choice was arbitrary as many articles were left out, and we instead elected to include these articles. Otherwise, we attempted to make WikiText(100w)* as close as possible to WikiText(100w).

Another noteworthy observation is that the corpus variants with tables, infoboxes, and lists, rows (4) and (6), have many more articles, but fewer average words per passage compared to their counterparts without the semi-structured data, rows (3) and (5). This was due to the added sentences being shorter in length, on average, compared to those from the bodies of the articles in the text-only variants.

As expected, the corpora with a passage size of 8 sentences, rows (5) and (6), have longer passages than their counterparts with a passage size of 6 sentences, rows (3) and (4). Naturally, the variants with longer passages have fewer passages overall.

3   Experimental Design

Our experiments adopted the standard retriever–reader architecture for open-domain question answering. We evaluated retrieval effectiveness on the same five datasets as Karpukhin et al. [10]: the open-domain version of Natural Questions (NQ) [11,12], the open-domain version of TriviaQA [9], WebQuestions [2], CuratedTREC [1], and SQuAD [20]. For end-to-end effectiveness, we followed Izacard and Grave [8] and only performed evaluations on the Natural Questions and the TriviaQA datasets.

3.1   Retriever Model

For each corpus variant, DPR models were trained on Google's TPU v3–8 using the Tevatron toolkit [6], with the same hyperparameters as those from Karpukhin et al. [10]. We fine-tuned the uncased variant of BERT-base with a batch size of 128 for 40 epochs, using a dropout rate of 10%, the ADAM optimizer with a learning rate of 10^{-5}, and linear scheduling with warm-up steps. We limited the question and passage lengths to 32 and 256 tokens, respectively, using the WordPiece tokenizer of BERT [5]. We considered two settings in terms of labeled data: first, only NQ; second, a multi-dataset approach that combined training data from NQ, TriviaQA, WebQuestions, and CuratedTREC.

One challenge we faced was the selection of positive and negative passages for fine-tuning the retrieval models. The original DPR paper implemented two methods: For NQ and SQuAD, each question has a corresponding span of text in Wikipedia from which the original annotators identified the answer. This span of text maps to a passage in WikiText(100w). These passages are referred to as "gold passages" and form the positive examples for each question to train the DPR model. Karpukhin et al. [10] prepared negative passages by first using BM25 for retrieval and then filtering out the positives. The authors considered these "hard negatives" because although the passages rank highly in terms of BM25 scores, they do not contain answers to the questions.

Instead of using the gold passages as the positive examples, positive passages can be selected using distant supervision. For the TREC, WebQuestions, and TriviaQA datasets, since there are no gold passages from Wikipedia, the highest-ranked passage from BM25 that contains the answer (where the search query is the concatenation of the question and the answer) is used as the single positive passage for training. Karpukhin et al. [10] prepared negative passages the same way as when gold passages are used.

To avoid matching answer spans with gold passages for each corpus, we adopted a distant supervision approach when preparing the training set for each QA dataset. In our case, the Pyserini IR toolkit [13] was used to perform BM25 retrieval using the parameters $k_1 = 0.9$ and $b = 0.4$.

After training a DPR model for each corpus, we then performed a second iteration of DPR training, inspired by Xiong et al. [21] and Oğuz et al. [16]. Again, with distant supervision, but instead of using BM25, we leveraged the already-trained DPR model to retrieve 100 passages for each query, and those that did not contain the answer were selected as the negative passages. The top-ranked passage that contained the answer was selected as the single positive passage. Using these positive and negative passages, we trained a new DPR model from scratch to obtain the second iteration DPR model.

In all these experiments, we performed dense retrieval with the Tevatron toolkit [6] and evaluated retrieval effectiveness in Pyserini [13] using the top-k accuracy metric.

3.2   Reader Model

For each corpus, we trained a FiD reader model [8] with T5-Large [19] using code made available by the authors. Hyperparameters were chosen to be consistent with Izacard and Grave [8] when computationally possible. Specifically, we used a batch size of 64, a dropout rate of 10%, and an ADAMW optimizer with a peak learning rate of 5×10^{-5}. The length of the concatenation of the question and the passage was limited to 250 tokens using the SentencePiece tokenizer of T5 [19]. Given hardware restrictions, we trained the models using 4× A100 40GB GPUs with gradient accumulation to achieve a batch size of 64. Model training was performed in mixed precision using the `bfloat16` datatype for computational efficiency. We trained the model with 10K gradient steps, which we found to be adequate, and selected the best model based on the exact match score on the validation set.

A FiD reader model was trained separately on the Natural Questions dataset and the TriviaQA dataset. On NQ, the target answer was sampled randomly from the list of answers. However, for training on TriviaQA, the unique human-generated answer was used as the target (after normalization by capitalizing the first letter of every word and leaving everything else as lowercase).

4   Results

4.1   Retrieval Results

We begin with a focus on the Natural Questions (NQ) dataset. Table 2 explores the retrieval effectiveness of DPR models trained on NQ with the different corpora discussed. Broadly, they are divided into DPR leveraging gold passages, DPR leveraging distant supervision, and DPR leveraging both distant supervision and a second round of fine-tuning.

Karpukhin et al. [10] stated that choosing positive passages using distant supervision instead of using the gold passages results in only a minor decrease in top-k accuracy for retrieval; see rows (1a) and (2f) in Table 2. We observe in row (2e) that training a DPR model on our WikiText(100w)* corpus using distantly supervised passages achieves similar top-100 effectiveness but lower top-20 effectiveness compared to WikiText(100w), row (2f). Nonetheless, using distantly supervised passages is promising and we explored other variants with this setting.

Table 2. Retrieval effectiveness comparing the use of distant supervision vs. gold passages for selecting training data, choice of segmentation, and inclusion of semi-structured data on the Natural Questions dataset.

Method	NQ-Dev		NQ-Test	
	top20	top100	top20	top100
DPR 1st round—Gold Passages				
(1a) WikiText(100w) [10]	78.1	85.0	78.4	85.4
DPR 1st round—Distantly Supervised Passages				
(2a) WikiText(6, 3)	75.9	85.0	77.0	86.5
(2b) WikiAll(6, 3)	78.1	88.2	79.3	89.1
(2c) WikiText(8, 4)	75.8	85.1	77.6	86.9
(2d) WikiAll(8, 4)	79.7	88.7	81.1	90.0
(2e) WikiText(100w)*	75.1	84.5	76.6	85.4
(2f) WikiText(100w) [10]	77.1	84.4		
DPR 2nd round—Distantly Supervised Passages				
(3a) WikiText(6, 3)	80.8	87.3	81.6	88.5
(3b) WikiAll(6, 3)	84.7	90.9	85.2	92.3
(3c) WikiText(8, 4)	81.0	87.3	82.5	89.2
(3d) WikiAll(8, 4)	85.2	91.2	86.4	92.4
(3e) WikiText(100w)*	79.6	87.1	81.2	87.8

When we trained a DPR model for the second iteration with positive and negative passages chosen using the first iteration DPR model, we observe that top-k retrieval accuracy results were higher on our WikiText(100w)* corpus than the results from both the gold and the distant supervision settings from Karpukhin et al. [10], row 3(e) vs. rows (1a) and (2f). The reason could be that better hard negatives from the first iteration DPR model (compared to BM25 hard negatives) produce a more effective final model.

Turning our attention to all the open-domain QA datasets in our study, Table 3 shows retrieval effectiveness using a DPR model trained on the amalgamation of the NQ, TriviaQA, WQ, and CuratedTREC datasets. The table begins with results across the corpus variants from BM25, standard DPR (DPR$_1$), and the hybrid of BM25 and DPR$_1$ retrieval, rows 1(a)–3(e). In the hybrid conditions, the passages are retrieved using reciprocal rank fusion (RRF) between the rankings from the DPR model and BM25. Then we have results from the two-round refined DPR (DPR$_2$) and the hybrid of BM25 and DPR$_2$ retrieval, rows 4(a)–5(e).

Except for the SQuAD dataset, the DPR models achieve higher top-k accuracy scores than BM25 retrieval, shown in rows 1(a)–(f) vs. rows 2(a)–(f) and 4(a)–(e). This observation is consistent with results from previous work [10,14]. Across the board, the second iteration DPR model seems to outperform the first iteration DPR model, shown in rows 2(a)–(e) vs. rows 4(a)–(e) of this table and rows 2(a)–(e) vs. rows 3(a)–(e) in Table 2. This finding confirms the importance of selecting negatives when fine-tuning DPR (and related) models.

Table 3. Retrieval effectiveness comparing the different corpora across the NQ, TriviaQA, WQ, CuratedTREC, and SQuAD datasets. We trained the DPR models on a combination of the NQ, TriviaQA, WQ, and CuratedTREC datasets.

Method	NQ		TriviaQA		WQ		Curated		SQuAD	
	top20	top100	top20	top100	top20	top100	top20	top100	top20	top100
BM25										
(1a) WikiText(6, 3)	64.3	78.9	77.5	84.2	62.8	76.5	81.4	91.2	74.1	84.4
(1b) WikiAll(6, 3)	66.7	81.7	78.3	84.8	64.0	78.7	80.6	91.4	72.7	83.3
(1c) WikiText(8, 4)	66.7	79.6	78.6	84.7	65.2	78.2	82.7	92.2	74.8	85.0
(1d) WikiAll(8, 4)	69.6	82.9	79.5	85.5	66.1	80.2	82.7	92.1	73.5	84.1
(1e) WikiText(100w)*	63.8	78.0	76.3	83.4	61.2	75.4	80.3	90.8	70.4	81.4
(1f) WikiText(100w) [14]	62.9	78.3	76.4	83.2	62.4	75.5	80.7	89.9	71.1	81.8
DPR_1										
(2a) WikiText(6, 3)	76.4	85.8	77.9	85.0	70.8	81.2	85.7	92.9	53.6	69.6
(2b) WikiAll(6, 3)	78.7	89.0	78.3	85.6	73.1	82.0	88.5	94.4	53.2	68.5
(2c) WikiText(8, 4)	76.9	86.3	78.5	85.4	71.8	81.3	88.0	94.7	55.1	70.7
(2d) WikiAll(8, 4)	80.4	89.8	79.0	85.7	74.1	83.1	87.9	94.0	54.4	70.2
(2e) WikiText(100w)*	75.4	85.1	78.0	84.8	70.1	80.9	88.2	94.1	50.5	67.2
(2f) WikiText(100w) [14]	79.4	87.0	78.5	84.5	75.3	83.0	88.2	94.4	58.3	72.4
$RRF(DPR_1, BM25)$										
(3a) WikiText(6, 3)	79.6	87.9	82.9	87.0	76.1	83.9	89.5	94.5	76.8	85.8
(3b) WikiAll(6, 3)	83.3	91.3	83.8	87.8	77.1	85.8	92.2	95.2	75.6	85.1
(3c) WikiText(8, 4)	81.2	88.3	83.5	87.6	75.6	84.5	91.1	95.7	77.1	86.0
(3d) WikiAll(8, 4)	84.7	91.9	84.2	88.1	78.7	85.8	91.5	95.1	76.0	85.8
(3e) WikiText(100w)*	79.6	87.8	82.4	87.0	73.8	83.5	90.1	94.7	73.3	83.1
DPR_2										
(4a) WikiText(6, 3)	81.1	88.3	81.2	86.2	76.5	84.2	90.3	94.2	60.6	75.3
(4b) WikiAll(6, 3)	85.5	91.8	81.9	87.0	80.1	87.4	91.9	95.8	60.8	75.0
(4c) WikiText(8, 4)	82.2	88.6	81.1	86.4	77.8	85.3	91.6	95.2	60.3	74.5
(4d) WikiAll(8, 4)	85.9	92.7	82.4	87.4	80.2	86.8	90.6	95.4	61.1	74.8
(4e) WikiText(100w)*	80.0	88.1	80.2	85.8	75.0	84.0	91.3	95.1	57.7	72.7
$RRF(DPR_2, BM25)$										
(5a) WikiText(6, 3)	81.6	89.2	83.3	87.4	77.4	84.7	91.6	95.4	77.9	86.5
(5b) WikiAll(6, 3)	85.3	93.0	84.2	88.0	80.3	87.5	91.5	96.4	76.9	85.9
(5c) WikiText(8, 4)	82.6	89.2	84.0	87.7	77.9	85.2	91.9	95.1	78.1	86.7
(5d) WikiAll(8, 4)	86.0	93.2	84.6	88.5	80.5	87.5	91.5	95.8	77.5	86.4
(5e) WikiText(100w)*	81.2	88.9	82.5	87.1	76.0	84.4	91.2	94.7	74.4	84.1

Also, much like in Ma et al. [14], retrieval accuracy seems to benefit from hybrid retrieval. Reciprocal rank fusion [4] between DPR ranked lists and BM25 ranked lists tend to achieve higher top-k accuracy scores than either ranking method alone, shown in rows 1(a)–(e), 2(a)–(e) vs. rows 3(a)–(e) and rows 1(a)–(e), 4(a)–(e) vs. rows 5(a)–(e).

Between the corpus variants we considered, retrieval on those with the addition of tables, infoboxes, and lists, WikiAll(6, 3) and WikiAll(8, 4), generally results in higher retrieval accuracy, shown in rows (∗a) vs. (∗b) and rows (∗c) vs. (∗d), in both Tables 2 and 3. This observation is consistent in all except for the CuratedTREC and the SQuAD datasets. Nonetheless, including these semi-structured data provides value.

Furthermore, retrieval accuracy is usually higher with WikiAll(8, 4) compared to WikiAll(6, 3), perhaps because of the longer passages, shown in rows (∗d) vs.

Table 4. End-to-end QA effectiveness in terms of the exact match score. In the "DPR Ranking" condition, the passages are retrieved using the second iteration DPR model. In the "Hybrid Retrieval" condition, retrieval applied Reciprocal Rank Fusion between the rankings from the second iteration DPR model and BM25 retrieval.

Condition	NQ		TriviaQA	
	Dev	Test	Dev	Test
DPR ranking				
(1a) WikiText(6, 3)	51.0	52.4	70.6	70.7
(1b) WikiText(8, 4)	51.3	53.2	70.2	70.3
(1c) WikiAll(6, 3)	**54.3**	54.8	**71.8**	71.8
(1d) WikiAll(8, 4)	54.0	**55.0**	71.1	**72.3**
(1e) WikiText(100w)*	50.4	51.1	70.4	70.4
Hybrid ranking				
(2a) WikiText(6, 3)		53.0		72.9
(2b) WikiText(8, 4)		53.3		72.5
(2c) WikiAll(6, 3)		**55.8**		**73.7**
(2d) WikiAll(8, 4)		55.7		73.1
(2e) WikiText(100w)*		51.4		72.5

(*b). Similarly, retrieval accuracy is usually higher with WikiText(8, 4) over Wiki-Text(6, 3), again perhaps because of the difference in average passage lengths, shown in rows (*c) vs. (*a). Retrieval accuracy tends to be the lowest on the WikiText(100w)* corpus, shown in rows (∗a, ∗b, ∗c, ∗d) vs. (∗e). These observations are generally consistent in all datasets except for CuratedTREC, where there are some exceptions.

4.2   Reader Results

Arguably, the more important measure to compare the different corpus variants is end-to-end question answering effectiveness. Retrieval evaluation is insufficient, because scores may be higher simply due to some corpora having longer passages. We evaluated end-to-end QA effectiveness in the PyGaggle toolkit using the exact match score.

Of the different corpora studied, we observe in Table 4 that exact match scores are highest in those with the inclusion of semi-structured data, seen in rows (1c), (2c), (1d), (2d). Comparing rows (∗a), (∗b), and (∗e), we see that using the text-only corpora, WikiText(6, 3) or WikiText(8, 4), results in an improvement in terms of exact match scores over the WikiText(100w)* corpus for the Natural Questions dataset, suggesting a potential superiority in segmenting articles as discussed in Sect. 2.1. Nonetheless, results for the different corpus variants are much closer on TriviaQA. It is crucial to note that answers for questions in the Natural Questions dataset were sourced from Wikipedia, whereas the same does not hold for TriviaQA, and such collection biases could be the reason for this inconsistent finding.

Table 5. Comparison to previous work for end-to-end QA effectiveness. Original results reported for FiD [8], DKRR [7], and UniK-QA [16].

Condition	NQ-test	TriviaQA-test
(1) FiD (Original) [8]	51.4	67.6
(2) DKRR [7]	54.4	72.5
(3) UniK-QA [16]	54.1	65.1
(4) WikiAll(6, 3)	**55.8**	**73.7**

Comparing the passage size and stride configurations of $(6, 3)$ vs $(8, 4)$, seen in rows (*a) vs (*b) and (*c) vs (*d), we see that results are often similar with the configurations taking turns leading.

Retrieving passages using hybrid retrieval leads to improved exact match scores, seen in rows 1(a)–(e) vs. 2(a)–(e). That is, hybrid retrieval improves not just retrieval scores but end-to-end QA effectiveness as well.

Our best experimental setting, which uses hybrid retrieval on the WikiAll(6, 3) corpus followed by a FiD-large reader model [8] to extract the answer, achieves improved effectiveness over comparable prior work also based on the FiD-large reader model. Table 5 presents these comparisons. Our best experimental setting, displayed in row (4), shows an improvement of over one point on both Natural Questions and TriviaQA.

5 Conclusion

Our paper replicates and builds on a line of work that uses a retriever–reader pipeline to process passages from the WikiText(100w) corpus to answer open-domain questions. We more thoroughly examined various techniques and strengthened findings from previous work.

We showed that not only does including tables, infoboxes, and lists in the Wikipedia corpus improve retrieval effectiveness for the Natural Questions dataset, a finding reported in Oğuz et al. [16], but that the addition clearly improves scores for also the TriviaQA and the WQ datasets. We also demonstrated that including linearized semi-structured data improves end-to-end effectiveness for both of the datasets considered: Natural Questions and TriviaQA. Splitting Wikipedia into passages in a fashion where we preserve complete sentences and allow for overlapping passages also benefits both retrieval and end-to-end effectiveness compared to the previous segmentation approach.

In terms of modeling, we find that training a DPR model for an additional iteration improves retrieval accuracy for QA, confirming Oğuz et al. [16]. Finally, following Ma et al. [14], we replicated the finding that hybrid retrieval offers better effectiveness across these new corpora. As a result of all this careful study, we have achieved the best scores that we know of for this class of models that leverages a T5-large variant of a Fusion-in-Decoder reader model to answer questions from the Natural Questions and TriviaQA datasets. Together, these improvements demonstrate the importance of careful corpus preparation and thoroughly re-examining previous work to make sure that all design variants have been explored.

Acknowledgements. This research was supported in part by the Natural Sciences and Engineering Research Council (NSERC) of Canada. Computational resources were provided in part by Compute Ontario and Compute Canada. In addition, thanks to Google Cloud and the TPU Research Cloud Program for credits to support some of our experimental runs.

References

1. Baudiš, P., Šedivý, J.: Modeling of the question answering task in the YodaQA system. In: Mothe, J., et al. (eds.) CLEF 2015. LNCS, vol. 9283, pp. 222–228. Springer, Cham (2015). https://doi.org/10.1007/978-3-319-24027-5_20
2. Berant, J., Chou, A., Frostig, R., Liang, P.: Semantic parsing on Freebase from question-answer pairs. In: Proceedings of the 2013 Conference on Empirical Methods in Natural Language Processing, Seattle, Washington, USA, pp. 1533–1544, October 2013
3. Chen, D., Fisch, A., Weston, J., Bordes, A.: Reading Wikipedia to answer open-domain questions. In: Proceedings of the 55th Annual Meeting of the Association for Computational Linguistics (Volume 1: Long Papers), Vancouver, Canada, pp. 1870–1879, July 2017
4. Cormack, G.V., Clarke, C.L.A., Buettcher, S.: Reciprocal rank fusion outperforms condorcet and individual rank learning methods. In: Proceedings of the 32nd International ACM SIGIR Conference on Research and Development in Information Retrieval, SIGIR 2009, pp. 758–759. Association for Computing Machinery, New York (2009)
5. Devlin, J., Chang, M.W., Lee, K., Toutanova, K.: BERT: pre-training of deep bidirectional transformers for language understanding. In: Proceedings of the 2019 Conference of the North American Chapter of the Association for Computational Linguistics: Human Language Technologies, Volume 1 (Long and Short Papers), pp. 4171–4186 (2019)
6. Gao, L., Ma, X., Lin, J., Callan, J.: Tevatron: an efficient and flexible toolkit for dense retrieval. arXiv:2203.05765 (2022)
7. Izacard, G., Grave, E.: Distilling knowledge from reader to retriever for question answering. In: International Conference on Learning Representations (2021)
8. Izacard, G., Grave, E.: Leveraging passage retrieval with generative models for open domain question answering. In: Proceedings of the 16th Conference of the European Chapter of the Association for Computational Linguistics: Main Volume, pp. 874–880. Online, April 2021
9. Joshi, M., Choi, E., Weld, D., Zettlemoyer, L.: TriviaQA: a large scale distantly supervised challenge dataset for reading comprehension. In: Proceedings of the 55th Annual Meeting of the Association for Computational Linguistics (Volume 1: Long Papers), Vancouver, Canada, pp. 1601–1611, July 2017
10. Karpukhin, V., et al.: Dense passage retrieval for open-domain question answering. In: Proceedings of the 2020 Conference on Empirical Methods in Natural Language Processing (EMNLP), pp. 6769–6781, November 2020. Online
11. Kwiatkowski, T., et al.: Natural questions: a benchmark for question answering research. Trans. Assoc. Comput. Linguist. **7**, 452–466 (2019)

12. Lee, K., Chang, M.W., Toutanova, K.: Latent retrieval for weakly supervised open domain question answering. In: Proceedings of the 57th Annual Meeting of the Association for Computational Linguistics, Florence, Italy, pp. 6086–6096, July 2019

13. Lin, J., Ma, X., Lin, S.C., Yang, J.H., Pradeep, R., Nogueira, R.: Pyserini: a Python toolkit for reproducible information retrieval research with sparse and dense representations. In: Proceedings of the 44th International ACM SIGIR Conference on Research and Development in Information Retrieval, SIGIR 2021, pp. 2356–2362. Association for Computing Machinery, New York (2021)

14. Ma, X., Sun, K., Pradeep, R., Li, M., Lin, J.: Another look at DPR: reproduction of training and replication of retrieval. In: Hagen, M., et al. (eds.) ECIR 2022. LNCS, vol. 13185, pp. 613–626. Springer, Cham (2022). https://doi.org/10.1007/978-3-030-99736-6_41

15. Nogueira, R., Jiang, Z., Pradeep, R., Lin, J.: Document ranking with a pretrained sequence-to-sequence model. In: Findings of the Association for Computational Linguistics: EMNLP 2020, pp. 708–718. Online, November 2020

16. Oguz, B., et al.: UniK-QA: unified representations of structured and unstructured knowledge for open-domain question answering. In: Findings of the Association for Computational Linguistics: NAACL 2022, Seattle, United States, pp. 1535–1546, July 2022

17. Pradeep, R., Li, Y., Wang, Y., Lin, J.: Neural query synthesis and domain-specific ranking templates for multi-stage clinical trial matching. In: Proceedings of the 45th International ACM SIGIR Conference on Research and Development in Information Retrieval, SIGIR 2022, pp. 2325–2330. Association for Computing Machinery, New York (2022)

18. Pradeep, R., Nogueira, R., Lin, J.J.: The expando-mono-duo design pattern for text ranking with pretrained sequence-to-sequence models. arXiv:2101.05667 (2021)

19. Raffel, C., et al.: Exploring the limits of transfer learning with a unified text-to-text transformer. J. Mach. Learn. Res. **21**(140), 1–67 (2020)

20. Rajpurkar, P., Zhang, J., Lopyrev, K., Liang, P.: SQuAD: 100,000+ questions for machine comprehension of text. In: Proceedings of the 2016 Conference on Empirical Methods in Natural Language Processing, Austin, Texas, pp. 2383–2392, November 2016

21. Xiong, L., et al.: Approximate nearest neighbor negative contrastive learning for dense text retrieval. In: Proceedings of the 9th International Conference on Learning Representations (ICLR 2021) (2021)

From Baseline to Top Performer: A Reproducibility Study of Approaches at the TREC 2021 Conversational Assistance Track

Weronika Lajewska$^{(\boxtimes)}$ (iD) and Krisztian Balog (iD)

University of Stavanger, Stavanger, Norway
{weronika.lajewska,krisztian.balog}@uis.no

Abstract. This paper reports on an effort of reproducing the organizers' baseline as well as the top performing participant submission at the 2021 edition of the TREC Conversational Assistance track. TREC systems are commonly regarded as reference points for effectiveness comparison. Yet, the papers accompanying them have less strict requirements than peer-reviewed publications, which can make reproducibility challenging. Our results indicate that key practical information is indeed missing. While the results can be reproduced within a 19% relative margin with respect to the main evaluation measure, the relative difference between the baseline and the top performing approach shrinks from the reported 18% to 5%. Additionally, we report on a new set of experiments aimed at understanding the impact of various pipeline components. We show that end-to-end system performance can indeed benefit from advanced retrieval techniques in either stage of a two-stage retrieval pipeline. We also measure the impact of the dataset used for fine-tuning the query rewriter and find that employing different query rewriting methods in different stages of the retrieval pipeline might be beneficial. Moreover, these results are shown to generalize across the 2020 and 2021 editions of the track. We conclude our study with a list of lessons learned and practical suggestions.

Keywords: Conversational search · Query rewriting · Dense retrieval · Passage re-ranking · TREC CAsT · Reproducibility

1 Introduction

The last few years have seen an acceleration of research on multi-turn, natural language, and long-term user modeling capabilities of search systems with an attempt to make them more conversational [33]. The Conversational Assistance Track (CAsT) at the Text Retrieval Conference (TREC) [7–9] has been a key enabler of progress in this area, by providing a reusable test collection for conversational search. The task at TREC CAsT is to identify relevant content from a collection of passages, "for conversational queries that evolve through a trajectory of a discussion on a topic" [9]. Over the years, query rewriting,

J. Kamps et al. (Eds.): ECIR 2023, LNCS 13982, pp. 177–191, 2023.
https://doi.org/10.1007/978-3-031-28241-6_12

passage retrieval, and passage reranking have emerged as the main components, which are combined in a pipeline architecture. Clearly, the ranking components can directly benefit from advances in dense/hybrid passage retrieval [17], and are indeed critical to overall system performance. However, what makes the task interesting from a conversational perspective, and different from passage retrieval, is the problem of query rewriting [14, 16, 18, 23, 24, 26, 31].

It has been shown that the best performing systems at TREC form a very competitive reference point for effectiveness comparison [2]. This means that, even if one's ultimate research interest lies in query rewriting, demonstrating strong absolute performance for conversational search requires a high degree of effectiveness from all system components. Our main objective in this paper is to reproduce (1) the best performing baseline method provided by the track organizers [9] and (2) the top performing (documented) system [28] from the latest (2021) edition of TREC CAsT. These two approaches are seen as representatives of a strong baseline and the state of the art, respectively. It is worth noting that the system description papers accompanying TREC submissions are not peer-reviewed and there is no explicit or implicit reproducibility requirement. This can make reproducibility particularly challenging and a study such as this particularly insightful.

Both selected systems follow a two-stage *retrieve-then-rerank* pipeline architecture with queries rewritten based on conversational context. Specifically, the baseline system [9] uses a T5-based query rewriting model fine-tuned on CANARD [11], first-pass retrieval based on BM25, and a pointwise T5 re-ranker. The top participating system [28] uses a different dataset for fine-tuning the query rewriting model (QReCC [1]) and employs more advanced ranking components: a combination of sparse-dense retrieval with pseudo relevance feedback for first-pass retrieval (ANCE/BM25/PRF), and pointwise/pairwise (mono/duoT5) re-ranking. We find that these complex multi-step architectures are challenging to reproduce due to the numerous components involved. Neither the baseline nor the top participating system can be fully reproduced due to key information missing about model choices, parameters, and various input preparation and collection preprocessing steps. With a best-effort attempt, the results we obtain are within 12% and 21% relative margins for the baseline and top performing systems, respectively, with regards to all evaluation measures reported in the track overview paper [9]. However, the relative differences between the two shrink from 18% in NDCG@3 and 37% in recall, according to the track overview, to 5% and 7%, respectively, according to our reproduced systems.

Since the two selected systems follow the same basic two-stage retrieval pipeline, we perform additional experiments in order to better understand how each pipeline component contributes to overall effectiveness. To shed light on the generalizability of findings, we report results on both the 2020 and 2021 editions of TREC CAsT. Since the query rewriter influences the effectiveness of both first-pass retrieval and re-ranking, we also perform experiments using a different retrieval pipeline, which can utilize different query rewriting methods for the two ranking stages. We find that final performance is indeed influenced by the position of the query rewriting component in the retrieval pipeline: T5 fine-tuned on CANARD gives better results than fine-tuning on QReCC in terms of first-pass retrieval (higher recall), whereas the best overall results (NDCG@3) are achieved by the system using QReCC for first-pass retrieval and CANARD for re-ranking. This suggests that employing different query rewriting methods for the different stages might be beneficial.

In summary, the main contributions of this paper are twofold. First, we attempt to reproduce two approaches from the latest edition of TREC CAsT, the organizers' baseline and the top performing submission, and report results and lessons learned. Second, we present additional experiments on two-stage retrieval pipelines and query rewriting models to provide insights into the potential contributions of various components. All resources developed within this study (i.e., source code, runfiles, evaluation results) are made publicly available.[1]

2   Related Work

We briefly introduce the TREC Conversational Assistance Track, discuss query rewriting approaches, and review ranking architectures used at TREC CAsT.

2.1   TREC Conversational Assistance Track

The Conversational Assistance Track at TREC has started in 2019 with the aim to facilitate research on conversational information seeking, by creating a large-scale reusable test collection [7]. The task is to identify relevant passages (in 2019 and 2020) or documents (in 2021) from a collection comprising MS MARCO [3], Wikipedia [20], TREC CAR [10] and the Washington Post v4.[2]

In TREC CAsT'19, user utterances may only refer to the information mentioned in previous user utterances. Since 2020 [8], utterances may refer to previous responses given by the system as well, which significantly extends the scope of contextual information that the system needs to use to understand a request. TREC CAsT'21 [9] is characterized by the increased dependence on previous system responses, as well as simple forms of user revealment, reformulation, and explicit feedback introduced in users' utterances.

By TREC CAsT'21, a two-step passage ranking architecture has emerged. A first-pass passage retrieval is usually performed using an unsupervised sparse model (e.g., BM25), which is followed by re-ranking using a neural model trained for passage retrieval (e.g., T5 trained on MS MARCO [6]). Additionally, most systems employ a query rewriting step, where the original query is de-contextualized to be independent of the previous turns.

2.2   Query Rewriting

The goal of query rewriting is to handle common conversational phenomena such as omission, coreference [7], zero anaphora, topic change, and topic return [26].

---

[1] https://github.com/iai-group/ecir2023-reproducibility.
[2] https://trec.nist.gov/data/wapost/.

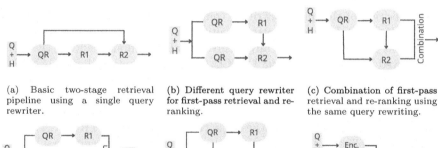

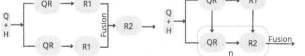

(a) Basic two-stage retrieval pipeline using a single query rewriter.

(b) Different query rewriter for first-pass retrieval and re-ranking.

(c) Combination of first-pass retrieval and re-ranking using the same query rewriting.

(d) **Re-ranking of fused first-pass results that use different query rewriters.**

(e) Fusion of multiple passage re-rankings using different rewrites.

(f) Few-shot conversational dense retrieval.

Fig. 1. Pipeline architectures for conversational search (**Q+H**: raw query and conversational history; **QR**: query rewriter; **R1**: first-pass retriever; **R2**: re-ranker; **Enc.**: encoder; **Docs**: document collection; **Dot prod.**: dot product).

Approaches can be broadly categorized into unsupervised, supervised feature-based, and (weakly-)supervised neural methods. Unsupervised query rewriting methods expand the original query with terms from the conversation history, for example, from previous utterances based on BM25 score [30], cosine similarity [25], or other frequency-based signals [16]. Supervised feature-based methods use linguistic features based on dependency parsing, coreference resolution, named entity resolution, or part-of-speech tagging [18]. Supervised neural query rewriting approaches utilize large pre-trained language models, and in particular generative models such as GPT-2 [24] or T5 model [4,13,28]. These models are fine-tuned on a conversational dataset, such as CANARD [4,13,16,23,24] or QReCC [28]. The generated query reformulations may further be expanded with terms from conversation history [24], with paraphrases [13], or related sentences from semantically related documents [4]. Weakly supervised neural query rewriting methods aim to fine-tune large pre-trained language models [31] or term selection classifiers [14] with weak supervision data that is created using rule-based or self-supervised approaches. The best results are reported using a combination of term-based query expansion with generative models for query reformulation [14,16,24].

2.3 Pipeline Architectures

Systems participating in TREC CAsT exhibit a wide variety of approaches, not only in terms of component-level choices but also in terms of the overall architectures of their ranking pipelines. The most common choice is a two-stage retrieval pipeline with a query rewriting module. Different variants of this cascade

Table 1. Overview of approaches reproduced in this paper.

	Query rewriting	First-pass retrieval	Re-ranking
BaselineOrganizers	T5 fine-tuned on CANARD	BM25	monoT5
WaterlooClarke	T5 fine-tuned on QReCC	BM25 with PRF + ANCE	mono/duoT5

architecture include systems with the same rewriting method used for both first-pass retrieval and re-ranking [4,13,18,22–24,28,31] (Fig. 1a), different query rewriting modules for both stages [29] (Fig. 1b), or using rewriting only for first-pass retrieval [12,29].

More advanced architectures may use a two-stage retrieval pipeline with the same query rewriter for each stage, but combine the scores obtained from retrieval and re-ranking to produce a final ranking [25] (Fig. 1c) or use two different versions of the rewritten query for first-pass retrieval and a fusion of the ranked lists for the re-ranking stage [16] (Fig. 1d). Another architecture variant consists of first-pass retrieval using the rewritten query, followed by a fusion of multiple contextualized passage re-ranking of several different rewrites [14] (Fig. 1e). An alternative to the retrieve-then-rerank approach is a few-shot conversational dense retrieval system that learns contextualized embeddings of utterances and documents in the collection, and scores documents solely using the dot product of the embeddings [32] (Fig. 1f).

3   Selected Approaches

We present the two approaches from TREC CAsT'21 that we aim to reproduce in this paper: (1) the best performing official baseline provided by the track organizers' and (2) the top performing system submitted by participants.[3] These approaches may be regarded as representatives of a strong baseline and of the state of the art, respectively. Both may be seen as instantiations of the basic two-stage retrieval pipeline approach (cf. Fig. 1a), with query rewriting, first-pass retrieval, and re-ranking components, as shown in Table 1. In this section, we focus on a high-level description of these approaches, based on the corresponding TREC papers; specific implementation details are discussed in Sect. 4.

3.1   Organizers' Baseline

Of the several baselines provided by the track organizers, org_auto_bm25_t5 was the best performing run [9]; this will be referred to as the **BaselineOrganizers** approach henceforth. The query rewriting component is using T5 fine-tuned on CANARD for generative query rewriting. The rewriter uses all previous queries and the three previous canonical responses as context. For first-pass retrieval, BM25 is used to collect the top 1000 documents from the collection. The documents are re-ranked with a pointwise (mono) T5 model trained on MS MARCO.

---

[3] More specifically, this is the best performing system that is accompanied by a system description and can thus be (attempted to be) reproduced.

3.2   Top Performer: WaterlooClarke

The top-performing documented system was the `clarke-cc` run by Yan et al.
[28]; this will be referred to as the **WaterlooClarke** approach henceforth. The
query rewriting component is based on a T5 model that is fine-tuned on the
QReCC dataset [1]. For context, the rewriter uses previously rewritten utterances
and the last canonical result. First-pass retrieval comprises two sub-components:
a sparse and a dense retriever. The sparse retriever utilizes a BM25 with pseudo-
relevance feedback (PRF), with the parameters tuned to maximize recall. PRF
is run over both the target corpus and the C4 corpus.[4] The dense retriever is
based on the ANCE approach [27]. Both retrieval systems return the top 1000
documents that are merged into one final ranking. Re-ranking is performed using
a pointwise T5 re-ranker, followed by another re-ranking of the top 50 documents,
using pairwise duoT5 [21].

4   Reproducibility Experiments

In this section, we answer our first research question: Can the organizers' baseline
and the best performing system at the TREC CAsT'21 be reproduced? We
describe the implementation details of the two systems and discuss their end-to-
end performance with respect to the results reported in the track overview [9].

4.1   Baseline Implementation

We base the implementation solely on the description of the track organizers'
`org_auto_bm25_t5` baseline in the overview paper [9], without resorting to additional communication with the authors.

The passage collection is indexed using Elasticsearch, using the built-in analyzer for tokenization, stopwords removal, and KStem stemming. For query rewriting, we use a pre-existing T5 model that has been fine-tuned on the CANARD
dataset (`castorini/t5-base-canard`).[5] Our implementation is based on the
Hugging Face transformers library.[6] According to [9], the context for the query
rewriter is of the form: $q_1, q_2, \ldots, q_{i-3}, r_{i-3}, q_{i-2}, r_{i-2}, q_{i-1}, r_{i-1}, q_i$, where q_i and
r_i are the ith raw query and canonical response, respectively. Contexts exceed-
ing the allowed model input length are not handled. This, however, can result
in trimming the input in a way that the raw query that is to be rewritten
is removed. To increase the quality of the rewriting by ensuring the correct
form of the input and benefiting from previous rewrites, we alternatively use:
$\hat{q}_1, \hat{q}_2, \ldots, \hat{q}_{i-1}, trim(r_{i-1}), q_i$, where $\hat{q}_i$ is the ith rewritten query and $trim$ is a
function that cuts the canonical response if the length of the input is longer than
the capacity of the model. For first-pass retrieval, the passages are ranked using
BM25 on a `catch_all` field (concatenating the `title` and `body` fields) in the 2021
index and on the `body` field for the 2020 index. We initially used the parameters

---

[4] https://huggingface.co/datasets/allenai/c4.

[5] https://huggingface.co/castorini/t5-base-canard.

[6] https://github.com/huggingface/transformers.

reported by the organizers (k1 = 4.46, b = 0.82), but then achieved better results with the default parameters (k1 = 1.2, b = 0.75). The top 1000 candidates for each turn are re-ranked using the T5 model introduced by Nogueira et al. [19], which has been published on Hugging Face (`castorini/monot5-base-msmarco`).[7]

4.2   WaterlooClarke Implementation

We base our implementation on the WaterlooClarke group's TREC paper [28]. Additional information on specific details was obtained from the authors via email communication and inferred from the implementation made available.[8]

The approach requires two indices: an approximate nearest neighbor (ANN) index for ANCE dense retrieval and an inverted index for BM25. The authors use ANCE's own implementation[9] and a publicly released model checkpoint (passage ANCE(FirstP)) for the ANN index.[10] We use Pyterrier's plugin[11] for creating the ANN index, which is based on the original paper, and allows for easier integration with other modules in our pipeline. For building the ANN index we use MS MARCO Passage and TREC CAR collections provided by the ir_datasets package,[12] and implement our own generator for the WaPo 2020, MS MARCO Documents, and KILT collections. No additional preprocessing is performed when building the dense retrieval index. The inverted index used by BM25 is the same as in Sect. 4.1.

The query reformulation step in WaterlooClarke is based on a T5 model trained on the QReCC dataset [1]. All the previous rewritten utterances and the canonical response for the last utterance are used as context to reformulate the current question (i.e., the input is given as: $\hat{q}_1, \hat{q}_2, \ldots, \hat{q}_{i-1}, trim(r_{i-1}), q_i$). If the length of the input sentence exceeds 512, the answer passage is cut off. (See footnote 10) The authors fine-tune a pretrained `t5-base` model[13] with the training partition of the QReCC dataset for 3 epochs, using the original test partition as a validation set. (See footnote 10) The train batch size is equal to 2 and the learning rate is 5×10^{-5}. (See footnote 10) We use the Simple Transformers library[14] for the fine-tuning procedure (as opposed to PyTorch Lightning[15] and Hugging Face transformers used by the authors (See footnote 10)).

There are two first-pass rankers involved: (1) sparse retrieval using BM25 with pseudo relevance feedback (PRF) and (2) dense retrieval using ANCE [27]. The final sparse retrieval ranking is a fusion of two rankings. (See footnote 10) PRF is applied on the top 17 documents to expand the query with the top 26 terms; the expanded query is then scored using BM25 to generate the first sparse

---

[7] https://huggingface.co/castorini/monot5-base-msmarco.

[8] https://github.com/claclark/Cottontail/blob/main/apps/treccast21.cc.

[9] https://github.com/microsoft/ANCE.

[10] Missing information provided by the authors in personal communication.

[11] https://github.com/terrierteam/pyterrier_ance.

[12] https://github.com/allenai/ir_datasets.

[13] https://huggingface.co/t5-base.

[14] https://simpletransformers.ai/.

[15] https://www.pytorchlightning.ai/.

ranking. Additionally, the authors use the top 16 weighted answer candidates generated by a statistical question-answering method ran against the C4 corpus to create the second ranking (answer candidates are used by BM25). (See footnote 10) The first and the second ranking produced by the sparse retrieval are fused with Reciprocal Rank Fusion (RRF) [5]. (See footnote 10) There is no further information disclosed about the question-answering system used (neither in the paper nor in the GitHub repository). Therefore, we skip the second ranking in reproducibility and focus on standard BM25 with PRF. The BM25 parameters are tuned to maximize recall over manually rewritten questions from previous years. The exact details of this remain unclear. We tune BM25 parameters on our 2020 and 2021 indices and take the average of the best parameters found for each year (b = 0.45, k1 = 0.95), since the parameters used in their code (b = 0.45, k1 = 1.18) gave worse results on our indices. For query expansion, since the choice of PRF algorithm could not be resolved, we opted for RM3 [15], which we implemented from scratch.

The results of sparse and dense retrieval are fused to generate the final set of 1000 candidate passages for re-ranking. Since the fusion method is not stated in the paper, we assume that this step also employs RRF; we utilize the TrecTools library,[16] which implements a RRF as defined in [5].

The re-ranking stage in this approach is based on a pointwise monoT5 re-ranker (on all candidate passages), followed by a pairwise duoT5 re-ranker (on the top 50 passages re-ranked by monoT5). The original re-ranking implementation is based on the Pyaggle library[17] with the default model checkpoints. Our implementation of duoT5 is based on the Hugging Face transformers library and the `castorini/duot5-base-msmarco` model published on Hugging Face.[18]

4.3 Results

Table 2 reports our results on the CAsT'21 collection. Following the official setup, we consider measures with both binary and graded relevance. The main measure is NDCG@3; other measures are computed with a rank cutoff of 500. For binary measures, we apply a relevance threshold of 2.

For the baseline, the results reported in the overview paper [9] are included verbatim and regarded as the reference, since the raw runfile (`org_auto_bm25_t5`) is not available in the TREC archive. We include results using the original query rewriting method and reported BM25 parameters (BaselineOrganizers-QR-BM25), using the improved query rewriter while keeping the reported BM25 parameters (BaselineOrganizers-BM25), and finally using the improved query rewriter with default BM25 parameters (BaselineOrganizers). We find that the latest variant performs best; it is still 9% below the reference result in terms of NDCG@3, but 2% better in terms of Recall@500.

Regarding WaterlooClarke, the performance of our reproduced system is 19% lower in terms of NDCG@3 and 20% lower in terms of Recall@500 than the

---

[16] https://github.com/joaopalotti/trectools.

[17] https://github.com/castorini/pygaggle.

[18] https://huggingface.co/castorini/duo5-base-msmarco.

Table 2. Reproducibility experiments on the TREC CAsT'21 dataset.

Approach	R@500	MAP	MRR	NDCG	NDCG@3
BaselineOrganizers@TREC'21 (in [9])	0.636	0.291	0.607	0.504	0.436
BaselineOrganizers-QR-BM25	0.5632	0.2268	0.4947	0.4317	0.3457
BaselineOrganizers-BM25	0.5894	0.2546	0.5405	0.4672	0.3966
BaselineOrganizers	0.6472	0.2628	0.5354	0.4885	0.3968
WaterlooClarke@TREC'21 (in [9])	0.869	0.362	0.684	0.640	0.514
WaterlooClarke@TREC'21 (runfile)	0.8534	0.3494	0.6626	0.6240	0.4950
WaterlooClarke reproduced by us	0.6915	0.2864	0.5712	0.5176	0.4151

official results reported for this approach. The discrepancy in the results is most likely caused by the lack of the C4-based question-answering step performed in first-pass retrieval. This element of the system is not sufficiently described in the paper nor has been resolved via personal email communication. Surprisingly, we observe discrepancies between the official results reported in the overview paper and a direct evaluation of the `clarke-cc` runfile taken from the TREC archive (cf. rows 5 vs. 6 in Table 2). The latter results are lower, with a relative drop of almost 4% in NDCG@3, which is a non-negligible difference. We cannot explain this discrepancy; however, it also puts into question the results reported in the track overview. When comparing our reproduced results against their runfile, the relative differences are under 16% and 19% in terms of NDCG@3 and Recall@500, respectively.

Overall, according to the track overview paper, the relative differences between BaselineOrganizers and WaterlooClarke are 18% and 37% in terms of NDCG@3 and Recall@500, respectively (cf. rows 1 vs. 5 in Table 2). However, the respective differences in our reproduced approaches are 5% and 7% (cf. rows 4 vs. 7 in Table 2). Moreover, these differences are no longer statistically significant, based on a paired t-test with $p < 0.05$. The same test does indicate significant differences when performed against the WaterlooClarke runfile.

4.4 Summary

In summary, neither approach could be fully reproduced due to key information missing. In the case of BaselineOrganizers, the specifics of the models used for query rewriting and re-ranking were lacking, and the formulation of input sequences for query rewriting was underspecified (esp. with regards to exceeding the length limits of the model). As for WaterlooClarke, the complexity of the system and shortages in technical details made it impossible to fully implement the system. Most notably, the involvement of a question-answering system for sparse retrieval is not even mentioned in the paper. We do want to acknowledge the kind, helpful, and open communication by the authors via email, which allowed us to resolve questions around the query rewriting model and its parameters, the BM25 and PRF parameters used, and the rank fusion method employed. Nevertheless, after several rounds of email exchanges, we are still missing details

Table 3. Variants of a two-stage retrieval pipeline on TREC CAsT'20 and '21.

Approach	R@1000	MAP	MRR	NDCG	NDCG@3
TREC CAsT 2020					
T5_CANARD + BM25 + monoT5	0.5276	0.2191	0.5457	0.4353	0.3789
T5_QReCC + BM25 + monoT5	0.5100	0.2056	0.5106	0.4065	0.3618
T5_CANARD + ANCE/BM25 + mono/duoT5	0.6781	0.2540	0.5512	0.5027	0.4052
T5_QReCC + ANCE/BM25 + mono/duoT5	0.6449	0.2443	0.5357	0.4804	0.4061
T5_CANARD + ANCE/BM25/PRF + mono/duoT5	**0.6878**	**0.2555**	**0.5541**	**0.5063**	**0.4086**
T5_QReCC + ANCE/BM25/PRF + mono/duoT5	0.6608	0.2451	0.5355	0.4840	0.4052
Approach	R@500	MAP	MRR	NDCG	NDCG@3
TREC CAsT 2021					
T5_CANARD + BM25 + monoT5	0.6472	0.2628	0.5354	0.4885	0.3968
T5_QReCC + BM25 + monoT5	0.6018	0.2530	0.5369	0.4670	0.3933
T5_CANARD + ANCE/BM25 + mono/duoT5	0.7259	0.2886	0.5575	0.5316	0.4068
T5_QReCC + ANCE/BM25 + mono/duoT5	0.6799	0.2843	0.5702	0.5135	**0.4159**
T5_CANARD + ANCE/BM25/PRF + mono/duoT5	**0.7306**	**0.2915**	0.5573	**0.5330**	0.4061
T5_QReCC + ANCE/BM25/PRF + mono/duoT5	0.6915	0.2864	**0.5712**	0.5176	0.4151

about the PRF algorithm, the question-answering system employed, the exact approach used for tuning the BM25 parameters, the preprocessing employed for the inverted index, and the method used for combining sparse and dense rankings. It is also worth noting that while BM25 parameters were shared for both approaches, those parameters were not the optimal ones for us, which is likely due to differences in document preprocessing. It, however, means that BM25 parameters alone, without further details on preprocessing or collection statistics, are only moderately useful. We shall reflect more generally on some of these challenges and possible remedies in Sect. 6.

5   Additional Experiments

We have reproduced two approaches, BaselineOrganizers and WaterlooClarke, which follow the same basic two-stage retrieval pipeline (cf. Fig. 1a), but differ in each of the query rewriting, first-pass retrieval, and re-ranking components. We experiment with different configurations of this basic pipeline to understand which changes contribute most to overall performance (Sect. 5.1). Additionally, we consider a different pipeline architecture (Sect. 5.2). In both sets of experiments, we are interested in the generalizability of findings, therefore we also report results on the TREC CAsT'20 dataset. (Note that the rank cut-off for 2020 collection is 1000, while for 2021 it is 500.)

5.1   Variants of a Two-Stage Retrieval Pipeline

In this experiment, we gradually switch out the components of a baseline system (BaselineOrganizers) with components of a state-of-the-art system (WaterlooClarke). The results are presented in Table 3; the first and last rows within each block correspond to BaselineOrganizers and WaterlooClarke, respectively.

Table 4. Performance of query rewriting approaches with different variants of the two-stage pipeline on the TREC CAsT'20 and '21 datasets. Highest scores for each year are in boldface.

R1	R2			
	Recall	NDCG@3	Recall	NDCG@3
	T5_CANARD		T5_QReCC	
T5_CANARD	2020: **0.6878**	2020: **0.4086**	2020: **0.6878**	2020: 0.3923
	2021: **0.7306**	2021: 0.4061	2021: 0.7267	2021: 0.4166
T5_QReCC	2020: 0.6608	2020: **0.4086**	2020: 0.6608	2020: 0.4052
	2021: 0.6879	2021: **0.4176**	2021: 0.6915	2021: 0.4151

Our observations are as follows. First, when changing the dataset used for training the T5-based query rewriter from CANARD to QReCC (rows 1 vs. 2, 3 vs. 4, and 5 vs. 6 in Table 3) we observe a noticable drop (3%–7%) in terms of recall, with smaller differences in NDCG@3 (below 2%, with one exception). Second, using more advanced retrieval methods (ANCE/BM25 instead of BM25 for first-pass ranking and mono/duoT5 instead of monoT5 for re-ranking; rows 1 vs. 3 and 2 vs. 4 in Table 3) does yield consistent improvements across metrics and datasets: +12%–29% in recall and +3%–12% in NDCG@3. Finally, using pseudo relevance feedback for first-pass retrieval (rows 3 vs. 5 and 4 vs. 6 in Table 3) results in small but consistent improvements in terms of recall (1%–2%) with negligible differences in NDCG@3 (<1%). It should be noted that none of the above differences are statistically significant, thereby the results are merely indicative. However, in terms of overall trends, our results are in line with the tendencies reported by Yan et al. [28]. Namely, that adding PRF and combining sparse and dense retrieval methods for first-pass retrieval improves performance.

5.2 Using a Different Pipeline Architecture

It is clear that query rewriting has a direct impact on both ranking steps: first-pass retrieval (R1) and re-ranking (R2). Still, it remains to be seen whether the two stages are impacted the same way. The basic two-stage retrieval pipeline (cf. Fig. 1a) uses the same query rewriter for both ranking stages and therefore cannot be used to answer this question. We thus switch to a different pipeline architecture—one that uses a different query rewriter component for R1 and R2, but is identical to the basic pipeline in the ranking components (cf. Fig. 1b).

Table 4 presents the results for the possible four-way combinations of query rewriters, T5_CANARD and T5_QReCC, and ranking stages, R1 and R2. The ranking components follow the WaterlooClarke approach (i.e., using T5_QReCC for both R1 and R2 corresponds to the last row in Table 3). The results reveal some interesting tendencies that generalize across both datasets (even though the differences are not statistically significant). Using T5_CANARD for first-pass retrieval results in the highest recall. However, the overall best combination in terms of final ranking (NDCG@3) is when T5_QReCC is employed in first-pass retrieval and T5_CANARD is used in re-ranking. Overall, we observe meaningful relative improvements for recall (up to 6%) and negligible improvements for NDCG@3 ($\leq$1%) on both datasets over the WaterlooClarke approach.

6   Reflections and Conclusions

In this work, we have attempted to reproduce approaches for the task of conversational passage retrieval, in the context of TREC CAsT. TREC papers can range anywhere between vague system descriptions to full-fledged research papers, which can make reproducibility a real challenge; this has certainly been the case for this study. We acknowledge that reproducibility is not a requirement for TREC submissions. Still, since they are often used for reference comparison in terms of absolute system performance on a given test collection (cf. [2]), it is worth considering how easy or difficult it is to reproduce them. Specifically, we have selected two approaches for our study: the best performing baseline by the track organizers and the best performing participant submission (that was accompanied by a paper) from the 2021 edition of TREC CAsT. We have decided against personal communication with the track organizers (thus implicitly subjecting them to a higher virtual bar-of-standard) while making a best effort to resolve any missing details with the participant team over email.

Overall, our reproducibility efforts have met with moderate success. Surprisingly, we have managed to come closer to reproducing the participant's submission (WaterlooClarke) than the organizers' baseline. In the case of the former, there is a missing sparse retrieval component that can well explain the difference. As for the organizers' results, the discrepancies between the reported results in the track overview paper and the actual runfiles found in the TREC archive would be worth a follow-up investigation. Generally, key missing information includes the names of specific algorithms and models used, and detailed-enough descriptions of procedures of constructing inputs to neural models and ways of obtaining models' parameters. We wish to note that sharing model parameters in some cases is not enough; consider, e.g., the simple case of BM25, where the length normalization parameter alone is not meaningful if collection statistics markedly differ due to how the collection is preprocessed. Given that multi-stage ranking architectures are common at TREC CAsT, but also beyond that, sharing intermediate results from the different components would be immensely valuable. These could include the rewritten or expanded queries, set of candidate document IDs, and intermediate document rankings. Sharing them would not only support reproducibility but also facilitate component-level evaluation.

Since the two reproduced systems follow the same basic two-stage retrieval pipeline, we have also performed additional experiments to study different configurations of this pipeline and have made some observations regarding the contributions made by the various components. Moreover, we have reported on experiments with different combinations of query rewriting methods using a different retrieval pipeline, which have yielded some novel findings. Further comparisons of different pipeline architectures would be an especially interesting direction for future work.

Post-acceptance communication with TREC CAsT organizers. Upon acceptance of this paper, we attempted to clarify the discrepancies between the results in this paper and those reported in the track overview via email communication with the track organizers. There is a difference in tooling: they used Pyserini[19] for building the index, while we used Elasticsearch. Differences in collection preprocessing (tokenization, stemming, stopword removal, etc.) may contribute to the gap in the results. Regarding the runfile, we were pointed to the track's GitHub repository[20] containing the raw runfile (`org_automatic_results_1000.v1.0.run`). However, evaluating this runfile against the official qrels still yields results different from those reported in the track overview paper (in parentheses): Recall@500 is 0.623 (vs. 0.636), MAP is 0.282 (vs. 0.291), and NDCG@3 is 0.424 (vs. 0.436). This is "in alignment" with the case of the WaterlooClarke (`clarke-cc`) runfile, in the sense that there is a mismatch between the numbers reported in the track overview paper and the evaluation of the actual runfiles (with the latter being lower). At the time of writing, this issue has not been resolved. We plan to update our online repository if new findings become available.

Acknowledgments. This research was supported by the Norwegian Research Center for AI Innovation, NorwAI (Research Council of Norway, project number 309834). We thank the members of the WaterlooClarke group (School of Computer Science, University of Waterloo, Canada), Xinyi Yan and Charlie Clarke for supporting our efforts to reproduce their TREC CAsT'21 submission. We also thank the TREC CAsT organizers for their efforts in coordinating the track and for providing us with additional technical details regarding their baseline.

References

1. Anantha, R., Vakulenko, S., Tu, Z., Longpre, S., Pulman, S., Chappidi, S.: Open-domain question answering goes conversational via question rewriting. In: Proceedings of the 2021 Conference of the North American Chapter of the Association for Computational Linguistics: Human Language Technologies, NAACL 2021, pp. 520–534 (2021)
2. Armstrong, T.G., Moffat, A., Webber, W., Zobel, J.: Improvements that don't add up: Ad-hoc retrieval results since 1998. In: Proceedings of the 18th ACM Conference on Information and Knowledge Management, CIKM 2009, pp. 601–610 (2009)
3. Campos, D.F., et al.: MS MARCO: a human generated machine reading comprehension dataset. arXiv, cs.CL/1611.09268 (2016)
4. Chang, C.-Y., et al.: Query expansion with semantic-based ellipsis reduction for conversational IR. In: The Twenty-Ninth Text REtrieval Conference Proceedings, TREC '20 (2020)
5. Cormack, G.V., Clarke, C.L.A., Büttcher, S.: Reciprocal rank fusion outperforms condorcet and individual rank learning methods. In: Proceedings of the 32nd International ACM SIGIR Conference on Research and Development in Information Retrieval, SIGIR 2009 (2009)

---

[19] https://github.com/castorini/pyserini.
[20] https://github.com/daltonj/treccastweb/tree/master/2021/baselines.

6. Craswell, N., Mitra, B., Yilmaz, E., Campos, D.: Overview of the TREC 2020 deep learning track. In: The Twenty-Ninth Text REtrieval Conference Proceedings, TREC 2020 (2020)

7. Dalton, J., Xiong, C., Callan, J.: TREC CAsT 2019: the Conversational Assistance track overview. In: The Twenty-Eighth Text REtrieval Conference Proceedings, TREC 2019 (2019)

8. Dalton, J., Xiong, C., Callan, J.: TREC CAsT 2020: the Conversational Assistance track overview. In: The Twenty-Ninth Text REtrieval Conference Proceedings, TREC 2020 (2020)

9. Dalton, J., Xiong, C., Callan, J.: TREC CAsT 2021: the Conversational Assistance track overview. In: The Thirtieth Text REtrieval Conference Proceedings, TREC 2021 (2021)

10. Dietz, L., Verma, M., Radlinski, F., Craswell, N.: TREC complex answer retrieval overview. In: The Twenty-Seventh Text REtrieval Conference Proceedings, TREC 2018 (2018)

11. Elgohary, A., Peskov, D., Boyd-Graber, J.: Can you unpack that? Learning to rewrite questions-in-context. In: Proceedings of the 2019 Conference on Empirical Methods in Natural Language Processing and the 9th International Joint Conference on Natural Language Processing, EMNLP-IJCNLP 2019, pp. 5918–5924 (2019)

12. Gemmell, C., Dalton, J.: Glasgow representation and information learning lab (GRILL) at the conversational assistance track 2020. In: The Twenty-Ninth Text REtrieval Conference Proceedings, TREC 2020 (2020)

13. Ju, J.-H., et al.: An exploration study of multi-stage conversational passage retrieval: Paraphrase query expansion and multi-view point-wise ranking. In: The Thirtieth Text REtrieval Conference Proceedings, TREC 2021 (2021)

14. Kumar, V., Callan, J.: Making information seeking easier: an improved pipeline for conversational search. In: Findings of the Association for Computational Linguistics: EMNLP 2020, EMNLP 2020, pp. 3971–3980 (2020)

15. Lavrenko, V., Croft, W.B.: Relevance based language models. In: Proceedings of the 24th Annual International ACM SIGIR Conference on Research and Development in Information Retrieval, SIGIR 2001, pp. 120–127 (2001)

16. Lin, S.-C., Yang, J.-H., Nogueira, R., Tsai, M.-F., Wang, C.-J., Lin, J.: Multi-stage conversational passage retrieval: an approach to fusing term importance estimation and neural query rewriting. ACM Trans. Inf. Syst. **39**(4), 1–29 (2021)

17. Luan, Y., Eisenstein, J., Toutanova, K., Collins, M.: Sparse, dense, and attentional representations for text retrieval. Trans. Assoc. Comput. Linguist. **9**, 329–345 (2021)

18. Mele, I., Muntean, C.I., Nardini, F.M., Perego, R., Tonellotto, N., Frieder, O.: Topic propagation in conversational search. In: Proceedings of the 43rd International ACM SIGIR Conference on Research and Development in Information Retrieval, SIGIR 2020, pp. 2057–2060 (2020)

19. Nogueira, R., Jiang, Z., Lin, J.: Document ranking with a pretrained sequence-to-sequence model. In: Findings of the Association for Computational Linguistics: EMNLP 2020, EMNLP 2020, pp. 708–718 (2020)

20. Petroni, F., et al.: KILT: a benchmark for knowledge intensive language tasks. arXiv, cs.CL/2009.02252 (2020)

21. Pradeep, R., Nogueira, R., Lin, J.: The expando-mono-duo design pattern for text ranking with pretrained sequence-to-sequence models. arXiv, cs.IR/2101.05667 (2021)

22. Vakulenko, S., Voskarides, N., Tu, Z., Longpre, S.: Leveraging query resolution and reading comprehension for conversational passage retrieval. In: The Twenty-Ninth Text REtrieval Conference Proceedings, TREC 2020 (2020)
23. Vakulenko, S., Longpre, S., Tu, Z., Anantha, R.: Question rewriting for conversational question answering. In: Proceedings of the 14th ACM International Conference on Web Search and Data Mining, WSDM 2021, pp. 355–363 (2021)
24. Vakulenko, S., Voskarides, N., Tu, Z., Longpre, S.: A comparison of question rewriting methods for conversational passage retrieval. In: European Conference on Information Retrieval, ECIR 2021, pp. 418–424 (2021)
25. Voskarides, N., Li, D., Panteli, A., Ren, P.: ILPS at TREC 2019 conversational assistant track. In: The Twenty-Eighth Text REtrieval Conference Proceedings, TREC 2019 (2019)
26. Voskarides, N., Li, D., Ren, P., Kanoulas, E., de Rijke, M.: Query resolution for conversational search with limited supervision. In: Proceedings of the 43rd International ACM SIGIR Conference on Research and Development in Information Retrieval, SIGIR 2020, pp. 921–930 (2020)
27. Xiong, L., et al.: Approximate nearest neighbor negative contrastive learning for dense text retrieval. In: International Conference on Learning Representations, ICLR 2020 (2020)
28. Yan, X., Clarke, C.L., Arabzadeh, N.: WaterlooClarke at the TREC 2021 conversational assistant track. In: The Thirtieth Text REtrieval Conference Proceedings, TREC 2021 (2021)
29. Yang, J.-H., Lin, S.-C., Wang, C.-J., Lin, J., Tsai, M.-F.: Query and answer expansion from conversation history. In: The Twenty-Eighth Text REtrieval Conference Proceedings, TREC 2019 (2019)
30. Yilmaz, Z.A., Wang, S., Lin, J.: H2oloo at TREC 2019: combining sentence and document evidence in the deep learning track. In: The Twenty-Eighth Text REtrieval Conference Proceedings, TREC 2019 (2019)
31. Yu, S., et al.: Few-shot generative conversational query rewriting. In: Proceedings of the 43rd International ACM SIGIR Conference on Research and Development in Information Retrieval, SIGIR 2020, pp. 1933–1936 (2020)
32. Yu, S., Liu, Z., Xiong, C., Feng, T., Liu, Z.: Few-shot conversational dense retrieval. In: Proceedings of the 44th International ACM SIGIR Conference on Research and Development in Information Retrieval, SIGIR 2021, pp. 829–838 (2021)
33. Zamani, H., Trippas, J.R., Dalton, J., Radlinski, F.: Conversational information seeking. arXiv, cs.IR/2201.08808 (2022)

Demonstration Papers

Exploring Tabular Data Through Networks

Aleksandar Bobic[1,2](✉) [ID], Jean-Marie Le Goff[1] [ID], and Christian Gütl[2] [ID]

[1] IPT Department, CERN, Geneva, Switzerland
{aleksandar.bobic,jean-marie.le.goff}@cern.ch
[2] CoDiS Lab ISDS, Graz University of Technology, Graz, Austria
c.guetl@tugraz.at

Abstract. Representing and visualizing data as networks is a widely spread approach to analyzing highly connected data in domains such as medicine, social sciences, and information retrieval. Investigating data as networks requires pre-processing, retrieval or filtering, conversion of data into networks, and application of various network analysis approaches. These processes are usually complex and hard to perform without some programming knowledge and resources. To the best of our knowledge, most solutions attempting to make these functionalities accessible to users focus on particular processes in isolation without exploring how these processes could be further abstracted or combined in a real-world application to assist users in their data exploration and knowledge extraction journey. Furthermore, most applications focusing on such approaches tend to be closed-source. This paper introduces a solution that combines the approaches above as part of Collaboration Spotting X (CSX), an open-source network-based visual analytics tool for retrieving, modeling, and exploring or analyzing data as networks. It abstracts the concepts above through the use of multiple interactive visualizations. In addition to being an easily accessible open-source platform for data exploration and analysis, CSX can also serve as a real-world evaluation platform for researchers in related computer science areas who wish to test their solutions and approaches to machine learning, visualizations, interactions, and more in a real-world system.

Keywords: Information retrieval · Network modeling · Network visual analytics

1 Introduction and Background

In recent years with the growth and variety of generated data, the needs of domain experts to extract insights and draw conclusions from it became apparent. To accomplish these objectives, experts tend to use a variety of analysis and visual analytics tools [7,8,13,14]. However, to the best of our knowledge, most tools are either not flexible enough to adequately support a wide range of use cases since they focus on a single domain and can't be generalized, lack abstractions that would facilitate insight extraction, or are not open-source and therefore limit their use-case [2,8,12,17,19]. These limitations entail that most domain

J. Kamps et al. (Eds.): ECIR 2023, LNCS 13982, pp. 195–200, 2023.
https://doi.org/10.1007/978-3-031-28241-6_13

experts have to either find an appropriate tool for each particular dataset, know how to code or have resources to develop their solutions, or rely on third parties to help them conduct their research.

To meet domain experts' expectations in terms of data analysis goals, a tool should provide iterative expression, and modification of their information needs, the capability of reshaping their retrieved data into formats that will enhance their ability to answer research questions and offer means to visualize their data in a way that will reveal hidden patterns and connections [6,15–17]. An effective approach for providing domain experts with means to explore implicit or hidden patterns and connections in their data is to use the concepts of networks, and network visualizations [1,2,4,5,9,11,18,19]. Although networks provide helpful insight into implicit connections, their usually large size makes them challenging to visualize and interpret without appropriate filtering, and aggregation tools or accompanying visualizations [2,20].

As part of this paper, we introduce Collaboration Spotting X (CSX), a network-based information retrieval and visual-analytics application in order to fulfill the above requirements and provide a broad spectrum of domain experts with an open-source, extensible tool that can be used with diverse datasets. It is built based on knowledge found in literature and experience with previous experimental versions of the tool [3,10]. It combines a visual information need builder, multiple juxtaposed views of node-link visualizations, tabular representations and charts, and multiple tools for molding, filtering, and exploring data abstracted through interactions and visualizations.

2   Collaboration Spotting X

Collaboration Spotting X, in most simple terms, enables users to upload their data in a CSV format and explore it through dynamic network schemas, statistics widgets, and list views. It abstracts many processes needed for expressing retrieval needs, visualizing the retrieval results, modifying the visualization of the retrieval results and modifying the retrieval results themselves, and finally analyzing the results through interactive visualizations. It is built using React.js, FastAPI, and Docker as the core technologies. It can either be run from a local machine using Docker or deployed on an institution server and be used by multiple users. The project is available on GitHub[1]. Additionally, as part of this paper, a demonstration video is provided highlighting CSX's main features and their usage[2].

2.1   Simplified User Workflow

Users can upload their datasets in the CSV file format with a simple drag-and-drop mechanism on the CSX homepage. Users can select the default dataset

---

[1] https://github.com/aleksbobic/csx.
[2] https://youtu.be/TwSA6nVkdec.

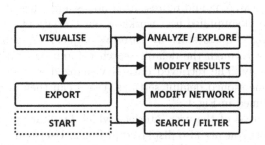

Fig. 1. Collaboration Spotting X user workflow. Users first perform a search or filter action or a set of actions in order to get their data. The data is initially visualized using the default settings. Users can then perform one of the actions on the right side of the diagram.

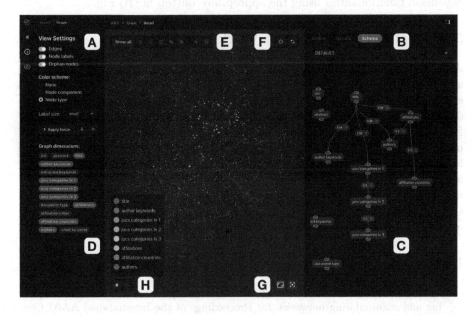

Fig. 2. Collaboration Spotting X detail network and detail schema. The displayed network represents the user's search results. The individual nodes represent cell values in the dataset and are connected if two cell values appear in the same row and are connected in the network schema. Users can modify the network's layout and visual properties using the view settings panel (A). They can build various statistics, view the data corresponding to the part of the network currently visualized in a tabular or list format, or modify the network schema using the data panel (B). Users can also modify the network topology using the schema designer (C) and select which network node types should be visible using the node type list (D). In addition, multiple network-related features for exploring direct connections (E), switching between network types (F), interacting with the entire view (G), and modifying retrieved results (H) are at users' disposal.

columns used for searching and representing nodes and edges during the upload process. The general simplified user workflow once a dataset has been uploaded is depicted in Fig. 1. Users start by entering a query or defining a workflow used for searching through the default search column or multiple user-defined columns respectively. The retrieved data is visualized as a node-link diagram using the above mentioned default settings. Users can then perform one of the other actions outlined in Fig. 1.

The relevant areas of the user interface which facilitate parts of the above-described user workflow are also displayed in Fig. 2. Users can interact with the visualization and explore their data either through elements of the view settings panel (A) Fig. 2, through stats in the details tab of the data panel (B) Fig. 2 or using built-in exploration functionalities (E) Fig. 2. They can modify the network topology using the interactive network schema (C) Fig. 2 or the node type list (D) Fig. 2. Finally, they may modify their search results using the trimming and expansion functionalities using the appropriate button in (H) Fig. 2.

3 Conclusion and Future Work

As part of this work, we introduce and demonstrate Collaboration Spotting X, an open-source network-based visual analytics tool for retrieving, modeling, and exploring or analyzing data as networks. We describe the main features and introduce the general user workflow in CSX. It can analyze a broad set of datasets from different domains. It can also be the foundation for advanced scientific evaluations and studies in information visualization, information retrieval, machine learning, and more. As part of our future work, we want to extend CSX with features such as history and data annotation. Once these features are implemented, we will perform an extensive evaluation focusing on multiple aspects of CSX such as the usability, help with task completion and comparison with other existing tools in various environments and use cases.

References

1. Bastian, M., Heymann, S., Jacomy, M.: Gephi: an open source software for exploring and manipulating networks. In: Proceedings of the International AAAI Conference on Web and Social Media, vol. 3, pp. 361–362 (2009)
2. Bigelow, A., Nobre, C., Meyer, M., Lex, A.: Origraph: interactive network wrangling, pp. 81–92, October 2019. https://doi.org/10.1109/VAST47406.2019.8986909
3. Bobic, A., Le Goff, J.M., Gütl, C.: Collaboration spotting x-a visual network exploration tool. In: Proceedings of the The Eighth International Conference on Social Networks Analysis, Management and Security: SNAMS 2021 (2021)
4. Cashman, D., et al.: Cava: a visual analytics system for exploratory columnar data augmentation using knowledge graphs. IEEE Trans. Visual Comput. Graph. **27**(2), 1731–1741 (2021). https://doi.org/10.1109/TVCG.2020.3030443
5. Chau, D.H., Kittur, A., Hong, J.I., Faloutsos, C.: Apolo: making sense of large network data by combining rich user interaction and machine learning. In: CHI 2011, pp. 167–176. Association for Computing Machinery, New York (2011). https://doi.org/10.1145/1978942.1978967

6. De Bie, T., De Raedt, L., Hernández-Orallo, J., Hoos, H.H., Smyth, P., Williams, C.K.I.: Automating data science. Commun. ACM **65**(3), 76–87 (2022). https://doi.org/10.1145/3495256
7. Demšar, J., et al.: Orange: data mining toolbox in python. J. Mach. Learn. Res. **14**, 2349–2353 (2013). http://jmlr.org/papers/v14/demsar13a.html
8. Dimara, E., Zhang, H., Tory, M., Franconeri, S.: The unmet data visualization needs of decision makers within organizations. IEEE Trans. Visualization Comput. Graph., 1 (2021). https://doi.org/10.1109/TVCG.2021.3074023
9. Heer, J., Perer, A.: Orion: A system for modeling, transformation and visualization of multidimensional heterogeneous networks. In: 2011 IEEE Conference on Visual Analytics Science and Technology (VAST), pp. 51–60 (2011). https://doi.org/10.1109/VAST.2011.6102441
10. Le Goff, J.M., Dardanis, D., Rattinger, A., Agocs, A., Forster, R., Ouvrard, X.: Collaboration spotting: a visual analytics platform to assist knowledge discovery. ERCIM News, pp. 46–48 (2017)
11. Nobre, C., Meyer, M., Streit, M., Lex, A.: The state of the art in visualizing multivariate networks. Comput. Graph. Forum **38**, 807–832 (2019). https://doi.org/10.1111/cgf.13728
12. Pienta, R., Tamersoy, A., Endert, A., Navathe, S., Tong, H., Chau, D.H.: Visage: interactive visual graph querying. In: Proceedings of the International Working Conference on Advanced Visual Interfaces, AVI 2016, pp. 272–279. Association for Computing Machinery, New York (2016). https://doi.org/10.1145/2909132.2909246
13. Polychronidou, E., Kalamaras, I., Votis, K., Tzovaras, D.: Health vision: an interactive web based platform for healthcare data analysis and visualisation. In: 2019 IEEE Conference on Computational Intelligence in Bioinformatics and Computational Biology (CIBCB), pp. 1–8 (2019). https://doi.org/10.1109/CIBCB.2019.8791462
14. Randles, B.M., Pasquetto, I.V., Golshan, M.S., Borgman, C.L.: Using the jupyter notebook as a tool for open science: an empirical study. In: 2017 ACM/IEEE Joint Conference on Digital Libraries (JCDL), pp. 1–2 (2017). https://doi.org/10.1109/JCDL.2017.7991618
15. Ruotsalo, T., Jacucci, G., Myllymäki, P., Kaski, S.: Interactive intent modeling: Information discovery beyond search. Commun. ACM **58**(1), 86–92 (2014). https://doi.org/10.1145/2656334
16. Russell, D.M., Stefik, M.J., Pirolli, P., Card, S.K.: The cost structure of sensemaking. In: Proceedings of the INTERACT '93 and CHI '93 Conference on Human Factors in Computing Systems, CHI 1993, pp. 269–276. Association for Computing Machinery, New York (1993). https://doi.org/10.1145/169059.169209
17. Russell-Rose, T., Chamberlain, J., Shokraneh, F.: A visual approach to query formulation for systematic search. In: Proceedings of the 2019 Conference on Human Information Interaction and Retrieval, CHIIR 2019, p. 379–383. Association for Computing Machinery, New York (2019). https://doi.org/10.1145/3295750.3298919
18. Shannon, P., Markiel, A., Ozier, O., Baliga, N.S., Wang, J.T., Ramage, D., Amin, N., Schwikowski, B., Ideker, T.: Cytoscape: a software environment for integrated models of biomolecular interaction networks. Genome Res. **13**(11), 2498–2504 (2003). https://doi.org/10.1101/gr.1239303

19. Valdivia, P., Buono, P., Plaisant, C., Dufournaud, N., Fekete, J.D.: Analyzing dynamic hypergraphs with parallel aggregated ordered hypergraph visualization. IEEE Trans. Visual Comput. Graphics **27**(1), 1–13 (2021). https://doi.org/10.1109/TVCG.2019.2933196
20. Yoghourdjian, V., Yang, Y., Dwyer, T., Lawrence, L., Wybrow, M., Marriott, K.: Scalability of network visualisation from a cognitive load perspective. IEEE Trans. Visual Comput. Graphics **27**(2), 1677–1687 (2021). https://doi.org/10.1109/TVCG.2020.3030459

InfEval: Application for Object Detection Analysis

Kirill Bogomasov$^{(\boxtimes)}$ (ID), Tim Geuer, and Stefan Conrad (ID)

Heinrich Heine University, Universitätsstr. 1, 40225 Düsseldorf, Germany
{bogomasov,tim.geuer,stefan.conrad}@uni-duesseldorf.de

Abstract. Object Detection is one of the most fundamental and challenging areas in computer vision. A detailed analysis and evaluation is key to understanding the performance of custom Deep Learning models. In this contribution, we present an application which is able to run inference on custom data for models created in different machine learning frameworks (e.g. TensorFlow, PyTorch), visualize the output and evaluate it in detail. Both, the Object Detection models and the data sets, are uploaded and executed locally without leaving the application. Numerous filtering options, for instance filtering on *mAP*, on *NMS* or on *IoU*, are provided.

Keywords: Object detection · Evaluation · Visualization

1 Introduction

Object Detection is one of the most fundamental and challenging areas in computer vision. By this reason, it has received considerable attention from the research community in recent years [9,12]. The variety of its application scenarios is huge and provides processing capability to all kinds of images which range from medical images, to natural images, and up to large-scale images. The idea to analyze and automatically evaluate increasingly complex imaging is part of many computer vision challenges. One of those is the ImageCLEF Coral task [4], hosted by the ImageCLEF [5] conference, which aims to detect different substrate types in large-scale underwater images. The particular difficulty of the data set is that the individual images may contain, to the extent of a three-digit number, objects belonging to different classes and largely varying in shape. Some of these objects are inlaying or overlapping. In our previous contributions [2,3,8] we have dealt extensively with the detection of corals in underwater images, but we have always lacked of tools for a comfortable visualization and comparison of our models on selected data. An all-in-one solution that would take care of the data management, call the trained models independently and allow the visualization of results including all useful filtering by a simple mouse click, was missing. Knowing that the results of a system based on a deep learning model can only be as good as the performance of the deep learning model behind it,

J. Kamps et al. (Eds.): ECIR 2023, LNCS 13982, pp. 201–205, 2023.
https://doi.org/10.1007/978-3-031-28241-6_14

makes clear that in-depth analysis and monitoring are necessary. Commonly, the evaluation is limited to numerical values, such as the widely used *mAP*.

In order to achieve good results in Object Detection, it is important to understand the data by uncovering potential pitfalls prior to finalizing the Object Detection module. Therefore, a closer look at the predictions the model performs poorly on is required. Furthermore, an overview over the ground truth objects in relation to the corresponding predictions, as well as in general the view on the impacts of selected parameters on the overall performance, is highly beneficial. The goal of visualization of the prediction results is a step towards getting a feeling for the strengths and weaknesses of the system. This insight may further on be used for model optimization.

Several platforms allow visualization of inference output in images. First and foremost to be named, is the fee-based platform Roboflow[1]. Besides of simple visualization of bounding boxes, it allows filtering at the confidence - and overlap threshold. However, the filtering by overlap and highlighting of boxes that can be considered as a *true positive* is not well implemented and by this reason the visualization is limited. In addition, it must be taken into account that commercially operated platforms are mostly client-server architectures, hosted on a commercial hardware. Meaning that the data needs to be transferred from client to server. When working with sensitive data, legal terms commonly restrict the data to be uploaded to third-party servers. Although privacy is often granted in the handling of one's own data, misuse or data leakage is a common issue. For this reason, it is often desirable to restrict the processing of the data on his local computer. Therefore, we provide our own solution, which is exclusively hosted locally and because of that safe to use. Another platform that we might consider as related is called Voxel51[2]. Although the platform is able to run locally, it does not support uploading own Object Detection models to get the inference output in the expected format. Additionally, useful filtering, particularly necessary on coral data such as filtering on *NMS*, is not provided. In the following subchapters, we present a brief overview of the functionality and the structure of the program.

2   Functionality

Usability and simple mouse-click control offer full control over the evaluation process. In the first step, the data (e.g. image data, class information and optionally validation labels) needs to be uploaded to an internal private user folder and saved temporarily. The entire control of the application, including the transfer of the data, runs via the frontend. In the backend the data is validated, stored and evaluated. Each uploaded file is validated in terms of correctness. Therefore, the given format of ground truth and prediction files are tested for the expected structure. For further processing, user trained Object Detection models are required. The application supports models written in TensorFlow 1, TensorFlow 2 and PyTorch. In addition, several YOLO implementations [7] [6],

---

[1]   https://roboflow.com/.
[2]   https://voxel51.com/.

as such proved to be useful, are supported. The uploaded models are validated by running inference on a blank image. The underlying security mechanism informs the user about unexpected properties of the uploaded data. Given that the verification is run successfully, inference on the selected data can be executed. The result can be further analyzed. Both, the ground truth bounding boxes and the predictions can be visualized. This visualization is highly customizable and allows e.g. leaving out certain classes, changing the colors of the bounding boxes or leaving out all class labels. Subsequently, output predictions can by easily compared to the ground truth bounding boxes. The whole amount of predictions can be filtered by a chosen confidence interval, by *NMS* and by an *IoU* threshold as well. Filtering predictions is very important if the model generates many predictions that overlap to a high degree. Dealing with the least promising predictions requires fine tuning of different parameters. Finetuning can be challenging as it is difficult to anticipate whether it improves the results on a given metric. In this scenario it is very useful to have an immediate visualization of the impact of the parameters. Experimenting with different threshold values helps to highlight whether a match with ground truth boxes exists. This is very useful in order to get a well-founded understanding of the predictions generated by the model. Since an elaboration of a models performance in terms of matching the ground truth bounding boxes is difficult, a detailed visualization is useful. All adjustments can be run on individual images or the entire data set. Besides of the visualization, a calculation of *mAP* for the current selection is performed and displayed. This calculation can be done for either just the current image or the entire data set. Finally, the images containing all visualizations with selected settings can by downloaded.

3   System Architecture

The architecture of InfEval is based on the server-client principle, in addition a distinction between frontend and backend is made. The communication between those two main components entirely relies on HTTP requests. The user authentication and authorization is realized by PyJWT as it protects the application against various attacks. A feature that might be useful for hosting the server remotely.

The frontend, as visualized in Fig. 1, is implemented in TypeScript using the Angular framework. In order to run inference tasks using the uploaded models, the application provides support to the most common frameworks. Along with TensorFlow 1 [1], Tensorflow 2 [1] and PyTorch [11], the application also allows YOLOv3 [7] and YOLOv5 [6] models. For all predictions, the COCO and PascalVOC metrics are evaluated [10]. Individual components created in Angular have a corresponding unit in the backend to contact with using the RESTful API. The GUI is designed with Bootstrap 5 and Angular Material UI.

The backend is implemented in Python 3.9 using the Django REST framework. As the most machine learning frameworks are written in Python, this choice allows us to easily include those into our project. The implementation supports multiple different users, each with its personal folder structure for uploads.

Fig. 1. Screenshot section presents an overview of the UI and the available filtering option applied on an image from the ImageCLEF coral task including inference results alongside the ground truth values colored in red. The objects with green backgrounds are overlapped by at least one prediction with the required mAP value. The objects with a red background were not hit accordingly. (Color figure online)

A brief description of the software architecture, divided into front- and backend, can be found on GitHub[3].

4   Concluding Remarks

In short, the InfEval application simplifies the inference evaluation of any Deep Learning model for Object Detection (limited to the selected frameworks) on any image data set and makes it convenient and easy to handle with just a few clicks. The evaluation results are visualized for the user selected image data without the necessity to upload sensitive data to a third-party server. Meaningful filtering options (e.g. by class labels, by confidence, by IoU, etc.) are provided, enabling besides listings of common metrics, also graphical highlighting. The application proves to be extremely useful for the evaluation of model performance on high-resolution images with two-digit numbers of objects in each file or even more. It is recently released on GitHub (See footnote 3), is publicly available and free to use. Continuous improvement is planned for the future. Any contribution of creative ideas and custom requirements as well as programmatic cooperation are welcome.

---

[3] https://github.com/tigeu/InfEval.

References

1. Abadi, M., et al.: TensorFlow: large-scale machine learning on heterogeneous systems (2015). https://www.tensorflow.org/, software available from tensorflow.org
2. Bogomasov, K., Grawe, P., Conrad, S.: A two-staged approach for localization and classification of coral reef structures and compositions. In: CLEF (Working Notes) (2019)
3. Bogomasov, K., Grawe, P., Conrad, S.: Enhanced localization and classification of coral reef structures and compositions. In: CLEF (Working Notes) (2020)
4. Chamberlain, J., Garcia Seco de Herrera, A., Campello, A., Clark, A.: ImageCLEF-coral task: coral reef image annotation and localisation. In: Experimental IR Meets Multilinguality, Multimodality, and Interaction. Proceedings of the 13th International Conference of the CLEF Association (CLEF 2022), LNCS Lecture Notes in Computer Science, Italy, 5–8 September 2022. Springer, Bologna (2022). https://doi.org/10.1007/978-3-031-13643-6
5. Ionescu, B., et al.: Overview of the imageclef 2022: Multimedia retrieval in medical, social media and nature applications. In: International Conference of the Cross-Language Evaluation Forum for European Languages, pp. 541–564. Springer, Heidelberg (2022). https://doi.org/10.1007/978-3-031-13643-6_31
6. Jocher, G., et al.: ultralytics/yolov5: v7.0 - YOLOv5 SOTA realtime instance segmentation (2022). https://doi.org/10.5281/zenodo.7347926
7. Jocher, G., et al.: ultralytics/yolov3: v9.6.0 - YOLOv5 v6.0 release compatibility update for YOLOv3 (2021). https://doi.org/10.5281/zenodo.5701405
8. Kerlin, F., Bogomasov, K., Conrad, S.: Monitoring coral reefs using faster r-cnn. In: CLEF (Working Notes) (2022)
9. Liu, L., et al.: Deep learning for generic object detection: a survey. Int. J. Comput. Vision 128(2), 261–318 (2020)
10. Padilla, R., Passos, W.L., Dias, T.L.B., Netto, S.L., da Silva, E.A.B.: A comparative analysis of object detection metrics with a companion open-source toolkit. Electronics 10(3) (2021). https://doi.org/10.3390/electronics10030279. https://www.mdpi.com/2079-9292/10/3/279
11. Paszke, A., et al.: Pytorch: an imperative style, high-performance deep learning library. In: Advances in Neural Information Processing Systems 32, pp. 8024–8035. Curran Associates, Inc. (2019). http://papers.neurips.cc/paper/9015-pytorch-an-imperative-style-high-performance-deep-learning-library.pdf
12. Zou, Z., Shi, Z., Guo, Y., Ye, J.: Object detection in 20 years: a survey. arXiv preprint arXiv:1905.05055 (2019)

The System for Efficient Indexing and Search in the Large Archives of Scanned Historical Documents

Martin Bulín[(✉)] , Jan Švec , and Pavel Ircing

Department of Cybernetics, University of West Bohemia, Pilsen, Czech Republic
{bulinm,honzas,ircing}@kky.zcu.cz

Abstract. The paper introduces software capable of indexing and searching large archives of scanned historical documents. The system capabilities are demonstrated on the collection containing documents from the archives of the post-Soviet security services. The backend of the system was designed with a focus on flexibility (it is actually already being used for other related tasks) and scalability to larger volumes of data. The graphical user interface design has been consulted with historians interested in using the archived documents and was developed in several iterations, gradually including the changes induced both by the user's requests and by our improving knowledge about the nature of the processed data.

Keywords: Indexing · GUI design · Scanned documents

1 Introduction

Many institutions aiming to preserve historical documents (primarily written and graphical ones, but recently also audiovisual) have successfully adopted modern means of digitizing and storing those materials. However, making these documents accessible for scholars and even the general public still remains a serious challenge. The issues that have to be addressed are actually multifold: (a) rather sophisticated machine learning methods have to be employed to extract searchable metadata (or, at least, a machine-readable text) from the digitized material, (b) an efficient backend solution needs to be implemented for indexing and storing both the data and the metadata and (c) there is a need to design a comprehensible Graphical User Interface (GUI) to make the search feasible for both experts and general users. The paper presents our solutions to (b) and (c) as implemented in the joint research project with the Institute for Study of the Totalitarian Regimes (ISTR), Czech Republic [4]. This institution has collected and scanned over 0.5 million pages of materials from the Ukrainian and other

The work described herein has been supported by the Ministry of Education, Youth and Sports of the Czech Republic project LINDAT/CLARIAH-CZ and by the grant of the University of West Bohemia, project No. SGS-2022-017.

J. Kamps et al. (Eds.): ECIR 2023, LNCS 13982, pp. 206–210, 2023.
https://doi.org/10.1007/978-3-031-28241-6_15

post-Soviet archives to document the persecution of people by the Soviet regime. The nature of the archive presents specific challenges for the machine learning techniques mentioned above – but those are out of the scope of this paper and are extensively described elsewhere [2,3]. On the other hand, the backend solution was implemented to be very universal and has already proven its efficiency in other related tasks [1,7]. The same is true for the GUI, although naturally, to a smaller extent—effectively presenting scanned documents, of course, has different requirements than presenting a video recording. The presented demo should be interesting for all archivists and historians struggling with efficient indexation and presentation of their large, diverse (and potentially multi-medial) digitized archive materials.

2   Backend Server Implementation

The backend server provides an HTTP interface for the graphical user interface described in Sect. 3. Along with this interface, it controls and executes the data processing pipeline and interacts with the database. We designed the backend server using the experience of building our prototype system for searching audiovisual archives [5]. Unlike in this prototype, we focused on the ability to automatically process the input documents in the end-to-end pipeline and provide the web browser user interface accessible over the Internet. Two kinds of components control the overall processing of the document collection:

- *Workers* – A worker represents an atomic processing step. The workers have a consistent interface and can be executed as stand-alone processes or as part of the pipeline. An example of a stand-alone worker is an importing worker, which loads the source documents with their metadata and stores them in the document database. Other workers for converting media formats, performing OCR, or indexing OCR results are executed from the pipeline.[1]
- *Pipelines* – The pipeline is a basic principle which allows reusing different workers in different scenarios. It interconnects the workers and creates dependencies between the outputs of one worker and the inputs of another.

In other words, the workers define atomic units of work and pipelines represent the workflows. The backend stores its state entirely in the shared document database; therefore, the processing can be executed in parallel on multiple machines. Such deployment is simplified using modern containerized technologies such as Docker and Kubernetes.

The backend uses the MongoDB document database in the current implementation, storing both the processed and presented documents and the interim processing results. The backend server is implemented in Python and uses the Tornado web server for implementing the HTTP interface.

---

[1] The core worker in the presented demo is an OCR worker who internally uses the Tesseract OCR engine [6]. The OCR engine is wrapped with additional functionalities improving the performance of the engine in the domain of scanned documents – see the details in [2,3].

The search functionality in the pre-indexed results is implemented as part of the indexation worker. In other words, the indexation worker indexes the results (of, for example, OCR) in a specific format of an index, and at the same time, it allows to search through the index using textual queries.

The architecture of the backend server is very flexible. We use it not only for indexing the scanned documents but also for multilingual audiovisual archives [7,8] and interactive archives [1].

3   Graphical User Interface

The topmost part of the processing pipeline is the graphical user interface based on the latest web technologies, including React and Typescript. The interface is hosted on a separate Next.js server, which is independent of the backend server and the mutual communication between these two is established over the HTTP protocol. This design allows us to exploit the principal features of the Next.js framework, especially working with the client-side and server-side rendering options and optimizing the interface load time. When a user accesses the website, their web browser, in the role of the client, starts communicating with the backend server and exchanges the dynamic data only. In contrast, the static data structures are processed server-side in parallel. The website is responsive and capable of detecting the type of device - a mobile version is also provided. The application is currently accessible at https://cechoslovacivgulagu.kky.zcu.cz/en.

The crucial function of the interface is to provide a user-friendly browser of the backend data and to allow the user to search in them. Therefore, the functional design provides two fundamental modes:

Data Browser. Data are loaded from the backend server in pre-defined structures gradually, depending on the user's requests. Rendering the title page requires only the metadata of available archives, while the list of documents and more detailed data are loaded after selecting the corresponding archive. As there are many documents in total, we use pagination and load the documents in chunks. Once loaded, all the data are saved in the cache memory of the user's web browser, making them instantly accessible when requested repeatedly. The viewer of scanned documents (Fig. 1, right side) is capable of image rotation, zooming, listing documents metadata, and entering the full-screen mode.

Browser of Search Results. The application's core feature is searching for a textual phrase in a vast database of scanned documents, namely with no or a reasonable time delay. The GUI provides a search frame accepting a Unicode phrase with additional settings of the search scope with options: 1) the whole database, 2) the currently opened archive, 3) the currently viewed document. Additionally, a Russian virtual keyboard is available for entering special characters.

After sending the search query to the backend, the results are returned and listed in the lefthand side of the interface (see Fig. 1). Every single occurrence of

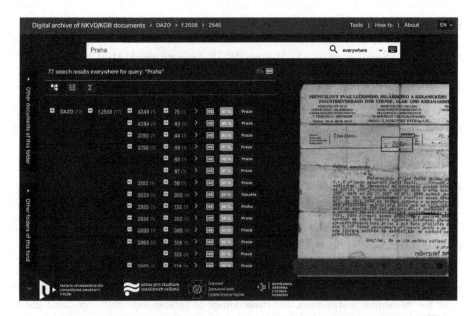

Fig. 1. GUI - showing results for search phrase "Praha" (**Lit.** *Prague*). The system found various Czech inflexion forms (*Praze, Prahy*) in many documents. Left: A tree structure of sorted results. Right: A multi-tool document viewer.

the requested phrase is returned with a confidence score. The documents then can be sorted by the highest score of contained occurrences, by the number of occurrences or hierarchically into a structured tree, which is the default option. Furthermore, results from unwanted archives can be filtered out when using the whole database scope. The list of occurrences on the lefthand side (Fig. 1) is connected to the bounding boxes in the document viewer; hence we can easily find the correct place in the image by clicking on the desired result in the list. Search results can also be downloaded into a well-arranged .csv file.

4   Conclusion

The presented demo system currently includes only a relatively small portion of the data collected by the cooperating historical institute ISTR. However, the backend solution has been proven to scale up in other tasks with much larger data sets. It is also important to stress that the current GUI design results from several rounds of graphical and functional design changes that were consulted with—and often initiated by—the actual target users, mainly the historians from the ISTR. Such changes include the addition of the Russian virtual keyboard or automated cropping of scanned documents.

References

1. Chýek, A., Šmídl, L., Švec, J.: Multimodal Dialog with the MALACH Audiovisual Archive. In: Proceedings of Interspeech 2019, pp. 3663–3664 (2019)
2. Gruber, I.: OCR improvements for images of multi-page historical documents. In: Karpov, A., Potapova, R. (eds.) SPECOM 2021. LNCS (LNAI), vol. 12997, pp. 226–237. Springer, Cham (2021). https://doi.org/10.1007/978-3-030-87802-3_21
3. Gruber, I., et al.: An automated pipeline for robust image processing and optical character recognition of historical documents. In: Karpov, A., Potapova, R. (eds.) SPECOM 2020. LNCS (LNAI), vol. 12335, pp. 166–175. Springer, Cham (2020). https://doi.org/10.1007/978-3-030-60276-5_17
4. Institute for Study of the Totalitarian Regimes (2022). https://www.ustrcr.cz/en/
5. Psutka, J., et al.: System for fast lexical and phonetic spoken term detection in a Czech cultural heritage archive. EURASIP J. Audio Speech Music Process. **2011**(1), 10 (2011). https://doi.org/10.1186/1687-4722-2011-10
6. Smith, R.: An overview of the tesseract ocr engine. In: Ninth International Conference on Document Analysis and Recognition (ICDAR 2007), vol. 2, pp. 629–633 (2007). https://doi.org/10.1109/ICDAR.2007.4376991
7. Stanislav, P., Švec, J., Ircing, P.: An engine for online video search in large archives of the holocaust testimonies. In: Proceedings of Interspeech 2016, pp. 2352–2353 (2016)
8. Zajíc, Z., et al.: Towards processing of the oral history interviews and related printed documents. In: Proceedings of the Eleventh International Conference on Language Resources and Evaluation (LREC 2018). European Language Resources Association (ELRA), Miyazaki (2018). https://aclanthology.org/L18-1331

Public News Archive: A Searchable Sub-archive to Portuguese Past News Articles

Ricardo Campos[1,2](✉) ⓘ, Diogo Correia[2], and Adam Jatowt[3] ⓘ

[1] LIAAD – INESCTEC, Porto, Portugal
ricardo.campos@ipt.pt
[2] Polytechnic Institute of Tomar, Ci2 – Smart Cities Research Center, Tomar, Portugal
aluno81470@ipt.pt
[3] University of Innsbruck, Innsbruck, Austria
adam.jatowt@uibk.ac.at

Abstract. Over the past few decades, the amount of information generated turned the Web into the largest knowledge infrastructure existing to date. Web archives have been at the forefront of data preservation, preventing the losses of significant data to humankind. Different snapshots of the web are saved everyday enabling users to surf the past web and to travel through this overtime. Despite these efforts, many people are not aware that the web is being preserved, often finding these infrastructures to be unattractive or difficult to use, when compared to common search engines. In this paper, we give a step towards making use of this preserved information to develop *"Public Archive"* an intuitive interface that enables end-users to search and analyze a large-scale of 67,242 past preserved news articles belonging to a Portuguese reference newspaper (*"Jornal Público"*). The referred collection was obtained by scraping 10,976 versions of the homepage of the *"Jornal Público"* preserved by the Portuguese web archive infrastructure (Arquivo.pt) during the time-period of 2010 to 2021. By doing this, we aim, not only to mark a stand in what respects to make use of this preserved information, but also to come up with an easy-to-follow solution, the Public Archive python package, which creates the roots to be used (with minor adaptations) by other news source providers interested in offering their readers access to past news articles.

Keywords: Web archives · Past news articles search engine · Data analysis

1 Introduction

Many web documents are preserved everyday by web archive infrastructures preventing the losses of ephemeral content. Such content preservation raises awareness of viewing web archives not only as archiving infrastructures, but also as a repository for data exploration and analysis. News outlets play an important role in this process by generating a large stream of data [9], yet making it challenging for journalist and readers alike, who have to deal with a large amount of information. This turns out to be even more difficult when the temporal dimension is added into the equation. Aware of this, several systems [1–3, 12] have been proposed over the last few years to make sense of archived data and

J. Kamps et al. (Eds.): ECIR 2023, LNCS 13982, pp. 211–216, 2023.
https://doi.org/10.1007/978-3-031-28241-6_16

to make a connection between present and past realm [13]. Despite clear advances in the literature, there is still an evident failure in the use and wide adoption of web archiving infrastructures [7] with people often privileging the recent web, even when searching for past information. In this paper, we want to raise awareness to the value of this kind of infrastructures and how they can be used, not only in an Information Retrieval perspective, but also in a Data Science dimension, particularly in data exploration and analysis. *Public Archive*[1] brings together a set of Natural Language Processing (NLP), Information Retrieval and Data Science technologies to automatically collect, extract and index valuable information from news articles belonging to a Portuguese reference newspaper ("*Jornal Público*[2]"). The interface developed allows users to explore, analyze and query a large volume of news articles preserved by the Portuguese web archive infrastructure (Arquivo.pt[3]) [8] during the time-period of 2010 to 2021. While a few reference newspapers have already developed their own archive solutions, thus offering users the chance to query past news articles, most of them do not offer a large spectrum of search, as older digital contents have been lost, and can only be found in the Arquivo.pt. This adds to the fact that small-scale media outlets, particularly those with a regional scope do not have the required infrastructures to meet the needs of developing and maintaining an in-house system. Our project lays the foundations for one such solution, by making available a system that allows newspapers to generate a dedicated searchable sub-archive collection on demand, even though news articles were not preserved by them beforehand. Anchored on Arquivo.pt, media outlets can thus offer their readers access to a searchable interface, where users can query for past news articles and look for more detailed information. Mentions to locations in the text, play here an important role by enabling geo-tagging news stories [6, 10]. Such is an important aspect, not only for readers, who can quickly access an article of their interest based on a specific location (and a particular year), but also for journalists or even historians, who have here a tool to explore geo-location aspects regarding newspaper articles coverage. To cope with this, we make available a python package[4] that eases the process of collecting past news articles and related information whenever available (title, snippet, url, author, date, related locations, organizations, people and keywords). The work described here stems from our participation at the Arquivo.pt 2022 competition, where we have been awarded the third place.

2 Architecture

Our system involves a three-tier architecture. The data layer is responsible for the process of data acquisition, data extraction, information extraction and data analysis. The application layer involves the process of indexing the information. Finally, the presentation layer consists of the web application. The architecture of the system is shown in Fig. 1 and consists of five main steps. In **Step 1: Data Acquisition**, we resort to the Text Search API to obtain the 10,976 preserved versions of the "*Jornal Público*" website within the period of 2010–2021 (contents from 2022 were not collected as Arquivo.pt

---

[1] http://arquivopublico.ipt.pt.

[2] http://www.publico.pt.

[3] http://arquivo.pt.

[4] https://github.com/diogocorreia01/PublicNewsArchive/.

has a one year embargo policy). Each of these versions is then processed in **Step 2: Data Extraction**. In this step, we develop and apply a web scraping function to automate the process of obtaining the set of news articles that are part of each collected version. In particular, we make use of the *BeautifulSoup* python library to extract the title, description, date, link and author of each news article, whenever each of these items is available. We make this function available in our python package, to be as general as possible, so as to encourage future contributions. As a rule-of-thumb, users of the python package, only have to specify, in the large majority of the cases, the HTML class of the corresponding item (title, snippet, etc.), easily available upon inspection of the webpage. While *"Public Archive"* demo has been built on top of the *"Jornal Público"*, we leveraged the opportunity to test our python package in the process of collecting news articles from further three national reference newspapers, thus confirming that our solution can be used by interested users to collect past news articles from different newspapers with minor adaptations, and from there, deepen subsequent procedures, such as named entity recognition, keyword extraction, geo-tagging and search capabilities. It is worth mentioning that, similarly to what is usually done when collecting or making use of Twitter data, we are not publishing the dataset, but the script to recreate it. Next, we move onto **Step 3: Information Extraction and Data Analysis**. In this final step, of the data layer, we conduct an Information Extraction process, where a diverse set of NLP techniques are applied. In particular we resort to Spacy (for the automatic detection of entities), Yake! [4, 5] (for the automatic extraction of relevant keywords) and to Geopy (for the mapping of locations identified by spacy in geodetic coordinates). With this information at hand, we move onto the application layer, in particular to **Step 4: Indexing**, where we make use of Elastic Search and the BM25 IR model [11] as our search engine (anchored on a Docker machine). Finally, we proceed to the presentation layer, and for our next **Step 5: Website**, where we resort to Flask to have our demo.

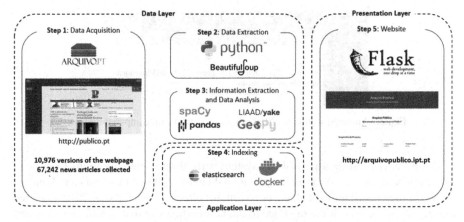

Fig. 1. Public archive architecture.

3   Demonstration

When navigating in the demo, users will have the chance to experience several features and deepen their knowledge about the collected news articles, in particular the
organizations, locations and persons most mentioned in the selected year (see Fig. 2).

Fig. 2. Top-2 organizations, locations and persons mentioned in 2019.

By clicking on any of the named-entities therein available, users will be given access
to all the corresponding news articles from that particular year. For instance, clicking on
Donald Trump image, will list, in a google-search results type interface, a total of 233
news articles (from 2019) with their corresponding title, snippet, author and publishing
date. Each news article is a clickable element that redirects the user to the archived
webpage (see Fig. 3). Dated from 09/01/2019, this snapshot reports to a time when
"Former Trump campaign director was accused of sharing information with Russians".
Further other interesting insights can be obtained from each of the 233 news articles
listed in the interface.

Fig. 3. Archived webpage (2019) about former Trump campaign director.

A further interesting aspect of our system is its ability detect locations from news
articles (whenever they exist) and to geo-map them on a google map instance. Figure 4
shows the map distribution of all the news article collected from the year 2010. Each red
icon redirects the user of the interface to the news articles of that specific location and
year.

For instance, clicking on the *"Irlanda"* white box, lists the 20 news articles written
about *"Ireland"* in the *"Jornal Publico"* throughout 2010. Figure 5 shows the top-4 results
of that list. Looking at the results, gives us valuable information about Ireland financial
crises (*"Moody's cuts Ireland's rating"*, *"Aid to Ireland implies drastic conditions"*,

Fig. 4. News articles geo-map plot 2010.

"Ireland is negotiating a request for assistance from Brussels and the IMF", *"Ireland on the verge of needing foreign aid"*). As in the previous example (Fig. 3), users may opt to click on each of the news articles to have access to the archived webpage.

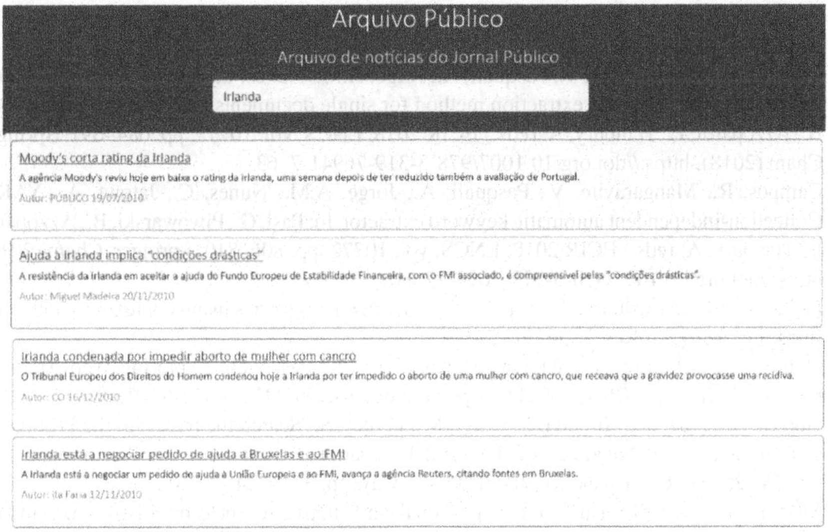

Arquivo Público

Arquivo de notícias do Jornal Público

Irlanda

Moody's corta rating da Irlanda
A agência Moody's reviu hoje em baixa o rating da Irlanda, uma semana depois de ter reduzido também a avaliação de Portugal.
Autor: PÚBLICO 19/07/2010

Ajuda à Irlanda implica "condições drásticas"
A resistência da Irlanda em aceitar a ajuda do Fundo Europeu de Estabilidade Financeira, com o FMI associado, é compreensível pelas "condições drásticas".
Autor: Miguel Madeira 20/11/2010

Irlanda condenada por impedir aborto de mulher com cancro
O Tribunal Europeu dos Direitos do Homem condenou hoje a Irlanda por ter impedido o aborto de uma mulher com cancro, que receava que a gravidez provocasse uma recidiva.
Autor: CO 16/12/2010

Irlanda está a negociar pedido de ajuda a Bruxelas e ao FMI
A Irlanda está a negociar um pedido de ajuda à União Europeia e ao FMI, avança a agência Reuters, citando fontes em Bruxelas.
Autor: Ba Faria 12/11/2010

Fig. 5. Top-4 results of Ireland news articles.

Users also have access to a yearly word cloud and the opportunity to issue a query in our search engine feature in a similar interface as to the one that can be observed in Fig. 5. Finally, we have a specific study concerning the way the newspaper covered news articles regarding the covid-19 during the year 2020. A demonstration video of our system can be found here. As future research, we plan to retrieve search results alongside with images of the named entities therein determined. To accomplish this objective, we will make use of the Arquivo.pt Image API.

Acknowledgments. This work is financed by National Funds through the Portuguese funding agency, FCT – Fundação para a Ciência e a Tecnologia, within project LA/P/0063/2020 and by the project Text2Story, financed by the ERDF – European Regional Development Fund through the Norte Portugal Regional Operational Programme – NORTE 2020 under the Portugal 2020 Partnership Agreement and by National Funds through the FCT – Fundação para a Ciência e a Tecnologia, I.P. (Portuguese Foundation for Science and Technology) within project Text2Story, with reference PTDC/CCI-COM/31857/2017 (NORTE-01-0145-FEDER-031857).

References

1. AlNoamany, Y., Weigle. M.C., Nelson. M.L.: Generating stories from archived collections. In: Proceedings of the 2017 ACM Conference on Web Science, WebSci 2017, pp. 309–318 (2017)
2. Alonso, O., Berberich, K., Bedathur, S., Weikum, G.: NEAT: news exploration along time. In: Gurrin, C., et al. (eds.) ECIR 2010. LNCS, vol. 5993, pp. 667–667. Springer, Heidelberg (2010). https://doi.org/10.1007/978-3-642-12275-0_72
3. Campos, R., Pasquali, A., Jatowt, A., Mangaravite, V., Jorge, A.: Automatic generation of timelines for past-web events. In: Gomes, D., Demidova, E., Winters, J., Risse, T. (eds.) The Past Web. Exploring Web Archives, pp. 225–242. Springer, Cham (2021). https://doi.org/10.1007/978-3-030-63291-5_18
4. Campos, R., Mangaravite, V., Pasquali, A., Jorge, A.M., Nunes, C., Jatowt, A.: A text feature based automatic keyword extraction method for single documents. In: Pasi, G., Piwowarski, B., Azzopardi, L., Hanbury, A. (eds.) ECIR 2018. LNCS, vol. 10772, pp. 684–691. Springer, Cham (2018). https://doi.org/10.1007/978-3-319-76941-7_63
5. Campos, R., Mangaravite, V., Pasquali, A., Jorge, A.M., Nunes, C., Jatowt, A.: YAKE! Collection-independent automatic keyword extractor. In: Pasi, G., Piwowarski, B., Azzopardi, L., Hanbury, A. (eds.) ECIR 2018. LNCS, vol. 10772, pp. 806–810. Springer, Cham (2018). https://doi.org/10.1007/978-3-319-76941-7_80
6. Ferdous, Md., Chowdhury, S., Jose, J.: Geo-tagging news stories using contextual modelling. Int. J. Inf. Retrieval Res. **7**, 50–71 (2017)
7. Gomes, D., Demidova, E., Winters, J., Risse, T.: The Past Web: Exploring Web Archives, pp. 1–297. Springer, Cham (2021). https://doi.org/10.1007/978-3-030-63291-5
8. Gomes, D., Cruz, D., Miranda, J., Costa, M., Fontes, S.: Search the past with the Portuguese web archive. In: Proceedings of the 22nd International Conference on World Wide Web (WWW 2013), Rio de Janeiro, Brazil, 13–17 May, pp. 321–324 (2013)
9. Martinez-Alvarez, M., et al.: First international workshop on recent trends in news information retrieval (NewsIR'16). In: Ferro, N., et al. (eds.) ECIR 2016. LNCS, vol. 9626, pp. 878–882. Springer, Cham (2016). https://doi.org/10.1007/978-3-319-30671-1_85
10. Rafiei, J., Rafiei, D.: Geotagging named entities in news and online documents. In: Proceedings of the 25th ACM International Conference on Information and Knowledge Management (CIKM 2016), USA, 24–28 October, pp. 1321–1330 (2016)
11. Robertson, S., Zaragoza, H.: The probabilistic relevance framework: BM25 and beyond. In Foundations and Trends in Information Retrieval. **3**(4), 333–389 (2009)
12. Saleiro, P., Teixeira, J., Soares, C., Oliveira, E.: TimeMachine: entity-centric search and visualization of news archives. In: Ferro, N., et al. (eds.) ECIR 2016. LNCS, vol. 9626, pp. 845–848. Springer, Cham (2016). https://doi.org/10.1007/978-3-319-30671-1_78
13. Sato, M., Jatowt, A., Duan, Y., Campos, R., Yoshikawa, M.: Estimating contemporary relevance of past news. In: Proceedings of the ACM/IEEE Joint Conference on Digital Libraries (JCDL 2021), 27–30 September, pp. 70–79 (2021)

TweetStream2Story: Narrative Extraction from Tweets in Real Time

Mafalda Castro[1,2]($\boxtimes$), Alípio Jorge[1,2], and Ricardo Campos[1,3]

[1] INESC TEC, Porto, Portugal
mafalda.r.castro@inesctec.pt
[2] FCUP, University of Porto, Porto, Portugal
amjorge@fc.up.pt
[3] Polytechnic Institute of Tomar, Ci2 - Smart Cities Research Center,
Tomar, Portugal
ricardo.campos@ipt.pt

Abstract. The rise of social media has brought a great transformation to the way news are discovered and shared. Unlike traditional news sources, social media allows anyone to cover a story. Therefore, sometimes an event is already discussed by people before a journalist turns it into a news article. Twitter is a particularly appealing social network for discussing events, since its posts are very compact and, therefore, contain colloquial language and abbreviations. However, its large volume of tweets also makes it impossible for a user to keep up with an event. In this work, we present TweetStream2Story, a web app for extracting narratives from tweets posted in real time, about a topic of choice. This framework can be used to provide new information to journalists or be of interest to any user who wishes to stay up-to-date on a certain topic or ongoing event. As a contribution to the research community, we provide a live version of the demo, as well as its source code.

Keywords: Narrative extraction · Natural language processing · Twitter

1 Introduction

Social media is a powerful tool that can provide great insights into a variety of topics. Using Twitter posts as a source for extracting narratives may bring us different information than a news article does, from the people who experience an event first-hand. The Twitter platform is a very helpful tool for journalists [4,7], however, its colloquial language and the large volume of *tweets* (6000 *tweets* are posted every second, on average [12]) makes it impractical to keep up with an event. For this reason, obtaining the most relevant tweet posts turns out to be of the utmost importance. To achieve this, researchers have presented a variety of methods regarding the automatic summarization of *tweet* streams [1,3,6,9,10,15], although none of these had narrative extraction in mind [11]. Recently, Campos et al. [2] have proposed the Tweet2Story

J. Kamps et al. (Eds.): ECIR 2023, LNCS 13982, pp. 217–223, 2023.
https://doi.org/10.1007/978-3-031-28241-6_17

framework[1], which performs the automatic narrative extraction from a bundle of *tweets*. However, this framework doesn't work in real time, requiring users to previously collect and process the *tweets* that will be given as input. In this paper, we present TweetStream2Story[2], an extension of Tweet2Story that fills this gap, by incorporating the real-time collection of *tweets* on a given topic, as well as the automatic extraction of narratives from these *tweets*. As a further contribution to the research community, we make the source code of our project available, thus challenging researchers to use and expand it[3].

2   TweetStream2Story

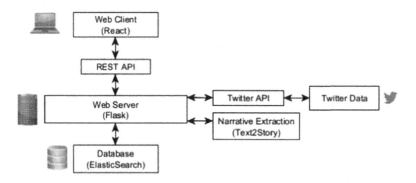

Fig. 1. Architecture overview of TweetStream2Story

Figure 1 depicts the architecture of the TweetStream2Story framework. The first step of this pipeline is to issue a topic, a query (e.g. Denmark Shooting) in the user interface (the web client) to search for related *tweets*. The user must also provide the time period for the collection of *tweets* (e.g. July 4 2022, 4:30 pm to July 5 2022, 12:30 am), which will be divided into time windows of a specified duration (e.g. 2 h). The narrative will be generated in two modes: in the global mode, each time window uses *tweets* since the start of the topic ([4:30 pm–6.30 pm], [4:30 pm - 8:30 pm], and on and so forth). In the interval mode, instead, each time window uses *tweets* posted strictly during that time window ([4:30 pm–6:30 pm], [6:30 pm–08:30 pm], and so on and so forth). Once this information is defined, we proceed by obtaining the collection of related *tweets* using either the Twitter API's Filtered stream, in case the user wants to follow up *tweets* posted in real-time, or the Full-archive search, to look for events in the past. For every collected *tweet*, a preprocessing stage, involving hashtags removal, hyperlinks and emojis is applied. Similar tweets, with a term-frequency cosine similarity higher

---

[1] http://tweet2story.inesctec.pt/.

[2] http://tweetstream2story.inesctec.pt/.

[3] https://github.com/LIAAD/tweetstream2story.

than 80% are also removed. The resulting set is then stored in Elasticsearch, a flexible document-oriented database. To reduce the amount of *tweets*, we then proceed with a summarization-like step where only the most relevant tweets are taken into account. To do this, we use, as in Rishab S. et al. [14], the Okapi BM25 function [8] as our IR model, a function that estimates the relevance of a document to a given search query, and by that, retrieve the top-X most relevant *tweets* belonging to a particular time window, where X equals 50 (a trade-off between the number of *tweets* and their Precision). Following, we proceed to use these *tweets* as input to the Text2Story narrative extraction pipeline. In the coming section, we demonstrate how such pipeline is used to create a visual representation of the topics narrative.

3   Demo

In this section, we describe the main features of this demo. Its live version can be used by anyone who wishes to extract the narrative of a specified topic from *tweets* posted either in real time or in the past. The first step for generating a narrative requires the user to input a topic of their interest. After typing in a topic and clicking on the Extract Narrative button, a modal opens where the user can specify parameters such as the desired language, the duration of each time window (e.g. 2 h), and the mode for collecting tweets (e.g. streaming, past *tweets*). Currently, the only languages supported are English and Portuguese. Although the focus of this work is the retrieval of *tweets* posted in real time, our framework also allows retrieving past *tweets*. In this case, however, the user must provide their Twitter API credentials, which will not be stored, but discarded as soon as they're used. Topics are automatically added to a private list of topics, owned by the user, allowing them to keep track of its status, visualize the corresponding narrative or perform actions such as stopping the retrieval of tweets or deleting the topic from the list. Figure 2 shows the interface for the list of topics therein presented.

#	Topic	Language	Start Date	End Date	Status	Options		
1	seleção portugal espanha	PT	2022-06-02 19:45	2022-06-02 22:15	✓ Completed	Open	🗑	
2	Queen Elizabeth death	EN	2022-09-10 15:27	2022-09-11 22:00	C In progress...	Open	Stop	

Fig. 2. Topics list

By clicking on a topic, users are offered the chance to visualize its narrative through a timeline, as shown in Fig. 3. Added, they can choose between the two modes mentioned before: global view or interval view. In each mode, the timeline is divided into time windows with the duration previously specified by the user, where each one shows its respective narrative and information. The default visualization of a narrative is the knowledge graph, which shows actors as nodes and semantic relationships as the edges between the actors. It also highlights in yellow the nodes that weren't present in the previous time window, as a way for the user to quickly see new information. Other modes of visualization include the list of tweets that were used to generate the narrative, as well as the list of actors, as can be seen in Fig. 4. Further advanced analysis can also be performed by viewing and downloading the information in formal representations, as is the case of DRS annotations [5] or the Text2Story annotation [13].

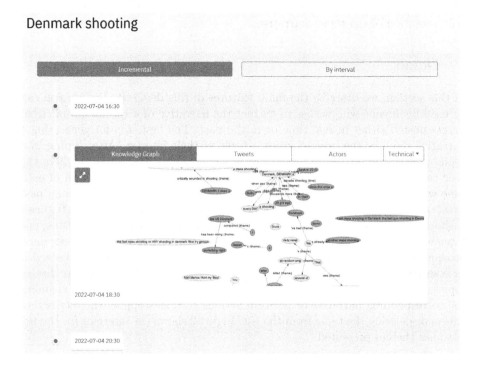

Fig. 3. Timeline representation of a topic

As a means to demonstrate Twitter's potential for narrative extraction, some examples of topics in both Portuguese and English, are pre-loaded in the interface. Figure 4 shows a visual representation of the topic *Denmark Shooting*, an event that occurred in Copenhagen in 2022. This knowledge graph is able to capture information about the number of deaths, critically wounded people, and previous shootings in the country. These examples are able to demonstrate Twitter's usefulness as a news source, as the information contained in some of the extracted actors and relations is able to complement a news article.

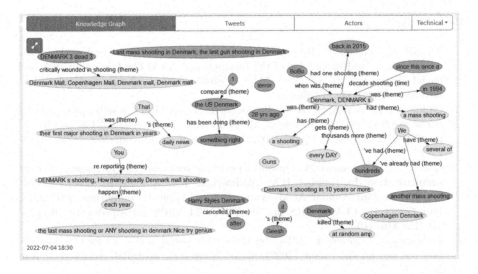

Fig. 4. Narrative representation of a time window

4   Conclusions and Future Work

In this paper, we have presented a framework that allows the automatic collection of *tweets* and extraction of their narrative elements, TweetStream2Story. This tool can be beneficial not only for journalists, but also for users interested in an ongoing event. Some of its limitations are the requirement for a user to enter their Twitter API credentials when generating narratives from events in the past, and the long computational time to extract the narrative. In the future, we would like to improve the quality of the results by incorporating techniques such as irony detection and offensive speech, as a way to filter out some *tweets*. We also plan on improving the user-system interactions, as well as implementing an abstractive summarization approach, in order to use original content as the source of the narratives.

Acknowledgements. This work is financed by National Funds through the Portuguese funding agency, FCT - Fundação para a Ciência e a Tecnologia, within project LA/P/0063/2020. The authors Alípio Jorge and Ricardo Campos are financed by the project Text2Story, financed by the ERDF - European Regional Development Fund through the Norte Portugal Regional Operational Programme - NORTE 2020 under the Portugal 2020 Partnership Agreement and by National Funds through the FCT - Fundação para a Ciência e a Tecnologia, I.P. (Portuguese Foundation for Science and Technology) within project Text2Story, with reference PTDC/CCI-COM/31857/2017 (NORTE-01-0145-FEDER-031857).

References

1. Alsaedi, N., Burnap, P., Rana, O.: Automatic summarization of real world events using twitter. In: Proceedings of the International AAAI Conference on Web and Social Media, vol. 10, no. 1, pp. 511–514 (2021). https://ojs.aaai.org/index.php/ICWSM/article/view/14766

2. Campos, V., Campos, R., Mota, P., Jorge, A.: Tweet2Story: a web app to extract narratives from twitter. In: Hagen, M., et al. (eds.) ECIR 2022. LNCS, vol. 13186, pp. 270–275. Springer, Cham (2022). https://doi.org/10.1007/978-3-030-99739-7_32

3. Chellal, A., Boughanem, M.: Optimization framework model for retrospective tweet summarization. In: Haddad, H.M., Wainwright, R.L., Chbeir, R. (eds.) Proceedings of the 33rd Annual ACM Symposium on Applied Computing, SAC 2018, Pau, France, 09–13 April 2018, pp. 704–711. ACM (2018). https://doi.org/10.1145/3167132.3167210

4. Jurkowitz, M., Gottfried, J.: Twitter is the go-to social media site for U.S. journalists, but not for the public (2022). https://pewrsr.ch/3yqfpRP. Accessed 31 Aug 2022

5. Kamp, H., Reyle, U.: From Discourse to Logic: Introduction to Model-theoretic Semantics of Natural Language, Formal Logic and Discourse Representation Theory. Springer, Dordrecht (1993). https://doi.org/10.1007/978-94-017-1616-1

6. Li, Q., Zhang, Q.: Twitter event summarization by exploiting semantic terms and graph network. In: Proceedings of the AAAI Conference on Artificial Intelligence, vol. 35, no. 17, pp. 15347–15354 (2021). https://ojs.aaai.org/index.php/AAAI/article/view/17802

7. MuckRack: The state of journalism 2022 (2022). https://muckrack.com/blog/2022/05/18/2022-state-of-journalism-on-twitter. Accessed 31 Aug 2022

8. Robertson, S., Zaragoza, H.: The probabilistic relevance framework: Bm25 and beyond. Found. Trends Inf. Retr. **3**, 333–389 (2009). https://doi.org/10.1561/1500000019

9. Rudra, K., Ghosh, S., Ganguly, N., Goyal, P., Ghosh, S.: Extracting situational information from microblogs during disaster events: a classification-summarization approach. In: Proceedings of the 24th ACM International Conference on Information and Knowledge Management, CIKM 2015, Melbourne, VIC, Australia, 19–23 October 2015, pp. 583–592. ACM (2015). https://doi.org/10.1145/2806416.2806485

10. Rudra, K., Goyal, P., Ganguly, N., Imran, M., Mitra, P.: Summarizing situational tweets in crisis scenarios: an extractive-abstractive approach. IEEE Trans. Comput. Social Syst. **6**(5), 981–993 (2019). https://doi.org/10.1109/TCSS.2019.2937899

11. Santana, B., Campos, R., Amorim, E., Jorge, A., Silvano, P., Nunes, S.: A survey on narrative extraction from textual data. Artif. Intell. Rev. (2023). https://doi.org/10.1007/s10462-022-10338-7

12. Sayce, D.: The number of tweets per day in 2022 (2022). https://www.dsayce.com/social-media/tweets-day/. Accessed 25 Sept 2022

13. Silvano, P., Leal, A., Cantante, I., Oliveira, F., Mario Jorge, A.: Developing a multilayer semantic annotation scheme based on ISO standards for the visualization of a newswire corpus. In: Proceedings of the 17th Joint ACL - ISO Workshop on Interoperable Semantic Annotation, pp. 1–13. Association for Computational Linguistics, Groningen, The Netherlands (online) (2021). https://aclanthology.org/2021.isa-1.1

14. Singla, R., Modha, S., Majumder, P., Mandalia, C.: Information extraction from microblog for disaster related event. In: Proceedings of the First International Workshop on Exploitation of Social Media for Emergency Relief and Preparedness co-located with European Conference on Information Retrieval, SMERP@ECIR 2017, Aberdeen, UK, 9 April 2017. CEUR Workshop Proceedings, vol. 1832, pp. 85–92. CEUR-WS.org (2017). http://ceur-ws.org/Vol-1832/SMERP-2017-DC-DAIICT-IR-LAB-Retrieval.pdf
15. Wang, Z., Shou, L., Chen, K., Chen, G., Mehrotra, S.: On summarization and timeline generation for evolutionary tweet streams. IEEE Trans. Knowl. Data Eng. **27**(5), 1301–1315 (2015). https://doi.org/10.1109/TKDE.2014.2345379

SimpleRad: Patient-Friendly Dutch Radiology Reports

Koen Dercksen[1,2(✉)] ⓘ, Arjen P. de Vries[1] ⓘ, and Bram van Ginneken[2] ⓘ

[1] Radboud University, Nijmegen, The Netherlands
{koen,arjen}@cs.ru.nl
[2] Radboud University Medical Center, Nijmegen, The Netherlands
bram.vanginneken@radboudumc.nl

Abstract. Patients increasingly have access to their electronic health records. However, much of the content therein is not specifically written for them; instead it captures communication about a patient's situation between medical professionals. We present SimpleRad, a prototype application to explore patient-friendly explanations of radiology terminology. In this demonstration paper, we describe the various modules currently included in SimpleRad such as an entity linker, summarizer, search page, and observation frequency estimator.

Keywords: Natural language processing · Information retrieval · Medical

1 Motivation and Background

Radiology reports are technical medical texts describing a radiologists interpretation of a particular radiology image (e.g. an x-ray or a CT-scan). The report is meant strictly for medical professionals, and is not intended to be read by patients. However, as a result of regulation like the European GDPR and similar legal developments, patients increasingly do have access to these reports. The technical nature of the reports can cause unnecessary anxiety and stress due to a lack of understanding. This potentially leads to extra unnecessary interactions with medical doctors that could be avoided.

When looking for technological solutions to overcome this problem, we run into complications due to a lack of resources. The reports we work with are written in Dutch, and most available tooling for medical texts is mainly aimed at English [5,9]. Ondov et al. conclude that the vast majority of research on medical text simplification uses English resources [7]. Ideally, we want to offer easily understandable information in Dutch to clarify the reports. While there are Dutch websites with patient-friendly medical information like Thuisarts ("home doctor")[1], Gezondheidsplein ("health square")[2] etc., there is a disconnect with

---

[1] https://thuisarts.nl, accessed 2023-01-12.
[2] https://gezondheidsplein.nl, accessed 2023-01-12.

© The Author(s), under exclusive license to Springer Nature Switzerland AG 2023
J. Kamps et al. (Eds.): ECIR 2023, LNCS 13982, pp. 224–229, 2023.
https://doi.org/10.1007/978-3-031-28241-6_18

radiology terminology. Explanations for frequently used radiology terms are often not available on these websites, which tend to aim more at general medical concepts (like diseases, procedures and body parts). Qenam et al. conclude the same for some English patient-oriented medical vocabularies, stating that radiology-specific terms should be added to the vocabulary they used in order for it to be applicable to radiology reports [8].

In order to address this situation, we develop SimpleRad: a tool for patient-friendly presentation of Dutch radiology reports. The main goals of SimpleRad are to provide Dutch patient-friendly information about technical jargon in the reports, and to give some sense of importance of the findings contained in the reports.

In terms of technological methods to accomplish this, we have focused on entity recognition, summarization and search-related techniques. We describe each of these below. We also provide a video demonstration with English captions for clarity.[3]

2 SimpleRad

SimpleRad is a web application that allows people to examine various parts of their radiology reports. It is made up of a separate back-end and front-end application, as well as a database containing medical concepts and their patient-friendly descriptions. In this section, we describe the various parts that make up SimpleRad.

2.1 Architecture

The SimpleRad codebase is split up into a front-end and back-end repository. The back-end is implemented with FastAPI[4], and is generally configurable in terms of parameters, models etc. used for various endpoints. This means that the back-end can also be used standalone, for example to do batch requests on only one endpoint (e.g. to obtain entity predictions for a set of reports). Each back-end module is configurable using a simple YAML file, and usually has multiple different *flavours* (see the respective module sections).

The frontend is implemented using ReactJS[5], primarily to have a single interface to test the different SimpleRad modules. A general overview of the interface can be viewed in Fig. 1.

The front- and back-end code is available on Github, and is actively worked on.[6]

---

[3] https://youtu.be/2316wQTboVg, accessed 2023-01-12.

[4] https://fastapi.tiangolo.com/, accessed 2023-01-12.

[5] https://reactjs.org/, accessed 2023-01-12.

[6] https://github.com/KDercksen/simplerad-frontend,
https://github.com/KDercksen/simplerad-backend, accessed 2023-01-12.

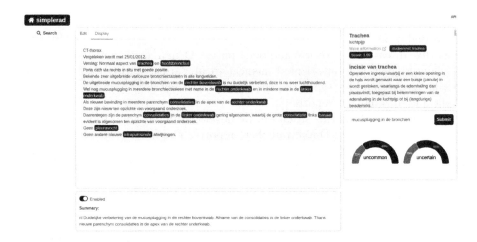

Fig. 1. SimpleRad interface. The radiology report with highlighted entities (top left), summarization (bottom left), entity information (top right), and observation frequency estimation (bottom right). See the video link in Footnote 3 for a demonstration with English captions.

2.2   Data

Radboud University Medical Center (Radboudumc) has a large set of approximately 250K anonymized radiology reports describing CT thorax/abdomen scans done between 2000–2021. We use this set of reports to train embedding models and create derived datasets for summarization and entity linking.

In order to construct the concept database, we pull information from various sources. The largest source is Nictiz' Dutch translation of SNOMEDCT. [7] We also pull data from sites like Thuisarts, Gezondheidsplein and others. We manually created a set of approximately 120 concepts that are most frequent in the Radboudumc collection, but not included in any of the other sources. This set will increase over time as we add more missing concepts. When a data source exposes synonyms for a particular concept, we include those as well. For example, we extract synonyms for many concepts from UMLS [3]. At the time of writing, the database contains around 7000 unique concepts with patient-friendly descriptions.

Data and indexing scripts are available upon request to the authors (some sources have their own distribution licensing).

2.3   Entity Recognition and Search

One of the most important usecases of SimpleRad is the ability to find clarifying information for medical concepts. We have implemented various entity recognition methods to tag medical concepts in Dutch radiology reports. The interface

---

[7] https://nictiz.nl/, accessed 2023-01-12.

highlights concepts—clicking on a concept lists the best matches from the concept database. Each match is displayed with a title and description, as well as its source, score and a link to the original article (available only for some data sources like Thuisarts).

Entity recognition and linking is divided into two separate back-end calls, in order to easily swap methods. SimpleRad also provides a search-only page, where you simply search the concept database with custom queries (skipping the entity recognition step altogether).

Currently implemented back-ends include a SimString-based fuzzy matching algorithm [6] and Flair-compatible neural models [1].

2.4 Summarization

The summarization module provides an interface to (abstractive) summarization models. The ultimate goal is a summarization model that outputs simplified summaries of the radiology reports. However, we do not yet have the data necessary to train such a model. Instead, we have models that automatically generate conclusions based on the findings in the entered radiology report. The back-end can be configured to use any text-to-text generation model compatible with the HuggingFace transformers library.[8]

Since these summarization models can be resource intensive, it is possible to disable this module from the front-end (i.e., the summarization back-end will not receive requests).

2.5 Observation Frequency Estimation

This final module situates observations (or *findings*) from the report within the collection. The goal is to assign noteworthiness to specific findings. For example, "degenerative changes in skeleton" may sound disconcerting, but once you know that this finding occurs in 90% of radiology reports there is probably nothing to worry about.

For any particular finding query, SimpleRad presents the estimated frequency of that finding occurring in the collection (based on embedding similarity), as well as an indication of the certainty of the given frequency estimation. The certainty estimation is based on the number of similar findings that are found within an acceptable range of similarity scores (i.e., on how many samples is the frequency estimation based). The acceptable range is determined automatically based on a sampling strategy that models the average finding similarity distribution over the collection. Any findings outside of this range are typically too different from the finding query to be included.

Additionally, returned estimations are discretized into easily understable terms like "very certain" or "rare" to help with interpretation of the values.

---

[8] https://huggingface.co/transformers, accessed 2023-01-12.

3 Final Remarks

We believe that, while still under development, SimpleRad is already well on its way to becoming a good tool to evaluate various parts of the patient-friendly report pipeline. Its modularity allows for easy changes to both front-end and back-end, making it trivial to add extra components. For example, if we were to integrate imaging information alongside the reports, it would be trivial to add visualization of e.g. segmentation maps or other model outputs as well.

The standalone front-end also opens up possibilities for user experience researchers to experiment with various presentations of data. Presentation is vital in our imagined usecase, and up until now SimpleRad has primarily been developed from a technical perspective.

To evaluate the use of SimpleRad for laymen, we plan to conduct a patient survey to investigate if respondents understand reports better when using the tool. Once we move on to offering simplified summaries, there are various quantitative metrics available to evaluate the summary quality alongside human judgment [2,4,10].

Other future work includes expanding the size of the concept database, and improving the quality of models used for the various tasks. We are also looking into adding functionality for collecting user feedback on summarization quality, missed entities, incorrect descriptions and more. With such functionality, SimpleRad could eventually be used for patient studies. We are working on publishing more detailed evaluations on the various models that underpin SimpleRad.

References

1. Akbik, A., Bergmann, T., Blythe, D., Rasul, K., Schweter, S., Vollgraf, R.: FLAIR: an easy-to-use framework for state-of-the-art NLP. In: Proceedings of the 2019 conference of the North American chapter of the Association for Computational Linguistics (demonstrations), pp. 54–59 (2019)
2. Van den Bercken, L., Sips, R.J., Lofi, C.: Evaluating neural text simplification in the medical domain. In: The World Wide Web Conference, pp. 3286–3292 (2019)
3. Bodenreider, O.: The unified medical language system (UMLS): integrating biomedical terminology. Nucleic Acids Res. **32**(suppl_1), D267–D270 (2004)
4. Devaraj, A., Marshall, I.J., Wallace, B.C., Li, J.J.: Paragraph-level simplification of medical texts. arXiv preprint arXiv:2104.05767 (2021)
5. Kraljevic, Z., et al.: MedCAT-medical concept annotation tool. arXiv preprint arXiv:1912.10166 (2019)
6. Okazaki, N., Tsujii, J.: Simple and efficient algorithm for approximate dictionary matching. In: Proceedings of the 23rd International Conference on Computational Linguistics (Coling 2010), pp. 851–859 (2010)
7. Ondov, B., Attal, K., Demner-Fushman, D.: A survey of automated methods for biomedical text simplification. J. Am. Med. Inf. Assoc. **29**(11), 1976–1988 (2022)
8. Qenam, B., Kim, T.Y., Carroll, M.J., Hogarth, M., et al.: Text simplification using consumer health vocabulary to generate patient-centered radiology reporting: translation and evaluation. J. Med. Internet Res. **19**(12), e8536 (2017)

9. Soldaini, L., Goharian, N.: QuickUMLS: a fast, unsupervised approach for medical concept extraction. In: MedIR Workshop, SIGIR, pp. 1–4 (2016)
10. Trienes, J., Schlötterer, J., Schildhaus, H.-U., Seifert, C.: Patient-friendly clinical notes: towards a new text simplification dataset. In: TSAR workshop, EMNLP (2022)

Automated Extraction of Fine-Grained Standardized Product Information from Unstructured Multilingual Web Data

Alexander Flick[1]([✉])[iD], Sebastian Jäger[1][iD], Ivana Trajanovska[1][iD], and Felix Biessmann[1,2][iD]

[1] Berlin University of Applied Sciences and Technology, Berlin, Germany
alexander.flick@bht-berlin.de
[2] Einstein Center Digital Future, Berlin, Germany

Abstract. Extracting structured information from unstructured data is one of the key challenges in modern information retrieval applications, including e-commerce. Here, we demonstrate how recent advances in machine learning, combined with a recently published multilingual data set with standardized fine-grained product category information, enable robust product attribute extraction in challenging transfer learning settings. Our models can reliably predict product attributes across online shops, languages, or both. Furthermore, we show that our models can be used to match product taxonomies between online retailers.

Keywords: Product information extraction · E-commerce

1 Introduction

Recent research achievements in the field of machine learning (ML) [1,13] have the potential to improve automated information extraction in applications such as e-commerce. However, the translation of these ML innovations into real-world application scenarios is impeded by the lack of publicly available data sets. Here we demonstrate that recent advances in ML can be translated into automated information extraction applications when leveraging carefully curated data. To better assess the contribution of this study, we first highlight some relevant data sets and methods that aim at the automated extraction of structured data in the field of e-commerce.

Public E-commerce Data Sets. We summarize publicly e-commerce data sets used for the automated extraction of product information in Table 1. To leverage the potential of ML, large and diverse data sets that follow a fine-grained product taxonomy are favorable. A common and detailed taxonomy is the Global Product Classification (GPC) standard, which "classifies products by grouping them into categories based on their essential properties as well as their relationships to

Table 1. Comparison of e-commerce data sets used for product attribute extraction and classification. Column *GPC* means whether or not the data set follows the GPC taxonomy.

	Regular	Multi-			GPC	Size
	Updated	Lingual	Shop	Family		
Farfetch product meta data [9]	✗	✗	✗	✗	✗	400 K
Product details on Flipkart [3]	✗	✗	✗	✓	✗	20 K
Amazon browse node classification [2]	✗	✗	✗	✓	✗	3 M
Amazon product-question answering [16]	✗	✗	✗	✓	✗	17.3 GB
Rakuten data challenge [10]	✗	✗	✗	✓	✗	1 M
MAVE [18]	✗	✗	✗	✓	✗	2.2 M
Innerwear from victoria's secret & co [15]	✗	✗	✓	✗	✗	600 K
WDC-MWPD [19]	✗	✗	✓	✗	✓	16 K
WDC-25 gold standard [14]	✗	✗	✓	✓	✓	24 K
GreenDB [7]	✓	✓	✓	✓	✓	>576 K

other products" [4]. For example, multiple *Brick*s (shirts and shorts) can belong to the same *Family* (clothing) but are different *Classes* (upper and lower body wear)[1].

Multilingual Fine-Grained Product Classification. There are few recent studies investigating automated extraction of standardized product information in text corpora. Brinkmann et al. [1] study how hierarchical product classification benefits from domain-specific language modeling. They report an improvement of 0.012 weighted F1 score by using schema.org product[2] annotations for pre-training. Peeters et al. [12] study cross-language learning for entity matching and demonstrate that multilingual transformers outperform single-language models (German BERT) by 0.143 F1 when trained on a single language (German) and tested on multiple (German and English). Furthermore, using additional training data for the second language (English) improves the performance by another 0.038 weighted F1.

These studies highlight the potential of modern ML methods for automated product attribute extraction. In this work, we show that transfer learning helps to extract structured information (product category) from unstructured data (product name and description) and to find reliable taxonomy mappings.

2 Experiments

We evaluate three transfer learning scenarios for product classification:

1. **Language Transfer:** training on data of one language, test on other language data.

---
[1] See the GPC Browser for more examples: https://gpc-browser.gs1.org/.
[2] Website: https://schema.org/Product.

2. **Shop Transfer:** training on data of one shop, test on other shop data.
3. **Language and Shop Transfer:** training on data of one shop and one language, test on data of different shops and languages.

Furthermore, we study whether ML methods can be used to find reliable taxonomy mappings. For this, we apply a model trained for a *target taxonomy* to data that uses a *source taxonomy*. For each source category, the majority of predicted target categories define the mapping from source to target taxonomy.

Data Sets. In our experiments, we use two data sets, the GreenDB [6] and the Farfetch data set [9]. The GreenDB[3] is a multilingual data set covering 5 European shops with about 576k unique products of the 37 most important product categories following the GPC taxonomy. It covers categories from the GPC segments Clothing, Footwear, Personal Accessories, Home Appliances, Audio Visual/Photography, and Computing. A recent publication [8] presents the GreenDB's high quality and usefulness for information extraction tasks. The Farfetch data set has about 400k unique products from a single shop. It does not follow a public taxonomy and covers only fashion products.

ML Model. The experiment implementation is based on autogluon's [17] TextPredictor and uses *mDeBERTav3* [5] as the backbone model. For training, we use the GreenDB and apply Cleanlab [11] to find and remove miss-classified products (211 were found). Our models use the product's name and description to predict their product category. $model_{baseline}$ is trained on the entire GreenDB (all shops), $model_{ZaDE}$ on the German, $model_{ZaFR}$ on the French, and $model_{ZaALL}$ on the German, French, and English Zalando products contained in the GreenDB.

Online Demo. To demonstrate the transfer capabilities, we published an online demo available: https://product-classification.demo.calgo-lab.de. As shown in Fig. 1, it automatically downloads the HTML of a given URL, extracts the products' name and description, and uses $model_{baseline}$ to predict its GPC category.

3   Results

The baseline performance ($model_{baseline}$) shows a strong 0.99 weighted F1 score on a GreenDB test set.

Transfer Tasks. $model_{ZaDE}$ demonstrates language transfer when it is applied to other languages of the same shop. It achieves weighted F1 scores of 0.898 for English and 0.873 for French. Applying $model_{ZaFR}$ and $model_{ZaDE}$ on other shops demonstrates shop transfer with weighted F1 scores from 0.648 to 0.836. If the model is fine-tuned on multi-lingual data ($model_{ZaALL}$), almost all shops benefit, see Table 2 for details. The language and shop transfer is even more challenging and performs worse for all shops. Transferring across data sets, i.e., applying $model_{baseline}$ to Farfetch data, achieves a 0.924 weighted F1 score.

---

[3] We use GreenDB version 0.2.2 available at https://zenodo.org/record/7225336.

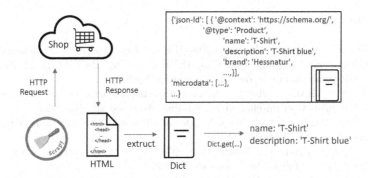

Fig. 1. Online demo overview. Automated extraction of schema.org information (product name and description) from HTML, used for product classification.

Table 2. Weighted F1 scores for shop transfer experiments. Scores from 0.648 to 0.836 demonstrate robust shop transfer. Shop transfer profits from additional data in other languages.

	Model	FR		DE	
		Asos	H&M	Otto	Amazon
Shop Transfer	$model_{ZaFR}$	0.836	0.678	–	–
	$model_{ZaDE}$	–	–	0.777	0.648
	$model_{ZaALL}$	0.842	0.717	0.762	0.739
Shop & Language Transfer	$model_{ZaFR}$	–	–	0.614	0.449
	$model_{ZaDE}$	0.795	0.666	–	–

Taxonomy Matching. Using $model_{baseline}$ to map products' categories from Far-fetch to GreenDB (GPC taxonomy) results in 41 out of 46 (>89%) correctly mapped categories.

4 Conclusion

We demonstrate that combining rich multilingual data sets and modern ML methods enables fine-grained standardized product information extraction from unstructured data. We investigate several transfer learning settings when training and testing on data from different shops and languages, even in zero-shot scenarios when no data from another shop and language was available in the training data.

Acknowledgements. This research was supported by the Federal Ministry for the Environment, Nature Conservation and Nuclear Safety based on a decision of the German Bundestag.

References

1. Brinkmann, A., Bizer, C.: Improving Hierarchical Product Classification using Domain-specific Language Modelling. IEEE Data Eng. Bull. **44**(2), 14–25 (2021). http://sites.computer.org/debull/A21june/p14.pdf
2. Challenge, A.M.: (2022). https://www.hackerearth.com/en-us/challenges/competitive/amazon-ml-challenge/, [Online; Accessed 23 May 2022]
3. Flipkart: (2022). https://www.kaggle.com/PromptCloudHQ/flipkart-products, [Online; Accessed 23 May 2022]
4. GS1: Global Product Classification (GPC) — GS1. https://www.gs1.org/standards/gpc, [Online; Accessed Oct 20 2022]
5. He, P., Gao, J., Chen, W.: Debertav 3: Improving deberta using electra-style pre-training with gradient-disentangled embedding sharing. CoRR abs/2111.09543 (2021). https://doi.org/10.48550/arxiv.2111.09543
6. Jäger, S., Greene, J., Jakob, M., Korenke, R., Santarius, T., Biessmann, F.: GreenDB: Toward a Product-by-Product Sustainability Database. Tech. rep., arXiv (May 2022). https://doi.org/10.48550/arXiv.2205.02908
7. Jäger, S., Bießmann, F., Flick, A., Sanchez Garcia, J.A., von den Driesch, K., Brendel, K.: GreenDB: A Product-by-Product Sustainability Database (Feb 2022). https://doi.org/10.5281/zenodo.6576662,Supported by the Federal Ministry for the Environment, Nature Conservation and Nuclear Safety based on a decision of the German Bundestag. Förderkennzeichen: 67KI2022B
8. Jäger, S., Flick, A., Garcia, J.A.S., Driesch, K.v.d., Brendel, K., Biessmann, F.: GreenDB - A Dataset and Benchmark for Extraction of Sustainability Information of Consumer Goods (Aug 2022). https://doi.org/10.48550/arXiv.2207.10733
9. Kale, A., Kallumadi, S., King, T.H., Malmasi, S., de Rijke, M., Tagliabue, J.: Ecom'22: The sigir 2022 workshop on ecommerce. In: Proceedings of the 45th International ACM SIGIR Conference on Research and Development in Information Retrieval. pp. 3485–3487. SIGIR '22, Association for Computing Machinery, New York, NY, USA (2022). https://doi.org/10.1145/3477495.3531701
10. Lin, Y., Das, P., Datta, A.: Overview of the SIGIR 2018 ecom rakuten data challenge. In: Degenhardt, J., Fabbrizio, G.D., Kallumadi, S., Kumar, M., Trotman, A., Lin, Y., Zhao, H. (eds.) The SIGIR 2018 Workshop On eCommerce co-located with the 41st International ACM SIGIR Conference on Research and Development in Information Retrieval (SIGIR 2018), Ann Arbor, Michigan, USA, July 12, 2018. CEUR Workshop Proceedings, vol. 2319. CEUR-WS.org (2018), http://ceur-ws.org/Vol-2319/ecom18DC_paper_13.pdf
11. Northcutt, C.G., Jiang, L., Chuang, I.L.: Confident learning: estimating uncertainty in dataset labels. J. Artif. Intell. Res. **70**, 1373–1411 (2021). https://doi.org/10.1613/jair.1.12125
12. Peeters, R., Bizer, C.: Cross-Language Learning for Entity Matching. In: Companion Proceedings of the Web Conference 2022, pp. 236–238 (Apr 2022). https://doi.org/10.1145/3487553.3524234
13. Peeters, R., Bizer, C., Glavas, G.: Intermediate training of BERT for product matching. In: Piai, F., Firmani, D., Crescenzi, V., Angelis, A.D., Dong, X.L., Mazzei, M., Merialdo, P., Srivastava, D. (eds.) Proceedings of the 2nd International Workshop on Challenges and Experiences from Data Integration to Knowledge Graphs co-located with 46th International Conference on Very Large Data Bases, DI2KG@VLDB 2020, Tokyo, Japan, August 31, 2020. CEUR Workshop Proceedings, vol. 2726. CEUR-WS.org (2020), http://ceur-ws.org/Vol-2726/paper1.pdf

14. Primpeli, A., Peeters, R., Bizer, C.: The wdc training dataset and gold standard for large-scale product matching. In: Companion Proceedings of The 2019 World Wide Web Conference, pp. 381–386. WWW '19, Association for Computing Machinery, New York, NY, USA (2019). https://doi.org/10.1145/3308560.3316609

15. PromptCloud: Innerwear Data from Victoria's Secret and Others. https://www.kaggle.com/datasets/PromptCloudHQ/innerwear-data-from-victorias-secret-and-others (2022). [Online; Accessed Oct. 20 2022]

16. Rozen, O., Carmel, D., Mejer, A., Mirkis, V., Ziser, Y.: Answering product-questions by utilizing questions from other contextually similar products. In: Proceedings of the 2021 Conference of the North American Chapter of the Association for Computational Linguistics: Human Language Technologies, pp. 242–253. Association for Computational Linguistics, Online (Jun 2021). https://doi.org/10.18653/v1/2021.naacl-main.23

17. Shi, X., Mueller, J., Erickson, N., Li, M., Smola, A.J.: Benchmarking multimodal automl for tabular data with text fields. CoRR abs/2111.02705 (2021). https://doi.org/10.48550/arxiv.2111.02705

18. Yang, L., et al.: MAVE: A Product Dataset for Multi-source Attribute Value Extraction. In: Proceedings of the Fifteenth ACM International Conference on Web Search and Data Mining, pp. 1256–1265. WSDM '22, Association for Computing Machinery, New York, NY, USA (Feb 2022). https://doi.org/10.1145/3488560.3498377

19. Zhang, Z., Bizer, C., Peeters, R., Primpeli, A.: MWPD2020: semantic web challenge on mining the web of html-embedded product data. In: Zhang, Z., Bizer, C. (eds.) Proceedings of the Semantic Web Challenge on Mining the Web of HTML-embedded Product Data co-located with the 19th International Semantic Web Conference (ISWC 2020), Athens, Greece, November 5, 2020. CEUR Workshop Proceedings, vol. 2720. CEUR-WS.org (2020), http://ceur-ws.org/Vol-2720/paper1.pdf

Continuous Integration for Reproducible Shared Tasks with `TIRA.io`

Maik Fröbe[1(✉)], Matti Wiegmann[2], Nikolay Kolyada[2], Bastian Grahm[3],
Theresa Elstner[3], Frank Loebe[3,4], Matthias Hagen[1], Benno Stein[2],
and Martin Potthast[3,4]

[1] Friedrich-Schiller-Universität Jena, Jena, Germany
`maik.froebe@uni.jena.de`
[2] Bauhaus-Universität Weimar, Weimar, Germany
[3] Leipzig University, Leipzig, Germany
[4] ScaDS.AI, Leipzig, Germany

Abstract. A major obstacle to the long-term impact of most shared tasks is their lack of reproducibility. Often only the test collections and the papers of the organizers and participants are published. Third parties who want to independently evaluate the state of the art for a task on other data must re-implement the participants' software. The tools developed to collect software from participants in shared tasks only partially verify its reliability at the time of submission, much less long-term, and do not enable third parties to reuse it later. We have overhauled the TIRA Integrated Research Architecture to address all of these issues. The new version simplifies task setup for organizers and software submission for participants, scales from a local computer to the cloud, supports on-demand resource allocation up to parallel CPU and GPU processing, and enables export for local reproduction with just a few lines of code. This is achieved by implementing the TIRA protocol with an industry-standard continuous integration and deployment (CI/CD) pipeline using Git, Docker, and Kubernetes.

1 Introduction

A shared task is a collaborative laboratory experiment to evaluate state-of-the-art computational solutions to a problem, the task. A reproducible shared task gathers the resources needed by third parties to reproduce the evaluation results. Reproducibility is only guaranteed if the datasets of the shared tasks and the software of the individual participants are available. However, most participants in shared tasks do not publish their software, and most organizers do not have the time or resources to collect it. Therefore, the results of shared tasks are usually difficult for third parties to reproduce after a shared task has been completed.

J. Kamps et al. (Eds.): ECIR 2023, LNCS 13982, pp. 236–241, 2023.
https://doi.org/10.1007/978-3-031-28241-6_20

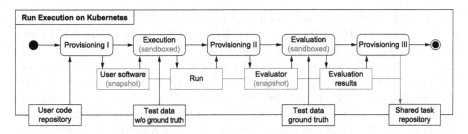

Fig. 1. UML activity diagram of the control flow (rounded boxes) and the data flow (rectangular boxes). All run artifacts (highlighted middle "row") are persisted through the provisioning steps in the shared task repository. The user software and evaluator are "untrusted" software and therefore run in a sandbox without network access.

In information retrieval, many initiatives have been launched to promote the reproducibility of experiments and software. These include, among others, the reproducibility and resource tracks at the IR conferences, the OSIRRC workshop [2], as well as the CENTRE lab held in close succession at CLEF [4], TREC [13], and NTCIR [12]. Moreover, the ACM SIGIR Artifact Badging Board [3] awards badges of honor to especially reproducible papers. With respect to reproducible experiments and shared tasks, a requirements catalog has been developed at the Evaluation-as-a-Service (EaaS) workshop [7], drawing inspiration from existing tools [5,6,14,15], as a guide to future ones [1,8,10,11,16].

In this paper, we present a new approach to make shared tasks reproducible based on continuous integration and deployment (CI/CD), a widely used industry standard in software engineering. In the context of research software engineering for shared tasks, CI refers to the reproducible evaluation of each software version (cf. Sect. 2), while CD refers to the automatic archiving of each evaluated software version in a centralized shared task repository (cf. Sect. 3). Our new approach is based on modern, widely used industry-standard tools, including GitHub and/or GitLab, Docker, and the cluster computing framework Kubernetes. This minimizes the effort for organizers and participants alike and allows for instant reproduction, all with just a few lines of code.

2 Snapshotting Software on Every Run

Reproducibility is always a trade-off between cost and benefit: Perfect reproducibility may, in extreme cases, require persistence of the hardware on which a piece of software was executed. However, implementing a practical form of reproducibility is already quite a challenge. Originally, TIRA persisted software through virtual machine snapshots [5], which was prone to scaling errors due to the VirtualBox implementation, sometimes requiring costly manual intervention. Recent progress in DevOps and continuous integration and deployment, well integrated with Git, provide all the components for automating reproducible shared tasks with mature and cloud-native tools. Consequently, our new backend for TIRA is cloud-native and based on a privately hosted GitLab, its container registry (12.4 PB HDD storage), and a Kubernetes cluster (1,620 CPU cores, 25.4 TB RAM, 24 GeForce GTX 1080 GPUs).

Figure 1 provides an overview of TIRA's new workflow for shared tasks. The workflow is based on a Git repository for shared tasks and uses the tools built into Git platforms: (1) a user code repository with a dedicated container registry where participants upload Docker images, (2) continuous integration runners that execute the five phases of the pipeline, (3) Kubernetes to orchestrate (provision, scale, and distribute) the runners in the cluster, and (4) a storage platform that provides access to the datasets.

First, users upload their software (in a Docker image) to the container registry in their private code repository, which is maintained on the Git platform. Access is granted at the time of login to TIRA via an automatically generated authentication token. Then, users can add their software to TIRA by specifying the command to execute the software in the Docker image. Any software added in this way is immutable: the command and Docker image cannot be changed afterwards (participants can upload as many pieces of software as they like). The container registry has a lower memory footprint compared to virtual machines. It is also private during the task (only the user and TIRA have access) to avoid premature publication [9].

The workflow for executing and evaluating software is specified via the declarative continuous integration API in the shared task's Git repository, which also contains a registry for the evaluator images. The workflow is triggered by a special commit in the shared task repository (specifying the software to be executed; done via the TIRA website or the command line) and consists of five steps:

1. *Provisioning I*: Prepares the execution environment by branching and cloning the shared task repository and copying the test data to the execution environment; all operations are trusted.
2. *Execution*: In the prepared execution environment with the test data, this step moves the user software into a sandbox and then executes it to generate its output as a so-called run file. Sandboxing cuts off the internet connection (using egress and ingress rules in Kubernetes) to ensure that (untrusted) user software does not leak test data. This enforces blind evaluation and ensures that the software is mature enough to run unattended in its Docker image (i.e., the software cannot download any data while it is running).
3. *Provisioning II*: Persists the run files and logs, and copies the test ground truth to the execution environment for evaluation; all operations are trusted.
4. *Evaluation*: In the prepared execution environment with the run files and the test truth, the evaluator is executed in a sandbox to generate the evaluation results. The evaluator is not trusted since it is an external software.
5. *Provisioning III*: Persists the evaluation results and logs and merges the executed software's branch into the main branch (all operations are trusted).

This workflow ensures that only the data of successfully executed and evaluated software is in the main branch of the Git repository of the shared task. Branches indicate queued or running software. Since the software itself is immutable, every software is snapshotted on every run.

```
import tira
df = tira.load_data('<dataset-name>')
# df can be manipulated for ablation/replicability/reproducibility studies
predictions, evaluation = tira.run(
    '<task-name>/<user-name>/<software-name>',
    data=df, evaluate='<evaluator-name>'
)
```

Listing 1: The software <software-name> by user <user-name> submitted in the <task-name> shared task is executed and evaluated on a pandas DataFrame df of dataset <dataset-name>. Demo available at tira.io/t/post-hoc-experimentation.

3   Repeat, Replicate, and Reproduce in One Line of Code

Through continuous deployment (CD), our new version of TIRA provides organizers with a self-contained Git repository that contains all shared tasks artifacts and is ready to be published. It contains all datasets, runs, evaluation results, logs, metadata and software snapshots. This "shared task repository" also contains utility scripts that allow all software submitted to the shared task to be run on the existing datasets, but also on other datasets as long as their formatting is the same. Listing 1 illustrates this by loading a (new or additional) dataset into a Pandas DataFrame, which then serves as input to software submitted to the shared task while the run output is evaluated directly. The utility script tira that enables this replication is in the repository; Docker and Python 3 are the only external dependencies. When archiving the shared task, all pieces of software are published to Docker Hub so that they can be loaded ad-hoc during replications (TIRA also maintains a local archive).

The final shared task repository can be easily published and serves as a natural entry point for a variety of follow-up studies. All researchers can fork such a repository and make contributions that can increase the impact of a shared task through additional material (e.g., additional datasets, evaluations, ablations, software, etc.). None of this was previously possible with the old version of TIRA or any other related tool to support reproducible experiments.

4   Conclusion

Our new version of TIRA enables reproducible shared tasks with software submissions in a cloud-native environment and is currently being used in two shared tasks at SemEval 2023. Cloud-native orchestration reduces the burden of organizing shared tasks with software submissions. Therefore, we plan to spread TIRA further, to collect more shared tasks on the platform for which post-hoc experiments are then possible, and to further encourage the submission of software in shared tasks.

In the future, we will port our pipeline to support more and also proprietary vendors (GitHub, AWS/Azure), which will make TIRA more accessible. In addition, we aim for one-click deployments that use private repositories in GitHub or (self-hosted instances of) GitLab as the backend for shared tasks.

Acknowledgments. This work has received funding from the European Union's Horizon Europe research and innovation programme under grant agreement No 101070014 (OpenWebSearch.EU, https://doi.org/10.3030/101070014).

References

1. Breuer, T., Schaer, P., Tavakolpoursaleh, N., Schaible, J., Wolff, B., Müller, B.: STELLA: Towards a Framework for the Reproducibility of Online Search Experiments. In: Proceedings of the Open-Source IR Replicability Challenge OSIRRC@SIGIR 2019, pp. 8–11 (2019)
2. Clancy, R., Ferro, N., Hauff, C., Lin, J., Sakai, T., Wu, Z.Z.: The SIGIR 2019 Open-Source IR Replicability Challenge (OSIRRC 2019). In: Proceedings of SIGIR 2019, pp. 1432–1434 (2019)
3. Ferro, N., Kelly, D.: SIGIR initiative to implement ACM artifact review and badging. SIGIR Forum **52**(1), 4–10 (2018)
4. Ferro, N., Maistro, M., Sakai, T., Soboroff, I.: Overview of CENTRE@CLEF 2018: A First Tale in the Systematic Reproducibility Realm. In: CLEF, pp. 239–246 (2018)
5. Gollub, T., Potthast, M., Beyer, A., Busse, M., Rangel, F., Rosso, P., Stamatatos, E., Stein, B.: Recent Trends in Digital Text Forensics and its Evaluation. In: Proceedings of CLEF 2013, pp. 282–302 (2013)
6. Hopfgartner, F., et al.: Benchmarking news recommendations: the CLEF NewsREEL use case. SIGIR Forum **49**(2), 129–136 (2015)
7. Hopfgartner, F., et al.: Evaluation-as-a-Service for the Computational Sciences: Overview and Outlook. ACM J. Data Inf. Qual. **10**(4), 15:1–15:32 (2018)
8. Jagerman, R., Balog, K., de Rijke, M.: OpenSearch: Lessons Learned from an Online Evaluation Campaign. ACM J. Data Inf. Qual. **10**(3), 13:1–13:15 (2018)
9. Lin, J., Campos, D., Craswell, N., Mitra, B., Yilmaz, E.: Fostering Coopetition While Plugging Leaks: The Design and Implementation of the MS MARCO Leaderboards. In: Proceedings of SIGIR 2022, pp. 2939–2948 (2022)
10. Pavao, A.: CodaLab Competitions: An Open Source Platform to Organize Scientific Challenges. Université Paris-Saclay, France, Tech. rep. (2022)
11. Potthast, M., Gollub, T., Wiegmann, M., Stein, B.: TIRA integrated research architecture. In: Information Retrieval Evaluation in a Changing World. TIRS, vol. 41, pp. 123–160. Springer, Cham (2019). https://doi.org/10.1007/978-3-030-22948-1_5
12. Sakai, T., Ferro, N., Soboroff, I., Zeng, Z., Xiao, P., Maistro, M.: Overview of the NTCIR-14 CENTRE Task. In: Proceedings of NTCIR-14 (2019)
13. Soboroff, I., Ferro, N., Sakai, T.: Overview of the TREC 2018 CENTRE Track. In: Proceedings of TREC 2018 (2018)
14. Tsatsaronis, G., et al.: An Overview of the BIOASQ Large-scale Biomedical Semantic Indexing and Question Answering Competition. BMC Bioinform. **16**, 138:1–138:28 (2015)

15. Vanschoren, J., van Rijn, J.N., Bischl, B., Torgo, L.: OpenML: networked science in machine learning. SIGKDD Explor. **15**(2), 49–60 (2013)
16. Yadav, D., et al.: EvalAI: Towards Better Evaluation Systems for AI Agents. arXiv 1902.03570 (2019)

Dynamic Exploratory Search
for the Information Retrieval Anthology

Tim Gollub[1]([⊠]), Jason Brockmeyer[2], Benno Stein[2], and Martin Potthast[1,3]

[1] Leipzig University, Leipzig, Germany
{tim.gollub,martin.potthast}@uni-leipzig.de
[2] Bauhaus-Universität Weimar, Weimar, Germany
{jason.brockmeyer,benno.stein}@uni-weimar.de
[3] ScaDS.AI, Leipzig, Germany

Abstract. This paper presents dynamic exploratory search technology for the analysis of scientific corpora. The unique dynamic features of the system allow users to analyze quantitative corpus statistics beyond document counts, and to switch between corpus exploration and corpus filtering. To demonstrate the innovation of our approach, we apply our technology to the IR Anthology, a comprehensive corpus of information retrieval publications. We showcase, among others, how to query for potential PC members and the "Salton number" of an author.

Keywords: IR anthology · Exploratory search · Faceted search

1 Introduction and Related Work

The Information Retrieval Anthology[1] compiles a comprehensive collection of publications on information retrieval [6]. At the time of this writing, it includes the bibliographic information of 57,330 IR publications that have appeared since 1963, and indexes the full text of about 88%. It is available online as a search and browsing tool with the goals of (1) providing the information retrieval community with a comprehensive overview of its publications, (2) facilitating scholarly search in a closed-world environment, and (3) enabling community introspection through exploratory and quantitative publication analysis.

While major achievements have already been made and published towards the first two goals, for the first time, this paper reports on our efforts towards the third goal: With "IR Anthology Analytics", we develop a unique exploratory search experience for corpus-based community introspection. Our system design is driven by the assumption that users of the search engine are not merely interested in relevant documents but, beyond that and foremost, in the *analytical statements* that can be made about them.

Existing exploratory search engines, like Relation Browser [1], SearchLens [2], Querium [4], gFacet [5], mSpace Explorer [7], or Flamenco Browser [10], do not embrace this way of thinking, in our view. Though all systems employ, as we

---

[1] https://ir.webis.de.

© The Author(s), under exclusive license to Springer Nature Switzerland AG 2023
J. Kamps et al. (Eds.): ECIR 2023, LNCS 13982, pp. 242–247, 2023.
https://doi.org/10.1007/978-3-031-28241-6_21

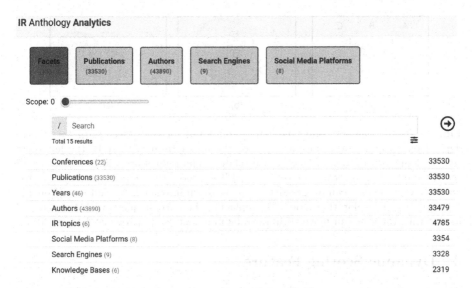

Fig. 1. Screenshot of the presented exploratory search engine. The user interface consists of two main elements. The facet pipe at the top and the facet view below, which shows the terms and term scores of the currently selected facet "Facets" (the root facet, highlighted in blue). The scope of a facet can be set by moving the "scope" slider below the facet pipe. (Color figure online)

do, the concept of faceted search [8] for exploration, facets are meant to serve as document filters much more than they are meant as carrier of analytical statements about the search results.

As a response to this shortcoming, our exploratory search system implements the following three dynamic features: (1) Any facet can be selected as the center of the search results page and hence becomes the target of the search. (2) Relation scores between arbitrary facets can be requested. (3) Both filtering and exploration of the current search results are supported.

The features are described in detail in the following sections. Our current prototype is available at https://ir-analytics.web.webis.de.

2   Dynamic Target Facet Feature

As mentioned above, to facilitate a convenient browsing through the terms of any facet, we do not reserve the center of the search results page for the display of the relevant documents. Rather, any facet can be selected to be the current target facet, which is then prominently shown in the center. A screenshot of our user interface, where the root facet "Facets" is the current target facet, is shown in Fig. 1. To enable target facet selection, we divide the search results page of our exploratory search engine into two elements, (1) the facet pipe, and (2) the facet view. In the facet pipe, which is displayed at the top of the page, users can add facets relevant to their investigations. Selecting a facet in the pipe shows respective facet terms in a facet view, which is displayed below the facet pipe at the center of the page. For the root facet, the facet view shows all available facets (see Fig. 1).

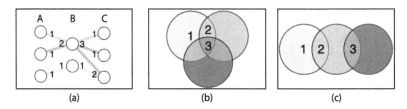

Fig. 2. (a) The numeric value indicators computed for each facet term (= circles) depend on its neighboring facets. E.g., for the upper term of facet B, the left score will be 2 since two terms of facet A are related. The right score will be 3 because of the three relations to terms of facet C. (b) Filter mode. By adding facet terms to the search query, the user can progressively narrow down into smaller result sets (from set 1 to set 3). (c) Exploration mode. By replacing the current query with the selected facet terms, the user can move within overlapping result sets (from set 1 to set 3).

3   Dynamic Scoring Feature

As an innovation to support the statistical analysis of facet term relations, the scores we compute for each facet term of the target facet (called numeric volume indicators in [9]) are not always document counts but depend on the position of the target facet in the facet pipe. The principle is illustrated in Fig. 2a. If Facet B is the current target facet, then each of its facet terms displayed in the facet view will feature on the left, the number of related facet terms from Facet A, and likewise, on the right, the number of related faceted terms from Facet C. By moving the position of a facet in the facet pipe, the computation of any term-relationships can be requested. As demonstrated in Sect. 5, this way, even with a small number of bibliographic facets, interesting statements about the IR community can be made.

4   Dynamic Facet Scope Feature

As pointed out by Gollub et al. in [3], the selection of a facet term by the user can be handled in two different ways: (1) by adding the selected term to the current search query, or (2) by replacing the current search query with the selected term. In the first case, the selected query term is used to filter current search results as illustrated in Fig. 2b. In the second case, the selected query term is explored, since the search results feature the relations that this term has to the other facets (see Fig. 2c). As both methods have their use cases (see Sect. 5), by implementing the idea of facet scopes (visible as "scope" slider in Fig. 1), the user can decide which method is applied to any facet of the facet pipe.

5   Selected Insights into the IR Anthology

This section illustrates the presented features of our exploratory search engine by linking and discussing the search engine result pages obtained for three selected queries.

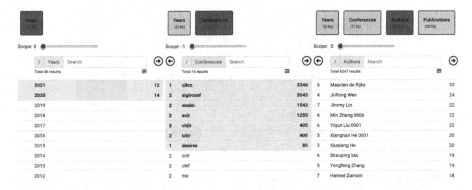

Fig. 3. Three screenshots showing how to construct a search query that reveals the authors (right) who published at any of seven selected IR conferences (middle) in the last two years (left), ordered by number of publications.

5.1 Scoring Module

A common task of conference organizers is to compile a list of active IR researchers which could serve as part of the program committee. To this end, in Fig. 3, a collage of three screenshots demonstrating how to query for the authors who published at a major IR conference in the last two years is shown[2] . First, the desired publication years and conferences have been selected in the first two facets. Adding then the authors facet reveals the list of matching authors. By adding a final publication facet, this authors list can be ranked either with respect to number of conferences (left score) or publications (right score).

5.2 Exploration with Scoped Facets

To demonstrate the difference between filtering and exploration, Fig. 4 shows a collage of screenshots from the co-author graph of Gerard Salton[3] . Taking inspiration from the Erdős number, which describes the collaborative distance between mathematician Paul Erdős and other persons, the distance of an author to IR pioneer Gerard Salton can be obtained by first adding an authors facet and selecting Gerard Salton, and to then add further authors facets with a reduced scope of -1 (=exploration) to the pipe until the desired author appears in the result. Due to the reduced scope, each authors facet reveals the co-authors of the authors in the previous facet. Note that from the 43 890 authors in the IR Anthology, 34 390 have a Salton number ($<= 12$).

5.3 Content Facets

In order to support explorations of the IR Anthology also with respect to custom facets, we implemented the integration of custom content facets via full-text

---

[2] https://ir-analytics.web.webis.de/pipes/scoringmodule.
[3] https://ir-analytics.web.webis.de/pipes/saltonnumber.

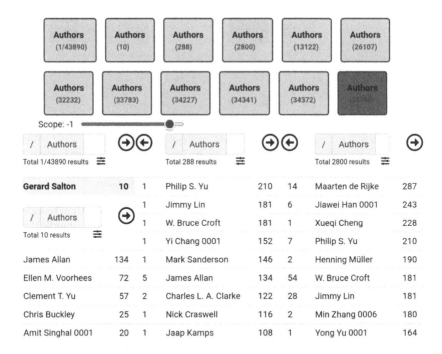

Fig. 4. Collage of screenshots illustrating the co- author graph of Gerard Salton which can be obtained by a series of Authors facets with a scope of -1. The facet views show the result of the first four Authors facets (from top left to bottom right).

retrieval. To demonstrate the potential of content facets, we provide a query showing the facet term distribution over publications for a selection of content facets, which we compiled by searching the Web for lists of IR related concepts[4]. The results reveal that Yahoo, Twitter, Wikipedia, and TREC are the most referenced entities in their respective facets Search Engines, Social Media Platforms, Knowledge Bases and Evaluation Forums.

6   Conclusion

This paper reports on our current prototype of an exploratory search engine for the IR Anthology, which we aim to publicly provide to the IR community for introspection. The prototype excels by providing a unique faceted search experience, which is designed especially for investigation tasks. As next steps, among others, we will further extend our scoring module to support relative term counts and non-binary relevance scores. Moreover, we are in the process of crowd-sourcing annotations in the available full-text which pertain to syntactical as well as semantic features. We invite the reader to explore the IR anthology with our current prototype available at https://ir-analytics.web.webis.de.

---

[4] https://ir-analytics.web.webis.de/pipes/customfacets.

References

1. Capra, R.G., Marchionini, G.: The relation browser tool for faceted exploratory search. In: Proceedings of the 8th ACM/IEEE-CS Joint Conference on Digital Libraries, pp. 420–420. JCDL '08, ACM, New York, NY, USA (2008). https://doi.org/10.1145/1378889.1378967, http://doi.acm.org/10.1145/1378889.1378967

2. Chang, J.C., Hahn, N., Perer, A., Kittur, A.: Searchlens: composing and capturing complex user interests for exploratory search. In: Proceedings of the 24th International Conference on Intelligent User Interfaces, IUI 2019, Marina del Ray, CA, USA, March 17–20, 2019, pp. 498–509 (2019). https://doi.org/10.1145/3301275.3302321

3. Gollub, T., Hutans, L., Jami, T.A., Stein, B.: Exploratory Search Pipes with Scoped Facets. In: 2019 ACM SIGIR International Conference on the Theory of Information Retrieval (ICTIR 2019). ACM (Oct 2019). https://doi.org/10.1145/3341981.3344247, http://doi.acm.org/10.1145/3341981.3344247

4. Golovchinsky, G., Diriye, A., Dunnigan, T.: The future is in the past: Designing for exploratory search. In: Proceedings of the 4th Information Interaction in Context Symposium, pp. 52–61. IIIX '12, ACM, New York, NY, USA (2012). https://doi.org/10.1145/2362724.2362738, http://doi.acm.org/10.1145/2362724.2362738

5. Heim, P., Ertl, T., Ziegler, J.: Facet graphs: complex semantic querying made easy. In: Aroyo, L., et al. (eds.) The Semantic Web: Research and Applications, pp. 288–302. Springer, Berlin Heidelberg, Berlin, Heidelberg (2010)

6. Potthast, M., et al.: The Information Retrieval Anthology. In: Diaz, F., Shah, C., Suel, T., Castells, P., Jones, R., Sakai, T. (eds.) 44th International ACM Conference on Research and Development in Information Retrieval (SIGIR 2021), pp. 2550–2555. ACM (Jul 2021). https://doi.org/10.1145/3404835.3462798, https://dl.acm.org/doi/10.1145/3404835.3462798

7. schraefel, M., Wilson, M., Russell, A., Smith, D.A.: mspace: Improving information access to multimedia domains with multimodal exploratory search. Commun. ACM **49**(4), 47–49 (Apr 2006). https://doi.org/10.1145/1121949.1121980, http://doi.acm.org/10.1145/1121949.1121980

8. Tunkelang, D.: Faceted Search, Synthesis Lectures on Information Concepts, Retrieval, and Services, vol. 1. Morgan & Claypool Publishers (2009). https://doi.org/10.2200/S00190ED1V01Y200904ICR005

9. Wilson, M.L., schraefel, m.c.: mspace: What do numbers and totals mean in a flexible semantic browser (2006), https://eprints.soton.ac.uk/262666/

10. Yee, K.P., Swearingen, K., Li, K., Hearst, M.: Faceted metadata for image search and browsing. In: Proceedings of the SIGCHI Conference on Human Factors in Computing Systems, pp. 401–408. CHI '03, ACM, New York, NY, USA (2003). https://doi.org/10.1145/642611.642681, http://doi.acm.org/10.1145/642611.642681

Text2Storyline: Generating Enriched Storylines from Text

Francisco Gonçalves[3]([✉]) [iD], Ricardo Campos[1,2] [iD], and Alípio Jorge[1,3] [iD]

[1] LIAAD – INESCTEC, Porto, Portugal
[2] Polytechnic Institute of Tomar, Ci2 - Smart Cities Research Center, Tomar, Portugal
`ricardo.campos@ipt.pt`
[3] FCUP, University of Porto, Porto, Portugal
`{up201604505,amjorge}@fc.up.pt`

Abstract. In recent years, the amount of information generated, consumed and stored has grown at an astonishing rate, making it difficult for those seeking information to extract knowledge in good time. This has become even more important, as the average reader is not as willing to spare more time out of their already busy schedule as in the past, thus prioritizing news in a summarized format, which are faster to digest. On top of that, people tend to increasingly rely on strong visual components to help them understand the focal point of news articles in a less tiresome manner. This growing demand, focused on exploring information through visual aspects, urges the need for the emergence of alternative approaches concerned with text understanding and narrative exploration. This motivated us to propose Text2Storyline, a platform for generating and exploring enriched storylines from an input text, a URL or a user query. The latter is to be issued on the Portuguese Web Archive (Arquivo.pt), therefore giving users the chance to expand their knowledge and build up on information collected from web sources of the past. To fulfill this objective, we propose a system that makes use of the Time-Matters algorithm to filter out non-relevant dates and organize relevant content by means of different displays: '*Annotated Text*', '*Entities*', '*Storyline*', '*Temporal Clustering*' and '*Word Cloud*'. To extend the users' knowledge, we rely on entity linking to connect persons, events, locations and concepts found in the text to Wikipedia pages, a process also known as Wikification. Each of the entities is then illustrated by means of an image collected from the Arquivo.pt.

Keywords: Temporal information extraction · Entity recognition · Wikification · Web archiving · Temporal narratives

1 Introduction

Recent years have shown a clear trend towards the consumption of information from different formats [11], especially in the younger generations. Driven by this new paradigm, different stakeholders have made an effort in an attempt to adapt their content to the consumption habits of an increasingly digital audience. In this context, the representation of texts from timelines appears as an alternative to the presentation of data made

© The Author(s), under exclusive license to Springer Nature Switzerland AG 2023
J. Kamps et al. (Eds.): ECIR 2023, LNCS 13982, pp. 248–254, 2023.
https://doi.org/10.1007/978-3-031-28241-6_22

solely from textual structures, offering users the possibility to become familiar with a given event in a short space of time [2, 16]. Several news outlets have been making efforts in this regard. One illustrative example of this is the manually created storyline óf a Portuguese reference newspaper (Jornal Público), which documents Portugal's participation in World War I. Despite the growing importance of timelines in the context of summarizing data from multiple documents [1, 9, 10, 21], little is known about their use and application in the context of individual documents and visual narratives. On the other hand, the immense volume of data existent in web documents also makes it prohibitive to manually build and make this type of interface available. This work is motivated by the concept of exploring more appealing and innovative ways to represent narratives and provide creative tools and features to enhance the user experience. Based on this, we intend to propose an alternative to the availability of purely textual structures or common timelines, offering users the possibility to create automatic visual narratives with a temporal focus. Each visual narrative is complemented with a set of related images automatically obtained from a collection of 584 million images preserved by the Arquivo.pt [15], the Portuguese web archive infrastructure. As a way of complementing the information obtained from text documents, we propose to identify a set of relevant keywords and named entities. The detailed information about each entity is obtained by connecting our system to an external knowledge database, Wikipedia, through a Wikification process, thus expanding the information initially obtained from the text.

To accomplish this objective, we adapted a previously introduced version of the Time-Matters online demo [http://time-matters.inesctec.pt/] [4], which was limited by only accepting text as input and it exclusively annotated the temporal expressions and keywords in the text. Our efforts produced a new **online platform** known as Text2Storyline [http://text2storyline.inesctec.pt/], that distinguished itself from its predecessor in two main ways. Firstly, by allowing other types of inputs, namely queries, in addition to the previously available URLs and single texts. Secondly, by enabling users to expand their knowledge with the identification of potentially relevant persons, events, locations, objects or concepts. For the former, we resort to the Arquivo.pt infrastructure [8, 18] and to the text search API to search for documents that are deemed relevant to the search term provided and, from there, generate a concise temporal narrative, which is performed by our in-lab Time-Matters algorithm whose purpose is to score temporal expressions identified in the text. A fully detailed description of the Time-Matters scientific approach and the evaluation methodology for the study of queries and multiple documents can be found in Campos et al. [3]. Readers are also recommended to refer to our wiki documentation for an in-depth understanding of the multiple document version explored in this demo. As a means to connect the present to the past [19], we also make use of the Google Trends' API, thus suggesting present trending terms for which getting past knowledge may be interesting. For the latter (expanding the user's knowledge), we use an API called Wikifier [20] that performs Entity Linking [7], more specifically a task called Wikification [17, 22], as it assigns a Wikipedia page to the entities found in the text. In both cases, each important moment of the narrative and each entity found in the text, is complemented with a related image and a description. Text2Storyline is a first step towards understanding and creating narratives automatically in a large scale environment. This may be beneficial not only for journalists looking for complementary aspects that can support their written narratives, or in need of fact-checking information from past stories, but also for large-scale infrastructures, such as web archives, for which displaying enriched visually outdated articles from the past, may be beneficial as a way to reach new audiences.

Few projects exist with similar objectives regarding temporal narratives for proper comparisons. The closest example is Digital Libraries' Narratives [12–14], which shares some similarities with Text2Storyline. Their Narrative Building and Visualising Tool rivals our Time-Matters algorithm. However, their system extracts content from Europeana while illustrations are obtained from Wikimedia Commons. In Text2Storyline both information is collected from the Arquivo.pt infrastructure which gives us the chance of collecting historical information, though it also raises questions as to the quality of the data collected, when compared to Europeana. These distinctions result in different storylines in each system, even when centered on the same subject.

2 Text2Storyline Demonstration

In order to allow the user to understand a given narrative, five unique components are on display: '*Annotated Text*', '*Storyline*', '*Entities*', '*Temporal Clustering*' and '*Word Cloud*'. In this paper, we put an emphasis on '*Storyline*', '*Entities*' and the '*Search Results*' components, due to space reasons. We demonstrate our approach by using, as input, the query "*Síndrome Respiratória do Médio Oriente*", which translates to "*Middle East Respiratory Syndrome*" or "*MERS*" for short. It should be noted that the relevant texts are obtained from the Portuguese archive Arquivo.pt and, as such, queries provided in Portuguese will result in more complete search results and timelines.

2.1 "Storyline"

Figure 1 shows the storyline interface, which explores the most relevant moments for that specific query through a temporal lens. For each moment, it highlights the relevant date ("*2002*"), its Time-Matters score ("*0.806*"), the sentence where the date occurs and the title ("*Síndrome Respiratória Aguda*" which translates to "*Acute Respiratory Syndrome*") obtained from the sentence by applying YAKE! [5, '6] keyword extractor. The story is also automatically illustrated with images, which are automatically obtained from the Portuguese web archive Arquivo.pt [8] image search API. While this web service can obtain results for any language, it naturally works better for its native language, Portuguese.

Users can then interactively navigate between the different time-periods. The bottom timeline bar displays a temporal overview of the entire story, containing all of its moments. At the top, a toggle is available which displays the moments of the least relevant dates when activated. Dates are considered non-relevant when scored lower than 0.35 by Time-Matters. At the end of the description, a clickable element ([+]) redirects the user to the source that the respective moment was extracted from. This particular behavior is only available for narratives generated from queries.

Fig. 1. "Storyline" interface

2.2 'Entities'

Figure 2 shows part of the 'Entities' component. Each entity is accompanied by an adequate image, description and primary class (e.g., "*geographic region*" for "*Médio Oriente*" or "*Middle East*"). The first is provided by Arquivo.pt's image search API, the second is extracted from Wikipedia through the MediaWiki API, and the last one is obtained with Wikifier [20]. A maximum of 50 entities can be displayed in this component. All of these are clickable, redirecting the user to the respective Wikipedia page. The language of the texts is taken into account. In this case, all texts were in Portuguese, thus all entities and respective descriptions are in the same dialect.

Fig. 2. "Entities" interface.

2.3 Search Results

Figure 3 shows the interface of the '*Search Results*' feature for the same query stated above. This component is only available when a query is supplied. In this feature, multiple documents deemed relevant will be obtained by issuing the query into Arquivo.pt and displayed accordingly to a google search-type interface. Relevant dates are determined by the Time-Matters algorithm, while irrelevant ones are discarded (or omitted). The user may then choose to sort these results chronologically according to the dates identified within each text and the scores determined by Time-Matters. In addition to this, users can choose to create a narrative from one of the results obtained (by clicking on the blue book icon), thus being able to visualize its text in an annotated, clustering, entities, word cloud and timeline fashion.

Fig. 3. "Search Results" interface (Color figure online).

The remaining features of Text2Storyline offer another look at the generated narrative. The '*Temporal Clustering*' component offers a similar view as the '*Storyline*' component, but in a vertical and more condensed display, without illustrations of events. The '*Word Cloud*' feature displays the top-30 most relevant keywords found in the text, obtained with YAKE!, for a summarized preview of what the user can expect to read from the sources. The '*Annotated Text*' feature is only available when looking at a single document, which does not occur when generating a narrative from a query. A video of the demo is available at: https://youtu.be/Rh1GmGA9n2U. In addition to this, we have also made the entire source code of our project fully available at: https://github. com/1Skkar1/Text2Storyline. As future research, we plan to improve its platform, more notably the '*Storyline*' and '*Entities*' components, by implementing a more accurate image selection method.

3 Conclusions

In current times, the increasing amount of information that is generated, consumed and stored has made it difficult for those seeking information to extract knowledge in reasonable time. Additionally, information has slightly shifted towards types of content that are easier and faster to digest, with a strong visual component. Text2Storyline emerge in this context, as a platform for generating and exploring enriched storylines from an input text, a URL or a user query, which distinguished itself from its predecessor by enabling the users to explore the generated narratives in a more in-depth manner.

Acknowledgements. This work is financed by the ERDF – European Regional Development Fund through the Norte Portugal Regional Operational Programme – NORTE 2020 under the Portugal 2020 Partnership Agreement and by National Funds through the FCT – Fundação para a Ciência e a Tecnologia, I.P. (Portuguese Foundation for Science and Technology) within project Text2Story, with reference PTDC/CCI-COM/31857/2017 (NORTE-01-0145-FEDER-031857).

References

1. Alonso, O., Shiells, K.: Timelines as summaries of popular scheduled events. In: Proceedings of the 22nd International Conference on World Wide Web (WWW 2013), Rio de Janeiro, Brazil, 13–17 May, pp. 1037–1044. ACM (2013)
2. Ansah, J., Liu L., Kang, W., Kwashie, S., Li, J., Li, J.: A graph is worth a thousand words: Telling event stories using timeline summarization graphs. In: Proceedings of the 30th International Conference on World Wide Web (WWW 2019), San Francisco, USA, 13–17 May, pp. 2565–2571. ACM (2019)
3. Campos, R., Dias, G., Jorge, A., Nunes, C.: Identifying top relevant dates for implicit time sensitive queries. Inf. Retrieval J. **20**(4), 363–398 (2017). ISSN 1386-4564
4. Campos, R., Duque, J., Cândido, T., Mendes, J., Dias, G., Jorge, A., Nunes, C.: Time-matters: temporal unfolding of texts. In: Hiemstra, D., Moens, M.-F., Mothe, J., Perego, R., Potthast, M., Sebastiani, F. (eds.) ECIR 2021. LNCS, vol. 12657, pp. 492–497. Springer, Cham (2021). https://doi.org/10.1007/978-3-030-72240-1_53
5. Campos, R., Mangaravite, V., Pasquali, A., Jorge, A.M., Nunes, C., Jatowt, A.: A text feature based automatic keyword extraction method for single documents. In: Pasi, G., Piwowarski, B., Azzopardi, L., Hanbury, A. (eds.) ECIR 2018. LNCS, vol. 10772, pp. 684–691. Springer, Cham (2018). https://doi.org/10.1007/978-3-319-76941-7_63
6. Campos, R., Mangaravite, V., Pasquali, A., Jorge, A.M., Nunes, C., Jatowt, A.: YAKE! collection-independent automatic keyword extractor. In: Pasi, G., Piwowarski, B., Azzopardi, L., Hanbury, A. (eds.) ECIR 2018. LNCS, vol. 10772, pp. 806–810. Springer, Cham (2018). https://doi.org/10.1007/978-3-319-76941-7_80
7. Collobert, R., Weston, J., Bottou, L., Karlen, M., Kavukcuoglu, K., Kuksa, P.P.: Natural language processing (almost) from scratch. J. Mach. Learn. Res. **12**, 2493–2537 (2011). ISSN 1532-4435
8. Gomes, D., Cruz, D., Miranda, J.A., Costa, M., Fontes, S.A.: Search the past with the portuguese web archive. In: Proceedings of the 22nd International Conference on World Wide Web (WWW 2013), Rio de Janeiro, Brazil, 13–17 May, pp. 321–324. ACM (2013)
9. La Quatra, M., Cagliero, L., Baralis, E., Messina, A,. Montagnuolo, M.: Summarize dates first: a paradigm shift in timeline summarization. In: Proceedings of the 44th International ACM SIGIR Conference on Research and Development in Information Retrieval (SIGIR 2021), 11–15 July, pp. 418–427 (2021)
10. Li, J., Cardie, C.: Timeline generation: Tracking individuals on twitter. In: Proceedings of the 23rd International Conference on World Wide Web (WWW 2014), Seoul, South Korea, 7–11 April, pp. 643–652. ACM (2014)

11. Martinez-Alvarez, M., et al.: First international workshop on recent trends in news information retrieval (NewsIR'16). In: Ferro, N., et al. (eds.) ECIR 2016. LNCS, vol. 9626, pp. 878–882. Springer, Cham (2016). https://doi.org/10.1007/978-3-319-30671-1_85

12. Meghini, C., Bartalesi, V., Metilli, D.: Steps towards accessing digital libraries using narratives. In: Proceedings of the 10th International Workshop on Artificial Intelligence for Cultural Heritage co-located with the 15th International Conference of the Italian Association for Artificial Intelligence (AI*IA 2016), Genoa, Italy, 28 November–1 December, pp. 10–17. CEUR (2016)

13. Meghini, C., Bartalesi, V., Metilli, D.: Using formal narratives in digital libraries. In: Grana, C., Baraldi, L. (eds.) IRCDL 2017. CCIS, vol. 733, pp. 83–94. Springer, Cham (2017). https://doi.org/10.1007/978-3-319-68130-6_7

14. Meghini, C., Bartalesi, V., Metilli, D.: Representing narratives in digital libraries: the narrative ontology. Seman. Web **12**, 241–264 (2020)

15. Mourão, A., Gomes, D.: The anatomy of a web archive image search engine, Technical report, Arquivo.pt, Lisboa, Portugal (2021)

16. Pasquali, A., Mangaravite, V., Campos, R., Jorge, A.M., Jatowt, A.: Interactive system for automatically generating temporal narratives. In: Azzopardi, L., Stein, B., Fuhr, N., Mayr, P., Hauff, C., Hiemstra, D. (eds.) ECIR 2019. LNCS, vol. 11438, pp. 251–255. Springer, Cham (2019). https://doi.org/10.1007/978-3-030-15719-7_34

17. Roth, D., Ji, H., Chang, M.-W., Cassidy, T.: Wikification and beyond: the challenges of entity and concept grounding. In: Proceedings of the 52nd Annual Meeting of the Association for Computational Linguistics: Tutorials, Baltimore, USA, p. 7. ACL (2014)

18. Santana, B., Campos, R., Amorim, E., Jorge, A., Purificação, S., Nunes, S.: A survey on narrative extraction from textual data. Artif. Intell. Rev. (2023)

19. Sato, M., Jatowt, A., Duan, Y., Campos, R., Yoshikawa, M.: Estimating contemporary relevance of past news. In: Proceedings of the 2021 ACM/IEEE Joint Conference on Digital Libraries (JCDL 2021), 27–30 September, pp. 70–79 (2021)

20. Schonhofen, P.: Annotating documents by wikipedia concepts. In: Proceedings of the 2008 IEEE/WIC/ACM International Conference on Web Intelligence and Intelligent Agent Technology (WI-IAT 2008), Sydney, Australia, 9–12 December, vol. 1, pp. 461–467. IEEE (2008)

21. Tran, G., Alrifai, M., Herder, E.: Timeline summarization from relevant headlines. In: Hanbury, A., Kazai, G., Rauber, A., Fuhr, N. (eds.) ECIR 2015. LNCS, vol. 9022, pp. 245–256. Springer, Cham (2015). https://doi.org/10.1007/978-3-319-16354-3_26

22. Tsai, C.-T., Mayhew, S., Roth, D.: Cross-lingual named entity recognition via wikification. In: Proceedings of the 20th SIGNLL Conference on Computational Natural Language Learning, Berlin, Germany, 11–12 August, pp. 219–228. ACL (2016)

Uptrendz: API-Centric Real-Time Recommendations in Multi-domain Settings

Emanuel Lacic[1], Tomislav Duricic[1,2], Leon Fadljevic[1], Dieter Theiler[1],
and Dominik Kowald[1,2(✉)]

[1] Know-Center GmbH, Graz, Austria
{elacic,tduricic,lfadljevic,dtheiler,dkowald}@know-center.at
[2] Graz University of Technology, Graz, Austria

Abstract. In this work, we tackle the problem of adapting a real-time recommender system to multiple application domains, and their underlying data models and customization requirements. To do that, we present Uptrendz, a multi-domain recommendation platform that can be customized to provide real-time recommendations in an API-centric way. We demonstrate (i) how to set up a real-time movie recommender using the popular MovieLens-100 k dataset, and (ii) how to simultaneously support multiple application domains based on the use-case of recommendations in entrepreneurial start-up founding. For that, we differentiate between domains on the item- and system-level. We believe that our demonstration shows a convenient way to adapt, deploy and evaluate a recommender system in an API-centric way. The source-code and documentation that demonstrates how to utilize the configured Uptrendz API is available on GitHub.

Keywords: Uptrendz · API-centric recommendations · Multi-domain recommendations · Real-time recommendations

1 Introduction

Utilizing recommender systems is nowadays recognized as a necessary feature to help users discover relevant content [14,15]. Most industry practitioners [3], when they build a recommender system, adapt existing algorithms to the underlying data and customization requirements of the respective application domain (e.g., movies, music, news, etc.). However, the focus of the research community has recently shifted towards building recommendation systems that simultaneously support multiple application domains [4,7,16] in an API-centric way.

In this work, we demonstrate Uptrendz[1], an API-centric recommendation platform, which can be configured to simultaneously provide real-time recommendations in an API-centric way to multiple domains. Uptrendz supports popular recommendation algorithms, e.g., Collaborative Filtering (CF), Content-based Filtering (CBF, or Most Popular (MP), that are applied across different

---

[1] https://uptrendz.ai/.

J. Kamps et al. (Eds.): ECIR 2023, LNCS 13982, pp. 255–261, 2023.
https://doi.org/10.1007/978-3-031-28241-6_23

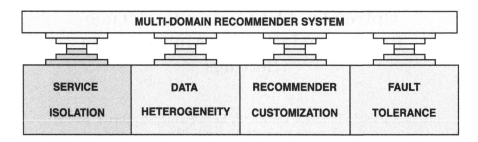

Fig. 1. Aspects that need to be addressed when building a recommender system for a multi-domain environment [10].

application domains. The focus of this demonstration is to show how domain-specific data-upload APIs can be created to support the customization of the respective recommendation algorithms. Using the MovieLens-100k dataset [6] and a real-world use-case of entrepreneurial start-up founding[2], we show how such an approach allows for a highly customized recommendation system that can be used in an API-centric way. The source-code and documentation for this demonstration is available via GitHub[3].

2   The Uptrendz Platform

The Uptrendz platform is built on top of the ScaR recommendation framework [11]. As shown in [10] and Fig. 1, the microservice-based system architecture addresses four distinctive requirements of a multi-domain recommender system, i.e., (i) service isolation, (ii) data heterogeneity, (iii) recommender customization, and (iv) fault tolerance. Uptrendz provides a layer on top of the framework to dynamically configure an application domain and to instantly provide an API to (i) upload item, user and interaction data, and (ii) request recommendations.

Domain-Specific Data Model. As discussed by [1], different domains may employ the same recommender algorithm but can differ with respect to what kind of data is utilized to build the model (e.g., interaction types, context, etc.). Given an API-centric approach, we show that in order to support the customization of recommender algorithms with domain-specific parameters, the underlying platform needs to unambiguously know which source of information should be used to calculate the recommendations. To do that, the Uptrendz platform first allows generating customized data upload APIs for multiple item and user entities (see Table 1). Second, with respect to interaction data, both user-item and user-user interactions can be configured. The interaction API is further customized in accordance to what kind of interactions the respective application domain actually supports, i.e., (i) registered users, anonymous sessions or

---

[2] https://cogsteps.com/.
[3] https://github.com/lacic/ECIR2023Demo.

Table 1. Supported attributes to configure the data upload API for items and users.

Type	Sub-Type	Description
Categorical Text	Single Value	String value, which usually represents a category. Used for **post-filtering** recommendation results.
	Multiple Values	List of string values, which usually represent an array of categories. Used for **post-filtering** recommendation results.
Free Text	English	**English text**, which is processed and utilized for **content-based** recommendations.
	German	**German text**, which is processed and utilized for **content-based** recommendations.
Numeric	Integer	Used for **post-filtering** recommendations (e.g., user age).
	Real	Used for **post-filtering** recommendations (e.g., price).
Date	–	Date information for the respective entity (e.g., creation date)

both, (ii) interaction timestamp tracking, and (iii) type of interaction (explicit or implicit).

Recommender Customization. The Uptrendz platform fosters the notion of defining personalization scenarios (i.e., use-cases) when creating recommendation APIs. The available selection of real-time recommendation models [11] for a given scenario depends on (i) what should be recommended (e.g., item or user entities), (ii) for whom the recommendations are targeted (e.g., registered or anonymous users) and, (iii) what kind of context is given [2] (e.g., item ID to recommend relevant content for). As we adopt a non-restricted configuration with respect to the number of freely defined user interaction types, algorithms that use this kind of data (e.g., Collaborative Filtering) can be customized to utilize any subset of the list of available interactions as well as to define how much weight a particular interaction type should have. With respect to post-filtering recommendation results, each model can use categorical (e.g., tags [12] or other semantic representations [8]) or numerical data attributes to ensure that the resulting recommendations either contain or exclude a particular value (see Table 1 for complete list of attributes).

3   Multi-domain Support

In order to provide a multi-domain recommender platform, we support the notions of a system-level and item-level domain in accordance with [5]. For the former, items and users belong to distinct systems (e.g., Netflix and Amazon).

Fig. 2. Example of supporting multiple domains on the item-level (up) and configuring a hybrid recommendation algorithm (below) with previously created APIs.

For the latter, individual domains have different types of items and users which may share some common attributes (e.g., movies and books).

Demo Walkthrough: System-level Domain. When a domain is created on a system level, the underlying data is physically stored in a different location than the data of other domains. Hence, domains do not share any data between themselves and the underlying services are isolated so that the performance of one domain does not impact the performance of another domain (e.g., during request load peaks). We demonstrate how to create a movie recommender on a system level. To utilize the MovieLens-100k dataset [6], we first need to configure the respective data services to upload (i) movie, (ii) user, and (iii) interaction data. Each entity needs to be separately created in the Uptrendz platform in order to generate an API that can be used to upload the MovieLens-specific data attributes. This allows creating recommendation scenarios for (i) similar

Fig. 3. Uptrendz requires the specification of (i) the item types that should be recommended (e.g., products or users, depending on the domain - left figure), and (ii) the user types for which recommendations should be generated (e.g., registered users or session users - right figure).

movies (CBF), (ii) popular horror movies (MP with post-filtering), (iii) movies based on ratings (CF), (iv) their weighted hybrid combination (e.g., for cold-start settings [13], and (v) a user recommender for a given movie.

Demo Walkthrough: Item-level Domain. To showcase how to configure Uptrendz to support multiple-domains on an item-level, we present the use-case of entrepreneurial start-up founding. Here, we recommend experts that can provide feedback to an innovation idea, support co-founder matching, help incubators, innovation hubs and accelerators to discover innovations but also provide relevant educational materials until the innovation idea matures enough to form a start-up. In this case, each recommendable entity has a separate data model and can be viewed as part of a standalone application domain. Figure 2 depicts how adding multiple item entities in the data catalog allows customizing data attributes for the respective domain. While configuring a recommendation algorithm, the respective item-level domain can be selected to be recommended. Here, via the example of a hybrid algorithm, only pre-configured algorithms can be utilized that belong to the same domain (i.e., innovation recommendations).

Finally, in Fig. 3, we show how Uptrendz allows the specification of (i) different item types that can be recommended, and (ii) different user types for which recommendations should be generated. Our demo application includes different specification examples.

4 Conclusion

In this paper, we present Uptrendz, an API-centric recommendation platform that can be customized to provide real-time recommendations for multiple domains. To do that, we support the notions of a system-level and item-level domain. We demonstrate Uptrendz using the popular MovieLens-100k dataset and the use-case of entrepreneurial start-up founding.

In future work, we plan to support even more use cases from other domains, e.g., music recommendations [9]. Here, we also want to integrate fairness-aware recommendation algorithms for mitigating e.g., popularity bias effects.

Acknowledgements. This research was funded by CogSteps and the "DDAI" COMET Module within the COMET - Competence Centers for Excellent Technologies

Programme, funded by the Austrian Federal Ministry for Transport, Innovation and Technology (bmvit), the Austrian Federal Ministry for Digital and Economic Affairs (bmdw), the Austrian Research Promotion Agency (FFG), the province of Styria (SFG) and partners from industry and academia.

References

1. Adomavicius, G., Tuzhilin, A.: Context-aware recommender systems. In: Proceedings of the 2008 ACM Conference on Recommender Systems, pp. 335–336. RecSys '08, ACM (2008). https://doi.org/10.1145/1454008.1454068, http://doi.acm.org/10.1145/1454008.1454068

2. Adomavicius, G., Tuzhilin, A.: Context-aware recommender systems. In: Ricci, F., Rokach, L., Shapira, B., Kantor, P.B. (eds.) Recommender Systems Handbook, pp. 217–253. Springer, Boston, MA (2011). https://doi.org/10.1007/978-0-387-85820-3_7

3. Amatriain, X., Basilico, J.: Past, present, and future of recommender systems: An industry perspective. In: Proceedings of the 10th ACM Conference on Recommender Systems, pp. 211–214 (2016)

4. Bonab, H., Aliannejadi, M., Vardasbi, A., Kanoulas, E., Allan, J.: Cross-market product recommendation. In: Proceedings of the 30th ACM International Conference on Information & Knowledge Management, pp. 110–119 (2021)

5. Cantador, I., Fernández-Tobías, I., Berkovsky, S., Cremonesi, P.: Cross-domain recommender systems. In: Ricci, F., Rokach, L., Shapira, B. (eds.) Recommender Systems Handbook, pp. 919–959. Springer, Boston, MA (2015). https://doi.org/10.1007/978-1-4899-7637-6_27

6. Harper, F.M., Konstan, J.A.: The movielens datasets: History and context. Acm Trans. Interact. Intell. Syst. (tiis) 5(4), 1–19 (2015)

7. Im, I., Hars, A.: Does a one-size recommendation system fit all? the effectiveness of collaborative filtering based recommendation systems across different domains and search modes. ACM Trans. Inform. Syst. (TOIS) 26(1), 4-es (2007)

8. Kowald, D., Dennerlein, S.M., Theiler, D., Walk, S., Trattner, C.: The social semantic server a framework to provide services on social semantic network data. In: Proceedings of I-SEMANTICS 2013), pp. 50–54 (2013)

9. Kowald, D., Muellner, P., Zangerle, E., Bauer, C., Schedl, M., Lex, E.: Support the underground: characteristics of beyond-mainstream music listeners. EPJ Data Sci. 10(1), 1–26 (2021). https://doi.org/10.1140/epjds/s13688-021-00268-9

10. Lacic, E., Kowald, D., Lex, E.: Tailoring recommendations for a multi-domain environment. In: Workshop on Intelligent Recommender Systems by Knowledge Transfer & Learning (RecSysKTL'2017) co-located with the 11th ACM Conference on Recommender Systems (RecSys'2017) (2017)

11. Lacic, E., Kowald, D., Parra, D., Kahr, M., Trattner, C.: Towards a scalable social recommender engine for online marketplaces: The case of apache solr. In: Workshop Proceedings of WWW'2014, pp. 817–822 (2014)

12. Lacic, E., Kowald, D., Seitlinger, P., Trattner, C., Parra, D.: Recommending items in social tagging systems using tag and time information. In: Proceedings of the 1st International Workshop on Social Personalisation (SP'2014) co-located with Hypertext'2014 (2014)

13. Lacic, E., Kowald, D., Traub, M., Luzhnica, G., Simon, J.P., Lex, E.: Tackling cold-start users in recommender systems with indoor positioning systems. In: Poster Proceedings of the 9th {ACM} Conference on Recommender Systems. Association of Computing Machinery (2015)

14. McNee, S.M., Riedl, J., Konstan, J.A.: Being accurate is not enough: how accuracy metrics have hurt recommender systems. In: CHI'06 Extended Abstracts on Human Factors in Computing Systems, pp. 1097–1101. ACM (2006)

15. Resnick, P., Varian, H.R.: Recommender systems. Commun. ACM **40**(3), 56–58 (1997)

16. Roitero, K., Carterette, B., Mehrotra, R., Lalmas, M.: Leveraging behavioral heterogeneity across markets for cross-market training of recommender systems. In: Companion Proceedings of the Web Conference 2020, pp. 694–702 (2020)

Clustering Without Knowing How To: Application and Evaluation

Daniil Likhobaba⬛, Daniil Fedulov⬛, and Dmitry Ustalov$^{(\boxtimes)}$⬛

Toloka, Belgrade, Serbia
{likhobaba-dp,mr-fedulow,dustalov}@toloka.ai

Abstract. Clustering plays a crucial role in data mining, allowing convenient exploration of datasets and new dataset bootstrapping. However, it requires knowing the distances between objects, which are not always obtainable due to the formalization complexity or criteria subjectivity. Such problems are more understandable to people, and therefore human judgements may be useful for this purpose. In this paper, we demonstrate a scalable crowdsourced system for image clustering, release its code at https://github.com/Toloka/crowdclustering under a permissive license, and also publish demo in an interactive Python notebook. Our experiments on two different image datasets, dresses from Zalando's FEIDEGGER and shoes from the Toloka Shoes Dataset, confirm that one can yield meaningful clusters with no machine learning purely with crowdsourcing. In addition, these two cases show the usefulness of such an approach for domain-specific clustering process in fashion recommendation systems or e-commerce.

Keywords: Clustering · Human-in-the-loop · Data mining · Crowdsourcing

1 Introduction

Clustering is a task of grouping objects in such a way that objects in the same group (called a *cluster*) are more similar to each other than to those in other groups [23]. This is important process in machine learning and arises in many applications, such as text [14] and image [8] segmentation, information retrieval [7], data mining [16] and pattern recognition [13]. In most cases, clustering is an unsupervised task [12] that requires knowing the distances between objects [15]. As the distances are often unknown or the clustering rules cannot be clearly defined [2], crowdsourcing may help to cope with these problems as such tasks often are trivial for humans [26]. It is known that people can apply their life experience to solve creative tasks [17], such as toxicity detection [1], relative rankings [20], fashion recommendation [3], etc.

The first idea for obtaining human judgements is to request labels from multiple humans, which allows computational quality control [24]. However, this

J. Kamps et al. (Eds.): ECIR 2023, LNCS 13982, pp. 262–268, 2023.
https://doi.org/10.1007/978-3-031-28241-6_24

approach is unscalable, unlike crowdsourcing. Although a proper use of crowd-sourcing requires a careful task design and quality control setup, recent studies show that it can approximate the distance function between the objects using crowd judgments [5,6,25]. Some of these papers are theoretical [21,22], evaluate performance on synthetic datasets [18], or require a prohibitively large number of human tasks to converge [11].

In this demonstration paper, we build a system for clustering with crowds, and evaluate it with crowds without involving any machine learning algorithms. We run our experiments on Toloka with two real world datasets, dresses from Zalando's FEIDEGGER [19] and shoes from the Toloka Shoes Dataset [9], and confirm the reproducibility of this method. Also, we release the source code of the built hybrid human-computer system under a free license and publish demo in an interactive Python notebook on Google Colab. We picked the *clustering by style* task as it is difficult to formalize as an algorithm, yet the task itself is relatively easy for humans: each of us can tell whether the style of clothes is similar or not.

2 Task Design and Worker Selection

To cluster objects, it is necessary to know how similar they are to each other; in the classical formulation, the pairwise matrix of distances is given. If the matrix is not known, it can trivially be approximated with crowds by running a pairwise comparison of all object pairs [11]. Unfortunately, it generates $\mathcal{O}(N^2)$ tasks for N objects, which is very expensive, i.e., 1000 objects would require roughly 500,000 comparisons, quickly impacting the annotation budget. Hence, there is a need for a task sampling method that is sufficient to divide the objects into meaningful groups as cheaply as possible.

2.1 Object Sampling

For clustering, we used Crowdclustering approach [10]. For each task, we show M objects and ask the crowd workers to combine them into groups based on the similarity of style. During prototyping, we found that the optimal choice of M is between 3 and 8 as clustering a large number of objects seems to require additional concentration from the workers, resulting in mistakes, such as failing to group all similar objects. In our setup, every task is completed by three differ-ent workers. We sampled each object for $V = \log_2 N \times \log_M N$ times to gather enough information on inter- and intra-relationships of the groups, allowing us to approximate the clustering.

2.2 Worker Training

Before starting, workers have to pass a training and a qualification test. The training consists of five pages of tasks, each have more pictures and requires more complex actions than the previous one. Starting with two images on the page and a step-by-step guide and ending with six pictures with more complex instructions. We made instruction and training tasks to familiarise workers with the interface and give them our understanding of how to group clothes by style, while leaving room for their subjective opinion. In the training and exam, workers receive a numerical skill value from 0 to 100, equal to their fraction of correct responses. Only those who achieved the skill value of at least 80 get access to the next step. Training tasks are obvious to everyone, so attentive workers do everything right, and we filter out those who did not understand the task at all. An example of an obvious task is "label all the high-heeled shoes with red color from palette."

Fig. 1. Clustering task. The worker groups similar clothes by highlighting similar ones with same color from the color palette; similar to [10].

2.3   Task Design

The task is formulated as *Group the objects by labeling similar ones using color palette*, and its interface is shown in Fig. 1. Workers should choose one color and label similar images with it, then choose another color and make another group, etc. This color palette simply serves as labels that are convenient for a worker, and *it has nothing in common with the colors of objects*. We then treat these colors simply as numbers assigned to each object on the page, placing objects with the same numbers in the same list for each page. Since each item is completely unique, the workers are told not to pay attention to the small details when grouping clothes, but to look at the style as a whole. There is a brief instruction on each page with the main points that should be kept in mind during grouping.

3 Clustering with Crowds

After labeling, we have information about how the pictures on each page have been grouped together, and which worker made each page. Since we have a sparse dataset of noisy labels for the objects, we need a special aggregation method to recover the clusters. For this, we applied and re-implemented using Python a probabilistic model called *Crowdclustering* [10]. This approach represents the objects as points in Euclidean space and models the workers as a plurality of binary classifiers. Each task is considered as a binary classification of the objects pairs regarding their membership in the same group or in a different group. The main advantage of Crowdclustering algorithm is that it allows each worker to group objects by any attribute (e.g. color, material, shape) and works with these groups called *atomic clusters*. Then, atomic clusters are assembled into resulting clusters, the number of which is not a fixed hyper-parameter and is estimated dynamically.

Fig. 2. Cluster visualization produced by the Crowdclustering method; dots are objects, colors are clusters.

4 Quality Evaluation

Having annotated and aggregated the clusters, we need to evaluate the quality of them. A common approach is to compare result with ground truth answers. However, we are solving a problem without a strict criterion for grouping objects that is difficult to formalise, i.e. *clustering clothes by style*. It leads to difficulties with making the only correct ground truth answers. And since we use subjective human judgements for clustering, we want to use them for evaluation as well. For this purpose we apply an approach called *Intruders* [4]. For each cluster, we sample from another cluster a random incorrect object called an intruder.

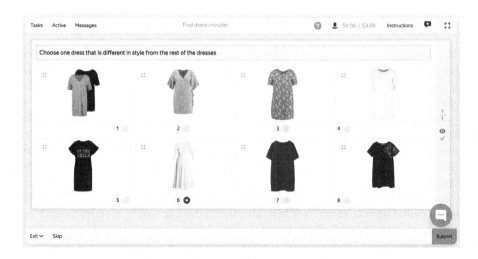

Fig. 3. Evaluation task. The worker has to select the artificially-inserted intruder element (no. 6); similar to [4].

Then, we run another crowdsourcing task, in which we ask the workers to select the out-of-style object. This task's interface is shown in Fig. 3. The clustering quality is a fraction of times the workers selected the intruder correctly and it is considered the better, the more often the workers choose this obviously incorrect object.

We ran the experiments on FEIDEGGER and Toloka Shoes in the same above-described configuration. Visualization of result is shown in Fig. 2, where dots in Euclidean space are projected on a 2D plane for clarity. We found that for the 2,000 dress images in FEIDEGGER the quality is 0.83, and for the 87 shoes images in Toloka Shoes Dataset the quality is 0.88. It means that in the clustering tasks crowd workers divided images into clusters according to their subjective opinion, which are consistent with the subjective opinion of other workers who evaluated these clusters.

5   Conclusion

We found that crowdsourcing allows to obtain a reasonable clustering of objects when distances between the objects could not be measured at all due to informal criteria. It allows using a human-understandable textual instruction instead of metric learning, while being more cost-efficient than the entire distance matrix annotation. This may be important when the problem is difficult to formalize for machine learning algorithms. We release a Python implementation of the system at https://github.com/Toloka/crowdclustering. It includes the labeling pipeline, a Crowdclustering algorithm, and quality evaluation algorithm.

References

1. Aroyo, L., Dixon, L., Thain, N., Redfield, O., Rosen, R.: Crowdsourcing Subjective Tasks: The Case Study of Understanding Toxicity in Online Discussions. In: Companion Proceedings of The 2019 World Wide Web Conference. pp. 1100–1105. WWW '19, Association for Computing Machinery, New York, NY, USA (2019). https://doi.org/10.1145/3308560.3317083
2. Ben Ayed, A., Ben Halima, M., Alimi, A.M.: Survey on clustering methods: Towards fuzzy clustering for big data. In: 2014 6th International Conference of Soft Computing and Pattern Recognition (SoCPaR), pp. 331–336 (2014). https://doi.org/10.1109/SOCPAR.2014.7008028
3. Burton, M.A., Brady, E., Brewer, R., Neylan, C., Bigham, J.P., Hurst, A.: Crowdsourcing Subjective Fashion Advice Using VizWiz: Challenges and Opportunities. In: Proceedings of the 14th International ACM SIGACCESS Conference on Computers and Accessibility. pp. 135–142. ASSETS '12, Association for Computing Machinery, New York, NY, USA (2012). https://doi.org/10.1145/2384916.2384941
4. Chang, J., Boyd-Graber, J., Gerrish, S., Wang, C., Blei, D.M.: Reading Tea Leaves: How Humans Interpret Topic Models. In: Advances in Neural Information Processing Systems 22, pp. 288–296. NIPS 2009, Curran Associates Inc, Vancouver, BC, Canada (2009), https://papers.nips.cc/paper/3700-reading-tea-leaves-how-humans-interpret-topic-models.pdf
5. Chang, J.C., Kittur, A., Hahn, N.: Alloy: Clustering with crowds and computation. In: Proceedings of the 2016 CHI Conference on Human Factors in Computing Systems, pp. 3180–3191 (2016)
6. Chen, J., Chang, Y., Castaldi, P., Cho, M., Hobbs, B., Dy, J.: Crowdclustering with partition labels. In: International Conference on Artificial Intelligence and Statistics, pp. 1127–1136. PMLR (2018)
7. Chen, Y., Wang, J.Z., Krovetz, R.: Content-Based Image Retrieval by Clustering. In: Proceedings of the 5th ACM SIGMM International Workshop on Multimedia Information Retrieval, pp. 193–200. MIR '03, Association for Computing Machinery, New York, NY, USA (2003). https://doi.org/10.1145/973264.973295
8. Coleman, G., Andrews, H.: Image segmentation by clustering. Proc. IEEE **67**(5), 773–785 (1979). https://doi.org/10.1109/PROC.1979.11327
9. Drutsa, A., Fedorova, V., Ustalov, D., Megorskaya, O., Zerminova, E., Baidakova, D.: Crowdsourcing Practice for Efficient Data Labeling: Aggregation, Incremental Relabeling, and Pricing. In: Proceedings of the 2020 ACM SIGMOD International Conference on Management of Data, pp. 2623–2627. SIGMOD '20, Association for Computing Machinery, New York, NY, USA (2020). https://doi.org/10.1145/3318464.3383127
10. Gomes, R., Welinder, P., Krause, A., Perona, P.: Crowdclustering. In: Shawe-Taylor, J., Zemel, R., Bartlett, P., Pereira, F., Weinberger, K. (eds.) Advances in Neural Information Processing Systems, vol. 24. Curran Associates, Inc. (2011), https://proceedings.neurips.cc/paper/2011/file/c86a7ee3d8ef0b551ed58e354a836f2b-Paper.pdf
11. Green Larsen, K., Mitzenmacher, M., Tsourakakis, C.: Clustering with a Faulty Oracle. In: Proceedings of The Web Conference 2020, pp. 2831–2834. WWW '20, Association for Computing Machinery, New York, NY, USA (2020). https://doi.org/10.1145/3366423.3380045
12. Grira, N., Crucianu, M., Boujemaa, N.: Unsupervised and semi-supervised clustering: a brief survey. A Rev. Mach. Learn. Tech. Process. Multimed. Content **1**, 9–16 (2004)

13. Hamerly, G., Elkan, C.: Alternatives to the K-Means Algorithm That Find Better Clusterings. In: Proceedings of the Eleventh International Conference on Information and Knowledge Management, pp. 600–607. CIKM '02, Association for Computing Machinery, New York, NY, USA (2002). https://doi.org/10.1145/584792.584890

14. Jain, A.K., Bhattacharjee, S.: Text segmentation using gabor filters for automatic document processing. Mach. Vision Appl. **5**(3), 169–184 (1992). https://doi.org/10.1007/BF02626996

15. Jain, A.K., Murty, M.N., Flynn, P.J.: Data Clustering: a review. ACM Comput. Surv. **31**(3), 264–323 (1999). https://doi.org/10.1145/331499.331504

16. Judd, D., McKinley, P., Jain, A.: Large-scale parallel data clustering. IEEE Trans. Patt. Anal. Mach. Intell. **20**(8), 871–876 (1998). https://doi.org/10.1109/34.709614

17. Kittur, A.: Crowdsourcing. Collab. Creativity. XRDS **17**(2), 22–26 (2010). https://doi.org/10.1145/1869086.1869096

18. Korlakai Vinayak, R., Hassibi, B.: Crowdsourced Clustering: Querying Edges vs Triangles. In: Advances in Neural Information Processing Systems. vol. 29. Curran Associates, Inc. (2016), https://proceedings.neurips.cc/paper/2016/file/82f2b308c3b01637c607ce05f52a2fed-Paper.pdf

19. Lefakis, L., Akbik, A., Vollgraf, R.: FEIDEGGER: A Multi-modal Corpus of Fashion Images and Descriptions in German. In: LREC 2018, 11th Language Resources and Evaluation Conference (2018)

20. Luon, Y., Aperjis, C., Huberman, B.A.: Rankr: a mobile system for crowdsourcing opinions. In: Zhang, J.Y., Wilkiewicz, J., Nahapetian, A. (eds.) MobiCASE 2011. LNICST, vol. 95, pp. 20–31. Springer, Heidelberg (2012). https://doi.org/10.1007/978-3-642-32320-1_2

21. Mazumdar, A., Saha, B.: Clustering with Noisy Queries. In: Advances in Neural Information Processing Systems, vol. 30. Curran Associates, Inc. (2017), https://proceedings.neurips.cc/paper/2017/file/db5cea26ca37aa09e5365f3e7f5dd9eb-Paper.pdf

22. Raman, R.K., Varshney, L.R.: Budget-optimal clustering via crowdsourcing. In: 2017 IEEE International Symposium on Information Theory (ISIT), pp. 2163–2167. IEEE (2017)

23. Rokach, L., Maimon, O.: Clustering Methods, pp. 321–352. Springer, US, Boston, MA (2005). https://doi.org/10.1007/0-387-25465-X_15

24. Ustalov, D., Pavlichenko, N., Losev, V., Giliazev, I., Tulin, E.: A General-Purpose Crowdsourcing Computational Quality Control Toolkit for Python. In: The Ninth AAAI Conference on Human Computation and Crowdsourcing: Works-in-Progress and Demonstration Track. HCOMP 2021 (2021), https://www.humancomputation.com/2021/assets/wips_demos/HCOMP_2021_paper_85.pdf

25. Yi, J., Jin, R., Jain, S., Yang, T., Jain, A.: Semi-crowdsourced clustering: Generalizing crowd labeling by robust distance metric learning. In: Advances in Neural Information Processing Systems, vol. 25 (2012)

26. Yuen, M.C., King, I., Leung, K.S.: A Survey of Crowdsourcing Systems. In: 2011 IEEE Third International Conference on Privacy, Security, Risk and Trust and 2011 IEEE Third International Conference on Social Computing, pp. 766–773 (2011). https://doi.org/10.1109/PASSAT/SocialCom.2011.203

Enticing Local Governments to Produce FAIR Freedom of Information Act Dossiers

Maarten Marx(✉) , Maik Larooij , Filipp Perasedillo, and Jaap Kamps

University of Amsterdam, Amsterdam, The Netherlands
{maartenmarx,larooij,kamps}@uva.nl

Abstract. Government transparency is central in a democratic society, and increasingly governments at all levels are required to publish records and data either proactively, or upon so-called Freedom of Information (FIA) requests. However, public bodies who are required by law to publish many of their documents turn out to have great difficulty to do so. And what they publish often is in a format that still breaches the requirements of the law, stipulating principles comparable to the FAIR data principles. Hence, this demo is addressing a timely problem: the FAIR publication of FIA dossiers, which is obligatory in The Netherlands since May 1st 2022.

Keywords: IR data collection · FAIR data · Governments records · Transparency

1 Introduction

Freedom of Information Act (FIA), sometimes called *Access to Information* or *sunhine* laws are effective in over 100 countries. They give citizens the right to request previously unreleased documents on policy issues from governmental bodies. In the Netherlands, more than 1000 such bodies exist and they are obliged to also publish the requested dossiers to the general public, for instance through their own website. For a good functioning of democracy it is important that these documents are easy to retrieve and to discover relevant information in them [11]. Together with democracy watchdog *OpenState* and the platform for investigative journalism *Follow The Money*, we decided to create a vertical search engine for these FIA dossiers, bringing them together in one portal.

Finding and harvesting the published documents was not difficult and possible with standard crawling technology. Somewhat to our surprise, the main

---

This research was supported in part by the Netherlands Organization for Scientific Research (NWO) through the ACCESS project grant CISC.CC.016, and by the University of Amsterdam through Humane AI.

Demo material is available at https://ecir2023.wooverheid.nl.

J. Kamps et al. (Eds.): ECIR 2023, LNCS 13982, pp. 269–274, 2023.
https://doi.org/10.1007/978-3-031-28241-6_25

problem was that the released data violated all 3 core assumptions behind Information Retrieval: that a collection consists of *documents* containing *words*, and having basic *metadata*. The predominant strategy to release FIA dossiers (containing the request, the argued decision, the list of relevant documents, and the released documents) is to print it out completely, scan the pile of pages and publish it as one (often huge) PDF file. Metadata is hidden inside the decision letter, individual documents are often published as one big PDF document without boundaries between the different documents and if the PDF contains characters, they should be obtained via OCR. This type of data is known in IR from the TREC legal and total recall tracks [3,4] and from IR for due diligence [6].

Of course, we can try to repair the data, using OCR, page stream segmentation [9] and knowledge extraction [7], and we have done so [5], but these processes are never error free and by nature frustrating, *as the process is most often reverse engineering what was originally in digital format and well structured available.* So we decided to repair this problem at the source and this demo describes our solution. Hence, our research question is

> *How can we entice, especially small local, governments to publish their FIA dossiers according to the FAIR data principles [10].*

We targeted municipal governments because they form the vast majority of FIA publishers, have little IT infrastructure and lack resources. So our solution had to be robust, simple and cheap with large, directly visible, concrete gains. We developed this FIA publication software in cooperation with the association of Dutch municipalities VNG.

At the demo we present our solution to the problem. Here, we first describe the model that best fits these FIA dossiers, then the requirements for us and for the users of the software, followed by the chosen implementation. We evaluate it by checking the desiderata and requirements. We use the extra page in the Proceedings to describe the state of our system in January 2023.

2   The Data: FIA Dossiers

A FIA dossier can be seen as an argued response to an uttered information need. It contains a request for information, and the response consisting of 1) a list of all relevant documents, 2) those documents from the list that the government wants to (partly) release, and 3) a decision letter, typically drafted by a lawyer, explaining and motivating the response. The resemblance to a TREC topic and its corresponding set of relevant documents is remarkable. The decision letter is drafted as text, but can better be seen as a set of attribute-value pairs, containing a number of required and optional attributes. The values can be of free type, like the text describing the information need, or constrained, like the dates of the request and decision or the applied articles of law to withheld information.

Such a dossier is best modeled as semi-structured data combining metadata with raw text, with many optional attributes and unbounded cardinality constraints. In addition, the released documents are currently all in PDF format, but obviously this is most often not their original technical format. Requested documents are often about *communication* (mail, social media like WhatsApp messages) and *structured data* (maps, spreadsheets), which are far more valuable in their original technical format than as a PDF produced to print.

To summarize, the data model of a FIA dossier consists of metadata on the level of the dossier, metadata for each released document, and the documents themselves. We chose to store a dossier as a zip file containing the released files, together with all metadata, including the text extracted from the documents, in XML.

The small municipality Waalwijk used our system to publish their data, resulting in this publication page with complete dossiers. Our datamodel is visible in the XML metadata file added to the zip file created for downloading a dossier.

3   Requirements

Our final aim thus is being able to harvest FIA dossiers of as high as possible technical (FAIR data) quality, in order to maintain a well functioning FIA search engine. This translated into 3 requirements: an API to harvest all data in validated XML format, OCR and full text extraction at the source, and checks and services to increase the availability and quality of the metadata.

To entice municipalities to use our system, data entry had to be almost effortless and the system had to have direct, visible and easy implementable advantages *for them*. This translated to an easy data entry interface with many prefilled values; automatic attractive publishing on a website of the municipality; automatic pushing to the obliged central government API; an internal search system for the FIA lawyers, and a data dashboard for managers/annual reports.

The FAIR data principles of course also had to apply to our system, so we required only open data standards and free open source software, both preferably top quality and with a large user base and community.

4   Implementation

Given these requirements, and the nature of the data, XML or JSON seemed the best option for data representation. We choose XML because we wanted extensive validation with good error messages, which is available through the XML constraint language Relax NG and the Jing validation software [8].

MySQL was well suited as our database backend because it is open source, has good full-text search mainly based on TF-IDF and BM25, solid security, and a wide active user base. We stored all metadata, the full text of all documents, and the original PDF files in the database. The relational schema is a straightforward implementation of the XML model of a dossier, witnessed by obvious translations back and forth. We used Python as the scripting language tying the different components together. The different components were implemented as follows:

- the translations from XML to the relational DB schema and back were done in Python using the etree module;
- publishing a list of dossiers as a webpage was done through an SQL query generating an answer which was processed by Python into an HTML page;
- publishing to the government API using Python, Flask, and SQL;
- search engine using MySQL full-text `MATCH AGAINST` queries on several text fields, with ranking by internal BM25 and TF-IDF relevancy ranking;
- data dashboard using MySQL, Pandas, Seaborn and Plotly.

The application runs on a server. Municipalities can choose to use the server or run it on their own platform by using a Docker image.

5   FAIRiscore and Full Text score

To encourage municipalities to create data of high FAIRness quality, we created a 5 point A–E *FAIRiscore* scale, reminiscent of the Nutriscore [2]. Using imported and overruled RelaxNG schemas it is easy to define and maintain 5 schemas of monotonically increasing tightness, one for each fairiscore value. After data entry, the data is validated immediately and the user receives not only the score but also suggestions for improving the score. As the score is monotonic it can be explained and visualized easily by coloring the XML tree using the 5 Nutriscore colors, as in this figure on the web. It must be read as follows: in order to reach Nutriscore say B, one needs to have score C, and fill in all fields coloured light green (the color of score B).

Similarly we provide a 5-point A-E score indicating the quality of the full text. Many documents are released as PDFs without underlying text, but the civil servants doing this are often not aware of this. We estimate the amount of real text per page using OCR and redacted text recognition, compare that to the text inside the PDF and give a score based on the overlap.

6   Evaluation

As the system is very new, we have no user studies yet, but expect to have them at the time of the conference. The system works well as a filter for data entry into our large Freedom of Information Act search engine https://woogle.wooverheid. nl. Even if the dossiers have the (lowest) FAIRiscore E, we have the full text (either the original or through OCR), and a tiny bit of metadata. This was our main goal, and thus reached. How well our data quality encouragment methods worked need to be seen, and will be reported on at the time of the conference.

7   Conclusion

Our system is simple but it contains for municipalties very desirable functionality. If the uploaded data is already slightly better than the worst FAIRiscore

E, a FIA search engine can both rank better and present much richer search snippets, making it easier for users to judge relevance already before they have to open a document.

At the time of writing we cannot yet answer the implicit question in the title of this demo. We expect to be able to do that at the time of the conference.

8 Update January 2023

We use the extra page in the proceedings to describe the state of our system in January 2023. On December 28, 2023, the Dutch central government announced that they stopped developing a platform having similar functionality as the system described here, after very negative advice from an independent ICT in government review committee [1].

After publishing our initiative we received over 30 reactions from interested municipalities, from which only 2 went further and 1 succeeded, Waalwijk. We decided to collect the large bulk of available data by dedicated crawling and around Christmas 2022 our search engine was complete for the published FIA dossiers of all Dutch ministeries, all provinces (10 of the 12 publish), and the top 80 largest municipalities (only 6 of them publish), see the overview on https://woogle.wooverheid.nl/overview.

The quality of the data turned out to be as bad as described in this paper. To summarize:

Publish seperate documents None of the ministeries, only 2 of the 10 provinces, and only 1 of the 6 municipalities.

Documents containing text From all 41K documents in our system, 46% contained not a single machine readable character (30% of all 1M pages). The amount of non machine readable pages ranges from 2% with the municipality Waalwijk and one ministery (publishing 40K pages) to 85%. It seems reasonable to assume that with 2% all documents are born digital, sometimes also called native PDFs.

Metadata Not a single publisher provided even these 4 basic properties as metadata: the dates of the request and the decision, the text of the request (the information need), and the decision.

Positive News. The single publisher Waalwijk using our system showed that it is possible to deliver what we asked for: release native PDF documents separately in a zip file instead of one concatenated PDF; each page machine readable and all basic metadata nicely in order; see https://doi.wooverheid.nl/?doi=nl.gm0867.

Negative News. Obviously with only one participant, we did not entice the municipalities. This seems partly due tp the unclear situation regarding the platform of the central government, but there is also another reason. From interviews with several publishers we learned that we simply asked too much, especially with regard to metadata. So, we changed strategy, focusing on just getting the raw data, preferably machine readable, using an API that can directly connect to the IT infrastructure of the data publisher.

Next Steps. We hope to fill our search engine using this API. We will highlight data publishers which produce FAIR data and hope that seeing good examples in action (besides their own not so good example) entices municipalities to change their publishing habits. On https://woogle.wooverheid.nl the reader can track our progress.

References

1. Adviescollege ICT-toetsing: BIT-advies Plooi. https://www.adviescollegeicttoet sing.nl/onderzoeken/documenten/publicaties/2022/11/28/bit-advies-plooi Accessed Dec 28 2022
2. Chantal, J., Hercberg, S., Organization, W.H., et al.: Development of a new front-of-pack nutrition label in france: the five-colour nutri-score. Public Health Panorama **3**(04), 712–725 (2017)
3. Grossman, M.R., Cormack, G.V.: Technology-assisted review in e-discovery can be more effective and more efficient than exhaustive manual review. Rich. JL Tech. **17**, 1 (2010)
4. Grossman, M.R., Cormack, G.V., Roegiest, A.: Trec 2016 total recall track overview. In: TREC (2016)
5. van Heusden, R., Kamps, J., Marx, M.: Wooir: A new open page stream segmentation dataset. In: Proceedings of the 2022 ACM SIGIR International Conference on Theory of Information Retrieval, pp. 24–33 (2022)
6. Roegiest, A., Hudek, A.K., McNulty, A.: A dataset and an examination of identifying passages for due diligence. In: The 41st International ACM SIGIR Conference on Research & Development in Information Retrieval, pp. 465–474 (2018)
7. Shi, P., Lin, J.: Simple bert models for relation extraction and semantic role labeling. arXiv preprint arXiv:1904.05255 (2019)
8. Van der Vlist, E.: Relax ng: A simpler schema language for xml. " O'Reilly Media, Inc." (2003)
9. Wiedemann, G., Heyer, G.: Multi-modal page stream segmentation with convolutional neural networks. Lang. Resour. Eval. **55**(1), 127–150 (2021)
10. Wilkinson, M.D., et al.: The FAIR guiding principles for scientific data management and stewardship. Sci. Data **3**(1), 1–9 (2016)
11. Worthy, B.: More open but not more trusted? the effect of the Freedom of Information Act 2000 on the United Kingdom Central Government. Governance **23**(4), 561–582 (2010). https://doi.org/10.1111/j.1468-0491.2010.01498.x

Which Country Is This? Automatic Country Ranking of Street View Photos

Tim Menzner[1], Florian Mittag[1], and Jochen L. Leidner[1,2]

[1] Coburg University of Applied Sciences and Arts, Friedrich-Streib-Str. 2,
96450 Coburg, Germany
tim.menzner@hs-coburg.de
[2] University of Sheffield, Regents Court, 211 Portobello, Sheffield S1 4DP, UK
leidner@acm.org

Abstract. In this demonstration, we present Country Guesser, a live system that guesses the country that a photo is taken in. In particular, given a Google Street View image, our federated ranking model uses a combination of computer vision, machine learning and text retrieval methods to compute a ranking of likely countries of the location shown in a given image from Street View. Interestingly, using text-based features to probe large pre-trained language models can assist to provide cross-modal supervision. We are not aware of previous country guessing systems informed by visual and textual features.

Keywords: Country identification · Content meta-data enrichment · Content-based image analysis · Cross-modal classification · Software demonstration

1 Introduction

If someone looked at a physical or electronic photo, one of the natural questions one may ask is "Where was this taken?". For instance, the Geographical Magazine of the Royal Geographical Society features a monthly competition showing a photo and asking its readers to identify the place depicted [10].

In this demonstration, we present a new system that addresses an easier sub-problem, namely: given a Google Street View image in specific, which *country* is shown on the image? Our approach is to use a set of individual classifiers, such that the individual signal evidence is combined to create a ranked list of countries. Interestingly, both visual and textual features turn out to be helpful in the task.

The bulk of this work was undertaken by the first author during his Master's thesis. We gratefully acknowledge the funding by the *High Tech Agenda Bavaria* (https://www.bayern.de/wp-content/uploads/2019/10/Regierungserklaerung_101019_engl.pdf) to the third author that supported part of this research.

J. Kamps et al. (Eds.): ECIR 2023, LNCS 13982, pp. 275–280, 2023.
https://doi.org/10.1007/978-3-031-28241-6_26

2   Related Work

One early notable contribution for solving the problem of determining the geolocation of images was the winning entry of the "where am I" contest carried out during the 2005 International Conference on Computer Vision. This approach featured a huge database of city street scenes tagged with GPS locations and used SIFT to find correspondences [15]: classical feature matching techniques were combined with landscape classification the classical feature matching by IM2GPS, which was, according to its authors, "the first to be able to extract geographic information from a single image" [3]. Later, the topic was addressed in a wide range of papers (see [2] for a detailed survey). One of the first papers that used machine learning for geolocating images was PlaNet [13], which divided earth's surface into thousands of grid cells. Then, 126 million images with geo information were assigned to the respective cells. A convolutional neuronal network was trained with these images to output a probability score for each and every cell. Zamir et al. [14] focused specifically on Google Street View images. They used trees of indexed SIFT descriptors and 100,000 Street View images from Pittsburgh, PA and Orlando, FL to find GPS coordinates for (not necessarily Street View) pictures from these cities. Also using Street View and inspired by the game GeoGuessr, [12] combined Google Street View imagery and machine learning to make educated guesses about photos from Street View's coverage of the United States. In contrast, our system provides a ranked list of countries and has world wide geographic scope.[1]

3   Graphical User Interface

The system's graphical user interface is shown in Fig. 1. The system can operate in two basic modes: in the first mode, users can load a photo either from the local hard drive or pulled via Google Street View's online API for an automated guess, while additional information about the results of each individual module are provided. There is also the possibility to switch into a game mode, where the user can directly compete against the system.

4   Method and Implementation

External Data. We used Geo-JSON country boundary polygons from the Natural Earth, public domain dataset. We identify countries based on cues, a database maintains this evidence, such as what language is spoken where; information about the language of text in an image is not much of use without such background knowledge. Therefore, every featured country also gets a fact sheet in the form of an external JSON file. The information for this fact sheet was

---

[1] Between acceptance of this paper and our preparation of the publication version, we found another work, [1] which attempts country guessing of photos that are not necessarily from Street View, which is very much in the spirit of this paper.

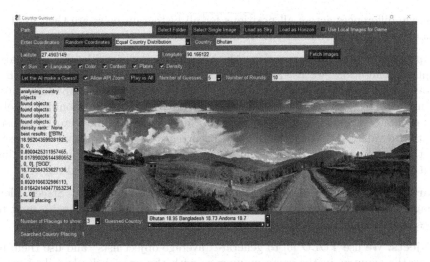

Fig. 1. Country Guesser recognizes a Google Street View image as located in Bhutan.

collected manually. Overall, 110 countries were found to have sufficient Street View coverage to be included in this project.

Solar Position. We narrow down the list of candidate countries, leveraging the sun's position via celestial direction. As the earth orbits the sun with an axial tilt of about 23.4°, the relative position of the sun throughout the day is different between the two hemispheres. With a southern sun position only possible on the Northern Hemisphere and northern sun position only possible on the Southern Hemisphere (except for the tropics), this concept can be used to determine the hemisphere where the panorama image was taken. For detecting the sun position, our approach is to split the panorama, which covers a 360° field of view, into separate images pitched towards the sky, to search for the brightest image.

Text and Languages. Street View, as the name suggests, mainly consist of images taken from streets, so often street signs, billboards etc. showing some kind of text are visible. We combine optical character recognition (OCR) using *EasyOCR* [6] to obtain text before checking the most likely language using the Python *langdetect* [9] and *lingua* [11] libraries and, finally, we scan text for place names mentioned and knowledge of the country the place is located in [7].

Coloration. Differences in climate and vegetation lead to different colors being associated with the "look" of a country: Ireland for example is well known to be very green (Fig. 2). To objectively model this relationship between color palette and country, we use one histogram for each RGB channel. The average histograms for a country can be generated by looking at the histograms of a number of images from that country, summing up the total number of pixels for each intensity value and respectively dividing it by the number of images used. The average differences across all positions for all histogram pairs are then used to calculate an overall similarity score and to rank all countries accordingly.

Captions. A powerful feature was the descriptive text provided by automatic caption generation (we used the *ClipCap* model [8]). One more concept that proved to be useful was analysing differences in the frequency of individual words being featured in the image descriptions generated by image captioning models. We generated average "word lists" for each country as an offline step. These lists contain a value indicating the average occurrence of all words that were used by the model when describing an image from the country during its generation (with non-descriptive words like articles and adverbs being filtered out).

Car License Plates. Despite license plates being blurred on Street View imagery for privacy reasons, their colours are still visible and can be used to guess the country.

Objects. We use *YOLO* [5] to generate "average object lists" and use them to guess the country, in the same manner as with the previously described word lists.

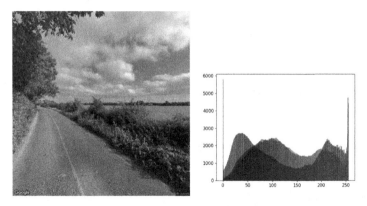

Fig. 2. A photo taken from Google Street View in Ireland (left) and the Associated Color Histogram (right). (Left image: Copyright by Google Inc. – Used Under Fair Use/Academic Research)

5   Evaluation

A test data set was created by downloading two panoramas of randomly selected coordinates for each of the included 110 countries, resulting in a total number of 4,620 individual images. For each panorama, the system produced a probability ranking including all countries. The weightings for each individual module were optimized with another (unseen) data set beforehand, where color and captions proved to be most useful among the indicators. For the 220 guesses made, the right country landed on rank 14.7 on average, with a standard deviation of 19.1 and a median of 7. In 35 cases, the searched county was successfully ranked first (Table 1).

Table 1. Feature/Sub-Model Ablation Study: Each row describes a system variant with a feature or sub-model removed from the setup in the row one above

System version	Quality (avg. rank)	Evidence exploited	Feature always available?
Full system	14.695	Image + Text	N/a
As above minus *Car License Plates*	14.655	Image (*text blurred*)	No
As above minus *Text & Languages*	14.927	Text	No
As above minus *Object Lists*	15.172	Text	No
As above minus *Solar Position*	15.736	Image	No
As above minus *Coloration*	19.727	Image	Yes
As above minus *Caption*	0	Text	Yes

6   Applications and Impact

Identifying countries can be the first step for increasingly fine-grained location identification. A system like our Country Guesser can serve multiple functions:

- entertainment: our system can be used as a quiz game to test one's ability to identify visual geographic cues about an image's whereabouts;
- education: as a teaching aid to create awareness for subtle geographical hints of country differences;
- investigative journalism: to pinpoint locations of events under investigation by reporters and to fact-check the authenticity of potential "fake news" by probing for potential re-use of accompanying images [4]; and
- law enforcement: to assist crime investigations that involve visual evidence.

Ethics and Privacy Note. Because the system makes use of data from Google Street View, it inherits any potential privacy issues from it. The system by design has a bias towards views that are visible from car-accessible roads, which includes only portions of each country.

7   Summary, Conclusions and Future Work

We have presented a system demonstration comprising a country guesser that attempts to use a range of evidence, visual features from the image itself and textual features via automatic caption generation, to compute a ranking of most likely countries shown in an arbitrary Google Street View image. The system's source code is available on GitLab[2], and our repository also contains a demo video.

Acknowledgements. The authors wish to thank the three anonymous reviewers, whose comments improved the quality of this paper.

---

[2] https://gitlab.com/Timperator/which-country-is-this (cited 2022-01-10).

References

1. Alamayreh, O., Dimitri, G.M., Wang, J., Tondi, B., Barni, M.: Which country is this picture from? New data and methods for DNN-based country recognition, September 2022. Unpublished Manuscript, Cornell University, ArXiv Pre-Print Server. https://arxiv.org/abs/2209.02429
2. Brejcha, J., Čadík, M.: State-of-the-art in visual geo-localization. Pattern Anal. Appl. **20**(3), 613–637 (2017). https://doi.org/10.1007/s10044-017-0611-1
3. Hays, J., Efros, A.A.: IM2GPS: estimating geographic information from a single image. In: 2008 IEEE Conference on Computer Vision and Pattern Recognition, pp. 1–8. IEEE (2008)
4. Higgins, E.: We Are Bellingcat: Global Crime, Online Sleuths, and the Bold Future of News. Bloomsbury, London (2021)
5. Jocher, G., et al.: ultralytics/yolov5: v6.1 - TensorRT, TensorFlow edge TPU and OpenVINO export and inference (2022). https://doi.org/10.5281/ZENODO.6222936. https://zenodo.org/record/6222936
6. Kittinaradorn, R.: Easy OCR. Software (2022). https://github.com/JaidedAI/EasyOCR
7. Leidner, J.L.: Toponym Resolution in Text: Annotation, Evaluation and Applications of Spatial Grounding of Place Names. Universal Press, Boca Raton (2008)
8. Mokady, R., Hertz, A., Bermano, A.H.: ClipCap: CLIP prefix for image captioning. arXiv preprint arXiv:2111.09734 (2021)
9. Nakatani, S.: Language detection library for Java (2010). https://github.com/shuyo/language-detection
10. Royal Geographical Society (ed.): Where in the World – Castles & Walled Cities. Geographical (2022). https://geographical.co.uk/crossword-and-quizzes/where-in-the-world-castles-walled-cities. Accessed 22 Aug 2022
11. Stahl, P.M.: Lingua (2022). https://github.com/pemistahl/lingua-py
12. Theethira, N.S.P., Ravindranath, D.: GeoguessrLSTM. Technical report, University of Colorado (unpublished). https://github.com/Nirvan66/geoguessrLSTM/blob/master/documentation/CSCI5922_ProjectReport.pdf. Accessed 14 Oct 2022
13. Weyand, T., Kostrikov, I., Philbin, J.: PlaNet - photo geolocation with convolutional neural networks. In: Leibe, B., Matas, J., Sebe, N., Welling, M. (eds.) ECCV 2016. LNCS, vol. 9912, pp. 37–55. Springer, Cham (2016). https://doi.org/10.1007/978-3-319-46484-8_3
14. Zamir, A.R., Shah, M.: Accurate image localization based on Google Maps Street view. In: Daniilidis, K., Maragos, P., Paragios, N. (eds.) ECCV 2010. LNCS, vol. 6314, pp. 255–268. Springer, Heidelberg (2010). https://doi.org/10.1007/978-3-642-15561-1_19
15. Zhang, W., Kosecka, J.: Image based localization in urban environments. In: Third International Symposium on 3D Data Processing, Visualization, and Transmission (3DPVT 2006), pp. 33–40. IEEE (2006)

Automatic Videography Generation from Audio Tracks

Debasis Ganguly[✉], Andrew Parker, and Stergious Aji

University of Glasgow, Glasgow, Scotland
Debasis.Ganguly@glasgow.ac.uk, {2389622P,2546916A}@student.gla.ac.uk

Abstract. This paper describes a prototype of an automatic videography generation system. Given any YouTube video of a song, a set of images are retrieved corresponding to each line of the song which are automatically inserted and aligned into a video track.

Keywords: Forced alignment · Keyword extraction · Image retrieval

1 Introduction

Multimodal sources of information are conducive to increased user engagement and satisfaction for tasks such as information seeking [12,19,20] and learning [5]. This is also true in the context of entertainment, where it has been shown that audio tracks of songs accompanied with video tracks lead to a more engrossing user experience [13]. However, there exists a plethora of music videos on online streaming platforms, such as YouTube [23], Dailymotion [14] etc., where the audio tracks are not accompanied with content-rich video tracks. In most cases, the video track consists of a single or a few static images, e.g., the video track of the official music video of the song 'A Horse with No Name' [16]. In such situations, music fans would often create video tracks by using a sequence of static images appropriate to the mood and the content of the song. In fact, several fan-made video tracks have been made for America's song 'A Horse with No Name', one of which [22] is shown in Fig. 1. While this is common for popular songs, there are many more tracks that are in absence of any videographic content. Specifically with vinyl tracks that have been digitised and posted on the platform, as well as songs in languages other than English.

The motivation of this demonstration paper is to automate the process of videographic content creation - namely retrieving a list of images to match the content from various segments of a song, then inserting them sequentially at appropriate time intervals. Such an automated system not only saves the manual effort but also contributes towards an enriched user experience, which in turn could also help popularise unknown artists/bands.

2 System Overview

Figure 2 shows a schematic overview of our system. The input to the system is a YouTube URL of a song, which is used to download the content. Firstly, the audio

J. Kamps et al. (Eds.): ECIR 2023, LNCS 13982, pp. 281–287, 2023.
https://doi.org/10.1007/978-3-031-28241-6_27

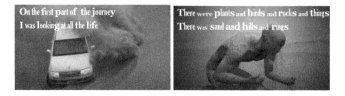

Fig. 1. First two images (left to right) from the manually created video track of the song 'A Horse with No Name'. The entire video track is constituted of a sequence of images that are topically related to the content of the current segment of the song being played.

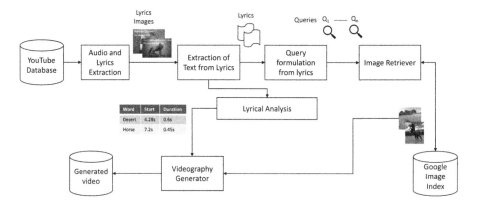

Fig. 2. Schematic workflow diagram of the videography generation system.

and lyrics are extracted from the YouTube metadata of the song (Sect. 3.1). In situations where the lyrics are not present, the '**Audio and Lyrics Extraction**' component will output a list of images that have the lyrics embedded within them (see Fig. 2).

The second component - the '**Text Extractor**' takes as input these images and extracts the text (i.e. lyrical content) for each (Sect. 3.2). Specifically, it leverages a standard OCR technique [11]. The output is then passed into the next component in the pipeline.

The third component - the '**Query Formulator**' (Sect. 3.3) partitions the entire lyrics into segments. For each segment, this component employs an unsupervised keyword extraction algorithm to formulate a query comprising at most 2 informative words or phrases.

The fourth component - the '**Image Retriever**' (Sect. 3.5) then retrieves a set of candidate images for these queries from the third-party index of Google Images using their API. While the images are being retrieved, a parallel thread of execution, namely the lyric analyser (Sect. 3.6), processes the lyrics in combination with the audio track to construct a timings list (a keyword to time interval map).

Finally, the video generator component (Sect. 3.7) takes as inputs both the candidate set of images and the timings list to generate a video track comprised of potentially relevant images which are then inserted in the video during appropriate time intervals.

3 Implementation Details

The system constitutes a web service developed with the Django web framework [4]. The reason to employ Django is due to its convenience in usage and deployment.

Functionality-wise, the system takes as input in a link to a YouTube [23] content, the name of the accredited artist, the song title and the method in which to analyse the video. The Github source code repository of this prototype system is available here[1]. In the subsequent sections, we now describe the components of the system and the processing steps involved in each.

3.1 Audio and Lyrics Extraction

The first major processing step in our system is extracting out the audio track from a YouTube content. In particular, we use the 'youtube_dl' package [6] for this step. In situations where the text metadata is absent, we need to extract the text from the images corresponding to the video frames. This workflow of the system requires processing the video in a frame by frame manner, for which we needed to download them locally. To download the videos, we used the Pytube [18] API.

3.2 Extracting the Text of Lyrics

We use the Python API lyricsgenius [21] to access Genius, which is a popular database of songs and artists. Using this API we formulate a query using the name of the song and the accredited artist that were supplied as a part of the input. The query is then used to retrieve a song transcript.

Alternatively, the transcript can be extrapolated from the video's captions. This method relies on the 'youtube-transcript-api' package [3], which returns text segments in JSON format. These are then collated to form the entire transcript.

3.3 Query Formulation from Lyrics Segments

Each method of extracting the lyrics in the form of text from an input YouTube content produces a text file of the transcript partitioned into multiple lines. Our system treats each line as a segment or a fine-grained unit of the song. Each segment is treated as a verbose query for which the objective is to retrieve a set of candidate images that are topically relevant to its content.

---

[1] https://tinyurl.com/d5x32aet.

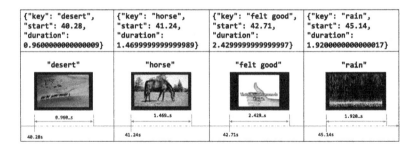

Fig. 3. An illustration of the automatic videography generation phase for 'A Horse with No Name'. The system uses the collated timings list to determine the start time and the duration for each image. A pre-generated version can be viewed on the 'Collection' tab of the system. This is also uploaded to the GitHub repository (https://tinyurl. com/d5x32aet).

3.4 Unsupervised Extraction of Keywords

For the process of query formulation, we extract a set of informative keywords from each segment. In particular, we employ the YAKE (Yet Another Keyword Extractor) tool [10], which implements an unsupervised keyword extraction algorithm. The algorithm itself makes use of a number of different features, such as the position of a word within a sentence, a term's relatedness to its context, its informativeness etc. to score and select a candidate set of terms [1]. From each segment, we extract a total of 2 keywords or key phrases.

3.5 Image Retrieval

To retrieve a candidate set of images, we programmatically invoke the Google Images search with each query formulated from the lyric segment. A number of filters are included in the search process to exclude inappropriate images like emotion/action words that could be associated with other entities, such as movie names.

The HTML page of the retrieved resultlist is parsed with the BeauitfulSoup [17] API to extract the image URLs, following which these images are downloaded. A filtering step is included to ensure that the dimensions of the image files are compatible with the package responsible for rendering the video, namely moviepy [24].

3.6 Lyrical Analysis

As a first step, the Opencv [8] tool is used to determine the frame rate of an extracted MP4 video file. This is then used to pull out a frame every half a second, i.e. every 14th frame is saved as a PNG file for a 28 frames per second (fps) video.

These frames are then analysed using the OCR tool - Pytesseract [11], to extract any textual information. A set of informative words along with their spans are obtained. A span starts from the frame where the words first appear and ends on the frame where they no longer exist. This can then be translated into a time interval to aid the system later in displaying retrieved images in time with the lyrics and audio.

In situations where no textual data can be extracted from the YouTube content, we employ the process of *forced lyrical alignment*. This is done using Selenium [9] to access the lyrics text and audio files submitted to a web-based form. This then automatically provides word-level alignment to a given polyphonic audio file [2,7].

The JSON dictionary returned from the caption extrapolation method also contains the start and duration of each corresponding text segment. Assuming reliable captions, the system requires little processing to generate a timings list with this method. However, it is impossible to know the exact times of individual words in the phrase, as a result of the image rendering occurs at the phrase-level, i.e., our system renders an image for the entire duration of a phrase.

3.7 Videography Generation

The output of the lyrical analysis step (Sect. 3.6) is a word: time-interval map, which we refer to as the timings list. This timings list is a sequence of objects, each comprising of a keyword, corresponding to a candidate image, and the start time and duration for which the image should be displayed. The list is sorted by the start timestamps. Figure 3 shows an example. This timings list is then used to create the video track using the pymovie package.

Since the opening lines of a song are often repeated along its entire duration, the system collects a list of retrieved images (5 in total) for each concept (word or phrase), and chooses an image at random to insert during a time interval corresponding to that concept.

4   Concluding Remarks

We described a prototype system that, given an audio track of a song, automatically creates a video track by retrieving images that match the content of the audio. Moreover, the video generation process ensures that the images are aligned to the appropriate segments of the lyrics. This potentially provides an enriched user experience and enables the users to better connect with the mood of a song.

There are different ways in which we can extend this system. The system currently works for audio in English. In future, we plan to work towards making this system multilingual, which may, in fact, contribute to popularising less known songs from resource-poor languages.

Additionally, we plan to use an open-source, static image collection rather than retrieving them from a Google Images search, which may not always be copyright-free. This will improve the reproducibility of results and allow fair evaluations against other automatic videography systems run on the same collection.

Another way in which this work could be extended is to make use of the multi-modal semantics between text and images to improve the retrieval quality. In particular, we plan to make use of OpenAI's Contrastive Language-Image Pre-training (CLIP) [15] which is the state-of-the-art multimodal embedding technique.

References

1. Campos, R., Mangaravite, V., Pasquali, A., Jorge, A., Nunes, C., Jatowt, A.: YAKE! Keyword extraction from single documents using multiple local features. Inf. Sci. **509**, 257–289 (2020). https://doi.org/10.1016/j.ins.2019.09.013. https://www.sciencedirect.com/science/article/pii/S0020025519308588
2. Gupta, C., Yilmaz, E., Li, H.: AutolyrixAlign (2020). https://github.com/chitralekha18/AutoLyrixAlign. Accessed 17 Feb 2022
3. Depoix, J.: youtube-transcript-api (2021). https://pypi.org/project/youtube-transcript-api/. Accessed 17 Feb 2022
4. Django: the web framework for perfectionists with deadlines (2005). https://www.djangoproject.com/. Accessed 17 Feb 2022
5. Doumanis, I., Economou, D., Sim, G.R., Porter, S.: The impact of multimodal collaborative virtual environments on learning: a gamified online debate. Comput. Educ. **130**, 121–138 (2019). https://doi.org/10.1016/j.compedu.2018.09.017. https://www.sciencedirect.com/science/article/pii/S0360131518302537
6. Garcia, R.: youtube-dl (2021). https://github.com/ytdl-org/youtube-dl. Accessed 17 Feb 2022
7. Gupta, C., Yılmaz, E., Li, H.: Automatic lyrics alignment and transcription in polyphonic music: does background music help? In: ICASSP 2020–2020 IEEE International Conference on Acoustics, Speech and Signal Processing (ICASSP), pp. 496–500 (2020). https://doi.org/10.1109/ICASSP40776.2020.9054567
8. Heinisuo, O.P.: opencv-python (2012). https://pypi.org/project/opencv-python/. Accessed 17 Feb 2022
9. Huggins, J.: selenium (2004). https://pypi.org/project/selenium/. Accessed 17 Feb 2022
10. Laboratory of Artificial Intelligence and Decision Support: yake (2021). https://github.com/LIAAD/yake. Accessed 17 Feb 2022
11. Lee, M.: pytesseract (2021). https://pypi.org/project/pytesseract/. Accessed 17 Feb 2022
12. Liao, L., Long, L.H., Zhang, Z., Huang, M., Chua, T.S.: MMConv: an environment for multimodal conversational search across multiple domains. In: Proceedings of the SIGIR 2021, pp. 675–684 (2021). https://doi.org/10.1145/3404835.3462970
13. Liikkanen, L.A., Salovaara, A.: Music on YouTube: user engagement with traditional, user-appropriated and derivative videos. Comput. Hum. Behav. **50**, 108–124 (2015). https://doi.org/10.1016/j.chb.2015.01.067. https://www.sciencedirect.com/science/article/pii/S0747563215000953
14. Daily Motion (2005). https://dailymotion.com. Accessed 17 Feb 2022
15. OpenAI: CLIP: Connecting Text and Images (2021). https://openai.com/blog/clip/. Accessed 14 Jan 2023

16. RHINO: America - a horse with no name (official audio) (2019). https://www.youtube.com/watch?v=na47wMFfQCo. Accessed 19 Oct 2022
17. Richardson, L.: beautifulsoup4 (2021). https://pypi.org/project/beautifulsoup4/. Accessed 17 Feb 2022
18. Ghose, R., Dahlin, T.F., Ficano, N.: pytube (2022). https://github.com/pytube/pytube. Accessed 17 Feb 2022
19. Sen, P., Ganguly, D., Jones, G.J.F.: Tempo-lexical context driven word embedding for cross-session search task extraction. In: NAACL-HLT, pp. 283–292. Association for Computational Linguistics (2018)
20. Sen, P., Ganguly, D., Jones, G.J.F.: I know what you need: investigating document retrieval effectiveness with partial session contexts. ACM Trans. Inf. Syst. **40**(3), 53:1–53:30 (2022)
21. Lehman, T., Zechory, I., Moghadam, M.: Genius (2009). https://pypi.org/project/lyricsgenius/. Accessed 17 Feb 2022
22. Verysweetify: A horse with no name - America (lyrics) (2012). https://www.youtube.com/watch?v=CpSdePGgVyQ. Accessed 19 Oct 2022
23. YouTube (2005). https://www.youtube.com/. Accessed 17 Feb 2022
24. Zulko: moviepy (2017). https://pypi.org/project/moviepy/. Accessed 17 Feb 2022

Ablesbarkeitsmesser: A System for Assessing the Readability of German Text

Florian Pickelmann[1], Michael Färber[2], and Adam Jatowt[1(✉)]

[1] University of Innsbruck, Innsbruck, Austria
`florian.picklemann@student.uibk.ac.at, adam.jatowt@uibk.ac.at`
[2] Karlsruhe Institute of Technology (KIT), Karlsruhe, Germany
`michael.faerber@kit.edu`

Abstract. While several approaches have been proposed for estimating the readability of English texts, there is much less work for other languages. In this paper, we present an online service, available at https://readability-check.org/, that provides five well-established statistical methods and two machine learning models for measuring the readability of texts in German. For the machine learning methods, we train two BERT models. To bring all the measures together, we provide an interactive website that allows users to evaluate the readability of German texts at the sentence level. Our research can be useful for anyone who wants to know whether the text content at hand is easy or difficult and therefore can be used in certain situations or rather needs to be adapted and improved. In education, for example, it can help to assess the suitability of a particular teaching material for a particular grade.

Keywords: Readability · German language · Online service

1 Introduction

When we communicate information in text form, how do we ensure that the recipient can understand it? For example, an explanation of how to use a product might be written in a language that is too difficult for average users. What might help mitigate this problem is to assess the readability of the text before using it. However, manual assessment of readability is difficult, especially for long texts, and requires a certain level of expertise. Moreover, a written text typically consists of many complex relations or hidden assumptions as for the knowledge of the reader that make a sentence or paragraph difficult or easy to read and understand. Hence, dedicated and sometimes complex measures need to be used to properly estimate the reading level of texts. However, most studies on readability have focused on English, while fewer studies have focused on other languages such as German, French, or Polish [5].

In this paper, we present an online service, available at https://readability-check.org/, that allows to assess the readability of German texts based on various

© The Author(s), under exclusive license to Springer Nature Switzerland AG 2023
J. Kamps et al. (Eds.): ECIR 2023, LNCS 13982, pp. 288–293, 2023.
https://doi.org/10.1007/978-3-031-28241-6_28

metrics. The goal is to provide a platform where you can paste a text and then easily assess how readable it is and which parts of the text cause readability problems. This would help people who need to publish or hand over written work that needs to be easily readable, such as legal documents, texts on difficult topics, or texts for non-native speakers. While we focus on German, similar solutions can be used to estimate readability in other languages. Our work demonstrates a feasible model that also has an explanation-oriented function in a sense that the difficulty of a document can be explained by the set of its difficult sentences indicated in the output.

Related Work. Automatically assessing text by means of readability metrics started in the early 1900s when readability formulas based on statistics (e.g., word frequency, word length, sentence length) were proposed. However, these measures have been criticized for their weak statistical bases and inability to capture more complex aspects of a language. With the development of state-of-the-art natural language processing methods, attempts where made to improve the measures created. The next leap in readability assessment was made when machine learning (ML) became mainstream [6,10]. The ML models were generally able to produce better results than the traditional formulas, but require more effort, such as creating data sets and training the model until they are ready to use. Specifically, for the German language, methods that use traditional language models [3,8] or which use semantic networks in comparison to simple surface-level indicators to calculate text readability [4] were created. However, to our knowledge there is no research work describing the process of building of a working tool for assessment of text readability, especially for German.

Statistical Readability Measures. Traditionally, readability of texts was measured by statistical readability formulas, which try to construct a simple human-comprehensible formula with a good correlation to what humans perceive as the degree of readability. The simplest of them is the average sentence length (ASL); others they take into account various other statistical factors, such as word length and word frequency. Most of these formulas were originally developed for English but could be also applicable to other languages with some modifications [10]. For our system we have chosen four formulas, the first three are among the most popular ones used [7] and the fourth one was designed for the German language [2].

(1) The *Gunning fog index* (GFI) [12] calculates the years of education a person needs to understand a text on its first reading. It is calculated with the following expression:

$$GFI = (\frac{numberOfWords}{numberOfSentences} + 100 * \frac{numberOfPolysyllables}{numberOfWords}) * 0.4,$$

where *numberOfPolysyllables* indicates the number of words with three or more syllables.

(2) The *Flesch reading ease* (FRE) [9] indicates the United States (US) grade level a reader needs in order to be able to understand the text. It is calculated

in the following way:

$$FRE = 206.835 - (1.015 * ASL) - (84,6 * ASW),$$

where ASL is the average sentence length and ASW is average number of syllables per word. For the use in German language, Amstad [1] proposed the formula

$$FRE_{deutsch} = 180 - ASL - (58.5 * ASW).$$

(3) The *Simple Measure of Gobbledygook* (SMOG grade) [11] is a readability formula originally used for checking health messages. It roughly corresponds to the years of education needed to understand the text. It is calculated as follows:

$$SMOG = 1.0430\sqrt{numberOfPolysyllables * \frac{30}{totalSentences}} + 3.1291,$$

(4) The *Wiener Sachtextformel* (WSTF) [2] calculates readability in terms of Austrian/German grade levels. It ranges from four to fifteen with four being easy and 15 being very hard.

$$WSTF = 0.1935 * MS + 0.1672 * SL + 0.1297 * IW - 0.0327 * ES - 0.875.$$

MS is the percentage of words with more than three syllables, SL represents the mean of the number of words, IW denotes the percentage of words with more than six letters and ES is the percentage of words with one syllable.

2   Datasets

We use two datasets for training our machine learning components.

TextcomplexityDE Dataset.[1] This dataset consists of 1,000 sentences in German taken from 23 Wikipedia articles. It was created to be used for generating models for text-complexity prediction and text simplification. Sentences were rated in three scales (complexity, understandability, and lexical difficulty) by 267 non-native German learners.

Deutsche Welle Dataset. This dataset was created to assist in a Deep Text Evaluation Project[2]. It was created by using different articles of the online magazine "Deutsche Welle" (DW) to create a classification for the reading level zero for B1 and one for B2 and C1. After splitting all the sentences and classifying them to the corresponding readability classes of their documents, we obtained 17,395 sentences (249,206 words) with readability class 0, being easy, and 43,814 sentences (685,577 words) with readability class 1, being hard.

Training BERT Models. We used the pre-trained BERT model $BERT_{BASE}$ and fine-tuned models for both the datasets described above. A user can select either of the models to be used as well as obtain readability scores by the statistical readability metrics discussed in Sect. 1.

---

[1] https://github.com/babaknaderi/TextComplexityDE.
[2] https://github.com/shlomihod/deep-text-eval.

3   Demonstration System

All the implementation code together with the processed datasets can be found on our GitLab[3]. The service can be accessed online[4] and the screenshots of the website are shown in Fig. 1a and 1b. A short video presentation of the website can be also found online.[5] To create a platform where everyone can check input text for readability we used the python package `streamlit`.[6] To let users input text we chose a simple textbox with a submit button. Having a text input, we then incorporated the classic measures into the website, showing their results as an output table.

We used the BERT models to highlight sentences with different complexities. To check different complexities, we decided to implement a slider that defines the lower threshold for displaying the predictions. A user is able to choose with the slider the minimal value that should be highlighted. Furthermore, the model output was normalized to be from zero to one using a min-max normalization. For highlighting a range of values we decided to color-code them from green (zero) to red (one). A range of colours was created using the `Color` objects from the `colours` package. Note that since the model trained on Deutsche Welle displays just two distinct values (0 or 1), we vary the color depending on the confidence of the prediction.

Fig. 1. (a) Screenshot of the website displaying the text input field and the output table of the classic measures (explained in Sect. 1). (b) Screenshot of the text highlighting using the BERT model trained on TextComplexityDE.

4   Evaluation

We calculated the mean square error (MSE) and the mean absolute error (MAE) of our models with 5-fold cross-validation. The average results over all the five

---

[3] https://git.uibk.ac.at/csaw3616/readability-detection.
[4] https://readability-check.org/.
[5] https://youtu.be/jtYJH2XxLl4.
[6] https://streamlit.io/.

Table 1. MSE and MAE for the TextComplexityDE and the Deutsche Welle models.

TextComplexityDE		Deutsche Welle	
MSE	MAE	MSE	MAE
0.172	0.413	0.197	0.435

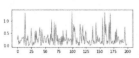
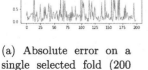
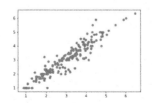
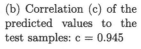
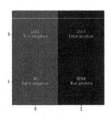

(a) Absolute error on a single selected fold (200 sentences). Sentences are arranged on axis x.

(b) Correlation (c) of the predicted values to the test samples: c = 0.945

(c) True positive, false positive, true negative and false negative

Fig. 2. Evaluation of the TextComplexityDE models test set ((a), (b)) and the confusion matrix of the classification using the Deutsche Welle model (c).

folds are shown in Table 1 for the TextComplexityDE model and Deutsche Welle model. Figure 2a depicts also the error of TextComplexityDE model. As can be seen, there are few outliers with high absolute error values. Furthermore we also measure the Pearson correlation coefficient of the predicted values and the ground truth scores of the test samples, as shown in Fig. 2b. The predicted values appear to be quite correlated (c = 0.945), which further proves the reliability of the model. Finally for Deutsche Welle model, as seen in Fig. 2c, 10,645 of 12,242 samples where correctly predicted, leading to an accuracy of 86.95%. The number of false positives is way bigger than the false negatives. This was to be expected, because the dataset in general is positively biased.

5   Conclusions

While the readability of English texts has been relatively well researched, much less work has been done on the assessment of the readability of texts in other languages. Furthermore, there is no work describing ready-made, workable systems for automatically estimating the readability of texts. The goal of this work is to develop a system that can estimate the readability of German texts using two different methods: classical statistical approaches and machine learning approaches. The proposed system not only allows the estimation of the readability of an entire document, but also indicates which parts might cause difficulties for the reader, thus providing an explainable result that can assist the user in text correction. For the future, we plan to estimate the readability of text at a sub-sentence level (clauses or multi-word expressions).

References

1. Amstad, T.: Wie verständlich sind unsere Zeitungen? Abhandlung: Philosophische Fakultät I. Zürich. 1977, Studenten-Schreib-Service (1978). https://books.google.at/books?id=kiI7vwEACAAJ
2. Bamberger, R., Vanecek, E.: Lesen-Verstehen-Lernen-Schreiben: die Schwierigkeitsstufen von Texten in deutscher Sprache. Jugend und Volk (1984). https://books.google.at/books?id=TElTAAAACAAJ
3. Blaneck, P.G., Bornheim, T., Grieger, N., Bialonski, S.: Automatic readability assessment of German sentences with transformer ensembles. In: Proceedings of the GermEval 2022 Workshop on Text Complexity Assessment of German Text, Potsdam, Germany, pp. 57–62, September 2022
4. vor der Brück, T., Hartrumpf, S.: A semantically oriented readability checker for German, January 2007
5. Collins-Thompson, K.: Computational assessment of text readability: a survey of current and future research. ITL-Int. J. Appl. Linguist. **165**(2), 97–135 (2014)
6. Crossley, S.A., Skalicky, S., Dascalu, M., McNamara, D.S., Kyle, K.: Predicting text comprehension, processing, and familiarity in adult readers: new approaches to readability formulas. Discourse Process. **54**(5–6), 340–359 (2017)
7. Dubay, W.: The principles of readability, pp. 631–3309, January 2004
8. Hancke, J., Vajjala, S., Meurers, D.: Readability classification for German using lexical, syntactic, and morphological features. In: Proceedings of COLING 2012, Mumbai, India, pp. 1063–1080. The COLING 2012 Organizing Committee, December 2012. https://aclanthology.org/C12-1065
9. Kincaid, J.P., Fishburne Jr., R.P., Rogers, R.L., Chissom, B.S.: Derivation of new readability formulas (automated readability index, fog count and flesch reading ease formula) for navy enlisted personnel. Technical report, Naval Technical Training Command Millington TN Research Branch (1975)
10. Martinc, M., Pollak, S., Robnik-Šikonja, M.: Supervised and unsupervised neural approaches to text readability. Comput. Linguist. **47**(1), 141–179 (2021)
11. Mc Laughlin, G.H.: Smog grading-a new readability formula. J. Read. **12**(8), 639–646 (1969)
12. Robert, G.: The Technique of Clear Writing, Revised edn. McGraw-Hill (1968)

FACADE: Fake Articles Classification and Decision Explanation

Erasmo Purificato[1,2](✉) [iD], Saijal Shahania[1,3] [iD], Marcus Thiel[1] [iD], and Ernesto William De Luca[1,2] [iD]

[1] Otto von Guericke University Magdeburg, Magdeburg, Germany
{erasmo.purificato,saijal.shahania,marcus.thiel,ernesto.deluca}@ovgu.de
[2] Leibniz Institute for Educational Media | Georg Eckert Institute,
Brunswick, Germany
{erasmo.purificato,deluca}@gei.de
[3] German Centre for Higher Education Research and Science Studies,
Hanover, Germany
shahania@dzhw.eu

Abstract. The daily use of social networks and the resulting dissemination of disinformation over those media have greatly contributed to the rise of the *fake news* phenomenon as a global problem. Several manual and automatic approaches are currently in place to try to tackle and defuse this issue, which is becoming nearly uncontrollable. In this paper, we propose *Facade*, a fake news detection system that aims to provide a complete solution for classifying news articles and explain the motivation behind every prediction. The system is designed with a cascading architecture composed of two classification pipelines dealing with either low-level or high-level descriptors, with the overall goal of achieving a consistent confidence score on each outcome. In addition, the system is equipped with an explainable user interface through which fact-checkers and content managers can visualise in detail the features leading to a certain prediction and have the possibility for manual cross-checking.

Keywords: Fake news detection · Feature engineering · Explainability

1 Introduction

Over the last few years, the term *fake news* has become extremely popular to the point of making this phenomenon a worldwide issue [15,17]. This concept gained traction following the emblematic 2016 US elections, in which the diffusion of misinformation on social networks has been used as a form of propaganda to get substantial political advantages [8]. The main characteristics of fake news, i.e. *volume*, *variety* and *velocity* [22], are sustained by the rapid spread of *web bots* [12] that make fabricated articles easy to publish and disinformation sources even more difficult to recognise and control. In this scenario, attention is being paid by fact-checkers [10] and content managers [18] in automatic detection systems for two main motivations: 1) manual detection by experts and organisations

J. Kamps et al. (Eds.): ECIR 2023, LNCS 13982, pp. 294–299, 2023.
https://doi.org/10.1007/978-3-031-28241-6_29

is a time-consuming and expensive process, with a huge human-resources involvement to maintain it [11]; 2) the nature and composition of fake news are not the same for every fabricated article. Indeed, some news entries are blatant lies, while others hide their disinformation content among the facts. Furthermore, the outcomes have to be transparent to increase trust in such systems since the results must be cross-checked to be deemed false.

In this paper, we propose **Facade**[1], an automatic system for *fake articles classification and decision explanation*. The system is designed with a cascading architecture composed of two classification pipelines. For each document to analyse, the detection process starts with a first classifier which exploits basic linguistic features (*low-level descriptors*) previously extracted from several fake news datasets. The second pipeline makes use of more complex features (*high-level descriptors*), such as sentiment, emotion, and attribution to known real or fake sources, computed by additional algorithms. We further present an explainable user interface (UI) which can help end users understand what parts of the investigated article are likely to be fake and for what reasons through the implementation of feature importance and post-hoc methods.

2   Existing Fake News Detection Systems

The early-stage detection systems started with manual fact-checking initiatives, and despite the enormous human effort required, some of them are nowadays still hugely reliable, such as *Truth-o-Meter* [1] and *Snopes* [2]. On the automatic detection front, many works, such as [19], shape their systems around the notion of linguistic similarity of the analysed content with known real or fake articles. Nevertheless, the state of the art is unsurprisingly dominated by machine learning and deep learning models, which usually rely on a supervised learning approach (e.g. [24,26]). In a recent publication in this field, Zhang et al. [23] leveraged the relationship between the emotions portrayed in the news content and the end users' emotions expressed in the related comments. In most of the existing systems, however, the component of interpretability is almost overlooked. Due to the coexistence of fake and real news, it is necessary to incorporate the vision of experts and the audience in general [14,25], and this can be achieved through an effective explainable UI. Only a few works, such as *dEFEND* [20] and *Xfake* [21], presented a solution having explainability as a fundamental part of the system.

3   The Facade System

The **Facade** system is designed with a cascading architecture composed of three main phases: 1) **Feature extraction**: low-level and high-level features are extracted from the adopted fake news datasets: *ISOT Fake News Dataset* [3],

---

[1] https://github.com/dtdh/facade (links to **demo video** and **live webapp** inserted inside the repository).

Fake News Dataset [4], *Fake News Corpus* [5], *Multi-Perspective Question Answering Dataset* (MPQA) [6] and *Myers-Briggs Personality Type Dataset* (MBTI) [7]. 2) **Classification**: leveraging the low-level features, a first classifier is executed to the documents to produce the probability of how likely the analysed news is fake or real. 3) **Filtering**: based on the resulting probability and the related confidence level of the classifier that receives **low-level descriptors** (i.e. basic linguistic features extracted from the article texts and headlines, such as size, number of grammatical errors, parts of speech and term frequencies), each news is filtered and marked as *fake, real* or *uncertain.* For the latter group, a second classification is applied, making use of **high-level descriptors** (i.e. complex features detected from the news content with additional algorithms, like sentiment, entailment, attribution, syntactical structure, tones and latent topics). Both pipelines have different classifiers catering to the inputted features. The classifiers were selected based on specific evaluation metrics such as accuracy, recall, precision, F1 scores and other customised metrics, whose detailed discussion is out of the scope of this paper.

The explainability methods included in the system, constituting the basis for the UI, are *feature importance, partial dependence plots* and *SHAP.* **Feature importance** [9] is a widely used method for finding the attributes that contribute the most towards the classifier's predictions. **Partial dependence plots** (PDP) [13] is a model-agnostic and global method, aiming to create a link between the target label (in our case, fake or real) and the attributes utilised by the classifiers (i.e. low-level and high-level descriptors). **SHAP** (SHapley Additive exPlanations) [16] is a state-of-the-art explainability technique and it is mainly used to figure out the effect of each attribute of a classifier's prediction.

4   Demonstration

In this section, we will guide our readers in the exploration of the functionalities of Facade, whose UI has been designed with a *Harry Potter* style, resembling a wizard revealing the truth or the falsehoods of an investigated article.

The initial page (Fig. 1) shows a welcome message (Fig. 1a) with two possible input options (Fig. 1b): insert the URL of a public article and manually type a custom text to analyse, useful to evaluate only a piece of news.

After the execution of the two pipelines, we land on the result page (Fig. 2a), where we can view the prediction (top left corner) and the related confidence score to the right. Optionally, we can highlight the sentences attributed by the system to real or fake sources, coloured in green and red, respectively. The colour gradient relates to the similarity score between a sentence and the attributed source. As displayed in Fig. 2b, we can also check the detailed explanations for attributions and features by hoovering over the specific contributions. The list of the most influential features is on the right-hand side of the result page. The arrows next to each feature name indicate how strong the contribution of that feature towards the prediction is through their number.

Additionally, by browsing the *Explainer Dashboard*, we can visualise all the SHAP values and the partial dependence plots for a single prediction. With the

(a) Welcome page   (b) Input options

Fig. 1. Initial page of Facade

"What If" module, we can adjust the feature importance scores to see how the prediction changes accordingly in a counterfactual scenario.

(a) Sentences highlights   (b) Feature explanations

Fig. 2. Result page of Facade with explanation details (Color figure online)

To summarise, the system is designed to deal with the needs of computer scientists and non-expert audiences. The more specific aspects, such as entering the URL or directly the news text to be questioned, and highlighting parts of the articles considered fake or real by the system in the second pipeline, mainly cater to the non-expert audiences. The highlighting is done in a realistic colour scheme so that it is easier for everyone to follow, irrespective of their background or technical knowledge. The red and green colours are commonly used as a convention for wrong and right, respectively. Hence the same idea translates to them being associated with the fakeness or realness of the news articles. Moreover, highlighting and pop-up boxes are standard methods in UI design and might help in external validation by the user, who can check the reasoning behind the decision and be used to further improve the system in case of incorrect tagging. The design of the explainer dashboard is mainly done to ensure the technical information is communicated with accuracy and clarity, and it is openly addressed to computer scientists.

5 Conclusion

We presented a novel fake news detection system which includes a set of capabilities able to overcome the limitations of the existing systems by exploiting both linguistic features extracted from benchmarking fake news datasets to analyse an article's text and complex features (e.g. sentiment, topic, attribution) computed for enriching the range of descriptors and enhance the classification performance. In addition, through the implementation of an explainable UI, we aim to provide fact-checkers and content managers with a reliable tool for cross-checking the validity of the system results. In the next steps, we plan to improve the system's response time and perform a user study to evaluate the overall user satisfaction in interacting with Facade and its UI.

References

1. https://www.politifact.com/truth-o-meter/
2. https://www.snopes.com/fact-check/
3. https://www.uvic.ca/ecs/ece/isot/datasets/fake-news/index.php
4. https://www.kaggle.com/datasets/jruvika/fake-news-detection
5. https://github.com/several27/FakeNewsCorpus
6. https://mpqa.cs.pitt.edu/corpora/mpqacorpus/
7. https://www.kaggle.com/datasets/datasnaek/mbti-type
8. Allcott, H., Gentzkow, M.: Social media and fake news in the 2016 election. J. Econ. Perspect. **31**(2), 211–36 (2017)
9. Altmann, A., Toloşi, L., Sander, O., Lengauer, T.: Permutation importance: a corrected feature importance measure. Bioinformatics **26**(10), 1340–1347 (2010)
10. Clayton, K., et al.: Real solutions for fake news? Measuring the effectiveness of general warnings and fact-check tags in reducing belief in false stories on social media. Polit. Behav. **42**(4), 1073–1095 (2020). https://doi.org/10.1007/s11109-019-09533-0
11. Dale, R.: NLP in a post-truth world. Nat. Lang. Eng. **23**(2), 319–324 (2017)
12. Ferrara, E., Varol, O., Davis, C., Menczer, F., Flammini, A.: The rise of social bots. Commun. ACM **59**(7), 96–104 (2016)
13. Goldstein, A., Kapelner, A., Bleich, J., Pitkin, E.: Peeking inside the black box: visualizing statistical learning with plots of individual conditional expectation. J. Comput. Graph. Stat. **24**(1), 44–65 (2015)

14. Ha, L., Andreu Perez, L., Ray, R.: Mapping recent development in scholarship on fake news and misinformation, 2008 to 2017: disciplinary contribution, topics, and impact. Am. Behav. Sci. **65**(2), 290–315 (2021)

15. Loomba, S., de Figueiredo, A., Piatek, S.J., de Graaf, K., Larson, H.J.: Measuring the impact of COVID-19 vaccine misinformation on vaccination intent in the UK and USA. Nat. Hum. Behav. **5**(3), 337–348 (2021)

16. Lundberg, S.M., Lee, S.I.: A unified approach to interpreting model predictions. In: Advances in Neural Information Processing Systems, vol. 30 (2017)

17. McGonagle, T.: "Fake news" False fears or real concerns? Neth. Q. Hum. Rights **35**(4), 203–209 (2017)

18. Molina, M.D., Sundar, S.S., Le, T., Lee, D.: "Fake news" is not simply false information: a concept explication and taxonomy of online content. Am. Behav. Sci. **65**(2), 180–212 (2021)

19. Pérez-Rosas, V., Kleinberg, B., Lefevre, A., Mihalcea, R.: Automatic detection of fake news. arXiv preprint arXiv:1708.07104 (2017)

20. Shu, K., Cui, L., Wang, S., Lee, D., Liu, H.: defend: explainable fake news detection. In: Proceedings of the 25th ACM SIGKDD International Conference on Knowledge Discovery & Data Mining, pp. 395–405 (2019)

21. Yang, F., et al.: XFake: explainable fake news detector with visualizations. In: The World Wide Web Conference, pp. 3600–3604 (2019)

22. Zhang, X., Ghorbani, A.A.: An overview of online fake news: characterization, detection, and discussion. Inf. Process. Manag. **57**(2), 102025 (2020)

23. Zhang, X., Cao, J., Li, X., Sheng, Q., Zhong, L., Shu, K.: Mining dual emotion for fake news detection. In: Proceedings of the Web Conference 2021, pp. 3465–3476 (2021)

24. Zhou, X., Jain, A., Phoha, V.V., Zafarani, R.: Fake news early detection: a theory-driven model. Digit. Threats Res. Pract. **1**(2), 1–25 (2020)

25. Zhou, X., Zafarani, R., Shu, K., Liu, H.: Fake news: fundamental theories, detection strategies and challenges. In: Proceedings of the Twelfth ACM International Conference on Web Search and Data Mining, pp. 836–837 (2019)

26. Zuo, C., Karakas, A., Banerjee, R.: A hybrid recognition system for check-worthy claims using heuristics and supervised learning. In: CEUR Workshop Proceedings, vol. 2125 (2018)

PsyProf: A Platform for Assisted Screening of Depression in Social Media

Anxo Pérez$^{(\boxtimes)}$, Paloma Piot-Pérez-Abadín , Javier Parapar ,
and Álvaro Barreiro

Information Retrieval Lab, CITIC, Universidade da Coruña, Campus de Elviña s/n,
15071 A Coruña, Spain
{anxo.pvila,paloma.piot,javier.parapar,alvaro.barreiro}@udc.es

Abstract. Depression is one of the most prevalent mental disorders.
For its effective treatment, patients need a quick and accurate diagnosis.
Mental health professionals use self-report questionnaires to serve that
purpose. These standardized questionnaires consider different depression
symptoms in their evaluations. However, mental health stigmas heavily
influence patients when filling out a questionnaire. In contrast, many
people feel more at ease discussing their mental health issues on social
media. This demo paper presents a platform for assisted examination
and tracking of symptoms of depression for social media users. In order
to bring a broader context, we have complemented our tool with user
profiling. We show a platform that helps professionals with data labelling,
relying on depression estimators and profiling models.

Keywords: Depression estimation · Author profiling · BDI-II

1 Motivation and Background

Social media platforms are channels people tend to consider comfortable for
expressing their honest feelings and concerns [12], where factors such as the
anonymity status may influence people on a sincere manifestation of their
thoughts [3]. Computational methods have obtained promising results in detecting mental health states by exploiting this user-generated data. There is a large
body of prior work in assessing users at risk from different mental disorders,
such as suicidal ideation [20], eating disorders [13] or pathological gambling [16].
In this context, Major Depressive Disorder (MDD), also known as depression,
attracted the attention of many researchers, as it is one of the most common
and debilitating mental illnesses [11]. We can find rich bodies of work identifying
indicators that characterize depression based on user texts from different social
platforms, such as Twitter, Reddit and Facebook [6,7,22,23]. Traditional studies
on depression detection focused on extracting engineered features with the use
of standard machine learning classifiers.

Previous works considered a different set of features as linguistic markers.
For example, the ones from the LIWC analysis tool [17], which covers psychological categories, or relied on a different emotion and sentiment lexicons [1,22].

J. Kamps et al. (Eds.): ECIR 2023, LNCS 13982, pp. 300–306, 2023.
https://doi.org/10.1007/978-3-031-28241-6_30

Other studies focused on analyzing the metadata of the user's activity, revealing significant differences between depressive and control users [2]. These features include the frequency and time gap of their writings or the average number of words per publication.

The solutions mentioned above obtained remarkable results in a variety of datasets [4,25] and benchmark evaluations [13,14,16] that consider depression and control groups. However, the integration of these methods in clinical settings faces several challenges. Health professionals favour models that base their decisions on interpretable features, as they need to be inspected and validated [24]. However, the exclusive use of engineered features does not provide enough context to be interpretable indicators [5]. Diverse studies have also shown that the performance of mental health models is not stable across different social media platforms [9,10]. To overcome these limitations, a new line of work focused on developing solutions that integrate symptoms from different clinical questionnaires as reliable markers. In this regard, recent works demonstrated the potential of symptom-based models in terms of performance, interpretability and generalization [15,19,26].

In this demo paper, we present PsyProf, a monitoring platform for assisted screening of depression in social media. To measure the depression severity level of the individuals, we use models that estimate the presence of recognized symptoms. For this purpose, we use the symptoms of a validated clinical questionnaire, the Beck Depression Inventory (BDI-II) [8]. The BDI-II covers 21 depressive symptoms, such as agitation, energy loss or pessimism. These symptoms have four alternative options, which range in severity from 0 to 3, and the models estimate the severity level of each one. PsyProf also reports the BDI score, which is the sum of the option responses to the 21 symptoms. The BDI score is associated with four depression levels. Moreover, we have complemented our tool with user profiling methods that can bring wither context when measuring at-risk users. We use Reddit as the target platform due to its wide acceptance in previous studies [4,16,25], and our tool provides scalability to process the large amount of data coming from this platform. Finally, the data from the social media users can be downloaded to CSV format and can help create symptom-based datasets with the inspection and labels coming from health professionals.

2 PsyProf

We conceived PsyProf both as a demonstration platform for models for the task of severity estimation of depression and for being used by professionals for doing automated user screening and validation of the results. This dual motivation is the reason for designing the platform's use cases that show the results to the professionals for validation and correction. In this way, PsyProf is not only a proof-of-concept of the utility of the automated models for massive screening but also a tool for obtaining insights from the corrections or validations that the professionals make based on the provided evidence.

Architecture and Implementation. Figure 1 illustrates PsyProf's overall architecture, which consists of (1) a web-based front-end and back-end built with the web framework Django (Python). (2) A scalable system for processing user publications from Reddit, built with Celery and Redis. The API calls to the Reddit API can be made asynchronously. Therefore, the models can infer user estimations without the need of waiting to process all the remaining data. (3) Two different REST API regarding to the depression and profiling models that consume the calls from the Celery workers. Furthermore, PsyProf is portable since all the components run in a different Docker container orchestrated with Docker compose.

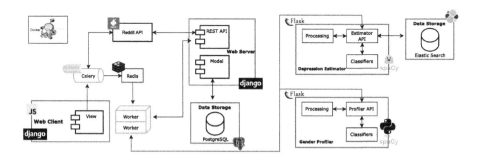

Fig. 1. PsyProf architecture overview.

User Interface and Interaction. PsyProf is a web application intended to be used exclusively by clinical professionals where administrators can run experiments related to profiling and estimating depression severity levels from Reddit users. Figure 2 shows screenshots regarding the platform's main functionalities. The clinicians have two ways of monitoring: (1) In the upper left corner, we can see the form obtain a pool of Reddit users estimations. It contains five fields: the subreddit from which to obtain users (e.g., *mentalhealth*), the number of users, the number of threads and comments to inspect. After filling in these fields, the application will obtain estimates of users that meet these characteristics. (2) Moreover, PsyProf includes the feature to obtain the estimations of specific users by introducing the Reddit username. The application also allows the clinician to export all this data (e.g., for research purposes such as creating new unsupervised/supervised collections) that can be organized in different corpus. It allows exporting to JSON and CSV formats.

Inspection and Validation of the Data. Once the clinician has requested to process users, PsyProf displays the estimations from our models in the Profiles view (bottom of Fig. 2). In this view, we have two categories of users: (1) The users who have not been validated yet appear with the tag *Not processed*. (2) On the other hand, the tag of validated users will be (*Depression/No depression*). For both categories, the profiling attributes always appear (*age, gender, country, personality*) if they surpass the confidence threshold assigned by the

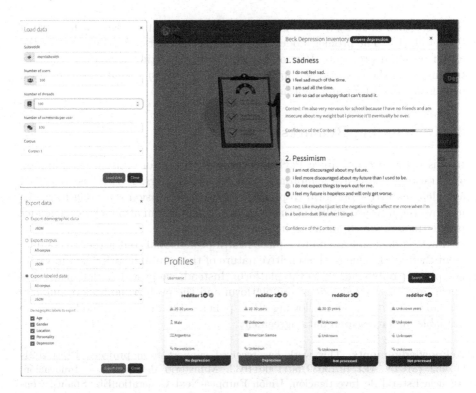

Fig. 2. User interfaces related to the main functionalities of PsyProf.

models. Otherwise, the tag *Unknown* is shown. Finally, the main PsyProf functionality is related to the estimation of symptoms as clinical markers, displayed in the top of Fig. 2. The clinicians can access the predicted responses of all the BDI symptoms and the total severity level associated with each user. Moreover, the models also highlight the most relevant user comment on the symptoms as the decision context. The predicted responses are not definitive. The application allows the clinician to manually change the responses for every decision, if considered necessary.

Models. To perform the user estimations, we rely on two different models. (1) The depression estimator automatically fills in the BDI by predicting the response to the 21 symptoms of the questionnaire. For this purpose, we use the Sense2Vec model from Pérez et al. [19], which is based on the use of word embeddings to capture the semantics of the symptom options. The predict these options, Sense2Vec compares the similarity of each option with the user embedding representation. To do so, we index the embeddings from the training texts corresponding to each option in an Elasticsearch index and perform a vector search using the test user embedding as query. Following this approach, we can compare test users with the options of the BDI. Sense2vec training data corresponds to the eRisk2019 collections [13], obtained from Reddit. (2) The gender profiler, which includes a set of models for author profiling based on the work of Piot et al. [18]. The profiler models exploit a different set of features, such as the use of personal pronouns, Part-of-Speech (PoS) or sentiment analysis using as training data the collections from *PAN Author Profiling* shared task. This task belong to the CLEF campaign in 2019 [21].

3 Concluding Remarks

In this paper, we introduce PsyProf, a web platform for assisted examination and tracking depressive symptoms in social media users. PsyProf is conceived as a demonstrative platform to produce effective depression screening tools. To improve the interpretability of the decisions, the platform also includes a gender profiler model, which allows to improve the context.

PsyProf does not intend to replace health professionals but rather to complement their work. Due to the sensitive nature of the mental health domain, we do not provide public access to the platform. Instead, we provide a demonstration video[1] and the source code of the platform and the profiler models are publicly available[2,3]. Finally, and following eRisk policies, the depression models will be available under research data agreement.

Acknowledgements. This work has received support from projects: PLEC2021-007662 (MCIN/AEI/10.13039/501100011033, Ministerio de Ciencia e Innovación, Agencia Estatal de Investigación, Unión Europea-Next GenerationEU, Spain); Consellería de Educación, Universidade e Formación Profesional, Spain (accreditation 2019–2022 ED431G/01 and GPC ED431B 2022/33) and the European Regional Development Fund, which acknowledges the CITIC Research Center, an ICT of the University of A Coruña as a Research Center of the Galician University System.

References

1. Aragón, M.E., López-Monroy, A.P., González-Gurrola, L.C., Montes-y Gómez, M.: Detecting depression in social media using fine-grained emotions. In: Proceedings of the 2019 Conference of the North American Chapter of the Association for Computational Linguistics: Human Language Technologies, Volume 1 (Long and Short Papers), Minneapolis, Minnesota, pp. 1481–1486. Association for Computational Linguistics, June 2019. https://doi.org/10.18653/v1/N19-1151. https://aclanthology.org/N19-1151
2. Cacheda, F., Fernandez, D., Novoa, F.J., Carneiro, V., et al.: Early detection of depression: social network analysis and random forest techniques. J. Med. Internet Res. **21**(6), e12554 (2019)
3. Chancellor, S., De Choudhury, M.: Methods in predictive techniques for mental health status on social media: a critical review. NPJ Digit. Med. **3**(1), 1–11 (2020)

---

[1] https://irlab.org/psyprof.mp4.

[2] https://github.com/palomapiot/early.

[3] https://github.com/palomapiot/profiler-buddy.

4. Cohan, A., Desmet, B., Yates, A., Soldaini, L., MacAvaney, S., Goharian, N.: SMHD: a large-scale resource for exploring online language usage for multiple mental health conditions. In: Proceedings of the 27th International Conference on Computational Linguistics, Santa Fe, New Mexico, USA, pp. 1485–1497. Association for Computational Linguistics, August 2018. https://aclanthology.org/C18-1126

5. Coppersmith, G., Leary, R., Crutchley, P., Fine, A.: Natural language processing of social media as screening for suicide risk. Biomed. Inform. Insights **10**, 1–11 (2018). 1178222618792860

6. Couto, M., Pérez, A., Parapar, J.: Temporal word embeddings for early detection of signs of depression. In: Proceedings of the 2nd Joint Conference of the Information Retrieval Communities in Europe (CIRCLE 2022), Samatan, Gers, France, 4–7 July 2022. CEUR Workshop Proceedings, vol. 3178. CEUR-WS.org (2022)

7. De Choudhury, M., Counts, S., Horvitz, E.: Social media as a measurement tool of depression in populations. In: WebSci 2013, pp. 47–56. Association for Computing Machinery, New York (2013). https://doi.org/10.1145/2464464.2464480

8. Dozois, D.J., Dobson, K.S., Ahnberg, J.L.: A psychometric evaluation of the Beck Depression Inventory-II. Psychol. Assess. **10**(2), 83 (1998)

9. Ernala, S.K., et al.: Methodological gaps in predicting mental health states from social media: triangulating diagnostic signals. In: Proceedings of the 2019 CHI Conference on Human Factors in Computing Systems, CHI 2019, pp. 1–16. Association for Computing Machinery, New York (2019). https://doi.org/10.1145/3290605.3300364

10. Harrigian, K., Aguirre, C., Dredze, M.: Do models of mental health based on social media data generalize? In: Findings of the Association for Computational Linguistics: EMNLP 2020, pp. 3774–3788. Association for Computational Linguistics, November 2020. https://doi.org/10.18653/v1/2020.findings-emnlp.337

11. Hollon, S.D., Thase, M.E., Markowitz, J.C.: Treatment and prevention of depression. Psychol. Sci. Public Interest **3**(2), 39–77 (2002)

12. Kauer, S.D., Mangan, C., Sanci, L.: Do online mental health services improve help-seeking for young people? A systematic review. J. Med. Internet Res. **16**(3), e3103 (2014)

13. Losada, D.E., Crestani, F., Parapar, J.: Overview of eRisk 2019 early risk prediction on the internet. In: Crestani, F., et al. (eds.) CLEF 2019. LNCS, vol. 11696, pp. 340–357. Springer, Cham (2019). https://doi.org/10.1007/978-3-030-28577-7_27

14. Losada, D.E., Crestani, F., Parapar, J.: eRisk 2020: self-harm and depression challenges. In: Jose, J.M., et al. (eds.) ECIR 2020. LNCS, vol. 12036, pp. 557–563. Springer, Cham (2020). https://doi.org/10.1007/978-3-030-45442-5_72

15. Nguyen, T., Yates, A., Zirikly, A., Desmet, B., Cohan, A.: Improving the generalizability of depression detection by leveraging clinical questionnaires. In: Muresan, S., Nakov, P., Villavicencio, A. (eds.) Proceedings of the 60th Annual Meeting of the Association for Computational Linguistics (Volume 1: Long Papers), ACL 2022, Dublin, Ireland, 22–27 May 2022, pp. 8446–8459. Association for Computational Linguistics (2022). https://doi.org/10.18653/v1/2022.acl-long.578

16. Parapar, J., Martín-Rodilla, P., Losada, D.E., Crestani, F.: Overview of eRisk 2021: early risk prediction on the internet. In: Candan, K.S., et al. (eds.) CLEF 2021. LNCS, vol. 12880, pp. 324–344. Springer, Cham (2021). https://doi.org/10.1007/978-3-030-85251-1_22

17. Pennebaker, J.W., Mehl, M.R., Niederhoffer, K.G.: Psychological aspects of natural language use: our words, our selves. Ann. Rev. Psychol. **54**(1), 547–577 (2003)

18. Piot-Perez-Abadin, P., Martín-Rodilla, P., Parapar, J.: Experimental analysis of the relevance of features and effects on gender classification models for social media author profiling. In: ENASE, pp. 103–113 (2021)

19. Pérez, A., Parapar, J., Barreiro, Á.: Automatic depression score estimation with word embedding models. Artif. Intell. Med. **132**, 102380 (2022). https://doi.org/10.1016/j.artmed.2022.102380

20. Ramírez-Cifuentes, D., et al.: Detection of suicidal ideation on social media: multimodal, relational, and behavioral analysis. J. Med. Internet Res. **22**(7), e17758 (2020)

21. Rangel, F., Rosso, P.: Overview of the 7th author profiling task at PAN 2019: bots and gender profiling in Twitter. In: Proceedings of the CEUR Workshop, Lugano, Switzerland, pp. 1–36 (2019)

22. Ríssola, E.A., Losada, D.E., Crestani, F.: A survey of computational methods for online mental state assessment on social media. ACM Trans. Comput. Healthc. **2**(2), 1–31 (2021). https://doi.org/10.1145/3437259

23. Trotzek, M., Koitka, S., Friedrich, C.: Utilizing neural networks and linguistic metadata for early detection of depression indications in text sequences. IEEE Trans. Knowl. Data Eng. **32**, 588–601 (2018). https://doi.org/10.1109/TKDE.2018.2885515

24. Walsh, C.G., et al.: Stigma, biomarkers, and algorithmic bias: recommendations for precision behavioral health with artificial intelligence. JAMIA open **3**(1), 9–15 (2020)

25. Yates, A., Cohan, A., Goharian, N.: Depression and self-harm risk assessment in online forums. In: Proceedings of the 2017 Conference on Empirical Methods in Natural Language Processing, Copenhagen, Denmark, pp. 2968–2978. Association for Computational Linguistics, September 2017. https://doi.org/10.18653/v1/D17-1322

26. Zhang, Z., Chen, S., Wu, M., Zhu, K.Q.: Psychiatric scale guided risky post screening for early detection of depression. In: Raedt, L.D. (ed.) Proceedings of the Thirty-First International Joint Conference on Artificial Intelligence, IJCAI 2022, Vienna, Austria, 23–29 July 2022, pp. 5220–5226. ijcai.org (2022). https://doi.org/10.24963/ijcai.2022/725

SOPalign: A Tool for Automatic Estimation of Compliance with Medical Guidelines

Luke van Leijenhorst[1,3](✉) [iD], Arjen P. de Vries[1] [iD], Thera Habben Jansen[2] [iD],
and Heiman Wertheim[3] [iD]

[1] Institute for Computing and Information Sciences, Radboud University,
Nijmegen, Netherlands
{luke.vanleijenhorst,arjen.devries}@ru.nl

[2] Department of Infection Prevention and Control, Amphia Hospital,
Breda, Netherlands
thabbenjansen@amphia.nl

[3] Department of Medical Microbiology, Radboudumc, Nijmegen, Netherlands
heiman.wertheim@radboudumc.nl

Abstract. SOPalign is a tool designed for hospitals and other healthcare providers in the Netherlands to automatically estimate the compliance of internal standard operating procedures (SOPs) for employees with the national guidelines. In this tool, users can upload the SOPs of their hospital and the recommendations from the most recent guidelines. SOPalign will then link the individual recommendations from the guidelines to the relevant passages of text in the SOPs and determine whether these passages are compliant with the recommendations. To link the SOP passages to the recommendations from the guideline, we make use of a Semantic Textual Similarity (STS) model based on the siamese BERT-network architecture. For efficiency reasons, we only apply the STS model to sentences that exceed a threshold in n-gram cosine similarity. To estimate compliance of SOPs with guideline recommendations, we have fine-tuned pre-trained language models using two different Dutch Natural Language Inference (NLI) datasets.

Keywords: Information retrieval · Natural language inference · Semantic textual similarity

1 Introduction

In healthcare, medical knowledge is captured into guidelines that support healthcare professionals to deliver the best possible quality of care. In hospitals and other healthcare settings, these guidelines are implemented and integrated in local *standard operating procedures (SOPs)*. When guidelines are updated, it is important to also update all SOPs that are based on this guideline. In most hospitals, this is a manual process in which employees periodically adjust the

J. Kamps et al. (Eds.): ECIR 2023, LNCS 13982, pp. 307–312, 2023.
https://doi.org/10.1007/978-3-031-28241-6_31

SOPs associated with the guideline in the quality system. Given the number of guidelines and the number of SOPs in a hospital, this is a vulnerable process that may lead to SOPs that are (partially) outdated.

SOPalign is a tool that uses natural language processing techniques to assess where SOPs are compliant with the applicable guidelines, and where they differ. The tool can support healthcare facilities to keep their procedures updated with the guidelines. It also helps to discover bottlenecks in the guideline recommendations, when recommendations in guidelines do not appear in SOPs, or the SOPs are not compliant with the guideline. This is an important step in the improvement cycle of guidelines and supports the perspective of 'living guidelines'. The authors foresee a transition from a collection of independent documents into a hyperlinked network of guidelines and SOPs. Our tool aims to enable this transition at a low cost.

2   Tool Design

An overview of SOPalign can be found in Fig. 1. The input consists of a file with a list of recommendations drawn from the guidelines we want to validate, and the PDFs of the SOPs we want to analyse. Before we analyse the SOPs, we extract the text fully automatically from the PDFs and attempt to merge these lines of text into their original paragraphs. These passages of text, together with the recommendations from the guidelines are then passed to the guidelines linker.

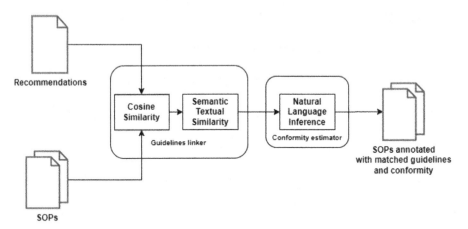

Fig. 1. Architecture of the SOPalign tool.

Guidelines Linker. The goal of the guidelines linker is to link every recommendation of the guidelines to their corresponding passages of text in the SOPs to identify where this recommendation is implemented. Because a guideline can contain up to a hundred recommendations, and every recommendation gets coupled to every passage of every SOP, this generates a very large candidate set.

Before we use our large Semantic Textual Similarity (STS) language model to assess whether passages correspond to the recommendations from the guideline, we calculate the character n-gram based cosine similarity between the SOP passage and the guideline recommendations, and only keep pairs above a certain threshold so the candidate set becomes substantially smaller. By doing so, we make the assumption that at least some character n-grams of the recommendation will also be in the SOP passage where it is implemented.

Now we feed the smaller dataset to our multilingual STS language model. We make use of the *paraphrase xlm-r-multilingual-v1* model which is constructed from sentence embeddings using siamese networks [1]. We only keep the instances where this STS score exceeds a certain threshold. All the SOP passage and recommendation pairs that exceeded both the cosine similarity and STS thresholds are linked together and are sent to the Compliance estimator. In our application of SOPalign, for a set of 329 SOPs and 104 guidelines, there were roughly 5.200.000 candidate pairs. Of this number, 7611 ($\approx 0.15\%$) exceeded the cosine threshold, and of those pairs 543 ($\approx 7.13\%$) exceeded both thresholds.

Compliance Estimator. The compliance estimator determines whether the passage in the SOP that was coupled to a guideline, is compliant with that guideline. It consists of a single Natural Language Inference (NLI) model. The output of the model is a list of probabilities for the following three NLI labels, of which the maximum probability is selected:

- **Entailment:** The SOP passage is compliant with the guideline;
- **Neutral:** The guideline is not implemented in the given SOP passage;
- **Contradiction:** The SOP passage is not compliant with the guideline.

An example of this compliance evaluation can be found in Table 1.

Table 1. Example of compliance evaluation using the three NLI labels.

Recommendation	SOP passage	NLI label
A COVID-19 patient should wear a mask	The patient wears a mask	Entailment
A COVID-19 patient should wear a mask	The patient does not wear a mask	Contradiction
A COVID-19 patient should wear a mask	The patient wears gloves	Neutral

Because SOPalign has been developed for healthcare in the Netherlands, we could not simply rely on a high-quality English NLI model, and had to localize our approach at this step in the pipeline by training a Dutch NLI model.

To train our Dutch NLI model, we experimented with fine-tuning three different pre-trained large language models for Dutch: RobBERT [2], BERTje [3] and mBERT [4].

We experimented with two different datasets for fine-tuning. First, we used the SICK-NL dataset [5], which is a Dutch semi-automatically translated version of the original SICK dataset [6]. The sentences in this dataset are generated from descriptions of images. These sentences are short and not of the clinical domain, which renders these quite different from the sentences we encounter in the SOPs and guidelines. Perhaps not so surprisingly then, the model did not generalise to the sentences in our application domain.

Therefore, we translated the medNLI dataset [7] to Dutch using machine translation. This dataset was specifically created for medical texts and thus, it was a much better fit. For translation, we experimented with both Google Translate and DeepL. Table 2 shows the results on the medNLI test set for the different settings where the best model was selected after training for 20 epochs. The combination of DeepL and BERTje yields the best results for the Dutch language. The performance loss caused by the translation and the use of Dutch pre-trained models seems limited, when compared to their English counterparts.

Table 2. Results on the medNLI test set for the different translators on the three different pre-trained models.

Pre-trained model	Data language	Translator	F1-score
mBERT	English	None	79.87
mBERT	Dutch	Google	74.44
mBERT	Dutch	DeepL	75.68
BERT	English	None	79.57
BERTje	Dutch	Google	73.86
BERTje	Dutch	DeepL	76.34
RoBERTa	English	None	80.96
RobBERT	Dutch	Google	74.33
RobBERT	Dutch	DeepL	73.84

The linker and compliance modules are used as follows. We automatically generate annotations inside the SOP PDF for the guideline recommendations linked to each passage, together with the corresponding NLI label. The list of annotated PDFs is shown to the user, along with two tables, one containing all the annotations made and another with annotation counts per recommendation along with the NLI labels. The latter helps the user to find mismatches between guidelines and SOPs. The user can adapt the strictness of the system using a slider to control the threshold and influence the precision/recall trade-off. A screenshot of a hospital SOP annotated with matched guidelines is shown in Fig. 2.

The tool will be made available for hospitals and long-term care facilities to look for improvements on user experience. A built-in possibility to provide feedback on the (mis)matches identified by the tool will be used to generate new

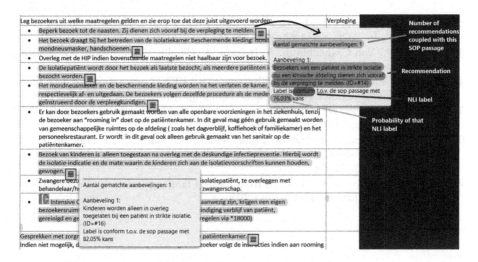

Fig. 2. Screenshot of annotated output PDF for a SOP.

and realistic labeled data to improve performance. A limitation of our current approach is that our model is trained on machine-translated NLI data, resulting in noisy sentences. This feedback addition will overcome this. We will also calculate the Inter-Annotator Agreement (IAA) between the tool and at least two human annotators to evaluate the tool, and we will perform an error analysis to look for systematic errors.

3 Demonstration

Users can access our code through GitHub[1]. The repository contains the code for the tool, the API that is used, the training of the models, the translation of the data, and the translated medNLI dataset. To showcase the important features of our tool we have created a short video[2] demonstrating an example usage of the tool where a compliance check is done of two hospital SOPs with the recommendations of the Dutch hospital MRSA guideline. The tool can easily be adapted to allow for different languages by changing the underlying STS and NLI models.

Acknowledgements. This work is funded by ABR Zorgnetwerk Utrecht and Zorgnetwerk GAIN.

References

1. Reimers, N., Gurevych, I.: Sentence-BERT: sentence embeddings using siamese BERT-networks. In: EMNLP, pp. 3973–3983 (2019)

---

[1] https://github.com/Lukevanl/SOPalign.
[2] https://youtu.be/vVPHGCMsmvY.

2. Delobelle, P., Winters, T., Berendt, B.: RobBERT: a Dutch RoBERTa-based Language model. In: ACL, pp. 3255–3265 (2020)
3. De Vries, W., Van Cranenburgh, A, Bisazza, A. Caselli, T. Van Noord, G. Nissim, M.: BERTje: a dutch BERT model (2019). arXiv:1912.09582
4. Devlin, J., Chang, M. -W, Lee, K., Toutanova, K.: BERT: pre-training of deep bidirectional transformers for language understanding. In: ACL, pp. 4171–4186 (2019)
5. Wijnholds, G., Moortgat, M.: SICKNL: a dataset for Dutch natural language inference. In: EACL, pp. 1474–1479 (2021)
6. Marelli, M., Menini, S., Baroni, M., Bentivogli, L., Bernardi, R., Zamparelli, R.: A SICK cure for the evaluation of compositional distributional semantic models. In: LREC, pp. 216–223 (2014)
7. Romanov, A., Shivade, C.: Lessons from natural language inference in the clinical domain. In: EMNLP, pp. 1586–1596 (2018)

Tutorials

Understanding and Mitigating Gender Bias in Information Retrieval Systems

Amin Bigdeli[1]([✉]), Negar Arabzadeh[2], Shirin Seyedsalehi[1], Morteza Zihayat[1],
and Ebrahim Bagheri[1]

[1] Toronto Metropolitan University, Toronto, Canada
{abigdeli,shirin.seyedsalehi,bagheri}@torontomu.ca
[2] University of Waterloo, Waterloo, Canada
narabzad@uwaterloo.ca

Abstract. Recent studies have shown that information retrieval systems may exhibit stereotypical gender biases in outcomes which may lead to discrimination against minority groups, such as different genders, and impact users' decision making and judgements. In this tutorial, we inform the audience of studies that have systematically reported the presence of stereotypical gender biases in Information Retrieval (IR) systems and different pre-trained Natural Language Processing (NLP) models. We further classify existing work on gender biases in IR systems and NLP models as being related to (1) relevance judgement datasets, (2) structure of retrieval methods, (3) representations learnt for queries and documents, (4) and pre-trained embedding models. Based on the aforementioned categories, we present a host of methods from the literature that can be leveraged to measure, control, or mitigate the existence of stereotypical biases within IR systems and different NLP models that are used for down-stream tasks. Besides, we introduce available datasets and collections that are widely used for studying the existence of gender biases in IR systems and NLP models, the evaluation metrics that can be used for measuring the level of bias and utility of the models, and de-biasing methods that can be leveraged to mitigate gender biases within those models.

1 Motivation and Overview

There have been both qualitative and quantitative studies that have effectively shown that societal biases have become prevalent in various Natural Language Processing (NLP) and Information Retrieval (IR) techniques, models, and datasets [2,3,10,11,13,15,17,18,24,28,35,37]. Given these tools are often deployed at scale, such biases have the potential to directly impact the lives of many people. More specifically within the context of IR systems, biased methods can exacerbate biases by exposing users to a set of biased documents in response to user queries. Such biases can have a potentially harmful impact on the users' judgments when exposed to unfair and biased search results. This is concerning since not only do a large number of search engine users heavily rely on retrieval systems on a daily basis but also the search results often constitute a major component of important practical systems such as recommendation systems, question answering systems, intelligent assistants, to name a few.

ⓒ The Author(s), under exclusive license to Springer Nature Switzerland AG 2023
J. Kamps et al. (Eds.): ECIR 2023, LNCS 13982, pp. 315–323, 2023.
https://doi.org/10.1007/978-3-031-28241-6_32

Researchers such as Draws et al. [14] have most recently shown that when search results are biased, the users who are exposed to the biases results will tend to favor the biased viewpoint. This aligns very well with several forms of cognitive bias identified by Azzopardi [1] including *Availability bias*, which points to user biases towards content that are more easily accessible, and *Anchoring Bias* that reports that users are more likely to focus on the first piece of information that they receive. Thus, it is important to systematically control the degree of biases that are exhibited by such retrieval systems to avoid their detrimental effects on the users' beliefs and decisions. To systematically address such biases, various researchers have proposed methods that can help control and/or mitigate biases, such as gender biases, in information retrieval systems. In this tutorial, we will provide a classification of existing work in the literature [2,5,7–10,12,16,19–21,23,30,32,33,35,38,39] and introduce the state-of-the-art methods that are available for managing gender biases within IR systems. The structure of this tutorial can be summarized as follows:

1. The tutorial will present concrete evidence, using real-world examples of cases where gender biases are introduced and intensified in natural language processing and IR systems;
2. The tutorial will draw inspiration from and provide adequate contextual information from experience reports and methodological work in natural language processing that have already explored gender biases [34,35];
3. A systematic classification of possible sources for gender biases will be presented and details of how biases can be transferred from these sources will be provided. These sources include relevance judgement collections, ranker characteristics, objective functions, and neural embeddings, to name a few.
4. The tutorial will review existing methods that have attempted to control or mitigate gender biases and will also provide an in-depth treatment of the retrieval effectiveness-bias tradeoff. This tradeoff is concerned with the right balance between maximizing retrieval effectiveness and minimizing gender bias, which may not always be synergistic;
5. A clear description of evaluation methodology, datasets, and metrics that have been used in the literature for investigating gender bias will be provided.

Our tutorial will build on our recent tutorial at SIGIR 2022 [6] (https://bit.ly/3TfDMss) as well as four invited talks that we have delivered at Microsoft Research, Center for Intelligent Information Retrieval at UMass Amherst, the keynote talk at the BIAS workshop at ECIR 2022, and Radboud University. The central focus of the tutorial and these talks have been on methods for controlling and mitigating gender biases, which can broadly be classified as follows:

1. *Relevance Judgement Collections*: Relevance judgment documents are often considered as gold standard benchmark datasets used for training and evaluating ranking models. Researchers have already introduced methodological processes for studying possible traces of gender bias in relevance judgment collections [9,27], and show that stereotypical gender biases can be observable in these collections, which are capable of making their way into the algorithmic aspects of ranking models that are trained and evaluated based on them.

We will also introduce those approaches that have been introduced in the literature for de-biasing relevance judgment collections. We will report on the findings from these studies that when neural ranking models are trained based on de-biased relevance judgments, the level of gender biases may be reduced while retrieval effectiveness is maintained.

2. *Neural Representations*: Neural embeddings have been widely adopted in IR systems for different tasks such as document retrieval [4]. Since neural representations have often been pre-trained on large corpora, they may have picked up existing gender stereotypes and biases. Many research works [5,10,12,35,39] have investigated gender biases within these neural representations, and have proposed methods for mitigating the levels of bias using different approaches such as data augmentation and embeddings de-biasing techniques [10,17,25,29]. We will cover how such techniques can be adopted in practice to manage gender biases.

3. *Query Representation*: The query submitted by the user can itself be highly influential on the retrieved list of documents. For instance, Kulshrestha et al. [22] examine the impact of such biases in the context of political search queries. Therefore, we will report on studies that explore the gendered nature of search queries [36], as well as those that present query reformulation mechanisms that attempt to revise an initial query in a way that will lead to a less biased list of documents while maintaining retrieval effectiveness [8].

4. *Retrieval Methods*: Recent studies show that neural-based retrieval methods can intensify the level of gender biases within the retrieved list of documents [16,31]. Therefore, it is important to manage the level of gender bias at the ranker level. Researchers have already looked into how rankers can be made less biased (or in other terms more fair) through approaches such as introducing bias-aware loss functions or bias-aware negative sampling strategies. In the tutorial, we will cover various existing work in this space. For instance, we will introduce methods such as ADVBERT [30] that leverages adversarial components within the BERT reranker loss function for decreasing the level of bias in neural ranking models. We will also introduce the bias-aware neural ranker [32], which explicitly incorporates a notion of gender bias and hence control how bias is expressed in documents that are retrieved. We will also cover bias-aware negative sampling strategies that consider the degrees of gender bias when sampling documents to be used for training neural rankers [7].

We highlight that this tutorial will build on but significantly expand the scope of our talks by providing comprehensive information about evaluation strategies, available datasets, and bias measurement techniques. Most important, we will discuss the limitations of existing work from both technical as well as conceptual perspectives. For instance, we will highlight the following two limitations: (1) existing work in the literature have focused primarily on the notion of sex as a binary construct and assumed that search queries and results can be analyzed from their association with the male or female gender. This is a major limitation that needs to be addressed in future work; and (2) Most existing work assume that gender bias can be measured based on the frequency of gendered terms. This overlooks the complexity associated with the stance and position of documents with regards to different gender identities in favor of simplifying computation.

2   Objectives

The objectives of this tutorial can be enumerated as follows: 1) Show the presence of gender biases in IR systems and large scale corpora relevance judgments; 2) Introduce bias measurement metrics used for calculating the level of gender biases within the retrieved list of documents; 3) Present datasets used for investigating gender biases in IR results; 4) Introduce de-biasing methods for reducing the level of bias in relevance judgment datasets; 5) Describe existing methods for mitigating the level of bias within neural ranking models; 6) Present existing methods for the exploration and mitigation of bias in neural embeddings; 7) Highlight important theoretical and conceptual limitations of existing work when dealing with the concept of gender.

The aforementioned topics will give participants a thorough understanding of existing datasets and bias measurement metrics used for investigating gender biases within information retrieval results. Besides, they become familiar with methods used for reducing gender biases within IR systems. As a result, they can take advantage of these techniques to release models that are aware of gender biases and expose users to a less biased list of documents without being worried about the retrieval effectiveness of their model. In addition, these topics can be beneficial for researchers who are conducting research in a similar area in terms of applying introduced de-biasing methods for other *types of societal biases* and can serve as a useful starting point.

3   Format and Schedule

The length of this tutorials will be half day, i.e., 3 h plus breaks and will be delivered in-person by the presenters. This tutorial covers the following sections:

Introduction to the Topic of Gender Biases in IR [30 min]. The tutorial will begin by covering the foundations of IR methods as well as the datasets, which will be referred to throughout the tutorial. We will provide evidence to show the footprints of various forms of gender bias in IR systems and will introduce bias measurement methods that will be used for measuring the level of bias in retrieval results.

Exploration of Gender Biases in IR Relevance Judgments and Retrieval Methods [30 min]. We discuss the presence of stereotypical gender biases within various IR methods and compare the level of gender bias among their retrieved results. Subsequently, we explore the possibility of gender biases within relevance judgement datasets, also known as gold standard datasets. Through a methodological approach, we discuss that such datasets could be a potential source of bias.

Mitigation of Gender Biases in IR Methods [60 min]. In this session, we review existing methods for reducing gender bias through different classes of retrieval methods, namely, term-frequency-based methods as well as neural ranking models. These de-biasing methods can be classified based on four different strategies, namely (1) Adversarial Training, (2) Regularizing the Loss Function,

(3) Data Augmentation, and (4) Query Representation. Additionally, we show the effectiveness of each of the proposed methods in reducing the level of bias within the retrieved results and their utility. We will demonstrate that leveraging these methods allows for maintaining utility and at the same time mitigating the level of bias. Finally, we demonstrate how each of the proposed methods can be applied for other societal attributes other than gender.

Exploration and Mitigation of Gender Biases in Neural Embeddings [40 min]. In this session, we will explore the existence of gender biases within the representation and algorithmic aspects of different classes of neural embeddings, namely (1) static word embeddings and (2) dynamic word embeddings. In addition, we will cover the proposed methods used for de-basing neural embeddings and will show their impact on both reducing gender biases and performance of down-stream tasks.

Limitations and Future Work [20 min]. This session will discuss major theoretical and conceptual limitations of existing work and will present avenues for future work.

4   Audience and Relevance

Fairness and ethical issues surrounding the practice of IR has become a major topic of concern among IR researchers [8–10,30,31,39]. The existence of gender stereotypes in IR systems can influence an individual's judgments, leading to unfair treatments and outcomes. In an ideal world, the expectation from IR systems is to be fair towards different gender identities and avoid reflecting unfair prejudices that may exist within society. We hope that our work contributes to the growing body of knowledge in this area, and helps the IR community to become familiar with the datasets, metrics, and methods that can be used for reducing the level of such biases in retrieval methods.

It is worth mentioning that there have been many attempts by industrial entities to address biases from a practical sense. For instance, we can point to the investigation of fairness in neural-based models by Microsoft, the responsible machine learning initiative at Twitter, which tackles gender and racial biases, or the PAIR group at Google Brain that explores responsible AI in Google systems.

We note that while there have been similar tutorials related to investigating fairness issues in IR systems in other venues, the topics proposed in this tutorial distinguish themselves by focusing on proposing systematic and well-validated methods for reducing gender biases in retrieval results. The following tutorials can be considered complementary and synergistic to the theme of ours:

1. *Addressing Bias and Fairness in Search Systems* by Ruoyuan Gao and Chirag Shah at SIGIR 2021. Similar to our topic, this tutorial focuses on introducing the issue of bias in data, algorithms, and search process.
2. *Fairness of Machine Learning in Recommender Systems* by Yunqi Li, Yingqiang Ge, Yongfeng Zhang at CIKM 2021. This tutorial introduces and describes fairness definitions as well as evaluation metrics in recommender systems.

3. *Fair Graph Mining* by Jian Kang, Hanghang Tong at CIKM 2021. The purpose of this tutorial is to introduce state-of-the-art techniques for increasing fairness on graph mining and describe challenges as well as future directions.

There are many other similar tutorials presented at major venues similar to the above. Our goal in this tutorial is to provide comprehensive knowledge about the methods and techniques that can be used for reducing gender bias in IR systems, while past tutorials are not related to retrieval tasks.

The target audience for this tutorial will be those who have interest in IR methods especially neural ranking models and well-known datasets. The tutorial will provide an overview of some of the IR concepts and components for those who are new to the field of IR. As such, sufficient details will be provided as appropriate so that the content will be accessible and understandable to those who only have a basic understanding of IR principles. This tutorial will only assume that the audiences is familiar with different topics included in an undergraduate IR course such as those covered in [26].

5   Presenters

Amin Bigdeli is a Data Scientist at Warranty Life and a Research Associate at Toronto Metropolitan University. His research work focuses on issues of fairness in information retrieval systems. Amin has published multiple research papers in this area in top IR venues such as SIGIR, CIKM, EDBT, and ECIR.

Negar Arabzadeh is a Research intern at Google brain working on fairness evaluation of text to image generation models. She is also completing her Ph.D. at the University of Waterloo. Her research is aligned with Ad hoc Retrieval and Conversational search in IR and NLP. Negar has published relevant papers in prestigious conferences and journals such as SIGIR, CIKM, ECIR, and IP&M. She previously interned at Microsoft Research and Spotify research and is one of the lead organizers of NeurIPS 2022 IGLU competition on NLP task.

Shirin Seyedsalehi is a Ph.D. student at Toronto Metropolitan University. Her research so far is focused on fairness in Information Retrieval and Neural Rankers. She has published papers in well known conferences such as SIGIR, CIKM and EDBT. She previously interned at Microsoft Research.

Morteza Zihayat is an Associate Professor and co-founder of the centre for Digital Enterprise Analytics & Leadership (DEAL) at Toronto Metropolitan University. His research concerns user modeling, applied machine learning and bias, debiasing, and fairness in NLP and IR. He has published in various well-respected journals and conferences in Information Retrieval, Machine Learning, and Information Systems such as IEEE TKDE, Information Processing and Management, ACM SIGKDD, SIGIR, ECIR, PKDD, SIAM, and SDM.

Ebrahim Bagheri is a Professor and the Director for the Laboratory for Systems, Software and Semantics (LS^3) at Toronto Metropolitan University. He holds a Canada Research Chair (Tier II) in Social Information Retrieval as well as an NSERC Industrial Research Chair in Social Media Analytics. He currently leads the NSERC Program on Responsible AI (http://responsible-ai.ca). He is an Associate Editor for ACM Transactions on Intelligent Systems and Technology (TIST) and Wiley's Computational Intelligence.

6 Type of Support Materials

As for the supporting materials, we will publicly share a Github repository several weeks prior to the conference so the participants of the tutorial can familiarize themselves with the content. The repository will include a comprehensive slide deck, links to code, models, datasets, and run files.

References

1. Azzopardi, L.: Cognitive biases in search: a review and reflection of cognitive biases in information retrieval. In: Proceedings of the 2021 Conference on Human Information Interaction and Retrieval, pp. 27–37 (2021)
2. Baeza-Yates, R.: Bias on the web. Commun. ACM **61**, 54–61 (2018)
3. Baeza-Yates, R.: Bias in search and recommender systems. In: Fourteenth ACM Conference on Recommender Systems, p. 2 (2020)
4. Bagheri, E., Ensan, F., Al-Obeidat, F.: Neural word and entity embeddings for ad hoc retrieval. Inf. Proc. Manag. **54**(4), 657–673 (2018)
5. Basta, C., Costa-Jussà, M.R., Casas, N.: Evaluating the underlying gender bias in contextualized word embeddings. arXiv preprint arXiv:1904.08783 (2019)
6. Bigdeli, A., Arabzadeh, N., SeyedSalehi, S., Zihayat, M., Bagheri, E.: Gender fairness in information retrieval systems. In: Proceedings of the 45th International ACM SIGIR Conference (2022)
7. Bigdeli, A., Arabzadeh, N., Seyedsalehi, S., Zihayat, M., Bagheri, E.: A light-weight strategy for restraining gender biases in neural rankers. In: Hagen, M., et al. (eds.) ECIR 2022. LNCS, vol. 13186, pp. 47–55. Springer, Cham (2022). https://doi.org/10.1007/978-3-030-99739-7_6
8. Bigdeli, A., Arabzadeh, N., Seyersalehi, S., Zihayat, M., Bagheri, E.: On the orthogonality of bias and utility in ad hoc retrieval. In: Proceedings of the 44rd International ACM SIGIR Conference (2021)
9. Bigdeli, A., Arabzadeh, N., Zihayat, M., Bagheri, E.: Exploring gender biases in information retrieval relevance judgement datasets. In: Hiemstra, D., Moens, M.-F., Mothe, J., Perego, R., Potthast, M., Sebastiani, F. (eds.) ECIR 2021. LNCS, vol. 12657, pp. 216–224. Springer, Cham (2021). https://doi.org/10.1007/978-3-030-72240-1_18
10. Bolukbasi, T., Chang, K.W., Zou, J.Y., Saligrama, V., Kalai, A.T.: Man is to computer programmer as woman is to homemaker? Debiasing word embeddings. In: Advances in Neural Information Processing Systems, vol. 29 (2016)
11. Bordia, S., Bowman, S.R.: Identifying and reducing gender bias in word-level language models (2019)
12. Brunet, M.E., Alkalay-Houlihan, C., Anderson, A., Zemel, R.: Understanding the origins of bias in word embeddings. In: International Conference on Machine Learning, pp. 803–811. PMLR (2019)
13. Caliskan, A., Bryson, J.J., Narayanan, A.: Semantics derived automatically from language corpora contain human-like biases. Science **356**(6334), 183–186 (2017)

14. Draws, T., Tintarev, N., Gadiraju, U., Bozzon, A., Timmermans, B.: This is not what we ordered: exploring why biased search result rankings affect user attitudes on debated topics (2021)
15. Ekstrand, M.D., Das, A., Burke, R., Diaz, F.: Fairness in information access systems. arXiv preprint arXiv:2105.05779 (2021)
16. Fabris, A., Purpura, A., Silvello, G., Susto, G.A.: Gender stereotype reinforcement: measuring the gender bias conveyed by ranking algorithms. Inf. Proc. Manag. **57**(6), 102377 (2020)
17. Font, J.E., Costa-Jussa, M.R.: Equalizing gender biases in neural machine translation with word embeddings techniques. arXiv preprint arXiv:1901.03116 (2019)
18. Gerritse, E.J., Hasibi, F., de Vries, A.P.: Bias in conversational search: the double-edged sword of the personalized knowledge graph. In: Proceedings of the 2020 ACM SIGIR on International Conference on Theory of Information Retrieval (2020)
19. Klasnja, A., Arabzadeh, N., Mehrvarz, M., Bagheri, E.: On the characteristics of ranking-based gender bias measures. In: 14th ACM Web Science Conference 2022, pp. 245–249 (2022)
20. Krieg, K., Parada-Cabaleiro, E., Medicus, G., Lesota, O., Schedl, M., Rekabsaz, N.: Grep-BiasIR: a dataset for investigating gender representation-bias in information retrieval results. arXiv preprint arXiv:2201.07754 (2022)
21. Krieg, K., Parada-Cabaleiro, E., Schedl, M., Rekabsaz, N.: Do perceived gender biases in retrieval results affect relevance judgements. In: Boratto, L., Faralli, S., Marras, M., Stilo, G. (eds.) BIAS 2022. Communications in Computer and Information Science, vol. 1610, pp. 104–116. Springer, Cham (2022). https://doi.org/10.1007/978-3-031-09316-6_10
22. Kulshrestha, J., et al.: Quantifying search bias: investigating sources of bias for political searches in social media. In: Proceedings of the 2017 ACM Conference on Computer Supported Cooperative Work and Social Computing, pp. 417–432 (2017)
23. Liu, H., Dacon, J., Fan, W., Liu, H., Liu, Z., Tang, J.: Does gender matter? Towards fairness in dialogue systems. arXiv preprint arXiv:1910.10486 (2019)
24. Liu, H., Wang, W., Wang, Y., Liu, H., Liu, Z., Tang, J.: Mitigating gender bias for neural dialogue generation with adversarial learning (2020)
25. Lu, K., Mardziel, P., Wu, F., Amancharla, P., Datta, A.: Gender bias in neural natural language processing. In: Nigam, V., et al. (eds.) Logic, Language, and Security. LNCS, vol. 12300, pp. 189–202. Springer, Cham (2020). https://doi.org/10.1007/978-3-030-62077-6_14
26. Manning, C.D., Raghavan, P., Schütze, H.: Introduction to Information Retrieval. Cambridge University Press, Cambridge (2008)
27. Nguyen, T., et al.: MS MARCO: a human generated machine reading comprehension dataset. In: CoCo@ NIPS (2016)
28. Olteanu, A., et al.: FACTS-IR: fairness, accountability, confidentiality, transparency, and safety in information retrieval. In: ACM SIGIR Forum, vol. 53, pp. 20–43. ACM New York, NY, USA (2021)
29. Prost, F., Thain, N., Bolukbasi, T.: Debiasing embeddings for reduced gender bias in text classification. arXiv preprint arXiv:1908.02810 (2019)
30. Rekabsaz, N., Kopeinik, S., Schedl, M.: Societal biases in retrieved contents: measurement framework and adversarial mitigation for BERT rankers (2021)
31. Rekabsaz, N., Schedl, M.: Do neural ranking models intensify gender bias?. In: Proceedings of the 43rd International ACM SIGIR Conference (2020)

32. SeyedSalehi, S., Bigdeli, A., Arabzadeh, N., Mitra, B., Zihayat, M., Bagheri, E.: Bias-aware fair neural ranking for addressing stereotypical gender biases. In: EDBT, pp. 2–435 (2022)
33. Seyedsalehi, S., Bigdeli, A., Arabzadeh, N., Zihayat, M., Bagheri, E.: Addressing gender-related performance disparities in neural rankers. In: Proceedings of the 45th International ACM SIGIR Conference on Research and Development in Information Retrieval, pp. 2484–2488 (2022)
34. Stanczak, K., Augenstein, I.: A survey on gender bias in natural language processing. arXiv preprint arXiv:2112.14168 (2021)
35. Sun, T., et al.: Mitigating gender bias in natural language processing: literature review. arXiv preprint arXiv:1906.08976 (2019)
36. Wang, J., Liu, Y., Wang, X.E.: Are gender-neutral queries really gender-neutral? mitigating gender bias in image search. arXiv preprint arXiv:2109.05433 (2021)
37. Yang, Z., Feng, J.: A causal inference method for reducing gender bias in word embedding relations. In: Proceedings of the AAAI Conference (2020)
38. Zhao, J., Mukherjee, S., Hosseini, S., Chang, K.W., Awadallah, A.H.: Gender bias in multilingual embeddings and cross-lingual transfer (2020)
39. Zhao, J., Wang, T., Yatskar, M., Cotterell, R., Ordonez, V., Chang, K.W.: Gender bias in contextualized word embeddings. arXiv preprint arXiv:1904.03310 (2019)

ECIR 23 Tutorial: Neuro-Symbolic Approaches for Information Retrieval

Laura Dietz[1]([⊠])(iD), Hannah Bast[2](iD), Shubham Chatterjee[1,3](iD), Jeff Dalton[3](iD), Edgar Meij[4](iD), and Arjen de Vries[5](iD)

[1] University of New Hampshire, Durham, USA
`dietz@cs.unh.edu`
[2] University of Freiburg, Freiburg im Breisgau, Germany
`bast@cs.uni-freiburg.de`
[3] University of Glasgow, Glasgow, Scotland
{`shubham.chatterjee,jeff.dalton`}`@glasgow.ac.uk`
[4] Bloomberg, New York, UK
`emeij@bloomberg.net`
[5] Radboud University, Nijmegen, The Netherlands
`arjen@acm.org`

Abstract. This tutorial will provide an overview of recent advances on neuro-symbolic approaches for information retrieval. A decade ago, knowledge graphs and semantic annotations technology led to active research on how to best leverage symbolic knowledge. At the same time, neural methods have demonstrated to be versatile and highly effective.

From a neural network perspective, the same representation approach can service document ranking or knowledge graph reasoning. End-to-end training allows to optimize complex methods for downstream tasks.

We are at the point where both the symbolic and the neural research advances are coalescing into neuro-symbolic approaches. The underlying research questions are how to best combine symbolic and neural approaches, what kind of symbolic/neural approaches are most suitable for which use case, and how to best integrate both ideas to advance the state of the art in information retrieval.

Keywords: Neural networks · Semantics · IR

1 Motivation

Being able to reason on what is relevant for an information need is important for all kinds of information retrieval tasks: web search, question answering, dialogues, image search, task assistance, or e-commerce. As traditional keyword-matching approaches are successively being replaced with neural-representation approaches [15,17], the question is whether symbolic approaches still have merit.

A decade ago, advances in knowledge graphs and semantic annotations, such as via entity linking, led to significant improvements in text ranking tasks [6,22]. These, in turn, set new standards for entity-oriented downstream tasks like question answering [1,8]. Now, neural representations for semantic annotations or

© The Author(s), under exclusive license to Springer Nature Switzerland AG 2023
J. Kamps et al. (Eds.): ECIR 2023, LNCS 13982, pp. 324–330, 2023.
https://doi.org/10.1007/978-3-031-28241-6_33

other kinds of symbols have taken hold in the field of knowledge management. The information retrieval community is split into research that solely relies on neural representations (abandoning symbols altogether) and research that integrates neural and symbolic approaches.

Symbolic approaches have been studied in information retrieval over the years. The IR community has had a continued interest in entity retrieval tasks [2,3,9]. Sometimes, information needs are best answered using knowledge from external databases [24]—other times text can contextualize knowledge [20]. Furthermore, effective query expansion via pseudo-relevance feedback relies upon approaches that analyze retrieved documents—and reason about why these are relevant.

The goal of this tutorial is to consolidate findings and initiate a synergistic transfer across different IR-relevant use cases with respect to neuro-symbolic approaches.

2 Format and Target Audience

In this full-day tutorial, we will provide different perspectives on neural-symbolic methods, provide different perspectives on the topic, and discuss customizations for different use cases.

Our goal is to provide useful information to a *wide variety* of audiences. The first part of the tutorial will be *introductory*, designed to bring audience members up to speed who have only basic knowledge in neural representations and/or symbolic approaches, such as knowledge graphs and entity linking.

The second part will of interest to both a *beginners and intermediate audience*, where different speakers provide their own perspective on the topic and look at different use cases where we need to reason on what is relevant.

To conclude the tutorial, we will invite all speakers and some additional guests to a panel discussion. Our goal is to spur a discussion of what works where, when, and why.

3 Topics Covered

1. **Foundational Topics**

    Neural Text Representations. BERT and other large neural language models (LLMs) of text have led to tremendous increases in performance improvements. LLM-based document re-ranking models are either based on Siamese-models like the Duet Model [16], or transformers, such as mono-BERT or duo-BERT [15]. We also cover neural query expansion [21] and query rewriting [14].

    Symbols and Knowledge Graphs. Several repositories of symbolic knowledge are readily available: Entities derived from Wikipedia pages, or nodes in a knowledge graph such as DBpedia or Wikidata. Word-oriented knowledge bases for Common Sense Reasoning are COMET [12] as well as ConceptNet. Neural representations of such symbols are provided in E-BERT

[19], Wikipedia2Vec [25], or BERT-ER [3]. Graph neural networks such as HOPE [18], allow to reason across the graph structure.

Text-Symbol Alignment and Semantic Annotations. The task of Entity Linking [11] (aka Wikification) is to annotate unstructured text with detected and disambiguated entity identifiers. Such entity links in the text can serve as a set of logical symbols to reason with. The entity links also provide a means to align the text with nodes in a knowledge graph to perform inferences in. For some tasks, such as for conversations, specialized entity linking methods obtain better performance [13]. Other fine-grained information can be extracted from text with relation extraction, semantic role labeling, type predictions, and entity-aspect linking. Neural alignment methods allow the utilization of information in text and symbols for better ranking quality, such as EM-BERT [9].

2. **Perspectives**

Reasoning about Relevance. Retrieval models aim to reason about what is relevant. Hence, we summarize related ideas from other areas, such as logic-based reasoning in knowledge graphs as well as Natural Language Inference (NLI). Some systems include retrieval into their neural inference models, such as REALM [10]. Following on research on probabilistic reasoning with neural approaches for logic-based reasoning in knowledge bases, such as FuzzQE [5]. Chain-of-Thought reasoners [23] are leveraging neural few-shot learners to generate well-reasoned arguments.

Ranking Wikipedia Entities/Aspects. Given a query and a knowledge graph, the entity ranking task is to retrieve and rank entities from the knowledge graph according to their relevance to the query. Entity ranking has also been shown to be useful for tasks that require an explicit semantic understanding of text [4]. Two broad directions for entity ranking are (1) Non-neural approaches that leverage symbols and semantic annotations in text, and (2) Neural approaches that leverage dense representations of entities learnt using neural networks. Finally, we discuss future directions for learning better entity representations for IR.

Explainability for Pseudo-relevance Feedback. Traditionally, pseudo-relevance feedback (PRF/RM3) is a technique to identify relevant terms for query expansion. This idea has been generalized to identify relevant entities for expanding queries with expansion entities [6], or augmenting neural representations [21].

In "Explainability" the focus shifts from making correct predictions to explaining why a prediction was made. One explainability approach is to analyze model gradients to approximate input importance [7]. In a PRF setting, such information can be used to glean information on why a document was deemed relevant, with the goal to augment and refine the search query.

3. **Different Use Cases**

Use Case: Question Answering on Knowledge Graphs. This task accepts a question in arbitrary natural language, which should then be translated to a corresponding structured query (for example, in SPARQL) on a given knowledge graph. The currently best approaches for solving this problem [1] are all inherently neuro-symbolic: the knowledge is given in strongly structured (symbolic) form, yet the learning is neural. Correspondingly, the challenges are twofold. The *symbolic* challenge is to understand the nature of the structured queries, which are often surprisingly complex and non-trivial—even for seemingly simple questions. The *neural* challenge is to learn a high-quality translation model that can handle also complex questions and requires little supervision.

Use Case: Task-based Assistance. Information agents support complex real-world tasks and must not only retrieve relevant information, but also perform complex tasks using external symbolic tools and computation. This requires grounded reasoning about information and world state that is multimodal across text, images, video, and structured knowledge. Further, they must support the user in explainable and controllable fashion that involves eliciting structured information and storing it in personal knowledge graphs to incorporate structured symbolic constraints ("make it vegan") as well as being adaptable to mood, situation, and skill level.

Use Case: Generating Relevant Articles. Some usage scenarios ask for long, multi-faceted answers without the need for a user to interact. The goal is to foresee obvious next questions, and be forthcoming with such information without being explicitly asked. To satisfy this use case requires to solve a range of inter-dependent tasks: (1) high-recall retrieval with broad coverage, (2) query-specific clustering for subtopic-detection, (3) organization of content into a sequential structure, and (4) summarization and natural language generation.

4. **Discussion Panel:** The goal is to identify synergistic opportunities across different use cases. We are discussing approaches that (according to the literature) are supposed to work, but do not yet yield satisfactory results, leaving ample room for improvement. We also debate some controversial questions, such as "Since we have neural text representations, do we really need symbolic approaches?" The tutorial presenters and panel speakers are selected because they represent a broad spectrum of expert opinions on the topic.

4   Presenters

Laura Dietz, Associate Professor, University of New Hampshire (Main Contact). Dr. Dietz focuses on integrating relevant-oriented tasks, using full-text search, Wikipedia knowledge, and fine-grained semantic annotations, along with subtopic extraction, content organization, and natural language generation. She organized the KG4IR Tutorial and Workshop Series (ICTIR 2016, WSDM 2017, SIGIR 2017, SIGIR 2018) and the TREC Complex Answer Retrieval track (2017–2019).

Hannah Bast, Full Professor, University of Freiburg. Dr. Bast is interested in all aspects of information retrieval, with a focus on efficiency, ease of use, and fully functional systems. Her search systems power DBLP, Google Maps, and maybe soon Wikidata. Her work combines indexing and search in full text and structured knowledge for downstream applications such as question answering.

Shubham Chatterjee, Research Associate, University of Glasgow. Dr. Chatterjee focuses on neural entity-oriented information retrieval and extraction, particularly text understanding using entities and entity ranking. His goal is to build an intelligent search system that can answer open-ended information needs.

Jeff Dalton, Senior Lecturer, University of Glasgow. Dr. Dalton focuses on methods for effectively leveraging knowledge for complex information-seeking tasks. His work on entity-based query feature expansion published at SIGIR in 2014 is one of the first to demonstrate the effectiveness of using general-purpose knowledge graphs for search. He is a Turing AI Acceleration Fellow at the Turing Institute with a prestigious UKRI fellowship, and the lead organizer of the TREC Conversational Assistance track.

Edgar Meij, Head of Search and Discovery, Bloomberg AI. Dr. Meij leads several teams of researchers and engineers that work on information retrieval, semantic parsing, question answering, and smart contextual suggestions under severe latency constraints. Together, these researchers and engineers build, maintain, and leverage the company's search, autocomplete, and question-answering systems. He has taught several tutorials and organized various workshops on knowledge graphs, entity linking, and semantic search at top-tier conferences.

Arjen de Vries, Full Professor, Radboud University. Dr. de Vries uses structured and unstructured information to improve information access. He works on entity linking as well as entity retrieval, demonstrating that having knowledge of the entities in the query can help improve retrieval performance for entity-oriented search tasks. Dr. de Vries organized the first information retrieval evaluation campaigns that looked beyond documents into entities—the Enterprise Search track at TREC and later, the Entity Ranking track at INEX.

References

1. Bast, H., Haussmann, E.: More accurate question answering on freebase. In: Proceedings of the 24th ACM International on Conference on Information and Knowledge Management, pp. 1431–1440 (2015)
2. Cao, N.D., Izacard, G., Riedel, S., Petroni, F.: Autoregressive entity retrieval. CoRR abs/2010.00904 (2020). https://arxiv.org/abs/2010.00904
3. Chatterjee, S., Dietz, L.: BERT-ER: query-specific BERT entity representations for entity ranking. In: Proceedings of the 45th International ACM SIGIR Conference on Research and Development in Information Retrieval, SIGIR 2022, pp. 1466–1477. Association for Computing Machinery, New York (2022). https://doi.org/10.1145/3477495.3531944

4. Chatterjee, S., Dietz, L.: Predicting guiding entities for entity aspect linking. In: Proceedings of the 31st ACM International Conference on Information and Knowledge Management, CIKM 2022. Association for Computing Machinery, New York (2022). https://doi.org/10.1145/3511808.3557671
5. Chen, X., Hu, Z., Sun, Y.: Fuzzy logic based logical query answering on knowledge graphs. In: Proceedings of the AAAI Conference on Artificial Intelligence, vol. 36, pp. 3939–3948 (2022)
6. Dalton, J., Dietz, L., Allan, J.: Entity query feature expansion using knowledge base links. In: Proceedings of the 37th International ACM SIGIR Conference on Research and Development in Information Retrieval, SIGIR 2014, pp. 365–374. Association for Computing Machinery, New York (2014). https://doi.org/10.1145/2600428.2609628
7. Funke, T., Khosla, M., Rathee, M., Anand, A.: Zorro: valid, sparse, and stable explanations in graph neural networks. IEEE Trans. Knowl. Data Eng. (2022)
8. Gerritse, E.J., Hasibi, F., de Vries, A.P.: Graph-embedding empowered entity retrieval. In: Jose, J.M., Yilmaz, E., Magalhães, J., Castells, P., Ferro, N., Silva, M.J., Martins, F. (eds.) ECIR 2020. LNCS, vol. 12035, pp. 97–110. Springer, Cham (2020). https://doi.org/10.1007/978-3-030-45439-5_7
9. Gerritse, E.J., Hasibi, F., de Vries, A.P.: Entity-aware transformers for entity search. In: Proceedings of the 45th International ACM SIGIR Conference on Research and Development in Information Retrieval, SIGIR 2022, pp. 1455–1465. Association for Computing Machinery, New York (2022). https://doi.org/10.1145/3477495.3531971
10. Guu, K., Lee, K., Tung, Z., Pasupat, P., Chang, M.W.: Realm: retrieval-augmented language model pre-training. In: Proceedings of the 37th International Conference on Machine Learning, pp. 3929–3938 (2020)
11. van Hulst, J.M., Hasibi, F., Dercksen, K., Balog, K., de Vries, A.P.: Rel: an entity linker standing on the shoulders of giants. In: Proceedings of the 43rd International ACM SIGIR Conference on Research and Development in Information Retrieval, pp. 2197–2200 (2020)
12. Hwang, J.D., Bhagavatula, C., Le Bras, R., Da, J., Sakaguchi, K., Bosselut, A., Choi, Y.: (comet-) atomic 2020: on symbolic and neural commonsense knowledge graphs. In: Proceedings of the AAAI Conference on Artificial Intelligence, vol. 35, pp. 6384–6392 (2021)
13. Joko, H., Hasibi, F., Balog, K., de Vries, A.P.: Conversational entity linking: problem definition and datasets. In: Proceedings of the 44th International ACM SIGIR Conference on Research and Development in Information Retrieval, pp. 2390–2397 (2021)
14. Lewis, M., et al.: Bart: denoising sequence-to-sequence pre-training for natural language generation, translation, and comprehension. In: Proceedings of the 58th Annual Meeting of the Association for Computational Linguistics, pp. 7871–7880 (2020)
15. Lin, J., Nogueira, R., Yates, A.: Pretrained transformers for text ranking: BERT and beyond. CoRR abs/2010.06467 (2020). https://arxiv.org/abs/2010.06467
16. Mitra, B., Craswell, N.: An updated duet model for passage re-ranking. arXiv preprint arXiv:1903.07666 (2019)
17. Mitra, B., Craswell, N., et al.: An introduction to neural information retrieval. Found. Trends® Inf. Retrieval 13(1), 1–126 (2018)
18. Ou, M., Cui, P., Pei, J., Zhang, Z., Zhu, W.: Asymmetric transitivity preserving graph embedding. In: Proceedings of the 22nd ACM SIGKDD International Conference on Knowledge Discovery and Data Mining, pp. 1105–1114 (2016)

19. Poerner, N., Waltinger, U., Schütze, H.: E-BERT: efficient-yet-effective entity embeddings for BERT. In: Findings of the Association for Computational Linguistics: EMNLP 2020, pp. 803–818. Association for Computational Linguistics, Online, November 2020. https://doi.org/10.18653/v1/2020.findings-emnlp.71

20. Ponza, M., Ceccarelli, D., Ferragina, P., Meij, E., Kothari, S.: Contextualizing trending entities in news stories. In: Proceedings of the 14th ACM International Conference on Web Search and Data Mining, pp. 346–354 (2021)

21. Pradeep, R., Nogueira, R., Lin, J.: The expando-mono-duo design pattern for text ranking with pretrained sequence-to-sequence models. arXiv e-prints pp. arXiv-2101 (2021)

22. Reinanda, R., Meij, E., de Rijke, M., et al.: Knowledge graphs: an information retrieval perspective. Found. Trends® Inf. Retrieval **14**(4), 289–444 (2020)

23. Saparov, A., He, H.: Language models are greedy reasoners: a systematic formal analysis of chain-of-thought. arXiv preprint arXiv:2210.01240 (2022)

24. Xiong, C., Liu, Z., Callan, J., Hovy, E.: Jointsem: combining query entity linking and entity based document ranking. In: Proceedings of the 2017 ACM SIGIR Conference on Information and Knowledge Management, CIKM 2017, pp. 2391–2394. Association for Computing Machinery, New York (2017). https://doi.org/10.1145/3132847.3133048

25. Yamada, I., et al.: Wikipedia2Vec: an efficient toolkit for learning and visualizing the embeddings of words and entities from Wikipedia. In: Proceedings of the 2020 Conference on Empirical Methods in Natural Language Processing: System Demonstrations, pp. 23–30. Association for Computational Linguistics, Online, October 2020. https://doi.org/10.18653/v1/2020.emnlp-demos.4

Legal IR and NLP: The History, Challenges, and State-of-the-Art

Debasis Ganguly[1]($\boxtimes$), Jack G. Conrad[2], Kripabandhu Ghosh[3],
Saptarshi Ghosh[4], Pawan Goyal[4], Paheli Bhattacharya[4],
Shubham Kumar Nigam[5], and Shounak Paul[4]

[1] University of Glasgow, Glasgow, UK
Debasis.Ganguly@glasgow.ac.uk
[2] Thomson Reuters Labs, Minneapolis, MN, USA
[3] Indian Institute of Science Education and Research Kolkata, Mohanpur, India
[4] Indian Institue of Technology Kharagpur, Kharagpur, India
[5] Indian Institute of Technology Kanpur, Kanpur, India

Abstract. Artificial Intelligence (AI), Machine Learning (ML), Information Retrieval (IR) and Natural Language Processing (NLP) are transforming the way legal professionals and law firms approach their work. The significant potential for the application of AI to Law, for instance, by creating computational solutions for legal tasks, has intrigued researchers for decades. This appeal has only been amplified with the advent of Deep Learning (DL). It is worth noting that working with legal text is far more challenging as compared to the other subdomains of IR/NLP, mainly due to the typical characteristics of legal text, such as considerably longer documents, complex language and lack of large-scale annotated datasets.

In this tutorial, we introduce the audience to these characteristics of legal text, and with it, the challenges associated with processing the legal documents. We touch upon the history of AI and Law research, and how it has evolved over the years from relatively simpler approaches to more complex ones, such as those involving DL. We organize the tutorial as follows. First, we provide a brief introduction to state-of-the-art research in the general domain of IR and NLP. We then discuss in more detail IR/NLP tasks specific to the legal domain. We outline the methodologies (both from an academic and industry perspective), and the available tools and datasets to evaluate the methodologies. This is then followed by a hands-on coding/demo session.

Keywords: AI & Law · Legal data analytics · Natural language processing · Legal information retrieval

1 Goals/Objectives of the Tutorial

AI for Law is one of the most promising areas in IR/NLP research, having become immensely popular in the research community over the past decade, as is evident from the increasing volume of research works in this field [36,43].

© The Author(s), under exclusive license to Springer Nature Switzerland AG 2023
J. Kamps et al. (Eds.): ECIR 2023, LNCS 13982, pp. 331–340, 2023.
https://doi.org/10.1007/978-3-031-28241-6_34

Additionally, the critical need for AI for Law solutions in many countries with highly overburdened legal systems and where access to justice is difficult and costly for the common citizen, is motivating researchers to contribute to this field, given its potential practical utility.

Due to the compelling needs cited above, countries across the world are making efforts in digitizing legal records and funding research in AI for Law. Thus, the future is promising in terms of the availability of resources and funds, and the potential scope of automating solutions to practical problems in the legal domain is significant. Furthermore, applications of AI for Law are increasingly being adopted in industrial settings as well; law firms and startups are investing significant effort and resources to introduce cutting-edge solutions to many such problems [16,36]. Thus, research in IR/NLP based AI for Law is likely to be a lucrative opportunity for both academia and industry alike.

Through this tutorial, the participants are likely to gain some understanding of the challenges and opportunities in this field, which are quite distinct as compared to other sub-domains of IR/NLP. There have been some tutorials in this field in recent years, such as at ICAIL 2017[1] and IJCAI-ECAI 2018[2], and some workshops like those at SIGIR 2020[3] and annual legal workshops like AI4Legal[4] and NLLP[5]. Yet, the topics to be covered in this tutorial are significantly different. Moreover, there has not been any tutorial on this important domain in prominent IR/NLP conferences to date.

Target Audience. Anyone interested in AI & Law would benefit from this tutorial. Practitioners who are interested in the applications of IR/NLP/ML to sub-domains such as law, can also benefit from this tutorial.

Prerequisites. A knowledge of basic IR, NLP and ML techniques would be helpful for attendees to grasp the tasks and tools discussed in this tutorial; hence we view the tutorial to be at an 'intermediate' level of difficulty.

2   Tutorial Outline

While there has been lot of work on legal text in other European languages [25] and in Chinese [39], this tutorial primarily focuses on application of AI to English legal text. The brief outline of the tutorial is provided in Table 1.

Topic 1 – Background on Legal Text: The tutorial starts with a brief introduction to the basics of legal systems used across the world (Civil and Common Law Systems), and different types of legal documents (e.g., court case documents, statutes, patents, contracts, etc.). There is also a brief discussion on the *legal citation networks* that exist between court case documents, statutes, etc.

---

[1] https://www.andrew.cmu.edu/user/mgrabmai/ainlawtutorial2017/.

[2] https://www.ijcai-18.org/wp-content/uploads/2018/05/T04-AI-and-the-Law-IJCAI-ECAI-18.pdf.

[3] https://legalai2020.github.io/.

[4] https://ai4legal.linkeddata.es/.

[5] https://nllpw.org/workshop/.

Table 1. Plan of the tutorial

Topic	Presenter(s)
Background on Legal Text	Saptarshi, Kripa, Debasis
Challenges in processing legal text	
A brief history of AI & Law research, important milestones	Jack
Background on recent approaches in NLP & IR	Pawan
Survey on the state-of-the-art in Legal IR/NLP problems; available tools and datasets	Saptarshi, Paheli, Debasis
The industry perspective	Jack
Future directions in IR/NLP/ML research in legal domain	Jack, Kripa
Hands-on Coding / demo session	Debasis, Pawan, Paheli, Shounak, Shubham

Topic 2 – Challenges in Processing Legal Text: Legal documents are characteristically largely different from those of the general type, e.g., legal documents are lengthy and usually written in a formal and complex language. Tasks considered relatively simple in the open domain, such as detecting sentence boundaries or Named Entity Recognition (NER), are more challenging for the legal domain [34]. Moreover, in many countries, legal documents are unstructured (do not have demarcations such as headings, etc.), and do not follow any common pattern.

Although it is now possible to gain access to unlabelled legal corpora relatively easily (via resources [21] or api-based tools [22]), large-scale, task-specific annotated datasets are still difficult to find for many countries and languages. Moreover, the process of manual annotation requires the assistance of legal experts, which proves to be expensive both in terms of time and effort. These challenges are discussed in this part of the tutorial.

Topic 3 – A Brief History of AI & Law Research, and Important Milestones: This part examines the roots of AI and Law, noting that this field has been around for much longer than the recent focus would suggest [20,33]. We explore the field in terms of its development and expansion starting in the 1980s s and study how seminal research was conducted and reported on in conference proceedings such as ICAIL and publications such as the AI and Law journal [2, 12–14,20,33,38]. After this look at its foundations, we look at the more recent history and examine use-cases and AI-based applications that have been created to address them.

Topic 4 – Brief Background of Recent Approaches in NLP and IR: This section of the tutorial briefly discusses the recent advances in NLP and IR that are particularly useful in the legal domain, such as architectures like transformers [37], and resources such as pre-trained embeddings and language models, such as BERT [15], RoBERTa [27] and XLNet [40]. We also touch upon advances in practical use cases of NLP such as low resource settings [17,24,32], and similar.

Topic 5 – A Survey on the State-of-the-Art in Some Specific Problems in Legal IR/NLP, Available Tools and Datasets: In this section we discuss some common tasks and approaches in the legal domain. These include tasks like summarization [5,35], legal judgment prediction [28] and semantic segmentation [6,7], which are typically solved through approaches such as retrieval (recall-oriented search), text classification (both sentence and document-level), and so on. In the legal domain, often alternative sources of knowledge (such as citation networks) can be combined with classical text processing for tasks like document similarity [4] and legal statute identification [30]. Also, recently, many pre-trained models for the legal domain have been developed [10,21,31,42], which can be applied with good results on many end tasks. Table 2 presents some of the datasets and tools to be covered. We also briefly cover popular shared tasks in this domain, such as TREC Legal, FIRE, CLEF, NTCIR, SemEval, and so on.

Topic 6 – The Industry Perspective: Here we discuss the problems that are particularly important from an industry point of view. We start by briefly covering the three key task-focused areas that legal knowledge workers pursue: finding, analyzing, and making decisions about information [12]. These are critical areas where legal professionals develop refined capabilities in order to address their stakeholders' legal needs. Today, many industries focus on injecting technology (data mining, natural language processing, and machine learning) into the last two tasks: analyzing and deciding. One finds that, within frequent legal issues, certain patterns repeat and practitioners benefit from seeing such patterns consisting of facts, claims, counter-claims, legal principles applied, analysis and decisions. However, a statistical discovery of these relevant patterns is only possible in large datasets. In today's operational legal settings, one finds ML models, trained on large amounts of data, making predictions in order to assist practitioners with their analysis and decision making. Increasingly such models deploy deep learning to support advanced legal workflows. Some of the challenges that arise around these AI models involve issues such model transparency, explainability, bias, and accountability. We conclude this section by examining some of these evolving challenges.

Topic 7 – Future Directions in IR/NLP/ML Research in Legal Domain: We discuss the current problems plaguing the field of AI & Law, and the future research directions. These include topics like fairness and bias related concerns [18,23], and the need for explainibility in DL models [1,26]. Such issues are of utmost importance to prevent undesirable consequences when such technologies are incorporated into real judicial and legal systems of countries [9,19,41].

Topic 8 – Hands-on Coding/Demo Session: The tutorial concludes with a hands-on demo session to introduce the basics of the coding and modeling strategies to implement and run contemporary models on common legal tasks, such as semantic segmentation [6,7], summarization [5] and court judgment prediction [28]. We also have a look at how citation networks can be used in conjunction with text for tasks such as legal statute identification [30]. This will also cover some existing state-of-the-art resources such as tools and datasets for such tasks.

3   Biography of the Tutorial Presenters

- **Debasis Ganguly**[6], *Lecturer (Assistant Professor), School of Computing Science, University of Glasgow, Glasgow, Scotland.* Formerly, he was a research staff member at IBM Research Europe, Dublin, Ireland. Generally speaking, his research activities span topics on IR and NLP. More specifically, he is interested in semantic search, neural retrieval models, explainable

Table 2. Brief description of some of the resources to be discussed under Topic 5 of the tutorial

Datasets	
Dataset	Description
Semantic Segmentation [7]	Corpus of 150 Indian and 50 U.K. Supreme Court Case Documents annotated for the Semantic Segmentation Task
ILSI [30]	Corpus of 65k Indian criminal court case documents for the Legal Statute Identification Task
ILDC [28]	Multiple corpora of Indian Supreme Court case documents for the Court Judgment Prediction and Explanation Task
Summarization [35]	A collection of 3 legal case document summarization datasets
LexGLUE Benchmark [11]	A collection of multiple datasets (mostly EU, UK or US-based) for different legal tasks
Pre-trained Language Models	
Model	Description
LegalBERT [10]	Pre-trained Language Model over EU, UK and US legal text
CaseLawBERT [42]	Pre-trained Language Model over US case documents
PoLBERT [21]	Pre-trained Language Model over many types of legal documents
InLegalBERT & InCaseLawBERT [31]	Pre-trained Language Model over Indian court case documents

---

[6] https://gdebasis.github.io/ (contact person) Email: debasis.ganguly@glasgow.ac.uk.

search and recommendation, fair and trustworthy search, and privacy preserving AI. Apart from this, he is interested in automatically constructing knowledge bases from legal documents for structured and explainable search. He is a part of the organization committee of the Symposium on Artificial Intelligence and Law (SAIL).

– **Jack G. Conrad**[7], *Director of Applied Research, Thomson Reuters Labs, Minneapolis, MN USA.* Jack Conrad is Director of Applied Research at Thomson Reuters Labs where he focuses on a broad range of technical application areas involving AI, ML and textual data processing. For over two decades, he has delivered critical artifacts and infrastructure for research and business directed projects across a diverse spectrum of domains that have included legal, tax and news. Jack has published more than 50 peer reviewed research papers and has eight patents. He is passionate about the power of AI transformation in enterprise environments. Jack is past president of the International Association for Artificial Intelligence and Law (IAAIL.org) and has served on the IAAIL Executive Committee for 8 years. Jack's areas of expertise include research in the fields of information retrieval, question answering, NLP, machine learning, data mining, and system evaluation.

– **Kripabandhu Ghosh**[8], *Assistant Professor, Department of Computational & Data Sciences, IISER Kolkata, West Bengal, India.* He completed his Ph.D. from Indian Statistical Institute, Kolkata on "Information Retrieval in the Legal Domain". He has been working on AI-Law topics for the last 13 years. His papers have received awards at the two most recognized international conferences in AI and Law - Best Paper award at JURIX 2019 and Best Student Paper award at ICAIL 2021. He has also published papers in AI-Law in journals such as AI and Law, Springer (the most reputed AI-Law journal) and conferences such as SIGIR, CIKM, ACL, ECIR etc. In addition, he has organized several AI-Law events, including a workshop with an international conference (ACM CIKM), shared tasks, and international symposiums such as SAIL'21[9] and SAIL'22,[10] that hosted talks by reputed researchers in the domain. He has recently been inducted to the Editorial Board of the AI and Law journal, Springer.

– **Saptarshi Ghosh**[11], *Assistant Professor, Department of Computer Science & Engineering, IIT Kharagpur, West Bengal, India.* His research interests include Legal analytics, Social media analytics, and Algorithmic bias and fairness. His works on AI & Law have been published at premier conferences including SIGIR, AAAI, CIKM, ECIR, COLING, and have been awarded at top AI & Law conferences, including the *Best Paper Award* at the International Conference on Legal Knowledge and Information Systems (JURIX) 2019, and the *Best Student Paper Award* at the International Conference on

---

[7] http://www.conradweb.org/~jackg/.

[8] https://www.iiserkol.ac.in/web/en/people/faculty/cds/kripaghosh.

[9] https://sites.google.com/view/sail-2021/.

[10] https://sites.google.com/view/sail-2022/.

[11] http://cse.iitkgp.ac.in/~saptarshi.

Artificial Intelligence and Law (ICAIL) 2021. He is presently the Section Editor on Legal Information Retrieval for the journal 'Artificial Intelligence and Law' (Springer).

- **Pawan Goyal**[12], *Associate Professor, Deptt. of Computer Science & Engineering, IIT Kharagpur, West Bengal, India.* He received his B. Tech. degree in Electrical Engineering from IIT Kanpur in 2007 and his PhD degree from University of Ulster, UK in 2011. He was then a post doctoral fellow at INRIA Paris Rocquencourt. His research interests include Natural Language Understanding, Information Retrieval and Sanskrit Computational Linguistics. He has published his research work in various top conferences including ACL, EMNLP, NAACL, KDD, SIGIR, WSDM, AAAI, IJCAI, as well as journals such as Computational Linguistics, ACM and IEEE Transactions. He was the recipient of Google India AI/ML research awards 2020 and INAE Young Engineers Award 2020.

- **Paheli Bhattacharya**[13], *PhD, Dept. of Computer Science & Engineering, IIT Kharagpur, West Bengal, India.* Her PhD thesis was based on developing text and graph-based methods for mining legal court case documents. She has received best paper awards from the two most prestigious conferences in AI and Law - JURIX 2019 [6] and ICAIL 2021 [8]. She actively co-organized shared tasks at FIRE from 2019–2021 and symposiums - SAIL from 2021–2022 and IGLAIS (an Indo-German initiative) in 2021 – all of which were aimed at fostering research in the application of AI techniques in the legal domain. During her MS (by Research) from the same department, she worked on cross-lingual IR on Indian languages. She has publications at SIGIR [3], ECIR [5], AACL-IJCNLP [35], IPM [4], AI & Law [7] and TALLIP. Her research interests are in the area of natural language processing, deep learning and information retrieval.

- **Shubham Kumar Nigam**[14], *Senior Research Fellow, Department of Computer Science &, Engineering, IIT Kanpur, Uttar Pradesh, India.* His research interests include legal data analytics and applications of NLP and IR in the legal domain. Before joining IIT Kanpur, he worked as a project assistant at the Aeronautical Development Agency (ADA), Ministry of Defence, India. He is on the Symposium on Artificial Intelligence and Law (SAIL) organization committee. He actively participates in shared tasks like competitions on legal information extraction and entailment (COLIEE) and semantic evaluation (SemEval). His work has been presented at conferences such as ACL 2021, JURISIN 2022, SemEval 2022, and NLLP 2022.

- **Shounak Paul**[15], *Senior Research Fellow, Department of Computer Science &, Engineering, IIT Kharagpur, West Bengal, India.* His research interests mainly include legal data analytics and applications of NLP in the legal domain. His works on AI & Law for Indian use cases have been published

---

[12] http://cse.iitkgp.ac.in/~pawang.

[13] https://sites.google.com/site/pahelibh/.

[14] https://sites.google.com/view/shubhamkumarnigam.

[15] https://sites.google.com/view/shounakpaul95.

in premier conferences and journals such as: semantic segmentation [6,7] (JURIX 2019, best paper award; AI & Law Journal 2021), charge identification [29] (COLING 2020) and legal statute identification using citation networks [30] (AAAI 2022).

References

1. Alammar, J.: Ecco: an open source library for the explainability of transformer language models. In: Proceedings of ACL-IJCNLP (2021)
2. Bench-Capon, T., et al.: A history of AI and Law in 50 papers: 25 years of the international conference on AI and Law. AI & Law (2012)
3. Bhattacharya, P., Ghosh, K., Pal, A., Ghosh, S.: Hier-SPCNet: a legal statute hierarchy-based heterogeneous network for computing legal case document similarity. In: Proceedings of SIGIR (2020)
4. Bhattacharya, P., Ghosh, K., Pal, A., Ghosh, S.: Legal case document similarity: You need both network and text. Information Processing & Management (2022)
5. Bhattacharya, P., Hiware, K., Rajgaria, S., Pochhi, N., Ghosh, K., Ghosh, S.: A comparative study of summarization algorithms applied to legal case judgments. In: Proceedings of ECIR (2019)
6. Bhattacharya, P., Paul, S., Ghosh, K., Ghosh, S., Wyner, A.: Identification of rhetorical roles of sentences in Indian legal judgments. In: Proceedings of JURIX (2019)
7. Bhattacharya, P., Paul, S., Ghosh, K., Ghosh, S., Wyner, A.: DeepRhole: deep learning for rhetorical role labeling of sentences in legal case documents. AI & Law (2021)
8. Bhattacharya, P., Poddar, S., Rudra, K., Ghosh, K., Ghosh, S.: Incorporating domain knowledge for extractive summarization of legal case documents. In: Proceedings of ICAIL (2021)
9. Branting, K., et al.: Semi-supervised methods for explainable legal prediction. In: Proceedings of ICAIL (2019)
10. Chalkidis, I., Fergadiotis, M., Malakasiotis, P., Aletras, N., Androutsopoulos, I.: LEGAL-BERT: the muppets straight out of law school. In: Proceedings of EMNLP (2020)
11. Chalkidis, I., et al.: LexGLUE: a benchmark dataset for legal language understanding in English. In: Proceedings of ACL (2022)
12. Conrad, J.G., Al-Kofahi, K.: Scenario analytics: analyzing jury verdicts to evaluate legal case outcomes. In: Proceedings of ICAIL (2017)
13. Conrad, J.G., Zeleznikow, J.: The Significance of Evaluation in AI and Law: A case study re-examining ICAIL proceedings. In: Proceedings of ICAIL (2013)
14. Conrad, J.G., Zeleznikow, J.: The Role of Evaluation in AI and Law: an examination of its different forms in the AI and Law Journal. In: Proceedings of ICAIL (2015)
15. Devlin, J., Chang, M.W., Lee, K., Toutanova, K.: BERT: pre-training of deep bidirectional transformers for language understanding. In: Proceedings of NAACL (2019)
16. Dhani, J.S., Bhatt, R., Ganesan, B., Sirohi, P., Bhatnagar, V.: Similar cases recommendation using legal knowledge graphs. CoRR (2021)
17. Diao, S., Xu, R., Su, H., Jiang, Y., Song, Y., Zhang, T.: Taming pre-trained language models with n-gram representations for low-resource domain adaptation. In: Proceedings of ACL-IJCNLP (2021)

18. Garrido-Muñoz, I., Montejo-Ráez, A., Martínez-Santiago, F., Ureña-López, L.A.: A survey on bias in deep NLP. Applied Sciences (2021)
19. Górski, Ł., Ramakrishna, S.: Explainable artificial intelligence, lawyer's perspective. In: Proceedings of ICAIL (2021)
20. Governatori, G., Bench-Capon, T., Verheij, B., Araszkiewicz, M., Francesconi, E., Grabmair, M.: Thirty years of Artificial Intelligence and Law: the first decade. AI & Law (2022)
21. Henderson, P., Krass, M.S., Zheng, L., Guha, N., Manning, C.D., Jurafsky, D., Ho, D.E.: Pile of Law: Learning Responsible Data Filtering from the Law and a 256GB Open-Source Legal Dataset. arXiv (2022)
22. Iyengar, P.: Case study-indiankanoon (2011)
23. Joshi, P., Santy, S., Budhiraja, A., Bali, K., Choudhury, M.: The state and fate of linguistic diversity and inclusion in the NLP world. arXiv (2020)
24. Kann, K., Cho, K., Bowman, S.R.: Towards realistic practices in low-resource natural language processing: the development set. arXiv (2019)
25. Leitner, E., Rehm, G., Moreno-Schneider, J.: A dataset of German legal documents for named entity recognition. In: Proceedings of LREC (2020)
26. Lertvittayakumjorn, P., Toni, F.: Explanation-based human debugging of NLP models: a survey. Trans. Assoc. Comput. Linguist. (2021)
27. Liu, Y., et al.: Roberta: a robustly optimized BERT pretraining approach. arXiv (2019)
28. Malik, V., et al.: ILDC for CJPE: Indian Legal Documents Corpus for Court Judgment Prediction and Explanation. In: Proceedings of ACL-IJCNLP (2021)
29. Paul, S., Goyal, P., Ghosh, S.: Automatic charge identification from facts: a few sentence-level charge annotations is all you need. In: Proceedings of COLING (2020)
30. Paul, S., Goyal, P., Ghosh, S.: LeSICiN: a heterogeneous graph-based approach for automatic legal statute identification from Indian legal documents. In: Proceedings of AAAI (2022)
31. Paul, S., Mandal, A., Goyal, P., Ghosh, S.: Pre-training Transformers on Indian Legal Text. arXiv (2022)
32. Şahin, G.G.: To augment or not to augment? a comparative study on text augmentation techniques for low-resource NLP. Computational Linguistics (2022)
33. Sartor, G., et al.: Thirty years of Artificial Intelligence and Law: the second decade. AI & Law (2022)
34. Savelka, J., Walker, V., Grabmair, M., Ashley, K.: Sentence boundary detection in adjudicatory decisions in the United States. TAL (2017)
35. Shukla, A., et al.: Legal case document summarization: extractive and abstractive methods and their evaluation. In: Proceedings of AACL (2022)
36. Sil, R., Roy, A., Bhushan, B., Mazumdar, A.: Artificial intelligence and machine learning based legal application: the state-of-the-art and future research trends. In: 2019 International Conference on Computing, Communication, and Intelligent Systems (ICCCIS) (2019)
37. Vaswani, A., et al.: Attention is all you need. In: Proceedings of NeurIPS (2017)
38. Villata, S., et al.: Thirty years of Artificial Intelligence and Law: the third decade. AI & Law (2022)
39. Xiao, C., Hu, X., Liu, Z., Tu, C., Sun, M.: Lawformer: a pre-trained language model for Chinese legal long documents. AI Open (2021)
40. Yang, Z., Dai, Z., Yang, Y., Carbonell, J., Salakhutdinov, R.R., Le, Q.V.: XLNet: generalized autoregressive pretraining for language understanding. In: Proceedings of NeurIPS (2019)

41. Yu, W., et al.: Explainable legal case matching via inverse optimal transport-based rationale extraction. In: Proceedings of SIGIR (2022)
42. Zheng, L., Guha, N., Anderson, B.R., Henderson, P., Ho, D.E.: When does pre-training help? assessing self-supervised learning for law and the CaseHOLD dataset of 53,000+ legal holdings. In: Proceedings of ICAIL (2021)
43. Zhong, H., Xiao, C., Tu, C., Zhang, T., Liu, Z., Sun, M.: How does NLP benefit legal system: A summary of legal artificial intelligence. In: Proceedings of ACL (2020)

Deep Learning Methods for Query Auto Completion

Manish Gupta[1(✉)], Meghana Joshi[2], and Puneet Agrawal[1]

[1] Microsoft, Hyderabad, India
{gmanish,punagr}@microsoft.com
[2] Microsoft, Vancouver, Canada
mejoshi@microsoft.com

Abstract. Query Auto Completion (QAC) aims to help users reach their search intent faster and is a gateway to search for users. Everyday, billions of keystrokes across hundreds of languages are served by Bing Autosuggest in less than 100 ms. The expected suggestions could differ depending on user demography, previous search queries and current trends. In general, the suggestions in the AutoSuggest block are expected to be relevant, personalized, fresh, diverse and need to be guarded against being defective, hateful, adult or offensive in any way. In this tutorial, we will first discuss about various critical components in QAC systems. Further, we will discuss details about traditional machine learning and deep learning architectures proposed for four main components: ranking in QAC, personalization, spell corrections and natural language generation for QAC.

Keywords: Query auto completion · Autosuggest · Deep learning

1 Motivation

Query Auto Completion (QAC) is the first service that search users interact with. Thus, it is critical to ensure that QAC is highly accurate and efficient. While web search suffers from intent gap (gap between user's information need and the actual query typed by user), the intent gap for QAC is larger since the system needs to guess the intent with just a partially typed query. QAC is challenging – showing fresh relevant suggestions across languages and regions in a time sensitive and personalized manner involves a very complex orchestration of multiple sub-systems. Advances in deep learning have improved several areas in Natural Language Processing (NLP) and information retrieval including QAC. Given so much work on deep learning for QAC, we think this is a good time to summarize the work in this area in an organized manner. The tutorial should be relevant to researchers and data scientists working in web search companies, domain specific search portals (like Monster, apartments.com, etc.), enterprise search companies, etc.

Learning objectives are: (1) establish an understanding of important components in QAC systems, (2) understand basic machine learning approaches for ranking and personalization for QAC, and (3) understand state-of-the-art deep learning approaches for ranking, personalization, defect removal and natural language generation for QAC systems.

J. Kamps et al. (Eds.): ECIR 2023, LNCS 13982, pp. 341–348, 2023.
https://doi.org/10.1007/978-3-031-28241-6_35

2   Relevance to the Information Retrieval Community

Every web-facing company supports search and hence QAC. The tutorial team has vast experience in experimenting and deploying various methods for QAC. Thus, we believe that this tutorial will be extremely relevant to the information retrieval and web mining community at ECIR. While traditional methods for QAC have used feature engineering and machine learning methods, recent methods have leveraged deep learning architectures. Thus, folks interested in the broad area of applications of deep learning for information retrieval will also find this tutorial interesting. We strongly believe that a tutorial on this topic at ECIR is very timely and will attract a lot of interest from a large section of conference participants.

3   Tutorial Format

This tutorial will be for half-day duration, i.e., 3 h plus breaks. Detailed schedule for various parts of the tutorial is as follows.

1. Components in Query Auto Completion systems [40 min]
2. Ranking [30 min]
3. Personalization [30 min]
4. Handling defective suggestions and prefixes [30 min]
5. Natural Language Generation [30 min]
6. Summary and Future Trends [20 min]

4   Detailed Outline

Here is a detailed outline of the tutorial with relevant references.

- Components in Query Auto Completion systems
  - Ranking suggestions: Most popular completion [3], Time sensitive suggestions [20,25], Location sensitive suggestions [1,26], Personalization
  - Ghosting [18]
  - Session co-occurrences [2]
  - Online spell correction [6]
  - Defect handling
  - Non-prefix matches [8]
  - Generating suggestions
  - Mobile QAC [29]
  - Enterprise QAC [9]
- Suggestion Ranking
  - Traditional Machine Learning methods for ranking suggestions [5,10,22]
  - Convolutional Latent Semantic Model [14]
  - LSTM encoder [24]
  - BERT and BART [15]

- Personalization
  - Traditional Machine Learning methods [19]
  - Hierarchical RNN Encoder-decoder [21] with pointer generator [4]
  - GRUs with user and time representations [7]
  - Transformer-based hierarchical encoder [28]
- Handling defective suggestions and prefixes
  - LSTMs for inappropriate query suggestion detection [27]
  - Online Spell Correction: A* search with spell correction pairs [6]; using character RNNs [23]
  - Offline Spell Correction [30]
- Natural Language Generation
  - RNNs with character and word embeddings [17]
  - LSTMs with subword embeddings [12]
  - Hierarchical RNN Encoder-decoder [11,22]
  - Next Phrase Prediction with T5 [13]
  - Problems with NLG [16]
- Summary and Future Trends

5   Target Audience and Prerequisites

Practitioners and people from the search industry will clearly benefit from the discussions both from the methods perspective, as well from the real challenges and experiences that the team had during deployment of such solutions. This tutorial will give them a systematic overview of recent work on QAC using deep learning. A lot of recent research happens based on close collaboration between industry and academia. Thus, this tutorial could be very useful for Masters/PhD students looking for interesting problems in the search industry.

This tutorial can be considered an intermediate level tutorial where we assume the folks in audience to know some basic deep learning architectures. Prerequisite knowledge includes introductory level knowledge in deep learning, specifically recurrent neural networks models, and transformers. Also, basic understanding of natural language processing and machine learning concepts is expected.

6   Tutorial History

We offered a similar 1.5 h tutorial in-person at IJCAI 2022. It was well received by 40+ participants. https://aka.ms/dl4qac

We are not aware of any other recent tutorial at ECIR or related conferences (including SIGIR, WSDM, WWW, KDD, ACL, RecSys, ICML, etc.) on this important topic.

7   Type of Support Materials to Be Supplied to Attendees

The tutorial is lecture-style and will be delivered using PowerPoint slides. The slides will be made available online to the attendees before the tutorial.

8   Contact Information

Contact details of the presenters are as follows.

Manish Gupta (Main contact person):

   Homepage: http://research.microsoft.com/en-us/people/gmanish/

   Email: gmanish@microsoft.com

   Address: Microsoft Campus, Gachibowli, Hyderabad, India.

Meghana Joshi:

   LinkedIn: https://www.linkedin.com/in/meghanajoshi154/

   Email: mejoshi@microsoft.com

   Address: Microsoft Campus, 858 Beatty St, Vancouver, BC V6B 1C1, Canada.

Puneet Agrawal:

   LinkedIn: https://www.linkedin.com/in/puneet-agrawal-2291808/

   Email: punagr@microsoft.com

   Address: Microsoft Campus, Gachibowli, Hyderabad, India.

9   Presenter Biographies

Manish, Meghana and Puneet lead the Bing Autosuggest team which is responsible for powering query auto completion across Bing and other Microsoft endpoints. The presenters have developed a unique perspective having worked on live product while innovating with deep learning methods in this space, and the audience will benefit from this perspective.

9.1   Manish Gupta

Manish Gupta (Homepage Link) is a Principal Applied Researcher at Microsoft AI and Research at Hyderabad, India. He is also an Adjunct Faculty at a premier engineering school in India - International Institute of Information Technology, Hyderabad (IIIT-H) and a visiting faculty at the Indian School of Business (ISB). He received his Masters in Computer Science from IIT Bombay in 2007 and his Ph.D. from the University of Illinois at Urbana-Champaign in 2013. With research interests in the areas of deep learning, web mining and information retrieval, he has published more than 100 research papers in referred journals and conferences, including WWW, SIGIR, ECIR, ICDE, KDD, WSDM conferences. He has also co-authored two books: one on Outlier Detection for Temporal Data and another one on Information Retrieval with Verbose Queries. Currently, along with Puneet, he works in the Bing Autosuggest team which is responsible for powering query auto completion across Bing and other Microsoft endpoints.

Manish has a strong track record of teaching. He taught a full credit course on *Web Mining* at IIIT-H, India in 2013 and in 2014. He currently teaches a course on Information Retrieval and Extraction (with Prof. Vasudeva Varma) at IIIT-H and courses on Text Analytics and Deep Learning at ISB, Hyderabad. He has an extensive experience in offering tutorials at top conferences. Following is a list of selected tutorials he has offered.

- **Manish Gupta**, Puneet Agrawal. Deep Learning Methods for Query Auto Completion. *IJCAI 2022*.
- Subba Reddy Oota, Jashn Arora, **Manish Gupta**, Raju S. Bapi, Mariya Toneva. Deep Learning for Brain Encoding and Decoding. *CogSci 2022*.
- Sarah Masud, Pinkesh Badjatiya, Amitava Das, **Manish Gupta**, Preslav Nakov, Tanmoy Chakraborty. Combating Online Hate Speech: Roles of Content, Networks, Psychology, User Behavior, etc. *WSDM 2022*.
- Sarah Masud, Pinkesh Badjatiya, Amitava Das, **Manish Gupta**, Vasudeva Varma, Tanmoy Chakraborty. Combating Online Hate Speech: Roles of Content, Networks, Psychology, User Behavior and Others. *ECML-PKDD 2021*.
- **Manish Gupta**, Vasudeva Varma, Sonam Damani, Kedhar Nath Narahari. Compression of Deep Learning Models for NLP. *CIKM 2020*.
- **Manish Gupta**, Vasudeva Varma, Sonam Damani, Kedhar Nath Narahari. Compression of Deep Learning Models for NLP. *IJCAI-PRICAI 2020*.
- **Manish Gupta**, Michael Bendersky. Information Retrieval with Verbose Queries. *SIGIR 2015*.
- **Manish Gupta**, Rui Li, Kevin C. Chang. Towards a Social Media Analytics Platform: Event and Location Detection for Microblogs. *WWW 2014*.
- **Manish Gupta**, Jing Gao, Charu Aggarwal, Jiawei Han. Outlier Detection for Temporal Data. *CIKM 2013*.
- **Manish Gupta**, Jing Gao, Charu Aggarwal, Jiawei Han. Outlier Detection for Graph Data. *ASONAM 2013*.
- **Manish Gupta**, Jing Gao, Charu Aggarwal, Jiawei Han. Outlier Detection for Temporal Data. *SDM 2013*.

9.2 Meghana Joshi

Meghana Joshi (LinkedIn Link) is a seasoned Applied Scientist at Microsoft at Vancouver, Canada, who has built scalable AI services that have served billions of requests across 235 Bing regions. In her current role, she is Principal Architect of Bing Autosuggest and deals with interesting cross section of compute heavy modern AI models and low latency, high throughput requirements of Autosuggest. Her experience has endowed her with unique perspective on practical challenges faced by large scale AI systems in real word. She is not only passionate about delivering deep learning models to users but also optimizing them to allow these models to be deployed at web-scale. Earlier, she has been a founding member of the core team for Ruuh, Microsoft's AI-powered chatbot, and has worked on several projects including novel art generation using AI, Bing Search Relevance, among others. Meghana has actively shared her learnings at various forums including WWW, ACL and at several top engineering schools in India in past.

9.3 Puneet Agrawal

Puneet Agrawal (LinkedIn Link) is a Principal Software Engineering Manager at the Bing team in Microsoft Artificial Intelligence and Research, India R&D Private Limited at Hyderabad, India. He heads the Bing Autosuggest team which is responsible for powering query auto completion across Bing and other Microsoft endpoints. He received his BTech in Computer Science from Indian Institute of Technology (IIT) Delhi in 2008. He interned at Yahoo! in 2007, and has been with Microsoft for the past 14 years. His research interests include web search, natural language processing, deep learning, and conversational agents. During his stint at Microsoft, he has actively led and developed several AI powered products and features that have reached millions of users. He is especially passionate about creating products with a human-like personality and was the co-creator of Cortana's personality. He is the founder of Microsoft's Ruuh.ai, an AI-powered chatbot. He also co-organized an ACL task: "SemEval-2019 Task 3: EmoContext – Contextual Emotion Detection in Text" which received 311 submissions, and co-organized a successful series of workshops on "Humanizing AI" at IJCAI 2018 and IJCAI 2019.

References

1. Backstrom, L., Kleinberg, J., Kumar, R., Novak, J.: Spatial variation in search engine queries. In: Proceedings of the 17th International Conference on World Wide Web, pp. 357–366 (2008)
2. Bar-Yossef, Z., Kraus, N.: Context-sensitive query auto-completion. In: Proceedings of the 20th International Conference on World Wide Web, pp. 107–116 (2011)
3. Citeseer: Exploring Real-Time Temporal Query Auto-Completion (2013)
4. Dehghani, M., Rothe, S., Alfonseca, E., Fleury, P.: Learning to attend, copy, and generate for session-based query suggestion. In: Proceedings of the 2017 ACM on Conference on Information and Knowledge Management, pp. 1747–1756 (2017)
5. Di Santo, G., McCreadie, R., Macdonald, C., Ounis, I.: Comparing approaches for query autocompletion. In: Proceedings of the 38th International ACM SIGIR Conference on Research and Development in Information Retrieval, pp. 775–778 (2015)
6. Duan, H., Hsu, B.J.: Online spelling correction for query completion. In: Proceedings of the 20th International Conference on World Wide Web, pp. 117–126 (2011)
7. Fiorini, N., Lu, Z.: Personalized neural language models for real-world query auto completion. In: Proceedings of NAACL-HLT, pp. 208–215 (2018)
8. Gog, S., Pibiri, G.E., Venturini, R.: Efficient and effective query auto-completion. In: Proceedings of the 43rd International ACM SIGIR Conference on Research and Development in Information Retrieval, pp. 2271–2280 (2020)
9. Hawking, D., Griffiths, K.: An enterprise search paradigm based on extended query auto-completion: do we still need search and navigation? In: Proceedings of the 18th Australasian Document Computing Symposium, pp. 18–25 (2013)
10. Jiang, J.Y., Ke, Y.Y., Chien, P.Y., Cheng, P.J.: Learning user reformulation behavior for query auto-completion. In: Proceedings of the 37th International ACM SIGIR Conference on Research & Development in Information Retrieval, pp. 445–454 (2014)
11. Jiang, J.Y., Wang, W.: Rin: reformulation inference network for context-aware query suggestion. In: Proceedings of the 27th ACM International Conference on Information and Knowledge Management, pp. 197–206 (2018)

12. Kim, G.: Subword language model for query auto-completion. In: Proceedings of the 2019 Conference on Empirical Methods in Natural Language Processing and the 9th International Joint Conference on Natural Language Processing (EMNLP-IJCNLP), pp. 5022–5032 (2019)
13. Lee, D.H., Hu, Z., Lee, R.K.W.: Improving text auto-completion with next phrase prediction. In: Findings of the Association for Computational Linguistics: EMNLP 2021, pp. 4434–4438 (2021)
14. Mitra, B., Craswell, N.: Query auto-completion for rare prefixes. In: Proceedings of the 24th ACM International on Conference on Information and Knowledge Management, pp. 1755–1758 (2015)
15. Mustar, A., Lamprier, S., Piwowarski, B.: Using bert and bart for query suggestion. In: Joint Conference of the Information Retrieval Communities in Europe, vol. 2621. CEUR-WS. org (2020)
16. Olteanu, A., Diaz, F., Kazai, G.: When are search completion suggestions problematic? Proc. ACM Hum.-Comput. Inter. 4(CSCW2), 1–25 (2020)
17. Park, D.H., Chiba, R.: A neural language model for query auto-completion. In: Proceedings of the 40th International ACM SIGIR Conference on Research and Development in Information Retrieval, pp. 1189–1192 (2017)
18. Ramachandran, L., Murthy, U.: Ghosting: contextualized query auto-completion on amazon search. In: Proceedings of the 42nd International ACM SIGIR Conference on Research and Development in Information Retrieval, pp. 1377–1378 (2019)
19. Shokouhi, M.: Learning to personalize query auto-completion. In: Proceedings of the 36th International ACM SIGIR Conference on Research and Development in Information Retrieval, pp. 103–112 (2013)
20. Shokouhi, M., Radinsky, K.: Time-sensitive query auto-completion. In: Proceedings of the 35th International ACM SIGIR Conference on Research and Development in Information Retrieval, pp. 601–610 (2012)
21. Song, J., Xiao, J., Wu, F., Wu, H., Zhang, T., Zhang, Z.M., Zhu, W.: Hierarchical contextual attention recurrent neural network for map query suggestion. IEEE Trans. Knowl. Data Eng. 29(9), 1888–1901 (2017)
22. Sordoni, A., Bengio, Y., Vahabi, H., Lioma, C., Grue Simonsen, J., Nie, J.Y.: A hierarchical recurrent encoder-decoder for generative context-aware query suggestion. In: Proceedings of the 24th ACM International on Conference on Information and Knowledge Management, pp. 553–562 (2015)
23. Wang, P.W., Zhang, H., Mohan, V., Dhillon, I.S., Kolter, J.Z.: Realtime query completion via deep language models. In: eCOM@ SIGIR (2018)
24. Wang, S., Guo, W., Gao, H., Long, B.: Efficient neural query auto completion. In: Proceedings of the 29th ACM International Conference on Information & Knowledge Management, pp. 2797–2804 (2020)
25. Wang, Y., Ouyang, H., Deng, H., Chang, Y.: Learning online trends for interactive query auto-completion. IEEE Trans. Knowl. Data Eng. 29(11), 2442–2454 (2017)
26. Welch, M.J., Cho, J.: Automatically identifying localizable queries. In: Proceedings of the 31st Annual International ACM SIGIR Conference on Research and Development in Information Retrieval, pp. 507–514 (2008)
27. Yenala, H., Chinnakotla, M., Goyal, J.: Convolutional bi-directional LSTM for detecting inappropriate query suggestions in web search. In: Kim, J., Shim, K., Cao, L., Lee, J.-G., Lin, X., Moon, Y.-S. (eds.) PAKDD 2017. LNCS (LNAI), vol. 10234, pp. 3–16. Springer, Cham (2017). https://doi.org/10.1007/978-3-319-57454-7_1

28. Yin, D., Tan, J., Zhang, Z., Deng, H., Huang, S., Chen, J.: Learning to generate personalized query auto-completions via a multi-view multi-task attentive approach. In: Proceedings of the 26th ACM SIGKDD International Conference on Knowledge Discovery & Data Mining, pp. 2998–3007 (2020)
29. Zhang, A., et al.: Towards mobile query auto-completion: an efficient mobile application-aware approach. In: Proceedings of the 25th International Conference on World Wide Web, pp. 579–590 (2016)
30. Zhang, S., Huang, H., Liu, J., Li, H.: Spelling error correction with soft-masked bert. In: Proceedings of the 58th Annual Meeting of the Association for Computational Linguistics, pp. 882–890 (2020)

Trends and Overview: The Potential of Conversational Agents in Digital Health

Tulika Saha[1]([✉]), Abhishek Tiwari[2], and Sriparna Saha[2]

[1] University of Liverpool, Liverpool, UK
sahatulika15@gmail.com
[2] Indian Institute of Technology Patna, Daulatpur, India
abhisek_1921cs16@iitp.ac.in

1 Motivation

With the COVID-19 pandemic serving as a trigger, 2020 saw an unparalleled global expansion of tele-health [23]. Tele-health successfully lowers the need for in-person consultations and, thus, the danger of contracting a virus. While the COVID-19 pandemic sped up the adoption of virtual healthcare delivery in numerous nations, it also accelerated the creation of a wide range of other different technology-enabled systems and procedures for providing virtual healthcare to patients. Rightly so, the COVID-19 has brought many difficulties for patients[1] who need continuing care and monitoring for mental health issues and/or other chronic diseases.

One important technological advancement is the increasing use of Conversational Agents (CAs) or Virtual Assistants (VAs) in people's life, which now have numerous health applications. Chatbots or CAs communicate with users using a text-based or speech-based interface and consequently can make their services and applications available to a large segment of the population. Due in part with the on-set of COVID-19, CAs are being used more frequently in the healthcare industry as a promising tool to enhance delivery and quality. Numerous healthcare surveys conducted over the past years have revealed a worrying shortage of doctors [14] compared to the doctor-to-population ratio in physical health and even more severe for mental health. Thus, CAs in healthcare is becoming more and more popular, driven by the need to assist the doctors and utilize their time effectively. The usage of CAs by patients and medical personnel has also shown to be well accepted, with high ratings for perceived utility, convenience, and participation in overcoming service and logistical constraints. The recent advancement in messaging services amongst leading social media firms and the latest rally to develop automated systems has driven novel research in this area. Thus, it has become imperative, more than ever to focus on understanding and analysing growing trends of human-computer interfaces, i.e., VAs which will further pave way for developing robust computational models in healthcare including mental

---

[1] https://www.who.int/news/item/02-03-2022-covid-19-pandemic-triggers-25-increase-in-prevalence-of-anxiety-and-depression-worldwide.

J. Kamps et al. (Eds.): ECIR 2023, LNCS 13982, pp. 349–356, 2023.
https://doi.org/10.1007/978-3-031-28241-6_36

health. The motivation behind this tutorial is to analyze the growing trend of VAs in healthcare and provide the IR researchers with an overall perspective of where the AI and NLP communities are heading which can further pave way for ground-breaking novelties benefitting the research community and the society at large.

2   Objective

By attending this tutorial, participants will learn about: **(i)** the existing limitations in digital health in the realms of NLP specifically VAs and the recent upcoming frameworks which alleviate these limitations and enhance their capabilities; **(ii)** the basics of a Dialogue System aka Conversational AI and the recent most emerging branches of communication with the end user; **(iii)** the most successful techniques addressing multiple sub-modules of a VA for both physical and mental health; **(iv)** for practical exposure, the tutorial will also provide a live demonstration of the recently published *symptom investigation and disease diagnosis VA* and **(v)** some pressing challenges to the adoption of AI in healthcare, such as Reliability, Explainability, and Safety (RES) as well as discuss possible remedies to accelerate progress in this area.

3   Relevance to the IR Community

In the early 20s, research in digital health specifically VAs was unheard of, largely because of the communities disbelief of whether an automated system can be accurate, knowledgeable, intelligent and human enough to make decisions in the sensitive domain of health. With the advent of Deep Learning and massive expansion of e-commerce in the mid 20s through recommender systems, CAs etc. paved the way for novel ideas and digitisation of health goals.

Initially, the research in digital or tele-health were widely focused on social media analysis or support forums due to the lack of real-time data available owing to privacy concerns. Soon enough the NLP conferences were flooded with tele-health oriented papers related to social media. However, in the recent times there has been a massive shift towards real-time data processing and inference where researchers realized that the overall health of the society can be benefitted while only dealing with real-time data between patients and doctors. During this shift, the IR community remained consumers but not key innovators in healthcare VAs, largely falling in the track of AI for Social Good.

It is important for the IR community to resume greater focus on digital health, guided by the need of CAs in healthcare including mental health. In the context of IR, there are two main threads of work in an end-to-end CA: conversational Q&A and conversational recommendation. Currently, these are viewed as two distinct systems with unique objectives, architectures, and evaluation standards. Instead, the two should be seamlessly integrated into CAs to better help users, changing the perspective from one that is isolated to one that is more unified. In order to actively encourage good engagement, the multi-modality of interactions also needs to be more thoroughly acknowledged.

4   Format and Detailed Schedule

The tutorial will interleave slide-based presentation, scribbling on a whiteboard, screen-sharing short demos, and Q&A sessions (at least every 25 min). The tutorial organisers may set up Slack or Piazza for attendees to interact with them and with each other. The schedule is for 165 min of presentation and 15 min of Q&A suitable for introductory to advanced target audience.

4.1   Preliminaries (20 Min)

- **Dialogue System.** The tutorial will introduce the basics of a typical Dialogue System (DS). A DS is primarily known to comprise of three prime modules: (i) Natural Language Understanding; (ii) Dialogue Policy Learning also known as Dialogue Management (DM) Strategy and (iii) Natural Language Generation also known as Response Generation in dialogues. Lately, two branch of DSs have emerged focused on varying ways of communicating with the user: (i) Modularized DS and (ii) End-to-end Generation Framework. Each of these variation and its workflow will be inspected in details.
- **Need and Utility of VAs in Healthcare.** The range of CAs in healthcare has been discussed in a number of recent review studies. A significant majority of CAs in healthcare have been categorised as goal-oriented agents since they have been built to assist patients and healthcare workers in certain tasks. For CAs in health, a taxonomy has been created in recent times which categorizes six health-related purposes into training, education, aid, prevention, diagnostic, and assistance for the elderly. The tutorial will cover in details the different use cases concerning the above mentioned six purposes which best describes the role of VAs in healthcare.
- **Challenges in Healthcare VAs.** Although AI has been a tremendous success for healthcare in recent years, there are a few pressing challenges for healthcare VAs which needs to be addressed in the upcoming years. For e.g., adequate amount of data due to privacy concerns, risk associated with the failure of AI models, explainable decision support system etc. will be discussed in details.

4.2   Natural Language Understanding (45 Min)

Natural Language Understanding (NLU) module is responsible for making sense of the user input. It extracts various information from the user utterance (coarse to fine-grained) which is used further in the conversational framework. NLU module is typically framed as a classification task. Formally, consider a training set, $\{x_i, y_i\}_{i=1}^{N}$, where x_i is the i-th input sample, y_i is the label vector for the classification task, respectively and N is the number of training instances. $f(x, w)$ denotes the classifier and w is its parameters. The task is to find the optimal parameter, w^*, by minimizing the training loss, $1/N \sum_{i=1}^{N} L^{train}(w)$, where $L^{train}(w) = l(y_i, f(x_i, w))$. In healthcare VAs, different type of NLU modules have been developed in the recent times covering both physical and mental health ranging from multi-modality to multi-tasking frameworks.

- **Physical Health.** Lately, disease diagnosis VA is an upcoming research area. Understanding patients' concerns from their utterances are critical to diagnosis and treatment outcomes [4]. The recent works on medical dialogue understanding can broadly be grouped into two categories: (a) pre-trained transformer-based joint intent and entity model and (b) Multi-label entity classification [20]. The community is also slowly advancing towards multimodal signal processing in digital health. Primarily, when we consult with doctors, we often show our signs/symptoms through visuals. Thus, an symptom image identification is an integral part of medical disease diagnosis DS. Authors of [21] introduce a context aware image identification model, which incorporates conversation history for identifying an image adequately.
- **Mental Health.** Mental health dialogue understanding is somewhat an implicit task as opposed to disease diagnosis VAs [21] where users are expected to provide task information explicitly. Thus, these mental health VAs [16] also utilize semantic features such as sentiment and emotion for understanding users concern accurately and to serve more appropriately. Lately, researchers are focused on identifying mental health disorder [17], symptom investigation [12], gender prediction [11] etc. from conversations either collected from support forums [5] or from real counselling settings [1]. Multi-modality has also been explored in the recent times to identify mental health diseases such as depression [13].

4.3   Dialogue Policy Learning (40 Min)

Dialogue Policy Learning also known as DM strategy is responsible for deciding the flow of conversation in any DS. It takes as input the information extracted by the NLU module in its state space and outputs an appropriate action based on a policy to communicate with the user so as to maximize a reward (short or long term goals). DM is often viewed as a sequential decision problem and is formulated as a Markov Decision Process (MDP) which optimizes the dialogue policy through a Reinforcement Learning (RL) algorithm. Formally, MDP can be represented as a five-tuple (S, A, P, R, π), where S is the dialogue state, A denotes the set of possible actions for the VA, P signifies transition probability $(P(S_{t+1}|S_t))$ and R denotes reward model.

- **Physical Health.** For a large number of non-fatal diseases, doctors typically identify patients' diseases by conducting a symptom examination through conversations only. Inspired by such real-world scenarios, the researchers have formulated automatic disease diagnosis as a task-oriented dialogue framework [22]. Authors of [24] further proposed a knowledge routed relational dialogue system that incorporates an external rich medical knowledge graph into dialogue policy learning for knowledge grounded topic transition. To overcome the rule-based dependency of medical department identification, Liao et al. [8] proposed a novel policy learning framework for symptom investigation and disease diagnosis using Hierarchical Reinforcement Learning (HRL). The community is also advancing towards multi-modal VAs to provide end-users with a human-like experience. Motivated by the importance of visual form of symptom reporting, researchers proposed a HRL based multi-modal disease diagnosis VA [21].

- **Mental Health.** When talking to someone, our physical presence is usually appreciated, and it becomes more crucial if it is related to a serious concern like anxiety or depression. In [3], the authors have developed an animated virtual interviewer SimSensei Kiosk to create engaging face-to-face interactions with mental health support seekers.

4.4 Generation Frameworks (45 Min)

The generation framework typically skips the DM module and is focused on generating the next response of the VA aligned with the user context. Formally, given a user utterance, $X_t = (x_{t,1}, x_{t,2}, ..., x_{t,n})$, a conversational context/history, $C = (c_1, c_2, ..., c_{t-1})$, where $c_i = (X_i, Z_i)$, the task is to generate next textual response of the VA, $Z_t = (z_{t,1}, z_{t,2}, ..., z_{t,n''})$. In the literature, three popular approaches have been employed for generating VAs response: (a) template-based response modeling [24] (b) Seq2Seq [7] and (c) pre-trained generation frameworks [2]. *The tutorial will also demonstrate Seq2Seq and fine-tuning of GPT models for dialogue response generation task.*

- **Physical Health.** In real-time, doctors' investigation also depends on patients' personal information, such as age and gender, in addition to patients' reported major difficulties. Inspired by such scenarios, the authors have proposed a context-aware HRL-based dialogue system [6] for symptom investigation followed by disease prediction. Authors of [9] have proposed a graph-based dialogue generation framework that utilizes commonsense knowledge for identifying new diseases. In the diagnostic process, external medical knowledge aids clinicians in narrowing their investigation space and efficiently utilize the gathered information. Motivated by the observation, the authors [10] have proposed medical knowledge graph guided response generation model.
- **Mental Health.** Extensive research has been carried out in mental health addressing the aspects of empathy and motivation which have been identified as key affective factors providing positive outcome in support based conversations. In [19], authors presented a computational approach to understand empathy based on three communication mechanisms: *Emotional Reactions*, *Explorations* and *Interpretations*. Furthermore, in [18], authors developed an empathetic rewriting framework, named *PARTNER* that transforms lower empathetic responses into higher empathetic content. Authors of [16] proposed a VA acting as the first point of contact for mentally distressed support seekers afflicted with some form of mental illness. Authors of [15] went ahead and combined both these aspects of empathy and motivation in a unified end-to-end system for online mental health support.

4.5 Overall Summary and Scope for Future Work (15 Min)

CAs are an emerging technology for use in healthcare that has not yet undergone a thorough evaluation. Future studies should concentrate on evaluating the viability, acceptability, safety, and efficacy of various CA forms that are in line with the requirements and preferences of the target audience. Additionally, there is a need for more in-depth research on the function of CAs in the current health systems as well as clearer guidelines for the creation and evaluation of CAs connected to healthcare.

5   Supporting Material

The tutorial will provide the following material: (1) lecture video recording, (2) annotated slides, (3) assignments, exams and projects from related courses taught, (4) extended bibliography, and (5) compendium of public software and data sets.

6   Contact Information

- Dr. Tulika Saha (corresponding author) is a Lecturer of Computer Science at the University of Liverpool, United Kingdom (UK). Her current research interests include ML, DL, NLP typically Dialogue Systems, AI for Social Good, Social Media Analysis etc. She was a postdoctoral research fellow at the National Centre for Text Mining, University of Manchester, UK. Previously she earned her Ph.D. from Indian Institute of Technology Patna, India. Her research articles are published in top-tier conferences such as ACL, ACM SIGIR etc. and peer-reviewed journals.
- Abhisek Tiwari is a research scholar (Prime Minister Research Fellow) in Computer science and Engineering, Indian Institute of Technology, Patna. His research interest includes AI for Social Good, NLP, typically Conversational AI, and RL. He is also serving as a guest lecturer at NSIT Bihta, India. His research works have been published in reputable conferences, such as CIKM, IJCNLP, and peer-reviewed journals. Abhisek has delivered several tutorials including the GIAN Course on DL Techniques for Conversational AI, conducted birds-of-a-feather sessions at top-tier conferences such as ACL, ICLR, and NeurIPS.
- Dr. Sriparna Saha is currently serving as an Associate Professor (h5-index:33, total citations: 6201 as per Google Scholar), Head of Department in Computer Science and Engineering, Indian Institute of Technology Patna, India (https://www.iitp.ac.in/~sriparna/). Her current research interests include ML, DL, NLP, AI for Social Good, Information Retrieval. She has published more than 400 papers in reputed journals and conferences including ACL, SIGIR, AAAI, EMNLP, ECIR, COLING, ACM MM etc. Her tutorial on "Summarization Systems: From Text to Multimodal" is accepted to be delivered in ICONIP 2022 to be held in New Delhi, India. She is one of the special session organizers of ICONIP 2021 on the topic of "Smart Home Technologies & Services for the Wellbeing and Sustainability of Society". She was one of the special session organizers of IEEE SSCI 2021 on the topic of "Computational Intelligence for Natural Language Processing". She has delivered a

tutorial session on "Multi-modality Helps in Solving Biomedical Problems: Theory and Applications" in IEEE WCCI 2020.

References

1. Althoff, T., Clark, K., Leskovec, J.: Large-scale analysis of counseling conversations: an application of natural language processing to mental health. Trans. Assoc. Comput. Linguist. **4**, 463–476 (2016)
2. Chen, X., et al.: Bridging the gap between prior and posterior knowledge selection for knowledge-grounded dialogue generation. In: Proceedings of the 2020 Conference on Empirical Methods in Natural Language Processing (EMNLP), pp. 3426–3437 (2020)
3. DeVault, D., et al.: Simsensei kiosk: a virtual human interviewer for healthcare decision support. In: Proceedings of the 2014 International Conference on Autonomous Agents and Multi-agent Systems, pp. 1061–1068 (2014)
4. Enarvi, S., et al.: Generating medical reports from patient-doctor conversations using sequence-to-sequence models. In: Proceedings of the First Workshop on Natural Language Processing for Medical Conversations, pp. 22–30 (2020)
5. Ji, S., Li, X., Huang, Z., Cambria, E.: Suicidal ideation and mental disorder detection with attentive relation networks. Neural Comput. Appl. **34**(13), 10309–10319 (2022)
6. Kao, H.C., Tang, K.F., Chang, E.: Context-aware symptom checking for disease diagnosis using hierarchical reinforcement learning. In: Proceedings of the AAAI Conference on Artificial Intelligence, vol. 32 (2018)
7. Li, D., Ren, Z., Ren, P., Chen, Z., Fan, M., Ma, J., de Rijke, M.: Semi-supervised variational reasoning for medical dialogue generation. In: Proceedings of the 44th International ACM SIGIR Conference on Research and Development in Information Retrieval, pp. 544–554 (2021)
8. Liao, K., et al.: Task-oriented dialogue system for automatic disease diagnosis via hierarchical reinforcement learning. arXiv preprint arXiv:2004.14254 (2020)
9. Lin, S., et al.: Graph-evolving meta-learning for low-resource medical dialogue generation. In: Proceedings of the 35th AAAI Conference on Artificial Intelligence, pp. 13362–13370. AAAI Press (2021)
10. Liu, W., Tang, J., Liang, X., Cai, Q.: Heterogeneous graph reasoning for knowledge-grounded medical dialogue system. Neurocomputing **442**, 260–268 (2021)
11. Lokala, U., et al.: A computational approach to understand mental health from reddit: knowledge-aware multitask learning framework. In: Proceedings of the International AAAI Conference on Web and Social Media, vol. 16, pp. 640–650 (2022)
12. Patra, B.G., Kar, R., Roberts, K., Wu, H.: Mental health severity detection from psychological forum data using domain-specific unlabelled data. AMIA Summits Transl. Sci. Proc. **2020**, 487 (2020)
13. Qureshi, S.A., Saha, S., Hasanuzzaman, M., Dias, G.: Multitask representation learning for multimodal estimation of depression level. IEEE Intell. Syst. **34**(5), 45–52 (2019)
14. Rasmussen, K., et al.: Offline elearning for undergraduates in health professions: a systematic review of the impact on knowledge, skills, attitudes and satisfaction. J. Global Health **4**(1) (2014)

15. Saha, T., Gakhreja, V., Das, A.S., Chakraborty, S., Saha, S.: Towards motivational and empathetic response generation in online mental health support. In: Proceedings of the 45th International ACM SIGIR Conference on Research and Development in Information Retrieval, pp. 2650–2656 (2022)

16. Saha, T., Reddy, S., Das, A., Saha, S., Bhattacharyya, P.: A shoulder to cry on: towards a motivational virtual assistant for assuaging mental agony. In: Proceedings of the 2022 Conference of the North American Chapter of the Association for Computational Linguistics: Human Language Technologies, pp. 2436–2449 (2022)

17. Saha, T., Reddy, S.M., Saha, S., Bhattacharyya, P.: Mental health disorder identification from motivational conversations. IEEE Trans. Comput. Soc. Syst. (2022)

18. Sharma, A., Lin, I.W., Miner, A.S., Atkins, D.C., Althoff, T.: Towards facilitating empathic conversations in online mental health support: a reinforcement learning approach. In: Leskovec, J., Grobelnik, M., Najork, M., Tang, J., Zia, L. (eds.) WWW '21: The Web Conference 2021, Virtual Event/Ljubljana, Slovenia, April 19-23, 2021, pp. 194–205. ACM/IW3C2 (2021). https://doi.org/10.1145/3442381.3450097, https://doi.org/10.1145/3442381.3450097

19. Sharma, A., Miner, A.S., Atkins, D.C., Althoff, T.: A computational approach to understanding empathy expressed in text-based mental health support. In: Webber, B., Cohn, T., He, Y., Liu, Y. (eds.) Proceedings of the 2020 Conference on Empirical Methods in Natural Language Processing, EMNLP 2020, Online, 16–20 November, 2020, pp. 5263–5276. Association for Computational Linguistics (2020). https://doi.org/10.18653/v1/2020.emnlp-main.425, https://doi.org/10.18653/v1/2020.emnlp-main.425

20. Shi, X., Hu, H., Che, W., Sun, Z., Liu, T., Huang, J.: Understanding medical conversations with scattered keyword attention and weak supervision from responses. In: Proceedings of the AAAI Conference on Artificial Intelligence, vol. 34, pp. 8838–8845 (2020)

21. Tiwari, A., Manthena, M., Saha, S., Bhattacharyya, P., Dhar, M., Tiwari, S.: Dr. can see: towards a multi-modal disease diagnosis virtual assistant. In: Proceedings of the 31st ACM International Conference on Information & Knowledge Management, pp. 1935–1944 (2022)

22. Wei, Z., et al.: Task-oriented dialogue system for automatic diagnosis. In: Proceedings of the 56th Annual Meeting of the Association for Computational Linguistics (Volume 2: Short Papers), pp. 201–207 (2018)

23. Wosik, J., et al.: Telehealth transformation: Covid-19 and the rise of virtual care. J. Am. Med. Inform. Assoc. **27**(6), 957–962 (2020)

24. Xu, L., Zhou, Q., Gong, K., Liang, X., Tang, J., Lin, L.: End-to-end knowledge-routed relational dialogue system for automatic diagnosis. In: Proceedings of the AAAI Conference on Artificial Intelligence, vol. 33, pp. 7346–7353 (2019)

Crowdsourcing for Information Retrieval

Dmitry Ustalov[1]([⊠])(ID), Alisa Smirnova[2](ID), Natalia Fedorova[1](ID),
and Nikita Pavlichenko[1](ID)

[1] Toloka, Belgrade, Serbia
{dustalov,natfedorova,pavlichenko}@toloka.ai
[2] Toloka, Lucerne, Switzerland
zero@toloka.ai
https://toloka.ai/research

Abstract. In our tutorial, we will share more than six years of our crowdsourced data labeling experience and bridge the gap between crowdsourcing and information retrieval communities by showing how one can incorporate human-in-the-loop into their retrieval system to gather the real human feedback on the model predictions. Most of the tutorial time is devoted to a hands-on practice, when the attendees will, under our guidance, implement an end-to-end process for information retrieval from problem statement and data labeling to machine learning model training and evaluation.

Keywords: Data labeling · Information retrieval ·
Human-in-the-loop · Crowdsourcing · Intent classification

1 Motivations, Learning Objectives, and Scope of the Tutorial

Although the impressive progress in deep learning and especially autoregressive models and diffusion models, we observe that predictions and generations made by these models still do not correlate with actual human preferences. This impacts downstream user-facing applications, like information retrieval, recommender systems, conversational agents, etc. In our tutorial, we will share more than six years of our crowdsourced data labeling experience and bridge the gap between crowdsourcing and information retrieval communities by showing how one can incorporate data labeling into their retrieval system to gather the real human feedback on the model predictions. Our learning objectives are:

- to teach how to apply human-in-the-loop pipelines with crowdsourcing to address information retrieval problems
- to improve the attendees data labeling and machine learning skills on a real problem related to information retrieval, namely, a query intent classification task, by annotating data and training a machine learning model
- to introduce the mathematical methods and their open-source implementations to increase the annotation quality and the accuracy of the learned model without additional labeling

J. Kamps et al. (Eds.): ECIR 2023, LNCS 13982, pp. 357–361, 2023.
https://doi.org/10.1007/978-3-031-28241-6_37

We will provide useful theoretical aspects of data labeling and learning from crowdsourced data and offer the necessary icebreakers, so even a beginner will be able to succeed. Most of the tutorial time is devoted to a hands-on practice, when the attendees will, under our guidance, implement an end-to-end process for information retrieval from problem statement and data labeling to machine learning model training and evaluation. We will use query intent classification task for our practice, but its scope can be simply generalized to other classification problems faced in information retrieval.

2   Tutorial Format, Length, and a Detailed Outline

We would like to organize our half-day (three hours) tutorial according to the following schedule:

- Introduction (15 min)
- Crowdsourcing Essentials (45 min)
- Hands-On Practice Session (60 min)
- Learning from Crowds (45 min)
- Conclusion (15 min)

During the Introduction, Crowdsourcing Essentials, and Conclusion parts we will teach the crowdsourcing essentials and answer the attendees' questions. However, most of the tutorial time (Hands-On Practice Session and Learning from Crowds) will be focused on hands-on learning of the corresponding methodology using the real systems, code, and data that we will provide in advance.

2.1   Introduction

We will start with an introduction that includes crowdsourcing and human-in-the-loop terminology and examples of tasks on crowdsourcing marketplaces. We will also demonstrate why crowdsourcing is becoming more popular in working with data on a large scale, show successful human-in-the-loop applications for information retrieval and e-commerce, and describe current industry trends of crowdsourcing use.

2.2   Crowdsourcing Essentials

In this part, we will discuss quality control techniques. We will talk about the approaches that are applicable before task performance (selection of annotators, training of annotators, and exam tasks), the ones applicable during task performance (golden tasks, motivation of annotators, tricks to remove bots and cheaters), and approaches applicable after task performance (post verification/acceptance, inter-annotator agreement). We will share best practices, including critical aspects and pitfalls when designing instructions and interfaces for annotators, vital settings in different types of templates, training and examination for annotators selection, pipelines for evaluating the labeling process.

2.3   Hands-on Practice Session

We will conduct *a hands-on practice session*, which is the vital and the longest part of our tutorial. We will encourage the attendees to apply the techniques and best practices learned during the first part of the tutorial. For this purpose, we will let the attendees run their own crowdsourcing project for intent classification for the conversational agents on real crowd annotators. As the *input* for the crowdsourcing project the attendees will have search queries from the Web. The *output* of the project will be intent classes for each search query. We will use the CLINC150[1] dataset for the practice.

The attendees will start with brainstorming the suitable data labeling pipeline for the given task. Next, they will set up and launch the query intent classification project online on the real crowd. Since creating a project from scratch might be time-consuming, we propose that our attendees choose from the most popular pre-defined templates (e.g., text or picture input). We will also provide the attendees with pre-allocated accounts and datasets for annotation. By the end of the practice session, the attendees will learn to construct an efficient pipeline for data collection and labeling, become familiar with crowdsourcing marketplaces, and launch projects on their own. As a result of this part, the attendees will obtain a labeled intent classification dataset that they will use in the next part of the tutorial.

2.4   Learning from Crowds

In this part, we will describe how to process the raw labels obtained from the crowdsourcing marketplace and transform them into knowledge suitable for a downstream human-in-the-loop application. Since in crowdsourcing every object has multiple labels provided by different annotators, we will consider a few popular answer aggregation models in crowdsourcing, including methods for aggregating categorical responses (Dawid-Skene, GLAD, etc.) and recent methods for deep learning from crowds (CrowdLayer, CoNAL, etc.).

We will present Crowd-Kit, an open-source Python library implementing all these methods. We will put a special attention to the Crowd-Kit Learning module for training a complete deep learning model from crowdsourced data without an extra aggregation step.[2]

This part of our tutorial will be done on the data collected by the attendees in the hands-on practice session described in the previous section. The attendees will use a Python notebook in an environment like Google Colab (or its alternative) for training the model, inspecting and evaluating its performance, and preparing it for user-facing deployment. We will prepare the corresponding documented and working notebook for the attendees. They will be able to re-use the code and data from this tutorial in their own practice. As a result of this part the attendees will have a machine learning model for intent classification, trained on the data they annotated during our tutorial.

---

[1]   https://github.com/clinc/oos-eval.
[2]   https://github.com/Toloka/crowd-kit.

2.5   Conclusion

We will conclude the tutorial with analysis of the results obtained from the launched projects. By doing so, we will teach the participants to verify the collected labels. Together with the attendees, we will analyze outcome label distribution, check annotators' quality and contribution, elaborate on budget control, detect possible anomalies and problems. Also, we will show the useful datasets, software, and references for further studies and experiments. We will then share practical advice, discuss pitfalls and possible solutions, ask the attendees for feedback on the learning progress, and answer the final questions.

3   Target Audience and Prerequisite Knowledge Required

Our tutorial doesn't require any prerequisites beyond minimal label collection knowledge (no knowledge of crowdsourcing or deep learning is needed). We ensure this by giving an introduction lecture that covers the main components of data labeling processes.

We aim to attract researchers and practitioners that develop a web service or a software product that is based on data and/or machine learning. To this end, we plan to share rich experiences of shipping large-scale data collection pipelines while highlighting the best practices and pitfalls. Each attendee will learn how to construct a label collection pipeline and obtain labels with high quality under a limited budget while avoiding common pitfalls.

Since the practical part will constitute a half of the tutorial timeline, each attendee — even a beginner — will be able to practice their skills in collecting data labels via a crowdsourcing marketplace and training a machine learning model on these labels.

Our tutorial pays a special attention to modern and advanced techniques on computational quality and pricing control, which allow advanced specialists to learn how to improve data labeling processes and, thus, to conduct them efficiently. The attendees will be able to re-use the code and data from this tutorial in their own practice.

In addition, during our tutorial, we will point out open research questions and current challenges useful for research scientists.

References

1. Chu, Z., Ma, J., Wang, H.: Learning from crowds by modeling common confusions. In: Proceedings of the AAAI Conference on Artificial Intelligence, vol. 35(7), 5832–5840 (2021). https://doi.org/10.1609/aaai.v35i7.16730
2. Daniel, F., Kucherbaev, P., Cappiello, C., Benatallah, B., Allahbakhsh, M.: Quality control in crowdsourcing: a survey of quality attributes, assessment techniques, and assurance actions. ACM Comput. Surv. **51**(1), 7:1–7:40 (2018)
3. Dawid, A.P., Skene, A.M.: Maximum likelihood estimation of observer error-rates using the EM algorithm. J. Roy. Stat. Soc. **28**(1), 20–28 (1979). https://doi.org/10.2307/2346806

4. Rodrigues, F., Pereira, F.: Deep Learning from Crowds. In: Proceedings of the AAAI Conference on Artificial Intelligence, vol. 32(1) (2018). https://doi.org/10.1609/aaai.v32i1.11506
5. Ustalov, D., Pavlichenko, N., Losev, V., Giliazev, I., Tulin, E.: A general-purpose crowdsourcing computational quality control toolkit for python. In: The Ninth AAAI Conference on Human Computation and Crowdsourcing: Works-in-Progress and Demonstration Track, HCOMP 2021 (2021). https://arxiv.org/abs/2109.08584
6. Zheng, Y., Li, G., Li, Y., Shan, C., Cheng, R.: Truth inference in crowdsourcing: is the problem solved? Proceedings VLDB Endow. **10**(5), 541–552 (2017)

Uncertainty Quantification for Text Classification

Dell Zhang[1](✉)(iD), Murat Sensoy[2], Masoud Makrehchi[3],
and Bilyana Taneva-Popova[4]

[1] Thomson Reuters Labs, London, UK
`dell.z@ieee.org`
[2] Amazon Alexa AI, London, UK
`drmuratsensoy@gmail.com`
[3] Thomson Reuters Labs, Toronto, Canada
`masoud.makrehchi@thomsonreuters.com`
[4] Thomson Reuters Labs, Zug, Switzerland
`bilyana.taneva-popova@thomsonreuters.com`

Abstract. This half-day tutorial introduces modern techniques for
practical uncertainty quantification specifically in the context of multi-
class and multi-label text classification. First, we explain the usefulness of
estimating aleatoric uncertainty and epistemic uncertainty for text classification models. Then, we describe several state-of-the-art approaches
to uncertainty quantification and analyze their scalability to big text
data: Virtual Ensemble in GBDT, Bayesian Deep Learning (including
Deep Ensemble, Monte-Carlo Dropout, Bayes by Backprop, and their
generalization Epistemic Neural Networks), as well as Evidential Deep
Learning (including Prior Networks and Posterior Networks). Next, we
discuss typical application scenarios of uncertainty quantification in text
classification (including in-domain calibration, cross-domain robustness,
and novel class detection). Finally, we list popular performance metrics for the evaluation of uncertainty quantification effectiveness in text
classification. Practical hands-on examples/exercises are provided to the
attendees for them to experiment with different uncertainty quantification methods on a few real-world text classification datasets such as
CLINC150.

Keywords: Uncertainty quantification · Text classification

1 Motivations and Objectives

Machine Learning (ML) models can only be *trustworthy* if they are able to
express the *uncertainty* about their predictions honestly, i.e., "they should know
what they don't know". Commonly used ML models (including deep neural
networks) tend to be overconfident in their predictions [8], especially when there
are *data distribution shifts* [26,31]. For example, novel classes never seen during
the training of the ML model may appear at inference time, and if they are not

J. Kamps et al. (Eds.): ECIR 2023, LNCS 13982, pp. 362–369, 2023.
https://doi.org/10.1007/978-3-031-28241-6_38

detected by the ML model they may lead to catastrophic mistakes in critical applications like autonomous driving, medical diagnosis, and automated legal document processing.

Although the problem of uncertainty quantification has received great attention from ML researchers and many new techniques for uncertainty quantification have emerged in recent years, existing tutorials/overviews/surveys [1,10,13, 14,19,20,25,29,32] on this topic are dominated by the use cases in regression or multi-class *image classification*. By contrast, we would like to present a tutorial about uncertainty quantification specifically in the context of *text classification* [2,3,33,38,41]. Both *multi-class* (single-label) classification and *multi-label* classification would be covered: the latter occurs much more frequently for the classification of text documents than images.

We hope that this tutorial will provide a convenient entry point for IR researchers and practitioners to get a grip on modern uncertainty quantification techniques (beyond traditional *probability calibration*) for their various text classification tasks.

2   Scope and Relevance

As mentioned above, in this tutorial about uncertainty quantification, we focus on the use cases in text classification (aka text categorization) which has always been an important topic for Information Retrieval (IR) conferences including ECIR.

The text classifications models used to illustrate uncertainty quantification techniques here could be either the classic *gradient boosted decision trees* (GBDT) or the latest pre-trained *large language models* like BERT. Between these two, the former is simple and fast to run which makes it ideal for online interactive demonstration during the tutorial, while the latter is more computationally-intensive and time-consuming which can be run offline afterwards.

In many situations [17,24,36], it would be useful to distinguish the reducible uncertainty caused by the model's lack of in-domain knowledge (called *epistemic uncertainty* or *model uncertainty*) from the irreducible uncertainty caused by the inherent stochasticity or noise in the input data (called *aleatoric uncertainty* or *data uncertainty*) [19,21].

When describing the following state-of-the-art approaches to uncertainty quantification, we put our emphasis on practicality and analyze their scalability to big text data: *Virtual Ensemble* in GBDT [28], Bayesian Deep Learning [29] (including *Deep Ensemble* [22], *Monte-Carlo Dropout* [9], *Bayes by Backprop* [4], and their generalization *Epistemic Neural Networks* [30]), as well as Evidential Deep Learning [37] (including *Prior Networks* [27] and *Posterior Networks* [7,18,35]).

For the application scenarios [38] where uncertainty quantification can help text classification, we will discuss (i) in-domain calibration, (ii) cross-domain robustness, and (iii) novel class detection. The corresponding performance metrics to evaluate the effectiveness of uncertainty quantification include Expected

Calibration Error (ECE) [12,39,40], Negative Log Likelihood (NLL), Brier Score, Area Under the Receiver Operating Characteristic curve (AUROC) [15], Area Under the Precision-Recall curve (AUPR) [15], and False Alarm Rate at 90% Recall (FAR90) [16]. The text classification datasets used in our examples/exercises are real-world ones such as CLINC150[1] [23,38].

3   Outline and Materials

The tentative outline of this tutorial is as follows.

- Introduction
  - Text classification
  - Why do we need uncertainty
  - Where does uncertainty come form: aleatoric vs epistemic [19,21]
  - How can we use uncertainty in text classification [17,24,36]
- Approaches
  - Virtual Ensemble in GBDT [28]
  - Bayesian Deep Learning [29]
    * Deep Ensemble [22]
    * Monte-Carlo Dropout [9]
    * Bayes by Backprop [4]
    * Generalization to Epistemic Neural Networks [30]
  - Evidential Deep Learning [37]
    * The Dirichlet distribution
    * Prior Networks [27]
    * Posterior Networks [7,18,35]
- Application scenarios [38]
  - In-domain calibration
  - Cross-domain robustness
  - Novel class detection
- Evaluation metrics [18,38]
  - ECE [12,39,40]
  - NLL and Brier Score
  - AUROC/AUPR [15] and FAR90 [16]
- Take-home messages
  - Summary of techniques and practical recommendations
  - Recent developments and outlook for research

The tutorial website[2] will provide the following materials to the attendees.

- Presentation slides (PDF).
- Practical *hands-on* examples/exercises in the form of Python Jupyter notebooks.
- A curated webpage/repository with the pointers to relevant resources (such as datasets) and references.

---

[1] https://paperswithcode.com/dataset/clinc150.

[2] https://sites.google.com/view/uq-tutorial.

4   Format and Length

This tutorial will be delivered *in person* (with *remote support*) on Sunday 2nd April 2023. It contains a mix of presentations and short practical sessions with examples/exercises to experiment with.

It is a *half-day* tutorial, i.e., three hours plus a half-hour break in the middle.

5   Target Audience

This tutorial is intended for IR researchers and practitioners who have some experience in text classification and would like to study modern techniques for uncertainty quantification in a practical hands-on setting. To really benefit from the tutorial, the participants should be familiar with basic IR concepts[3] and be comfortable with Python[4] programming using Jupyter[5] notebooks.

6   Tutorial History

This tutorial is developed as a by-product of a research project at Thomson Reuters Labs. It has not been presented in any conference before.

There was a virtual tutorial about uncertainty quantification for deep learning in general with a focus on *out-of-distribution robustness* in the NeurIPS-2020 conference[6]. In addition, a tutorial specifically about "Uncertainty Quantification 360" (UQ360)[7]—an open-source Python package developed by IBM Research—has been given in the CODS-COMAD conference in 2022 [11]. In the related research field of NLP, a tutorial on uncertainty estimation has been given in the COLING-2022 conference[8]. However, to the best of our knowledge, so far there has not been any tutorial in major IR conferences dedicated to "uncertainty quantification for text classification". Moreover, this tutorial distinguishes itself from the above related tutorials by its stress on the combination of theoretical concepts and practical hands-on examples/exercises.

7   Tutorial Presenters

Dell Zhang (dell.z@ieee.org.) is the *main contact person* for this tutorial. He currently leads the Applied Research team at Thomson Reuters Labs in London, UK. Prior to this role, he was a Tech Lead Manager at ByteDance AI Lab and TikTok UK, a Staff Research Scientist at Blue Prism AI Labs, and a Reader in Computer Science at Birkbeck College, University of London. He

---

[3] https://nlp.stanford.edu/IR-book/information-retrieval-book.html.
[4] https://www.python.org/.
[5] https://jupyter.org/.
[6] https://neurips.cc/Conferences/2020/Schedule?showEvent=16649.
[7] https://github.com/IBM/UQ360; https://uq360.mybluemix.net/.
[8] https://sites.google.com/view/uncertainty-nlp.

is a Senior Member of ACM, a Senior Member of IEEE, and a Fellow of RSS. He got his PhD from the Southeast University (SEU) in Nanjing, China, and then worked as a Research Fellow at the Singapore-MIT Alliance (SMA) until he moved to the UK in 2005. His main research interests include Machine Learning, Information Retrieval, and Natural Language Processing. He has published 110+ papers, graduated 11 PhD students, received multiple best paper awards, and won several prizes from international data science competitions. He was also involved in the founding of 3 software startups. He has been giving lectures to both undergraduate and postgraduate students in Birkbeck and UCL. He presented tutorials in ECIR and SIGIR conferences before.

Murat Sensoy (drmuratsensoy@gmail.com.) is an Applied Machine Learning Scientist at Amazon Alexa AI in London, UK. He was previously a Senior Research Scientist at Blue Prism AI Labs, a Visiting Scholar at UCL, an Associate Professor at Ozyegin University, and a Postdoctoral Research Fellow at the University of Aberdeen. He received his PhD degree in Computer Engineering at Bogazici University in 2008. He developed semantic reasoning mechanisms for sensor networks, which are used by US Army Research Lab and IBM Research. His work on uncertainty quantification has been published in top conferences like NeurIPS [35], AAAI [34], and IJCAI [6] as well as the Machine Learning journal [5]. He presented tutorials in ECIR and SIGIR conferences before.

Masoud Makrehchi (masoud.makrehchi@thomsonreuters.com.) is a Director of Research at Thomson Reuters Labs in Toronto, Canada. He received his PhD in Electrical and Computer Engineering from the University of Waterloo in 2007, and then worked as a postdoctoral research associate at the Center for Pattern Analysis and Machine Intelligence for a year. From May 2008 to June 2012, he was with Thomson Reuters R&D as (Senior) Research Scientist. In July 2012, he accepted an associate professor position at the University of Ontario Institute of Technology (OntarioTech University). From January 2018, his research areas are Natural Language Processing, Machine Learning, Artificial Intelligence, Social Computing and domain-specific NLP such as Legal Analytics. Since Oct 2019, he is leading a team of scientists to develop a state-of-the-art contract analytics solution. He has taught a variety of computer science courses at OntarioTech over the past ten years.

Bilyana Taneva-Popova (bilyana.taneva-popova@thomsonreuters.com.) is a Senior Applied Research Scientist at Thomson Reuters Labs in Zug, Switzerland. She received her PhD from the Max-Planck Institute for Informatics in Saarbrücken, Germany. Her main research interests are in Natural Language Processing, Machine Learning, Knowledge Bases, Information Retrieval, and Data-centric AI. At TR Labs, she has been working on a variety of internal and external products. Prior positions of hers include two start-up companies, Telepathy Labs (developing conversational agents), and AVA women (developing health related products), as well a position at Nokia Bell Labs in Dublin and at the Grenoble Informatics Laboratory in France.

References

1. Abdar, M.: A review of uncertainty quantification in deep learning: techniques. Appli. Challen. Inf. Fusion **76**, 243–297 (2021). https://doi.org/10.1016/j.inffus.2021.05.008
2. Arora, G., Jain, C., Chaturvedi, M., Modi, K.: HINT3: Raising the bar for Intent Detection in the Wild. In: Proceedings of the First Workshop on Insights from Negative Results in NLP, pp. 100–105. Association for Computational Linguistics, Online (Nov 2020). https://doi.org/10.18653/v1/2020.insights-1.16
3. Arora, U., Huang, W., He, H.: Types of out-of-distribution texts and how to detect them. In: Proceedings of the 2021 Conference on Empirical Methods in Natural Language Processing, pp. 10687–10701. Association for Computational Linguistics, Online and Punta Cana, Dominican Republic (Nov 2021). https://doi.org/10.18653/v1/2021.emnlp-main.835
4. Blundell, C., Cornebise, J., Kavukcuoglu, K., Wierstra, D.: Weight Uncertainty in neural network. In: Proceedings of the 32nd International Conference on Machine Learning, pp. 1613–1622. PMLR (Jun 2015)
5. Cerutti, F., Kaplan, L.M., Kimmig, A., Şensoy, M.: Handling epistemic and aleatory uncertainties in probabilistic circuits. Mach. Learn. **111**(4), 1259–1301 (2022)
6. Cerutti, F., Kaplan, L.M., Sensoy, M.: Evidential reasoning and learning: a survey. In: Thirty-First International Joint Conference on Artificial Intelligence, vol. 6, pp. 5418–5425 (Jul 2022). https://doi.org/10.24963/ijcai.2022/760
7. Charpentier, B., Zügner, D., Günnemann, S.: Posterior network: uncertainty estimation without ood samples via density-based pseudo-counts. In: Advances in Neural Information Processing Systems, vol. 33, pp. 1356–1367. Curran Associates, Inc. (2020)
8. Foong, A.Y.K., Li, Y., Hernández-Lobato, J.M., Turner, R.E.: 'In-Between' Uncertainty in Bayesian Neural Networks (Jun 2019). https://doi.org/10.48550/arXiv.1906.11537
9. Gal, Y., Ghahramani, Z.: Dropout as a bayesian approximation: representing model uncertainty in deep learning. In: Proceedings of The 33rd International Conference on Machine Learning, pp. 1050–1059. PMLR (Jun 2016)
10. GGawlikowski, J., et al.: A Survey of Uncertainty in Deep Neural Networks (Jan 2022). https://doi.org/10.48550/arXiv.2107.03342
11. Ghosh, S., et al.: Uncertainty quantification 360: a hands-on tutorial. In: 5th Joint International Conference on Data Science & Management of Data (9th ACM IKDD CODS and 27th COMAD), CODS-COMAD 2022, pp. 333–335. Association for Computing Machinery, New York (Jan 2022). https://doi.org/10.1145/3493700.3493767
12. Guo, C., Pleiss, G., Sun, Y., Weinberger, K.Q.: On calibration of modern neural networks. In: Proceedings of the 34th International Conference on Machine Learning, ICML 2017, vol. 70, pp. 1321–1330. JMLR.org, Sydney, NSW, Australia (Aug 2017)
13. Guo, Z., et al.: A Survey on Uncertainty Reasoning and Quantification for Decision Making: Belief Theory Meets Deep Learning (Jun 2022). https://doi.org/10.48550/arXiv.2206.05675
14. Hariri, R.H., Fredericks, E.M., Bowers, K.M.: Uncertainty in big data analytics: survey, opportunities, and challenges. J. Big Data **6**(1), 1–16 (2019). https://doi.org/10.1186/s40537-019-0206-3

15. Hendrycks, D., Gimpel, K.: A baseline for detecting misclassified and out-of-distribution examples in neural networks. In: International Conference on Learning Representations (2017)
16. Hendrycks, D., Mazeika, M., Dietterich, T.G.: Deep anomaly detection with outlier exposure. In: 7th International Conference on Learning Representations, ICLR 2019, New Orleans, LA, USA, 6–9 May 2019. OpenReview.net (2019)
17. Houlsby, N., Huszár, F., Ghahramani, Z., Lengyel, M.: Bayesian Active Learning for Classification and Preference Learning (Dec 2011). https://doi.org/10.48550/arXiv.1112.5745
18. Hu, Y., Khan, L.: Uncertainty-aware reliable text classification. In: Proceedings of the 27th ACM SIGKDD Conference on Knowledge Discovery & Data Mining, KDD 2021, pp. 628–636. Association for Computing Machinery, New York (Aug 2021). https://doi.org/10.1145/3447548.3467382
19. Hüllermeier, E., Waegeman, W.: Aleatoric and epistemic uncertainty in machine learning: an introduction to concepts and methods. Mach. Learn. 110(3), 457–506 (2021). https://doi.org/10.1007/s10994-021-05946-3
20. Kabir, H.M.D., Khosravi, A., Hosen, M.A., Nahavandi, S.: Neural network-based uncertainty quantification: a survey of methodologies and applications. IEEE Access 6, 36218–36234 (2018). https://doi.org/10.1109/ACCESS.2018.2836917
21. Kendall, A., Gal, Y.: What uncertainties do we need in bayesian deep learning for computer vision? In: Advances in Neural Information Processing Systems, vol. 30. Curran Associates, Inc. (2017)
22. Lakshminarayanan, B., Pritzel, A., Blundell, C.: Simple and scalable predictive uncertainty estimation using deep ensembles. In: Advances in Neural Information Processing Systems, vol. 30. Curran Associates, Inc. (2017)
23. Larson, S., et al.: An evaluation dataset for intent classification and out-of-scope prediction. In: Proceedings of the 2019 Conference on Empirical Methods in Natural Language Processing and the 9th International Joint Conference on Natural Language Processing (EMNLP-IJCNLP), pp. 1311–1316. Association for Computational Linguistics, Hong Kong, China (Nov 2019). https://doi.org/10.18653/v1/D19-1131
24. Leibig, C., Allken, V., Ayhan, M.S., Berens, P., Wahl, S.: Leveraging uncertainty information from deep neural networks for disease detection. Sci. Rep. 7(1), 17816 (2017). https://doi.org/10.1038/s41598-017-17876-z
25. Li, Y., Chen, J., Feng, L.: Dealing with uncertainty: a survey of theories and practices. IEEE Trans. Knowl. Data Eng. 25(11), 2463–2482 (2013). https://doi.org/10.1109/TKDE.2012.179
26. Lu, X., et al.: Robustness of Epinets against Distributional Shifts (Jun 2022). https://doi.org/10.48550/arXiv.2207.00137
27. Malinin, A., Gales, M.: Predictive uncertainty estimation via prior networks. In: Advances in Neural Information Processing Systems, vol. 31. Curran Associates, Inc. (2018)
28. Malinin, A., Prokhorenkova, L., Ustimenko, A.: Uncertainty in gradient boosting via ensembles. In: International Conference on Learning Representations (Mar 2021)
29. Mena, J., Pujol, O., Vitrià, J.: A survey on uncertainty estimation in deep learning classification systems from a bayesian perspective. ACM Comput. Sur. 54(9), 193:1–193:35 (2021). https://doi.org/10.1145/3477140
30. Osband, I., et al.: Epistemic Neural Networks (Oct 2022). https://doi.org/10.48550/arXiv.2107.08924

31. Ovadia, Y., et al.: Can you trust your model's uncertainty? evaluating predictive uncertainty under dataset shift. In: Advances in Neural Information Processing Systems, vol. 32. Curran Associates, Inc. (2019)

32. Psaros, A.F., Meng, X., Zou, Z., Guo, L., Karniadakis, G.E.: Uncertainty Quantification in Scientific Machine Learning: Methods, Metrics, and Comparisons (Jan 2022). https://doi.org/10.48550/arXiv.2201.07766

33. Rawat, M., Hebbalaguppe, R., Vig, L.: PnPOOD : Out-Of-Distribution Detection for Text Classification via Plug and Play Data Augmentation (Oct 2021). https://doi.org/10.48550/arXiv.2111.00506

34. Sensoy, M., Kaplan, L., Cerutti, F., Saleki, M.: Uncertainty-Aware Deep Classifiers Using Generative Models. In: Proceedings of the AAAI Conference on Artificial Intelligence, vol. 34(04), pp. 5620–5627 (2020). https://doi.org/10.1609/aaai.v34i04.6015

35. Sensoy, M., Kaplan, L., Kandemir, M.: Evidential Deep Learning to Quantify Classification Uncertainty. In: Advances in Neural Information Processing Systems, vol. 31. Curran Associates, Inc. (2018)

36. Siddhant, A., Lipton, Z.C.: Deep bayesian active learning for natural language processing: results of a large-scale empirical study. In: Proceedings of the 2018 Conference on Empirical Methods in Natural Language Processing, pp. 2904–2909. Association for Computational Linguistics, Brussels, Belgium (Oct 2018). https://doi.org/10.18653/v1/D18-1318

37. Ulmer, D.: A Survey on Evidential Deep Learning For Single-Pass Uncertainty Estimation (Dec 2021). https://doi.org/10.48550/arXiv.2110.03051

38. Van Landeghem, J., Blaschko, M., Anckaert, B., Moens, M.F.: Benchmarking scalable predictive uncertainty in text classification. IEEE Access 10, 43703–43737 (2022). https://doi.org/10.1109/ACCESS.2022.3168734

39. Widmann, D., Lindsten, F., Zachariah, D.: Calibration tests in multi-class classification: a unifying framework. In: Advances in Neural Information Processing Systems, vol. 32. Curran Associates, Inc. (2019)

40. Widmann, D., Lindsten, F., Zachariah, D.: Calibration tests beyond classification. In: International Conference on Learning Representations (2021)

41. Zhang, H., Xu, H., Zhao, S., Zhou, Q.: Learning Discriminative Representations and Decision Boundaries for Open Intent Detection (Jul 2022). https://doi.org/10.48550/arXiv.2203.05823

Workshops

Fourth International Workshop on Algorithmic Bias in Search and Recommendation (Bias 2023)

Ludovico Boratto[1] , Stefano Faralli[2] , Mirko Marras[1(✉)] ,
and Giovanni Stilo[3]

[1] University of Cagliari, Cagliari, Italy
{ludovico.boratto,mirko.marras}@acm.org
[2] Sapienza University of Rome, Rome, Italy
stefano.faralli@uniroma1.it
[3] University of L'Aquila, L'Aquila, Italy
giovanni.stilo@univaq.it

Abstract. Creating search and recommendation models responsibly requires monitoring more than just effectiveness and efficiency. Before moving these models into production, it is imperative to audit training data and evaluate their predictions for bias. Prior work has uncovered and studied the effects of different types of bias that can manifest in search and recommendation results. Despite of the debiasing approaches only recently emerged, there is still a long way to develop trustworthy search and recommendation models. This workshop aims to collect the recent advances in this field and offer a fresh ground for interested scientists from academia and industry. More information about the workshop is available at https://biasinrecsys.github.io/ecir2023/.

Keywords: Bias · Algorithms · Search · Recommendation · Fairness

1 Introduction

Both *search* and *recommendation* models provide a user with a ranking of results that aims to match their needs and interests. Despite the (non) personalized perspective that characterizes each class of models, both learn patterns from historical data, which conveys biases in terms of *imbalances* and *inequalities*.

In most cases, the trained models and, as a consequence, the final ranking, capture and unfortunately strengthen these biases in the learned patterns [10,13]. When a bias impacts human beings as individuals or as groups with legally protected characteristics (e.g., their race, gender, or religion), the inequalities even lead to *severe societal consequences* as discrimination and unfairness [4,7,8].

Recent efforts in academia and industry are focused, among others, on controlling the effects of popularity bias to improve the user's perceived quality of the rankings [15,16], supporting stakeholders with fairer rankings [5,9,11,14], and explaining why a model provides a given result [6,12]. Overall, characterizing and mitigating bias while keeping high effectiveness is a prominent objective.

J. Kamps et al. (Eds.): ECIR 2023, LNCS 13982, pp. 373–376, 2023.
https://doi.org/10.1007/978-3-031-28241-6_39

With the rapidly-changing methods for search and recommendation and the increasing interest of the European Information Retrieval (IR) community, Bias 2023 represents a timely workshop on the recent advances in bias within search and recommendation models. Bias 2023 comes after three successful iterations of this event at ECIR [1–3]. Specifically, the goal is to favor a dialogue on new advances in this field through a workshop having the following objectives:

1. Raise awareness on the algorithmic bias problem within the IR community;
2. Identify social and human dimensions affected by algorithmic bias in IR;
3. Solicit contributions from researchers who are facing algorithmic bias in IR;
4. Get insights on existing approaches, recent advances, and open issues;
5. Familiarize the IR community with existing practices from the field;
6. Uncover gaps between academic research and real-world needs in the field.

Every year, our workshop is one of the events with the highest number of participants, considering tutorials as well. The event is also supported by the ACM Conference on Fairness, Accountability, and Transparency (FAccT) Network.

2   Contributions' Selection and Topics

This workshop is of interest for individuals studying and mitigating the effects generated by biases in search and recommendation, including among others *information retrieval*, *data mining*, and *machine learning* researchers and *practitioners*, from both *academic institutions*, *research centers*, and *companies*.

Submitted papers fall into one out of four categories (full papers, reproducibility papers, short papers, and position papers). Their topics are focused but not limited to data set collection and preparation (e.g., Interplay between bias and imbalanced data or rare classes, collection pipelines that lead to fair and less biased data sets), countermeasure design and development (e.g., Bias and fairness concepts formalization, exploratory analyses that uncover novel types of bias), evaluation protocol and metric formulation (e.g., Auditing studies with respect to bias and fairness, quantitative experimental studies on bias and unfairness), and case study exploration (e.g., In e-commerce and education).

The accepted contributions are selected through a peer-reviewing process with at least *three programme committee members*. Decisions consider the relevance for the workshop, novelty/originality, significance, technical correctness, clarity of presentation, quality of references, and reproducibility. The program committee is composed of scientists from academia, research centers, and the industry, to provide comments deriving from the different reviewers' backgrounds.

During the workshop, the accepted contributions are organized in groups of three/four papers, based on the topic or application domain. Each accepted paper is accompanied by a presentation talk, whose slides and video recordings are also disseminated. These paper sessions are paired with classic elements that characterize a workshop, including three keynote speakers. The insights coming from each paper session and the keynote talks inform a final discussion with the participants, moderated by the organizers, for setting up a road-map for shared initiatives (e.g., special issues and research funding opportunities).

3   Workshop Organizers

Ludovico Boratto is (https://www.ludovicoboratto.com) Assistant Professor at the Department of Mathematics and Computer Science of the University of Cagliari (Italy). His research interests focus on recommender systems and their impact on stakeholders, both considering accuracy and beyond-accuracy evaluation metrics. He has a wide experience in workshop organizations, with 10+ events organized at ECIR, IEEE ICDM, ECML-PKDD, and ACM EICS and has given tutorials on recommender systems at UMAP and ICDM 2020, and WSDM, ICDE, and ECIR 2021.

Stefano Faralli is Assistant Professor at the University of Rome Sapienza, Rome, Italy. His research interests include Ontology Learning, Distributional Semantics, Word Sense Disambiguation/Induction, Recommender Systems, and Linked Open Data. He co-organized the International Workshop: Taxonomy Extraction Evaluation (TexEval) Task 17 of Semantic Evaluation (SemEval-2015), the International Workshop on Social Interaction-based Recommendation (SIR 2018), and the ECIR 2020, 2021 and 2022 BIAS workshops.

Mirko Marras (https://www.mirkomarras.com/) is Assistant Professor at the Department of Mathematics and Computer Science of the University of Cagliari (Italy). His research interests focus on responsible machine learning, with particular attention to educational environments. He has a leading role in chairing the ECIR BIAS workshop editions (2020-22) and workshops held in conjunction with other top-tier venues, such as WSDM, ICCV, and EDM. He is currently giving tutorials on bias and explainability in recommender systems at UMAP and ICDM 2020, and WSDM, ICDE, ECIR 2021, and RecSys 2022.

Giovanni Stilo is Associate Professor and Head of the Master's Degree course in Applied Data Science at the Department of Information Engineering, Computer Science and Mathematics of the University of L'Aquila. His research interests focus on Data Science and Artificial Intelligence, more specifically on reinforcement learning, temporal mining, ranking models, network medicine, semantics-aware recommender systems, anomaly detection, and machine learning pipelines with quality attributes. He also has organized several international workshops, held in conjunction with top-tier conferences (ICDM, CIKM, and ECIR), with the ECIR 2020-2022 BIAS workshops being three of them.

References

1. Boratto, L., Faralli, S., Marras, M., Stilo, G. (eds.): Bias and Social Aspects in Search and Recommendation - First International Workshop, BIAS 2020, Proceedings of Communications in Computer and Information Science, vol. 1245. Springer (2020)
2. Boratto, L., Faralli, S., Marras, M., Stilo, G. (eds.): Advances in Bias and Fairness in Information Retrieval - Second International Workshop on Algorithmic Bias in Search and Recommendation, BIAS 2021, Proceedings, Communications in Computer and Information Science, vol. 1418. Springer (2021)

3. Boratto, L., Faralli, S., Marras, M., Stilo, G. (eds.): Advances in Bias and Fairness in Information Retrieval - Third International Workshop on Algorithmic Bias in Search and Recommendation, BIAS 2022, Proceedings, Communications in Computer and Information Science, vol. 1610. Springer (2022)

4. Boratto, L., Fenu, G., Marras, M., Medda, G.: Consumer fairness in recommender systems: contextualizing definitions and mitigations. In: Hagen, M., et al. (eds.) ECIR 2022. LNCS, vol. 13185, pp. 552–566. Springer, Cham (2022). https://doi.org/10.1007/978-3-030-99736-6_37

5. Boratto, L., Fenu, G., Marras, M., Medda, G.: Practical perspectives of consumer fairness in recommendation. Inf. Process. Manag. **60**(2), 103208 (2023)

6. Deldjoo, Y., Bellogín, A., Noia, T.D.: Explaining recommender systems fairness and accuracy through the lens of data characteristics. Inf. Process. Manag. **58**(5), 102662 (2021)

7. Ekstrand, M.D., Das, A., Burke, R., Diaz, F.: Fairness in information access systems. Found. Trends Inf. Retr. **16**(1–2), 1–177 (2022)

8. Fabbri, F., Bonchi, F., Boratto, L., Castillo, C.: The effect of homophily on disparate visibility of minorities in people recommender systems. In: Fourteenth International AAAI Conference on Web and Social Media, ICWSM 2020, Proceedings, pp. 165–175. AAAI Press (2020)

9. Gómez, E., Zhang, C.S., Boratto, L., Salamó, M., Marras, M.: The winner takes it all: Geographic imbalance and provider (un)fairness in educational recommender systems. In: SIGIR 2021: The 44th International ACM SIGIR Conference on Research and Development in Information Retrieval, Proceedings. pp. 1808–1812. ACM (2021)

10. Huang, J., Oosterhuis, H., de Rijke, M.: It is different when items are older: Debiasing recommendations when selection bias and user preferences are dynamic. In: WSDM 2022: The Fifteenth ACM International Conference on Web Search and Data Mining, Proceedings, pp. 381–389. ACM (2022)

11. Kirnap, Ö., Diaz, F., Biega, A., Ekstrand, M.D., Carterette, B., Yilmaz, E.: Estimation of fair ranking metrics with incomplete judgments. In: WWW 2021: The Web Conference 2021, Proceedings, pp. 1065–1075. ACM / IW3C2 (2021)

12. Li, R., Li, J., Mitra, B., Diaz, F., Biega, A.J.: Exposing query identification for search transparency. In: WWW 2022: The ACM Web Conference 2022, Proceedings, pp. 3662–3672. ACM (2022)

13. Liu, D., et al.: Mitigating confounding bias in recommendation via information bottleneck. In: RecSys 2021: Fifteenth ACM Conference on Recommender Systems, Proceedings, pp. 351–360. ACM (2021)

14. Oosterhuis, H.: Computationally efficient optimization of plackett-luce ranking models for relevance and fairness. In: SIGIR 2021: The 44th International ACM SIGIR Conference on Research and Development in Information Retrieval, Proceedings, pp. 1023–1032. ACM (2021)

15. Yalcin, E., Bilge, A.: Investigating and counteracting popularity bias in group recommendations. Inf. Process. Manag. **58**(5), 102608 (2021)

16. Zhang, Y., et al.: Causal intervention for leveraging popularity bias in recommendation. In: SIGIR 2021: The 44th International ACM SIGIR Conference on Research and Development in Information Retrieval, Proceedings, pp. 11–20. ACM (2021)

The 6th International Workshop on Narrative Extraction from Texts: Text2Story 2023

Ricardo Campos[1,2](✉) , Alípio Jorge[1,3] , Adam Jatowt[4] , Sumit Bhatia[5] , and Marina Litvak[6]

[1] LIAAD – INESCTEC, Porto, Portugal
[2] Ci2 – Smart Cities Research Center – Polytechnic Institute of Tomar, Tomar, Portugal
ricardo.campos@ipt.pt
[3] FCUP, University of Porto, Porto, Portugal
amjorge@fc.up.pt
[4] University of Innsbruck, Innsbruck, Austria
adam.jatowt@uibk.ac.at
[5] MDSR Lab, Adobe Systems, Noida, India
Sumit.Bhatia@adobe.com
[6] Shamoon College of Engineering, Beer Sheva, Israel
marinal@sce.ac.il

Abstract. Over these past five years, significant breakthroughs, led by Transformers and large language models, have been made in understanding natural language text. However, the ability to capture contextual nuances in longer texts is still an elusive goal, let alone the understanding of consistent fine-grained narrative structures in text. These unsolved challenges and the interest in the community are at the basis of the sixth edition of Text2Story workshop to be held in Dublin on April 2nd, 2023 in conjunction with the 45th European Conference on Information Retrieval (ECIR'23). In its sixth edition, we aim to bring to the forefront the challenges involved in understanding the structure of narratives and in incorporating their representation in well-established models, as well as in modern architectures (e.g., transformers) which are now common and form the backbone of almost every IR and NLP application. It is hoped that the workshop will provide a common forum to consolidate the multi-disciplinary efforts and foster discussions to identify the wide-ranging issues related to the narrative extraction and generation task. Text2Story includes sessions devoted to full research papers, work-in-progress, demos and dissemination papers, keynote talks and space for an informal discussion of the methods, of the challenges and of the future of this research area.

1 Motivation

The continuous growth of social networks such as Facebook, Instagram, TikTok, and Twitter, together with an ever-increasing presence of traditional news media outlets on the Web, has changed the way information is being generated and consumed. Rather than relying on a few sources of information about an event or a news item topic (e.g., US and China trade war), readers now have easy access to the content via multiple sources produced by disparate content creators such as journalists, subject matter experts, and social media influencers and users. Further, active reader participation also occurs in the comments section of news articles with discussions lasting over days, weeks or possibly, months. Such a stream of continuously evolving information makes it unmanageable and time-consuming for an interested reader to track and process, which is essential to keep up with all the developments in various aspects of the topic of interest.

Automated narrative extraction and generation from text offers a compelling approach to this problem. It involves identifying the sub-set of interconnected raw documents, extracting the critical narrative story elements, their semantic meaning and representing them in an adequate final form (e.g., timelines) that conveys the key points of the story in an easy-to-understand format.

Although, information extraction and natural language processing have made significant progress towards an automatic interpretation of texts, the problem of automated identification and analysis of the different elements of a narrative present in a document (set) still presents significant unsolved challenges [25]. While these large language models have shown promise in capturing various linguistic nuances and generating high quality text [10], it has been shown that these models fare poorly at capturing longer range dependencies and contextual information for longer texts such as chapters in fictional literary works [26]. Industries such as finance [2, 9, 11, 29], business [13], news outlets [19], and health care [21] may be the ultimate beneficiaries of the advances in this area. The ultimate goal is to offer users tools allowing to quickly understand the information conveyed in economic and financial reports, patient records, verify the extracted information, and to offer them more appealing and alternative formats of exploring common narratives through interactive visualizations [12]. Timelines [23] and infographics for instance, can be employed to represent in a more compact way automatically identified narrative chains in a cloud of news articles [18] or keywords [4], assisting human readers in grasping complex stories with different key turning points and networks of characters. Also, the automatic generation of text [28] shows impressive results towards computational creativity. However, it still needs to develop means for controlling the narrative intent of the output and a profound understanding of the applied methods by humans (explainable AI). There are various open problems and challenges in this area, such as hallucination in generated text [20], bias in text [14], transparency and explainability of the generation techniques [1], reliability of the extracted facts [24, 27], and efficient and objective evaluation of generated narratives [3, 8].

The Text2Story workshop, now in its sixth edition [https://text2story23.inesctec.pt], aims to provide a common forum to consolidate the multi-disciplinary efforts and foster discussions to identify the wide-ranging issues related to the narrative extraction task. In the first five editions [5–7, 15, 16], we had an approximate number of 270 participants in total, 50 of which took part in the last edition of the workshop. This adds to the fact that a Special Issue on IPM Journal [17] devoted to this matter has also been proposed in the past demonstrating the growing activity of this research area. In this year's edition, we welcomed contributions from interested researchers on all aspects related to narrative understanding, including the extraction and formal representation of events, their temporal aspects and intrinsic relationships, verification of extracted facts, generation and evaluation of the generated texts, and more. Research works submitted to the workshop should foster the scientific advance on all aspects of storyline generation from texts including but not limited to narrative and content generation, text simplification, formal representation, knowledge extraction, and visualization of narratives. In addition to this, we seek contributions related to alternative means of presenting the information and on the formal aspects of evaluation, including the proposal of new datasets, annotation schema, and evaluation metrics. Special attention will be given to multilingual approaches and resources. Finally, we challenge the interested researchers to consider submitting a paper that makes use of the tls-covid19 dataset [22], which consists of a number of curated topics related to the Covid-19 outbreak, with associated news articles from Portuguese and English news outlets and their respective reference timelines as gold-standard. While it was designed to support timeline summarization research tasks, it can also be used for other tasks including the study of news coverage about the COVID-19 pandemic. The full list of topics for workshop submissions is as follows:

- Narrative Representation Models
- Story Evolution and Shift Detection
- Temporal Relation Identification
- Temporal Reasoning and Ordering of Events
- Causal Relation Extraction and Arrangement
- Narrative Summarization
- Multi-modal Summarization
- Automatic Timeline Generation
- Storyline Visualization
- Comprehension of Generated Narratives and Timelines
- Big Data Applied to Narrative Extraction
- Personalization and Recommendation of Narratives
- User Profiling and User Behavior Modeling
- Sentiment and Opinion Detection in Texts
- Argumentation Analysis
- Bias Detection and Removal in Generated Stories
- Ethical and Fair Narrative Generation
- Misinformation and Fact Checking
- Bots Influence
- Narrative-focused Search in Text Collections
- Event and Entity importance Estimation in Narratives
- Multilinguality: Multilingual and Cross-lingual Narrative Analysis
- Evaluation Methodologies for Narrative Extraction
- Resources and Dataset Showcase
- Dataset Annotation for Narrative Generation/Analysis
- Applications in Social Media (e.g. narrative generation during a natural disaster)
- Language Models and Transfer Learning in Narrative Analysis
- Narrative Analysis in Low-resource Languages
- Text Simplification

2 Scientific Objectives

The main goal of the workshop is to bring together scientists conducting relevant research in the field of identifying and producing narratives/stories from textual sources, such as journalistic texts, scientific articles, or even social networks. Researchers will present their latest breakthroughs with an emphasis on the application of their findings across a wide range of areas including information extraction, information retrieval, natural language processing, text mining, artificial intelligence, machine learning, and natural language generation. Overall, the workshop has the following main objectives: (1) raise awareness within the IR community to the problem of narrative extraction and understanding; (2) shorten the gap between academic research, practitioners and industrial application; (3) obtain insight on new methods, recent advances and challenges, as well on future directions; (4) share experiences of research projects, case studies and scientific outcomes, (5) identify dimensions potentially affected by the automatization of the narrative process, (6) highlight tested hypotheses that did not result in the expected outcomes. Our topics are organized around the basic research questions related to narrative generation, which are as follows: How to efficiently extract/generate reliable and accurate narrative from a large multi-genre and multi-lingual data? How to annotate data and evaluate new approaches? Is a new approach transparent, explainable and easily reproducible? Is it adjustable to new tasks, genres and languages without much effort required?

3 Organizing Team

Ricardo Campos is an assistant professor at the Polytechnic Institute of Tomar. He is an integrated researcher of LIAAD-INESC TEC, the Artificial Intelligence and Decision Support Lab of U. Porto, and a collaborator of Ci2.ipt, the Smart Cities Research Center of the Polytechnic of Tomar. He is PhD in Computer Science by the University of Porto (U. Porto). He has over ten years of research experience in IR and NLP. He is an editorial board member of the IPM Journal (Elsevier), co-chaired international conferences and workshops, being also a PC member of several international conferences. More in http://www.ccc.ipt.pt/~ricardo.

Alípio M. Jorge works in the areas of data mining, ML, recommender systems and NLP. He is a PhD in Comp. Science (CS) by the University of Porto (UP). He is an Associate Professor of the dep. of CS of the UP since 2009 and is the head of that dep. Since 2017. He is the coordinator of the research lab LIAAD-INESC TEC. He has projects in NLP, web automation, recommender systems, IR, text mining and decision support for the management of public transport. He represents Portugal in the Working Group on Artificial Intelligence at the European Commission and was the coordinator for the Portuguese Strategy on Artificial Intelligence "AI Portugal 2030".

Adam Jatowt is Full Professor at the University of Innsbruck. He has received his Ph.D. in Information Science and Technology from the University of Tokyo, Japan in 2005. His research interests lie in an area of IR, knowledge extraction from text and in digital history. Adam has been serving as a PC co-chair of IPRES2011, SocInfo2013, ICADL2014,

JCDL2017 and ICADL2019 conferences and a general chair of ICADL2020, TPDL2019 and a tutorial co-chair of SIGIR2017. He was also a co-organizer of 3 NTCIR evaluation tasks and co-organizer of over 20 international workshops at WWW, CIKM, ACL, ECIR, IJCAI, IUI, SOCINFO, TPDL and DH conferences.

Sumit Bhatia is a Senior Machine Learning Scientist at Media and Data Science Research Lab, Adobe Systems, India. He received his Ph.D. from the Pennsylvania State University in 2013. His doctoral research focused on enabling easier information access in online discussion forums followed by a post-doc at Xerox Research Labs on event detection and customer feedback monitoring in social media. With primary research interests in the fields of Knowledge Management, IR and Text Analytics, Sumit is a co-inventor of more than a dozen patents. He has served on program committees of multiple conferences and journals including WWW, CIKM, ACL, EMNLP, NAACL, TKDE, TOIS, WebDB, JASIST, IJCAI, and AAAI.

Marina Litvak is a Senior Lecturer at the department of Software Engineering, Shamoon College of Engineering (SCE), Israel. Marina received her PhD degree from Information Sciences dept. at Ben Gurion University at the Negev, Israel in 2010. Marina's research focuses mainly on Multilingual Text Analysis, Social Networks, Knowledge Extraction from Text, and Summarization. Marina published over 70 academic papers, including journal and top-level conference publications. She constantly serves on program committees and editorial boards in multiple journals and conferences. Marina is a co-organizer of the MultiLing (2011, 2013, 2015, 2017, and 2019) and the FNP (2020, 2021, and 2022) workshop series.

Acknowledgements. Ricardo Campos and Alípio Jorge are financed by National Funds through the Portuguese funding agency, FCT – Fundação para a Ciência e a Tecnologia, within project LA/P/0063/2020 and by the project Text2Story, financed by the ERDF – European Regional Development Fund through the Norte Portugal Regional Operational Programme – NORTE 2020 under the Portugal 2020 Partnership Agreement and by National Funds through the FCT – Fundação para a Ciência e a Tecnologia, I.P. (Portuguese Foundation for Science and Technology) within project Text2Story, with reference PTDC/CCI-COM/31857/2017 (NORTE-01-0145-FEDER-031857).

References

1. Alonso, J.M., et al.: Interactive natural language technology for explainable artificial intelligence. In: TAILOR, pp. 63–70 (2020)
2. Athanasakou, V., et al.: Proceedings of the 1st Joint Workshop on Financial Narrative Processing and MultiLing Financial Summarisation (FNP-FNS 2020) Co-located to Coling 2020, Barcelona, Spain (Online), 12 December, pp. 1–245 (2020)
3. Ayed, A.B., Biskri, I., Meunier, J.-G.: An efficient explainable artificial intelligence model of automatically generated summaries evaluation: a use case of bridging cognitive psychology and computational linguistics. In: Sayed-Mouchaweh, M. (ed.) Explainable AI Within the Digital Transformation and Cyber Physical Systems, pp. 69–90. Springer, Cham (2021). https://doi.org/10.1007/978-3-030-76409-8_5

4. Campos, R., Mangaravite, V., Pasquali, A., Jorge, A.M., Nunes, C., Jatowt, A.: A text feature based automatic keyword extraction method for single documents. In: Pasi, G., Piwowarski, B., Azzopardi, L., Hanbury, A. (eds.) ECIR 2018. LNCS, vol. 10772, pp. 684–691. Springer, Cham (2018). https://doi.org/10.1007/978-3-319-76941-7_63

5. Campos, R., Jorge, A., Jatowt, A., Bhatia, S., Litvak, M.: The 5th international workshop on narrative extraction from texts: Text2Story 2022. In: Hagen, M., Verberne, S., Macdonald, C., Seifert, C., Balog, K., Nørvåg, K., Setty, V. (eds.) ECIR 2022. LNCS, vol. 13186, pp. 552–556. Springer, Cham (2022). https://doi.org/10.1007/978-3-030-99739-7_68

6. Campos, R., Jorge, A., Jatowt, A., Bhatia, S., Finlayson, M.: The 4th international workshop on narrative extraction from texts: Text2Story 2021. In: Hiemstra, D., Moens, M.-F., Mothe, J., Perego, R., Potthast, M., Sebastiani, F. (eds.) ECIR 2021. LNCS, vol. 12657, pp. 701–704. Springer, Cham (2021). https://doi.org/10.1007/978-3-030-72240-1_84

7. Campos, R., Jorge, A., Jatowt, A., Sumit, B.: Third international workshop on narrative extraction from texts (Text2Story 2020). In: Jose, J., et al. (eds.) ECIR 2020, LNCS, vol. 12036, pp. 648–653. Springer, Cham (2020)

8. Celikyilmaz, A., Clark, E., Gao, J.: Evaluation of text generation: a survey. arXiv preprint arXiv:2006.14799(2020)

9. El-Haj, M., Litvak, M., Pittaras, N., Giannakopoulos, G.: The financial narrative summarisation shared task (FNS 2020). In: Proceedings of the 1st Joint Workshop on Financial Narrative Processing and MultiLing Financial Summarisation, pp. 1–12 (2020)

10. Elkins, K., Chun, J.: Can GPT-3 pass a writer's turing test? J. Cult. Analytics 5(2), 17212 (2020)

11. El-Haj, M., Rayson, P., Zmandar, N.: Proceedings of the 3rd Financial Narrative Processing Workshop. ACL (2021)

12. Gonçalves, F., Campos, R., Jorge, A.: Text2Storyline: generating enriched storylines from text. In: Caputo, A., et al. (eds.) Advances in Information Retrieval, ECIR 2023, Dublin, Ireland, 02–06 April. LNCS. Springer, Cham (2023)

13. Grobelny, J., Smierzchalska, J., Krzysztof, K.: Narrative gamification as a method of increasing sales performance: a field experimental study. Int. J. Acad. Res. Bus. Soc. Sci. 8(3), 430–447 (2018)

14. Guo, W., Caliskan, A.: Detecting emergent intersectional biases: contextualized word embeddings contain a distribution of human-like biases. In: Proceedings of the 2021 AAAI/ACM Conference on AI, Ethics, and Society, pp. 122–133 (2021)

15. Jorge, A.M., Campos, R., Jatowt, A., Bhatia, S.: The 2nd international workshop on narrative extraction from text: Text2Story 2019. In: Azzopardi, L., Stein, B., Fuhr, N., Mayr, P., Hauff, C., Hiemstra, D. (eds.) ECIR 2019. LNCS, vol. 11438, pp. 389–393. Springer, Cham (2019). https://doi.org/10.1007/978-3-030-15719-7_54

16. Jorge, A., Campos, R., Jatowt, A., Nunes, S.: First international workshop on narrative extraction from texts (Text2Story 2018). In: Pasi, G., et al. (eds.) Advances in Information Retrieval, ECIR 2018, Grenoble, France, 26–29 March. LNCS, vol. 10772, pp. 833–834 (2018)

17. Jorge, A., Campos, R., Jatowt, A., Nunes, S.: Special issue on narrative extraction from texts (Text2Story): preface. IPM J. 56(5), 1771–1774 (2019)

18. Liu, S., et al.: TIARA: interactive, topic-based visual text summarization and analysis. ACM Trans. Intell. Syst. Technol. 3(2) (2012). Article 25, 28 pages

19. Martinez-Alvarez, M., et al.: First international workshop on recent trends in news information retrieval (NewsIR'16). In: Ferro, N., et al. (eds.) ECIR 2016. LNCS, vol. 9626, pp. 878–882. Springer, Cham (2016). https://doi.org/10.1007/978-3-319-30671-1_85

20. Maynez, J., Narayan, S., Bohnet, B., McDonald, R.: On faithfulness and factuality in abstractive summarization. In: Proceedings of the 58th Annual Meeting of the Association for Computational Linguistics, pp. 1906–1919 (2020)

21. Özlem, U., Amber, S., Weiyi, S.: Chronology of your health events: approaches to extracting temporal relations from medical narratives. Biomedical Inf. **46**, 1–4 (2013)
22. Pasquali, A., Campos, R., Ribeiro, A., Santana, B., Jorge, A., Jatowt, A.: TLS-Covid19: a new annotated corpus for timeline summarization. In: Hiemstra, D., Moens, M.-F., Mothe, J., Perego, R., Potthast, M., Sebastiani, F. (eds.) ECIR 2021. LNCS, vol. 12656, pp. 497–512. Springer, Cham (2021). https://doi.org/10.1007/978-3-030-72113-8_33
23. Pasquali, A., Mangaravite, V., Campos, R., Jorge, A.M., Jatowt, A.: Interactive system for automatically generating temporal narratives. In: Azzopardi, L., Stein, B., Fuhr, N., Mayr, P., Hauff, C., Hiemstra, D. (eds.) ECIR 2019. LNCS, vol. 11438, pp. 251–255. Springer, Cham (2019). https://doi.org/10.1007/978-3-030-15719-7_34
24. Saakyan, A., Chakrabarty, T., Muresan, S.: COVID-fact: fact extraction and verification of real-world claims on COVID-19 pandemic. arXiv preprint arXiv:2106.03794(2021)
25. Santana, B., Campos, R., Amorim, E., Alípio, J., Purificação, S., Nunes, S.: A survey on narrative extraction from textual data. Artif. Intell. Rev. (2023). https://doi.org/10.1007/s10462-022-10338-7
26. Sun, S., Krishna, K., Mattarella-Micke, A., Iyyer, M.: Do long-range language models actually use long-range context? In: Proceedings of the 2021 Conference on Empirical Methods in Natural Language Processing, pp. 807–822. ACL, Online and Punta Cana, Dominican Republic (2021)
27. Vo, N., Lee, K.: Learning from fact-checkers: Analysis and generation of fact-checking language. In: Proceedings of the 42nd International ACM SIGIR Conference on Research and Development in Information Retrieval, Paris, France, 21–25 July, pp. 335–344 (2019)
28. Wu, Y.: Is automated journalistic writing less biased? An experimental test of auto-written and human-written news stories. J. Pract. **14**(7), 1–21 (2019)
29. Zmandar, N., El-Haj, M., Rayson, P., Litvak, M., Giannakopoulos, G., Pittaras, N.: The financial narrative summarisation shared task FNS 2021. In: Proceedings of the 3rd Financial Narrative Processing Workshop, pp. 120–125 (2021)

2nd Workshop on Augmented Intelligence in Technology-Assisted Review Systems (ALTARS)

Giorgio Maria Di Nunzio[1(✉)] 🆔, Evangelos Kanoulas[2] 🆔,
and Prasenjit Majumder[3]

[1] Department of Information Engineering, University of Padova, Padua, Italy
`giorgiomaria.dinunzio@unipd.it`
[2] Faculty of Science, Informatics Institute University of Amsterdam,
Amsterdam, The Netherlands
`E.Kanoulas@uva.nl`
[3] DAIICT, Gandhinagar, India
`prasenjit_t@isical.ac.in`

Abstract. In this second edition of the workshop on Augmented Intelligence in Technology-Assisted Review Systems (ALTARS), we focus on the evaluation of High-recall Information Retrieval (IR) systems which tackle challenging tasks that require the finding of (nearly) all the relevant documents in a collection. In fact, despite the number of evaluation measures at our disposal to assess the effectiveness of a "traditional" retrieval approach, there are additional dimensions of evaluation for these systems. During the workshop, the organizers as well as the participants will discuss these issues and prepare a set of guidelines for the preparation of a correct evaluation of these kinds of systems.

Keywords: Technology-Assisted Review Systems · Augmented Intelligence · Evaluation · Systematic reviews · eDiscovery

1 Introduction

Technology-assisted review (TAR) systems use a kind of human-in-the-loop approach where classification and/or ranking algorithms are continuously trained according to the relevance feedback from expert reviewers, until a substantial number of the relevant documents are identified. This approach has been shown to be more effective and more efficient than traditional e-discovery and systematic review practices, which typically consists of a mix of keyword searches and manual review of the search results.

The first edition of the ALTARS workshop was successfully held at ECIR 2022[1] [4]. During (and after) that workshop, there was a lively discussion about open questions that still need to be addressed and clarified in this research area.

---

[1] https://altars2022.dei.unipd.it/.

J. Kamps et al. (Eds.): ECIR 2023, LNCS 13982, pp. 384–387, 2023.
https://doi.org/10.1007/978-3-031-28241-6_41

In this second edition of the workshop, we aim to study innovative approaches to fathom the effectiveness of these systems. In fact, despite the number of evaluation measures at our disposal to assess the effectiveness of a "traditional" retrieval approach, there are additional dimensions of evaluation for TAR systems. For example, it is true that an effective high-recall system should be able to find the majority of relevant documents using the least number of assessments. However, this type of evaluation discards the resources used to achieve this goal, such as the total time spent on those assessments, or the amount of money spent for the experts judging the documents.

2   Workshop Goals and Objectives

The main goal of this workshop is to focus on the evaluation of the different definitions of the effectiveness of TAR systems which is a research challenge itself. The idea is to go beyond a "traditional" retrieval approach and study the problem from different perspectives.

In the first edition of the workshop, we organized a special issue for the Intelligent Systems with Applications journal dedicated to the extended version of the papers presented at the workshop[2].

The desired outcome of this second workshop is having a collection of high-quality papers that will be published on CEUR-WS. Then, a selected number of papers will be invited in a new special issue and gather momentum in this interdisciplinary area together with researchers and stakeholders in the fields[3] as well as international projects[4].

The goals of the workshop are threefold:

- To foster cross-discipline collaborations between researchers with different perspectives and research backgrounds in the TAR systems;
- To combine and analyze existing theoretical and empirical contributions in order to determine shared issues, and novel research questions;
- To create a set of shared datasets dedicated towards the evaluation of TAR systems, thus enabling a wider research community to benefit from the outcomes of the workshop.

One last goal of the workshop is to discuss open issues and challenges and have as a final product a Horizon Europe/NSF proposal to strengthen the network of researchers in this topic.

3   Format and Structure

This workshop will be a full-day workshop and will be structured in four sessions: two sessions in the morning and two in the afternoon.

---

[2] https://www.journals.elsevier.com/intelligent-systems-with-applications/call-for-papers/technology-assisted-review-systems.

[3] Such as ICASR, https://icasr.github.io/about.html and Zylab https://www.zylab.com/en/.

[4] Such as DOSSIER https://dossier-project.eu.

The call for papers of the workshop will include both full and short papers. All the authors of the accepted papers will have the possibility to give a talk; moreover, during an afternoon coffee break, we plan to organize a poster session to have some additional time for extra discussions that may arise during the day.

There will be an invited keynote speaker in the morning (to be confirmed) and a panel/general discussion in the afternoon after all the paper presentation.

4   Organizing Team

All the three organizing committee members have been active participants in the past editions of the TREC, CLEF and FIRE evaluation forum for the Total Recall and Precision Medicine TREC Tasks, TAR in eHealth tasks, and AI for Legal Assistance. The committee members have strong research record with a total of more than 400 papers in international journals and conferences. They have been doing research in technology-assisted review systems and problems related to document distillation both in the eHealth and eDiscovery domain and made significant contributions in this specific research area.

Giorgio Maria Di Nunzio is Associate Professor at the Department of Information Engineering of the Universiryt of Padova. He has been the co-organizer of the ongoing Covid-19 Multilingual Information Access Evaluation forum,[5] in particular for the evaluation of high-recall systems and high-precision system tasks. He will bring to this workshop the perspective of alternative (to the standard) evaluation measures and multilingual challenges [2,3,5,7].

Evangelos Kanoulas is Full Professor at the Faculty of Science of the Informatics Institute at the University of Amsterdam. He has been the co-organizer CLEF eHealth Lab and of the Technologically Assisted Reviews in Empirical Medicine task.[6] He will bring to the workshop the perspective of the evaluation of the costs in eHealth TAR systems, in particular of the early stopping strategies [6,9].

Prasenjit Majumder is Associate Professor at the Dhirubhai Ambani Institute of Information and Communication Technology (DA-IICT), Gandhinagar and TCG CREST, Kolkata, India. He has been the co-organizer of the Forum for Information Retrieval Evaluation and, in particular, the Artificial Intelligence for Legal Assistance (AILA) task.[7] He will bring to the workshop the perspective of the evaluation of the costs of eDiscovery, in particular of the issues related to legal precedent findings [1,8].

---

[5] http://eval.covid19-mlia.eu.

[6] https://clefehealth.imag.fr.

[7] https://sites.google.com/view/aila-2021.

References

1. Bhattacharya, P., et al.: FIRE 2020 AILA track: Artificial intelligence for legal assistance. In: Majumder, P., Mitra, M., Gangopadhyay, S., Mehta, P. (eds.) FIRE 2020: Forum for Information Retrieval Evaluation, Hyderabad, India, 16-20 December 2020, pp. 1–3. ACM (2020). https://doi.org/10.1145/3441501.3441510
2. Clipa, T., Di Nunzio, G.M.: A study on ranking fusion approaches for the retrieval of medical publications. Inf. **11**(2), 103 (2020). https://doi.org/10.3390/info11020103
3. Di Nunzio, G.M., Faggioli, G.: A study of a gain based approach for query aspects in recall oriented tasks. Appli. Sci. **11**(19) (2021). https://doi.org/10.3390/app11199075, https://www.mdpi.com/2076-3417/11/19/9075
4. Di Nunzio, G.M., Kanoulas, E., Majumder, P.: Augmented intelligence in technology-assisted review systems (ALTARS 2022): Evaluation Metrics and Protocols for ediscovery and systematic review systems. In: Hagen, M., et al. (eds.) ECIR 2022. LNCS, vol. 13186, pp. 557–560. Springer, Cham (2022). https://doi.org/10.1007/978-3-030-99739-7_69
5. Di Nunzio, G.M., Vezzani, F.: Did I miss anything? A study on ranking fusion and manual query rewriting in consumer health search. In: Barrón-Cedeño, A., et al., (eds.) CLEF 2022. LNCS, vol. 13390, pp. 217–229. Springer, Cham (2022). https://doi.org/10.1007/978-3-031-13643-6_17
6. Li, D., Kanoulas, E.: When to stop reviewing in technology-assisted reviews: Sampling from an adaptive distribution to estimate residual relevant documents. ACM Trans. Inf. Syst. **38**(4), 41:1–41:36 (2020). https://doi.org/10.1145/3411755
7. Marchesin, S., Di Nunzio, G.M., Agosti, M.: Simple but effective knowledge-based query reformulations for precision medicine retrieval. Inf. **12**(10), 402 (2021). https://doi.org/10.3390/info12100402
8. Parikh, V., et al.: AILA 2021: Shared task on artificial intelligence for legal assistance. In: Ganguly, D., Gangopadhyay, S., Mitra, M., Majumder, P. (eds.) FIRE 2021: Forum for Information Retrieval Evaluation, Virtual Event, India, 13 - 17 December 2021.,pp. 12–15. ACM (2021). https://doi.org/10.1145/3503162.3506571
9. Zou, J., Kanoulas, E.: Towards question-based high-recall information retrieval: Locating the last few relevant documents for technology-assisted reviews. ACM Trans. Inf. Syst. **38**(3), 27:1–27:35 (2020). https://doi.org/10.1145/3388640

QPP++ 2023: Query-Performance Prediction and Its Evaluation in New Tasks

Guglielmo Faggioli[1]([✉]) [ID], Nicola Ferro[1] [ID], Josiane Mothe[2] [ID],
and Fiana Raiber[3]

[1] University of Padova, Padova, Italy
{faggioli,ferro}@dei.unipd.it
[2] INSPE, Université de Toulouse, IRIT UMR5505 CNRS, Toulouse, France
Josiane.Mothe@irit.fr
[3] Yahoo Research, Haifa, Israel
fiana@yahooinc.com

Abstract. Query-Performance Prediction (QPP) is currently primarily applied to ad-hoc retrieval tasks. The Information Retrieval (IR) field is reaching new heights thanks to recent advances in large language models and neural networks, as well as emerging new ways of searching, such as conversational search. Such advancements are quickly spreading to adjacent research areas, including QPP, necessitating a reconsideration of how we perform and evaluate QPP. This workshop sought to elicit discussion on three topics related to the future of QPP: exploiting advances in IR to improve QPP, instantiating QPP on new search paradigms, and evaluating QPP on new tasks.

Keywords: Query performance prediction · Neural IR · Conversational search · Evaluation

1 Introduction

The advent of large language models and the rise of new tasks, such as conversational search, semantic search and question answering, enabled by the availability of new powerful technological tools, have led to a previously unseen rapid growth in the variety and quality of Information Retrieval (IR) systems. Several ancillary research fields have also flourished due to the scientific uptake of new Natural Language Processing (NLP) methodologies, facilitating advancement in new IR tasks. The Query-Performance Prediction and Its Evaluation in New Tasks (QPP++ 2023) workshop aimed to further fuel such growth in the renowned and important area of Query-Performance Prediction (QPP).

The QPP task is defined as estimating search effectiveness in the absence of human relevance judgments [1]. Since its introduction at the beginning of the 21st century, QPP has established itself as an essential tool in numerous tasks, including model selection [1,13], query suggestion [1,13], and rank fusion [10].

J. Kamps et al. (Eds.): ECIR 2023, LNCS 13982, pp. 388–391, 2023.
https://doi.org/10.1007/978-3-031-28241-6_42

The QPP++ 2023 workshop was a collaborative effort of researchers to master the new tools made available by the NLP community and learn how to effectively use them for the QPP task. The workshop focused on applying QPP in traditional scenarios, such as ad-hoc retrieval, and in new domains, including conversational and semantic search, passage retrieval, and question answering. QPP++ 2023 also allowed the community to reexamine past weaknesses and challenges linked to the QPP task, such as its evaluation, while establishing a roadmap to organize and guide the community's future efforts to advance the QPP research field.

QPP and Novel Search Paradigms. Given the recent developments in IR, the prediction quality of existing QPP approaches may be significantly affected in new domains and scenarios for the following three reasons. First, some of the traditional predictors exploit statistics derived from the collection [5], while new IR models often use indexes of embeddings or apply machine learning to re-rank documents [7]. Second, the vast majority of the recently developed retrieval models in IR utilize semantic information that, with a few notable exceptions [8, 12], is rarely exploited by QPP models. This, in turn, impairs the performance of traditional QPP models applied on IR systems based on new paradigms [3]. Finally, QPP can be used for new processes such as selective query processing [2].

The QPP++ 2023 workshop aimed to provide a platform for the community to jointly discuss ways to address these challenges and create a better alignment between the latest technologies, retrieval models, and QPP approaches. Along with the challenges mentioned above, the recent advances in NLP present great opportunities for enhancing the state of the art in QPP. The workshop also sought to encourage collaboration between researchers to exploit these opportunities.

QPP and its Evaluation on New Tasks. The quality of QPP methods is typically evaluated by computing the correlation between the scores assigned to queries by a QPP method and the true performance values, e.g., Average Precision (AP), attained for these queries using relevance judgements. Previous research demonstrated the unreliability of this approach when multiple experimental factors (i.e., IR models, corpora, and predictors) are considered [4,6,11]. In addition, researchers demonstrated that high correlation does not necessarily translate to improved retrieval effectiveness [6,9]. These issues are further exacerbated in new domains, such as question answering or conversational search, where the evaluation of the retrieval models is often more challenging. The QPP++ 2023 workshop aimed at fostering discussion in the community regarding these challenges.

2 Workshop Topics and Goals

The workshop provided a forum for researchers and practitioners to discuss the following key research challenges emerging following the recent advances in IR:

- Can existing QPP techniques be exploited, or which new QPP theories and models need to be devised, for new tasks, such as passage-retrieval, question answering, and conversational search?
- How can new technologies, such as contextualized embeddings, large language models, and neural networks be exploited to improve QPP?
- How should QPP techniques be evaluated, including best practices, datasets, and resources?
- Should QPP be evaluated in the same manner for different IR tasks?
- What changes should we make to the QPP evaluation paradigm to accommodate new domains and IR techniques?

The workshop is expected to have two main outcomes:

- We intend to compile the workshop proceedings from the submitted papers. The proceedings will be published in the CEUR-WS.org proceedings series.
- We intend to draft a position paper describing the roadmap identified during the discussions and submit it to the SIGIR forum.

3 Workshop Organizers

Guglielmo Faggioli is a PhD student in Information Engineering at the University of Padua, Italy. His main research concerns information retrieval evaluation, focusing on performance modeling, query performance prediction models, and conversational search systems. His thesis concerns the modeling and prediction of information retrieval systems' performance. He contributed as co-editor to the Proceedings of the Twelfth and Thirteenth International Conference of the CLEF Association (CLEF 2021 and CLEF 2022).

Nicola Ferro is a full professor in computer science at the University of Padua, Italy. He works in information retrieval and its evaluation. He is the coordinator of CLEF (Conference and Labs of the Evaluation Forum) and has organized several evaluation tasks over the years. He has co-authored many papers on IR evaluation, and his current interests are reproducibility of IR experiments, IR system performance modeling and prediction, formal models, and properties of IR evaluation measures. He co-organized several workshops at major conferences, among which GLARE at CIKM 2018, acted as general chair of ECIR 2016, short paper co-chair of ECIR 2020, and resource paper co-chair at CIKM 2021.

Josiane Mothe is full professor in Computer Science at INSPE-Université Toulouse Jean-Jaurès and researcher at Institut de Recherche en Informatique de Toulouse, UMR 5505 CNRS. Her research focuses on information systems and information retrieval, including selective query processing and query performance prediction. She serves as a senior PC member in major conferences in IR, is co-editor of SIGIR Forum, associate editor at TOIS, board member of IR journal, SIGIR 2023 co-chair, ECIR 2021 short papers co-chair, and CIRCLE 2022 general chair.

Fiana Raiber is a senior manager at Yahoo Research. She earned her Ph.D. from the Technion - Israel Institute of Technology, where she currently holds a research associate position, collaborating with faculty members and graduate students. Fiana is a co-author of multiple conference and journal papers in information retrieval, including several publications on query-performance prediction. She served as the SIGIR 2018 workshops co-chair and SIGIR 2022 short papers co-chair. She is a (senior) program committee member of numerous conferences, including SIGIR, ICTIR, WSDM, and CIKM.

References

1. Carmel, D., Yom-Tov, E.: Estimating the query difficulty for information retrieval. Synthesis Lect. Inf. Concepts, Retrieval Serv. **2**(1), 1–89 (2010)
2. Deveaud, R., Mothe, J., Ullah, M.Z., Nie, J.Y.: Learning to adaptively rank document retrieval system configurations. ACM Trans. Inf. Syst. (TOIS) **37**(1), 1–41 (2018)
3. Faggioli, G., Formal, T., Marchesin, S., Clinchant, S., Ferro, N., Benjamin, P.: Query performance prediction for neural IR: are we there yet? In: Proceedings of ECIR (2023)
4. Faggioli, G., Zendel, O., Culpepper, J.S., Ferro, N., Scholer, F.: An enhanced evaluation framework for query performance prediction. In: Proceedings of ECIR, pp. 115–129 (2021)
5. Hauff, C.: Predicting the effectiveness of queries and retrieval systems. Ph.D. Dissertation. University of Twente, pp. 1–179 (2010)
6. Hauff, C., Azzopardi, L., Hiemstra, D.: The combination and evaluation of query performance prediction methods. In: Proceedings of ECIR, pp. 301–312 (2009)
7. Mitra, B., Craswell, N.: An introduction to neural information retrieval. Found. Trends Inf. Retr. **13**(1), 1–126 (2018)
8. Mothe, J., Tanguy, L.: Linguistic features to predict query difficulty. In: Proceedings of SIGIR, pp. 7–10 (2005)
9. Raiber, F., Kurland, O.: Query-performance prediction: setting the expectations straight. In: Proceedings of SIGIR, pp. 13–22 (2014)
10. Roitman, H.: Enhanced performance prediction of fusion-based retrieval. In: ICTIR 2018, pp. 195–198 (2018)
11. Scholer, F., Garcia, S.: A case for improved evaluation of query difficulty prediction. In: Proceedings of SIGIR, pp. 640–641 (2009)
12. Shtok, A., Kurland, O., Carmel, D.: Using statistical decision theory and relevance models for query-performance prediction. In: Proceedings of SIGIR, pp. 259–266 (2010)
13. Thomas, P., Scholer, F., Bailey, P., Moffat, A.: Tasks, queries, and rankers in pre-retrieval performance prediction. In: ADCS 2017, pp. 1–4 (2017)

Bibliometric-Enhanced Information Retrieval: 13th International BIR Workshop (BIR 2023)

Ingo Frommholz[1]([✉]), Philipp Mayr[2], Guillaume Cabanac[3,4], and Suzan Verberne[5]

[1] School of Engineering, Computing and Mathematical Sciences, University of Wolverhampton, Wolverhampton, UK
`ifrommholz@acm.org`
[2] GESIS – Leibniz-Institute for the Social Sciences, Cologne, Germany
`philipp.mayr@gesis.org`
[3] Computer Science Department, University of Toulouse, IRIT UMR 5505, Toulouse, France
`guillaume.cabanac@univ-tlse3.fr`
[4] Institut Universitaire de France (IUF), Paris, France
[5] LIACS, Leiden University, Leiden, The Netherlands
`s.verberne@liacs.leidenuniv.nl`

Abstract. The 13th iteration of the Bibliometric-enhanced Information Retrieval (BIR) workshop series will take place at ECIR 2023 as a full-day workshop. BIR tackles issues related to, for instance, academic search and recommendation, at the intersection of Information Retrieval, Natural Language Processing, and Bibliometrics. As an interdisciplinary scientific event, BIR brings together researchers and practitioners from the Scientometrics/Bibliometrics community on the one hand, and the Information Retrieval community on the other hand. BIR is an ever-growing topic investigated by both academia and the industry.

Keywords: Academic search · Information retrieval · Digital libraries · Bibliometrics · Scientometrics

1 Motivation and Relevance to ECIR

The aim of the Bibliometric-enhanced Information Retrieval (BIR)[1] workshop series and its 13th iteration at ECIR 2023[2] is to bring together researchers and practitioners from Scientometrics/Bibliometrics as well as Information Retrieval (IR). Scientometrics is a sub-field of Bibliometrics which, like IR, is in turn a sub-field of Information Science. Bibliometrics and Scientometrics are concerned with all quantitative aspects of information and academic literature [5], which naturally make them interesting for IR research, in particular when it comes to academic search, recommendation, and other domains in which citations play a

---

[1] https://sites.google.com/view/bir-ws/home?authuser=0.
[2] https://sites.google.com/view/bir-ws/bir-2023?authuser=0.

J. Kamps et al. (Eds.): ECIR 2023, LNCS 13982, pp. 392–397, 2023.
https://doi.org/10.1007/978-3-031-28241-6_43

central role, for example, legal and patent retrieval. In the early 1960 s,s, Salton was already striving to enhance IR by including clues inferred from bibliographic citations [6]. In the course of decades, both disciplines (Bibliometrics and IR) evolved apart from each other over time, leading to the two loosely connected fields we know today [7].

However, the exploding number of scholarly publications and the requirement to satisfy scholars' specific information needs led Bibliometric-enhanced IR to receive growing recognition in the IR as well as the Scientometrics communities. Challenges in academic search and recommendation became particularly apparent during the COVID-19 crisis as well as the ongoing trend of publishing on preprint servers first (and sometimes exclusively), for instance in the rapidly developing field of artificial intelligence and deep learning. This results, for instance, in the information overload researchers and practitioners are facing, as well as the need to ensure the timeliness and quality of published research. Tackling these challenges requires effective and efficient solutions for scholarly search, recommendation and discovery of high-quality publications and heterogeneous data. Bibliometric-enhanced IR tries to provide these solutions to the peculiar needs of scholars to keep on top of the research in their respective fields, utilising the wide range of suitable relevance signals that come with academic scientific publications, such as keywords provided by authors, topics extracted from the full-texts, co-authorship networks, citation networks, altmetrics, bibliometric figures, and various classification schemes of science. Bibliometric-enhanced IR systems must deal with the multifaceted nature of scientific information by searching for or recommending academic papers, patents, venues (i.e., conference proceedings, journals, books, manuals, grey literature), authors, experts (e.g., peer reviewers), references (to be cited to support an argument), and datasets. A further discussion of the various research directions in bibliometric-enhanced IR can be found in [3].

To this end, the BIR workshop series was founded in 2014 [4] to tackle these challenges by tightening up the link between IR and Bibliometrics. We strive to bring the 'retrievalists' and 'citationists' [7] active in both academia and industry together. The success of past BIR events, as shown in Table 1, evidences that BIR@ECIR is a much-needed interdisciplinary scientific event that attracts researchers and practitioners from IR, Bibliometrics, and Natural Language Processing alike.

2   Workshop Goals and Objectives

Our vision is to bring together researchers and practitioners from Scientometrics/Bibliometrics on the one hand and IR, on the other hand, to create better methods and systems for instance for academic search and recommendation. Our view is to expose people from one community to the work of the respective other community and to foster fruitful interaction across communities. Therefore, in the call for papers for the 2023 BIR workshop at ECIR, we address, but are not limited to, current research issues regarding 3 aspects of the academic search/recommendation process:

Table 1. Overview of the BIR workshop series and CEUR proceedings

Year	Conference	Venue	Papers	Proceedings
2014	ECIR	Amsterdam, NL	6	Vol-1143
2015	ECIR	Vienna, AT	6	Vol-1344
2016	ECIR	Padua, IT	8	Vol-1567
2016	JCDL	Newark, US	$10 + 10^a$	Vol-1610
2017	ECIR	Aberdeen, UK	12	Vol-1823
2017	SIGIR	Tokyo, JP	11	Vol-1888
2018	ECIR	Grenoble, FR	9	Vol-2080
2019	ECIR	Cologne, DE	14	Vol-2345
2019	SIGIR	Paris, FR	$16 + 10^b$	Vol-2414
2020	ECIR	Lisbon (Online), PT	9	Vol-2591
2021	ECIR	Lucca (Online), IT	9	Vol-2847
2022	ECIR	Stavanger, NO	5	Vol-3230

[a] with CL-SciSumm 2016 Shared Task; [b] with CL-SciSumm 2019 Shared Task

1. User needs and behaviour regarding scientific information, such as:
   - Finding relevant papers/authors for a literature review.
   - Filtering high-quality research papers, e.g. in preprint servers.
   - Measuring the degree of plagiarism in a paper.
   - Identifying expert reviewers for a given submission.
   - Flagging predatory conferences and journals, or other forms of scientific misbehaviour
   - Understanding information-seeking behaviour and HCI in academic search.
2. Mining the scientific literature, such as:
   - Information extraction, text mining and parsing of scholarly literature.
   - Natural language processing (e.g., citation contexts).
   - Discourse modelling and argument mining.
3. Academic search/recommendation systems, such as:
   - Modelling the multifaceted nature of scientific information.
   - Building test collections for reproducible BIR.
   - System support for literature search and recommendation.

3   Target Audience and Dissemination

The target audience of the BIR workshops is researchers and practitioners, junior and senior, from Scientometrics as well as IR and Natural Language Processing (NLP). These could be IR/NLP researchers interested in potential new application areas for their work as well as researchers and practitioners working with bibliometric data and interested in how IR/NLP methods can make use of such data. BIR 2022 will be open to anyone interested in the topic.

The BIR organisers are well established in the IR and the Bibliometrics community, respectively and have a years-long experience as workshop organizers.

We will send the call for papers to major professional mailing lists in IR (ACM SIGIR, IR-List, JISC IR) and Bibliometrics (ASIS&T Sigmetrics and ISSI). We will also send it to the former BIR and BIRNDL participants (in the range of a few hundred people) and scientists who publish in both IR and Bibliometrics venues, based on the mining of the DBLP. We will further advertise the call for papers through our social media channels.

The 10th-anniversary edition in 2020 ran online with an audience peaking at 97 online participants [1]. BIR 2022, the 12th edition, was run as a hybrid event with on-site and online participants. We were surprised by how successful this model went and how satisfied the speakers, audience, and organizers were with the hybrid workshop.

In December 2020, we published our third special issue emerging from the past BIR workshops [2].

4   Workshop Format, Structure and Peer Review Process

Our peer review process will be supported by Easychair. Each submission is assigned to 2 to 3 reviewers, preferably at least one expert in IR and one expert in Bibliometrics or NLP. The programme committee for 2023 will consist of peer reviewers from all participating communities. Accepted papers are either long papers (15-min talks) or short papers (5-min talks). Two interactive sessions close the morning and afternoon sessions with posters and demos, allowing attendees to discuss the latest developments in the field and opportunities (e.g., shared tasks). We also invite attendees to demonstrate prototypes during flash presentations (5 min).

These interactive sessions serve as ice-breakers, sparking interesting discussions that usually continue during lunch and the evening social event. The sessions also allow our speakers to further discuss their work. BIR has a friendly and open atmosphere where there is an opportunity for participants (including students) to share their ideas and current work and to receive feedback from the community.

5   Previous Workshops

The BIR workshop series has a long tradition of taking place along major IR conferences such as ECIR, SIGIR, and JCDL, as documented on the BIR overview page. Table 1 provides an overview of past events. BIR@ECIR2023 would be the continuation of a highly successful conference series.

6   Organizers

Ingo Frommholz is Reader (Associate Professor equivalent) in Data Science at the University of Wolverhampton, UK. He has been working on formal IR systems and models taking user aspects into account. He has been co-organizing several IR-related research events, for instance, the IR and Foraging Autumn School at Schloß Dagstuhl, Germany, the Future Directions in Information Access (FDIA) symposia, BCS Search Solutions, BIRDS 2020 and 2021 (at SIGIR and CHIIR, respectively) and the ongoing Bibliometrics and IR (BIR@ECIR) workshop series. He is managing editor of the *International Journal on Digital Libraries* (IJDL, Springer).

Philipp Mayr is a team leader at the GESIS – Leibniz-Institute for the Social Sciences department Knowledge Technologies for the Social Sciences. His research group *Information and Data Retrieval* is working on methods and techniques of interactive information and dataset retrieval and maintains and further develops information systems for the social sciences. He has been co-organizing several IR-related research events, for instance, the Workshops on Scholarly Document Processing at EMNLP/NAACL/COLING and the ongoing Bibliometrics and IR (BIR@ECIR) workshop series.

Guillaume Cabanac is a Professor of Computer Science at the University of Toulouse and holds a research chair at the Institut Universitaire de France. His interdisciplinary research on the quantitative study of science is at the crossroads between Information Retrieval, Digital Libraries, and Scientometrics. He serves on the editorial boards of the *Journal of the Association for Information Science and Technology (JASIST)* and *Scientometrics*. His current work on the Problematic Paper Screener contributes to the identification and reporting of algorithmically generated and fraudulent papers published-often sold-by academic publishers. Cabanac was nicknamed 'Deception sleuth' in the *Nature's 10* list of 'ten people who helped shape science in 2021.'

Suzan Verberne is an associate professor at the Leiden Institute of Advanced Computer Science (LIACS) at Leiden University and group leader of Text Mining and Retrieval Leiden. She obtained her PhD in 2010 on the topic of Question Answering and has since then been working on the edge between Natural Language Processing (NLP) and Information Retrieval (IR). Her recent work centers around interactive information access for specific domains, covering academic, biomedical, legal, and patent data. She is highly active in the NLP and IR communities, holding chairing positions in large worldwide conferences, and is an associate editor for *Transactions on Information Systems*.

References

1. Cabanac, G., Frommholz, I., Mayr, P.: Report on the 10th anniversary workshop on bibliometric-enhanced information retrieval (BIR 2020). SIGIR Forum **54**(1) (2020). https://doi.org/10.1145/3451964.3451974
2. Cabanac, G., Frommholz, I., Mayr, P.: Scholarly literature mining with information retrieval and natural language processing: Preface. Scientometrics **125**(3), 2835–2840 (2020). https://doi.org/10.1007/s11192-020-03763-4
3. Frommholz, I., Cabanac, G., Mayr, P., Verberne, S.: Report on the 11th bibliometric-enhanced information retrieval workshop (BIR 2021). SIGIR Forum **55**(1) (2021). https://doi.org/10.1145/3476415.3476426

4. Mayr, P., Scharnhorst, A., Larsen, B., Schaer, P., Mutschke, P.: Bibliometric-enhanced information retrieval. In: de Rijke, M., et al. (eds.) ECIR 2014. LNCS, vol. 8416, pp. 798–801. Springer, Cham (2014). https://doi.org/10.1007/978-3-319-06028-6_99
5. Pritchard, A.: Statistical bibliography or bibliometrics? [Documentation notes]. J. Document. **25**(4), 348–349 (1969). https://doi.org/10.1108/eb026482
6. Salton, G.: Associative document retrieval techniques using bibliographic information. J. ACM **10**(4), 440–457 (1963). https://doi.org/10.1145/321186.321188
7. White, H.D., McCain, K.W.: Visualizing a discipline: An author co-citation analysis of Information Science, 1972–1995. J. Am. Soc. Inf. Sci. **49**(4), 327–355 (1998). b57vc7

Geographic Information Extraction
from Texts (GeoExT)

Xuke Hu[1]([✉])(ID), Yingjie Hu[2](ID), Bernd Resch[3](ID), and Jens Kersten[1](ID)

[1] Institute of Data Science, German Aerospace Center (DLR), Jena, Germany
{xuke.hu,jens.kersten}@dlr.de
[2] Department of Geography, University at Buffalo, Buffalo, USA
yhu42@buffalo.edu
[3] Department of Geoinformatics, University of Salzburg, Salzburg, Austria
bernd.resch@plus.ac.at

Abstract. A large volume of unstructured texts, containing valuable
geographic information, is available online. This information – provided
implicitly or explicitly – is useful not only for scientific studies (e.g.,
spatial humanities) but also for many practical applications (e.g., geographic information retrieval). Although large progress has been achieved
in geographic information extraction from texts, there are still unsolved
challenges and issues, ranging from methods, systems, and data, to applications and privacy. Therefore, this workshop will provide a timely opportunity to discuss the recent advances, new ideas, and concepts but also
identify research gaps in geographic information extraction.

Keywords: Geographic information extraction · Document
geolocation · Geoparsing · Toponym recognition · Toponym resolution

1 Motivation and Relevance to ECIR

Huge and ever-increasing amounts of semi- and unstructured text data, like
news, scientific articles, historical archives, travel blogs, and social media posts
are available online and offline. These documents often refer to geographic regions
or specific places on earth, and contain valuable but hidden geographic information in the form of toponyms, location references, and more complex location
descriptions [10]. Scientists from many fields, like information retrieval, natural language processing, and geographic information science, have an increased
interest in researching and applying methods to infer the geographical focus of
documents [18] or to extract geographic references from unstructured and heterogeneous texts and finally resolve these references unambiguously to places [14]
or spaces (e.g., a geographic grid) [4] on the earth's surface. The information is
useful not only for scientific studies (e.g., sociolinguistics and spatial humanities) [20], but can also contribute to many practical applications [8], such as
geographic information retrieval [22], disaster management [23], urban planning
[21], disease surveillance [1], tourism planning [5], and crime prevention [2].

© The Author(s), under exclusive license to Springer Nature Switzerland AG 2023
J. Kamps et al. (Eds.): ECIR 2023, LNCS 13982, pp. 398–404, 2023.
https://doi.org/10.1007/978-3-031-28241-6_44

Despite the encouraging progress in geographic information extraction, there are still many unsolved challenges and issues, ranging from methods, systems, and data, to applications and privacy. For example, the approaches for geolocating documents at hyper-local levels and resolving fine-grained toponyms (e.g., POI and streets) in informal texts (e.g., tweets), complex location descriptions (e.g., 'intersection of North-South Road and Great East Road'), and historical or ancient toponyms need further improvements. Unified and well-generalizing solutions satisfying several application-related needs (e.g., correctness and speed) are still missing. Furthermore, there is a lack of open, sufficient, and high-quality datasets for evaluating different characteristics (e.g., considering different text and toponym types) of existing approaches. Gazetteers need to be updated, completed, and further enriched to support the resolving of both, the latest geospatial changes as well as ancient toponyms and toponym variants. Last but not least, more attention should be paid to protecting the location privacy of online users.

Geographic information extraction from texts is an important sub-task of information extraction and retrieval and has been investigated by many researchers from information retrieval. The topic of this proposal is thus relevant to ECIR.

2 Workshop Goals

2.1 Objectives and Overall Vision

In this workshop, we aim to foster the discussion and exchange on recent advances in geographic information extraction from unstructured texts. Special emphasis for this first workshop is on methods for document geolocation and toponym recognition and resolution. Our goal is to establish a common, long-term forum to consolidate multi-disciplinary efforts addressing both, research and practitioners in Europe and beyond. Accordingly, submissions addressing several different aspects of geographic information extraction are welcome. Our general topics of interest are related, but not limited, to the following:

- Document geocoding
- Toponym recognition and resolution
- Toponym matching
- Method generalizability (regions, languages, data sources)
- Multi-source data (e.g., text and image) fusion for method improvement
- Location description extraction and resolution
- Relation extraction and disambiguation
- Tweets geolocating at hyper-local levels
- Fine-grained toponym resolving
- Historical toponym resolving
- Historical archive geovisualization
- Gazetteer enrichment for resolving ancient and evolving toponyms
- Dataset annotation for method training and evaluation
- Platforms and metrics for approach evaluation

- Framework or service to support solution customizing
- Location privacy protection for online users
- Standards or unified interfaces for system development
- Novel applications of geographic information in texts
- Analysing and processing of geotagged social media.

Even though we seek to gather and address researchers and practitioners, contributions focusing on methods and data are highly welcome.

2.2   Desired Outcomes

We aim to support continued discussions and exchanges between researchers but also between researchers and practitioners. The workshop at ECIR'23 itself would be a great starting point for this, and we aim to build up a network for continuous exchange. Already during the preparation of the workshop, a fruitful discussion between the involved actors came up and we would like to support this, e.g. through accompanying interactive methods and tool demonstrations. In this regard, the following outcomes are desired:

- Collection of papers, posters, data sets, and tools submitted by participants
- Compilation of overall findings contributed by participants in a summarizing paper
- Establish a common shared task in the future
- Establish a network/interest group of researchers and practitioners
- Foster new collaborations in national and international projects
- Paving the road towards follow-up ECIR workshops on this topic.

3   Format and Structure

Since this would be the first joint workshop of the organising team at ECIR, we initially would like to allocate a half-day workshop. Intended to support discussions and dialogues, we would like to encourage the following types of submissions:

- Full paper + oral presentation
- Short paper + oral presentation
- Short paper + poster presentation
- Demonstration paper + corresponding hands-on demo.

The papers (original, unpublished contributions) will be included in an open-access proceedings volume of CEUR Workshop Proceedings (indexed by both Scopus and DBLP).

4   Organisers

Xuke Hu received his doctoral degree (Dr.-Ing.) in Geoinformatics from Universität Heidelberg, Germany in 2020. Before this, he obtained a Master's and Bachelor's Degree in Software Engineering at the China University of Geosciences (Wuhan). He is now a senior researcher at the Data Science Institute of the German Aerospace Center (DLR) and acting as the Secretary of the Working Group ('Global mapping for SDGs') of ISPRS Technical Commission IV/III (2023–2026). He has several years of working experience in geographic information extraction, especially toponym recognition and resolution in the context of disaster management. He recently proposed and published two approaches named GazPNE [6] and GazPNE2 [9] and thoroughly reviewed and compared [8] the approaches for extracting place names from texts, and currently researches the fusion of toponym disambiguation methods [7].

Yingjie Hu Yingjie Hu received his PhD in Geography from the University of California, Santa Barbara in 2016. He obtained his Master's and Bachelor's Degree in Cartography and Geographic Information Science from East China Normal University. He is an Assistant Professor in the Department of Geography at the University at Buffalo (UB) and an Affiliated Professor of the UB AI and Data Science Institute. He has been involved in research on extracting location descriptions and place names from texts since 2012 [10–12]. His recent work investigates the use of AI methods to recognize location descriptions from social media messages posted during natural disasters to help reach the people in need [13, 24].

Bernd Resch is an Associate Professor at the University of Salzburg's Department of Geoinformatics - Z_GIS and a Visiting Scholar at Harvard University (USA). He heads the Geo-social Analytics Lab and the iDEAS:lab. Bernd Resch did his PhD in the area of "Live Geography" (real-time monitoring of environmental geo-processes) together with University of Salzburg and MIT. His research interest revolves around understanding cities as complex systems through analysing a variety of digital data sources, focusing on developing machine learning algorithms to analyse human-generated data like social media posts and physiological measurements from wearable sensors. The findings are relevant to a number of fields including urban research, disaster management, epidemiology, and others. Bernd received the Theodor Körner Award for his work on "Urban Emotions". Amongst a variety of other functions, he is an Editorial Board Member of IJHG, IJGI and PLOS ONE, a scientific committee member of various international conferences (having chaired several conferences), speaker of the Faculty of Digital and Analytical Sciences at PLUS, and an Executive Board Member of Spatial Services GmbH.

Jens Kersten received his doctoral degree (Dr.-Ing.) in remote sensing and computer vision and his diploma in geodesy from the Technical University of Berlin, Germany. He is a senior researcher at the German Aerospace Center (DLR). His group at DLR's Institute of Data Science researches methods to acquire and analyze web- and social media-data for applications related to civil security, natural hazards and environmental impacts. His research focuses on machine learning methods for a robust [17,19] and flexible [15] identification of crisis-related social media content, as well as on the orchestration of ML methods for multi-facteted analyses of geospatial data [16]. Developing new methods for place name extraction [9] and resolution [7] are essential to address the groups overarching objective of supporting operational crisis response [3].

Kristin Stock is Director of the Massey Geoinformatics Collaboratory and Associate Professor in Computer Science and Information Technology. She has qualifications in surveying (and is a Registered Surveyor), urban and regional planning, computer science and geospatial science, and her PhD addressed the subject of semantic integration of geographic databases. She has worked as a mining, engineering and cadastral surveyor in private practice and in 1999 she designed a land information system for the Australian Capital Territory government that is still in use. Kristin ran a successful geospatial data consultancy in the UK for several years, led several large international projects and is currently leading the BioWhere and QuakeText projects. Her current research focuses on extracting geospatial knowledge from text sources using natural language processing techniques in application areas including disaster management and biological collections.

5   Potential PC Members and Invited Speakers

Prof. Ross Purves from the University of Zurich will give a keynote talk for the workshop.

PC members

- Beatrice Alex, Edinburgh University, UK
- Mariona Coll Ardanuy, The Alan Turing Institute, UK
- Andrea Ballatore, King's College London, UK
- Tao Cheng, University College London, UK
- Hongchao Fan, Norwegian University of Science and Technology, Norway
- Matthias Hagen, FSU Jena, Germany
- Krzysztof Janowicz, Universität Wien, Austria
- Christopher B. Jones, Cardiff University, UK
- Hao Li, Technische Universität München, Germany
- Nicolás José Fernández Martínez, University of Jaén, Spain
- Bruno Martins, University of Lisbon, Portugal
- Katherine McDonough, The Alan Turing Institute, UK
- Ludovic Moncla, INSA Lyon, France
- Ross Purves, University of Zurich, Switzerland
- Yeran Sun, University of Lincoln, UK
- Thora Tenbrink, Bangor University, UK
- Diedrich Wolter, University of Bamberg, Germany
- Zhiyong Zhou, University of Zurich, Switzerland
- Benjamin Adams, University of Canterbury, New Zealand
- Fuqiang Gu, Chongqing University, China

- Muhammad Imran, Qatar Computing Research Institute, Qatar
- Morteza Karimzadeh, University of Colorado Boulder, US
- Grant McKenzie, McGill University, Canada
- Qinjun Qiu, China University of Geosciences (Wuhan), China
- Stephan Winter, University of Melbourne, Australia.

References

1. Allen, T., et al.: Global hotspots and correlates of emerging zoonotic diseases. Nat. Commun. **8**(1), 1–10 (2017)
2. Arulanandam, R., Savarimuthu, B.T.R., Purvis, M.A.: Extracting crime information from online newspaper articles. In: Proceedings of the Second Australasian Web Conference, vol. 155, pp. 31–38 (2014)
3. Bongard, J., Kersten, J., Klan, F.: Searching and structuring the twitter stream for crisis response: a flexible concept to support research and practice. In: Proceedings of the 4th International Open Search Symposium, OSSYM (2022)
4. Gritta, M., Pilehvar, M.T., Limsopatham, N., Collier, N.: What's missing in geographical parsing? Lang. Resour. Eval. **52**(2), 603–623 (2018)
5. Haris, E., Gan, K.H.: Mining graphs from travel blogs: a review in the context of tour planning. Inf. Technol. Tourism **17**(4), 429–453 (2017). https://doi.org/10.1007/s40558-017-0095-2
6. Hu, X., et al.: Gazpne: annotation-free deep learning for place name extraction from microblogs leveraging gazetteer and synthetic data by rules. Int. J. Geogr. Inf. Sci. **36**(2), 310–337 (2021)
7. Hu, X., Sun, Y., Kersten, J., Zhou, Z., Klan, F., Fan, H.: How can voting mechanisms improve the robustness and generalizability of toponym disambiguation? arXiv preprint arXiv:2209.08286 (2022)
8. Hu, X., et al.: Location reference recognition from texts: a survey and comparison. arXiv preprint arXiv:2207.01683 (2022)
9. Hu, X., et al.: Gazpne2: a general place name extractor for microblogs fusing gazetteers and pretrained transformer models. IEEE Internet Things J. **9**(17), 16259–16271 (2022). https://doi.org/10.1109/JIOT.2022.3150967
10. Hu, Y., Adams, B.: Harvesting big geospatial data from natural language texts. In: Werner, M., Chiang, Y.-Y. (eds.) Handbook of Big Geospatial Data, pp. 487–507. Springer, Cham (2021). https://doi.org/10.1007/978-3-030-55462-0_19
11. Hu, Y., Janowicz, K.: Improving personal information management by integrating activities in the physical world with the semantic desktop. In: Proceedings of the 20th International Conference on Advances in Geographic Information Systems, pp. 578–581 (2012)
12. Hu, Y., Janowicz, K., Prasad, S.: Improving wikipedia-based place name disambiguation in short texts using structured data from dbpedia. In: Proceedings of the 8th Workshop on Geographic Information Retrieval, pp. 1–8 (2014)
13. Hu, Y., Wang, J.: How do people describe locations during a natural disaster: an analysis of tweets from hurricane harvey. arXiv preprint arXiv:2009.12914 (2020)
14. Karimzadeh, M., Pezanowski, S., MacEachren, A.M., Wallgrün, J.O.: Geotxt: a scalable geoparsing system for unstructured text geolocation. Trans. GIS **23**(1), 118–136 (2019)

15. Kersten, J., Bongard, J., Klan, F.: Gaussian processes for one-class and binary classification of crisis-related tweets. In: Proceedings of the 19th International Conference on Information Systems for Crisis Response and Management, ISCRAM (2022)

16. Kersten, J., Klan, F.: What happens where during disasters? a workflow for the multifaceted characterization of crisis events based on twitter data. J. Contingencies Crisis Manage. **28**(3), 262–280 (2020)

17. Kersten, J., Kruspe, A., Wiegmann, M., Klan, F.: Robust filtering of crisis-related tweets. In: Proceedings of the 16th International Conference on Information Systems for Crisis Response and Management, 19–22 May ISCRAM (2019)

18. Kinsella, S., Murdock, V., O'Hare, N.: " I'm eating a sandwich in glasgow" modeling locations with tweets. In: Proceedings of the 3rd International Workshop on Search and Mining User-Generated Contents, pp. 61–68 (2011)

19. Kruspe, A., Kersten, J., Klan, F.: Review article: detection of actionable tweets in crisis events. Nat. Hazards Earth Syst. Sci. **21**(6), 1825–1845 (2021)

20. Melo, F., Martins, B.: Automated geocoding of textual documents: a survey of current approaches. Trans. GIS **21**(1), 3–38 (2017)

21. Milusheva, S., Marty, R., Bedoya, G., Williams, S., Resor, E., Legovini, A.: Applying machine learning and geolocation techniques to social media data (twitter) to develop a resource for urban planning. PLOS ONE **16**(2), 1–12 (2021). https://doi.org/10.1371/journal.pone.0244317

22. Purves, R.S., Clough, P., Jones, C.B., Hall, M.H., Murdock, V.: Geographic information retrieval: Progress and challenges in spatial search of text. Found. Trendső Inf. Retrieval **12**(2–3), 164–318 (2018). https://doi.org/10.1561/1500000034

23. Scalia, G., Francalanci, C., Pernici, B.: Cime: context-aware geolocation of emergency-related posts. Geoinformatica **26**(1), 125–157 (2022). https://doi.org/10.1007/s10707-021-00446-x

24. Wang, J., Hu, Y., Joseph, K.: Neurotpr: a neuro-net toponym recognition model for extracting locations from social media messages. Trans. GIS **24**(3), 719–735 (2020)

ROMCIR 2023: Overview of the 3rd Workshop on Reducing Online Misinformation Through Credible Information Retrieval

Marinella Petrocchi[1] and Marco Viviani[2]

[1] Institute of Informatics and Telematics – National Research Council (IIT – CNR), Pisa, Italy
marinella.petrocchi@iit.cnr.it

[2] Department of Informatics, Systems, and Communication (UNIMIB – DISCo), University of Milano-Bicocca, Milan, Italy
marco.viviani@unimib.it
https://www.iit.cnr.it/marinella.petrocchi/ ,
https://ikr3.disco.unimib.it/people/marco-viviani/

Abstract. With the advent of the Social Web, we are constantly and more than ever assaulted by different kinds of information pollution, which may lead to severe issues for both individuals and society as a whole. In this context, it becomes essential to guarantee users access to genuine information that does not distort their perception of reality. For this reason, in recent years, numerous approaches have been proposed for the identification of misinformation, in different contexts and for different purposes. However, the problem has not yet been sufficiently addressed in the field of Information Retrieval, because it has been treated primarily as a classification task to identify information versus misinformation. Hence, the purpose of this Workshop is to address the IR community for solutions in which, among other issues, the genuineness of information is considered as one of the dimensions of relevance within search engines or recommender systems, early detection of misinformation can be achieved, the results obtained are explainable with respect to the users of Information Retrieval Systems, user's privacy is taken into consideration.

Keywords: Information retrieval · Information access · Information disorder · Information genuineness · Misinformation

1 Motivation and Relevance to ECIR

"Technology is so much fun but we can drown in our technology.
The fog of information can drive out knowledge".

This quote by American historian Daniel J. Boorstin, about the computerization of libraries, appeared in July 1983 in the New York Times. Some 40 years later, it is probably still more relevant than it was then. In fact, the lack of mediation that Web 2.0 technologies brought in the generation and dissemination of online

J. Kamps et al. (Eds.): ECIR 2023, LNCS 13982, pp. 405–411, 2023.
https://doi.org/10.1007/978-3-031-28241-6_45

content within the Social Web has led to the well-known problems of *information overload* [23] and the spread of *misinformation* [6], which make it difficult for users to find information that is truly useful for their purposes, leading to possible severe issues for both individuals and society as a whole [3,10,20,26]. Fake news can, for example, guide public opinion in political and financial choices [24, 30]; false reviews can promote or discredit economic activities [5,14]; unverified medical information can lead people to follow behaviors that can be harmful to their own health and to that of society as a whole (let the reader think, for example, to the set of unverified news stories that have been disseminated about Covid-19) [4,8,18,25].

Hence, the central topic of the third edition of the ROMCIR Workshop concerns providing access to users to both topically relevant and genuine information (without neglecting other relevant dimensions), to mitigate the *information disorder* phenomenon with respect to distinct domains. By "information disorder" we mean all forms of communication pollution, from misinformation made out of ignorance, to the intentional sharing of false content [27,28]. In this context, all those approaches that can serve to assess the genuineness of information circulating online and in social media find their place. This topic is very broad, as it concerns different contents (e.g., Web pages, news, reviews, medical information, online accounts, etc.), different Web and social media platforms (e.g., microblogging platforms, social networking services, social question-answering systems, etc.), and different purposes (e.g., identifying false information, accessing information based on its genuineness, retrieving genuine information, etc.). Each of these aspects is relevant to the European Conference on Information Retrieval.

2   Aim and Topics of Interest

The aim of the Workshop is to generate a discussion on, and possibly provide countermeasures to, the problem of online information disorder, by providing access to information that is genuine. Given that the problem in recent years has been addressed from various points of view (e.g., fake news detection, bot detection, information genuineness assessment, and news source reputability, to cite a few), the purpose of this Workshop is to consider these issues in the context of Information Access and Retrieval, also considering related Artificial Intelligence fields such as Natural Language Processing (NLP), Natural Language Understanding (NLU), Computer Vision, Machine and Deep Learning. Hence, the topics of interest of ROMCIR 2023 include, but are not limited to, the following:

- Access to genuine information
- Bias detection
- Bot/spam/troll detection
- Computational fact-checking
- Crowd-sourcing for information genuineness assessment
- Deep fakes

- Disinformation/misinformation detection
- Evaluation strategies to assess information genuineness
- Fake news/review detection
- Harassment/bullying/hate speech detection
- Information polarization in online communities, echo chambers
- Propaganda identification/analysis
- Retrieval of genuine information
- Security, privacy and information genuineness
- Sentiment/emotional analysis
- Stance detection
- Trust and reputation
- Societal reaction to misinformation

3 Past Editions

The first two editions of the Workshop, both co-located with the ECIR conference, led to fervent discussion and presentation of innovative work with respect to a variety of open issues related to information genuineness and IR. The first edition, namely ROMCIR 2021, took place in online mode on April 1, 2021;[1] the second edition, namely ROMCIR 2022 took place in hybrid mode on April 10, 2022.[2] Further details are provided here below.

- ROMCIR 2021 received 15 submissions, of which 6 were accepted, with an acceptance rate of 40%. The accepted articles, collected in CEUR Proceedings [22], proposed distinct solutions for distinct open issues. There were those tangentially related to Credible Information Retrieval, such as those of *authorship verification* [29] and *bias detection* in science evaluation [2]. Furthermore, the problems of opinion mining and misinformation identification were tackled, such as those of *hate speech detection* [11] and *claim verification* [13]. Finally, the problem of access to genuine information was considered, by proposing the definition of Information Retrieval Systems to support users in *retrieving genuine news* [12], and the study of new IR methods able to consider the *genuineness* of the data collected in the retrieval process [7].
- ROMCIR 2022 received 10 submissions, of which 6 were accepted, so with an acceptance rate of 60%. The articles, collected in CEUR Proceedings [21] and summarized in the June 2022 SIGIR Forum volume,[3] have primarily considered two issues from distinct and new points of view compared to those considered the previous year, reflecting the interest and new challenges related to the aim of the Workshop. The first theme concerned *genuine health* IR [9,15,16,19], an application domain that is generating much discussion among researchers; the second theme concerned *multi-modal genuine* IR [1,17], and papers discussed the challenges that multimedia content analysis were raising with respect to those of text content analysis in information genuineness assessment.

---

[1] https://romcir.disco.unimib.it/2021-edition/2021-workshop/.

[2] https://romcir.disco.unimib.it/2022-edition/2022-workshop/.

[3] https://sigir.org/forum/issues/june-2022/.

4   List of Organizers

The following people contributed in different capacities to the organization of the Workshop and to the verification of the quality of the submitted work.

4.1   Workshop Chairs

Marinella Petrocchi is a Senior Researcher at the Institute of Informatics and Telematics of the National Research Council (IIT-CNR) in Pisa, Italy, under the Trust, Security, and Privacy research unit. She also collaborates with the Sysma unit at IMT School for Advanced Studies, in Lucca, Italy. Her field of research lies between Cybersecurity, Artificial Intelligence, and Data Science. Specifically, she studies novel techniques for online fake news/fake accounts detection and automated methods to rank the reputability of online news media. She is the author of several international publications on these themes and she usually gives talks and lectures on the topic. She is in the core team of the TOFFEe project (TOols for Fighting FakEs), funded by IMT. *Website*: https://www.iit.cnr.it/marinella. petrocchi/

Marco Viviani is an Associate Professor at the University of Milano-Bicocca, Department of Informatics, Systems, and Communication (DISCo). He works in the Information and Knowledge Representation, Retrieval and Reasoning (IKR3) Lab. He has been co-chair of several special tracks and workshops at international conferences, and general co-chair of MDAI 2019. He is an Associate Editor of Social Network Analysis and Mining, an Editorial Board Member of Online Social Networks and Media, and a Guest Editor of several Special Issues in International Journals related to information disorder detection. His main research activities include Social Computing, Information Retrieval, Text Mining, Natural Language Processing, Trust and Reputation Management, and User Modeling. On these topics, he has written several international publications. *Website*: https://ikr3.disco. unimib.it/people/marco-viviani/

4.2   Proceedings Chair

Rishabh Upadhyay. is a Research Fellow at the University of Milano-Bicocca, Department of Informatics, Systems, and Communication (DISCo).

His research interests are related to Machine and Deep Learning, Information Retrieval, and Social Computing. He is currently working within the EU Horizon 2020 ITN/ETN DoSSIER project on Domain-Specific Systems for Information Extraction and Retrieval, in particular on the project: "Assessing Credibility, Value, and Relevance". He was one of the co-organizers of Task 2: Consumer Health Search, at CLEF 2021 eHealth Lab Series. He has recently published papers at International Conferences on the topic of health misinformation detection. *Website*: https://www.unimib.it/rishabh-gyanendra-upadhyay/

4.3 Program Committee Members

- Rino Falcone, ISTC – National Research Council (CNR), Italy;
- Carlos A. Iglesias, Universidad Politécnica de Madrid, Spain
- Petr Knoth, Research Studio Data Science (DSc), Austria
- Udo Kruschwitz, University of Regensburg, Germany
- Yelena Mejova, ISI Foundation, Italy
- Preslav Nakov, Hamad Bin Khalifa University, Qatar
- Symeon Papadopoulos, Information Technologies Institute (ITI), Greece
- Gabriella Pasi, University of Milano-Bicocca, Italy
- Marinella Petrocchi, IIT – National Research Council (CNR), Italy
- Francesco Pierri, Politecnico di Milano, Italy
- Manuel Pratelli, IMT School for Advanced Studies Lucca, Italy
- Fabio Saracco, Centro Ricerche Enrico Fermi, Italy
- Marco Viviani, University of Milano-Bicocca, Italy
- Arkait Zubiaga, Queen Mary University, UK

Acknowledgements. This work was supported by re-DESIRE: *DissEmination of ScIentific REsults* 2.0, funded by IIT–CNR, TOFFEe: *TOols for Fighting FakES*, funded by IMT Scuola Alti Studi Lucca, and DoSSIER: *Domain Specific Systems for Information Extraction and Retrieval*, EU Horizon 2020 ITN/ETN (H2020-EU.1.3.1., ID: 860721).

References

1. Aghada, K.A.: An alternative approach to ranking videos and measuring dissimilarity between video content and titles. In: Petrocchi, M., Viviani, M. (eds.) ROMCIR 2022 CEUR Workshop Proceedings, vol. 3138, pp. 89–99. CEUR-WS.org (2022)
2. Bethencourt, A.M., Luo, J., Feliciani, T.: Bias and truth in science evaluation: a simulation model of grant review panel discussions. In: ROMCIR 2021 CEUR Workshop Proceedings, vol. 2838, pp. 16–24 (2021)
3. Caldarelli, G., De Nicola, R., Petrocchi, M., Del Vigna, F., Saracco, F.: The role of bot squads in the political propaganda on Twitter. Commun Phys. **3**, 81 (2020)
4. Caldarelli, G., De Nicola, R., Petrocchi, M., Pratelli, M., Saracco, F.: Flow of online misinformation during the peak of the covid-19 pandemic in Italy. EPJ Data Sci. **10**(1), 34 (2021)
5. Crawford, M., Khoshgoftaar, T.M., Prusa, J.D., Richter, A.N., Al Najada, H.: Survey of review spam detection using machine learning techniques. J. Big Data **2**(1), 1–24 (2015). https://doi.org/10.1186/s40537-015-0029-9

6. Del Vicario, M., et al.: The spreading of misinformation online. Proc. Natl. Acad. Sci. **113**(3), 554–559 (2016)
7. Denaux, R., Gomez-Perez, J.M.: Sharing retrieved information using linked credibility reviews. In: ROMCIR 2021 CEUR Workshop Proceedings, vol. 2838, pp. 59–65 (2021)
8. Di Sotto, S., Viviani, M.: Health misinformation detection in the social web: an overview and a data science approach. Int. J. Environ. Res. Public Health **19**(4), 2173 (2022)
9. Fröbe, M., Günther, S., Bondarenko, A., Huck, J., Hagen, M.: Using keyqueries to reduce misinformation in health-related search results. In: Petrocchi, M., Viviani, M. (eds.) ROMCIR 2022 CEUR Workshop Proceedings, vol. 3138, pp. 1–10. CEUR-WS.org (2022)
10. Ginsca, A.L., Popescu, A., Lupu, M., et al.: Credibility in information retrieval. Found. Trends® Inf. Retrieval **9**(5), 355–475 (2015)
11. Gupta, S., Nagar, S., Nanavati, A.A., Dey, K., Barbhuiya, F.A., Mukherjea, S.: Consumption of hate speech on twitter: a topical approach to capture networks of hateful users. In: ROMCIR 2021 CEUR Workshop Proceedings, vol. 2838, pp. 25–34 (2021)
12. Gupta, V., Beckh, K., Giesselbach, S., Wegener, D., Wi, T.: Supporting verification of news articles with automated search for semantically similar articles. In: ROMCIR 2021 CEUR Workshop Proceedings, vol. 2838, pp. 47–58 (2021)
13. Hatua, A., Mukherjee, A., Verma, R.M.: On the Feasibility of using GANs for claim verification- experiments and analysis. In: ROMCIR 2021 CEUR Workshop Proceedings, vol. 2838, pp. 35–46 (2021)
14. Heydari, A., ali Tavakoli, M., Salim, N., Heydari, Z.: Detection of review spam: a survey. Expert Syst. Appl. **42**(7), 3634–3642 (2015)
15. Huang, Y., Xu, Q., Wu, S., Nugent, C., Moore, A.: Fight against COVID-19 misinformation via clustering-based subset selection fusion methods. In: Petrocchi, M., Viviani, M. (eds.) ROMCIR 2022 CEUR Workshop Proceedings, vol. 3138, pp. 11–26. CEUR-WS.org (2022)
16. Janzen, S., Orr, C., Terp, S.: Cognitive security and resilience: a social ecological model of disinformation and other harms with applications to COVID-19 vaccine information behaviors. In: Petrocchi, M., Viviani, M. (eds.) ROMCIR 2022 CEUR Workshop Proceedings, vol. 3138, pp. 48–88. CEUR-WS.org (2022)
17. Kirdemir, B., Adeliyi, O., Agarwal, N.: Towards characterizing coordinated inauthentic behaviors on YouTube. In: Petrocchi, M., Viviani, M. (eds.) ROMCIR 2022 CEUR Workshop Proceedings, vol. 3138, pp. 100–116. CEUR-WS.org (2022)
18. Mattei, M., Pratelli, M., Caldarelli, G., Petrocchi, M., Saracco, F.: Bow-tie structures of Twitter discursive communities. Sci. Rep. **12**, 12944 (2022)
19. Pankovska, E., Schulz, K., Rehm, G.: Suspicious sentence detection and claim verification in the COVID-19 domain. In: Petrocchi, M., Viviani, M. (eds.) ROMCIR 2022 CEUR Workshop Proceedings, vol. 3138, pp. 27–47. CEUR-WS.org (2022)
20. Pasi, G., Viviani, M.: Information credibility in the social web: contexts, approaches, and open issues. arXiv preprint arXiv:2001.09473 (2020)
21. Petrocchi, M., Viviani, M.: Overview of ROMCIR 2022: the 2nd workshop on reducing online misinformation through credible information retrieval. In: ROMCIR 2022 CEUR Workshop Proceedings, vol. 3138, pp. i–vii (2022)
22. Saracco, F., Viviani, M.: Overview of ROMCIR 2021: workshop on reducing online misinformation through credible information retrieval. In: ROMCIR 2021 CEUR Workshop Proceedings, vol. 2838, pp. i–vii (2021)

23. Schmitt, J.B., Debbelt, C.A., Schneider, F.M.: Too much information? Predictors of information overload in the context of online news exposure. Inf. Commun. Soc. **21**(8), 1151–1167 (2018)
24. Sharma, K., Qian, F., Jiang, H., Ruchansky, N., Zhang, M., Liu, Y.: Combating fake news: a survey on identification and mitigation techniques. ACM Trans. Intell. Syst. Technol. (TIST) **10**(3), 1–42 (2019)
25. Upadhyay, R., Pasi, G., Viviani, M.: Vec4cred: a model for health misinformation detection in web pages. Multimed. Tools Appl. **82**, 5271–5290 (2022)
26. Viviani, M., Pasi, G.: Credibility in social media: opinions, news, and health information-a survey. Wiley Interdisc. Rev. Data Min. Knowl. Discovery **7**(5), e1209 (2017)
27. Wardle, C., Derakhshan, H.: Information disorder: toward an interdisciplinary framework for research and policy making. Council of Europe, vol. 27 (2017)
28. Wardle, C., Derakhshan, H., et al.: Thinking about 'information disorder': formats of misinformation, disinformation, and mal-information. Ireton, Cherilyn; Posetti, Julie. Journalism, 'fake news' & disinformation. Unesco, Paris, pp. 43–54 (2018)
29. Zhang, Y., Boumber, D., Hosseinia, M., Yang, Mukherjee, A.: Improving authorship verification using linguistic divergence. In: ROMCIR 2021 CEUR Workshop Proceedings, vol. 2838, pp. 1–15 (2021)
30. Zhou, X., Zafarani, R.: A survey of fake news: fundamental theories, detection methods, and opportunities. ACM Comput. Surv. (CSUR) **53**(5), 1–40 (2020)

ECIR 2023 Workshop: Legal Information Retrieval

Suzan Verberne[1]([✉]), Evangelos Kanoulas[2], Gineke Wiggers[3], Florina Piroi[4], and Arjen P. de Vries[5]

[1] Leiden Institute of Advanced Computer Science, Leiden University, Leiden, The Netherlands
`s.verberne@liacs.leidenuniv.nl`
[2] Informatics Institute, University of Amsterdam, Amsterdam, The Netherlands
[3] eLaw Center for Law and Digital Technologies, Leiden University, Leiden, The Netherlands
[4] Institute of Information Systems Engineering, TU Wien, Vienna, Austria
[5] Institute for Computing and Information Sciences, Radboud University, Nijmegen, The Netherlands

Abstract. The full-day workshop on Legal Information Retrieval (LegalIR) takes place at ECIR 2023. Although this is the first legal IR workshop organized at ECIR, the topic has a long history of prior successful events and benchmark campaigns. In this workshop we will cover a broad variety of tasks, challenges, and methods in the legal domain. We have three invited speakers and presentations based on submissions of extended abstracts. There is also time for discussion with all participants.

Keywords: Legal information retrieval · Legal informatics · Domain-specific IR

1 Motivation and Relevance to ECIR

Legal professionals spend up to a third of their time doing research and investigation [12]. During this research legal information retrieval is important because "...the number of legal documents published online is growing exponentially, but accessibility and searchability have not kept pace with this growth rate." [18]

Foundational research into legal IR includes Blair and Marron [6], who demonstrate the gap between the percentage of relevant results retrieved and the percentage of relevant results thought to have been retrieved by legal professionals. Cole and Kuhlthau [7] investigate the impact of work experience on the notion of relevance of legal documents. Lastres [12] describes the results of a survey that asked legal professionals about their information seeking habits. Mart [15] extensively describes the working of the Lexis Nexis and Westlaw legal IR systems. Van Opijnen and Santos [18] apply the framework of relevance as described by Saracevic [17] to the legal domain, and describe the notion of domain relevance. Geist [9] extensively describes the difference between the completeness ideal and research reality in legal IR. Recent developments include research into the factors that impact the perception of relevance of legal professionals and the role of citations in the legal domain by Wiggers et al. [20,21].

J. Kamps et al. (Eds.): ECIR 2023, LNCS 13982, pp. 412–419, 2023.
https://doi.org/10.1007/978-3-031-28241-6_46

One specific legal task that has attracted the attention of the IR community in the past decades is eDiscovery. Grossman and Cormack [10] demonstrated in 2010 that technology-assisted eDiscovery can result in higher precision and F1 scores than human assessment alone. The TREC Total Recall track [11] has been an important benchmarking activity for legal IR research.

A second legal task that has led to important benchmarking activities, is case law retrieval [13,16]. Case law retrieval is a query-by-document retrieval task in which one legal case is a query and the goal is to retrieve relevant prior cases. The extremely long queries in this task provide challenges for Transformer-based rankers: Askari and Verberne [4] showed that lexical ranking with BM25 is hard to beat for case law retrieval, but automatic query generation [14] a promising technique to realize better results. In ECIR 2022, Althammer et al. proposed paragraph aggregation retrieval as an alternative solution for the long query documents [3], and Abolghasemi et al. showed the effectiveness of multi-task optimization for the same problem [1].

eDiscovery and case law retrieval are only two of many retrieval tasks in the legal domain; both are tasks that are strongly recall-oriented. Other legal IR tasks have received less attention, for example legal web search in commercial legal search engines [21], legal community question answering, or lawyer finding [5]. In this workshop, we aim to address the complete scope of legal IR tasks, challenges, and methods needed to address those challenges.

2 Workshop Goals/objectives and Overall Vision

The goal of the workshop is to make evident the many aspects of Legal IR in the different communities that work with legal data on a daily basis. In the call for papers for the workshop, we listed the following topics:

- Data collection and benchmark development for legal IR
- Evaluation in the legal domain
- Text processing for legal IR
- Ranking models tailored to the legal domain
- Search UI/UX for the legal domain
- eDiscovery
- Case law retrieval
- Expert finding in legal (e.g., finding the right lawyer for a case)
- Legal Question Answering
- Legal Knowledge Graphs
- Bibliometric-enhanced legal information retrieval
- Other IR tasks in the legal domain

We also encouraged discussions about the challenges of research on legal IR: How can we build a community around the legal domain with a strong focus on data-driven IR/NLP methods; where do we publish our works; how do we build benchmarks, datasets that are realistic but bypass all sensitivity issues; how do we open up industrial focus, industrial advances, and get industry become part of the research?

3   Format and Structure

The LegalIR workshop is a full-day workshop with talks by invited speakers, talks based on submitted extended abstracts, and a discussion session. The invited talks are 45 min; the regular talks 20 min including sufficient time for Q&A.

To make the workshop programme of high quality and interesting for the audience, we have ensured early commitment of three invited speakers, from industry and academia, junior and senior, male and female.

In our experience, workshops thrive on interaction between the participants. We will therefore have a slot of open discussion at the end of the workshop involving all participants. The organizers will collect the main topics for the discussion during the workshop day, based on the talks and common topics between the speakers.

3.1   Invited Speakers

Maura R. Grossman, J.D., Ph.D. is a Research Professor in the School of Computer Science at the University of Waterloo, an Adjunct Professor at Osgoode Hall Law School of York University, and an affiliate faculty member at the Vector Institute of Artificial Intelligence, all in Ontario, Canada. She also is Principal at Maura Grossman Law, an eDiscovery law and consulting firm in Buffalo, New York, U.S.A. Maura is most well known for her scholarly work on technology-assisted review ("TAR"), which has been widely cited in case law, both in the U.S. and abroad. She is also known for her appointments as a special master in multiple high-profile U.S. federal and state court cases.

Milda Norkute is a Lead Designer at Thomson Reuters Labs in Zug, Switzerland, one of several labs worldwide. She works closely with data scientists and engineers on enhancing products and services across Thomson Reuters product portfolio such as Westlaw Edge, Practical Law and others with Artificial Intelligence (AI) solutions. Milda's work is focused on user research and design of the products to figure out how and where to put the human in the loop in AI solutions taking into account, for example, their explainability needs when interacting with such systems and selecting suitable mechanisms. She is also currently co-leading the Human Centred AI research theme as part of the Labs research program. Before joining Thomson Reuters Milda worked at Nokia and CERN. Milda holds a masters in Human Computer Interaction and a bachelors degree in Psychology.

Tjerk de Greef is a Director of Advanced & Search Technology in the global technology organization of Wolters Kluwer. Tjerk (or TJ) is keen to merge the worlds of Data, Artificial Intelligence, and Search in the Legal & Regulatory domain. Many innovations in Information Retrieval rely on the application of artificial intelligence. He is passionate to merge the deep understanding of legal professionals with advanced technologies to drive and deliver customer value at scale. As a tech-savvy leader, he applies the state-of-the-art Agile DevOps practices following a User Centered Design methodology. In this role TJ is responsible for a centralized ring-fenced Advanced Technology team acting as a central hub for all Legal and Regulatory AI engagements in Wolters Kluwer, aiming to reduce technology sprawl.

3.2 Extended Abstract Submissions

We disseminated an open call for extended abstracts to the common mailing lists, in our own network and social media channels, through the professional networks of the invited speakers and the PC members, and through the Natural Legal Language Processing (NLLP) slack channel.

Extended abstracts can describe work in progress, or summarize previously published papers in high-quality venues. The main purpose of this submission format is that it brings together work from different communities related to the topic of legal IR.

We had a peer review process (in EasyChair) for the extended abstracts, mainly based on the relevance to the workshop. Each abstract was reviewed by at least 2 reviewers (see Sect. 5.2) Since extended abstracts might summarize previously published work, novelty is not the main criterion.

The purpose of the workshop is not to be a competitive venue, but to bring together international researchers working on IR in the legal domain.

4   Intended Audience

The target audience of the workshop consists of researchers and students working on legal IR problems, as well as professionals working on legal IR technology.

We expect a core group of around 30 active participants, including the speakers. In our experience, this is a good number to have interaction and engagement in a hybrid setting. In addition, we are confident that with three invited speakers, our workshop will be of interest to a broader range of ECIR participants, who might attend parts of the workshop.

We encourage the speakers to come to the conference in person.

5   Organizing Committee

The organizers are well established in the international IR community and all have a track record around domain-specific search, with a clear focus on the legal domain.

5.1   List of Organisers

Suzan Verberne is an associate professor at the Leiden Institute of Advanced Computer Science (LIACS) at Leiden University. She obtained her PhD in 2010 on the topic of Question Answering. Her current research focus is domain-specific

NLP and IR. She supervises three PhD projects about legal IR. She is highly active in the NLP and IR communities, holding chairing positions in large worldwide conferences, and is an associate editor for *Transactions on Information Systems*. She was chair of the First International Workshop on Professional Search in 2018, and co-chair of the BIR workshop series since 2020.

Evangelos Kanoulas is a professor of Information Retrieval and Evaluation at the University of Amsterdam. His research focuses on developing evaluation methods and algorithms for search and recommendation. Together with his students he is exploring ways for machines to effectively converse with humans so they better understand the searcher's need and effectively respond to it, especially in areas where retrieval is still difficult and require high recall, such as legal or medical search. Evangelos has participated in the coordination of the NIST TREC Million Query, Session, Tracks and Common Core tracks, as well as the CLEF e-Health track. Recently he organized the 1st Workshop on Augmented Intelligence for Technology-Assisted Reviews, while he has been a constant participant in the Competitions on Legal Information Extraction/Entailment (COLIEE).

Gineke Wiggers has recently completed her PhD thesis on legal information retrieval (awaiting defence). Her research focuses on bibliometric-enhanced information retrieval and the evaluation of domain-specific live IR systems.

Florina Piroi is a senior researcher at the Institute of Information Systems Engineering, Technische Universität Wien, Austria. She is an Information Retrieval researcher with experience in domain specific search (e.g. search in patent data, medical data, scientific and mathematical repositories), search engine evaluation and running evaluation campaigns. Florina was the main organiser of the CLEF-IP benchmarking activities in Information Retrieval for the Intellectual Property domain, and she has co-organized the TREC-Chem tracks. Since 2019 she is one of the PatentSemTech workshop series main organizers.

Arjen P. de Vries is professor of Information Retrieval at Radboud University Nijmegen in the Netherlands. He aims to resolve the question how users and systems may cooperate to improve information access, with a specific focus on the value of a combination of structured and unstructured information representations. Does a generic search system exist, or will it always be adapted to the user's specific situation and preferences? Can we involve the user into such adaptations of retrieval systems to task and domain? His research into these questions has led to a low-code solution for creating custom search engines, brought to market by Spinque BV in Utrecht. He is a member of the TREC PC, has been PC chair and general chair of SIGIR, and currently serves as the research director of the university's Institute for Computing and Information Sciences (iCIS).

5.2   Programme Committee

Since we ask for extended abstracts instead of full paper submissions, the workshop organizers will act as reviewers themselves, extended with people from our networks: Jeremy Pickens (expert in legalAIIA), Julien Rossi (expert from the COLIEE community), Gábor Recski (expert on legal NLP), Daniel Locke (expert on high-recall IR), and Procheta Sen (expert on socially responsible AI and IR).

6   Relation to Previous Workshops

Although this is the first legal IR workshop in this form with this focus, the workshop area has precedents that have been relevant in constituting our research community.

The First International Workshop on Professional Search[1] was co-located with SIGIR 2018, in Ann Arbor, Michigan. The workshop addressed the specific requirements and challenges of professional search, as opposed to web search. In the workshop report, the organizers – of whom three are also authors of the current proposal – concluded that "there are many open problems that potentially have an impact on the IR community beyond the professional domains" [19].

The International Workshop on AI and Intelligent Assistance for Legal Professionals in the Digital Workplace (LegalAIIA) has been organized twice, in 2019[2] and 2021[3], and will be organized again in 2023. LegalAIIA originated from the DESI workshop on eDiscovery. The focus of LegalAIIA is the relationship between AI developments and human-centered approaches (IA stands for intelligence augmentation). The goal is to emphasize comparisons in and between both an AI and an IA approach.

In the NLP community, the Natural Legal Language Processing (NLLP)[4] workshop has been successfully organized since 2019. The 2021 edition, co-located with EMNLP, received a record number of 48 submissions and accepted 28 papers [2]. We expect the legal IR community to be smaller than the legal NLP community, and ECIR is a smaller conference than EMNLP, but the success of the NLLP workshop indicates a growing interest in the legal domain.

Closely related to the legal domain is the patent domain, which also entails high-recall search and legal language use. The PatentSemTech[5] workshop series was co-located with SEMANTiCS 2019, and with SIGIR 2021 and 2022. PatentSemTech aims to establish a long-term collaboration and a two-way communication channel between the Intellectual Property industry and academia from relevant fields such as Information Retrieval (IR), natural-language processing (NLP), text and data mining (TDM) and semantic technologies (ST).

---

[1] https://jiyinhe.github.io/ProfS2018/.
[2] https://sites.google.com/view/legalaiia2019/home.
[3] https://sites.google.com/view/legalaiia-2021/home.
[4] https://nllpw.org/workshop/.
[5] http://ifs.tuwien.ac.at/patentsemtech/.

Benchmarking activities relevant to legal IR have been conducted in the TREC legal track 2006-2012 [8], the TREC total recall track [11] (both directed at eDiscovery), and COLIEE [16].

Finally, legal IR has a prominent place in the H2020 MSCA Training Network DoSSIER[6] on Domain Specific Systems for Information Extraction and Retrieval. Nine of the project's PhD students work on methods, evaluation and user studies in the legal domain. We expect involvement of these students (as author, audience, or local assistant) in the workshop.

7   Concluding Remarks

With the Legal IR workshop at ECIR 2023, we hope to bring together a group of people who have been working on a range of challenges in the legal domain over the past two decades, and learn from each other. This will hopefully bring new synergy and substantiates our research lines for the coming decade.

References

1. Abolghasemi, A., Verberne, S., Azzopardi, L.: Improving BERT-based query-by-document retrieval with multi-task optimization. In: Hagen, M., et al. (eds.) ECIR 2022. LNCS, vol. 13186, pp. 3–12. Springer, Cham (2022). https://doi.org/10.1007/978-3-030-99739-7_1

2. Aletras, N., Androutsopoulos, I., Barrett, L., Goanta, C., Preotiuc-Pietro, D. (eds.): Proceedings of the Natural Legal Language Processing Workshop 2021. Association for Computational Linguistics, Punta Cana, Dominican Republic, November 2021. https://aclanthology.org/2021.nllp-1.0

3. Althammer, S., Hofstätter, S., Sertkan, M., Verberne, S., Hanbury, A.: PARM: a paragraph aggregation retrieval model for dense document-to-document retrieval. In: Hagen, M., et al. (eds.) ECIR 2022. LNCS, vol. 13185, pp. 19–34. Springer, Cham (2022). https://doi.org/10.1007/978-3-030-99736-6_2

4. Askari, A., Verberne, S.: Combining lexical and neural retrieval with longformer-based summarization for effective case law retrieva. In: Proceedings of the Second International Conference on Design of Experimental Search & Information Retrieval Systems, pp. 162–170. CEUR (2021)

5. Askari, A., Verberne, S., Pasi, G.: Expert finding in legal community question answering. In: Hagen, M., et al. (eds.) ECIR 2022. LNCS, vol. 13186, pp. 22–30. Springer, Cham (2022). https://doi.org/10.1007/978-3-030-99739-7_3

6. Blair, D.C., Maron, M.E.: An evaluation of retrieval effectiveness for a full-text document-retrieval system. Commun. ACM **28**(3), 289–299 (1985)

7. Cole, C., Kuhlthau, C.: Information and information seeking of novice versus expert lawyers: how experts add value. Rev. Inf. Behav. Res. **1**, 103–115 (2000)

8. Cormack, G.V., Grossman, M.R., Hedin, B., Oard, D.W.: Overview of the trec 2010 legal track. In: TREC (2010)

9. Geist, A.C.J.: Rechtsdatenbanken und Relevanzsortierung. Doctoral dissertation, uniwien, Vienna, Austria (2016)

---

[6] https://dossier-project.eu/.

10. Grossman, M.R., Cormack, G.V.: Technology-assisted review in e-discovery can be more effective and more efficient than exhaustive manual review. Rich. JL Tech. **17**, 1 (2010)
11. Grossman, M.R., Cormack, G.V., Roegiest, A.: Trec 2016 total recall track overview. In: TREC (2016)
12. Lastres, S.A.: Rebooting legal research in a digital age (2013). https://www.lexisnexis.com/documents/pdf/20130806061418_large.pdf
13. Locke, D., Zuccon, G.: A test collection for evaluating legal case law search. In: Proceedings of the 41st International ACM SIGIR Conference on Research & Development in Information Retrieval (SIGIR), pp. 1261–1264 (2018)
14. Locke, D., Zuccon, G., Scells, H.: Automatic query generation from legal texts for case law retrieval. In: Asia Information Retrieval Symposium, vol. 10648, pp. 181–193. Springer, Cham (2017). https://doi.org/10.1007/978-3-319-70145-5_14
15. Mart, S.: The algorithm as a human artifact: implications for legal [re]search. Law Library J. **109**, 387 (2017)
16. Rabelo, J., Kim, M.-Y., Goebel, R., Yoshioka, M., Kano, Y., Satoh, K.: A summary of the COLIEE 2019 competition. In: Sakamoto, M., Okazaki, N., Mineshima, K., Satoh, K. (eds.) JSAI-isAI 2019. LNCS (LNAI), vol. 12331, pp. 34–49. Springer, Cham (2020). https://doi.org/10.1007/978-3-030-58790-1_3
17. Saracevic, T.: Relevance: a review of and framework for the thinking on the notion in information science. J. Am. Soc. Inf. Sci. **1975**, 321–343 (1975)
18. van Opijnen, M., Santos, C.: On the concept of relevance in legal information retrieval. Artif. Intell. Law **25**(1), 65–87 (2017). https://doi.org/10.1007/s10506-017-9195-8
19. Verberne, S., He, J., Kruschwitz, U., Wiggers, G., Larsen, B., Russell-Rose, T., de Vries, A.P.: First international workshop on professional search. In: ACM SIGIR Forum, vol. 52, pp. 153–162. ACM New York, NY, USA (2019)
20. Wiggers, G., Verberne, S., Zwenne, G.J., van Loon, W.: Exploration of domain relevance by legal professionals in information retrieval systems. Legal Inf. Manage. **22**, 49–67 (2022)
21. Wiggers, G., Verberne, S., Zwenne, G.J.: Citation metrics for legal information retrieval: scholars and practitioners intertwined? Legal Inf. Manage. **22**(2), 88–103 (2022)

Doctoral Consoritum

Building Safe and Reliable AI Systems for Safety Critical Tasks with Vision-Language Processing

Shuang Ao[✉] [iD]

The Open University, Walton Hall, Kents Hill, Milton Keynes MK7 6AA, England
shuang.ao@open.ac.uk

Abstract. Although AI systems have been applied in various fields and achieved impressive performance, their safety and reliability are still a big concern. This is especially important for safety-critical tasks. One shared characteristic of these critical tasks is their risk sensitivity, where small mistakes can cause big consequences and even endanger life. There are several factors that could be guidelines for the successful deployment of AI systems in sensitive tasks: (i) failure detection and out-of-distribution (OOD) detection; (ii) overfitting identification; (iii) uncertainty quantification for predictions; (iv) robustness to data perturbations. These factors are also challenges of current AI systems, which are major blocks for building safe and reliable AI. Specifically, the current AI algorithms are unable to identify common causes for failure detection. Furthermore, additional techniques are required to quantify the quality of predictions. All these contribute to inaccurate uncertainty quantification, which lowers trust in predictions. Hence obtaining accurate model uncertainty quantification and its further improvement are challenging. To address these issues, many techniques have been proposed, such as regularization methods and learning strategies. As vision and language are the most typical data type and have many open source benchmark datasets, this thesis will focus on vision-language data processing for tasks like classification, image captioning, and vision question answering. In this thesis, we aim to build a safeguard by further developing current techniques to ensure the accurate model uncertainty for safety-critical tasks.

Keywords: Deep learning · Model calibration · Uncertainty

1 Introduction

Despite the impressive performance of AI algorithm in various fields, their safety and reliability is still a concern. Recent studies have achieved successful performance in areas like image [5] and text classification [18], object detection [10], segmentation [9], image captioning [20], visual question answer [8] and graph scene generation [19], and some tasks obtain near-perfect results. However, AI

© The Author(s), under exclusive license to Springer Nature Switzerland AG 2023
J. Kamps et al. (Eds.): ECIR 2023, LNCS 13982, pp. 423–428, 2023.
https://doi.org/10.1007/978-3-031-28241-6_47

has not been fully deployed in sensitive fields like autonomous driving, medical diagnosing, or assistance for socially vulnerable groups. The major limitation lies in the lack of safeguard in these safety-critical tasks. One shared characteristic of these tasks is their risk sensitivity: it raises serious concern as the mistake by the AI algorithm can be expensive and even endanger human life. Guidelines from academic papers [2] and industry whitepapers [1] for deploying AI systems for safety-critical tasks include: identifying common causes of failure detection and out-of-distribution (OOD) detection, identifying overfitting in training data, quantifying uncertainty in prediction, and making the model robust to data perturbations. However, these recommendations are not fully satisfied in specific tasks, leading to the limitation of deployment of AI systems in these fields.

One serious limitation of current AI systems is that they tend to give the wrong prediction confidently [12,16]. Humans feel uncertain when their decision is potentially wrong or ambiguous in the decision-making process, and AI systems are supposed to have similar behaviors. In the recent decade, the quality of network architectures has significantly improved by utilizing deeper and wider networks such as VGG [15] and ResNet [5]. These state-of-the-art networks significantly improve feature learning for text and image but also raise the question of models with poor uncertainty quantification and being over-confident. It refers to when the overall confidence score is higher than the overall accuracy for testing data. Specifically, the model is supposed to show low confidence when the prediction is ambiguous or likely to be wrong and vice versa. The over-confident issue leads to the concern of inaccurate uncertainty quantification and trustworthiness of predictions.

The confidence and accuracy level of the system should match so that human experts can tell when the system tends to make mistakes. Hence addressing the over-confidence issue is essential to building a safe and reliable AI system. Even though a deep learning model can output the prediction for trained tasks, it cannot provide feedback about the quality of its prediction. For example, which class is poorly performed or if the overall output is reliable or not. In other words, such quality refers to how doubtful or uncertain the model for its prediction is, known as the model uncertainty. Ideally, when the model uncertainty is high, the model should suggest a second opinion and defer the task to human experts to re-examine it. With human intervention, the unexpected behaviour or wrong predictions from the AI system can be prevented. This process is crucial for safety-critical tasks as it can enhance failure detection, meaning a model to detect its own wrong predictions during the deployment or real-time applications without checking with the ground truth. Hence precise quantification and sufficient improvement of model uncertainty lie in the heart of building a safe and reliable AI.

In practice, many safety-critical tasks require multi-modality processing, especially with vision and language data. For example, autonomous driving systems process images, audio from user's input, and also signal data from sensors. Models for medical diagnosing deal with the image data such as MRI and text data of patients records. As real-world applications are more complicated than single modality data processing such as solely image or text classification, I will research reliable multi-modal data processing of image and text, and possibly other data sources.

2 Research Question

The main goal of my research is to build a safeguard for vision-language processing, by developing techniques and learning strategies to improve estimates of model uncertainty. RQ1.2 and RQ 2.1 will be discussed at the Doctoral Consortium.

RQ1: Can model uncertainty quantification be improved without adding additional computational complexity to build safe AI systems?

- **RQ1.1: Can we improve upon the uniform distribution in Label Smoothing by generalizing the soft label in a more reasonable way for different applications?**
  The traditional label smoothing clips the hard label into uniformly distributed soft labels, but not the case in practice. Hence we need to tackle this issue for a more accurate soft label generalization.
- **RQ1.2: How to efficiently conduct automatic failure detection (FD) for identify model's own wrong prediction during inference time?**
  FD is a significant criteria for the trustworthiness of a model.
- **RQ1.3: For curriculum learning, can we rank difficulty of samples more accurately to build the framework of curriculum learning?**
  We will use the model confidence as a proxy for ranking the difficulty of training samples.

RQ2: By integrating the techniques of safe AI systems developed in RQ1, can we improve the reliability and robustness of the model in the application of vision-language processing?

- **RQ2.1: For vision-language processing, how to build an end-to-end pipeline to reduce the dependency on the prior procedure, as well as the computational cost?**

  Tasks like image captioning and VQA include several processing steps such as object detection and feature extraction. Hence it is necessary to reduce the dependency of the prior step, to reduce its influence to latter processing.

3 Preliminary Result

We investigate 3 CNNs (ResNet34 [5], DenseNet121 [6] and VGG16 [15]) and 3 transformers models (ViT [4], SwinT [11] and Deit [17]) for ImageNet [14] and CIFAR100 [7] dataset. The evaluation metrics are accuracy and Expected Calibration Error (ECE) [13], and ECE is the metric to measure calibration.

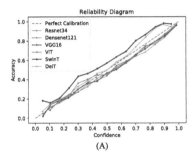

Models	ECE(%)	Accuracy(%)
ResNet34	4.0	71.2
DenseNet121	3.0	71.8
VGG16	3.3	69.9
ViT	2.2	83.3
SwinT	8.7	84.4
DeiT	6.4	81.3

(A)   (B)

Fig. 1. Left (A): Reliability diagram for ResNet34, DenseNet121, VGG16, ViT, SwinT, and DeiT models with ImageNet dataset. Right (B): Results of accuracy and ECE of various model architectures on ImageNet dataset. All results are shown in percentages. The lower the ECE the better.

In Fig. 1, the table shows that models that perform better in accuracy are not always better in ECE, such as VGG16 and SwinT, which is illustrated further in the reliability diagram in 1 (A). Hence it is essential to build a model with good calibration and high accuracy to improve the trustworthiness and reliability.

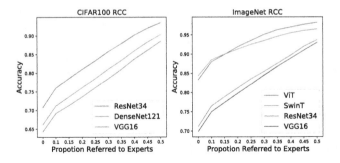

Fig. 2. Risk coverage curve for CIFAR100 and ImageNet dataset of CNNs and transformers. The x-axis indicates the percentage of data removed from the entire test set, and the accuracy is calculated on the remaining test set. The higher accuracy means better performance of automatic failure detection.

The Risk Coverage Curve (RCC) demonstrates the efficiency of automatic failure detection (AFD) as shown in Fig. 2. We utilize the predictive uncertainty to distinguish correct from incorrect samples by following [3]. The data belonging to "Referred to experts" are wrongly predicted. Comparing ViT and SwinTrans models in the ImageNet RCC, SwinTrans has higher accuracy with the entire test set, but ViT outperforms it with more wrong predictions removed. It suggests that ViT is more reliable than SwinT as the model can detect more wrong predictions.

4   Conclusion

In this thesis, I will explore the topic of safe and reliable AI, focusing on the applications in vision-language processing, such as image captioning and visual question answering. The main goal of my research is to build a safeguard for safety-critical tasks with multi-modal data processing.

References

1. Aptiv, A., Apollo, B., Continenta, D., FCA, H., Infineon, I.V.: Safety first for automated driving. In: Continental, Daimler, FCA, HERE, Infineon, Intel, and Volkswagen, pp. 1–157. White Paper (2019)
2. Ashmore, R., Calinescu, R., Paterson, C.: Assuring the machine learning lifecycle: desiderata, methods, and challenges. ACM Comput. Surv. (CSUR) 54(5), 1–39 (2021)
3. Corbière, C., Thome, N., Bar-Hen, A., Cord, M., Pérez, P.: Addressing failure prediction by learning model confidence. In: Advances in Neural Information Processing Systems, vol. 32 (2019)
4. Dosovitskiy, A., et al.: An image is worth 16×16 words: transformers for image recognition at scale. arXiv preprint arXiv:2010.11929 (2020)
5. He, K., Zhang, X., Ren, S., Sun, J.: Deep residual learning for image recognition. In: Proceedings of the IEEE Conference on Computer Vision and Pattern Recognition, pp. 770–778 (2016)
6. Huang, G., Liu, Z., Van Der Maaten, L., Weinberger, K.Q.: Densely connected convolutional networks. In: Proceedings of the IEEE Conference on Computer Vision and Pattern Recognition, pp. 4700–4708 (2017)
7. Krizhevsky, A., Hinton, G., et al.: Learning multiple layers of features from tiny images. University of Toronto, Technical Report (2009)
8. Li, W., Sun, J., Liu, G., Zhao, L., Fang, X.: Visual question answering with attention transfer and a cross-modal gating mechanism. Pattern Recogn. Lett. **133**, 334–340 (2020)
9. Li, X., Chen, H., Qi, X., Dou, Q., Fu, C.W., Heng, P.A.: H-DenseUNet: hybrid densely connected UNet for liver and tumor segmentation from CT volumes. IEEE Trans. Med. Imaging **37**(12), 2663–2674 (2018)
10. Liu, W., et al.: SSD: Single shot multibox detector. In: Leibe, B., Matas, J., Sebe, N., Welling, M. (eds.) ECCV 2016. LNCS, vol. 9905, pp. 21–37. Springer, Cham (2016). https://doi.org/10.1007/978-3-319-46448-0_2
11. Liu, Z., et al.: Swin transformer: hierarchical vision transformer using shifted windows. In: Proceedings of the IEEE/CVF International Conference on Computer Vision, pp. 10012–10022 (2021)
12. Müller, R., Kornblith, S., Hinton, G.E.: When does label smoothing help? In: Advances in Neural Information Processing Systems, vol. 32 (2019)
13. Naeini, M.P., Cooper, G., Hauskrecht, M.: Obtaining well calibrated probabilities using bayesian binning. In: Twenty-Ninth AAAI Conference on Artificial Intelligence (2015)
14. Russakovsky, O., et al.: Imagenet large scale visual recognition challenge. Int. J. Comput. Vis. **115**(3), 211–252 (2015)
15. Simonyan, K., Zisserman, A.: Very deep convolutional networks for large-scale image recognition. arXiv preprint arXiv:1409.1556 (2014)

16. Szegedy, C., Vanhoucke, V., Ioffe, S., Shlens, J., Wojna, Z.: Rethinking the inception architecture for computer vision. In: Proceedings of the IEEE Conference on Computer Vision and Pattern Recognition, pp. 2818–2826 (2016)
17. Touvron, H., Cord, M., Douze, M., Massa, F., Sablayrolles, A., Jégou, H.: Training data-efficient image transformers & distillation through attention. In: International Conference on Machine Learning, pp. 10347–10357. PMLR (2021)
18. Vaswani, A., et al.: Attention is all you need. In: Advances in Neural Information Processing Systems, vol. 30 (2017)
19. Yang, J., Lu, J., Lee, S., Batra, D., Parikh, D.: Graph r-cnn for scene graph generation. In: Proceedings of the European conference on computer vision (ECCV), pp. 670–685 (2018)
20. Zhou, L., Palangi, H., Zhang, L., Hu, H., Corso, J., Gao, J.: Unified vision-language pre-training for image captioning and VQA. In: Proceedings of the AAAI Conference on Artificial Intelligence, vol. 34, pp. 13041–13049 (2020)

Text Information Retrieval in Tetun

Gabriel de Jesus$^{(\boxtimes)}$ [ORCID]

INESC TEC and Faculty Engineering of the University of Porto (FEUP),
Rua Dr. Roberto Frias, 4200-465 Porto, Portugal
gabriel.jesus@inesctec.pt

Abstract. Tetun is one of Timor-Leste's official languages alongside Portuguese. It is a low-resource language with over 932,000 speakers that started developing when Timor-Leste restored its independence in 2002. Newspapers mainly use Tetun and more than ten national online news websites actively broadcast news in Tetun every day. However, since information retrieval-based solutions for Tetun do not exist, finding Tetun information on the internet and digital platforms is challenging. This work aims to investigate and develop solutions that can enable the application of information retrieval techniques to develop search solutions for Tetun using *Tetun INL* and focus on the ad-hoc text retrieval task. As a result, we expect to have effective search solutions for Tetun and contribute to the innovation in information retrieval for low-resource languages, including making Tetun datasets available for future researchers.

Keywords: Information retrieval · Tetun · Search · Ad-hoc retrieval · Low-resource language

1 Introduction and Motivation

Information retrieval (IR) deals with finding documents of unstructured nature that satisfy an information need from within large collections [1]. In the ad-hoc retrieval task, users typically formulate information needs as a query using natural language text and execute it through search. Then, the retrieval system retrieves documents that are relevant to the given query from large textual document collections and returns them to the user in an ordered list. One important IR application is keyword-based web search, as showcased by the Google search engine. Since specific information retrieval-based approaches for Tetun do not exist, finding documents that satisfy an information need written in Tetun is challenging.

Tetun is the language spoken in Timor-Leste. Timor-Leste is a multilingual country with two official languages, Tetun and Portuguese, two working languages, Indonesian and English [5], and more than 30 dialects [6]. Tetun is a dialect that was previously used as a trade language before Timor-Leste restored

This PhD research is financed by the Portuguese Foundation for Science and Technology (FCT) under the scholarship grant number SFRH/BD/151437/2021.

its independence and became a new sovereign state on 20 May 2002. In 2002, the Government of Timor-Leste designated Tetun as one of its official languages and since then, it has become a dominant language in public life.

There are two major varieties of Tetun: Tetun Dili (referred to as Tetun) and Tetun Terik [2]. Tetun Dili comprises Tetun INL and Tetun DIT. Tetun Terik is one of Timor-Leste's dialects. Tetun INL is Tetun for which the government of Timor-Leste produced standard orthography through the *Instituto Nacional de Linguísticas* (INL) [3]. Tetun INL has become the official Tetun dialect being used in the education system, official publications, and media [4]. Tetun DIT is produced by the linguists at Dili Institute of Technology (DIT) with a few standard difference with the Tetun INL in writing structures [2], such as Tetun INL uses "ñ" and "ll", while Tetun DIT uses "nh" and "lh", e.g., millaun (Tetun INL), milhaun (Tetun DIT).

By the 2015 census, the population of Timor-Leste was 1.17 million and the proportion of Tetun speakers was 79.04%, alongside Portuguese (2.56%), Indonesian (2.02%), and English (1.04%) [6]. Most of the Tetun verbs, nouns, and adjectives are Portuguese loanwords and Tetun shows much more influence from Portuguese noted in media [7–9]. Consequently, Tetun is still widely varied in writing mainly influenced by Portuguese and Indonesian.

To tackle the aforementioned problem, this work aims to develop search solutions for Tetun using *Tetun INL* and focus on the ad-hoc text retrieval task, including developing a text corpus and a test collection for evaluation. Since online news has played a crucial role in promoting Tetun in the last five years, from 2017 to 2022, we will use Timor News [10] as our case study. Timor News will be used to showcase Tetun search solutions to further evaluate the algorithms' performance and retrieval effectiveness. Timor News broadcasts national and international news only in Tetun INL and has over 4,000 news articles registered in the database. The portal is accessed by visitors from around the globe, with a total of 1,500 visitors on average per day, and as it was founded by the author of this paper [11], we have full access to the platform. The remainder of this paper is organized as follows. Section 2 presents the related work. We describe our main research questions in Sect. 3. Then, Sects. 4 and 5 present the research methodology and the specific research issues. Finally, Sect. 6 closes with a final remark.

2   Related Work

Low-resource languages (LRLs) can be understood as the languages that are less studied, less computerized, low density [12,13], that lack text corpora, or reduced accessibility [14]. Developing information retrieval tools for a LRL requires corpora, specific techniques and algorithms, and test collections. For the text corpora construction, Artetxe et al. [15] built a Basque corpus comprising 12.5 M documents and 423 M tokens employing tailored crawling and reported that it could be an effective alternative to obtain high-quality data for LRLs. Linder et al. [16] developed SwissCrawl, a Swiss German corpus composed of over half one million sentences, by crawling the web using a tool they developed.

Dovbnia and Wróblewska [17] developed a language identification (LID) model experimented with three Celtic language variations comprising 9,969 sentences corpus and reported that using n-gram characters as input enables building a robust LID model for LRLs in supervised and semi-supervised settings. Ferilli [18] described that term and document frequency (TDF) could be used to identify stopwords from a small number of corpora and stated that it outperformed the classic term frequency (TF) [19–21] and the normalized inverse document frequency (NIDF) from Lo et al. [22]. Tukeyev et al. [23] developed a Turkic stemmer using stopwords, affixes, and root lists and reported a ratio of 97% in the experiment tested with 1,000 Turkic words.

Chavula and Suleman [24] built a test collection for Chichewa, Citumbuka, and Cinyanja, where the topics formulation and relevance assessment of the document-query pairs were conducted by five external assessors. Esmaili et al. [25] built a test collection for Sorani Kurdish and Aleahmad et al. [26] created the Hamshahri test collection for Persian. The relevance of the query-document pairs for the first test collection was judged by the native Sorani speakers, and Persian students collaborated in the latter.

There are several search engines for low-resource languages, as showcased in the works for Bantu languages. Holy et al. [27] developed a search engine for nine Bantu languages that enable users to search using multilingual queries. Malumba et al. [28] developed a search engine for isiZulu and a search engine for IsiXhosa was developed by Kyeyune [29]. All these search engines were developed using a similar infrastructure - crawling the documents from the web using focused crawling [30], identifying documents using a language identification model, and indexing, retrieval, and ranking using the Solr platform.

3   Research Questions

This research will be focused on developing solutions for ad-hoc text retrieval for Tetun and our main research questions (RQs) are the following.

RQ1—*What retrieval strategies provide the most effective solutions for Tetun text-based search?* Different retrieval strategies for the ad-hoc retrieval task have been studied, from the classical probabilistic-based [31] to neural-based approaches [32,33]. Considering the language diversity and the fact that Tetun is a low-resource language, this question extends our knowledge to investigate the works in ad-hoc information retrieval on low-resource languages which can be used to support in developing text retrieval solutions for Tetun search.

RQ2—*Do query processing operations improve the text retrieval effectiveness in Tetun INL?* Tetun INL words contain accented letters (á, é, í, ó, ú, ñ), apostrophes ('), and hyphens (for mono-semantic). Moreover, a multilingual environment influences Tetun in writing. This research question intends to investigate how applying query and document preprocessing operations, such as normalizing the accented letters, removing stopwords, and stemming, can improve text retrieval effectiveness in Tetun text search.

RQ3—*What impact does the Tetun variant used in Tetun INL has on the text retrieval effectiveness in Tetun text search?* Since Tetun is still widely varied in writing, this research question aims to investigate whether properly written Tetun INL documents improve text retrieval effectiveness.

4   Research Methodology

The proposed steps for the research are described as follows. We will start by crawling the World Wide Web to collect documents and store them in a repository. Documents are preprocessed to remove unnecessary elements, apply the LID model to extract Tetun texts from the collections, and then build a text corpus for Tetun. Three individuals with proficiency in Tetun will be hired to evaluate the corpus quality using a similar approach to Artetxe et al. [15].

Documents and topics will be characterized according to TREC guidelines, where some topics will be extracted from search query logs of Timor News, and the team members will formulate the others. A TREC-style approach, based on the Cranfield approach [34], will be employed to build a test collection for evaluation. The open-source platforms for IR research, such as Solr, Elasticsearch, and Terrier IR, will be used to index and rank documents for each query and then create a pool of documents to judge their relevance. Five Tetun native speakers will be hired to conduct the relevance judgments.

After a corpus and a test collection are built, we will conduct experiments using the topics and documents developed to respond to the research questions outlined in Sect. 3. The most effective retrieval strategy for Tetun will be showcased in a search prototype using Timor News as a case study to further evaluate the algorithms' performance.

5   Specific Research Issue

Lin [35] stated that neural-based models had shown substantial improvements over traditional methods, even in the low-resource languages, referring to the work of Yang et al. [36]. However, the result reported by Yang et al. [36] was still quite far (0.3152 AP) from the best-known result on Robust04 (0.3686 AP). Therefore, since Tetun is a newly developed and LRL, we intend to discuss the gap between traditional and neural-based approaches in ad-hoc retrieval in our research context to get insightful feedback. So the question would be: is it possible to do transfer learning or create a multilingual aligned model from high-resource languages like English or partially similar languages like Portuguese?

6   Final Remark

This paper presents the idea of developing search solutions for Tetun which includes main research questions, methodologies to be adopted, and the specific research issues to be discussed. As a continuation of the work, we will follow the steps outlined in Sect. 4 to respond the research questions in Sect. 3.

References

1. Manning, C.-D., Raghavan, P., Schütze, H.: An Introduction to Information Retrieval. Cambridge University Press, Cambridg (2009)
2. van-Klinken, C.-W., Hajek, J., Nordlinger R.: Tetun Dili: a grammar of an East Timorese language, Pacific Linguistics, Canberra, Australia (2002)
3. The standard orthography of the tetum language. https://archive.org/details/the-standard-orthography-of-the-tetum-language. Accessed 31 Oct 2022
4. Government decree-law No. 1/2004 of 14 April 2004 - the standard orthography of the tetun language. https://mj.gov.tl/jornal/lawsTL/RDTL-Law/RDTL-Gov-Decrees/Gov-Decree-2004-01.pdf. Accessed 31 Oct 2022
5. Constitution of the democratic republic of timor-leste. https://timor-leste.gov.tl/wp-content/uploads/2010/03/Constitution_RDTL_ENG.pdf/. Accessed 31 Oct 2022
6. Timor-leste population and housing Census 2015. General directorate of statistics, ministry of finance, democratic republic of timor-leste. https://www.statistics.gov.tl/category/publications/census-publications Accessed 31 Oct 2022
7. Hajek, J., van-Klinken., C.-W.: language contact and gender in Tetun Dili: what happens when Austronesian meets romance?. University of Hawai'i Press **58**, 59–91 (2019). https://doi.org/10.1353/ol.2019.0003
8. Zuzana, G.: Tetun in Timor-Leste: The role of language contact in its development. PhD thesis, Universidade de Coimbra, Portugal (2018). https://hdl.handle.net/10316/80665
9. van-Klinken, C. W., Hajek, J.: Language contact and functional expansion in Tetun Dili: the evolution of a new press register. Multilingual **37**, 613–647 (2018)
10. Timor news: an online news agency based in Dili, Timor-Leste, https://www.timornews.tl
11. The registered and licensed social communication agencies in press council of timor-Leste. https://conselhoimprensa.tl/baze-de-dadus/registu-media. Accessed 31 Oct 2022
12. Magueresse, A., Carles, V., Heetderks, E.: Low-resource languages: a review of past work and future challenges. CoRR, abs/2006.07264 (2020). https://arxiv.org/abs/2006.07264
13. Cieri, C., Maxwell, M., Strassel, M.-S., Tracey, J.: Selection criteria for low resource language programs. In: Calzolari, N., et al. (eds.) Proceedings of the Tenth International Conference on Language Resources and Evaluation LREC 2016, Portorož, Slovenia, 23–28 May 2016. European Language Resources Association (ELRA) (2016). https://www.lrec-conf.org/proceedings/lrec2016/summaries/1254.html'
14. Hoenen, A., Koc, C., Rahn, M.-D.: A manual for web corpus crawling of low resource languages. Umanistica Digitale **4**(8), 342–344 (2020). https://doi.org/10.6092/issn.2532-8816/9931
15. Artetxe, M., Aldabe, I., Agerri, R., Perez-de-Viñaspre, O., Soroa, A.: Does corpus quality really matter for low-resource languages?. CoRR abs/2203.08111 (2022). https://doi.org/10.48550/arXiv.2203.08111
16. Linder, L., Jungo, M., Hennebert, J., Musat, C.-C., Fischer, A.: Automatic creation of text corpora for low-resource languages from the internet: the case of swiss German. In Béchet, F., et al. (eds.) Proceedings of The 12th Language Resources and Evaluation Conference, LREC 2020, Marseille, France, 11–16 May 2020, pp. 2706–2711, European Language Resources Association (2020). https://aclanthology.org/2020.lrec-1.329/

17. Dovbnia, O., Wróblewska, A.: Automatic language identification for celtic texts. CoRR abs/2203.04831 (2022). https://doi.org/10.48550/arXiv.2203.04831

18. Ferilli, S.: Automatic multilingual stopwords identification from very small corpora. Electron. **10**(17) (2021). https://doi.org/10.3390/electronics10172169

19. Ferilli, S., Izzi, G.L., Franza, T.: Automatic stopwords identification from very small corpora. In: Stettinger, M., Leitner, G., Felfernig, A., Ras, Z.W. (eds.) ISMIS 2020. SCI, vol. 949, pp. 31–46. Springer, Cham (2021). https://doi.org/10.1007/978-3-030-67148-8_3

20. Baeza-Yates, R., Ribeiro-Neto, B.-A.: Modern Information Retrieval - the Concepts and Technology Behind Search, 2nd edn. Pearson Education Ltd., Harlow (2011)

21. Croft, W.-B., Metzler, D., Strohman. T.: Search Engines - Information Retrieval in Practice. Pearson Education, London (2009). https://www.search-engines-book.com

22. Lo, R.-T., He, B., Ounis, T.: Automatically building a stopword list for an information retrieval system. J. Digital Inf. Manage. **3**(1), 3–8 (2005)

23. Tukeyev, U., Karibayeva, A., Turganbayeva, A., Amirova, D.: Universal programs for stemming, segmentation, morphological analysis of Turkic words. In: Nguyen, N.T., Iliadis, L., Maglogiannis, I., Trawiński, B. (eds.) ICCCI 2021. LNCS (LNAI), vol. 12876, pp. 643–654. Springer, Cham (2021). https://doi.org/10.1007/978-3-030-88081-1_48

24. Chavula, C., Suleman, H.: Ranking by language similarity for resource scarce southern bantu languages. In: International Conference on the Theory of Information Retrieval (ICTIR), Virtual Event, Canada, 2021, pp. 137–147. Association for Computing Machinery, New York, NY, USA (2021). https://doi.org/10.1145/3471158.3472251

25. Esmaili, K.-S., et al.: Building a test collection for Sorani Kurdish. In: ACS International Conference on Computer Systems and Applications, AICCSA 2013, Ifrane, Morocco, 27–30 May 2013, pp. 1–7, IEEE Computer Society (2013). https://doi.org/10.1109/AICCSA.2013.6616470

26. Aleahmad., A., Amiri, H., Darrudi, E., Rahgozar, M., Oroumchian, F.: Hamshahri: A standard Persian text collection. Knowl. Based Syst. **22**(5), 382–387 (2009). https://doi.org/10.1016/j.knosys.2009.05.002

27. von-Holy, A., Bresler, A., Shuman, O., Chavula, C., Suleman, H.: Bantuweb: a digital library for resource scarce South African languages. In: Masinde, M. Proceedings of the South African Institute of Computer Scientists and Information Technologists, SAICSIT 2017, Thaba Nchu, South Africa, 26–28 September 2017, pp. 36:1–36:10, Association for Computing Machinery (2017). https://doi.org/10.1145/3129416.3129446

28. Malumba, N., Moukangwe, K., Suleman, H.: AfriWeb: a web search engine for a marginalized language. In: Allen, R.B., Hunter, J., Zeng, M.L. (eds.) ICADL 2015. LNCS, vol. 9469, pp. 180–189. Springer, Cham (2015). https://doi.org/10.1007/978-3-319-27974-9_18

29. Kyeyune, M.-J.: IsiXhosa search engine development report. Department of Computer Science, University of Cape Town (2015). https://pubs.cs.uct.ac.za/id/eprint/1035/1/report.pdf. Accessed 10 Jul 2022

30. Chakrabarti, S., van-den-Berg, M., Dom, B.: Focused crawling: a new approach to topic-specific web resource. Comput. Netw. **31**(11–16), 1623–1640 (1999). https://doi.org/10.1016/S1389-1286(99)00052-3

31. Robertson, S., Zaragoza, H.: The probabilistic relevance framework: BM25 and Beyond. Foundations and Trends in Information Retrieval, April 2009. vol. 3, pp.

333–389, Now Publishers Inc., Hanover, MA, USA (2009). https://doi.org/10.1561/1500000019

32. Nogueira, R., Jiang, Z., Pradeep, R., Lin, J.: Document Ranking with a Pretrained Sequence-to-Sequence Model. In: Chon, T., He, Y, Liu, Y. Findings of the Association for Computational Linguistics: EMNLP 2020, Online Event, 16–20 November 2020, ACL, vol. EMNLP 2020, pp. 708–718. Association for Computational Linguistics (2020). https://doi.org/10.18653/v1/2020.findings-emnlp.63

33. Yang, W., Zhang, H., Lin, J.: Simple Applications of BERT for Ad hoc document retrieval. CoRR abs/1903.10972, (2019). https://arxiv.org/abs/1903.10972

34. Clough, P.-D., Sanderson, M.: Evaluating the performance of information retrieval systems using test collections. Inf. Res. 18(2) (2013). https://www.informationr.net/ir/18-2/paper582.html'

35. Lin, J.: The neural hype, justified!: a recantation. ACM SIGIR Forum 53(2), 88–93 (2019). https://doi.org/10.1145/3458553.3458563

36. Yang, W., Lu, K., Yang, P., Lin, J.: Critically examining the "Neural Hype": Weak baselines and the additivity of effectiveness gains from neural ranking models. In: Piwowarski, B., Gaussier, É., Maarek, Y., Nie, J., Scholer, F. (eds.) In: Proceedings of the 42nd International ACM SIGIR Conference on Research and Development in Information Retrieval, SIGIR 2019, Paris, France, 21–25 July 2019, pp. 1129–1132, ACM (2019). https://doi.org/10.1145/3331184.3331340

Identifying and Representing Knowledge Delta in Scientific Literature

Alaa El-Ebshihy[1,2,3](✉) [ID]

[1] Research Studio Austria, Vienna, Austria
alaa.el-ebshihy@researchstudio.at
[2] Technische Universität Wien, Vienna, Austria
[3] Alexandria University, Alexandria, Egypt

Abstract. The process of continuously keeping up to date with the state-of-the-art on a specific research topic is a challenging task for researchers not least due to the rapid increase of published research. In this research proposal, we define the term *Knowledge Delta* (KD) between scientific articles which refers to the differences between pairs of research articles that are similar in some aspects. We propose a three-phase research methodology to identify and represent the KD between articles. We intend to explore the effect of applying different text representations on extracted facts from scientific articles on the downstream task of KD identification.

Keywords: Knowledge delta · Text representation · Literature update

1 Motivation

For any research topic, there are various available scientific articles from conferences, journal publications, etc. It is crucially important for a researcher to continuously keep up to date with the evolution of new research in her area. The number of published scientific articles grows exponentially every year [6] and as a result, the task of searching and keeping up to date on scientific literature for a given topic is a challenging and time consuming one. Moreover, it may end up that some relevant papers are not considered by the researcher through this process.

It is important for researchers, with different levels of expertise, to keep up to date with literature for multiple reasons. Some researchers try to spot the gaps in a specific research point and so it is important that they make sure that the problem they are trying to solve is a novel one. Other researchers try to find out solutions to problems they are facing in their own research. On the other hand, young researchers or researchers new to specific research areas tend to build up knowledge in their research area in addition to the requirement of keeping up to date with the state-of-the-art in their area.

This process of keeping up to date with the state-of-the-art on a specific research area is a challenging one. For experienced researchers, it is some times difficult to prioritize this task beside other tasks. In addition to their research, they have other tasks like teaching, writing proposals, reviewing papers, giving talks, etc. For young researchers, it is challenging to do a dual task of building up knowledge in research area and keeping up to date with the newly published research in the same time.

© The Author(s), under exclusive license to Springer Nature Switzerland AG 2023
J. Kamps et al. (Eds.): ECIR 2023, LNCS 13982, pp. 436–442, 2023.
https://doi.org/10.1007/978-3-031-28241-6_49

There are online resources that help the researcher to search for related scientific articles to their areas, for example: CiteSeerX [36], Google Scholar [18], Semantic Scholar [17], Connected Papers[1] and others. However, these tools define the relation between articles based on the articles' citations. Furthermore, if a tool recommends a paper to the user, it doesn't state why and how a particular article is recommended [6].

If there is a tool that can: (1) point out the novelty in the research by taking into account the differences between research articles belongs to the same topic and (2) retrieve the articles based on these differences, could contribute to solve the problem of keeping up to date with the state-of-the-art on specific research topic. In our proposal, we refer to the differences between documents as *Knowledge Delta* (KD).

To sum up, identifying a method that allows us to determine the differences (*Knowledge Delta*) between articles can help the researchers in the following aspects: (1) making the process of updating the literature review, on a research topic, fast and efficient, (2) reducing the risk of missing citing relevant papers to the research area, (3) fastening the process of building knowledge in a new research area or topic, and (4) the process of building surveys for research fields.

2 Background and Related Work

We classify the previous research to help researchers in updating their literature review into the following categories: *Scientific article summarization*, *Scholarly citation recommendation*, and *Citation role understanding*.

Scientific Article Summarization. Approaches, to tackle the challenges of scientific article summarization, include: Argumentative Zoning and Citation Based Summarization approaches. Argumentative Zoning (AZ) refers to the examination of the argumentative status of sentences in scientific articles and their assignment to specific argumentative zones. Its main goal is to collect sentences that belong to predefined zones, such as "claim" or "method". Annotated AZ corpora has been created by Teufel et al. [31–33,35] with approaches to AZ identification reported by Liu et al. [26]. Argument mining is similar to AZ and refers to the identification of arguments, its components and relations in text [3–5]. In the Citation Based Summarization approach, citation sentences are used to produce summaries for cited articles. In this approach, it is claimed that the citation sentences usually cover the most important aspects of reference paper (e.g. method, results, drawbacks and limitations) [2]. Aggregating citation sentences from different papers can give an informative summary for the content of a reference paper. In addition to the citation based approach proposed by Abu-Jbara et al. [2], a series of pilot tasks known as CL-SciSumm [11,12,19–21] have emerged using the same approach to generate summaries. Beginning with 2020, the CL-SciSumm shared task creators have introduced two new challenges: the CL-LaySumm and the Long-Summ challenge. The former challenge target is to generate summaries of scientific article for lay persons while the target of the latter is to generate informative summaries for the articles. Some other approaches emerged to generate summaries for articles. For example, Guy et al. [25] proposed a summarization approach by utilizing video talks

---

[1] https://www.connectedpapers.com/.

presentation of papers and Michihiro et al. [38] presented a neural network model to generate scientific summaries.

Scholarly Citation Recommendation. We divide the approaches for recommending scientific articles to the following: Content-Based and Topic Modelling approaches. The content based approach, for example the work presented by Bhagavatula et al. [8] and Ding et al. [14], relies on building a user model consists of mostly textual features of items (e.g. emails and web pages) that the user interact with. Then, the user model is used to find the best candidate for recommendation, for example using the cosine similarity between documents representations in a vector space [7]. In other approaches topic modeling and citation relationships are used to build a model for paper recommendation [22,30,37]. In addition to that, Chakraborty et al. [9] proposed a citation recommendation framework using scientific articles facets and citation network. Recently, Bedi et al. [6] proposed a new task to recommend papers that can be used as baselines given a scientific article by developing a neural network classification model using a manually annotated corpus of scientific papers in which references are classified to one of two classes: *Baseline* or *Non-baseline* reference.

Citation Role Understanding. Stevens et al. [29] mentioned 15 reasons to provide reference citations, including: criticizing others work, supporting claims, providing background and identifying methodology. Teufel et al. [34] introduced the citation function, which is defined as *"the author's reason for citing a given paper"*. Some work has been done to create annotated datasets for citation reasons [24,27] with some approaches to automatically classify the reason of citation [9,13,15]. In addition to understanding the reasons for citation, some researchers are interested in studying the strength of references to citing papers. Chakraborty et al. [10] argued that references are not equally related to the citing article.

3   Proposed Research

In our research, we aim to investigate how to identify the differences between scientific articles which are similar in some aspects (e.g. identify the difference in approaches of a pair of articles which tackle the same research problem). We define these differences as *Knowledge Delta* between scientific articles. Therefore, we propose our high-level research question as follows:

> ***What is a good way to identify the differences in factual knowledge between two documents?***

The question refers to extracting the factual knowledge of an article, representing this knowledge, and identifying the differences between representations. The facts can be the components of the scientific articles (e.g. the claims, the methods, the results, the conclusions, etc.). They may also represent the topics and sub-topics which the article belongs to. Hence, this high-level research question is divided to three sub-research questions:

- **RQ1**: What is a good way to detect the regions of the scientific article that cover the article's research questions, claimed novel contributions, and other facts?
- **RQ2**: What is a good way to extract and represent the facts and the relationship between them?
- **RQ3**: What is a good way to determine and quantify the Knowledge Delta using different fact representations?

4   Research Methodology and Proposed Experiments

The methodology targets to answer the high-level research question stated in Sect. 3. Our approach is divided into three phases. In each phase, we investigate to answer each research question. First, we will build a model to identify the regions that cover factual knowledge from the scientific articles. Second, we will investigate different text representation methods on the extracted facts. Finally, we will use the facts representation to identify the Knowledge Delta (KD) between articles.

Phase 1 - Fact Extraction: We aim at finding regions from scientific articles that cover the main components of a scientific article. Any research article covers the following aspects: *the research questions, hypothesis, methodology, results and conclusions.* One of the approaches that can be used to identify the main components of the scientific article is the *Argumentative Zoning (AZ)* [35] which can be used to assign sentences of scientific articles to specific categories (i.e. zones). We are planning to use AZ automatic identification approaches to identify the main claims and contributions in scientific articles [3,26,32,33]. Also, we can use an annotation tool which we previously developed for the purpose of creating an AZ benchmark dataset to improve the AZ prediction [16].

Phase 2 - Knowledge Representation: We are interested in exploring different text representations for extracted facts in order to relate extracted facts to broader concepts. We are planning to explore different text extraction and representations techniques beginning with simple methods going through more complex ones. We will start by extracting keywords from text using noun phrase extraction methods [28]. Then, we will explore topic detection methods including conceptual embeddings [1]. Finally, we shall explore the usage of knowledge graphs to identify entities and relations within an article [23].

Phase 3 - Knowledge Delta Identification: We aim to identify the KD between articles using the articles representations. In other words, we are planning to build a model that identifies the similarities and differences between pairs of articles using the text representations from Phase 2. Chakraborty et al. [9] showed some of the relations between citing article and cited article where they defined four relations: *Background, Alternative Approaches, Methods* or *Comparisons.* In our work, we plan to start with the *Alternative Approaches* relation between pairs of articles to compare and find the differences between the proposed approaches of articles pairs. The differences in the detected approaches will be used in representing the KD between articles pairs. To evaluate the different KD models we will run user experiments where students/researchers will give feedback on the KD identified approaches. We plan to identify the KD as a list of differences in approaches between pairs of articles and ask them to verify its correctness.

References

1. Abdulahhad, K.: Concept embedding for information retrieval. In: Pasi, G., Piwowarski, B., Azzopardi, L., Hanbury, A. (eds.) ECIR 2018. LNCS, vol. 10772, pp. 563–569. Springer, Cham (2018). https://doi.org/10.1007/978-3-319-76941-7_45

2. Abu-Jbara, A., Radev, D.: Coherent citation-based summarization of scientific papers. In: Proceedings of the 49th Annual Meeting of the Association for Computational Linguistics: Human Language Technologies, pp. 500–509 (2011)

3. Accuosto, P., Neves, M., Saggion, H.: Argumentation mining in scientific literature: from computational linguistics to biomedicine. In: Frommholz, I., Mayr, P., Cabanac, G., Verberne, S. (eds.) BIR 2021: 11th International Workshop on Bibliometric-Enhanced Information Retrieval; 1 April 2021, CEUR 2021, Lucca, Italy, pp. 20–36. CEUR Workshop Proceedings, Aachen (2021)

4. Accuosto, P., Saggion, H.: Transferring knowledge from discourse to arguments: a case study with scientific abstracts. In: Stein, B., Wachsmuth, H. (eds.) Proceedings of the 6th Workshop on Argument Mining, 1 August 2019, Florence, Italy, pp. 41–51. Association for Computational Linguistics. ACL (Association for Computational Linguistics), Stroudsburg (2019)

5. Accuosto, P., Saggion, H.: Mining arguments in scientific abstracts with discourse-level embeddings. Data Knowl. Eng. **129**, 101840 (2020)

6. Bedi, M., Pandey, T., Bhatia, S., Chakraborty, T.: Why did you not compare with that? Identifying papers for use as baselines. In: Hagen, M., et al. (eds.) Advances in Information Retrieval, pp. 51–64. Springer, Cham (2022). https://doi.org/10.1007/978-3-030-99736-6_4

7. Beel, J., Gipp, B., Langer, S., Breitinger, C.: Research-paper recommender systems: a literature survey. Int. J. Digit. Libr. **17**(4), 305–338 (2015). https://doi.org/10.1007/s00799-015-0156-0

8. Bhagavatula, C., Feldman, S., Power, R., Ammar, W.: Content-based citation recommendation. In: Proceedings of the 2018 Conference of the North American Chapter of the Association for Computational Linguistics: Human Language Technologies, vol. 1 (Long Papers), pp. 238–251. Association for Computational Linguistics, New Orleans, Louisiana, June 2018. https://doi.org/10.18653/v1/N18-1022, https://aclanthology.org/N18-1022

9. Chakraborty, T., Krishna, A., Singh, M., Ganguly, N., Goyal, P., Mukherjee, A.: FeRoSA: a faceted recommendation system for scientific articles. In: Bailey, J., Khan, L., Washio, T., Dobbie, G., Huang, J.Z., Wang, R. (eds.) Advances in Knowledge Discovery and Data Mining, pp. 528–541. Springer, Cham (2016). https://doi.org/10.1007/978-3-319-31750-2_42

10. Chakraborty, T., Narayanam, R.: All fingers are not equal: intensity of references in scientific articles. In: Proceedings of the 2016 Conference on Empirical Methods in Natural Language Processing, pp. 1348–1358. Association for Computational Linguistics, Austin, Texas, November 2016. https://doi.org/10.18653/v1/D16-1142, https://aclanthology.org/D16-1142

11. Chandrasekaran, M.K., Feigenblat, G., Hovy, E., Ravichander, A., Shmueli-Scheuer, M., De Waard, A.: Overview and insights from scientific document summarization shared tasks 2020: CL-SciSumm, LaySumm and LongSumm. In: Proceedings of the First Workshop on Scholarly Document Processing (SDP 2020) (2020)

12. Chandrasekaran, M.K., Yasunaga, M., Radev, D., Freitag, D., Kan, M.Y.: Overview and results: CL-SciSumm shared task 2019. arXiv preprint arXiv:1907.09854 (2019)

13. Cohan, A., Ammar, W., van Zuylen, M., Cady, F.: Structural scaffolds for citation intent classification in scientific publications. In: Proceedings of the 2019 Conference of the North American Chapter of the Association for Computational Linguistics: Human Language Technologies, vol. 1 (Long and Short Papers), pp. 3586–3596. Association for Computational Linguistics, Minneapolis, Minnesota, June 2019. https://doi.org/10.18653/v1/N19-1361, https://aclanthology.org/N19-1361

14. Ding, Y., Zhang, G., Chambers, T., Song, M., Wang, X., Zhai, C.: Content-based citation analysis: the next generation of citation analysis. J. Am. Soc. Inf. Sci. **65**(9), 1820–1833 (2014)
15. Dong, C., Schäfer, U.: Ensemble-style self-training on citation classification. In: Proceedings of 5th International Joint Conference on Natural Language Processing, pp. 623–631. Asian Federation of Natural Language Processing, Chiang Mai, Thailand, November 2011. https://aclanthology.org/I11-1070
16. El-Ebshihy, A., Ningtyas, A.M., Andersson, L., Piroi, F., Rauber, A.: A platform for argumentative zoning annotation and scientific summarization. In: Proceedings of the 31st ACM International Conference on Information & Knowledge Management, CIKM 2022, pp. 4843–4847. Association for Computing Machinery, New York, NY, USA (2022). https://doi.org/10.1145/3511808.3557193
17. Fricke, S.: Semantic scholar. J. Med. Libr. Assoc. JMLA **106**(1), 145 (2018)
18. Jacsó, P.: Google Scholar: the pros and the cons. Online Inf. Rev. **29**(2), 208–214 (2005)
19. Jaidka, K., et al.: The computational linguistics summarization pilot task. In: Proceedings of Text Analysis Conference, Gaithersburg, USA (2014)
20. Jaidka, K., Chandrasekaran, M.K., Rustagi, S., Kan, M.Y.: Overview of the CL-SciSumm 2016 shared task. In: Proceedings of the Joint Workshop on Bibliometric-Enhanced Information Retrieval and Natural Language Processing for Digital Libraries (BIRNDL), pp. 93–102 (2016)
21. Jaidka, K., Yasunaga, M., Chandrasekaran, M.K., Radev, D., Kan, M.Y.: The CL-SciSumm shared task 2018: results and key insights. arXiv preprint arXiv:1909.00764 (2019)
22. Jeong, C., Jang, S., Park, E., Choi, S.: A context-aware citation recommendation model with BERT and graph convolutional networks. Scientometrics **124**(3), 1907–1922 (2020). https://doi.org/10.1007/s11192-020-03561-y
23. Ji, S., Pan, S., Cambria, E., Marttinen, P., Yu, P.S.: A survey on knowledge graphs: representation, acquisition, and applications. IEEE Trans. Neural Netw. Learn. Syst. **33**(2), 494–514 (2022). https://doi.org/10.1109/tnnls.2021.3070843
24. Jurgens, D., Kumar, S., Hoover, R., McFarland, D., Jurafsky, D.: Measuring the evolution of a scientific field through citation frames. Trans. Assoc. Comput. Linguist. **6**, 391–406 (2018). https://doi.org/10.1162/tacl_a_00028, https://aclanthology.org/Q18-1028
25. Lev, G., Shmueli-Scheuer, M., Herzig, J., Jerbi, A., Konopnicki, D.: TalkSumm: a dataset and scalable annotation method for scientific paper summarization based on conference talks. CoRR abs/1906.01351 (2019). http://arxiv.org/abs/1906.01351
26. Liu, H.: Automatic argumentative-zoning using Word2vec. CoRR abs/1703.10152 (2017). http://arxiv.org/abs/1703.10152
27. Pride, D., Knoth, P.: An authoritative approach to citation classification, pp. 337–340. Association for Computing Machinery, New York, NY, USA (2020). https://doi.org/10.1145/3383583.3398617
28. Siddiqi, S., Sharan, A.: Keyword and keyphrase extraction techniques: a literature review. Int. J. Comput. Appl. **109**, 18–23 (2015). https://doi.org/10.5120/19161-0607
29. Stevens, M.E., Giuliano, V.E., Garfield, E.: Can citation indexing be automated ? (1964)
30. Tang, J., Zhang, J.: A discriminative approach to topic-based citation recommendation. In: Theeramunkong, T., Kijsirikul, B., Cercone, N., Ho, T.B. (eds.) Advances in Knowledge Discovery and Data Mining, pp. 572–579. Springer, Cham (2009). https://doi.org/10.1007/978-3-642-01307-2_55
31. Teufel, S., Carletta, J., Moens, M.: An annotation scheme for discourse-level argumentation in research articles. In: Ninth Conference of the European Chapter of the Association for Computational Linguistics (1999)
32. Teufel, S., Moens, M.: Summarizing scientific articles: experiments with relevance and rhetorical status. Comput. Linguist. **28**(4), 409–445 (2002)

33. Teufel, S., Siddharthan, A., Batchelor, C.: Towards domain-independent argumentative zoning: evidence from chemistry and computational linguistics. In: Proceedings of the 2009 Conference on Empirical Methods in Natural Language Processing, pp. 1493–1502 (2009)
34. Teufel, S., Siddharthan, A., Tidhar, D.: Automatic classification of citation function. In: Proceedings of the 2006 Conference on Empirical Methods in Natural Language Processing, pp. 103–110. Association for Computational Linguistics, Sydney, Australia, July 2006. https://aclanthology.org/W06-1613
35. Teufel, S., et al.: Argumentative zoning: Information extraction from scientific text. Ph.D. thesis, Citeseer (1999)
36. Wu, J., et al.: CiteSeerX: AI in a digital library search engine. AI Mag. **36**(3), 35–48 (2015)
37. Yang, L., et al.: A LSTM based model for personalized context-aware citation recommendation. IEEE Access **6**, 59618–59627 (2018)
38. Yasunaga, M., et al.: ScisummNet: a large annotated corpus and content-impact models for scientific paper summarization with citation networks. In: Proceedings of AAAI 2019 (2019)

Investigation of Bias in Web Search Queries

Fabian Haak[1,2]([⊠])(iD)

[1] TH Köln, Gustav-Heinemann-Ufer 54, 50678 Cologne, Germany
`fabian.haak@th-koeln.de`
[2] University of Hildesheim, Universitätsplatz 1, Hildesheim, Germany

Abstract. The dissertation investigates the correlations and effects between biases in search queries and search query suggestions, search results, and users' states of knowledge. Search engines are an important factor in opinion formation, while search queries determine the information a user is exposed to in information search. Search query suggestions play a crucial role in what users search for [22]. Biased query suggestions can be especially problematic if a user's information need is not set and the interaction with query suggestions is likely. Only recently, research has started to investigate the general assumption that biased search queries lead to biased search results, focusing on political stance bias [17]. However, the correlation between biases in search queries and biases in search results has not been sufficiently investigated. Sparse context and limited data access pose challenges in detecting biases in search queries. This dissertation thus contributes datasets and methodological approaches that enable media bias research in the field of search queries and search query suggestions.

Keywords: Bias · Search queries · Search query suggestions · Information search

1 Motivation for the Proposed Research

Search engines are a popular and trusted means of finding information on political topics, not only online information sources considered [9,26]. Users communicate their information needs via search queries, while search query suggestions presented by the search engines to these queries often lead the searches' direction. Search query suggestions, the list of suggested search queries provided by search engines during the input of a search query, play a crucial role in what people search for [22]. It has been shown, that media bias can strongly impact the (public) perception of reported topics [7,14,19]. By impacting what users search for, biased query suggestions can indirectly induce biased opinions. This is especially problematic since it was demonstrated, that query suggestions can be manipulated and could therefore be used in malicious ways [33]. Bias in online information aggravates the problem known as filter bubble or echo chamber [29], i.e., readers consume only news that corresponds to their beliefs, views, or personal liking [18,19]. Research on bias in online search has shown, that "factors

J. Kamps et al. (Eds.): ECIR 2023, LNCS 13982, pp. 443–449, 2023.
https://doi.org/10.1007/978-3-031-28241-6_50

such as the topic of the query, the phrasing of query and the time at which a query is issued also impact the bias seen by the users" [17]. Despite the relevance and potential implications of bias in search queries and search query suggestions, not much research has been published on the topic. There is no concise knowledge of the types of bias and how to detect them. Research has focused primarily on the (U.S. American) political domain, investigating partisanship and other related bias concepts; other domains lack research severely. Furthermore, the mechanisms between query suggestions, search queries, and users' states of knowledge have not been researched sufficiently. It is still unclear if biased search queries and what types of inherent biases lead to more biased results. In these research gaps, further elaborated in Sects. 2 and 4, I see this dissertations' relevance founded.

2   Background and Related Work

Problems with bias in search queries and its detection. Search engines are seen as a trustworthy source of information on many topics, including political information [9,26]. Further, they have proven to significantly impact political opinion formation [10]. Trust in search engines is problematic because their results are prone to be biased [4,16,21]. Research on bias detection has been focused on news texts and the identification of bias at sentence- and document level. Published research on bias in search queries and search query suggestions is sparse and has not yet seen a universal problem formulation on the main challenges of the identification of bias in query suggestions. A problematic aspect is the severe lack of datasets for search queries. Collecting search queries from users is problematic due to privacy concerns, previously available datasets such as the AOL query log dataset [23] are no longer available and the application is morally debatable. Search query suggestions are based on real queries issued by users, but due to the black box characteristics of search engines, their use as proxies for "real" search queries is not ideal. However, they are generally unpersonalized, can easily be crawled, and have their own potential implications. Another possible clue for understanding bias in search queries lies in the strategies and mechanisms in query formulation [32]. How users perceive and interact with query suggestions has been investigated [8,15], the effect of information search strategies and the exposure to biased suggestions and content on users' query formulations still lacks research. However, for research on bias in search queries, these interactions should not be ignored.

Approaches to Bias Detection. The primarily non-phrasal structure of search queries [34] and their lack of contextual metadata make them a difficult subject for bias detection and have been previously addressed [2,11]. Thus, bias detection in query suggestions usually relies on context expansion based on search results [28]. Most studies dealing with bias in search query suggestions chose to focus on a specific domain, for example, person-related search [31] or political parties and politicians [17,28]. Most earlier studies focus on showing that certain searches lead to biased suggestions, by primarily performing manual analyses of queries [1]. So far, few types of bias in search query suggestions have been researched. Gender bias [20] and topical group bias [5,24] are the focus of the majority of the studies. For the identification of bias in search queries and query suggestions, topical modeling and identification of bias via linguistic and lexical features along with statistical methods is the most common methodology [5,12,25]. However, in the identification of media bias in other types of documents, transformer-based language models have risen in popularity and tend to outperform previous techniques significantly [3,30]. For bias in search queries, transformer-based techniques have not yet been utilized.

3 Description of Proposed Research

This dissertation investigates bias in and connections between search queries, query suggestions, search results, and users' states of knowledge. The goal of the research is to produce a set of methodological approaches, that enable the identification of problematic biases in search queries and search query suggestions. These will enable the promotion of transparency and fairness in the online search processes and the identification of dangers and mechanisms in online information search. The dissertation aims to leverage and develop approaches for bias identification and to answer the following research questions:

- What biases exist in search queries and search query suggestions and how can they be categorized?
- What methods and metrics are suited for identifying biases in search queries?
- How do bias in search queries and bias in search results correlate?
- What types of search queries and what information search strategies lead to more and which to less biased search results?

4 Research Methodology in Publications

Publication: Perception-Aware Bias Detection for Query Suggestions. In this first publication, we investigate perception-aware metrics in query suggestion bias detection [12]. Treating query suggestion datasets as lists of unique suggestions does not consider two critical aspects of how query suggestions are perceived: The frequency and order of the suggestions. To solve these issues, we include rank- and frequency-aware metrics, DCG and nDCG, to analyze topical group bias in search query suggestions for names of German politicians as search terms. This way, we measure topical group bias towards a selected set of meta-attributes including gender, age, and party affiliation. The results of our study suggest a gender bias, with significantly lower average DCG and nDCG scores for political suggestions for female politicians.

Publication: Auditing Search Query Suggestion Bias Through Recursive Algorithm Interrogation. In [13], we include aspects of the iterative approach, in which information search is performed. Online information search usually consists of iterative rephrasing and multiple searches until either the results represent the expected outcome or the information need is constituted [6]. Datasets used in previous research did not reflect this information search and -exposure.

This publication employs the technique of Recursive Algorithm Interrogation [27] along with systematic query alteration to retrieve an extensive set of query suggestions for each root query. The results of the study indicate a gender bias for two of three topical clusters. Contrary to what previous studies have shown (c.f. [5,12]), searches for female politicians return more suggestions that have a political or economics-related topic than for male politicians. Thus we conclude that perceived bias in search query suggestions for person-related search is more dependent on the employed search strategy than the effects of biased meta-attributes such as gender or age.

Work in Progress. **A: Media Bias Taxonomy and Literature Review of Computer Science Methods for media bias detection**. In cooperation with the media bias group, we currently perform a systematic literature review of methods used for media bias detection. Furthermore, the publication attempts to produce the first structured framework of media bias. This will aid future work, such as this dissertation, to find a common ground in naming investigated types of bias.

B: Do Biased Search Queries Lead to Biased Search Results? A second work in progress investigates the effects of biased search queries on search results. We first finetune transformer-based language models on a dataset of biased news, that contains linguistically biased news. These language models are then used to generate variations of queries collected via recursive algorithm interrogation for a set of keywords [27].

5   Specific Research for Discussion at the Doctoral Consortium

Aside from the general direction of research of this dissertation, the main methodological contribution of this dissertation could benefit most from feedback. The contribution for this dissertation ought to investigate users' exposure to biased information during information search. For this simulation study, artificial (or real) users information needs form the basis. These information needs incorporate different forms of levels of knowledge, cognitive biases by restricting the query in complexity and the vocabulary in general. Via simulation, search is carried out using the queries, and query suggestions and search results are collected. The simulated user's knowledge is then updated by perception-aware factors based on information search strategies. This forms a new state of knowledge and a reformulated information need. Based on that, a new query is formulated and these processes are repeated iteratively based on a stopping criterion based on the initial information need and user properties. We then investigate the properties, changes and bias in information exposures for various different user properties and behaviour patterns.

References

1. Baker, P., Potts, A.: 'Why do white people have thin lips?' Google and the perpetuation of stereotypes via auto-complete search forms, vol. 10, pp. 187–204. Routledge (2013). https://doi.org/10.1080/17405904.2012.744320

2. Balog, K.: Entity-Oriented Search, The Information Retrieval Series, vol. 39. Springer, Cham (2018). https://doi.org/10.1007/978-3-319-93935-3

3. Blackledge, C., Atapour-Abarghouei, A.: Transforming fake news: robust generalisable news classification using transformers. arXiv Version Number: 2 (2021). https://doi.org/10.48550/ARXIV.2109.09796

4. Bolukbasi, T., Chang, K.W., Zou, J., Saligrama, V., Kalai, A.: Man is to computer programmer as woman is to homemaker? Debiasing word embeddings. In: Proceedings of the 30th International Conference on Neural Information Processing Systems, NIPS 2016, pp. 4356–4364. Curran Associates Inc., Red Hook, NY, USA (2016)

5. Bonart, M., Samokhina, A., Heisenberg, G., Schaer, P.: An investigation of biases in web search engine query suggestions. Online Inf. Rev. 44(2), 365–381 (2019). https://doi.org/10.1108/oir-11-2018-0341

6. Cai, F., de Rijke, M.: A Survey of Query Auto Completion in Information Retrieval, vol. 10, pp. 273–363 (2016). https://doi.org/10.1561/1500000055

7. Dallmann, A., Lemmerich, F., Zoller, D., Hotho, A.: Media bias in German online newspapers. In: Proceedings of the 26th ACM Conference on Hypertext & Social Media, HT 2015, pp. 133–137. Association for Computing Machinery, New York, NY, USA (2015). https://doi.org/10.1145/2700171.2791057

8. Dean, B.: We Analyzed 5 Million Google Search Results. Here's What We Learned About Organic CTR, August 2019. https://backlinko.com/google-ctr-stats

9. Edelman: 2022 edelman trust barometer (2022). https://www.edelman.com/trust/2022-trust-barometer

10. Epstein, R., Robertson, R.E.: The search engine manipulation effect (SEME) and its possible impact on the outcomes of elections, vol. 112, pp. E4512–E4521. National Academy of Sciences Section: PNAS Plus, August 2015. https://doi.org/10.1073/pnas.1419828112

11. Feuer, A., Savev, S., Aslam, J.A.: Evaluation of phrasal query suggestions. In: Proceedings of the Sixteenth ACM Conference on Conference on Information and Knowledge Management, CIKM 2007, pp. 841–848. Association for Computing Machinery, New York, NY, USA (2007). https://doi.org/10.1145/1321440.1321556

12. Haak, F., Schaer, P.: Perception-aware bias detection for query suggestions. In: Boratto, L., Faralli, S., Marras, M., Stilo, G. (eds.) BIAS 2021. CCIS, vol. 1418, pp. 130–142. Springer, Cham (2021). https://doi.org/10.1007/978-3-030-78818-6_12

13. Haak, F., Schaer, P.: Auditing search query suggestion bias through recursive algorithm interrogation. In: WebSci 2022: 14th ACM Web Science Conference 2022, Barcelona, Spain, 26–29 June 2022, pp. 219–227. ACM (2022). https://doi.org/10.1145/3501247.3531567

14. Hamborg, F., Donnay, K., Gipp, B.: Automated identification of media bias in news articles: an interdisciplinary literature review. Int. J. Digit. Libr. 20(4), 391–415 (2018). https://doi.org/10.1007/s00799-018-0261-y

15. Hofmann, K., Mitra, B., Radlinski, F., Shokouhi, M.: An eye-tracking study of user interactions with query auto completion. In: Li, J., Wang, X.S., Garofalakis, M.N., Soboroff, I., Suel, T., Wang, M. (eds.) Proceedings of the 23rd ACM International Conference on Conference on Information and Knowledge Management, CIKM 2014, Shanghai, China, 3–7 November 2014, pp. 549–558. ACM (2014). https://doi.org/10.1145/2661829.2661922

16. Introna, L., Nissenbaum, H.: Defining the web: the politics of search engines, vol. 33, pp. 54–62. Institute of Electrical and Electronics Engineers (IEEE) (2000). https://doi.org/10.1109/2.816269

17. Kulshrestha, J., et al.: Search bias quantification: investigating political bias in social media and web search. Inf. Retrieval J. **22**(4), 188–227 (2018). https://doi.org/10.1007/s10791-018-9341-2

18. Jeong Lim, S., Jatowt, A., Yoshikawa, M.: DEIM forum 2018 C 1–3 towards bias inducing word detection by linguistic cue analysis in news articles (2018)

19. Jeong Lim, S., Jatowt, A., Yoshikawa, M.: Understanding characteristics of biased sentences in news articles. In: CIKM Workshops (2018)

20. Mertens, A., Pradel, F., Rozyjumayeva, A., Wäckerle, J.: As the tweet, so the reply? In: Proceedings of the 10th ACM Conference on Web Science - WebSci 2019. ACM Press (2019). https://doi.org/10.1145/3292522.3326013

21. Mitra, B., Shokouhi, M., Radlinski, F., Hofmann, K.: On user interactions with query auto-completion. In: Proceedings of the 37th International ACM SIGIR Conference on Research & Development in Information Retrieval. ACM (2014). https://doi.org/10.1145/2600428.2609508

22. Niu, X., Kelly, D.: The use of query suggestions during information search. Inf. Process. Manage. **50**, 218–234 (2014). https://doi.org/10.1016/j.ipm.2013.09.002

23. Pass, G., Chowdhury, A., Torgeson, C.: A picture of search. In: Proceedings of the 1st International Conference on Scalable Information Systems, p. 1-es. InfoScale 2006. Association for Computing Machinery, New York, NY, USA (2006). https://doi.org/10.1145/1146847.1146848

24. Pitoura, E., et al.: On measuring bias in online information. In: CoRR. vol. abs/1704.05730 (2017)

25. Pradel, F.: Biased representation of politicians in Google and Wikipedia search? Joint Effect Party Identity Gender Identity Elections **38**, 447–478 (2021). https://doi.org/10.1080/10584609.2020.1793846

26. Ray, L.: 2020 Google search survey: how much do users trust their search results? (2020). https://moz.com/blog/2020-google-search-survey

27. Robertson, R.E., Jiang, S., Lazer, D., Wilson, C.: Auditing autocomplete: suggestion networks and recursive algorithm interrogation. In: Proceedings of the 10th ACM Conference on Web Science, WebSci 2019, pp. 235–244. ACM, New York, NY, USA (2019). https://doi.org/10.1145/3292522.3326047

28. Robertson, R.E., Lazer, D., Wilson, C.: Auditing the personalization and composition of politically-related search engine results pages. In: Proceedings of the 2018 World Wide Web Conference, WWW 2018, pp. 955–965. International World Wide Web Conferences Steering Committee, Republic and Canton of Geneva, CHE (2018). https://doi.org/10.1145/3178876.3186143

29. Spinde, T., Jeggle, C., Haupt, M., Gaissmaier, W., Giese, H.: How do we raise media bias awareness effectively? Effects of visualizations to communicate bias. PLOS ONE **17**(4), 1–14 (2022). https://doi.org/10.1371/journal.pone.0266204

30. Spinde, T., Plank, M., Krieger, J.D., Ruas, T., Gipp, B., Aizawa, A.: Neural media bias detection using distant supervision with BABE - bias annotations by experts. In: Findings of the Association for Computational Linguistics: EMNLP 2021, pp. 1166–1177. Association for Computational Linguistics, Punta Cana, Dominican Republic, November 2021. https://doi.org/10.18653/v1/2021.findings-emnlp.101
31. Stier, S., et al.: Systematically monitoring social media: the case of the German federal election 2017 (2018). https://doi.org/10.17605/OSF.IO/5ZPM9
32. Ter, A., Proper, H., Weide, T.: Query formulation as an information retrieval problem. Comput. J. **39**, 255–274 (1996). https://doi.org/10.1093/comjnl/39.4.255
33. Wang, P., et al.: Game of missuggestions: semantic analysis of search-autocomplete manipulations, January 2018. https://doi.org/10.14722/ndss.2018.23071
34. Zhou, Q., Wang, C., Xiong, M., Wang, H., Yu, Y.: SPARK: adapting keyword query to semantic search. In: Aberer, K., et al. (eds.) The Semantic Web, pp. 694–707. Springer, Cham (2007). https://doi.org/10.1007/978-3-540-76298-0_50

Monitoring Online Discussions and Responses to Support the Identification of Misinformation

Xin Yu Liew[✉] [ID]

School of Computer Science, University of Nottingham, Jubilee Campus, Wollaton Road,
Nottingham NG8 1BB, UK
hcyxll@nottingham.ac.uk

Abstract. Misinformation prospers on online social networks and impacts society in various aspects. They spread rapidly online; therefore, it is crucial to keep track of any information that could potentially be false as early as possible. Many efforts have focused on detecting and eliminating misinformation using machine learning methods. Our proposed framework aims to leverage the strength of human roles engaging with a machine learning tool, providing a monitoring tool to identify the risk of misinformation on Twitter at an early stage. Specifically, this work is interested in a visualisation tool that prioritises popular Twitter topics and analyses the responses of the higher-risk topics through stance classification. Besides tackling the challenging task of stance classification, this work also aims to explore features within the information from Twitter that could provide further aspects of a response to a topic using sentiment analysis. The main objective is to provide an engaging tool for people who are also working towards the issue of online misinformation, i.e., fact-checkers in identifying and managing the risk of a specific topic at an early stage by taking appropriate actions towards it before the consequences worsen.

Keywords: Fake news · Misinformation · NLP · Machine learning · Online social networks

1 Motivation

Think about a scenario where a fact checker tries to identify or prioritise a topic worth looking into because of a higher risk of misinformation. Typically, it would take them only to notice it once the misinformation has become popular and damage has made it to the news. At this point, managing the misinformation would be more challenging as the stages of damage have progressed very quickly. With the amount of data shared on the public like Twitter, that allows public speech and engagement, how can a fact checker keep up to date on dangerous topics discussed online and react according to how the public has responded towards it? This research will focus on stance classification to understand whether the responses to misinformation (e.g., tweet replies) can support misinformation and therefore require further intervention to mitigate its harmful impacts.

The nature of online social networks (OSN) has become a norm where people easily connect, engage, and share their opinions publicly. Furthermore, this convenience of

© The Author(s), under exclusive license to Springer Nature Switzerland AG 2023
J. Kamps et al. (Eds.): ECIR 2023, LNCS 13982, pp. 450–455, 2023.
https://doi.org/10.1007/978-3-031-28241-6_51

communication has allowed sharing of information despite its falsehood, causing misinformation beliefs. Misinformation spreads rapidly on OSN and chasing this targeted issue has been challenging as they change dynamically over time. During a pandemic, different opinions occur when the government enforces preventive measures because people are confused and overwhelmed by information online [3]. Hence, it is crucial to manage information online at an early stage before the misinformation causes more damage to society. Studies have widely explored computational methods like machine learning in classifying truthfulness but evaluating the risk at an early stage also plays an essential role.

In this research, we are targeting the early stages of misinformation to support the control and management of opinions to reduce further engagements. This research will experiment on COVID-19 information on Twitter. Considering that vaccine-related misinformation that populated OSN during COVID-19 has become a massive factor for vaccine hesitancy, as WHO declared it as a top 10 threat to world health [16]. Besides, a now-retracted study from the previous measles, mumps, and rubella (MMR) falsely claimed vaccine relates to autism, contributing to the current false beliefs in COVID-19 vaccines [7, 14]. Other related health misinformation informs the public to perform actions that cause harmful consequences like manipulating the thoughts of an individual performing dangerous actions within society. For example, the conspiracy theory of COVID-19 linking to the 5G network drove the provocation in individuals burning the 5G towers in the UK and creating violent threats [1]. Another example is the misinformation that provokes individuals to consume bleach to kill the virus has led to many deaths [13].

2 Related Work

There are several attempts made to support the monitoring of public opinions towards vaccines and the use of the stance of replies for a machine learning model to identify misinformation. D'Andrea et al. proposed a real-time monitoring framework for Italian public opinion about vaccines, and their system gathers vaccine-related tweets and uses SVM to perform stance classification [4]. Lemmens et al. presented a tool named Vaccinpraat, which monitors the stance towards COVID-19 vaccinations and the most frequently used anti-vaccination argument by presenting statistics and insights [11]. Dungs et al. proposed to use the stance of replies towards pre-defined rumours as a feature to model the truthfulness of a rumour using Hidden Markov Models (HMM) [6].

Studies that perform stance of responses towards a tweet showed an improvement to support the identification of rumour [6]. Stance classifications, also known as opinion mining, are based on determining opinions, e.g., in favour, against, or neutral towards a pre-defined target of interest, regardless of the explicit mention in a text [4]. Therefore, the following studies on stance classification will be observed as benchmarks to achieve the objective of our research in supporting the monitoring of misinformation through the stance of replies. Cotfas et al. analysed one month of public opinion regarding COVID-19 vaccination using BERT and achieved 78.94% accuracy [3]. Limitations mentioned in their research are improving the robustness of the algorithm and extending the data collection period, as vaccinations last for an extended period [3]. Giovanni et al. proposed a semi-automated process to label a large dataset and consider a BERT base model's geographical, temporal, and linguistic distribution in classifying the stance towards vaccination [8]. Their study suggested that future directions could investigate limitations like overfitting, multi-classification and observe the model on daily basis datasets [8]. Tahir et al. proposed multimodal online and offline data for COVID-19 vaccine stance detection and explainable AI to analyse the results of the predictions on emerging samples [15]. Their study suggests investigating other offline data could further improve the model [15].

The work will also study the role of sentiment analysis in misinformation. Sentiment analysis classifies a text based on the general attitude, e.g., positive, negative, or neutral [4]. The following related work proposes methods that this work will adapt to analyse sentiments and topics. Xue et al. analysed the COVID-19 sentiments of public tweets related to 11 topics derived from LDA topic modelling and concluded that fear is the most dominant emotion [17]. Kaur et al. proposed to monitor the dynamics of emotions during the first few months of COVID-19 using IBM Watson Tome Analyzer, they also compared and analysed the sentiments of tweets [10].

3   Proposed Research Questions and Methodology

To the best of my knowledge, none of the previous work has investigated the stance of the public response towards a wide range of topics discussed during COVID-19. In addition, none has combined sentiment analysis to monitor misinformation during a pandemic with explainable machine learning. Hence, this work proposes to investigate the following research question in this doctoral thesis:

1. **How can we efficiently prioritise tweets worth looking into first to narrow down the topics at higher risk?** This work calculates priority tweets based on two factors (1) popular tweets prediction, as the "virality" of tweets with potential high engagement attracts public attention to populate them; (2) false or non-evident health-threatening topics, as they are topics that are not officially stated as fact with evidence. To answer RQ1, we will first investigate state-of-the-art methods to predict the popularity of tweets in estimating the potential spread, and topic modelling approaches to visualise top misinformation to be addressed. For predicting popularity, probabilistic models like the Bayesian approach will be used by observing temporal information like the number of retweets or followers. Then, a topic modelling approach should summarise the most discussed topics within the prediction of popular tweets. The topic modelling approach will look at the given corpus, i.e., tweets and retrieve the similarities for clustering; the common method is the Latent Dirichlet Allocation (LDA) [5, 9, 12]. At this point, human users, i.e., fact-checkers, will select a topic of interest to be addressed.

2. **How can we learn feature representations useful for stance classification in relation to misinformation?** To classify responses as, for instance, supporting versus refuting the misinformation, there are multiple aspects of an OSN discussion that could bring meaningful perspective. For the basis, textual context retrieved from tweets provides linguistic information to be analysed by NLP. For example, the

BERT model is trained on sentence and paragraph levels for understanding text-level semantics [3, 8]. Some studies have proven the improvement using the stance of replies towards a pre-defined topic for rumour classification [6]. To answer RQ2, different features will be examined on the filtered tweets from the previous experiment in RQ1. Within the stance of replies, the work will further explore features like sentiment analysis. Common approaches in sentiment analysis include VADER [2, 5], TextBlob [12], and ABSA [9]. To generalise the model, the study will feed an unknown perspective as an input to evaluate how the model performs in adversarial settings.

3. **How can we explain the result to assist fact-checkers in identifying misinformation at an early stage?** Machine learning methods are widely researched, and our framework to build a monitoring tool provides leverage to the limitations of human and machine learning. To answer RQ3, the research will investigate how the information gathered based so far can efficiently communicate to human users (e.g., the users of the monitoring tool). This part of the research will involve understanding human-computer interaction components to develop an interface to present machine learning results.

4. **Which factors enable the monitoring tool users to make better decisions?** This experiment aims to recruit a group of participants to feedback and evaluate the tool for additional perspectives to improve its functionality. For RQ4, the work plans to generate a set of evaluation rules based on the research, e.g., usability, and questionnaire, to analyse the final tool in an evaluative user study. Ideally, it would add additional insights to communicate with fact-checkers about the practicality of our work.

4 Research Issues for Discussion

1. Is the focus on exploring stance in replies reasonable in addressing and managing misinformation within OSN? This idea is based on providing an early warning on dangerous topics, and many current misinformation solutions tend to focus on detection and elimination solely based on the binary classification of true and false.

2. What metrics are advisable to be used to evaluate a monitoring tool that this work is proposing apart from standard machine learning evaluation metrics? How to design a quantifiable evaluation to examine the quality of the tool for fact-checkers?

5 Appendix

Note from Doctoral Student. I am currently in the second year of my doctoral thesis. My work is supervised by Prof. Joel Fischer, Dr. Jeremie Clos, and Dr. Nazia Hameed at the University of Nottingham, United Kingdom. I have presented my proposed work in detail based on my research findings and I would like to join this consortium to discuss my ideas and gather feedback for my research areas to support my thesis. Mainly to get a further evaluation of the research directions I am aiming for from an expert's perspective. Besides, I would like to network with other PhD students in related fields to get additional support and to bring my research ideas to the table. I am also hoping to understand if I have considered sufficient perspective in my proposed work for users like fact-checkers based on real-life challenges.

Note from Advisor. Xin Yu Liew's research focuses on dealing with the problem of information overload from the perspective of fact-checkers by modelling the risk of misinformation and dealing with it accordingly (varying degrees of automation). She would benefit from attending the doctoral consortium by being paired with a mentor with expertise in expert-driven information processing systems and information filtering related to misinformation or otherwise. She would also benefit by networking with other doctoral scholars who study similar systems. Xin is currently in the second year of her PhD and produced a survey paper currently under review. We expect that she will defend her PhD thesis in Spring 2025.

Acknowledgements. The author is supported by the Engineering and Physical Sciences Research Council [grant number EP/V00784X/1].

References

1. Wasim, A., Vidal-Alaball, J., Downing, J., López Seguí, F.: COVID-19 and the 5G conspiracy theory: social network analysis of Twitter data. J. Med. Internet Res. **22**, e19458 (2020). https://doi.org/10.2196/19458

2. Bahja, M., Safdar, G.A.: Unlink the link between COVID-19 and 5G networks: an NLP and SNA based approach. IEEE Access **8**, 209127–209137 (2020). https://doi.org/10.1109/ACCESS.2020.3039168

3. Cotfas, L.A., Delcea, C., Roxin, I., Ioanăş, C., Gherai, D.S., Tajariol, F.: The longest month: analyzing COVID-19 vaccination opinions dynamics from tweets in the month following the first vaccine announcement. IEEE Access **9**, 3320333223 (2021). https://doi.org/10.1109/ACCESS.2021.3059821

4. D'Andrea, E., Ducange, P., Bechini, A., Renda, A., Marcelloni, F.: Monitoring the public opinion about the vaccination topic from tweets analysis. Expert Syst. Appl. **116**, 209226 (2019). https://doi.org/10.1016/j.eswa.2018.09.009

5. Daradkeh, M.: Analysing sentiments and diffusion characteristics of COVID-19 vaccine misinformation topics in social media: a data analytics framework. Int. J. Bus. Analytics (IJBAN) **9**, 1–22 (2022). https://doi.org/10.4018/IJBAN.292056

6. Dungs, S., Aker, A., Fuhr, N., Bontcheva, K.: Can rumour stance alone predict veracity? In: Proceedings of the 27th International Conference on Computational Linguistics, COLING 2018 (2018)

7. Eggertson, L.: Lancet retracts 12-year-old article linking autism to MMR vaccines. Can. Med. Assoc. J. **182**, E199–E200 (2010). https://doi.org/10.1503/cmaj.109-3179

8. di Giovanni, M., Corti, L, Pavanetto, S., Pierri, F., Tocchetti, A., Brambilla, M.: A content-based approach for the analysis and classification of vaccine-related stances on Twitter: the Italian scenario introduction and related work. Association for the Advancement of Artificial Intelligence (2021)

9. Jang, H., Rempel, E., Roth, D., Carenini, G., Janjua, N.Z.: Tracking COVID-19 discourse on Twitter in North America: infodemiology study using topic modeling and aspect-based sentiment analysis. J. Med. Internet Res. **23**, e25431 (2021). https://doi.org/10.2196/25431

10. Kaur, S., Kaul, P., Zadeh, P.M.: Monitoring the dynamics of emotions during Covid-19 using twitter data. Procedia Comput. Sci. **177**, 423–430 (2020)
11. Lemmens, J., Dejaeghere, T., Kreutz, T., van Nooten, J., Markov, I., Daelemans, W.: Vaccinpraat: monitoring vaccine skepticism in Dutch Twitter and Facebook comments. Comput. Linguist. Neth. J. **11**, 173–188 (2021)
12. Melton, C.A., Olusanya, O.A., Ammar, N., Shaban-Nejad, A.: Public sentiment analysis and topic modeling regarding COVID-19 vaccines on the Reddit social media platform: a call to action for strengthening vaccine confidence. J. Infect. Public Health **14**, 1505–1512 (2021). https://doi.org/10.1016/j.jiph.2021.08.010
13. Christina, M.: $1 Million in toxic bleach sold as 'miracle' cure, officials say. In: The New York Times (2021). https://nyti.ms/3j8E7OJ. Accessed 20 Jun 2022
14. Romer, D., Jamieson, K.H.: Conspiracy theories as barriers to controlling the spread of COVID-19 in the US. Soc. Sci. Med. **263**, 113356 (2020). https://doi.org/10.1016/j.socscimed.2020.113356
15. Tahir, A., Cheng, L., Sheth, P., Liu, H.: Improving vaccine stance detection by combining online and offline data. arXiv arXiv:2208.04491 (2022)
16. World Health Organization: Ten threats to global health in 2019 (2019). https://www.who.int/news-room/spotlight/ten-threats-to-global-health-in-2019. Accessed 7 Jun 2022
17. Xue, J., Chen, J., Chen, C., Zheng, C., Li, S., Zhu, T.: Public discourse and sentiment during the COVID 19 pandemic: using latent dirichlet allocation for topic modeling on Twitter. PLoS ONE **15**, e0239441 (2020). https://doi.org/10.1371/journal.pone.0239441

User Privacy in Recommender Systems

Peter Müllner[(✉)] [ID]

Know-Center GmbH, Graz, Austria
pmuellner@know-center.at

Abstract. Recommender systems process abundances of user data to generate recommendations that fit well to each individual user. This utilization of user data can pose severe threats to user privacy, e.g., the inadvertent leakage of user data to untrusted parties or other users. Moreover, this data can be used to reveal a user's identity, or to infer very private information as, e.g., gender. Instead of the plain application of privacy-enhancing techniques, which could lead to decreased accuracy, we tackle the problem itself, i.e., the utilization of user data. With this, we aim to equip recommender systems with means to provide high-quality recommendations that respect users' privacy.

Keywords: Recommender systems · Differential privacy · Data minimization · Neighborhood reuse

1 Motivation

Recommender systems are quintessential tools that help users navigating through the overload of information prevalent in many applications. Typically, historic user interaction data is utilized to generate personalized recommendations. This, however, poses a privacy threat, since through the utilization of their data, users' private information could be disclosed to untrusted parties or other users.

To hinder the disclosure of private information, legal and technical solutions have been implemented. For example, the General Data Protection Regulation's principle of data minimization requires that only the necessary data shall be processed [24]. However, requiring the recommender system to use a minimal amount of user data leads to a drop in recommendation accuracy [5]. Therefore, we study *RQ1: How can recommender systems use fewer data to generate meaningful recommendations?* Also, homomorphic encryption [10,15], federated learning [17,27], or differential privacy [6] have been applied to ensure privacy. Especially differential privacy became widely-used in a broad body of information retrieval applications, e.g., recommender systems [9,31]. However, differential privacy negatively affects recommendation accuracy, since it adds random noise to the user data [4,20]. Thus, we study *RQ2: How can we improve the recommendation accuracy of differentially private recommender systems?* and *RQ3: In which ways does differential privacy impact personalized recommendations?*

J. Kamps et al. (Eds.): ECIR 2023, LNCS 13982, pp. 456–461, 2023.
https://doi.org/10.1007/978-3-031-28241-6_52

Overall, we investigate how recommender systems can process user data in a more responsible way and also, we explore how we can make the negative impacts, differential privacy can have on users, less impactful.

2   Related Work

Due to the increasing awareness of the potential disclosure of private information, numerous works identify several ways in which recommender systems jeopardize users' privacy [1,2,14,26]. For example, Strucks et al. [25] can infer users' gender based on parts of the user's rating data. Also, the utilization of users' rating data to generate recommendations poses a severe privacy threat [23]. Through the generated recommendations, the recommender system could leak information about the users, whose data has been used. Similarly, Zhang et al. [30] illustrate how to identify users, whose rating data was used in the recommendation process.

To make these privacy risks less serious, for example, Biega et al. [5] minimize a user's data that the recommender system is allowed to use and find that the accuracy loss is less serious than expected. Plus, the severity of the accuracy loss is different for different users (see our work on *RQ1* in Sect. 3).

Besides applying homomorphic encryption [15] and federated learning [18], especially differential privacy [6] has been used in many research works [8,9, 13,20]. For example, Zhu et al. [31] use differential privacy to privately select users, whose data is used to generate recommendations. Plus, the data itself is protected as well via the addition of noise. Through fine-tuning the level of noise, Zhu et al. can make the accuracy-privacy trade-off less serious. Similarly, also Liu et al. [19] vary the level of noise to ensure meaningful recommendations. Xin and Jaakola [28] protect only a subset of users with differential privacy, and this way, recommendation accuracy benefits from the data of unprotected users (see our work on *RQ2* in Sect. 3).

Moreover, Zemel et al. [29] and Ekstrand et al. [7] discuss the impact of differential privacy on fairness, which is an example of the different ways in which differential privacy might impact users (see our work on *RQ3* in Sect. 3).

3   Research Questions and Preliminary Results

RQ1: How can recommender systems use fewer data to generate meaningful recommendations?

Data minimization can aid user privacy, however, it results in a drop in recommendation accuracy [5]. Thus, we study how strong the amount of data can be minimized and how serious this impacts recommendation accuracy. Plus, we explore if meta learning [11] can help to make this accuracy drop less severe.

Our Work. In our work [21], we experiment with MetaMF [18] and test its robustness against small privacy budgets. We measure robustness via the relative accuracy loss ΔMAE, and privacy budget is the fraction β of each user's

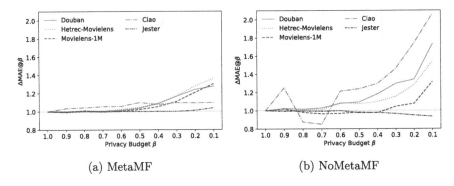

(a) MetaMF   (b) NoMetaMF

Fig. 1. Relative accuracy drop of MetaMF and NoMetaMF for different privacy budgets β, i.e., the fraction of a user's data that the recommender system can use. $\beta \geq 0.5$ is sufficient to keep recommendation accuracy. In case $\beta < 0.5$, meta learning is required to keep the accuracy loss at a moderate level.

data that the recommender system is allowed to use. Furthermore, we isolate the impact meta learning has on the recommendation accuracy via evaluating NoMetaMF (i.e., a variant without meta learning).

We find that for most datasets, approximately 50% of each user's data is sufficient to keep recommendation accuracy (see Fig. 1). In case fewer data is available, the relative accuracy loss increases much slower for MetaMF than for NoMetaMF. This shows that for small privacy budgets, meta learning helps to keep recommendation accuracy. Moreover, highly active users with lots of data experience a more severe accuracy loss than users that are less active, which is a sign of different accuracy-privacy trade-offs.

Open Issues. In our existing work, we assume that every piece of data is equally sensitive, i.e., how much it puts a user's privacy at risk when disclosed. In reality, there exists high-sensitive data, e.g., gender, as well as low-sensitive data, e.g., favorite color. Similarly, the recommender system can generate more or less accurate recommendations depending on the data it can use. Thus, we will study how to quantify the privacy-sensitivity of data and how important the data is for the recommender system to generate meaningful recommendations.

RQ2: How can we improve the recommendation accuracy of differentially private recommender systems?

Differential privacy [6] typically leads to a decrease in recommendation accuracy due to the addition of noise to the rating data [3]. Our idea is to minimize the set of users that need to be protected with differential privacy. This way, we limit the addition of noise, which leads to a better accuracy-privacy trade-off.

Our Work. In *KNN* recommender systems, neighbors' data is used to generate recommendations, which poses a privacy risk for these neighbors [23]. In our work [22], we develop the *ReuseKNN* recommender system, which reuses the same neighbors for many recommendations. This way, only a few users are used as neighbors and need to be protected with differential privacy, while most users do not need to be protected, since they are only rarely used as neighbors.

We find that *ReuseKNN* can substantially decrease the number of users that need to be protected with differential privacy. Depending on the dataset, down to 24% of users need to be protected, compared to 80% for traditional *UserKNN* [12,31]. Also, recommendation accuracy can be preserved, and in many cases improved over *UserKNN*, while the users' privacy risk decreases. Plus, we find that *ReuseKNN* does not exacerbate an existing popularity bias.

Open Issues. In our work, a user's privacy risk quantifies how often the user's data is utilized in the recommendation process. However, similar to our idea for future work on *RQ1*, this does not consider that the data of two users could differ in their sensitivity. Also, users with high-sensitive data could be protected differently than users with low-sensitive data.

RQ3: In which ways does differential privacy impact personalized recommendations?

In addition to the impact of differential privacy on recommendation accuracy, we are also interested in how beyond-accuracy objectives are impacted, e.g., popularity bias [16], and if different users are impacted differently. Also, we investigate the longitudinal effects on the users if differential privacy is applied.

Our Work. We apply differential privacy by adding noise to the users' ratings [6] and compare recommendation lists generated with and without differential privacy to quantify how severe differential privacy can change a user's recommendations. Plus, we study in which ways differential privacy impacts the recommendation lists by monitoring recommendation accuracy and popularity bias.

Our first experiments indicate that depending on the dataset and algorithm, between 54% and 80% of users are impacted by differential privacy. On average, these users' recommendation accuracy decreases by 5% to 26%. However, for 80% to 91% of these users, differential privacy leads to the recommendation of fewer popular items, which can help to alleviate popularity bias.

Open Issues. We identify two open issues. First, specific user behavior could influence the impact of differential privacy: for example, users with diverse preferences could be impacted in different ways than users with narrow preferences. Second, the impact of differential privacy could accumulate when many recommendations are generated for a user over time. For these open issues, we believe that a simulation study is a well-suited approach.

Acknowledgements. Thanks to my supervisors Dominik Kowald and Elisabeth Lex for their feedback on this work. This work is supported by the "DDAI" COMET Module within the COMET - Competence Centers for Excellent Technologies Programme, funded by the Austrian Federal Ministry for Transport, Innovation and Technology (bmvit), the Austrian Federal Ministry for Digital and Economic Affairs (bmdw), the

Austrian Research Promotion Agency (FFG), the province of Styria (SFG) and partners from industry and academia. The COMET Programme is managed by FFG.

References

1. Beigi, G., Liu, H.: "Identifying novel privacy issues of online users on social media platforms" by Ghazaleh Beigi and Huan Liu with Martin Vesely as coordinator. SIGWEB Newsl. (Winter) (2019). https://doi.org/10.1145/3293874.3293878
2. Beigi, G., Liu, H.: A survey on privacy in social media: identification, mitigation, and applications. ACM Trans. Data Sci. 1(1), 1–38 (2020)
3. Berkovsky, S., Kuflik, T., Ricci, F.: The impact of data obfuscation on the accuracy of collaborative filtering. Expert Syst. Appl. 39(5), 5033–5042 (2012)
4. Berlioz, A., Friedman, A., Kaafar, M.A., Boreli, R., Berkovsky, S.: Applying differential privacy to matrix factorization. In: Proceedings of the 9th ACM Conference on Recommender Systems, pp. 107–114 (2015)
5. Biega, A.J., Potash, P., Daumé, H., Diaz, F., Finck, M.: Operationalizing the legal principle of data minimization for personalization. In: Proceedings of the 43rd International ACM SIGIR Conference on Research and Development in Information Retrieval, pp. 399–408 (2020)
6. Dwork, C.: Differential privacy: a survey of results. In: Agrawal, M., Du, D., Duan, Z., Li, A. (eds.) TAMC 2008. LNCS, vol. 4978, pp. 1–19. Springer, Heidelberg (2008). https://doi.org/10.1007/978-3-540-79228-4_1
7. Ekstrand, M.D., Joshaghani, R., Mehrpouyan, H.: Privacy for all: ensuring fair and equitable privacy protections. In: Conference on Fairness, Accountability and Transparency, pp. 35–47. PMLR (2018)
8. Friedman, A., Berkovsky, S., Kaafar, M.A.: A differential privacy framework for matrix factorization recommender systems. User Model. User-Adap. Inter. 26(5), 425–458 (2016). https://doi.org/10.1007/s11257-016-9177-7
9. Gao, C., Huang, C., Lin, D., Jin, D., Li, Y.: DPLCF: differentially private local collaborative filtering. In: Proceedings of the 43rd International ACM SIGIR Conference on Research and Development in Information Retrieval, pp. 961–970 (2020)
10. Gentry, C.: A Fully Homomorphic Encryption Scheme. Stanford University (2009)
11. Ha, D., Dai, A., Le, Q.V.: Hypernetworks. arXiv preprint arXiv:1609.09106 (2016)
12. Herlocker, J.L., Konstan, J.A., Borchers, A., Riedl, J.: An algorithmic framework for performing collaborative filtering. In: Proceedings of the 22nd annual international ACM SIGIR Conference on Research and Development in Information Retrieval, pp. 230–237 (1999)
13. Hou, D., Zhang, J., Ma, J., Zhu, X., Man, K.L.: Application of differential privacy for collaborative filtering based recommendation system: a survey. In: 2021 12th International Symposium on Parallel Architectures, Algorithms and Programming (PAAP), pp. 97–101. IEEE (2021)
14. Jeckmans, A.J., Beye, M., Erkin, Z., Hartel, P., Lagendijk, R.L., Tang, Q.: Privacy in recommender systems. In: Ramzan, N., van Zwol, R., Lee, J.S., Clüver, K., Hua, X.S. (eds.) Social Media Retrieval, pp. 263–281. Springer, Cham (2013). https://doi.org/10.1007/978-1-4471-4555-4_12
15. Kim, S., Kim, J., Koo, D., Kim, Y., Yoon, H., Shin, J.: Efficient privacy-preserving matrix factorization via fully homomorphic encryption. In: Proceedings of the 11th ACM on Asia Conference on Computer and Communications Security, pp. 617–628 (2016)

16. Kowald, D., Muellner, P., Zangerle, E., Bauer, C., Schedl, M., Lex, E.: Support the underground: characteristics of beyond-mainstream music listeners. EPJ Data Sci. **10**(1), 1–26 (2021). https://doi.org/10.1140/epjds/s13688-021-00268-9

17. Li, Q., et al.: A survey on federated learning systems: vision, hype and reality for data privacy and protection. IEEE Trans. Knowl. Data Eng. (2021). https://ieeexplore.ieee.org/document/9599369

18. Lin, Y., et al.: Meta matrix factorization for federated rating predictions. In: Proceedings of the 43rd International ACM SIGIR Conference on Research and Development in Information Retrieval, pp. 981–990 (2020)

19. Liu, J., Hu, Y., Guo, X., Liang, T., Jin, W.: Differential privacy performance evaluation under the condition of non-uniform noise distribution. J. Inf. Secur. Appl. **71**, 103366 (2022)

20. Liu, X., et al.: When differential privacy meets randomized perturbation: a hybrid approach for privacy-preserving recommender system. In: Candan, S., Chen, L., Pedersen, T.B., Chang, L., Hua, W. (eds.) DASFAA 2017. LNCS, vol. 10177, pp. 576–591. Springer, Cham (2017). https://doi.org/10.1007/978-3-319-55753-3_36

21. Muellner, P., Kowald, D., Lex, E.: Robustness of meta matrix factorization against strict privacy constraints. In: Hiemstra, D., Moens, M.-F., Mothe, J., Perego, R., Potthast, M., Sebastiani, F. (eds.) ECIR 2021. LNCS, vol. 12657, pp. 107–119. Springer, Cham (2021). https://doi.org/10.1007/978-3-030-72240-1_8

22. Müllner, P., Lex, E., Schedl, M., Kowald, D.: ReuseKNN: neighborhood reuse for differentially-private KNN-based recommendations (2022). https://doi.org/10.48550/ARXIV.2206.11561

23. Ramakrishnan, N., Keller, B.J., Mirza, B.J., Grama, A.Y., Karypis, G.: When being weak is brave: privacy in recommender systems. IEEE Internet Comput. **5**(6), 54–62 (2001)

24. Parliament Regulation: Regulation (EU) 2016/679 of the European parliament and of the council. Regulation (EU) 679, 2016 (2016)

25. Strucks, C., Slokom, M., Larson, M.: BlurM(or)e: revisiting gender obfuscation in the user-item matrix (2019)

26. Wang, C., Zheng, Y., Jiang, J., Ren, K.: Toward privacy-preserving personalized recommendation services. Engineering **4**(1), 21–28 (2018)

27. Wang, Q., Yin, H., Chen, T., Yu, J., Zhou, A., Zhang, X.: Fast-adapting and privacy-preserving federated recommender system. VLDB J. **31**(5), 877–896 (2022)

28. Xin, Y., Jaakkola, T.: Controlling privacy in recommender systems. In: Proceedings of the 27th International Conference on Neural Information Processing Systems, NIPS 2014, vol. 2, pp. 2618–2626. MIT Press, Cambridge, MA, USA (2014)

29. Zemel, R., Wu, Y., Swersky, K., Pitassi, T., Dwork, C.: Learning fair representations. In: International Conference on Machine Learning, pp. 325–333. PMLR (2013)

30. Zhang, M., et al.: Membership inference attacks against recommender systems. In: Proceedings of the 2021 ACM SIGSAC Conference on Computer and Communications Security, pp. 864–879 (2021)

31. Zhu, T., Li, G., Ren, Y., Zhou, W., Xiong, P.: Differential privacy for neighborhood-based collaborative filtering. In: Proceedings of the 2013 IEEE/ACM International Conference on Advances in Social Networks Analysis and Mining, pp. 752–759 (2013)

Conversational Search for Multimedia Archives

Anastasia Potyagalova[(✉)]

ADAPT Centre, School of Computing, Dublin City University, Dublin 9, Ireland
anastasia.potyagalova2@mail.dcu.ie

Abstract. The growth of media archives (including text, speech, video and audio) has led to significant interest in developing search methods for multimedia content. An ongoing challenge of multimedia search is user interaction during the search process, including specification of search queries, presentation of retrieved content and user feedback. In parallel with this, recent years have seen increasing interest in conversational search methods enabling users to engage in a dialogue with an AI agent that supports their search activities. Conversational search seeks to enable users to find useful content more easily, quickly and reliably. To date, research in conversational search has focused on text archives. This project explores the integration of conversational search methods within multimedia search.

Keywords: Conversational search · Information retrieval · Multimedia retrieval

1 Introduction

My PhD research investigates the integration of conversational search into multimedia information retrieval (MIR) systems. MIR has been an area of investigation for many years; while recent years have seen significant advances in content-based indexing techniques, there remain challenges regarding user interaction with MIR systems. For example, specifying effect search queries, representing search output and user engagement with this output for the ongoing search processes. My PhD hypothesis is that user experience for MIR can be enhanced by integrating conversational search methods. Conversational search methods which enable users to engage in a dialogue with an AI agent to support their search activities have become a significant area of research investigation in recent years [17]. These methods seek to enable users to find useful content more easily, quickly and reliably. However, to date, research in conversational search has focused on text archives. My work explores the extension of the conversational search to their use in multimedia search.

Supported by the SFI Centre for Research Training (CRT) in Digitally-Enhanced Reality (d-real).

2   Background

There is a rapidly growing literature on conversational search, extensive existing work on MIR, and some work beginning to examine the use of conversational methods in multimedia systems. This section briefly introduces some of the most relevant work in these areas.

Concerning conversational search, this project focuses on a particular class of conversational systems - mixed-initiative systems - which involve the user and a conversational agent actively participating in dialogue and exchanging information through the interactive conversational process [1,17]. This approach is motivated by Radlinski and Craswell's [13] research which proposed that a conversational system should incorporate mixed-initiative features.

A multimodal conversational assistant, which uses images as a query, was introduced in Kim and Yoon's work [6]. Their research focused on the various image editing procedures and used a conversational mixed-initiative search assistant for more effective and quick query processing. They provided a convenient and efficient search framework for an image editing process. Nie and Jiao [10] presented research describing a contextual image search approach. They suggested a conversational model using an image as a search query. Moreover, Kaushik [5] introduced a basic multiview conversational image search system. This search concept included a multimedia search assistant which can find the intended image by proactively outputting a fixed number of relevant questions to clarify the intention of the user using a reinforcement learning algorithm.

MIR methods are being developed to enable more effective semantic analysis and objection recognition, which can potentially form an important part of a dialogue process in multimedia search. Meenaakshi and Shaveta [9] proposed an efficient content-based search using two approaches: text-based and feature vector-based ability. The work of Grycuk [3] presented a novel framework for retrieving images. Their application is based on content-based information retrieval and is designed to retrieve similar images from a large set of indexed images for a query image. The first step relies on automatically detecting objects, finding salient features from the images, and indexing them by database mechanisms. The study conducted by Pawaskar and Chaudhari [11] proposed a web image re-ranking application that learns the semantic meaning of images with numerous query keywords.

The key challenge of my research proposes a framework for using conversational engagement in MIR, including how this process is presented to the user during the search process.

3   Research Methodology and Main Research Questions

This research is being approached from an overall strategy of developing an experimental framework to investigate conversational engagement in MIR with state-of-the-art visual indexing methods. The goal of these investigations is to examine the following research question areas.

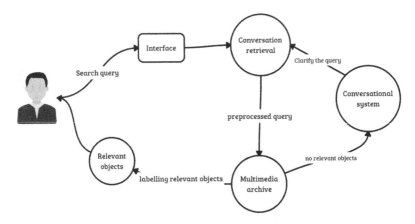

Fig. 1. Representation of conversational MIR workflow.

3.1 Research Questions

User experience in conversational MIR: How does user experience in MIR compare between a standard MIR system and an equivalent one integrating a conversational search agent, e.g. one the creation of clarification questions and query rewriting [15]?

Can clarifying questions be used effectively to resolve ambiguity and improve search effectiveness in conversational MIR? : Recent studies have highlighted the importance of clarifying questions in conversational search; generating them for open-domain search tasks still needs to be studied [2,16].

Can augmenting media views with text object labels be used to improve query construction in MIR : A similar strategy was introduced successfully in a search toolkit for MS the COCO image set [7]. This included the use of detected objects in the search query and the display of detected objects shown with corresponding images. Alternative work following similar principles is implemented in the FiftyOne solution [8].

3.2 Experimental Framework and Investigation

The first stage of an experimental conversational search application framework, including a dialogue web-search assistant and image and video archives preprocessing, has been completed. This currently includes a trained model for a simple dialogue chat with a user and processing of text user queries. Preprocessed queries are compared against text descriptions of the images or videos in a retrieval process. The functionality of the dialogue system includes various image and video retrieval techniques, including query expansion and clarification dialogue features. The prototype system will be further developed to include

Fig. 2. Initial conversational MIR interface

deep video analysis and its use in search. The CLIP [12] search techniques will be useful for the search among the video archives and construction of queries. Figure 1 shows a simple representation of the workflow of the conversational MIR system. The initial version of our conversational MIR search interface is shown in Fig. 2; note the agent's presence on the right-hand side of the image.

Initial pilot investigations are being conducted with users to examine the behaviour of our prototype framework. In particular, to identify the challenges of the user experience in MIR, which may be specifically addressed or improved using conversational engagement, and the opportunities for the use of augmented labelling of content in MIR. The entire web application, developed using the Python and Flask framework, will be deployed on a virtual machine server using one of the cloud technologies servers provided by the Rasa open library.

3.3 Datasets

Experiments will be performed on two public datasets: Flickr30k and MSR-VTT. This preprocessed and adapted for our MIR task.

Image Dataset. The Flickr30K Entities dataset is an extension of the Flickr30K dataset. It augments the original 158k captions with 244k coreference chains, linking mentions of the same entities across different captions for the same image and associating them with 276k manually annotated bounding boxes [4]. The Flickr dataset was chosen for this project because it contains many common photos of people, objects and places, so it is a good option for a common-purpose image search database.

Video Dataset. MSR-VTT (Microsoft Research Video to Text) is a large-scale open-domain video captioning dataset consisting of 10,000 video clips from 20 categories [14]. MSR-VTT dataset is well suited for searching video databases since it contains many videos with different visual objects.

Acknowledgement. This work was conducted with the financial support of the Science Foundation Ireland Centre for Research Training in Digitally-Enhanced Reality (d-real) under Grant No. 18/CRT/6224.

References

1. Aliannejadi, M., Azzopardi, L., Zamani, H., Kanoulas, E., Thomas, P., Craswell, N.: Analysing mixed initiatives and search strategies during conversational search. In: Proceedings of the 30th ACM International Conference on Information & Knowledge Management, pp. 16–26 (2021)
2. Aliannejadi, M., Zamani, H., Crestani, F., Croft, W.B.: Asking clarifying questions in open-domain information-seeking conversations. In: Proceedings of the 42nd International ACM SIGIR Conference on Research and Development in Information Retrieval, pp. 475–484 (2019)
3. Grycuk, R., Scherer, R.: Software framework for fast image retrieval. In: Proceedings of the 24th International Conference on Methods and Models in Automation and Robotics (MMAR 2019), pp. 588–593. IEEE (2019)
4. Hodosh, M., Young, P., Hockenmaier, J.: Framing image description as a ranking task: data, models and evaluation metrics. J. Artif. Intell. Res. **47**, 853–899 (2013)
5. Kaushik, A., Jacob, B., Velavan, P.: An exploratory study on a reinforcement learning prototype for multimodal image retrieval using a conversational search interface. Knowledge **2**(1), 116–138 (2022)
6. Kim, H., Kim, D., Yoon, S., Dernoncourt, F., Bui, T., Bansal, M.: CAISE: conversational agent for image search and editing. In: Proceedings of the Thirty-Sixth AAAI Conference on Artificial Intelligence (AAAI-2022) (2022)
7. Lin, T., et al: Microsoft COCO: common objects in context. CoRR abs/1405.0312 (2014). http://arxiv.org/abs/1405.0312
8. Moore, B.E., Corso, J.J.: Fiftyone (2020). https://github.com/voxel51/fiftyone
9. Munjal, M.N., Bhatia, S.: A novel technique for effective image gallery search using content based image retrieval system. In: 2019 International Conference on Machine Learning, Big Data, Cloud and Parallel Computing (COMITCon), pp. 25–29. IEEE (2019)
10. Nie, L., Jiao, F., Wang, W., Wang, Y., Tian, Q.: Conversational image search. IEEE Trans. Image Process. **30**, 7732–7743 (2021)
11. Pawaskar, S.K., Chaudhari, S.: Web image search engine using semantic of images's meaning for achieving accuracy. In: 2016 International Conference on Automatic Control and Dynamic Optimization Techniques (ICACDOT), pp. 99–103. IEEE (2016)
12. Radford, A., et al.: Learning transferable visual models from natural language supervision. CoRR abs/2103.00020 (2021). https://arxiv.org/abs/2103.00020
13. Radlinski, F., Craswell, N.: A theoretical framework for conversational search. In: Proceedings of the 2017 Conference on Conference Human Information Interaction and Retrieval, pp. 117–126 (2017)
14. Xu, J., Mei, T., Yao, T., Rui, Y.: MSR-VTT: a large video description dataset for bridging video and language. In: Proceedings of the IEEE Conference on Computer Vision and Pattern Recognition, pp. 5288–5296 (2016)
15. Zamani, H., Dumais, S., Craswell, N., Bennett, P., Lueck, G.: Generating clarifying questions for information retrieval. In: Proceedings of the Web Conference 2020, pp. 418–428 (2020)

16. Zamani, H., et al.: Analyzing and learning from user interactions for search clarification. In: Proceedings of the 43rd International ACM SIGIR Conference on Research and Development in Information Retrieval, pp. 1181–1190 (2020)
17. Zamani, H., Trippas, J.R., Dalton, J., Radlinski, F.: Conversational information seeking. arXiv preprint arXiv:2201.08808 (2022)

Disinformation Detection: Knowledge Infusion with Transfer Learning and Visualizations

Mina Schütz[1,2]([⊠])[iD]

[1] Darmstadt University of Applied Sciences, Darmstadt, Germany
[2] Austrian Institute of Technology GmbH, Vienna, Austria
`mina.schuetz@ait.ac.at`

Abstract. The automatic detection of disinformation has gained an increased focus by the research community during the last years. The spread of false information can be an issue for political processes, opinion mining and journalism in general. In this dissertation, I propose a novel approach to gain new insights on the automatic detection of disinformation in textual content. Additionally, I will combine multiple research domains, such as fake news, hate speech, propaganda, and extremism. For this purpose, I will create two novel and annotated datasets in German - a large multi-label dataset for disinformation detection in news articles and a second dataset for hate speech detection in social media posts, which both can be used for training the models in the listed domains via transfer learning. With the usage of transfer learning, an extensive data analysis and classification of the presented domains will be conducted. The classification models will be enhanced during and after training using a knowledge graph, containing additional information (i.e. named entities, relationships, topics), to find explicit insights about the common traits or lines of disinformative arguments in an article. Lastly, methods of explainable artificial intelligence will be combined with visualization techniques to understand the models predictions and present the results in a user-friendly and interactive way.

Keywords: Disinformation · NLP · Fake news · Transfer learning · XAI

1 Introduction

During the last years, news consumption shifted more and more to online articles and social media. Because of this, false information can spread more easily and is harder to distinguish by consumers [15]. Therefore, the need for an automatic disinformation detection system increased, which can help journalists with fact-checking, governmental institutions with hybrid threats, and private consumers with the information overload. Even though, there is a lot of research going on in this field, there is no universal solution found yet. Furthermore, textual classification approaches often only approach fake news in a binary or multi-class way (i.e. fake/real [47] or mostly true/mostly false/mixture of true and

J. Kamps et al. (Eds.): ECIR 2023, LNCS 13982, pp. 468–475, 2023.
https://doi.org/10.1007/978-3-031-28241-6_54

fake [40]). However, fake news is a much more complex domain that overlaps with other research fields [46]. This is often due to lack of time and resources - the annotation of large datasets is costly. This is especially important for deep learning models, which can overfit on specific words or syntax structures easily without enough diverse data [44]. Additionally, current machine learning models are often black-box systems. However, models have to go further than only to take an input and provide a label for a certain claim or news article [14]. This makes it essential to explain the models prediction and behavior to the experts and end-users with the usage of explainable artificial intelligence (XAI) methods and interactive visualizations. Some work has been done in this research area with systems such as exBERT [21], Verify2 [24], and XFake [54].

2   Related Work

One approach to detect disinformation is content-based with natural language processing (NLP). It focuses on linguistic features, such as writing style, syntactic and semantic features [56] either on character, word, sentence or document level [32]. For classification approaches, multiple datasets [10] and algorithms are compared regarding feature extraction and model architectures [2,3,22,36,37]. However, neural networks tend to be more effective than previous models [23,25]. Besides classification, other studies include emotional reactions [52], sentiment analysis [4], phrase detection and named entity recognition (NER) [5], or the influence of quotations used in articles [49]. However, there is a shift to rather use deep contextualized Transformer [51] models. Most often they outperform standard approaches and receive higher classification results for fake news [1,8,11,27,39,55] and hate speech detection [18,29,34,35].

Deep learning models are getting more complex and less understandable with the underlying mathematical calculations (comprehensibility of decisions), which is why they are also referred to as black boxes [33]. Due to their increased use in areas such as justice and terrorism [16] XAI is needed: explainability states why a prediction was made and interpretability deals with understanding the newly generated knowledge [17]. More even, when they are used in safety-critical systems XAI methods increase the users confidence in the predictions [17]. Additionally, the combination of graph and text analysis has proven itself promising in this area, compared to purely text-based analysis [6,19,20,30,50,53]. This "knowledge infusion" can be applied directly to the model (e.g. improvement of the error rate, context information in hidden layers) [17], as well as on the results and visualizations [16].

3   Main Contributions and Research Questions

The main scientific contributions of this dissertation are two-fold [43]: **we propose a novel approach that makes use of different machine learning methods and adapts itself through learning algorithms to the best fitting disinformation detection methods based on the underlying data for the German language.** This will be enhanced using a knowledge graph to find new relations between articles and their categories. Secondly, **we propose a model for explaining artificial intelligence through comprehensible visualization methods and automatic adaption methods**. The following research question and sub-questions are proposed:

To which extend can we obtain new linguistic knowledge about disinformation with natural language processing methods and visualization techniques?

- **RQ1:** To which extend can we classify the content and statements of articles with current state-of-the-art models in NLP?
- **RQ2:** To which extent can we extract statements and narratives (i.e. claims, storyline) with NLP approaches from corresponding articles and put them in a common context?
- **RQ3:** To which extend can transfer learning in combination with knowledge infusion be used to propose a better solution to understand the content behind disinformation?
- **RQ4:** To which extend are methods of XAI and the use of visualization techniques helpful to gain valuable insights on the decisions by models and to present them in a user-friendly way?

4   Methodology and Results

Data for disinformation detection is scarce, especially for the German language. Therefore, I propose to create two publicly available datasets as shown in Fig. 1. The first (Disinformation Dataset) will include general news articles for analyzing long texts and their specific characteristics. The second dataset (Hate Speech Dataset) is already published [12] and contains more detailed fine-granular labels for hate speech and toxicity for 10,278 short texts derived from Twitter.

To answer **RQ1 (Classification)**, I will extract linguistic patterns and features, additional to pre-training (unsupervised) and fine-tuning (supervised) Transformer models [42]. More features can be helpful for further training [42] and transparency in the final user interface. The models are trained (binary, multi-class, multi-label) depending on the dataset. First results show that the preprocessing steps and the content of articles (i.e. body text, titles, or both) can have an impact on the accuracy and F1-score in fake news detection [41,46]. I also found that multilingual models work better even for monolingual tasks [9], especially in combination with large domain-specific datasets for pre-training [42,44]. This can be enhanced with data augmentation strategies [26,28,45]. Next steps will include multi-task learning to make more robust predictions of the labels - especially for multi-label classification. To extract further information in a given text (**RQ2 - Narratives**), downstream NLP models - such as claim, entity, relationship, and event extraction - have been created. Those are also the basis for **RQ3 (Transfer Learning, Knowledge Infusion)**. After the extraction, the additional information will be used to create a knowledge graph in combination with the classification outputs. The graph will not only help to understand semantics in a given text, but can also open up the black-box

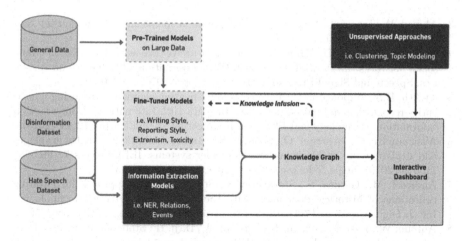

Fig. 1. Methodology overview.

machine learning models during training through knowledge infusion (combination of semantic and vector representations; enhanced loss functions). Finally, an interactive visual analytics dashboard will be implemented, which is used to make the models predictions interpretable (**RQ4 - XAI, Visualizations**). Additionally, initial experiments with local and global XAI methods, such as LIME [38] and SHAP [31], have shown good results, however they are not enough as a stand-alone approach for interpretability. Therefore, to explore possible correlations, similarity measures and clustering (K-means [48], DBSCAN [13]) has been done. To extend this, each topic of an article will be extracted via calculating the top terms that describe each document [7]. The combination of those approaches (classification, information retrieval, knowledge graph) will all be included in the multi-layered user interface to explore for end-users via interactive visualizations.

Acknowledgments. This work is enhanced by the Darmstadt University of Applied Sciences (h_da) within the research groups in Information Science (https://sis.h-da. de/) and HCI and Visual Analytics (https://vis.h-da.de/). In collaboration with the Austrian Institute of Technology GmbH (AIT) funded by the FFG projects "Defalsif-AI" (grant no. 879670) and "RAIDAR" (Austrian security research program KIRAS of the Federal Ministry of Finance (BMF)).

References

1. Aggarwal, A., Chauhan, A., Kumar, D., Mittal, M., Verma, S.: Classification of fake news by fine-tuning deep bidirectional transformers based language model. EAI Endorsed Trans. Scalable Inf. Syst. Online First (2020). https://doi.org/10. 4108/eai.13-7-2018.163973
2. Ahmed, H., Traore, I., Saad, S.: Detection of online fake news using N-gram analysis and machine learning techniques. In: Traore, I., Woungang, I., Awad, A. (eds.) ISDDC 2017. LNCS, vol. 10618, pp. 127–138. Springer, Cham (2017). https://doi. org/10.1007/978-3-319-69155-8_9

3. Ahmed, H., Traoré, I., Saad, S.: Detecting opinion spams and fake news using text classification. Secur. Priv. 1 (2018)

4. Ajao, O., Bhowmik, D., Zargari, S.: Sentiment aware fake news detection on online social networks. In: ICASSP 2019–2019 IEEE International Conference on Acoustics, Speech and Signal Processing (ICASSP), pp. 2507–2511 (2019)

5. Al-Ash, H.S., Wibowo, W.C.: Fake news identification characteristics using named entity recognition and phrase detection. In: 2018 10th International Conference on Information Technology and Electrical Engineering (ICITEE), pp. 12–17 (2018)

6. Alshammari, M., Nasraoui, O., Sanders, S.: Mining semantic knowledge graphs to add explainability to black box recommender systems. IEEE Access **7**, 110563–110579 (2019). https://doi.org/10.1109/ACCESS.2019.2934633

7. Andresel, M., Gordea, S., Stevanetic, S., Schütz, M.: An approach for curating collections of historical documents with the use of topic detection technologies. Int. J. Digit. Curation **17**(1), 12 (2022). http://www.ijdc.net/article/view/819

8. Antoun, W., Baly, F., Achour, R., Hussein, A., Hajj, H.: State of the art models for fake news detection tasks. In: 2020 IEEE International Conference on Informatics, IoT, and Enabling Technologies (ICIoT), pp. 519–524 (2020)

9. Böck, J., et al.: AIT_FHSTP at GermEval 2021: automatic fact claiming detection with multilingual transformer models. In: Proceedings of the GermEval 2021 Shared Task on the Identification of Toxic, Engaging, and Fact-Claiming Comments, pp. 76–82. Association for Computational Linguistics, Duesseldorf, September 2021. https://aclanthology.org/2021.germeval-1.11

10. Buntain, C., Golbeck, J.: Automatically identifying fake news in popular twitter threads. In: 2017 IEEE International Conference on Smart Cloud (SmartCloud), pp. 208–215 (2017)

11. Cruz, J.C.B., Tan, J.A., Cheng, C.: Localization of fake news detection via multitask transfer learning. In: Proceedings of The 12th Language Resources and Evaluation Conference, pp. 2596–2604. European Language Resources Association, Marseille, May 2020. https://www.aclweb.org/anthology/2020.lrec-1.316

12. Demus, C., Pitz, J., Schütz, M., Probol, N., Siegel, M., Labudde, D.: DeTox: a comprehensive dataset for German offensive language and conversation analysis. In: Proceedings of the Sixth Workshop on Online Abuse and Harms (WOAH), pp. 143–153. Association for Computational Linguistics, Seattle. July 2022. https://doi.org/10.18653/v1/2022.woah-1.14. https://aclanthology.org/2022.woah-1.14

13. Ester, M., Kriegel, H.P., Sander, J., Xu, X.: A density-based algorithm for discovering clusters in large spatial databases with noise. In: Proceedings of the Second International Conference on Knowledge Discovery and Data Mining, KDD 1996, pp. 226–231. AAAI Press (1996)

14. Figueira, Á., Guimarães, N., Torgo, L.: Current state of the art to detect fake news in social media: global trendings and next challenges. In: WEBIST (2018)

15. Figueira, Á., Oliveira, L.: The current state of fake news: challenges and opportunities. Procedia Comput. Sci. **121**, 817–825 (2017). https://doi.org/10.1016/j.procs.2017.11.106

16. Futia, G., Vetrò, A.: On the integration of knowledge graphs into deep learning models for a more comprehensible AI-three challenges for future research. Information **11**(2) (2020). https://doi.org/10.3390/info11020122. https://www.mdpi.com/2078-2489/11/2/122

17. Gaur, M., Faldu, K., Sheth, A.: Semantics of the black-box: can knowledge graphs help make deep learning systems more interpretable and explainable? IEEE Internet Comput. **25**(1), 51–59 (2021). https://doi.org/10.1109/MIC.2020.3031769

18. Gertner, A., Henderson, J., Merkhofer, E., Marsh, A., Wellner, B., Zarrella, G.: MITRE at SemEval-2019 task 5: transfer learning for multilingual hate speech detection. In: Proceedings of the 13th International Workshop on Semantic Evaluation, pp. 453–459. Association for Computational Linguistics, Minneapolis, Minnesota, June 2019. https://doi.org/10.18653/v1/S19-2080. https://aclanthology.org/S19-2080

19. Han, Y., Silva, A., Luo, L., Karunasekera, S., Leckie, C.: Knowledge enhanced multi-modal fake news detection. CoRR abs/2108.04418 (2021). https://arxiv.org/abs/2108.04418

20. Hojas-Mazo, W., Simón-Cuevas, A., De la Iglesia Campos, M., Romero, F., Olivas, J.: A concept-based text analysis approach using knowledge graph, vol. 854, pp. 696–708 (2018). https://doi.org/10.1007/978-3-319-91476-3_57

21. Hoover, B., Strobelt, H., Gehrmann, S.: exBERT: a visual analysis tool to explore learned representations in transformer models. In: Proceedings of the 58th Annual Meeting of the Association for Computational Linguistics: System Demonstrations, pp. 187–196. Association for Computational Linguistics, July 2020. https://www.aclweb.org/anthology/2020.acl-demos.22

22. Kaliyar, R.K., Goswami, A., Narang, P., Sinha, S.: FndNet - a deep convolutional neural network for fake news detection. Cogn. Syst. Res. **61**(C), 32–44 (2020). https://doi.org/10.1016/j.cogsys.2019.12.005

23. Kapil, P., Ekbal, A., Das, D.: Investigating deep learning approaches for hate speech detection in social media. CoRR abs/2005.14690 (2020). https://arxiv.org/abs/2005.14690

24. Karduni, A., et al.: Vulnerable to misinformation? Verifi! In: Proceedings of the 24th International Conference on Intelligent User Interfaces, IUI 2019, pp. 312–323. Association for Computing Machinery, New York (2019). https://doi.org/10.1145/3301275.3302320

25. Kumar, R., Ojha, A.K.: KMI-Panlingua at HASOC 2019: SVM vs BERT for hate speech and offensive content detection. In: FIRE (2019)

26. Köhler, J., Shahi, G.K., Struss, J.M., Wiegand, M., Siegel, M., Schütz, M.: Overview of the clef-2022 CheckThat! lab: task 3 on fake news detection. In: Faggioli, G., Ferro, N., Hanbury, A., Potthast, M. (eds.) Proceedings of the Working Notes of CLEF 2022 - Conference and Labs of the Evaluation Forum, Bologna, Italy, pp. 404–421 (2022). https://ceur-ws.org/Vol-3180/paper-30.pdf

27. Levi, O., Hosseini, P., Diab, M., Broniatowski, D.: Identifying nuances in fake news vs. satire: using semantic and linguistic cues. In: Proceedings of the Second Workshop on Natural Language Processing for Internet Freedom: Censorship, Disinformation, and Propaganda (2019). https://doi.org/10.18653/v1/d19-5004

28. Liakhovets, D., et al.: Transfer learning for automatic sexism detection with multilingual transformer models. In: y Gómez, M.M., et al. (eds.) Proceedings of the Iberian Languages Evaluation Forum (IberLEF 2022). CEUR-WS.org, A Coruna, Spain (2022). https://ceur-ws.org/Vol-3202/exist-paper1.pdf

29. Liu, P., Li, W., Zou, L.: NULI at SemEval-2019 task 6: transfer learning for offensive language detection using bidirectional transformers. In: Proceedings of the 13th International Workshop on Semantic Evaluation, pp. 87–91. Association for Computational Linguistics, Minneapolis, June 2019. https://doi.org/10.18653/v1/S19-2011

30. Liu, W., et al.: K-BERT: enabling language representation with knowledge graph. Proc. AAAI Conf. Artif. Intell. **34**(03), 2901–2908 (2020). https://doi.org/10.1609/aaai.v34i03.5681. https://ojs.aaai.org/index.php/AAAI/article/view/5681

31. Lundberg, S.M., Lee, S.: A unified approach to interpreting model predictions. CoRR abs/1705.07874 (2017). http://arxiv.org/abs/1705.07874

32. Mahid, Z.I., Manickam, S., Karuppayah, S.: Fake news on social media: brief review on detection techniques. In: 2018 Fourth International Conference on Advances in Computing, Communication Automation (ICACCA), pp. 1–5 (2018)

33. Mathew, B., Saha, P., Yimam, S.M., Biemann, C., Goyal, P., Mukherjee, A.: HateXplain: a benchmark dataset for explainable hate speech detection. CoRR abs/2012.10289 (2020). https://arxiv.org/abs/2012.10289

34. Mishra, S.: 3Idiots at HASOC 2019: fine-tuning transformer neural networks for hate speech identification in Indo-European languages. In: FIRE (2019)

35. Paraschiv, A., Cercel, D.C.: UPB at GermEval-2019 task 2: BERT-based offensive language classification of German Tweets. In: KONVENS (2019)

36. Pérez-Rosas, V., Kleinberg, B., Lefevre, A., Mihalcea, R.: Automatic detection of fake news. In: Proceedings of the 27th International Conference on Computational Linguistics, pp. 3391–3401. Association for Computational Linguistics, Santa Fe, August 2018. https://www.aclweb.org/anthology/C18-1287

37. Reis, J.C.S., Correia, A., Murai, F., Veloso, A., Benevenuto, F.: Explainable machine learning for fake news detection. In: Proceedings of the 10th ACM Conference on Web Science, WebSci 2019, pp. 17–26. Association for Computing Machinery, New York (2019). https://doi.org/10.1145/3292522.3326027

38. Ribeiro, M.T., Singh, S., Guestrin, C.: "why should I trust you?": explaining the predictions of any classifier. CoRR abs/1602.04938 (2016). http://arxiv.org/abs/1602.04938

39. Rodríguez, Á.I., Iglesias, L.L.: Fake news detection using deep learning (2019)

40. Santia, G.C., Williams, J.R.: BuzzFace: a news veracity dataset with Facebook user commentary and egos. In: ICWSM (2018)

41. Schütz, M.: Detection and identification of fake news: binary content classification with pre-trained language models. In: Information Between Data and Knowledge, Schriften zur Informationswissenschaft, vol. 74, pp. 422–431. Werner Hülsbusch, Glückstadt (2021). https://epub.uni-regensburg.de/44959/. gerhard Lustig Award Papers

42. Schütz, M., Demus, C., Pitz, J., Probol, N., Siegel, M., Labudde, D.: DeTox at GermEval 2021: toxic comment classification. In: Proceedings of the GermEval 2021 Shared Task on the Identification of Toxic, Engaging, and Fact-Claiming Comments, pp. 54–61. Association for Computational Linguistics, Duesseldorf, September 2021. https://aclanthology.org/2021.germeval-1.8

43. Schütz, M., Schindler, A., Siegel, M., Nazemi, K.: Automatic fake news detection with pre-trained transformer models. In: Del Bimbo, A., et al. (eds.) ICPR 2021. LNCS, vol. 12667, pp. 627–641. Springer, Cham (2021). https://doi.org/10.1007/978-3-030-68787-8_45

44. Schütz, M., et al.: Ait_fhstp at CheckThat! 2022: cross-lingual fake news detection with a large pre-trained transformer. In: Faggioli, G., Ferro, N., Hanbury, A., Potthast, M. (eds.) Proceedings of the Working Notes of CLEF 2022 - Conference and Labs of the Evaluation Forum, Bologna, Italy, pp. 660–670 (2022). https://ceur-ws.org/Vol-3180/paper-53.pdf

45. Schütz, M., et al.: Automatic sexism detection with multilingual transformer models. In: Montes, M., et al. (eds.) Proceedings of the Iberian Languages Evaluation Forum (IberLEF 2021), Málaga, Spain, pp. 346–355 (2021). https://ceur-ws.org/Vol-2943/exist_paper1.pdf

46. Schütz, M., Schindler, A., Siegel, M.: Disinformation detection: an explainable transfer learning approach (2021). https://www.unibw.de/code-events/code2021/04_schuetz.pdf

47. Shu, K., Mahudeswaran, D., Wang, S., Lee, D., Liu, H.: FakenewsNet: a data repository with news content, social context and dynamic information for studying fake news on social media. CoRR abs/1809.01286 (2018). http://arxiv.org/abs/1809.01286

48. Steinley, D., Brusco, M.J.: Initializing K-means batch clustering: a critical evaluation of several techniques. J. Classif. **24**(1), 99–121 (2007). https://doi.org/10.1007/s00357-007-0003-0

49. Traylor, T., Straub, J., Gurmeet, Snell, N.: Classifying fake news articles using natural language processing to identify in-article attribution as a supervised learning estimator. In: 2019 IEEE 13th International Conference on Semantic Computing (ICSC), pp. 445–449 (2019)

50. Tseng, Y.W., Yang, H.K., Wang, W.Y., Peng, W.C.: KAHAN: knowledge-aware hierarchical attention network for fake news detection on social media. In: Companion Proceedings of the Web Conference 2022, WWW 2022, pp. 868–875. Association for Computing Machinery, New York (2022). https://doi.org/10.1145/3487553.3524664

51. Vaswani, A., et al.: Attention is all you need (2017)

52. Vosoughi, S., Roy, D., Aral, S.: The spread of true and false news online. Science **359**(6380), 1146–1151 (2018). https://doi.org/10.1126/science.aap9559. https://science.sciencemag.org/content/359/6380/1146

53. Whitehouse, C., Weyde, T., Madhyastha, P., Komninos, N.: Evaluation of fake news detection with knowledge-enhanced language models. In: Proceedings of the International AAAI Conference on Web and Social Media, vol. 16, pp. 1425–1429 (2022)

54. Yang, F., et al.: XFake: explainable fake news detector with visualizations. In: The World Wide Web Conference on - WWW 2019, pp. 3600–3604. ACM Press, San Francisco (2019). https://doi.org/10.1145/3308558.3314119. http://dl.acm.org/citation.cfm?doid=3308558.3314119

55. Yang, K., Niven, T., Kao, H.: Fake news detection as natural language inference. CoRR abs/1907.07347 (2019). http://arxiv.org/abs/1907.07347

56. Zhou, X., Jain, A., Phoha, V.V., Zafarani, R.: Fake news early detection: a theory-driven model. Digit. Threats Res. Pract. **1**(2) (2020). https://doi.org/10.1145/3377478

A Comprehensive Overview of Consumer Conflicts on Social Media

Oliver Warke[✉]

School of Computing Science, University of Glasgow, Glasgow, UK
o.warke.1@research.gla.ac.uk

Abstract. The use of social media platforms is increasingly prevalent in society, providing brands with a multitude of opportunities to interact with consumers. However, literature has shown this increased usage has negative impacts for users who have experienced depression, anxiety, and stress and brands who see increasing volumes of hate within their communities such as bullying, conflicts, complaints, and harmful content. Existing research focuses on extreme forms of conflict, largely ignoring the lesser forms which still pose a significant threat to consumer and brand welfare. This research aims to capture the full spectrum of online conflict, providing a comprehensive overview of the problem from an interdisciplinary marketing and computer science perspective.

I propose a further investigation into online hate, utilising big data analysis to establish an understanding of triggers, consequences and brand responses to online hate. Initially, I will conduct a systematic literature review exploring the definitions and methodology used within the hate research domain. Secondly, I will conduct an investigation into state-of-the-art models and classification systems, producing an analysis on the prevalence of hate and its various forms on social media. Finally, I plan to establish the features of social media data which constitute triggers for online conflicts. Then, through a combination of user studies, sentiment analysis, and emotion detection I will examine the consequences of these conflicts.

This project represents a unique opportunity to combine cutting edge marketing theories with big data analysis, this collaborative approach will offer a considerable contribution to academic literature.

1 Motivation

Social media provides an unrivalled opportunity to users for communication, socialisation, and interaction. Due to these phenomena the use of social media is widespread across the globe [24]. Many brands are taking advantage of the large user base, exploiting the opportunity to interact with an ever growing range of consumers. These interactions between brands and consumers result in the formation of brand communities. Laroche et al. [19] find that 'brand communities established on social media have positive effects on customer/product, customer/brand, customer/company and customer/other customers relationships, which in turn have positive effects on brand trust, and trust has positive effects

J. Kamps et al. (Eds.): ECIR 2023, LNCS 13982, pp. 476–481, 2023.
https://doi.org/10.1007/978-3-031-28241-6_55

on brand loyalty'. Not all contents of these brand communities are positive, researchers have found increasing rates of bullying, conflicts, and other harmful content [12,26]. Whilst brands may be predominantly motivated by the impact on their brand image and culture, they should also take responsibility for their consumers well being. With consumers being exposed to increasingly toxic content within brand communities and the consequences that arise from said exposure. The rise of hate on social media and the severity of it's consequences is resulting in a crisis [29]. Despite this crisis there is a lack of intervention from social media platforms and brands. Researchers have long recognised this behaviour with Masud et al. [20] finding that; "Twitter... has a long history of accommodating hate speech, cyberbullying, and toxic behaviour." Whilst the Oxford Internet Institute [17] states that; "To develop effective responses to hate speech, including through education, it is essential to better monitor and analyse the phenomenon by drawing on clear and reliable data... this also means better understanding the occurrence, virulence and reach of online hate speech." In order to combat the phenomena of hate we must develop tools and methods to automatically detect it. The scale of social media, and the volume of traffic which flows through it means that manual detection through human operators is unfeasible. My work aims to provide a solution to this problem by employing a combination of data science and marketing techniques and theories. I aim to develop a suite of classification and analytic tools which can be utilised within the marketing community to reduce hate and conflict on social media.

2 Literature Review

Despite research confirming the severe impacts of social media hate for users [2], there is a division within Marketing literature on whether consumer conflicts within online brand communities have positive or negative consequences for the brand. Ilhan et al. [16] determine that brand bullying instances positively impact the brand's social media performance. Whilst Breitsohl et al. [5] and Dineva et al. [12] advocate for brand interaction when dealing with negative content on a brand page. Brand intervention involves brands responding to negative content on their brand page through various means; issuing warnings to the aggressive consumer, educating the consumer, or motivating the consumer to conduct themselves in a positive manner using constructive dialogue. From the brand's perspective this can mitigate the potentially harmful consequences of toxicity, improving the positive effects experienced by the brand as a result of the brand page. Marketing literature has attempted to answer this division using qualitative research, this therefore provides an opportunity to approach the problem from a quantitative data science perspective.

Within literature there are a multitude of datasets [9,14] and models available for the automatic classification of hate. However, the majority focus on the extreme forms of hate such as hate speech, abuse, and aggression. Those that do explore the finer nuances of hate tend to do so along the types of extreme hate such as racism, homophobia, and sexism. There is a lack of research into the full spectrum of conflict and hate [3]. There is therefore a gap within the domain for research which explores this spectrum; including conflict types such as teasing, sarcasm, and harassment, which all contribute to conflict on social media. Within the hate research domain there is a constant demand for accuracy and the generalisation of models. With the broad spectrum of hate on social media the ability to successfully differentiate between regular user generated content and hate is highly important [18]. As a result, SOTA models and classification techniques are used in order to improve performance. SOTA models include those such as BERT which has been used in various research within the domain [8,15,21]. As mentioned researchers have sought to improve upon the base SOTA models. Caselli et al. [7] produce HateBERT, a BERT model which has been pretrained on a large range of hateful language before being deployed for classification tasks. The authors test against other SOTA models and datasets, achieving superior results. Mutanga et al. [22] use distilBERT, a lightweight variation of BERT which has been proven to be competitive [27]. Researchers also investigate a variety of novel techniques such as; multi-faceted text representations [6], multi-task learning [1], and data augmentation [25]. Despite this development of novel techniques to improve hate detection there is still a lack of research surrounding the full spectrum of hate and it's intricacies [3]. There is a gap within literature for the investigation of fine gained hate.

3   Proposed Work and Methodology

- R1) How can we detect different shades of social media conflicts?
- R2) How prevalent are the different types of conflicts?

I wish to take a broad perspective of social media conflicts and hate, considering all types of consumer comments with potential to offend another consumer, including openly hostile (e.g. hate-speech, cyberbullying) and less hostile (e.g. criticism, joking) more subtle types of comments. To do this I will conduct a systematic literature review to investigate current definitions of the spectrum of conflict within marketing and computer science domains. Evaluating the current approaches to the problem in terms of definition and methodology. For the marketing community it is important that any system I develop meet the upmost degree of performance. Marketeers require this as any type of conflict is likely to have negative commercial consequences, such as customers not returning to the online community, making negative comments about it [23], and holding the company responsible for the conflicts [10]. To this end I will focus on developing models and techniques to surrounding multiclass hate classification. Additionally, literature within Computer Science has critiqued models within research due to their lack of generalisation across commonly used datasets [30]. I therefore want to ensure the robustness and generalisation of any models and techniques I create across a range of definitions and datasets.

- R3) What are the triggers of social media conflicts?
- R4) What are the consequences?
- R5) What is the impact when a brand intervenes?

These questions are important as we need to know why consumers engage in conflicts. Marketing stakeholders require knowledge surrounding these triggers in order to generate responses and incorporate any common topics such as gender equality and sustainability into their brand image and corporate culture. The marketing literature [4,12] shows that brands currently shy away from intervening in conflicts. However, since all major brands promote their social responsibility and values regarding equality, sustainability, etc. marketing managers are seeking to find scientific guidance on what kind of messages they can post in order to improve their customers' social media experience and stop the conflicts. I will quantify the impact that conflicts, and brand intervention within those conflicts has on the brand's customers through a combination of sentiment and emotional analysis. Whilst some studies in Marketing argue that the consequences are likely to be negative [5,11], there is little quantitative evidence on the exact negative consequences that conflicts have (e.g. emotional, social media engagement, etc), and whether the consequences differ depending on the type of conflict (e.g. hostile vs subtle). I am also considering a user study investigating users responses to a range of social media interactions such as those conducted by Breitsohl et al. [4], Ewing et al. [13], and Schmid et al. [28]. Investigating users' response to triggers, conflicts, and brand intervention. This would serve to form part of the answer to the above research questions.

4   Research Issues

Given my PhD is interdisciplinary I would appreciate advice on how to balance my research between computer science and marketing. From a computer science perspective I need to contribute novel technical research surrounding deep learning and data science. To the best of my knowledge no one has ever developed a detector for different shades of conflicts. From a marketing perspective the novelty originates from applying the data science techniques to the marketing theory, the techniques themselves don't need to be novel. Stemming from this I would appreciate advice on how to identify a technical area to work on which also has practical benefits, and how to integrate a technical contribution into my thesis proposal. Additionally, I would appreciate feedback on how I would include a user study within my work. Specifically how I would transition the results to computer science methodology. I believe a user study surrounding conflict triggers, responses, and consequences would be a valuable piece of research but I also want it to have practical use within my PhD. Guidance on how to develop a user study which both answers research questions and provides an avenue of research to explore would be much appreciated. Finally, I would like to gain some insight into the creation of a robust dataset. I have an understanding of the steps required in the formation of a dataset but have reservations on the content I would like it to contain. For example, what annotations would I like multi-label vs multi-class or what annotators should I use (small group of experts vs large group of novices).

References

1. Awal, M.R., Cao, R., Lee, R.K.-W., Mitrović, S.: AngryBERT: joint learning target and emotion for hate speech detection. In: Karlapalem, K., et al. (eds.) PAKDD 2021. LNCS (LNAI), vol. 12712, pp. 701–713. Springer, Cham (2021). https://doi.org/10.1007/978-3-030-75762-5_55
2. Best, P., Manktelow, R., Taylor, B.: Online communication, social media and adolescent wellbeing: a systematic narrative review. Child Youth Serv. Rev. 41, 27–36 (2014)
3. Bianchi, F., Hills, S.A., Rossini, P., Hovy, D., Tromble, R., Tintarev, N.: "it's not just hate": a multi-dimensional perspective on detecting harmful speech online. arXiv preprint arXiv:2210.15870 (2022)
4. Breitsohl, J., et al.: Bullying in online brand communities-exploring consumers' intentions to intervene. In: Hopfgartner, F., Jaidka, K., Mayr, P., Jose, J., Breitsohl, J. (eds.) International Conference on Social Informatics, pp. 436–443. Springer, Cham (2022). https://doi.org/10.1007/978-3-031-19097-1_30
5. Breitsohl, J., Roschk, H., Feyertag, C.: Consumer brand bullying behaviour in online communities of service firms. In: Service Business Development, pp. 289–312. Springer, Wiesbaden (2018). https://doi.org/10.1007/978-3-658-22424-0_13
6. Cao, R., Lee, R.K.W., Hoang, T.A.: DeepHate: hate speech detection via multi-faceted text representations. In: 12th ACM Conference on Web Science, pp. 11–20 (2020)
7. Caselli, T., Basile, V., Mitrović, J., Granitzer, M.: HateBERT: retraining BERT for abusive language detection in English. arXiv preprint arXiv:2010.12472 (2020)
8. Dai, X., Karimi, S., Hachey, B., Paris, C.: Cost-effective selection of pretraining data: a case study of pretraining BERT on social media. arXiv preprint arXiv:2010.01150 (2020)
9. Davidson, T., Warmsley, D., Macy, M., Weber, I.: Automated hate speech detection and the problem of offensive language. In: Proceedings of the International AAAI Conference on Web and Social Media, vol. 11, pp. 512–515 (2017)
10. Dineva, D., Breitsohl, J., Garrod, B., Megicks, P.: Consumer responses to conflict-management strategies on non-profit social media fan pages. J. Interact. Market. 52, 118–136 (2020). https://doi.org/10.1016/j.intmar.2020.05.002. https://www.sciencedirect.com/science/article/pii/S1094996820301006
11. Dineva, D., Breitsohl, J., Roschk, H., Hosseinpour, M.: Consumer-to-consumer conflicts and brand moderation strategies during covid-19 service failures: a framework for international marketers. Int. Market. Rev. (2022)
12. Dineva, D.P., Breitsohl, J.C., Garrod, B.: Corporate conflict management on social media brand fan pages. J. Mark. Manag. 33(9–10), 679–698 (2017)
13. Ewing, M.T., Wagstaff, P.E., Powell, I.H.: Brand rivalry and community conflict. J. Bus. Res. 66(1), 4–12 (2013)
14. Founta, A.M., et al.: Large scale crowdsourcing and characterization of Twitter abusive behavior. In: Twelfth International AAAI Conference on Web and Social Media (2018)
15. Heidari, M., Jones, J.H.: Using BERT to extract topic-independent sentiment features for social media bot detection. In: 2020 11th IEEE Annual Ubiquitous Computing, Electronics & Mobile Communication Conference (UEMCON), pp. 0542–0547. IEEE (2020)
16. Ilhan, B.E., Kübler, R.V., Pauwels, K.H.: Battle of the brand fans: impact of brand attack and defense on social media. J. Interact. Mark. 43, 33–51 (2018)

17. Institute, O.I., UNESCO, on Genocide Prevention, U.N.O., the responsibility to protect: addressing hate speech on social media: contemporary challenges (2021). https://unesdoc.unesco.org/ark:/48223/pf0000379177
18. Isaksen, V., Gambäck, B.: Using transfer-based language models to detect hateful and offensive language online. In: Proceedings of the Fourth Workshop on Online Abuse and Harms, pp. 16–27 (2020)
19. Laroche, M., Habibi, M.R., Richard, M.O.: To be or not to be in social media: how brand loyalty is affected by social media? Int. J. Inf. Manage. **33**(1), 76–82 (2013)
20. Masud, S., et al.: Hate is the new Infodemic: a topic-aware modeling of hate speech diffusion on Twitter. In: 2021 IEEE 37th International Conference on Data Engineering (ICDE), pp. 504–515. IEEE (2021)
21. Mozafari, M., Farahbakhsh, R., Crespi, N.: A BERT-based transfer learning approach for hate speech detection in online social media. In: Cherifi, H., Gaito, S., Mendes, J.F., Moro, E., Rocha, L.M. (eds.) COMPLEX NETWORKS 2019. SCI, vol. 881, pp. 928–940. Springer, Cham (2020). https://doi.org/10.1007/978-3-030-36687-2_77
22. Mutanga, R.T., Naicker, N., Olugbara, O.O.: Hate speech detection in twitter using transformer methods. Int. J. Adv. Comput. Sci. Appl. **11**(9) (2020)
23. Ounvorawong, N.: 'Brand victimisation': when consumers are bullied by fellow brand followers in online brand communities. University of Kent (United Kingdom) (2021)
24. Poushter, J., Bishop, C., Chwe, H.: Social media use continues to rise in developing countries but plateaus across developed ones. Pew Res. Center **22**, 2–19 (2018)
25. Rizos, G., Hemker, K., Schuller, B.: Augment to prevent: short-text data augmentation in deep learning for hate-speech classification. In: Proceedings of the 28th ACM International Conference on Information and Knowledge Management, pp. 991–1000 (2019)
26. Saha, K., Chandrasekharan, E., De Choudhury, M.: Prevalence and psychological effects of hateful speech in online college communities. In: Proceedings of the 10th ACM Conference on Web Science, pp. 255–264 (2019)
27. Sanh, V., Debut, L., Chaumond, J., Wolf, T.: DistilBERT, a distilled version of BERT: smaller, faster, cheaper and lighter. arXiv preprint arXiv:1910.01108 (2019)
28. Schmid, U.K., Kümpel, A.S., Rieger, D.: How social media users perceive different forms of online hate speech: a qualitative multi-method study. New Media Soc., 14614448221091185 (2022)
29. Walther, J.B.: Social media and online hate. Current Opin. Psychol. (2022)
30. Yin, W., Zubiaga, A.: Towards generalisable hate speech detection: a review on obstacles and solutions. CoRR abs/2102.08886 (2021). https://arxiv.org/abs/2102.08886

Designing Useful Conversational Interfaces for Information Retrieval in Career Decision-Making Support

Marianne Wilson(✉) ⓘ

Edinburgh Napier University, 10 Colinton Road, Edinburgh EH10 5DT, Scotland
m.wilson2@napier.ac.uk

Abstract. The proposal is an interdisciplinary problem-focused study to explore the usefulness of conversational information retrieval (CIR) in a complex domain. A research-through-design methodology will be used to identify the informational, practical, affective, and ethical requirements for a CIR system in the specific context of Career Education, Information, Advice & Guidance (CEIAG) services for young people in Scotland. Later phases of the research will use these criteria to identify appropriate techniques in the literature, and design and evaluate artefacts intended to meet these. This research will use an interdisciplinary approach to further understanding on the use and limitations of dialogue systems as intermediaries for information retrieval where there are a wide range of possible information tasks and specific users' needs may be ambiguous.

Keywords: Conversational information retrieval · Applied NLP · Research-through-design · Social impact · Evaluation · Ethics

1 Introduction

This project explores the role of text-based dialogue systems to support young people with career decision-making, in collaboration with Skills Development Scotland (SDS), Scotland's national Career Education, Information, Advice and Guidance (CEIAG) service. A two phase research-through-design (RtD) methodology will be used: exploration of the problem space with CEIAG experts and users to delineate the requirements for CIR in this domain, then identification and validation of appropriate approaches for the design and evaluation of a system that meets these requirements. This report provides an overview of the first phase of the research, including CEIAG-specific issues; relevant literature identified; research questions; and the methodology. The report concludes with a brief discussion of potential directions for the second phase.

CEIAG raises challenges for CIR. SDS's interventions include a range of activities and resources, that aim to support young people to effectively explore and integrate career-related information into their career decision-making processes [1, 2]. Diverse information is of relevance to career decisions, including information about personal preferences and skills, local job vacancies, training and qualification requirements, or

J. Kamps et al. (Eds.): ECIR 2023, LNCS 13982, pp. 482–488, 2023.
https://doi.org/10.1007/978-3-031-28241-6_56

macro-level labor market trends [3, 4]. SDS policies prioritize person-centered, social justice informed approach [5]. Therefore, their interventions aim to empower people to manage their own career decisions throughout their lives, in contrast with traditional interventions that seek to 'match' individuals to occupations [6, 7]. As such, the goal of CIR in this domain is to provide automated, interactive access to information, without inadvertently enacting a 'matching' paradigm.

Conversations with CEIAG practitioners may involve the discussion of complex, sensitive and emotive topics [8]. Therefore, CIR introduces an element of ethical risk, when compared to a static search-term based information retrieval approach. CEIAG has a significant impact on economic and social outcomes for both individuals and society, hence CEIAG services being a public policy issue in many countries, including Scotland [9–12]. Therefore, there is a strong ethical imperative to ensure that the design of CIR systems for this domain are aligned with the policies and professional standards of CEIAG services.

2 Related Work

Concerns regarding the use of automated systems have also prompted public policy response. Of relevance to this research are publications by the Scottish Government [13], the UK government [14], the OECD [15] and UNICEF's Policy guidance on AI for children [16]. Although these documents are currently advisory, a system designed for use by a public sector organization will be expected to adhere to these emerging standards. The core principles across all documents can be mapped to Dignum's [17] 'accountability, responsibility and transparency' [17–22]. In addition to general concerns regarding autonomous systems, conversational interfaces can enact representational and allocational harms to individuals and groups when the impacts of existing social hegemonies are not considered during system design [23–25]. Given the priorities of CEAIG, a foundational aim of this research is ensuring that these risks are adequately mitigated. This will extend beyond technical solution to include consideration of the role that people and processes may serve to mitigate or intensify any potential negative consequences [26].

CEIAG services already have a wide range of established methods and technologies to support the dissemination of career-related information [3, 27–30], and most intended users will be familiar with digital information consumption, including for careers support through SDS's existing digital services [31]. Therefore, for CIR to be of genuine benefit, a positive user experience (UX), that meets users' 'physical, cognitive and emotional…needs and expectations' [32] is required. As such, evaluation methods for applied CIR should consider of both the practical, task-orientated outcomes, and the social aspects of the interaction [33]. Several evaluation frameworks have been developed to address this issue [34–38], however, these are largely focused on customer service style tasks, where the problem and solution are more clearly bounded than a CEIAG intervention. Traditional approaches to designing UX do not readily transfer to conversational interfaces [33, 34]. Evaluation methods for dialogue systems largely focuses on utterance level measures rather than users' interaction with a system. These issues informed Moore & Arar's [39] proposed adoption of *recipient design, minimization* and *repair* from conversational analysis in the design and evaluation of conversational user interfaces. Recipient design strategies observed between human interlocutors aim to demonstrate their alignment with their conversational partner through both mirroring conversation style and lexicon and consideration of the recipients' understanding of the topic under discussion [39]. In addition to ensuring that the affective needs of users are met, this also serves the aim of minimization, the principle that the minimum number of utterances should be used to meet the conversational aims. Repair relates to utterances that are designed to correct issues arising from failures of the former strategies [39]. These conversational analysis structures have been identified in task-based human-computer dialogues, although the specific strategies deployed differ from human-human conversations [40, 41].

CIR research frequently focuses on a conversational, as opposed to utterance level, approach to the design and evaluation of systems [42, 43], modelled on human information seeking conversation [44]. Several interaction-focused CIR studies focus use approaches that are analogous to the conversational analysis concepts outlined above. The use of clarifying questions in mixed initiative CIR [45] and the elicitation of direct feedback on possible results to refine future responses [46] are examples of repair strategies. Minimization raises issues for CEIAG, as there is a need to balance the risk of cognitive overload with the requirement to avoid 'matching' users with too narrow a range of information. Vakulenko et al.'s [47] 'conversational browsing' approach has potential, as it addresses situations where the information needs of users are ambiguous and their knowledge of the information domain may be low. Document meta-information [48] may also have a role in avoiding information overload for users exploring across a dataset, while allowing users to focus on facets that are of most relevance to them.

The stylistic elements of recipient design are analogous to current research addressing problems of 'alignment', where the aim is to adjust system utterances to reflect the users' style [49] and lexicon [41]. However, issues of assessing and adjusting system responses based on users' knowledge in CEIAG are complex, given the breadth of both information needs and sources of information relevant to career decision-making.

3 Proposed Research Approach

SDS's services for young people present a well-defined use-case in which to explore the grounded development of CIR for ambiguous and complex information needs, while operating in accordance with defined ethical standards. Given the range and potential impact of CIR tasks for CEIAG discussed in Sect. 2, it is necessary to clearly delineate the goals and boundaries of the system. This leads to the following research questions to be addressed in the first phase of the research: RQ 1. In the context of SDS's CEIAG service, which career information tasks could CIR effectively meet users' practical and affective information needs? RQ 2. How can the principles of accountability, responsibility and transparency be assured in a CIR system in the CEIAG domain?

These will be addressed using a Research through Design (RtD) methodology. RtD is an approach to conducting scholarly research that leverages design practice to generate new knowledge [50]. Reflexivity and evaluation of the process of developing a solution, engenders deeper understanding of the problem space and the consideration of the potential impacts of proposed solutions [51]. In line with SDS and Scottish Government [1,

52] approaches to service design, the first stage of the research will focus on developing a detailed, understanding of the problems that CIR could address within CEIAG, based on the knowledge and experience of domain experts and system users. The outcome of this phase will be a delineation of the practical, interactional, and ethical requirements that a system would have to meet in order to be a useful addition to existing sources of support.

A Delphi study [53] is a method for canvassing a group of experts for their views on emerging technologies [54], and therefore, has been identified as suitable for addressing the RQs above. Three rounds of surveys are being conducted with CEIAG experts from across policy, service design and practice domains. The questions focus on identifying consensus on specific CIR tasks that could be usefully addressed in this domain, and appropriate information sources to incorporate. Feedback on the potential impacts of a CIR system on CEIAG professional standards and ethics, and mitigation strategies, will also be canvassed. This approach aims to leverage the knowledge and experience of domain experts to clearly map the problem space. These insights will be augmented through a Wizard-of-Oz study [55–57] with young people. This method involves a human researcher communicating with participants in a manner that implies that they are interacting with an automated agent. The study will provide evidence about young people's interaction preferences, through both their responses to direct questions and analysis of the transcripts. The dialogues will incorporate a range of conversational UX strategies [39, 58], in order to observe users' responses.

4   Conclusion and Future Work

By collecting data from both CEIAG experts and young people, the first phase of the research aims to develop a fully rounded view of the requirements for a dialogue system to be useful as an CEIAG information intermediary. This will inform the subsequent phase of the research, which will focus on identifying suitable CIR approaches for the design and evaluation of a system based on these requirements. In the course of designing and validating artefacts to resolve an established, complex, real-world CIR problem, the research aims to further current understanding of CIR in practice.

References

1. Career Review Programme Board: Careers by Design (2022)
2. Skills Development Scotland: Delivering Scotland's Career Service a Focus on Career Management Skills (2020)
3. Bimrose, J.: Labour market information for career development: pivotal or peripheral? In: Robertson, P.J., Hooley, T., McCash, P. (eds.) The Oxford Handbook of Career Development, pp. 282–296. Oxford University Press (2021). https://doi.org/10.1093/oxfordhb/978019006 9704.013.21
4. Law, B.: Career-learning space: New-DOTS thinking for careers education. Br. J. Guid. Counc. **27**, 35–54 (1999). https://doi.org/10.1080/03069889908259714
5. Skills Development Scotland: Annual review 2021/22 (2022)

6. Yates, J.: Career development theory: an integrated analysis. In: Robertson, P.J., Hooley, T., McCash, P. (eds.) The Oxford Handbook of Career Development. Oxford University Press, Oxford (2020). https://doi.org/10.1093/oxfordhb/9780190069704.013.10

7. Savickas, M.L., Savickas, S.: A history of career counselling. In: Athanasou, J.A., Perera, H.N. (eds.) International Handbook of Career Guidance, pp. 25–43. Springer, Cham (2019). https://doi.org/10.1007/978-3-030-25153-6_2

8. McMahon, M.: New trends in theory development in career psychology. In: Arulmani, G., Bakshi, A.J., Leong, F.T.L., Watts, A.G. (eds.) Handbook of Career Development. ICP, pp. 13–27. Springer, New York (2014). https://doi.org/10.1007/978-1-4614-9460-7_2

9. Watts, A.G., Sultana, R.G.: Career guidance policies in 37 countries: contrasts and common themes. Int. J. Educ. Vocat. Guid. **4**, 105–122 (2004). https://doi.org/10.1007/s10775-005-1025-y

10. Watts, A.G., Sultana, R.G., McCarthy, J.: The involvement of the European union in career guidance policy: a brief history. Int. J. Educ. Vocat. Guidance **10**, 89–107 (2010). https://doi.org/10.1007/s10775-010-9177-9

11. Varjo, J., Kalalahti, M., Hooley, T.: Actantial construction of career guidance in parliament of Finland's education policy debates 1967–2020. J. Educ. Policy **00**, 1–19 (2021). https://doi.org/10.1080/02680939.2021.1971772

12. Robertson, P.J., Melkumyan, A.: Career guidance and active labour market policies in the Republic of Armenia. Int. J. Educ. Vocat. Guidance **21**(2), 309–327 (2020). https://doi.org/10.1007/s10775-020-09443-2

13. Scottish Government: Scotland's Artificial Intelligence Strategy (2021)

14. Centre for Data Ethics and Innovation: The roadmap to an effective AI assurance ecosystem (2021). https://www.gov.uk/government/publications/the-roadmap-to-an-effective-ai-assurance-ecosystem

15. OECD: Recommendation of the Council on Artificial Intelligence, OECD/LEGAL/0449 (2021). https://legalinstruments.oecd.org/en/instruments/OECD-LEGAL-0449

16. UNICEF: Policy guidance on AI for children (2021)

17. Dignum, V.: Responsible Artificial Intelligence: How to Develop and Use AI in a Responsible Way. Springer, Cham (2019). https://doi.org/10.1007/978-3-030-30371-6

18. Koene, A., Clifton, C., Hatada, Y., Webb, H., Richardson, R.: A governance framework for algorithmic accountability and transparency, Brussels (2019). https://doi.org/10.2861/59990

19. OECD: State of Implementation of the OECD AI Principles: Insights from National AI Policies (2021)

20. Kerr, A., Barry, M., Kelleher, J.D.: Expectations of artificial intelligence and the performativity of ethics: Implications for communication governance. Big Data Soc. **7**, 205395172091593 (2020). https://doi.org/10.1177/2053951720915939

21. Rahwan, I.: Society-in-the-loop: programming the algorithmic social contract. Ethics Inf. Technol. **20**(1), 5–14 (2017). https://doi.org/10.1007/s10676-017-9430-8

22. Akhgar, B., et al.: CENTRIC: Accountability Principles for Artificial Intelligence (AP4AI) in the Internal Security Domain Summary Report on Expert Consultations (2022)

23. Blodgett, S.L., Barocas, S., Daumé III, H., Wallach, H.: Language (Technology) is power: a critical survey of "Bias" in NLP. arXiv preprint arXiv:2005.14050. (2020). https://doi.org/10.18653/v1/2020.acl-main.485

24. Kuziemski, M., Misuraca, G.: AI governance in the public sector: three tales from the frontiers of automated decision-making in democratic settings. Telecommun. Policy **44**, 101976 (2020). https://doi.org/10.1016/j.telpol.2020.101976

25. Bender, E.M.: English isn't generic for language, despite what NLP papers might lead you to believe (2019)

26. Wilson, M., Robertson, P., Cruickshank, P., Gkatzia, D.: Opportunities and risks in the use of AI in career development practice. J. Nat. Inst. Career Educ. Counselling **48**, 48–57 (2022). https://doi.org/10.20856/jnicec.4807

27. Hooley, T., Staunton, T.: The role of digital technology in career development. In: Robertson, P.J., Hooley, T., McCash, P. (eds.) The Oxford Handbook of Career Development, pp. 296–312. Oxford University Press, Oxford (2021). https://doi.org/10.1093/oxfordhb/9780190069704.013.22

28. Moore, N.: What has digital technology done for us and how can we evolve as a sector to make best use of what it has to offer? J. Nat. Inst. Career Educ. Counselling **46**, 25–31 (2021). https://doi.org/10.20856/jnicec.4605

29. Sampson, J.P., Kettunen, J., Vuorinen, R.: The role of practitioners in helping persons make effective use of information and communication technology in career interventions. Int. J. Educ. Vocat. Guidance **20**(1), 191–208 (2019). https://doi.org/10.1007/s10775-019-09399-y

30. Kettunen, J., Sampson, J.P.: Challenges in implementing ICT in career services: perspectives from career development experts. Int. J. Educ. Vocat. Guidance **19**(1), 1–18 (2018). https://doi.org/10.1007/s10775-018-9365-6

31. Skills Development Scotland: Skills Development Scotland Annual Review 2019/20 (2020)

32. International Organization for Standardization: BS EN ISO 9241–11:2018: Ergonomics of human-system interaction. Usability: Definitions and concepts (2018)

33. Clark, L., et al.: What makes a good conversation? Challenges in designing truly conversational agents. In: Proceedings of the Conference on Human Factors in Computing Systems. Association for Computing Machinery (2019). https://doi.org/10.1145/3290605.3300705

34. Følstad, A., Brandtzaeg, P.B.: Users' experiences with chatbots: findings from a questionnaire study. Qual. User Experience **5**(1), 1–14 (2020). https://doi.org/10.1007/s41233-020-00033-2

35. Smestad, T.L., Volden, F.: Chatbot personalities matters. In: Bodrunova, S.S., et al. (eds.) INSCI 2018. LNCS, vol. 11551, pp. 170–181. Springer, Cham (2019). https://doi.org/10.1007/978-3-030-17705-8_15

36. Weiss, A., Bartneck, C.: Meta analysis of the usage of the godspeed questionnaire series. In: Proceedings of the IEEE International Workshop on Robot and Human Interactive Communication, pp. 381–388, November 2015. https://doi.org/10.1109/ROMAN.2015.7333568

37. Skjuve, M., Brandzaeg, P.B.: Measuring user experience in chatbots: an approach to interpersonal communication competence. In: Bodrunova, S.S., et al. (eds.) INSCI 2018. LNCS, vol. 11551, pp. 113–120. Springer, Cham (2019). https://doi.org/10.1007/978-3-030-17705-8_10

38. Borsci, S., et al.: The chatbot usability scale: the design and pilot of a usability scale for interaction with ai-based conversational agents. Pers. Ubiquit. Comput. **26**(1), 95–119 (2021). https://doi.org/10.1007/s00779-021-01582-9

39. Moore, R.J., Szymanski, M.H., Arar, R., Ren, G.-J. (eds.): Studies in Conversational UX Design. HIS, Springer, Cham (2018). https://doi.org/10.1007/978-3-319-95579-7

40. Avgustis, I., Shirokov, A., Iivari, N.: "Please connect me to a specialist": scrutinising 'recipient design' in interaction with an artificial conversational agent. In: Ardito, C., et al. (eds.) INTERACT 2021. LNCS, vol. 12935, pp. 155–176. Springer, Cham (2021). https://doi.org/10.1007/978-3-030-85610-6_10

41. Spillner, L., Wenig, N.: Talk to me on my level - linguistic alignment for chatbots. In: Proceedings of 23rd ACM International Conference on Mobile Human-Computer Interaction (2021). https://doi.org/10.1145/3447526.3472050

42. Thomas, P., Czerwinksi, M., Mcduff, D., Craswell, N.: Theories of conversation for conversational IR. ACM Trans. Inf. Syst. **39**, 1–23 (2021). https://doi.org/10.1145/3439869

43. Zamani, H., Trippas, J.R., Dalton, J., Radlinski, F.: Conversational information seeking an introduction to conversational search, recommendation, and question answering (2022). https://doi.org/10.1561/XXXXXXXXX

44. Trippas, J.R., Spina, D., Cavedon, L., Joho, H., Sanderson, M.: Informing the design of spoken conversational search. In: Proceedings of the 2018 Conference on Human Information Interaction and Retrieval, CHIIR 2018, pp. 32–41. Association for Computing Machinery, Inc. (2018). https://doi.org/10.1145/3176349.3176387

45. Wang, Z., Ai, Q.: Simulating and modeling the risk of conversational search. ACM Trans. Inf. Syst. **40**, 1–33 (2022). https://doi.org/10.1145/3507357

46. Sun, Y., Zhang, Y.: Conversational recommender system. In: Proceedings of the 41st International ACM SIGIR Conference on Research and Development in Information Retrieval, SIGIR 2018, pp. 235–244. Association for Computing Machinery, Inc. (2018). https://doi.org/10.1145/3209978.3210002

47. Vakulenko, S., Savenkov, V., de Rijke, M.: Conversational browsing. arXiv preprint arXiv: 2012.03704 (2020)

48. Kiesel, J., Meyer, L., Potthast, M., Stein, B.: Meta-information in conversational search. ACM Trans. Inf. Syst. **39**, 1–44 (2021). https://doi.org/10.1145/3468868

49. Thomas, P., Czerwinski, M., McDuf, D., Craswell, N., Mark, G.: Style and alignment in information-seeking conversation. In: Proceedings of the 2018 Conference on Human Information Interaction and Retrieval, CHIIR 2018, March 2018, pp. 42–51 (2018). https://doi.org/10.1145/3176349.3176388

50. Zimmerman, J., Forlizzi, J.: Research through design in HCI. In: Olson, J.S., Kellogg, W.A. (eds.) Ways of Knowing in HCI, pp. 167–189. Springer, New York (2014). https://doi.org/10.1007/978-1-4939-0378-8_8

51. Eggink, W., Mulder-Nijkamp, M.: Research through design & research through education. In: Proceedings of the 18th International Conference on Engineering and Product Design Education: Design Education: Collaboration and Cross-Disciplinarity, pp. 216–221 (2016)

52. Scottish Government: The Scottish approach to service design how to design services for and with users (2019)

53. Linstone, H.A., Turoff, M.: The delphi method: techniques and applications (2002)

54. Custer, R.L., Scarcella, J.A., Stewart, B.R.: The modified delphi technique - a rotational modification. J. Vocat. Techn. Educ. **15** (1999). https://doi.org/10.21061/jcte.v15i2.702

55. Elsweiler, D., Frummet, A., Harvey, M.: Comparing Wizard of Oz & observational studies for conversational IR evaluation. Datenbank-Spektrum **20**(1), 37–41 (2020). https://doi.org/10.1007/s13222-020-00333-z

56. Medhi Thies, I., Menon, N., Magapu, S., Subramony, M., O'Neill, J.: How do you want your chatbot? An exploratory wizard-of-oz study with young, urban Indians. In: Bernhaupt, R., Dalvi, G., Joshi, A., Balkrishan, D.K., O'Neill, J., Winckler, M. (eds.) INTERACT 2017. LNCS, vol. 10513, pp. 441–459. Springer, Cham (2017). https://doi.org/10.1007/978-3-319-67744-6_28

57. Avula, S., Chadwick, G., Arguello, J., Capra, R.: Searchbots: User engagement with chatbots during collaborative search. In: Proceedings of the 2018 Conference on Human Information Interaction and Retrieval, CHIIR 2018, pp. 52–61. Association for Computing Machinery, Inc. (2018). https://doi.org/10.1145/3176349.3176380

58. Radlinski, F., Craswell, N.: A theoretical framework for conversational search. In: Proceedings of the 2017 Conference Human Information Interaction and Retrieval, CHIIR 2017, pp. 117–126. Association for Computing Machinery, Inc. (2017). https://doi.org/10.1145/3020165.3020183

CLEF Lab Descriptions

iDPP@CLEF 2023: The Intelligent Disease Progression Prediction Challenge

Helena Aidos[1], Roberto Bergamaschi[2], Paola Cavalla[3], Adriano Chiò[4],
Arianna Dagliati[2], Barbara Di Camillo[5], Mamede Alves de Carvalho[1],
Nicola Ferro[5(✉)], Piero Fariselli[4], Jose Manuel García Dominguez[6],
Sara C. Madeira[1], and Eleonora Tavazzi[7]

[1] University of Lisbon, Lisbon, Portugal
{haidos,sacmadeira}@fc.ul.pt, mamedemg@mail.telepac.pt
[2] University of Pavia, Pavia, Italy
roberto.bergamaschi@mondino.it, arianna.dagliati@unipv.it
[3] "Città della Salute e della Scienza", Turin, Italy
paola.cavalla@unito.it
[4] University of Turin,Turin, Italy
{adriano.chio,piero.fariselli}@unito.it
[5] University of Padua, Padua, Italy
{barbara.dicamillo,nicola.ferro}@unipd.it
[6] Gregorio Marañon Hospital in Madrid, Madrid, Spain
jgarciadominguez@salud.madrid.org
[7] IRCCS Foundation C. Mondino in Pavia, Pavia, Italy
eleonoratavazzi@gmail.com

Abstract. *Amyotrophic Lateral Sclerosis (ALS)* and *Multiple Sclerosis (MS)* are chronic diseases characterized by progressive or alternate impairment of neurological functions (motor, sensory, visual, cognitive). Patients have to manage alternated periods in hospital with care at home, experiencing a constant uncertainty regarding the timing of the disease acute phases and facing a considerable psychological and economic burden that also involves their caregivers. Clinicians, on the other hand, need tools able to support them in all the phases of the patient treatment, suggest personalized therapeutic decisions, indicate urgently needed interventions.

The goal of iDPP@CLEF is to design and develop an evaluation infrastructure for AI algorithms able to:

1. better describe disease mechanisms;
2. stratify patients according to their phenotype assessed all over the disease evolution;
3. predict disease progression in a probabilistic, time dependent fashion.

iDPP@CLEF run as a pilot lab in CLEF 2022, offering tasks on the prediction of ALS progression and a position paper task on explainability of AI algorithms for prediction; 5 groups submitted a total of 120 runs and 2 groups submitted position papers.

iDPP@CLEF will continue in CLEF 2023, focusing on the prediction of MS progression and exploring whether pollution and environmental data can improve the prediction of ALS progression.

J. Kamps et al. (Eds.): ECIR 2023, LNCS 13982, pp. 491–498, 2023.
https://doi.org/10.1007/978-3-031-28241-6_57

1   Introduction

Amyotrophic Lateral Sclerosis (ALS) and *Multiple Sclerosis (MS)* are severe chronic diseases characterized by a progressive but variable impairment of neurological functions, characterized by high heterogeneity both in presentation features and rate of disease progression. As a consequence patients' needs are different, challenging both caregivers and clinicians. Indeed, the time of relevant events is variable, which is associated with uncertainty regarding the opportunity of critical interventions, like non-invasive ventilation and gastrostomy in the case of ALS, with implications on the quality of life of patients and their caregivers. For this reason, clinicians need tools able to support their decision in all phases of disease progression and underscore personalized therapeutic decisions. Indeed, this heterogeneity is partly responsible for the lack of effective prognostic tools in medical practice, as well as for the current absence of a therapy able to effectively slow down or reverse the disease course. On the one hand, patients need support for facing the psychological and economic burdens deriving from the uncertainty of how the disease will progress; on the other, clinicians require tools that may assist them throughout the patient's care, recommending tailored therapeutic decisions and providing alerts for urgently needed actions.

We need to design and develop *Artificial Intelligence (AI)* algorithms to:

- stratify patients according to their phenotype all over the disease evolution;
- predict the progression of the disease in a probabilistic, time dependent way;
- better describe disease mechanisms.

The *Intelligent Disease Progression Prediction at CLEF (iDPP@CLEF)* lab[1] aims to deliver an evaluation infrastructure for driving the development of such AI algorithms. Indeed, in this context, it is fundamental, even if not so common yet, to develop shared approaches, promote the use of common benchmarks, foster the comparability and replicability of the experiments. Differently from previous challenges in the field, iDPP@CLEF addresses in a systematic way some issues related to the application of AI in clinical practice in ALS and MS. In addition to defining risk scores based on the probability of occurrence of an event in the short or long term period, iDPP@CLEF also addresses the issue of providing information in a more structured and understandable way to clinicians.

The paper is organized as follows: Section 2 discusses related works; Sect. 3 presents what has been done in iDPP@CLEF 2022 while Sect. 4 introduces the plans for iDPP@CLEF 2023; finally, Sect. 5 draws some conclusions.

2   Related Works

Within CLEF, there have been no other labs on this or similar topics before.

Outside CLEF, there have been a recent challenge on Kaggle[2] in 2021 and some older ones, the DREAM 7 ALS Prediction challenge[3] in 2012 and the

---

[1] https://brainteaser.health/open-evaluation-challenges/.

[2] https://www.kaggle.com/alsgroup/end-als.

[3] https://dreamchallenges.org/dream-7-phil-bowen-als-prediction-prize4life/.

DREAM ALS Stratification challenge[4] in 2015. The Kaggle challenge used a mix of clinical and genomic data to seek for insights about the mechanisms of ALS and difference between people with ALS who progress faster versus those who develop it more slowly. The DREAM 7 ALS Prediction challenge [8] asked to use 3 months of ALS clinical trial information (months 0–3) to predict the future progression of the disease (months 3–12), expressed as the slope of change in *ALS Functional Rating Scale Revisited (ALSFRS-R)* [4], a functional scale that ranges between 0 and 40. The DREAM ALS Stratification challenge asked participants to stratify ALS patients into meaningful subgroups, to enable better understanding of patient profiles and application of personalized ALS treatments. Differently from these previous challenges, iDPP@CLEF focuses on explainable AI and on temporal progression of the disease.

Finally, when it comes to MS, studies are mostly conducted on closed and proprietary datasets and iDPP@CLEF represents one of the first attempts to create a public and shared dataset.

3   iDPP@CLEF 2022

iDPP@CLEF run as a pilot lab for the first time in CLEF 2022[5] [6,7] and focused on pilot activities aimed both at an initial exploration of ALS progression prediction and at understanding of the challenges and limitations to refine and tune the labs itself for future iterations.

Tasks. iDPP@CLEF 2022 consisted of the following tasks:

- **Pilot Task 1 - Ranking Risk of Impairment**: it focused on ranking of patients based on the risk of impairment in specific domains. More in detail, we used the ALSFRS-R scale [4] to monitor speech, swallowing, handwriting, dressing/hygiene, walking and respiratory ability in time and asked participants to rank patients based on time to event risk of experiencing impairment in each specific domain.
- **Pilot Task 2 - Predicting Time of Impairment**: it refined Task 1 by asking participants to predict when specific impairments will occur (i.e. in the correct time-window). In this regard, we assessed model calibration in terms of the ability of the proposed algorithms to estimate a probability of an event close to the true probability within a specified time-window.
- **Position Paper Task 3 - Explainability of AI algorithms**: we evaluated proposals of different frameworks able to explain the multivariate nature of the data and the model predictions.

Participation. 43 participants registered for iDPP@CLEF 2022 and 5 participants successfully submitted a total of 120 runs for Task 1 and Task 2; moreover, 2 position papers were submitted for the explainability task, as detailed in Table 1. Submission of participants are openly available in git repositories[6].

[4] https://dx.doi.org/10.7303/syn2873386.
[5] https://brainteaser.health/open-evaluation-challenges/idpp-2022/.
[6] https://bitbucket.org/brainteaser-health/.

Table 1. Break-down of the runs submitted by participants for each task and sub-task. Participation in Task 3 does not involve submission of runs and it is marked just with a tick.

Team Name	Total	Task 1	Task 2	Task 3	Paper
BioHIT	18	9	9	–	–
CompBioMed	40	22	18	–	Pancotti et al. [12]
FCOOL	15	–	15	✓	Branco et al. [2] and Nunes et al. [11]
LIG GETALP	23	12	11	–	Mannion et al. [10]
SBB	24	12	12	–	Trescato et al. [13]
UNIPV	–	–	–	✓	Buonocore et al. [3]
Total	**120**	**55**	**65**		

Datasets. iDPP@CLEF 2022 created 3 datasets, for the prediction of specific events related to ALS, consisting of fully anonymized data from 2,250 real patients from medical institutions in Turin, Italy, and Lisbon, Portugal. The datasets contain both static data about patients, e.g. age, onset date, gender, ... and event data, i.e. 18,512 ALSFRS-R questionnaires and 4,015 spyrometries.

The following data are available for both the training and the test sets:

- the first available ALSFRS-R questionnaire at Time 0 (both single question scores and total score). , for example, time-of-onset and time-of-diagnosis are expressed as relative delta with respect to Time 0 in months (also fractions);
- the slope of the ALSFRS-R score between time-of-onset and Time 0 as:

$$slope = \frac{48 - \text{ALSFRS-R-score}\,(\texttt{Time0})}{\texttt{Time0} - \texttt{TimeOnset}}$$

- all the other static data, with a complete list available at http://brainteaser. dei.unipd.it/challenges/idpp2022/assets/other/static-vars.txt
- visits, containing either other ALSFRS-R questionnaires or Spirometry, i.e. *Forced Vital Capacity (FVC)*. The complete list of variables for each visit is available at http://brainteaser.dei.unipd.it/challenges/idpp2022/assets/ other/visits.txt.

Measures. iDPP@CLEF adopted several state-of-the-art evaluation measures to assess the performance of the prediction algorithms, among which:

- *ROC curve and/or the precision-recall curve (and area under the curve)* to show the trade-off between clinical sensitivity and specificity for every possible cut-off of the risk scores;
- *Concordance Index (C-index)* to summarize how well a predicted risk score describes an observed sequence of events.
- *E/O ratio and Brier Score* to assess whether or not the observed event rates match expected event rates in subgroups of the model population.

 - *Specificity and recall* to assess, for each interval, the ability of the models of correctly identify true positives and true negatives.
 - *Distance* to assess how far the predicted time interval was from the true time interval.

Approaches. BoiHIT explored the use of logistic regression, random forest classifiers, XGBoost, and LightGBM. Decision trees and boosting approaches were preferred due to their ability to deal with both categorical and numerical/continuous features and the interpretability they offer. Even if LightGBM was the model with the best performance, BoiHIT found out that this kind of approaches might not be appropriate for time dependent problems and that time to event analysis methods, such as survival analysis, might yield better results.

CompBioMed [12] considered three main approaches. The simplest one consisted on fitting a standard survival predictor separately for each event as outlined above for independent events, called Naive Multiple Event Survival (NMES). Another was the recently developed Deep Survival Machine (DSM), based on deep learning and capable of handling competing risks. Finally, they also proposed a time-aware classifier ensemble method, that also handles competing risks, called Time-Aware Classifier Ensemble (TACE). All the above approaches achieved comparable performance among them.

FCOOL [2] proposes a hierarchical approach, with a first-stage event prediction, followed by specialized models predicting the time window to a particular event. The procedure is three-fold: first, it creates patient snapshots based on clustering with constraints, thus organizing patient records in an efficient manner. Second, it uses a pattern-based approach that incorporates recent advances on temporal pattern mining to the context of classification. This approach performs end-stage event prediction while allowing the entire patient's medical history to be considered. Finally, exploiting the predictions from the previous step, specialized models are learned using the original features to predict the time window to an event. This two-stage prediction approach aimed to promote homogeneity and lessen the impact of class imbalance, in comparison to performing one single multilabel task.

LIG GETALP [10] employed Cox's proportional hazards model to the task of ranking the risk of impairment, using the gradient boosting learning strategy The output of the time-independent part of the survival function calculated by the gradient boosting survival analysis method is then mapped to the interval (0, 1), via a sigmoid function. To estimate the time-to-event, LIG GETALP used a regression model based on Accelerated Gradient Boosting (AGB). This being a standard regression model, it does not take censoring into account and [10] uses class predictions based on the Task 1 survival model to "censor" the time-to-event predictions.

SBB [13] considered three survival analysis methods, namely: Cox, SSVM, and RSF. They were chosen to represent a broad spectrum of baseline models including parametric (SSVM), semiparametric (Cox), linear (Cox, SSVM), and nonlinear (RSF) models.

For task 3, Nunes et al. [11] proposed a novel approach that generates semantic similarity-based explanations for patient-level predictions. The underlying idea is to explain the prediction for one patient by considering aspect-oriented semantic similarity with other relevant patients based on the most important features used by ML approaches or selected by users. To build rich and easy to understand semantic-similarity based explanations, [11] developed five steps: (1) the enrichment of the Brainteaser Ontology [1] through integration of other biomedical ontologies; (2) the semantic annotation of patients (if not already available); (3) the similarity calculation between patients; (4) selection of the set of patients to explain a specific prediction; and (5) the visualization of the generated similarity-based explanations.

For task 3, Buonocore et al. [3] trained a set of 4 well-known classifiers to predict death occurrence: Gradient Boosting (using XGB implementation), Random Forest, Logistic Regression and Multilayer perceptron. For the *eXplainable AI(XAI)* methods [3] focused our attention on three different methods for post-hoc, model-agnostic, local explainability, selecting SHAP, LIME and AraucanaXAI. Then, [3] evaluated and compared XAI approaches in terms of a set of metrics defined in previous research on XAI in healthcare: *identity*: if there are two identical instances, they must have the same explanations; *fidelity*: concordance of the predictions between the XAI surrogate model and the original ML model; *separability*: if there are 2 dissimilar instances, they must have dissimilar explanations; *time*: average time required by the XAI method to output an explanation across the entire test set.

4   iDPP@CLEF 2023

iDPP@CLEF 2023[7] will organize the following activities:

- **Task 1 – Predicting Risk of Disease Worsening (MS)**: It focuses on ranking subjects based on the risk of worsening, setting the problem as a survival analysis task. More specifically the risk of worsening predicted by the algorithm should reflect how early a patient experiences the event "worsening" and should range between 0 and 1. Worsening is defined on the basis of the *Expanded Disability Status Scale (EDSS)* [9], accordingly to clinical standards. In particular, e consider two different definitions of worsening corresponding to two different sub-tasks:
  - *Subtask 1a*: the patient crosses the threshold EDSS ≥ 3 at least twice within one year interval;
  - *Subtask 1b*: the second definition of worsening depends on the first recorded value accordingly to current clinical protocols. If Baseline EDSS < 1, worsening event occurs when and increase of EDSS by 1.5 points is first observed; if $1 \leq$ Baseline EDSS < 5.5, worsening event occurs when and increase of EDSS by 1 point is first observed; if baseline EDSS ≥ 5.5, worsening event occurs when and increase of EDSS by 0.5 points is first observed.

---

[7] https://brainteaser.health/open-evaluation-challenges/idpp-2023/.

In both cases the occurrence of the worsening event and the time of occurrence will be pre-computed by the challenge organizers.

- **Task 2 – Predicting Cumulative Probability of Worsening (MS)**: it refines Task 1 asking participants to explicitly assign the cumulative probability of worsening at different time windows, i.e. between years 2 and 4, 2 and 6, 2 and 8, 2 and 10. Worsening will be defined in two different ways in subtasks 2a and 2b similarly to Task 1.
- **Position Paper Task 3 – Impact of Exposition to Pollutants (ALS)**: we will evaluate proposals of different approaches to assess if exposure to different pollutants is a useful variable to predict time to Percutaneous Endoscopic Gastrostomy (PEG), Non-Invasive Ventilation (NIV) and death in ALS patients.

We will provide retrospective, fully anonymized MS and ALS clinical data including demographic and clinical characteristics, coming from clinical institutions in Italy, Portugal, and Spain.

For Task 1 and Task 2 we will release a brand new dataset with MS data consisting of about 1,800 patients. Accordingly to the survival analysis settings, for each subject in the training we will provide a label 0 or 1 indicating if the subject experienced the event "worsening" (label 1) or not (label 0) and the time-of-event, which indicates the time of the event for subjects experiencing it, or the time of censoring if the patient has not experience the event yet at the time of data dump.

For Position Paper Task 3 we will re-use the ALS dataset developed in iDPP@CLEF 2022, consisting of about 2,250 patients, and will extend it with environmental and pollution data.

5   Conclusions

iDPP@CLEF is a new shared tasks focusing on predicting the temporal progression of ALS and MS and on the explainability of the AI algorithms for such prediction. The first edition focused on ALS progression prediction and participation was satisfactory, hinting at the interest of the community concerning the task. More so, the solutions identified by participants range over several different techniques and provided valid input to such a highly relevant domain as the prediction of the ALS progression.

For the second iteration, iDPP@CLEF 2023 we plan to investigate MS progression prediction and how to exploit pollution and environmental data to improve progression prediction of ALS.

References

1. Bettin, M., et al.: Deliverable 9.1 - project ontology and terminology, including data mapper and RDF graph builder. BRAINTEASER, EU Horizon 2020, Contract N. GA101017598, December 2021. https://brainteaser.health/wp-content/uploads/2022/09/BRAINTEASER_D9.1_Final.pdf

2. Branco, R., et al.: Hierarchical modelling for ALS prognosis: predicting the progression towards critical events. In: [5], pp. 1211–1227
3. Buonocore, T.M., Nicora, G., Dagliati, A., Parimbelli, E.: Evaluation of XAI on ALS 6-months mortality prediction. In: [5], pp. 1236–1255
4. Cedarbaum, J.M., et al.: The ALSFRS-R: a revised ALS functional rating scale that incorporates assessments of respiratory function. J. Neurol. Sci. **169**(1–2), 13–21 (1999)
5. Faggioli, G., Ferro, N., Hanbury, A., Potthast, M. (eds.): CLEF 2022 Working Notes, CEUR Workshop Proceedings (CEUR-WS.org), ISSN 1613-0073 (2022). http://ceur-ws.org/Vol-3180/
6. Guazzo, A., et al.: Intelligent disease progression prediction: overview of iDPP@CLEF 2022. In: Barrón-Cedeño, A., et al. (eds.) Experimental IR Meets Multilinguality, Multimodality, and Interaction. Proceedings of the Thirteenth International Conference of the CLEF Association (CLEF 2022), pp. 395–422, Lecture Notes in Computer Science (LNCS), vol. 13390, pp. 395–422. Springer, Heidelberg (2022). https://doi.org/10.1007/978-3-031-13643-6_25
7. Guazzo, A., et al.: Overview of iDPP@CLEF 2022: the intelligent disease progression prediction challenge. In: [5], pp. 1130–1210
8. Küffner, R., et al.: Crowdsourced analysis of clinical trial data to predict amyotrophic lateral sclerosis progression. Nat. Biotechnol. **33**(1), 51–57 (2015)
9. Kurtzke, J.F.: Rating neurologic impairment in multiple sclerosis: an expanded disability status scale (EDSS). Neurology **33**(11), 1444–1452 (1983)
10. Mannion, A., Chevalier, T., Schwab, D., Goeuriot, L.: Predicting the risk of & time to impairment for ALS patients. In: [5]
11. Nunes, S., et al.: Explaining artificial intelligence predictions of disease progression with semantic similarity. In: [5], pp. 1256–1268
12. Pancotti, C., Birolo, G., Sanavia, T., Rollo, C., Fariselli, P.: Multi-event survival prediction for amyotrophic lateral sclerosis. In: [5], pp. 1269–1276
13. Trescato, I., et al.: Baseline machine learning approaches to predict amyotrophic lateral sclerosis disease progression. In: [5], pp. 1277–1293

LongEval: Longitudinal Evaluation of Model Performance at CLEF 2023

Rabab Alkhalifa[1,2], Iman Bilal[3], Hsuvas Borkakoty[4],
Jose Camacho-Collados[4], Romain Deveaud[6], Alaa El-Ebshihy[9],
Luis Espinosa-Anke[4,12], Gabriela Gonzalez-Saez[7], Petra Galuščáková[7],
Lorraine Goeuriot[7], Elena Kochkina[1,5], Maria Liakata[1,3,5],
Daniel Loureiro[4], Harish Tayyar Madabushi[8], Philippe Mulhem[7(✉)],
Florina Piroi[9], Martin Popel[10], Christophe Servan[6,11],
and Arkaitz Zubiaga[1]

[1] Queen Mary University of London, London, UK
[2] Imam Abdulrahman Bin Faisal University, Dammam, Saudi Arabia
[3] University of Warwick, Coventry, UK
[4] Cardiff University, Cardiff, UK
[5] Alan Turing Institute, London, UK
[6] Qwant, Paris, France
[7] Univ. Grenoble Alpes, CNRS, Grenoble INP,
Institute of Engineering Univ. Grenoble Alpes., LIG, Grenoble, France
Philippe.Mulhem@imag.fr
[8] University of Bath, Bath, UK
[9] Research Studios Austria, Data Science Studio, Vienna, Austria
[10] Charles University, Prague, Czech Republic
[11] Paris-Saclay University, CNRS, LISN, Gif-sur-Yvette, France
[12] AMPLYFI, Cardiff, UK

Abstract. In this paper, we describe the plans for the first LongEval CLEF 2023 shared task dedicated to evaluating the temporal persistence of Information Retrieval (IR) systems and Text Classifiers. The task is motivated by recent research showing that the performance of these models drops as the test data becomes more distant, with respect to time, from the training data. LongEval differs from traditional shared IR and classification tasks by giving special consideration to evaluating models aiming to mitigate performance drop over time. We envisage that this task will draw attention from the IR community and NLP researchers to the problem of temporal persistence of models, what enables or prevents it, potential solutions and their limitations.

Keywords: Evaluation · Temporal persistence · Temporal generalisability · Information retrieval · Text classification

© The Author(s), under exclusive license to Springer Nature Switzerland AG 2023
J. Kamps et al. (Eds.): ECIR 2023, LNCS 13982, pp. 499–505, 2023.
https://doi.org/10.1007/978-3-031-28241-6_58

1 Introduction

Recent research demonstrates that the performance of Text Retrieval and Classification systems drops over time as patterns observed in data change, due to linguistic and societal changes [2]. In classification systems, this drop is more pronounced when the testing data is further away in time from training data [1,5,7], a problem we refer to as the problem of *classifier temporal persistence*. Similarly, in Information Retrieval, it has been shown that a deep neural network-based IR is dependent on the consistency between the train and test data [8] Given that in most scenarios one has limited resources to continuously label new data to train models on, the aim of this shared task is to encourage the development of models that mitigate performance drop over time as the training data gets older. We do this by providing participants with training data distant in time from testing and un-annotated data from the testing time period. The challenges that come with such an evaluation setting are numerous, ranging from the definition and collection of the data on which the systems may be compared to the measures considered. As such, this lab focuses on two different tasks, both with a temporal axis in their design: (a) Task 1, Information Retrieval for the case in which Web documents evolve over the time, queries are not known a priori, relevance judgements are non-binary and submissions are required to provide ranked lists as results, and (b) Task 2, text classification in which the target classes are predefined while language usage associated with each class evolves rapidly over time, as in social media.

We encourage the development of novel approaches that can automatically adapt to possible temporal dynamics in textual data so as to progress towards time-insensitive computational methods. As such, the expected outcomes from this lab are threefold:

– to draw a deeper understanding of how time impacts IR and classification systems;
– to assess the effectiveness of different retrieval and classification approaches in achieving temporal persistence;
– to propose computational methods to leverage ageing labelled datasets, while minimising performance drop over time.

Given the prevalence of text classification in IR and NLP research across CLEF labs, as well as our objective to rank top models that provide high temporal persistence for NLP tasks, we propose our evolving sets which exabit natural language use change overtime (either over the short term or long term) compared to testing data from same time frame (within-time), LongEval is built on a common framework which add the temporal gap that define the distance between train and testing as a time-sensitivity measure [1]. As shown in Fig. 1, we compare the retrieval or classification temporal generalisability of a given IR or classification system when operating on data acquired at time t from same time as training, its persistence when operating on data acquired at time t' (occurring a short period after time t), and its persistence when operating on data acquired at time $t"$ (occurring a long period after time t). The system's

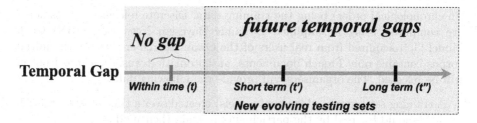

Fig. 1. Global framework for the LongEval Tasks.

ability to cope with dynamic data is thus evaluated using longitudinal datasets split at different temporal granularities, i.e. within-time, short and long time distances from the training data.

The remainder of the paper is structured as follows: LongEval-Retrieval is covered in Sect. 2.1, while LongEval-Classification is covered in Sect. 2.2. Both sections propose tasks and provide additional information about the data and the baseline to be used. Section 3 contains additional information and guidelines for participants.

2 Tasks

2.1 Task 1: LongEval-Retrieval

The goal of the Task 1 is to support the development of Information Retrieval systems that cope with temporal evolution. The retrieval systems evaluated in this task are expected to be persistent in their retrieval efficiency over time, as Web documents and Web queries evolve. To evaluate such features of systems, we rely on collections of documents and queries, corresponding to real data acquired from an actual Web search engine.

The *LongEval-Retrieval* organizes two sub-tasks.

Sub-task 1.A: Short-Term Persistence. In this task, participants will be asked to examine the retrieval effectiveness when the test documents are acquired shortly (typically within a range of few months distance) after the documents available in the train collection.

Sub-task 1.B: Long-Term Persistence. Here, participants will be asked to examine retrieval effectiveness on the documents published after a long period, at least 4 months after the documents in the train collection were published.

As mentioned above, **any participation in the sub-tasks 1.A or 1.B necessitate a "within time" run submission.**

Data. The data for this task is a sequence of Web document collections and queries, each containing a few million documents (e.g. 2.8 M for the training data) and hundreds of queries (e.g. almost 700 for the training data), provided by Qwant[1]. Each document set will have a release time stamp, with the first set

---

[1] Qwant search engine: https://www.qwant.com/.

(in chronological order) being the training data. Discrete relevance assessments are computed using a simplified Dynamic Bayesian Network (sDBN) Click Model [3,4] acquired from real users of the Qwant search engine. As the initial corpus contains only French documents, an automatic translation into English will be released. The organisers will provide the following data:

1. A training set (queries, documents, qrels) created over a time interval t. Such data should be used by the participants to train their models.
2. One "within time" test set (queries, documents) acquired at the same time frame as the training set. This test set will be used to assess the initial performance of the trained models, and will not be used to directly assess submissions;
3. Two test sets: one test set (queries, documents) acquired during a time interval t' occurring shortly after t (with no intersection between t and t') dedicated to evaluate **short-term persistence** sub-task 1.A, and one test set acquired long after t during a time interval t'' (with no intersection between t and t''), for **long-term persistence** evaluation, sub-task 1.B.

Evaluation. The submitted systems will be evaluated in two ways:

1. **nDCG** scores calculated on test set provided for the sub-tasks. Such a classical evaluation measure is consistent with Web search, for which the discount emphasises the ordering of the top results.
2. **Relative nDCG Drop (RnD)** measured by computing the difference between nDCG on "within time" test data vs short- or long-term testing sets. This measure relies on the "within time" test data, and supports the evaluation of the impact of the data changes on the system's results.

These measures will be used to assess the extent to which systems provide good results, but also the extent to which they are robust against the changes within the data (queries/documents) along time. Using these evaluation measures, a system that has good results using nDCG, and also good results according to the RnD measure is considered to be able to cope with the evolution over time of the Information Retrieval collection.

2.2   Task 2: LongEval-Classification

The first LongEval-Classification challenge focuses on systems that perform social media sentiment analysis, which is expressed as a binary classification task in English. By addressing this critical and widely studied task, we hope to attract attention and participation from the broader AI/NLP communities in order to better understand this emerging field and develop novel temporally persistent approaches.

The *LongEval-Classification* organizes two sub-tasks.

Sub-task 2.A: Short-Term Persistence. In this task participants will be asked to develop models which demonstrate performance persistence over short periods of time (test set within 1 year from the training data).

Sub-task 2.B: Long-Term Persistence. In this task participants will be asked to develop models which demonstrate performance persistence over longer period of time (test set over 1 year apart from the training data).

Data. The training data to be provided to the task participants will consist of the TM-Senti dataset[2] extended with a development set and three human-annotated novel test sets for submission evaluation. TM-Senti is a general large-scale Twitter sentiment dataset in English language, spanning a 9-year period from 2013 to 2021. Tweets are labelled for sentiment as either "positive" or "negative". The annotation is performed using distant supervision based on a manually curated list of emojis and emoticons [9] and, thus, can be easily extended to cover more recent years. We plan to release data in two phases:

1. In the **development phase**, participants will be given **(1) a distantly annotated training set** (tweet, label) created over a time interval t. Such data is dedicated model training, as well as **(2) human-annotated "within time" development set** (tweet, label) from the same time period t. This development set is intended to allow participants to develop their systems before the following phase, and will not be used to rank their submissions. For participant interested in data-centric approaches, we provide **(3) an un-labelled corpora** (timestamp, tweet) covering all periods of training, development and testing. All these resources, including python-based baseline code, evaluation scripts, and un-labelled temporal data, will be made available to participating teams upon data release in December 2022.
2. In the **evaluation phase**, participants will be provided with three human-annotated testing sets without their labels (id, tweet): **(1) "within time"** acquired during time period t, **(2) short-term** acquired during a time interval t' occurring shortly after t (with no intersection between t and t') dedicated to evaluate *short-term persistence (sub-task 2.A)*, and **(3) long-term** acquired long after t during a time interval t'' (with no intersection between t and t'') dedicated to evaluate *long-term persistence (sub-task 2.B)*. Similarly to Task 1, participating teams are required to provide a performance score for the "within time" test set, even if they are only interested in one of the sub-tasks to calculate persistence metrics, i.e. RPD.

Evaluation. The performance of the submissions will be evaluated using the following metrics:

1. **Macro-averaged F1-score** on the testing set of the corresponding sub-task
2. **Relative Performance Drop (RPD)** measured by computing the difference between performance on "within time" data vs short- or long-term distant testing sets.

The submissions for each sub-task will be ranked based on the first metric of macro-averaged F1. In order to identify the best submission, we will also calculate a unified score between the two sub-tasks as a **weighted average between the scores obtained for each sub-task (weighted-F1)**. This will encourage participants to contribute to both sub-tasks in order to be correctly placed on a joint leader board, as well as to enable better analysis of their system performance in both settings.

Baseline. Participants are expected to propose temporally persistent classifiers based on state-of-the-art data-centric or architecture-centric computational methods. The goal is to achieve high weighted-F1 performance across short and long temporally distant test sets while maintaining a reasonable RPD when compared to a test set from the same time period as training. We intend to use **RoBERTa**[3] [6] as a baseline classifier for our task because it has been demonstrated to be persistent over time [1].

3   LongEval Timeline

Information and updates about the LongEval Lab, the data and training/submission guidelines will be communicated mainly through the lab's website https://clef-longeval.github.io. The training data for both tasks will be released in December 2022, and the test data in February 2023. Participant submission deadline is planned for the end of April 2023, with the evaluation results to be released in June 2023.

During the CLEF 2023 conference, LongEval will organize a one-day workshop, with participant presentations as well as 2–3 invited speakers. The workshop will welcome other submissions on the topic of temporal persistence that were not part of the shared task.

Acknowledgements. This work is supported by the ANR Kodicare bi-lateral project, grant ANR-19-CE23-0029 of the French Agence Nationale de la Recherche, and by the Austrian Science Fund (FWF, grant I4471-N). This work is also supported by a UKRI/EPSRC Turing AI Fellowship to Maria Liakata (grant no. EP/V030302/1) and The Alan Turing Institute (grant no. EP/N510129/1) through project funding and its Enrichment PhD Scheme for Iman Bilal. This work has been using services provided by the LINDAT/CLARIAH-CZ Research Infrastructure (https://lindat.cz), supported by the Ministry of Education, Youth and Sports of the Czech Republic (Project No. LM2018101) and has been also supported by the Ministry of Education, Youth and Sports of the Czech Republic, Project No. LM2018101 LINDAT/CLARIAH-CZ.

---

[3] https://huggingface.co/roberta-base.

References

1. Alkhalifa, R., Kochkina, E., Zubiaga, A.: Building for tomorrow: assessing the temporal persistence of text classifiers. arXiv preprint arXiv:2205.05435 (2022)
2. Alkhalifa, R., Zubiaga, A.: Capturing stance dynamics in social media: open challenges and research directions. Int. J. Digit. Hum., 1–21 (2022)
3. Chapelle, O., Zhang, Y.: A dynamic Bayesian network click model for web search ranking. In: Proceedings of the 18th international conference on World Wide Web, WWW 2009, pp. 1–10. Association for Computing Machinery, New York (2009). https://doi.org/10.1145/1526709.1526711
4. Chuklin, A., Markov, I., Rijke, M.D.: Click models for web search. Synth. Lect. Inf. Concepts Retrieval Serv. **7**(3), 1–115 (2015). https://doi.org/10.2200/S00654ED1V01Y201507ICR043
5. Florio, K., Basile, V., Polignano, M., Basile, P., Patti, V.: Time of your hate: the challenge of time in hate speech detection on social media. Appl. Sci. **10**(12), 4180 (2020)
6. Liu, Y., et al.: RoBERTa: a robustly optimized BERT pretraining approach. arXiv preprint arXiv:1907.11692 (2019)
7. Lukes, J., Søgaard, A.: Sentiment analysis under temporal shift. In: Proceedings of the 9th Workshop on Computational Approaches to Subjectivity, Sentiment and Social Media Analysis, pp. 65–71 (2018)
8. Ren, R., et al.: A thorough examination on zero-shot dense retrieval (2022). arxiv:2204.12755. https://doi.org/10.48550/ARXIV.2204.12755
9. Yin, W., Alkhalifa, R., Zubiaga, A.: The emojification of sentiment on social media: collection and analysis of a longitudinal Twitter sentiment dataset. arXiv preprint arXiv:2108.13898 (2021)

The CLEF-2023 CheckThat! Lab: Checkworthiness, Subjectivity, Political Bias, Factuality, and Authority

Alberto Barrón-Cedeño[1]([⊠]) [iD], Firoj Alam[2] [iD], Tommaso Caselli[3], Giovanni Da San Martino[4], Tamer Elsayed[5] [iD], Andrea Galassi[1] [iD], Fatima Haouari[5], Federico Ruggeri[1] [iD], Julia Maria Struß[6] [iD], Rabindra Nath Nandi[7], Gullal S. Cheema[8], Dilshod Azizov[9], and Preslav Nakov[9] [iD]

[1] Università di Bologna, Bologna, Italy
a.barron@unibo.it
[2] Qatar Computing Research Institute, HBKU, Ar-Rayyan, Qatar
[3] University of Groningen, Groningen, The Netherlands
[4] University of Padova, Padua, Italy
[5] Qatar University, Doha, Qatar
[6] University of Applied Sciences Potsdam, Potsdam, Germany
[7] BJIT Limited, Dhaka, Bangladesh
[8] TIB – Leibniz Information Centre for Science and Technology, Hannover, Germany
[9] Mohamed bin Zayed University of Artificial Intelligence,
Abu Dhabi, United Arab Emirates

Abstract. The five editions of the `CheckThat!` lab so far have focused on the main tasks of the information verification pipeline: check-worthiness, evidence retrieval and pairing, and verification. The 2023 edition of the lab zooms into some of the problems and—for the first time—it offers five tasks in seven languages (Arabic, Dutch, English, German, Italian, Spanish, and Turkish): Task 1 asks to determine whether an item, text or a text plus an image, is check-worthy; Task 2 requires to assess whether a text snippet is subjective or not; Task 3 looks for estimating the political bias of a document or a news outlet; Task 4 requires to determine the level of factuality of a document or a news outlet; and Task 5 is about identifying authorities that should be trusted to verify a contended claim.

Keywords: Disinformation · Fact-checking · Check-worthiness · Subjectivity · Political bias · Factuality · Authority finding

1 Introduction

During its first five editions, the `CheckThat!` lab has focused on developing technology to assist the *journalist fact-checker* during the main steps of verification [7,8,18,19,47–49,51,52]. Figure 1 (top) shows the pipeline. First, a document (or a claim) is assessed for check-worthiness, i.e., whether a journalist should check its veracity. If this is so, the system needs to retrieve claims verified in the past that could be useful to fact-check the current one.

J. Kamps et al. (Eds.): ECIR 2023, LNCS 13982, pp. 506–517, 2023.
https://doi.org/10.1007/978-3-031-28241-6_59

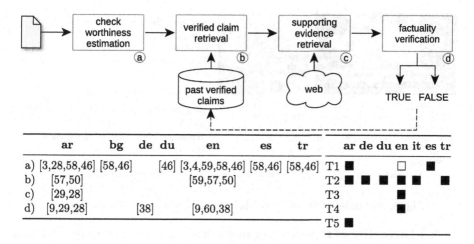

	ar	bg	de	du	en	es	tr		ar	de	du	en	it	es	tr
a)	[3,28,58,46]	[58,46]		[46]	[3,4,59,58,46]	[58,46]	[58,46]	T1	■			□		■	
b)		[57,50]			[59,57,50]			T2	■	■	■	■	■		■
c)		[29,28]						T3				■			
d)		[9,29,28]		[38]	[9,60,38]			T4				■			
								T5	■						

Fig. 1. Overview of the CheckThat! verification pipeline. The left table shows the core tasks addressed between the 2018 and the 2022 editions of the lab, including pointers to the relevant papers. The right table overviews the languages we target for the five tasks of the 2023 edition. Task 1 this year is the only one that belongs to the core tasks (a in the diagram), including multimodal data in English; a premier for the CheckThat! lab.

Further evidence to verify the claim is retrieved from the Web, if necessary. Finally, with the evidence gathered from the diverse sources, a decision can be made: whether the claim is factually true or not. The bottom-left table in Fig. 1 is the key to the technology developed for the tasks of the pipeline for all languages over the five editions of the CheckThat! lab so far.

Expert journalists consider that the most impactful technology in the verification process is check-worthiness, and that there are other aspects of news and social media that are relevant during analysis and verification, which have been overlooked.[1] With this in mind, the 2023 edition of the CheckThat! lab is organized around five tasks, four of which are run for the first time:

Task 1 Check-worthiness in tweets and political debates; the only task that has been organized during all the editions of the lab. It allows to reduce the workload of *listening* to social media for tweets that could be interesting. We introduce for the first time this year a multimodal track; cf. Sect. 2.

Task 2 Subjectivity in news articles to spot text that should be processed with specific strategies [56] (e.g., opinions may be filtered out and not checked, sarcasm and hyperboles might need further processing to extract the message they aim to convey); benefiting the fact-checking pipeline [33, 35, 64]; cf. Sect. 3.

Task 3 Political bias of news articles and news media to identify the political leaning of an article or media source, since a biased ones are more likely to make statements that are false or should be checked when they pursue the agenda and align with the bias of the author or the publisher; cf. Sect. 4.

---

[1] Private communication with organisations in various countries.

الان عودة الاشتباكات العنيفه في نهم والجيش الوطني يسحق الحوثيين في وادي حريب

Translation: Now the violent clashes are back in Nehm, and the national army is crushing the Houthis in Wadi Harib

(c) Checkworthy

The Malaysian Ministry of Education has introduced E-Skrol, an application built on the NEM blockchain to deal with the issue of certificate fraud through the use of blockchain technology.

Turns out I've been doing airborne precautions wrong my whole life. #coronavirus

واقعياً مليشيا الإصلاح بـ #مأرب هي خط الدفاع الأول عن الحوثيين

Translation: Realistically, the Islah militia in Marib is the first line of defense for the Houthis

(a) Checkworthy   (b) Not-checkworthy   (d) Not checkworthy

Fig. 2. Examples of tweets with their checkworthiness labels for Task 1.

Task 4 Factuality of reporting of news media is critical for media profiling; cf. Sect. 5.

Task 5 Authority finding in Twitter to help fact-checkers who aim to verify rumors propagating in social media find a trusted source (an authority that has "real knowledge" on the matter) that might help to confirm or to debunk a specific rumor. This task can be seen as a sub-problem of topical expert finding in Twitter [22,39,65]; cf. Sect. 6.

The bottom-right table in Fig. 1 gives an overview of the language coverage that we target for the five tasks this year.

2 Task 1: Check-Worthiness in Tweets

Task Definition. The aim of this task is to determine whether a claim is worth fact-checking. This year, we offer two kinds of data, which translate to the following two subtasks:

Subtask 1A (Multimodal – Tweets): The tweets to be judged include both a text snippet and an image.

Subtask 1B: Check-worthiness estimation from Multigenre (Unimodal) (text: A text snippet alone—from a tweet or a debate/speech transcription—has to be assessed for check-worthiness.

Subtask 1A is offered in Arabic and English, Subtask 1B is offered in Arabic, English and Spanish.

Data. For Task 1A, we use the annotation schema of [13]. Each tweet is annotated based on both the image and the text it contains for (i) the presence of a factual claim, (ii) check-worthiness, and (iii) visual relevance. The latter holds for two aspects: there is a piece of evidence (e.g., an event, an action, a situation, a person's identity, etc.) or illustration of certain aspects from the textual claim, or the image contains overlayed text that contains a claim in a textual form. The English data consists of 3k tweets. We also provide 82k unlabeled tweets that consist of text–image pairs and can be used for semi-supervised learning. The Arabic data consists of 3k tweets on topics such as COVID-19 and politics [1,47].

Table 1. Instances of subjective and objective sentences for Task 2.

	Instance	Class
1.	While it's misguided to put all focus or hope onto one section of the working class, we can't ignore this immense latent power that logistics workers possess	subj
2.	Taking refuge in public credit will cause that same infection to attack business, banking, industry, agriculture, the entire body of private enterprise	subj
3.	Workers would have a 24 percent wage increase by 2024, including an immediate 14 percent raise	obj
4.	University of Washington epidemiologist Ali Mokdad predicted a rise in reported COVID-19 cases	obj

The dataset for Subtask 1B consists of tweets in Arabic and Spanish. The Spanish tweets are collected from Twitter accounts and transcriptions from Spanish politicians and are manually annotated by professional journalists who are experts in fact-checking. The Arabic tweets for subtask 1B are collected using keywords related to COVID-19 and vaccines, using the annotation schema in [1]. The dataset for Subtask 1B (English) consists of political debates collected from U.S. general election presidential debates and annotated by human coders. Figure 2 shows examples of checkworthy and non-checkworthy tweets.

Evaluation. This is a binary classification task. The official evaluation measure is F_1 score for the positive class.

3   Task 2: Subjectivity in News Articles

Task Definition. The systems are challenged to distinguish whether a sentence from a news article expresses the subjective view of its author or presents an objective view of the covered topic. Given a list of sentences taken from a news article, the task asks to classify each of the sentences as subjective or objective. The task is offered in Arabic, Dutch, English, Italian, German, and Turkish.

Data. The focus is on sentences from newspaper articles. The data for Italian is partially derived from SubjectivITA [2] and consists of 2.2k examples, 25% of which are subjective. For English, we release a new dataset containing 1.2k sentences. The annotation process involved multiple annotators that labeled instances individually. Later on, annotators discussed and resolved the disagreements. We measured the Inter-Annotators Agreement (IAA) for the Italian dataset using Fleiss' kappa, and obtained a score of 0.61, which corresponds to substantial agreement. For the English dataset, we computed Krippendorf's alpha of 0.83. For the other languages, we plan to follow the same methodology to release datasets of comparable size. Table 1 shows examples of the English part of the dataset for Task 2.

Table 2. Examples of media with different biases for Task 3.

Name	URL	Bias
Loser.com	http://loser.com	Left
Die Hard Democrat	http://dieharddemocrat.com	Left
Democracy 21	http://www.democracy21.org	Center
Federal Times	http://www.federaltimes.com	Center
Gulf News	http://gulfnews.com	Center
Fox News	http://www.foxnews.com	Right

Evaluation. This is a binary classification task, and thus we use macro-averaged F_1 score as the official evaluation measure.

4   Task 3: Political Bias of News Articles and News Media

Task Definition. The goal of the task is to detect political bias of news reporting at the article and at the media level. This is an ordinal classification task and it is offered in English. It includes two subtasks:

Subtask 3A: Given an article, classify its leaning as left, center, or right.

Subtask 3B: Given the URL to a news outlet (e.g., www.cnn.com), predict the overall political bias of that news outlet as left, center, or right.

Data. We release a collection of 95k articles from 900 media sources annotated for bias at the article and at the media level, respectively. We used a subset of this data in previous research [6], but we have now crawled additional articles and sources for training and testing purposes.[2] Table 2 shows examples of news media with their political leaning. Note that we map the bias from a 7-point scale (Extreme-Left, Left, Center-Left, Center, Center-Right, Right, and Extreme-Right) to 3-point scale: left, center, and right.

Evaluation. This is an ordinal classification task, and thus we use mean absolute error as the official measure for both subtasks.

5   Task 4: Factuality of Reporting of News Media

Task Definition. We ask to predict the factuality of reporting at the media level, given the URL to a news outlet (e.g., www.cnn.com): low, mixed, and high. We offer the task in English.

Data. We use the same kind of data as for task 3, but with labels for factuality (again on an ordinal scale). We obtain the annotations and the analysis of the factuality of reporting and/or bias from mediabiasfactcheck.org, which are manually labeled by fact-checkers. The dataset consists of over 2k news media. Table 3 shows examples of news media and their factuality labels.

---

[2] The annotated labels for the articles are obtained from http://www.allsides.com/ and http://mediabiasfactcheck.org/.

Table 3. Examples of news media with different factuality labels for Task 4.

Name	URL	Factuality
Associated Press	http://apnews.com	High
NBC News	http://www.nbcnews.com/	High
Russia Insider	http://russia-insider.com	Mixed
Patriots Voice	https://www.patriotvoices.com	Low

Evaluation. This is an ordinal classification task, and we use mean absolute error as the official evaluation measure.

6   Task 5: Authority Finding in Twitter

Task Definition. The task asks systems to retrieve authority Twitter accounts for a given rumor that propagates in Twitter. Given a tweet spreading a rumor, the participating systems need to retrieve a ranked list of authority Twitter accounts that can help verify that rumor, as such accounts may tweet evidence that supports or denies the rumor [26]. This task is offered in Arabic.

Data. The training set comprises 150 rumors expressed in tweets associated with 1k authority Twitter accounts, a set of 400k Twitter accounts, 1.2M unique Twitter lists, and 878M timeline tweets. To construct the data, we selected rumors from Misbar[3], an Arabic fact-checking platform adopted by recent studies to construct datasets for Arabic rumor verification [27] and fake news detection [37]. For each rumor, two annotators were individually asked to find all possible authority Twitter accounts who can help confirm or deny that rumor following our detailed annotation guidelines. As part of the annotation process, the annotators were required to assign a grade for each authority to determine whether she is *highly relevant* or *relevant* to the rumor, i.e., having a higher priority to be contacted for verification or not. Finally, the annotators discussed their agreement on each others' selected authorities and their grades, and a third annotator helped resolve the disagreements. To evaluate the quality of the annotations, we considered the agreement both on whether the target Twitter account is labeled as authority with respect to the considered rumor as well as the graded relevance. The Cohen's Kappa inter-annotator agreement [14] was 0.78 and 0.71 for the former and for the latter, respectively, both scores corresponding to *substantial* agreement [40]. Table 4 shows an example rumor with authorities ranked according to relevance.

Evaluation. As this is a ranking task, we adopt P@5 as the official evaluation measure to evaluate how well the participating systems retrieve Twitter authorities at the top of a short retrieved list. We further report NDCG@5 to measure the ability of systems to retrieve highly relevant authority Twitter accounts higher up in that list.

---

[3] https://misbar.com/.

Table 4. An example of a rumor with corresponding authorities for Task 5.

Rumor Tweet: The Saudi Federation decided to ban the 'Al-Alamy' title from all league clubs and prohibit clubs from using it, whether banners inside stadiums or through clubs' websites	
Authority	Relevance
1 Saudi Arabian Football Federation	Highly relevant
2 President of the Saudi Arabian Football Federation	Highly relevant
3 Saudi Arabian Football Federation Media and Communications	Highly relevant
4 Ministry of Sport in Saudi Arabia	Relevant
5 Minister of Sports and President of the Saudi Olympic and Paralympic Committee	Relevant

7 Related Work

There has been a lot of research on checking the factuality of a claim, of a news article, or of an information source [5,6,34,41,45,67]. Given that misleading content is causing harm across different dimensions, a lot of attention has been paid to identifying disinformation and misinformation in social media [24,36,43, 61,66]. Check-worthiness estimation is still an understudied problem, especially in social media [21,30–32,63], and fake news detection for news articles is mostly approached as a binary classification problem [53].

CheckThat! is related to several tasks at SemEval: on determining rumor veracity [16,23], on stance detection [44], on fact-checking in community question answering forums [42], and on propaganda detection [15,17]. It is also related to the FEVER task [62] on fact extraction and verification, to the Fake News Challenge [25,55] and to the FakeNews task at MediaEval [54].

8 Conclusion

We presented the 2023 edition of the CheckThat! lab, which features complementary tasks to assist in the full fact-checking pipeline: from spotting check-worthy claims to identifying an authority that could help verify a rumor in social media. In line with one of the main missions of CLEF, we promote multi-linguality by offering tasks in seven languages: Arabic, Dutch, English, German, Italian, Spanish, and Turkish. Moreover, for the first time, we also promote a multimodal task.

Acknowledgments. The work of Tamer Elsayed was made possible by NPRP grant #NPRP-11S-1204-170060 from the Qatar National Research Fund (a member of Qatar Foundation). The work of Fatima Haouari is supported by GSRA grant #GSRA6-1-0611-19074 from the Qatar National Research Fund. The statements made herein are solely the responsibility of the authors.

References

1. Alam, F., et al.: Fighting the COVID-19 infodemic: modeling the perspective of journalists, fact-checkers, social media platforms, policy makers, and the society. In: Findings of EMNLP 2021, pp. 611–649 (2021)

2. Antici, F., Bolognini, L., Inajetovic, M.A., Ivasiuk, B., Galassi, A., Ruggeri, F.: SubjectivITA: an Italian corpus for subjectivity detection in newspapers. In: Candan, K.S., et al. (eds.) CLEF 2021. LNCS, vol. 12880, pp. 40–52. Springer, Cham (2021). https://doi.org/10.1007/978-3-030-85251-1_4

3. Atanasova, P., et al.: Overview of the CLEF-2018 CheckThat! lab on automatic identification and verification of political claims. Task 1: check-worthiness. In: Cappellato et al. [12]

4. Atanasova, P., Nakov, P., Karadzhov, G., Mohtarami, M., Da San Martino, G.: Overview of the CLEF-2019 CheckThat! lab on automatic identification and verification of claims. Task 1: check-worthiness. In: Cappellato et al. [11]

5. Ba, M.L., Berti-Equille, L., Shah, K., Hammady, H.M.: VERA: a platform for veracity estimation over web data. In: Proceedings of the 25th International Conference on World Wide Web, WWW 2016, pp. 159–162 (2016)

6. Baly, R., et al.: What was written vs. who read it: news media profiling using text analysis and social media context. In: Proceedings of the 58th Annual Meeting of the Association for Computational Linguistics, ACL 2020, pp. 3364–3374 (2020)

7. Barrón-Cedeño, A., et al.: CheckThat! at CLEF 2020: enabling the automatic identification and verification of claims in social media. In: Advances in Information Retrieval, ECIR 2020, pp. 499–507 (2020)

8. Barrón-Cedeño, A., et al.: Overview of CheckThat! 2020: automatic identification and verification of claims in social media. In: Arampatzis, A., et al. (eds.) CLEF 2020. LNCS, vol. 12260, pp. 215–236. Springer, Cham (2020). https://doi.org/10.1007/978-3-030-58219-7_17

9. Barrón-Cedeño, A., et al.: Overview of the CLEF-2018 CheckThat! lab on automatic identification and verification of political claims. Task 2: factuality. In: Cappellato et al. [12]

10. Cappellato, L., Eickhoff, C., Ferro, N., Névéol, A. (eds.): CLEF 2020 Working Notes. CEUR Workshop Proceedings (2020)

11. Cappellato, L., Ferro, N., Losada, D., Müller, H. (eds.): Working Notes of CLEF 2019 Conference and Labs of the Evaluation Forum. CEUR Workshop Proceedings (2019)

12. Cappellato, L., Ferro, N., Nie, J.Y., Soulier, L. (eds.): Working Notes of CLEF 2018-Conference and Labs of the Evaluation Forum. CEUR Workshop Proceedings (2018)

13. Cheema, G.S., Hakimov, S., Sittar, A., Müller-Budack, E., Otto, C., Ewerth, R.: MM-claims: a dataset for multimodal claim detection in social media. In: Findings of NAACL, pp. 962–979 (2022)

14. Cohen, J.: A coefficient of agreement for nominal scales. Educ. Psychol. Measur. **20**(1), 37–46 (1960)

15. Da San Martino, G., Barrón-Cedeno, A., Wachsmuth, H., Petrov, R., Nakov, P.: SemEval-2020 task 11: detection of propaganda techniques in news articles. In: Proceedings of the 14th Workshop on Semantic Evaluation, SemEval 2020, pp. 1377–1414 (2020)

16. Derczynski, L., Bontcheva, K., Liakata, M., Procter, R., Wong Sak Hoi, G., Zubiaga, A.: SemEval-2017 task 8: RumourEval: determining rumour veracity and support for rumours. In: Proceedings of the 11th International Workshop on Semantic Evaluation, SemEval 2017, pp. 69–76 (2017)

17. Dimitrov, D., et al.: SemEval-2021 task 6: detection of persuasion techniques in texts and images. In: Proceedings of the International Workshop on Semantic Evaluation, SemEval 2021, pp. 70–98 (2021)

18. Elsayed, T., et al.: CheckThat! at CLEF 2019: automatic identification and verification of claims. In: Azzopardi, L., Stein, B., Fuhr, N., Mayr, P., Hauff, C., Hiemstra, D. (eds.) ECIR 2019. LNCS, vol. 11438, pp. 309–315. Springer, Cham (2019). https://doi.org/10.1007/978-3-030-15719-7_41

19. Elsayed, T., et al.: Overview of the CLEF-2019 CheckThat! lab: automatic identification and verification of claims. In: Crestani, F., et al. (eds.) CLEF 2019. LNCS, vol. 11696, pp. 301–321. Springer, Cham (2019). https://doi.org/10.1007/978-3-030-28577-7_25

20. Faggioli, G., Ferro, N., Joly, A., Maistro, M., Piroi, F. (eds.): CLEF 2021 Working Notes. Working Notes of CLEF 2021-Conference and Labs of the Evaluation Forum (2021)

21. Gencheva, P., Nakov, P., Màrquez, L., Barrón-Cedeño, A., Koychev, I.: A context-aware approach for detecting worth-checking claims in political debates. In: Proceedings of the International Conference Recent Advances in Natural Language Processing, RANLP 2017, pp. 267–276 (2017)

22. Ghosh, S., Sharma, N., Benevenuto, F., Ganguly, N., Gummadi, K.: Cognos: crowdsourcing search for topic experts in microblogs. In: Proceedings of the 35th International ACM SIGIR Conference on Research and Development in Information Retrieval, SIGIR 2012, pp. 575–590 (2012)

23. Gorrell, G., et al.: SemEval-2019 task 7: RumourEval, determining rumour veracity and support for rumours. In: Proceedings of the 13th International Workshop on Semantic Evaluation, SemEval 2019, pp. 845–854 (2019)

24. Gupta, A., Kumaraguru, P., Castillo, C., Meier, P.: TweetCred: real-time credibility assessment of content on Twitter. In: Proceedings of the 6th International Social Informatics Conference, SocInfo 2014, pp. 228–243 (2014)

25. Hanselowski, A., et al.: A retrospective analysis of the fake news challenge stance-detection task. In: Proceedings of the 27th International Conference on Computational Linguistics, COLING 2018, pp. 1859–1874 (2018)

26. Haouari, F., Elsayed, T.: Detecting stance of authorities towards rumors in Arabic tweets: a preliminary study. In: Proceedings of the 45th European Conference on Information Retrieval (ECIR 2023) (2023)

27. Haouari, F., Hasanain, M., Suwaileh, R., Elsayed, T.: ArCOV19-Rumors: Arabic COVID-19 Twitter dataset for misinformation detection. In: Proceedings of the Arabic Natural Language Processing Workshop, WANLP 2021, pp. 72–81 (2021)

28. Hasanain, M., et al.: Overview of CheckThat! 2020 Arabic: automatic identification and verification of claims in social media. In: Cappellato et al. [10]

29. Hasanain, M., Suwaileh, R., Elsayed, T., Barrón-Cedeño, A., Nakov, P.: Overview of the CLEF-2019 CheckThat! lab on automatic identification and verification of claims. Task 2: evidence and factuality. In: Cappellato et al. [11]

30. Hassan, N., Li, C., Tremayne, M.: Detecting check-worthy factual claims in presidential debates. In: Proceedings of the 24th ACM International on Conference on Information and Knowledge Management, CIKM 2015, pp. 1835–1838 (2015)

31. Hassan, N., et al.: ClaimBuster: the first-ever end-to-end fact-checking system. Proc. VLDB Endow. **10**(12), 1945–1948 (2017)

32. Jaradat, I., Gencheva, P., Barrón-Cedeño, A., Màrquez, L., Nakov, P.: ClaimRank: detecting check-worthy claims in Arabic and English. In: Proceedings of the 2018 Conference of the North American Chapter of the Association for Computational Linguistics: Demonstrations, NAACL-HLT 2018, pp. 26–30 (2018)
33. Jerônimo, C.L.M., Marinho, L.B., Campelo, C.E.C., Veloso, A., da Costa Melo, A.S.: Fake news classification based on subjective language. In: Proceedings of the 21st International Conference on Information Integration and Web-based Applications & Services, pp. 15–24 (2019)
34. Karadzhov, G., Nakov, P., Màrquez, L., Barrón-Cedeño, A., Koychev, I.: Fully automated fact checking using external sources. In: Proceedings of the International Conference Recent Advances in Natural Language Processing, RANLP 2017, pp. 344–353 (2017)
35. Kasnesis, P., Toumanidis, L., Patrikakis, C.Z.: Combating fake news with transformers: a comparative analysis of stance detection and subjectivity analysis. Information 12(10), 409 (2021)
36. Kazemi, A., Garimella, K., Gaffney, D., Hale, S.: Claim matching beyond English to scale global fact-checking. In: Proceedings of the 59th Annual Meeting of the Association for Computational Linguistics and the 11th International Joint Conference on Natural Language Processing, ACL-IJCNLP 2021, pp. 4504–4517 (2021)
37. Khalil, A., Jarrah, M., Aldwairi, M., Jararweh, Y.: Detecting Arabic fake news using machine learning. In: Proceedings of the International Conference on Intelligent Data Science Technologies and Applications, IDSTA 2021, pp. 171–177 (2021)
38. Köhler, J., et al.: Overview of the CLEF-2022 CheckThat! lab task 3 on fake news detection. In: Working Notes of CLEF 2022–Conference and Labs of the Evaluation Forum, CLEF 2022 (2022)
39. Lahoti, P., De Francisci Morales, G., Gionis, A.: Finding topical experts in Twitter via query-dependent personalized PageRank. In: Proceedings of the 2017 IEEE/ACM International Conference on Advances in Social Networks Analysis and Mining, ASONAM 2017, pp. 155–162 (2017)
40. Landis, J.R., Koch, G.G.: The measurement of observer agreement for categorical data. Biometrics 159–174 (1977)
41. Ma, J., Gao, W., Mitra, P., Kwon, S., Jansen, B.J., Wong, K.F., Cha, M.: Detecting rumors from microblogs with recurrent neural networks. In: Proceedings of the International Joint Conference on Artificial Intelligence, IJCAI 2016, pp. 3818–3824 (2016)
42. Mihaylova, T., Karadzhov, G., Atanasova, P., Baly, R., Mohtarami, M., Nakov, P.: SemEval-2019 task 8: fact checking in community question answering forums. In: Proceedings of the 13th International Workshop on Semantic Evaluation, SemEval 2019, pp. 860–869 (2019)
43. Mitra, T., Gilbert, E.: CREDBANK: a large-scale social media corpus with associated credibility annotations. In: Proceedings of the Ninth International AAAI Conference on Web and Social Media, ICWSM 2015, pp. 258–267 (2015)
44. Mohammad, S., Kiritchenko, S., Sobhani, P., Zhu, X., Cherry, C.: SemEval-2016 task 6: detecting stance in tweets. In: Proceedings of the 10th International Workshop on Semantic Evaluation, SemEval 2016, pp. 31–41 (2016)
45. Mukherjee, S., Weikum, G.: Leveraging joint interactions for credibility analysis in news communities. In: Proceedings of the 24th ACM International Conference on Information and Knowledge Management, CIKM 2015, pp. 353–362 (2015)
46. Nakov, P., et al.: Overview of the CLEF-2022 CheckThat! lab task 1 on identifying relevant claims in tweets. In: Working Notes of CLEF 2022–Conference and Labs of the Evaluation Forum, CLEF 2022 (2022)

47. Nakov, P., et al.: Overview of the CLEF-2022 CheckThat! lab on fighting the COVID-19 infodemic and fake news detection. In: Proceedings of the 13th International Conference of the CLEF Association: Information Access Evaluation meets Multilinguality, Multimodality, and Visualization, CLEF 2022 (2022)

48. Nakov, P., et al.: The CLEF-2022 CheckThat! lab on fighting the COVID-19 infodemic and fake news detection. In: Hagen, M., et al. (eds.) ECIR 2022. LNCS, vol. 13186, pp. 416–428. Springer, Cham (2022). https://doi.org/10.1007/978-3-030-99739-7_52

49. Nakov, P., et al.: Overview of the CLEF-2018 lab on automatic identification and verification of claims in political debates. In: Working Notes of CLEF 2018 - Conference and Labs of the Evaluation Forum, CLEF 2018 (2018)

50. Nakov, P., Da San Martino, G., Alam, F., Shaar, S., Mubarak, H., Babulkov, N.: Overview of the CLEF-2022 CheckThat! lab task 2 on detecting previously fact-checked claims. In: Working Notes of CLEF 2022–Conference and Labs of the Evaluation Forum, CLEF 2022 (2022)

51. Nakov, P., et al.: Overview of the CLEF–2021 CheckThat! lab on detecting check-worthy claims, previously fact-checked claims, and fake news. In: Candan, K.S., et al. (eds.) CLEF 2021. LNCS, vol. 12880, pp. 264–291. Springer, Cham (2021). https://doi.org/10.1007/978-3-030-85251-1_19

52. Nakov, P., et al.: The CLEF-2021 CheckThat! lab on detecting check-worthy claims, previously fact-checked claims, and fake news. In: Hiemstra, D., Moens, M.-F., Mothe, J., Perego, R., Potthast, M., Sebastiani, F. (eds.) ECIR 2021. LNCS, vol. 12657, pp. 639–649. Springer, Cham (2021). https://doi.org/10.1007/978-3-030-72240-1_75

53. Oshikawa, R., Qian, J., Wang, W.Y.: A survey on natural language processing for fake news detection. In: Proceedings of the 12th Language Resources and Evaluation Conference, LREC 2020, pp. 6086–6093 (2020)

54. Pogorelov, K., et al.: FakeNews: corona virus and 5G conspiracy task at MediaEval 2020. In: Proceedings of the MediaEval 2020 Workshop, MediaEval 2020 (2020)

55. Pomerleau, D., Rao, D.: The fake news challenge: exploring how artificial intelligence technologies could be leveraged to combat fake news (2017). http://www.fakenewschallenge

56. Riloff, E., Wiebe, J.: Learning extraction patterns for subjective expressions. In: Proceedings of the 2003 Conference on Empirical Methods in Natural Language Processing, EMNLP 2003, pp. 105–112 (2003)

57. Shaar, S., et al.: Overview of the CLEF-2021 CheckThat! lab task 2 on detecting previously fact-checked claims in tweets and political debates. In: Faggioli et al. [20]

58. Shaar, S., et al.: Overview of the CLEF-2021 CheckThat! lab task 1 on check-worthiness estimation in tweets and political debates. In: Faggioli et al. [20]

59. Shaar, S., et al.: Overview of CheckThat! 2020 English: automatic identification and verification of claims in social media. In: Cappellato et al. [10]

60. Shahi, G.K., Struß, J.M., Mandl, T.: Overview of the CLEF-2021 CheckThat! lab: task 3 on fake news detection. In: Faggioli et al. [20]

61. Shu, K., Sliva, A., Wang, S., Tang, J., Liu, H.: Fake news detection on social media: a data mining perspective. SIGKDD Explor. Newsl. **19**(1), 22–36 (2017)

62. Thorne, J., Vlachos, A., Christodoulopoulos, C., Mittal, A.: FEVER: a large-scale dataset for fact extraction and VERification. In: Proceedings of the Conference of the North American Chapter of the Association for Computational Linguistics: Human Language Technologies, NAACL-HLT 2018, pp. 809–819 (2018)

63. Vasileva, S., Atanasova, P., Màrquez, L., Barrón-Cedeño, A., Nakov, P.: It takes nine to smell a rat: neural multi-task learning for check-worthiness prediction. In: Proceedings of the International Conference on Recent Advances in Natural Language Processing, RANLP 2019, pp. 1229–1239 (2019)
64. Vieira, L.L., Jerônimo, C.L.M., Campelo, C.E.C., Marinho, L.B.: Analysis of the subjectivity level in fake news fragments. In: Proceedings of the Brazillian Symposium on Multimedia and the Web, WebMedia 2020, pp. 233–240. ACM (2020)
65. Wei, W., Cong, G., Miao, C., Zhu, F., Li, G.: Learning to find topic experts in Twitter via different relations. IEEE Trans. Knowl. Data Eng. **28**(7), 1764–1778 (2016)
66. Zhao, Z., Resnick, P., Mei, Q.: Enquiring minds: early detection of rumors in social media from enquiry posts. In: Proceedings of the 24th International Conference on World Wide Web, WWW 2015, pp. 1395–1405 (2015)
67. Zubiaga, A., Liakata, M., Procter, R., Hoi, G.W.S., Tolmie, P.: Analysing how people orient to and spread rumours in social media by looking at conversational threads. PLoS ONE **11**(3), e0150989 (2016)

Overview of PAN 2023: Authorship Verification, Multi-author Writing Style Analysis, Profiling Cryptocurrency Influencers, and Trigger Detection
Extended Abstract

Janek Bevendorff[1], Mara Chinea-Ríos[7], Marc Franco-Salvador[7], Annina Heini[3], Erik Körner[1], Krzysztof Kredens[3], Maximilian Mayerl[4], Piotr Pęzik[3], Martin Potthast[5,6], Francisco Rangel[7], Paolo Rosso[2,3], Efstathios Stamatatos[8], Benno Stein[1], Matti Wiegmann[1(✉)], Magdalena Wolska[1], and Eva Zangerle[4]

[1] Bauhaus-Universität Weimar, Weimar, Germany
pan@webis.de
[2] Universitat Politècnica de València, Valencia, Spain
[3] Aston University, Birmingham, UK
[4] University of Innsbruck, Innsbruck, Austria
[5] Leipzig University, Leipzig, Germany
[6] ScaDS.AI, Leipzig, Germany
[7] Symanto Research, Valencia, Spain
[8] University of the Aegean, Mytilene, Greece

Abstract. The paper gives a brief overview of the four shared tasks organized at the PAN 2023 lab on digital text forensics and stylometry to be hosted at the CLEF 2023 conference. The general goal of the PAN lab is to advance the state-of-the-art in text forensics and stylometry while ensuring objective evaluation of new and established methods on newly developed benchmark datasets. PAN's tasks cover four areas of digital text forensics: author identification, multi-author analysis, author profiling, and content analysis. Some tasks follow up on past editions (cross-domain authorship verification, multi-author writing style analysis) and some explore novel ideas (profiling cryptocurrency influencers in social media and trigger detection). As with the previous editions, PAN invites software submissions rather than run submissions; more than 400 pieces of software have been submitted from PAN'12 through PAN'22 combined, with recent evaluations running on the TIRA experimentation platform. This proposal briefly outlines our goals for PAN as a lab and our contributions proposed for PAN'23.

J. Kamps et al. (Eds.): ECIR 2023, LNCS 13982, pp. 518–526, 2023.
https://doi.org/10.1007/978-3-031-28241-6_60

1   Introduction

PAN is a workshop series and a networking initiative for stylometry and digital text forensics. The workshop's goal is to bring together scientists and practitioners studying technologies that analyze texts with regard to originality, authorship, trust, and ethicality, among others. Since its inception 15 years back PAN has included shared tasks on specific computational challenges related to authorship analysis, computational ethics, and determining the originality of a piece of writing. Over the years, the respective organizing committees of the 64 shared tasks[1] have assembled evaluation resources for the aforementioned research disciplines that amount to 55 datasets[2] plus nine datasets contributed by the community. Each new dataset was compiled by the task's authors specifically for the given task and introduced new variants of author verification, profiling, or author obfuscation tasks as well as multi-author analysis and determining the morality, quality, or originality of a text. The tasks build incrementally on the experience and results of prior PAN shared tasks and extend them in meaningful ways by increasing complexity. The 2023 edition of PAN continues in the same vein, introducing new resources as well as previously unconsidered problems to the community. As in earlier editions, PAN is committed to reproducible research in IR and NLP therefore all shared tasks will ask for software submissions on our TIRA platform [9]. We briefly outline the upcoming tasks in the sections that follow.

2   Authorship Verification

Authorship verification is a fundamental task in author identification. All cases of questioned authorship can be decomposed into a series of verification instances, be it in a closed-set or open-set scenario [6]. The past editions of PAN considered the task of *cross-domain authorship verification*, where the texts of known and unknown authorship come from different domains [1,2,24]. In most of the examined cases, the domains corresponded to topics, thematic areas, or fandoms (non-professional fiction published online in significant quantities by fans of high-popularity authors or works, so-called fanfiction). The relatively high performance of the past submissions [1,2] demonstrates that authorship in most of these cases can be successfully verified. However, it is not clear yet how to handle more difficult authorship verification cases where texts of known and unknown authorship belong to different discourse types (DTs), especially when these DTs have few similarities (e.g., argumentative essays vs. text messages to family members). Hence, the most recent edition of PAN adopted a new and very challenging scenario: *cross-discourse type authorship verification*. Here, documents belong to different discourse types (i.e., essays, emails, text messages, business memos) whose style depends on the level of formality, intended audience, and communicative purpose [23]. The relatively low obtained evaluation results show that the task is still exceedingly difficult.

---

[1] Find PAN's past shared tasks at pan.webis.de/shared-tasks.html.
[2] Find PAN's datasets at pan.webis.de/data.html.

Cross-Discourse Type Author Verification at PAN'23

In its simplest form, authorship verification deals with determining whether two documents are written by the same author. In cross-discourse type authorship verification, introduced in the last edition of PAN [23], the two documents are of distinct DTs. Yet despite their differences, all documents in this and previous PAN editions are only forms of written language. At PAN'23, we will focus for the first time on (cross-discourse type) authorship verification where both written (e.g., essays, emails) and oral language (e.g., interviews, speech transcriptions) are represented in the set of discourse types. This will provide the opportunity to study the robustness and effectiveness of stylometric approaches in challenging and intriguing conditions. In addition, the ability of authorship verification methods to handle the different forms of expression in written and oral language will be highlighted. New training and evaluation datasets will be provided that cover DTs in both written and oral language. The same evaluation framework and measures as in the latest PAN editions of authorship verification tasks will be adopted [1,23]. The evaluation includes well-known measures like the area under the ROC curve, F_1 score, and Brier score, as well as more specialized measures that take into account non-answers, like c@1 (a variant of accuracy rewarding non-answers) and $F_{0.5u}$ (a variant of F-score rewarding correctly predicted same-author cases in addition to non-answers).

3   Author Profiling

Author profiling is the problem of distinguishing between classes of authors by studying how language is shared by people. Profiling can help to identify authors' individual characteristics, such as age, gender, or language variety, among others. During the years 2013–2022 we addressed several of these aspects in the shared tasks organized at PAN.[3] In 2013 the aim was to identify gender and age in social media texts for English and Spanish [16]. In 2014 we addressed age identification from a continuous perspective (without gaps between age classes) in the context of several genres, such as blogs, Twitter, and reviews (in Trip Advisor), both in English and Spanish [14]. In 2015, apart from age and gender identification, we addressed also personality recognition on Twitter in English, Spanish, Dutch, and Italian [18]. In 2016, we addressed the problem of cross-genre gender and age identification in English, Spanish, and Dutch [19]. The training data was gathered from Twitter and the test data was gathered from blogs and social media data. In 2017, we addressed gender and language variety identification in Twitter in English, Spanish, Portuguese, and Arabic [17]. In 2018, we investigated gender identification on Twitter from a multi-modal perspective, considering also the images linked within tweets; the dataset was composed of English, Spanish, and Arabic tweets [15]. From 2019 to 2022, we focused on a series of shared tasks related to profiling harmful information spreaders. In 2019 our focus was on profiling and discriminating bots from humans on the basis of textual data

---

[3] All our datasets comply with the EU General Data Protection Regulation [12].

only [13] and targeting both English and Spanish tweets. In 2020, we focused on profiling fake news spreaders [11], both in English and Spanish. The ease of publishing content on social media has also increased the amount of disinformation that is published and shared and our goal was to profile those authors who have shared some fake news in the past. In 2021, we focused on profiling hate speech spreaders in social media [10], both in English and Spanish. The goal was to identify Twitter users who can be considered haters, depending on the number of tweets with hateful content that they had spread. Finally, in 2022, we focused on profiling irony and stereotype spreaders on English tweets [20]. The goal was to profile highly ironic authors and those that employ irony to convey stereotypical messages, e.g. towards women or the LGTB community.

Profiling Cryptocurrency Influencers with Few-Shot Learning at PAN'23

Cryptocurrencies have massively increased their popularity in recent years [22]. The promise of independence from central authorities, the possibilities offered by the different projects, and the new, influencer-driven gold rush make cryptocurrencies a trendy topic in social media. Profiling research is particularly interested in the cryptocurrency ecosystem to identify influential actors that motivate others into action.

Producing sufficiently many high-quality annotations for author profiling is challenging. Profiling influencers in particular has high requirements in the economic and temporal cost, psychological and linguistic expertise needed by the annotator, and the congenital subjectivity involved in the annotation task [3,25]. Additionally, in a real environment, i.e. when traders want to leverage social media signals to forecast the market, profiling needs to be done in real-time in a few milliseconds. This difficult, expensive, and high-speed data collection process implies data scarcity: models need to work with as little data as possible and still perform.

In this shared task, we aim to profile cryptocurrency influencers in social media from a low-resource perspective, that is, using little data. Moreover, we propose to profile types of influencers also using a low-resource setting. Specifically, we focus on English Twitter posts for three different sub-tasks: (1) *Low-resource influencer profiling*: profile authors according to their degree of influence (null, nano, micro, macro, mega); (2) *Low-resource influencer interest profiling*: profile authors according to their main interests or areas of influence (technical information, price update, trading matters, gaming, other); (3) *Low-resource influencer intent profiling*: profile authors according to the intent of their messages (subjective opinion, financial information, advertising, announcement). Participants need to choose carefully which models to apply to this under-resourced setting. Concepts such as transfer learning [28] and few-shot learning [4,7,8,27] are key to succeed.

4   Multi-author Writing Style Analysis

Style change detection concerns itself with identifying positions within a given text document at which the writing style—and therefore, by extension, the author—changes. This task can be a constituent task of authorship identification and multi-author document analysis, and has applications in areas such as plagiarism detection. At PAN, the style change detection task has been studied since 2016, in various different forms. In 2016, participants had to identify the authors of fragments of a document, and group all fragments written by the same author together [21]. In 2017, the task was twofold [26]. First, participants had to determine whether a given document was written by one or by multiple authors. Second, for documents by multiple authors, they had to determine the exact positions within the documents where the author changes. In 2018, following feedback that the task posed in the previous year was too difficult, the problem was simplified to only identifying whether a document was written by one or more authors [5]. In the following years, we built on this and gradually made the task definitions more complex again. In 2019, participants had to first determine whether a document had a single or multiple authors, and, if it is multi-authored, determine the concrete number of authors involved in writing it [32]. In 2020, participants again had to determine whether a document is single- or multi-authored. For multi-authored documents, they also had to identify between which paragraphs in the document the author changes, and assign paragraphs to concrete authors [31]. The task posed in 2021 was very similar, but this time, we additionally provided participants with a simplified version of the task, where each document contained exactly one style change, and the participants had to determine between which paragraphs in the document this occurred [29]. Finally, in 2022, we added a more complex subtask where style changes could now occur not only between paragraphs but also between sentences [30].

Multi-author Writing Style Analysis at PAN'23

Traditionally, writing style analysis has focused on single-author documents. However, more recent research, including that conducted at previous editions of PAN, has shown that writing style analysis can effectively be employed for detecting author changes within a document. This can be used to partition a document into parts that have been written by different authors, which can be applied to areas such as plagiarism detection. In previous editions of PAN, our participants developed a range of different techniques for detecting author changes in documents. However, the datasets used in those editions of the task exhibited a large variety of topics, also within single documents. This allowed approaches to indirectly exploit topic changes to make the task easier.

In the 2023 edition, we have therefore paid special attention to developing datasets that do not exhibit this problem. We will provide participants with datasets of three difficulty levels: (1) "Easy dataset": The paragraphs of a document cover a variety of topics, allowing approaches to make use of topic information to detect authorship changes. (2) "Medium dataset": The topical variety in a document is small (though still present) forcing the approaches to focus more on style to solve the detection task effectively; (3) "Hard dataset": All paragraphs in a document are on the same topic. Similar to most tasks in recent editions, style changes are once again limited to occur between paragraphs (i.e., each paragraph belongs to a single author).

5   Trigger Detection

A trigger in psychology is a stimulus that elicits negative emotions or feelings of distress. In general, triggers include a broad range of stimuli, such as smells, tastes, sounds, textures, or sights, which may relate to possibly distressing acts or events of whatever type, such as violence, trauma, death, eating disorders, or obscenity. In order to proactively apprise the audience that a piece of media (writing, audio, video, etc.) contains potentially distressing material, the use of "trigger warnings" have become common. Trigger warnings are labels that indicate which type of triggering content is present. They are frequently used in online communities and in institutionalized education and allow a sensitive audience to prepare for the content to better manage their reactions. In the planned series of shared tasks on triggers, we propose a computational problem of identifying whether or not a given document contains triggering content, and if so, of what type.

Identifying Violent Content at PAN'23

In the pilot edition of the task at PAN'23, we will focus on a single trigger type: violence. As data we will use a corpus of fanfiction (millions of stories crawled from fanfiction.net and archiveofourown.org (Ao3)) in which trigger warnings have been assigned by the authors, that is, we do not define "violence" as a construct ourselves here, but rather rely on user-generated labels. We unify the set of label names where necessary and create a balanced corpus of positive and negative examples. The problem is formulated as binary classification at the document level as follows: Given a piece of fanfiction discourse, classify it as triggering or not triggering, that is, in the PAN'23 edition of the task, assign the trigger warning "violence" if appropriate. Standard measures of classifier quality will be used for evaluation.

Acknowledgments. The work from Symanto Research has been partially funded by the Pro²Haters - Proactive Profiling of Hate Speech Spreaders (CDTi IDI-20210776), the XAI-DisInfodemics: eXplainable AI for disinformation and conspiracy detection during infodemics (MICIN PLEC2021-007681), and the ANDHI - ANomalous Diffusion of Harmful Information (CPP2021-008994) R&D grants.

The work of Paolo Rosso was in the framework of the FairTransNLP research project (PID2021-1243610B-C31).

References

1. Bevendorff, J., et al.: Overview of PAN 2021: authorship verification, profiling hate speech spreaders on twitter, and style change detection. In: Experimental IR Meets Multilinguality, Multimodality, and Interaction - 12th International Conference of the CLEF Association, vol. 12880, pp. 419–431 (2021)
2. Bevendorff, J., et al.: Overview of PAN 2020: authorship verification, celebrity profiling, profiling fake news spreaders on twitter, and style change detection. In: Experimental IR Meets Multilinguality, Multimodality, and Interaction - 11th International Conference of the CLEF Association, vol. 12260, pp. 372–383 (2020)
3. Bobicev, V., Sokolova, M.: Inter-annotator agreement in sentiment analysis: machine learning perspective. In: Proceedings of the International Conference Recent Advances in Natural Language Processing (2017)
4. Chinea-Rios, M., Müller, T., Sarracén, G.L.D.l.P., Rangel, F., Franco-Salvador, M.: Zero and few-shot learning for author profiling. arXiv preprint arXiv:2204.10543 (2022)
5. Kestemont, M., et al.: Overview of the author identification task at PAN 2018: cross-domain authorship attribution and style change detection. In: CLEF 2018 Labs and Workshops, Notebook Papers (2018)
6. Koppel, M., Winter, Y.: Determining if two documents are written by the same author. J. Am. Soc. Inf. Sci. **65**(1), 178–187 (2014)
7. Mueller, T., Pérez-Torró, G., Franco-Salvador, M.: Few-shot learning with siamese networks and label tuning. In: Proceedings of the 60th Annual Meeting of the Association for Computational Linguistics (Volume 1: Long Papers), pp. 8532–8545 (2022)
8. Müller, T., Pérez-Torró, G., Basile, A., Franco-Salvador, M.: Active few-shot learning with FASL. arXiv preprint arXiv:2204.09347 (2022)
9. Potthast, M., Gollub, T., Wiegmann, M., Stein, B.: TIRA integrated research architecture. In: Ferro, N., Peters, C. (eds.) Information Retrieval Evaluation in a Changing World. TIRS, vol. 41, pp. 123–160. Springer, Cham (2019). https://doi.org/10.1007/978-3-030-22948-1_5
10. Rangel, F., De-La-Peña-Sarracén, G.L., Chulvi, B., Fersini, E., Rosso, P.: Profiling hate speech spreaders on Twitter task at PAN 2021. In: CLEF 2021 Labs and Workshops, Notebook Papers (2021)
11. Rangel, F., Giachanou, A., Ghanem, B., Rosso, P.: Overview of the 8th author profiling task at PAN 2019: profiling fake news spreaders on Twitter. In: CLEF 2020 Labs and Workshops, Notebook Papers. CEUR Workshop Proceedings (2020)
12. Rangel, F., Rosso, P.: On the implications of the general data protection regulation on the organisation of evaluation tasks. Lang. Law/Linguagem e Direito **5**(2), 95–117 (2019)
13. Rangel, F., Rosso, P.: Overview of the 7th author profiling task at pan 2019: bots and gender profiling. In: CLEF 2019 Labs and Workshops, Notebook Papers (2019)
14. Rangel, F., et al.: Overview of the 2nd author profiling task at PAN 2014. In: CLEF 2014 Labs and Workshops, Notebook Papers (2014)
15. Rangel, F., Rosso, P., Montes-y-Gómez, M., Potthast, M., Stein, B.: Overview of the 6th author profiling task at PAN 2018: multimodal gender identification in Twitter. In: CLEF 2019 Labs and Workshops, Notebook Papers (2018)

16. Rangel, F., Rosso, P., Moshe Koppel, M., Stamatatos, E., Inches, G.: Overview of the author profiling task at PAN 2013. In: CLEF 2013 Labs and Workshops, Notebook Papers (2013)

17. Rangel, F., Rosso, P., Potthast, M., Stein, B.: Overview of the 5th author profiling task at PAN 2017: gender and language variety identification in Twitter. Working Notes Papers of the CLEF (2017)

18. Rangel, F., Rosso, P., Potthast, M., Stein, B., Daelemans, W.: Overview of the 3rd author profiling task at PAN 2015. In: CLEF 2015 Labs and Workshops, Notebook Papers (2015)

19. Rangel, F., Rosso, P., Verhoeven, B., Daelemans, W., Potthast, M., Stein, B.: Overview of the 4th author profiling task at PAN 2016: Cross-genre evaluations. In: CLEF 2016 Labs and Workshops, Notebook Papers (2016). ISSN 1613-0073

20. Reynier, O.B., Berta, C., Francisco, R., Paolo, R., Elisabetta, F.: Profiling irony and stereotype spreaders on twitter (IROSTEREO) at pan 2022. In: CLEF 2021 Labs and Workshops, Notebook Papers (2022)

21. Rosso, P., Rangel, F., Potthast, M., Stamatatos, E., Tschuggnall, M., Stein, B.: Overview of PAN'16–new challenges for authorship analysis: cross-genre profiling, clustering, diarization, and obfuscation. In: Experimental IR Meets Multilinguality, Multimodality, and Interaction. 7th International Conference of the CLEF Initiative (CLEF 2016) (2016)

22. Sawhney, R., Agarwal, S., Mittal, V., Rosso, P., Nanda, V., Chava, S.: Cryptocurrency bubble detection: a new stock market dataset, financial task & hyperbolic models. In: Proceedings of the 2022 Conference of the North American Chapter of the Association for Computational Linguistics: Human Language Technologies, pp. 5531–5545 (2022)

23. Stamatatos, E., et al.: Overview of the authorship verification task at pan 2022. In: Faggioli, G., Ferro, N., Hanbury, A., Potthast, M. (eds.) CLEF 2022 Labs and Workshops, Notebook Papers. CEUR-WS.org (2022)

24. Stamatatos, E., Potthast, M., Pardo, F.M.R., Rosso, P., Stein, B.: Overview of the PAN/CLEF 2015 evaluation lab. In: Experimental IR Meets Multilinguality, Multimodality, and Interaction, vol. 9283, pp. 518–538 (2015)

25. Troiano, E., Padó, S., Klinger, R.: Emotion ratings: how intensity, annotation confidence and agreements are entangled. arXiv preprint arXiv:2103.01667 (2021)

26. Tschuggnall, M., et al.: Overview of the author identification task at PAN 2017: style breach detection and author clustering. In: CLEF 2017 Labs and Workshops, Notebook Papers (2017)

27. Wang, Y., Yao, Q., Kwok, J.T., Ni, L.M.: Generalizing from a few examples: a survey on few-shot learning. ACM Comput. Surv. (CSUR) 53, 1–34 (2020)

28. Weiss, K., Khoshgoftaar, T.M., Wang, D.D.: A survey of transfer learning. J. Big Data 3(1), 1–40 (2016). https://doi.org/10.1186/s40537-016-0043-6

29. Zangerle, E., Mayerl, M., Potthast, M., Stein, B.: Overview of the style change detection task at PAN 2021. In: Faggioli, G., Ferro, N., Joly, A., Maistro, M., Piroi, F. (eds.) CLEF 2021 Labs and Workshops, Notebook Papers. CEUR-WS.org (2021)

30. Zangerle, E., Mayerl, M., Potthast, M., Stein, B.: Overview of the style change detection task at PAN 2022. In: Faggioli, G., Ferro, N., Hanbury, A., Potthast, M. (eds.) CLEF 2022 Labs and Workshops, Notebook Papers. CEUR-WS.org (2022)
31. Zangerle, E., Mayerl, M., Specht, G., Potthast, M., Stein, B.: Overview of the style change detection task at PAN 2020. In: CLEF 2020 Labs and Workshops, Notebook Papers (2020)
32. Zangerle, E., Tschuggnall, M., Specht, G., Stein, B., Potthast, M.: Overview of the style change detection task at PAN 2019. In: CLEF 2019 Labs and Workshops, Notebook Papers (2019)

Overview of Touché 2023:
Argument and Causal Retrieval
Extended Abstract

Alexander Bondarenko[1]([✉]), Maik Fröbe[1], Johannes Kiesel[2],
Ferdinand Schlatt[3], Valentin Barriere[4], Brian Ravenet[5], Léo Hemamou[6],
Simon Luck[7], Jan Heinrich Reimer[3], Benno Stein[2], Martin Potthast[8],
and Matthias Hagen[1]

[1] Friedrich-Schiller-Universität Jena, Jena, Germany
alexander.bondarenko@uni-jena.de
[2] Bauhaus-Universität Weimar, Weimar, Germany
[3] Martin-Luther-Universität Halle-Wittenberg, Halle, Germany
[4] Centro Nacional de Inteligencia Artificial (CENIA), Macul, Chile
[5] Université Paris-Saclay, Gif-sur-Yvette, France
[6] Sanofi R&D France, Paris, France
[7] Alma Mater Studiorum - Università di Bologna, Bologna, Italy
[8] Leipzig University and ScaDS.AI, Leipzig, Germany
touche@webis.de, https://touche.webis.de/

Abstract. The goal of Touché is to foster and support the development of technologies for argument and causal retrieval and analysis. For the fourth time, we organize the Touché lab featuring four shared tasks: (a) argument retrieval for controversial topics, where participants retrieve web documents that contain high-quality argumentation and detect the argument stance, (b) causal retrieval, where participants retrieve documents that contain causal statements from a generic web crawl and detect the causal stance, (c) image retrieval for arguments, where participants retrieve images showing support or opposition to some stance from a focused web crawl, and (d) intra-multilingual multi-target stance classification, where participants detect the stance of comments on proposals from the multilingual participatory democracy platform CoFE. In this paper, we briefly summarize the results of Touché 2022 and describe the planned setup for the fourth lab edition at CLEF 2023.

1 Introduction

Making informed decisions and forming opinions on a matter often involves not only weighing pro and con arguments towards different options but also considering cause-effect relationships for one's actions [1]. Nowadays, everybody has the chance to acquire knowledge and find any kind of information on the Web on almost any topic for these tasks. However, conventional search engines are primarily optimized for returning *relevant* results and do not address the

---

L. Hemamou—Independent view, not influenced by Sanofi R&D France.

deeper analysis of arguments (e.g., argument quality and stance), or analysis of causal relationships. To close this gap, with the Touché lab's four shared tasks,[1] we intend to solicit the research community to develop respective approaches. In 2023, we organize the four following shared tasks:

1. Argumentative document retrieval from a generic web crawl to provide an overview of arguments and opinions on controversial topics.
2. Retrieval of web documents from a generic web crawl to understand whether a causal relationship between two events/actions exists (*new task*).
3. Image retrieval to corroborate and strengthen textual arguments and to provide a quick overview of public opinions on controversial topics.
4. Stance classification of comments on proposals from the multilingual participatory democracy platform CoFE,[2] written in different languages to support opinion formation on socially important topics (*new task*).

After having organized three successful Touché labs on argument retrieval at CLEF 2020–2022 [5–7], we propose a fourth lab edition to bring together researchers from the fields of information retrieval, natural language processing, and computational linguistics working on argumentation and causality. During the previous Touché labs, we received more than 210 runs from 64 participating teams. We manually labeled the relevance and argument quality of more than 27,000 argumentative texts, web documents, and images for 200 search topics; the topics and judgments are publicly available at https://touche.webis.de.

The previous three labs explored different granularities of argument retrieval and analysis: debates on various topics crawled from several online debating portals and their gist, complete web documents, and text passages; in the current lab iteration, we plan to investigate argument retrieval from the large web crawl corpus ClueWeb22-B [13] and stance detection of web documents and human-written comments in different languages. With the new task on evidence retrieval for causal questions, we aim for exploring effective approaches to retrieve web documents relevant to causality-related information needs and to analyze if a document supports or refutes the causal relationship specified in the question. Additionally, by repeating the task on image retrieval for arguments, we intend to collect new ideas that improve over the achieved results and to expand the test collection with additional manual judgments. Thus, we plan to investigate different retrieval modalities: text and images. As in the previous Touché editions, we will encourage participants to deploy their software in our cloud-based evaluation-as-a-service platform TIRA [14] for better reproducibility.

2   Task Definition

The first three Touché 2023 lab's shared tasks follow the classic TREC-style methodology: documents and search topics are provided to the participants who

---

[1] 'touché' is commonly "used to acknowledge a hit in fencing or the success or appropriateness of an argument, an accusation, or a witty point." [https://merriam-webster.com/dictionary/touche].

[2] https://futureu.europa.eu.

submit their ranked results (up to five runs) to be judged by human assessors. For the fourth task, the participants will submit the results with a predicted stance for respective data entries. The fourth Touché lab's edition will include the four shared tasks that are outlined below in detail.

Task 1: Argument Retrieval for Controversial Questions. Given a controversial topic and a collection of web documents, the task is to retrieve and rank documents by relevance to the topic, by argument quality, and to detect the document's stance. Participants of Task 1 will retrieve documents from the ClueWeb22-B crawl for 50 search topics. Our human assessors will label the ranked results both for their general topical relevance and for the rhetorical argument quality [16], i.e., "well-writtennes": (1) whether the document contains arguments and whether the argument text has a good style of speech, (2) whether the argument text has a proper sentence structure and is easy to follow, (3) whether it includes profanity, has typos, etc. Optionally, participants will detect the documents' stance: pro, con, neutral, or no stance.

Analogously to the previous Touché editions, our volunteer assessors will annotate the document's topical relevance with three levels: 0 (not relevant), 1 (relevant), and 2 (highly relevant). The argument quality will also be labeled with three labels: 0 (low quality, or no arguments in the document), 1 (average quality), and 2 (high quality). The annotators will be provided with detailed annotation guidelines, including examples, and will participate in a training phase with an initial kappa test and a follow-up discussion to clarify potential misinterpretations. Afterwards, each annotator will independently judge the results for disjoint subsets of the topics (i.e., each topic will be judged by one annotator only). We use this annotation policy due to a high annotation workload.

To lower the entry barrier for participants who cannot index the whole ClueWeb22-B corpus on their side, we provide a first-stage retrieval possibility via the API of the search engine ChatNoir [4] and a smaller version of the corpus that contains one million documents per topic. Additionally, participants are provided with a number of previously compiled resources that include the document-level relevance and quality judgments from the previous Touché editions.[3] For the identification of claims and premises in documents, participants can use any existing argument tagging tool such as the TARGER API [9] hosted on our own servers or develop their own tools if necessary. We will use nDCG@k[4] to evaluate rankings and accuracy to evaluate stance detection.

Topics. For the tasks on controversial questions (Task 1) and image retrieval (Task 3), we provide 50 search topics that represent various debated societal matters. The topics were chosen from the online debate portals (debatewise.org, idebate.org, debatepedia.org, and debate.org) having the largest number of user-generated comments, and thus representing the matters of the highest societal

---

[3] https://webis.de/data.html#touche-corpora.
[4] The value of k will depend on the number of result submissions and, thus, the annotation workload (nDCG@5 was used in the previous Touché editions).

interest. Each of these topics has a *title* (i.e., a question on a controversial issue), a *description* specifying the particular search scenario, and a *narrative* that serves as a guideline for the human assessors. The example topic is shown below:

```
<title> Should teachers get tenure? </title>
<description> A user has heard that some countries do give teachers
tenure and others don't. Interested in the reasoning for or against
tenure, the user searches for arguments [...] </description>
<narrative> Highly relevant statements clearly focus on tenure for
teachers in schools or universities. Relevant statements consider tenure
more generally, not specifically for teachers, or [...] </narrative>
```

Task 2: Evidence Retrieval for Causal Questions. Given a causality-related topic and a collection of web documents, the task is to retrieve and rank documents by relevance to the topic. For 50 search topics, participants of Task 2 will retrieve documents from the ClueWeb22-B crawl that contain relevant causal evidence. Optionally, participants will detect the document's *causal* stance. A document can provide supportive evidence (a causal relationship between the cause and effect from the topic holds), refutative (a causal relationship does not hold), neutral (in some cases holds and in some does not), or no evidence is entailed.

Our volunteer assessors will label the topical relevance documents according to three relevance levels: 0 (not relevant), 1 (relevant), and 2 (highly relevant). The direction of causality will be considered, e.g., a document stating that B causes A will be considered as off-topic (not relevant) for the topic 'Does A cause B?'. The document's stance will also be labeled to evaluate the optional stance detection task. In general, the labeling procedure will be analogous to Task 1, where volunteer assessors will participate in training and a discussion.

Like in Task 1, ChatNoir [4] can also be used for first-stage retrieval, and we will provide a smaller version of the corpus that contains one million documents per topic. Participants are free to use any additional existing tools or datasets and are encouraged to develop their own.

Topics. The 50 search topics for Task 2 describe scenarios, when users search for confirmation of whether some causal relationship holds, e.g., to know the possible reason for a current physical condition. Each of these topics has a *title* (i.e., a causal question), *cause* and *effect* entities, a *description* specifying the particular search scenario, and a *narrative* serving as a guideline for the assessors. The topics were manually selected from a corpus of causal questions [8] and a graph of causal statements [10] such that they span a diverse set of domains:

```
<title>Can eating broccoli lead to constipation?</title>
<cause>broccoli</cause>
<effect>constipation</effect>
<description> A young parent has a child experiencing constipation
after eating some broccoli for dinner and is wondering whether broccoli
could cause constipation [...] </description>
<narrative> Relevant documents will discuss if broccoli and other
high-fiber foods can cause or ease constipation [...] </narrative>
```

Task 3: Image Retrieval for Arguments. Given a controversial topic and a collection of web documents with images, the task is to retrieve for each stance (pro and con) images that show support for that stance. Participants of Task 3 should retrieve and rank images, possibly utilizing the corresponding web documents, from a focused crawl of 30,000 images and for a given set of 50 search topics (the same as in Task 1) [12]. Like in the last edition of this task, the focus is on providing users with an overview of public opinions on controversial topics, for which we envision a system that provides not only textual but also visual support for each stance in the form of images. Participants are able to use the approximately 6,000 relevance judgments from the last edition of the task for training supervised approaches [11].[5] Similar to the other tasks, participants are free to use any additional existing tools and datasets or develop their own.

Although rank-based metrics for single image grids exist [17], none have been proposed so far for a 'pro-con' layout. Therefore, like the last year, participants' submitted results will be evaluated by the simple ratio of relevant images among 20 retrieved images, namely 10 images for each stance (precision@10). We will again use three increasingly strict definitions of relevance, corresponding to three precision@10 evaluation measures: being on-topic, being in support of some stance (i.e., an image is "argumentative"), and being in support of the stance for which the image was retrieved.

Task 4: Intra-Multilingual Multi-Target Stance Classification. Given a proposal on a socially important issue, its title, and topic, the task is to classify whether a comment on the proposal is *in favor*, *against*, or *other* towards the proposal. The data used for the evaluation of the participants' approaches comes from the CoFE participatory democracy platform; the respective dataset was created by Barriere et al. [3] and contains about 4,000 proposals and 20,000 comments written in 26 languages. The participants will have to classify into 3 classes multilingual comments from 6 different languages.[6] We also provide an automatic English translation of the proposals and titles that are written in any of the 24 official EU languages (plus Catalan and Esperanto) since a proposal can be written in one language and its corresponding comment in another.

For training their classifiers, the participants are provided with three datasets from the same debating platform: (1) CF_{E-D}: a small set of comments annotated with three stance labels, (2) CF_S: a larger set of comments that are self-annotated in a binary way (*in favor* or *against* only), and (2) CF_U: a large set of unlabeled comments. Since the class *other* cannot be put on the same scale as *in favor* or *against*, we will use a non-ordinal metric for evaluation widely used for the evaluation of stance classifiers. To account for the class imbalance we will evaluate submitted approaches using the macro-averaged F1-score.

Within the task, we organize two subtasks: (1) *Cross-debate Classification*: the participants should not use comments from debates that are in the test set and (2) *All-data-available Classification*: the participants can use all the available

---

[5] https://webis.de/data.html#touche-corpora.
[6] German, English, Greek, French, Italian, and Hungarian.

data. Also, the participants can use any additional existing tools and datasets, e.g., the datasets that contain stance annotations created by Barriere et al. [2] and by Vamvas and Sennrich [15] for any of the subtasks.

Proposals. For Task 4, the participants are given a proposal and its title as well as the comment to classify its stance that are exemplified below:[7]

```
<title> Set up a program for returnable food packaging made from
recyclable materials </title>
    <proposal> The European Union could set up a program for returnable
food packaging made from recyclable materials (e.g. stainless steel,
glass). These packaging would be produced on the basis of open standards
and cleaned according to [...] </proposal>
    <comment> Ja, wir müssen den Verpackungsmül reduzieren. </comment>
    <label> In favor </label>
```

3   Touché at CLEF 2022: Brief Overview

For the Touché 2022 lab, we received 58 registrations, from which 23 teams actively participated in the tasks and submitted 84 results (runs; every team could submit up to 5 runs). Our evaluation of the submitted results showed that the most effective approaches to argument retrieval all share common characteristics. For instance, most use various strategies for query reformulation and expansion, such as using synonyms, relevance feedback, or generating new queries from scratch with pre-trained language models. For Task 1 (argument gist retrieval), the most challenging was identifying that a pair of sentences (premise and claim) is coherent. An interesting observation is that re-ranking first-stage retrieval results based on a quality assessment of arguments almost always improves the retrieval effectiveness. Specifically for Task 2 (comparative questions), re-ranking based on important terms such as comparison objects and aspects or argument units in documents (premises and claims) was successful. In Task 2, stance detection was a new subtask; and some participants included a re-ranking step based on the predicted stance in their retrieval pipelines, which had some promising effects on the overall retrieval effectiveness. However, the overall still rather low effectiveness of the stance detection approaches leaves room for future improvements. For Task 3 (image retrieval), the recognition of sentiment and emotion and the use of optical character recognition to analyze the text in images were particularly helpful. Stance detection for images was also very challenging. We also provide an online web service to visually explore the submitted runs to Task 3 (cf. Fig. 1). For more details about the Touché 2022 lab, refer to the overview paper [6]. We expect that the relevance and argument quality judgments and stance labels collected at the Touché 2022 lab will help participating teams achieve higher effectiveness in the new lab iteration.

---

[7] Example from https://futureu.europa.eu/en/processes/GreenDeal/f/1/proposals/ 83.

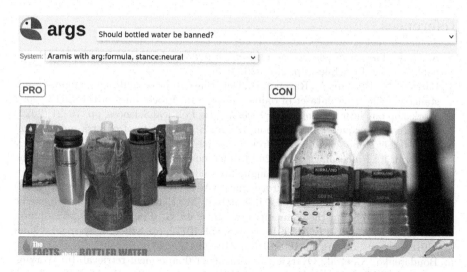

Fig. 1. Screenshot of the Task 3 run browser (see the task page at https://touche.webis.de).

4 Conclusion

At Touché, we continue our activities aimed for fostering research in argument and causal retrieval and analysis, building respective test collections, and bringing the research community together. During the previous three years of organizing Touché, we have observed the development of the participants' submitted approaches from sparse to dense retrieval to the deployment of zero-shot models combined with extensive approaches to assess document "argumentativeness," argument quality, stance detection, and sentiment analysis in images.

With the new Touché lab, we plan to investigate how argument retrieval and argument analysis approaches can be applied to a large collection of web documents and to better understand the evoked challenges. By repeating the image retrieval task, we expect to collect more ideas for understanding how argument analysis techniques, historically developed for text, can be used for visual argument representation. Moreover, with the two new shared tasks, we want to explore web document retrieval and analysis for causality-related information needs and multilingual multi-target stance classification.

Acknowledgments. This work has been partially supported by the Deutsche Forschungsgemeinschaft (DFG) in the project "ACQuA 2.0: Answering Comparative Questions with Arguments" (project 376430233) as part of the priority program "RATIO: Robust Argumentation Machines" (SPP 1999). V. Barriere's work was funded by the National Center for Artificial Intelligence CENIA FB210017, Basal ANID.

References

1. Ajzen, I.: The social psychology of decision making. In: Social Psychology: Handbook of Basic Principles, pp. 297–325 (1996)
2. Barriere, V., Balahur, A., Ravenet, B.: Debating europe: a multilingual multi-target stance classification dataset of online debates. In: Proceedings of the LREC 2022 Workshop on Natural Language Processing for Political Sciences, pp. 16–21 (2022)
3. Barriere, V., Jacquet, G., Hemamou, L.: CoFE: a new dataset of intra-multilingual multi-target stance classification from an online European participatory democracy platform. In: Proceedings of the 2nd Conference of the Asia-Pacific Chapter of the Association for Computational Linguistics and the 12th International Joint Conference on Natural Language Processing (AACL-IJCNLP), Taipei, Taiwan (2022)
4. Bevendorff, J., Stein, B., Hagen, M., Potthast, M.: Elastic ChatNoir: search engine for the ClueWeb and the common crawl. In: Pasi, G., Piwowarski, B., Azzopardi, L., Hanbury, A. (eds.) ECIR 2018. LNCS, vol. 10772, pp. 820–824. Springer, Cham (2018). https://doi.org/10.1007/978-3-319-76941-7_83
5. Bondarenko, A., et al.: Overview of Touché 2020: argument retrieval. In: Working Notes of CLEF 2020 - Conference and Labs of the Evaluation Forum, Thessaloniki, Greece, 22–25 September 2020, CEUR Workshop Proceedings, vol. 2696. CEUR-WS.org (2020). http://ceur-ws.org/Vol-2696/paper_261.pdf
6. Bondarenko, A., et al.: Overview of Touché 2022: argument retrieval. In: Proceedings of the Working Notes of CLEF 2022 - Conference and Labs of the Evaluation Forum, CEUR Workshop Proceedings, vol. 3180, pp. 2867–2903. CEUR-WS.org (2022). http://ceur-ws.org/Vol-3180/paper-247.pdf
7. Bondarenko, A., et al.: Overview of Touché 2021: argument retrieval. In: Proceedings of the Working Notes of CLEF 2021 - Conference and Labs of the Evaluation Forum, CEUR Workshop Proceedings, vol. 2936, pp. 2258–2284. CEUR-WS.org (2021). http://ceur-ws.org/Vol-2936/paper-205.pdf
8. Bondarenko, A., et al.: CausalQA: a benchmark for causal question answering. In: Calzolari, N., et al. (eds.) 29th International Conference on Computational Linguistics (COLING 2022), pp. 3296–3308. International Committee on Computational Linguistics (2022). https://aclanthology.org/2022.coling-1.291
9. Chernodub, A., et al.: TARGER: neural argument mining at your fingertips. In: Proceedings of the 57th Annual Meeting of the Association for Computational Linguistics, ACL 2019, pp. 195–200. ACL (2019). https://doi.org/10.18653/v1/p19-3031
10. Heindorf, S., Scholten, Y., Wachsmuth, H., Ngonga Ngomo, A.C., Potthast, M.: CauseNet: towards a causality graph extracted from the web. In: d'Aquin, M., et al. (eds.) 29th ACM International Conference on Information and Knowledge Management (CIKM 2020), pp. 3023–3030. ACM (2020). https://doi.org/10.1145/3340531.3412763
11. Kiesel, J., Potthast, M., Stein, B.: Dataset Touché22-Image-Retrieval-for-Arguments (2022). https://doi.org/10.5281/zenodo.6786948
12. Kiesel, J., Potthast, M., Stein, B.: Dataset Touché23-Image-Retrieval-for-Arguments (2023). https://doi.org/10.5281/zenodo.7497994
13. Overwijk, A., Xiong, C., Callan, J.: ClueWeb22: 10 billion web documents with rich information. In: Proceedings of the 45th International ACM SIGIR Conference on Research and Development in Information Retrieval, SIGIR 2022, pp. 3360–3362. ACM (2022). https://doi.org/10.1145/3477495.3536321

14. Potthast, M., Gollub, T., Wiegmann, M., Stein, B.: TIRA integrated research architecture. In: Ferro, N., Peters, C. (eds.) Information Retrieval Evaluation in a Changing World. TIRS, vol. 41, pp. 123–160. Springer, Cham (2019). https://doi.org/10.1007/978-3-030-22948-1_5

15. Vamvas, J., Sennrich, R.: X-stance: a multilingual multi-target dataset for stance detection. In: Proceedings of the 5th Swiss Text Analytics Conference and the 16th Conference on Natural Language Processing, SwissText/KONVENS 2020. CEUR-WS.org (2020). http://ceur-ws.org/Vol-2624/paper9.pdf

16. Wachsmuth, H., et al.: Computational argumentation quality assessment in natural language. In: Proceedings of the 15th Conference of the European Chapter of the Association for Computational Linguistics (EACL), pp. 176–187. Association for Computational Linguistics (2017). https://doi.org/10.18653/v1/e17-1017

17. Xie, X., et al.: Grid-based evaluation metrics for web image search. In: Liu, L., et al. (eds.) Proceedings of the 28th International World Wide Web Conference, WWW 2019, pp. 2103–2114. ACM (2019). https://doi.org/10.1145/3308558.3313514

CLEF 2023 SimpleText Track
What Happens if General Users Search Scientific Texts?

Liana Ermakova[1]([⊠]) [iD], Eric SanJuan[2] [iD], Stéphane Huet[2] [iD],
Olivier Augereau[3] [iD], Hosein Azarbonyad[4] [iD], and Jaap Kamps[5] [iD]

[1] Université de Bretagne Occidentale, HCTI, Brest, France
liana.ermakova@univ-brest.fr
[2] Avignon Université, LIA, Avignon, France
[3] ENIB, Lab-STICC UMR CNRS 6285, Brest, France
[4] Elsevier, Amsterdam, The Netherlands
[5] University of Amsterdam, Amsterdam, The Netherlands
https://simpletext-project.com

Abstract. The general public tends to avoid reliable sources such as scientific literature due to their complex language and lacking background knowledge. Instead, they rely on shallow and derived sources on the web and in social media – often published for commercial or political incentives, rather than the informational value. Can text simplification help to remove some of these access barriers? This paper presents the CLEF 2023 SimpleText track tackling technical and evaluation challenges of scientific information access for a general audience. We provide appropriate reusable data and benchmarks for scientific text simplification, and promote novel research to reduce barriers in understanding complex texts. Our overall use-case is to create a simplified summary of multiple scientific documents based on a popular science query which provides a user with an accessible overview on this specific topic. The track has the following three concrete tasks. Task 1 (*What is in, or out?*): selecting passages to include in a simplified summary. Task 2 (*What is unclear?*): difficult concept identification and explanation. Task 3 (*Rewrite this!*): text simplification - rewriting scientific text. The three tasks together form a pipeline of a scientific text simplification system.

Keywords: Scientific text simplification · (Multi-document) summarization · Terminology extraction · Keyword extraction · Contextualization · Background knowledge · Scientific information distortion · Information retrieval

1 Introduction

Scientific texts such as research publications are difficult to understand for the general public, or even for scientists outside the exact specialism. The CLEF 2023 SimpleText track is unique in its focus on text simplification for scientific texts, and its general public use-case naturally combining information retrieval and natural language processing aspects.

J. Kamps et al. (Eds.): ECIR 2023, LNCS 13982, pp. 536–545, 2023.
https://doi.org/10.1007/978-3-031-28241-6_62

Text complexity or reading levels and text simplification in general have been studied for long in linguistics, education science, and natural language processing. Simplified texts are more accessible for non-native speakers [28], young readers, people with reading disabilities [5,13,22] or needed for reading assistance (e.g. congenitally deaf people) [15] or lower level of education. Improving text comprehensibility remains a challenge as it is difficult to define a desirable output of simplification [14]. Traditional readability scores are limited to word or sentence length, while vocabulary overlap based metrics do not consider information distortion. Recently, text simplification is gaining interest. The workshop on Scholarly Document Processing[1] is targeting an NLP audience [4]. They hosted tasks on scientific document summarization including a Lay Summary task. At EMNLP 2022, the TSAR (Text Simplification, Accessibility, and Readability)[2] hosted a lexical simplification task, and TermEval 2020 ran a shared task on automatic term extraction [23]. In contrast to that, SimpleText is not limited to lexical and grammatical simplification.

SimpleText aims to improve information access to scientific knowledge for a general audience, by providing appropriate reusable data and benchmarks for text simplification, promoting novel research and tools to reduce barriers in understanding complex texts. In contrast to the previous work, we focus on (1) information selection which is suitable for a general public; (2) searching for difficult concepts, including words, abbreviations, etc. that need to be explained and can not be discard; and (3) evaluation of information distortion which might occur during the simplification process.

The track's setup is based on the following pipeline: (1) select the information to be included in a simplified summary; (2) decide whether the selected information is sufficient and comprehensible or provide some background knowledge if not; (3) improve the readability of the text [7]. This results in the following three tasks [11]:

- **Task 1: What is in, or out?** *Selecting passages to include in a simplified summary.*
- **Task 2: What is unclear?** *Difficult concept identification and explanation.*
- **Task 3: Rewrite this!** *Text simplification - rewriting scientific text.*

We also welcome any submission that uses our data in other ways as a fourth open task.

In the rest of this paper, we will first reflect on the CLEF 2022 edition of the track in Sect. 2, and then provide a detailed description of each task in Sect. 3.

---

[1] https://sdproc.org/2022/sharedtasks.html.
[2] https://taln.upf.edu/pages/tsar2022-st/.

2   Results and Lessons Learnt from SimpleText'22

In the first year of running SimpleText as a track at CLEF 2022, its counted a total of 62 registered teams [11]. A total of 40 teams downloaded data from the server. A total of 9 distinct teams submitted 24 runs, of which 10 runs were updated. For Task 1 (selecting passages/abstracts to include) [26], 6 runs were submitted. For Task 2 (identifying difficult terms) [9], we received 4 runs. For Task 3 (rewriting text) [10], a total of 14 submissions was made. We have seen several post-submission experiments, and with all 2022 data available, the track is expected to gain in participation in 2023.

For Task 1, we saw a clear difference in the reading level of journalistic and scientific articles. The 2022 topical relevance qrels provide a unique resource that can be reused and enriched with additional judgments. In 2023, we will extend relevance judgment with supplementary labels on text complexity and credibility of the source publication, based on large-scale automatic and small-scale manual judgments further enriching the 2022 qrels. As the recall base is small, we have to expand the test collection by increasing pooling depth and adding new subtopics and queries for the same set of popular science articles.

In the 2022 edition, the Task 2 was limited to difficult term spotting. However, several runs for Task 3 inserted some context or definition for difficult terms in additional to language simplification [11,25]. This shows a demand for a corpus with explanations of difficult terms integrated in a text. Thus, we will update Task 2 to provide further context for difficult terms. The evaluation stage allowed to increase the annotated data for term difficulty spotting. We will reuse these data as a first stage of the annotation for the corpus in 2023 and provide additional evaluation data.

Multiple SimpleText participants applied T5-based text simplification models which previously demonstrated strong performance [27]. However, we observed that direct application of large pre-trained models often keeps sentences unchanged as in case of the runs of PortLinguE with 36% of unchanged sentences [11,17]. We also observed that large pre-trained models tend to insert unnecessary and even false information in the simplification due to their generative nature. This is a general problem of generative models attracting massive attention in AI, and studying further safeguard against over-generation and gratuitous insertions feels necessary. Thus, we will continue to provide human evaluation results with regard to the errors produced during simplification, which distinguishes SimpleText from existing benchmarks using only automatic text simplification evaluation metrics. Our general observation is that state of the art text simplification systems perform well, but far below human simplifications in terms of the length and the complexity of the resulting simplifications.

Our shared tasks are interconnected. The corpus is based on abstracts in response to a popular science request. While some complex terms do not need to be explained as they will be further removed at the language simplification step, others must be kept even if they are too complex in order to avoid severe information distortion. In this case additional context or explanations could be inserted, integrating Tasks 2 and 3. And the other way around, information about the text complexity and the amount of revisions can inform the ranking stage of Task 1. For example, we can promote abstracts with more favorable reading levels in the ranking, and ensure our user is guided to relevant and already accessible abstracts first [18,19].

3   SimpleText 2023 Tasks

We will keep the three tasks for the 2023 edition. We will reuse data constructed in previous editions with additional topics and additional automatic and manual labels. We will also emphasize automatic evaluation and training using the 2022 data.

3.1   Task 1: Selecting Passages to Include in a Simplified Summary

Given a popular science article targeted to a general audience, this task aims at retrieving passages, that can help to understand this article, from a large corpus of academic abstracts and bibliographic metadata. Relevant passages should relate to any of the topics in the source article.

Data. We use the popular science articles as a source for the types of topics the general public is interested in, and as a validation of the reading level that is suitable for them. The main corpus is a large set of scientific abstracts plus associated metadata covering the field of computer science and engineering. We reuse the collection of academic abstracts from the Citation Network Dataset (12th version released in 2020)[3] [29]. This collection was extracted from DBLP, ACM, MAG (Microsoft Academic Graph), and other sources. It contains: 4,894,083 bibliographic references published before 2020, 4,232,520 abstracts in English, 3,058,315 authors with their affiliations, and 45,565,790 ACM citations. We provide an ElasticSearch index to allow participants to retrieve passages or abstracts using BM25 [24]. Through a simple API, queries can be done on the textual content of abstracts together with authorship. Thus, the shared dataset provides: document abstract content for LDA (Latent Dirichlet Allocation) or Word Embedding (WE); document authors for coauthoring analysis; citation relationship between documents for co-citation analysis; citations by author for author impact factor analysis.

On the other hand, press articles, targeted to a general audience, are drawn from two sources: *The Guardian*, a major international newspaper for a general audience with a tech section, and *Tech Xplore*,[4] a web site taking part in the Science X Network to provide a comprehensive coverage of engineering and technology advances. Each of these popular science article represents a general topic that has to be analyzed to retrieve relevant scientific information from the corpus. We provide the URLs to original articles, the title and the textual content of each popular science article as a general topic. Each general topic was

---

[3] https://www.aminer.cn/citation.
[4] https://techxplore.com/.

also enriched with one or more specific keyword queries manually extracted from their content, creating a familiar information retrieval task ranking passages or abstracts in response to a query. In the last year's edition, 40 articles, 20 of each source, were made available [11]. We plan to expand it with 10 other topics used as a test set. The 2022 qrels cover many topics (31) and queries (67) but with a limited pooling depth. In 2023, we will increase the pooling depth with at least 50 judged documents per query.

Evaluation. Topical relevance was only evaluated last year with a 0-5 score on the relevance degree towards the content of the original article [11]. Whereas this large scale can measure how close the retrieved abstract to the topic, the title or the textual content is, other facets, yet important in the context of text simplification, were missing. In 2023, we will continue evaluating on topical relevance, but also on text complexity (using readability measures and comparison to manually attributed scores), and source authoritativeness (using academic impact measures). The provided test collection will be simplified to three scores on a 0-2 scale:

- **Topic relevance**: Not relevant (0), relevant (1), highly relevant (2);
- **Text complexity**: Easy (0), difficult (1), very difficult (2);
- **Source credibility**: Low (0), medium (1), high credibility (2).

While these criteria can provide different levels of comparison between systems, we will still compute a unique ranking score using NDCG (as well as other measures) based on the fusion of the various criteria [21].

3.2 Task 2: Difficult Concept Identification and Explanation for a General Audience

The goal of this task is to decide which concepts in scientific abstracts require explanation and contextualization in order to help a reader to understand the scientific text. Complex Word Identification (CWI) and Lexical Simplification (LS) are the most popular approaches to assess and reduce the complexity [6,16, 31]. In the context of a query, some key concepts need to be contextualized with a definition, example and/or use-case that are easier to understand for a reader. There is ongoing research on this by generating definitions with a controllable complexity [1].

In 2023, we ask participants to identify such concepts and to provide useful and understandable explanations for them. Thus, the task has two steps:

1. to retrieve up to 5 difficult terms in a given passage from a scientific abstract;
2. to provide an explanation of these difficult terms (e.g. definition, abbreviation deciphering, example etc.).

Data. The corpus of Task 2 is based on the sentences in high-ranked abstracts to the requests of Task 1. For the first step of the task, i.e. retrieving difficult terms, we will use the train data collected in 2022 [11]. As for the test data, we will provide additional passages coming from the DBLP abstracts as in Task 1. For the second step of the task we will provide additional training data for definition generation, extracted from a much larger corpus of full text articles. This training data contains pairs of $<sentence, concept>$ and a label per pair is provided. The binary label indicates whether the sentence provides a good definition for the concept or not. Samples in this dataset are extracted from books and articles published in ScienceDirect[5]. This dataset contains 43,368 samples distributed across 8 different domains and all pairs in this dataset are annotated by subject matter experts. There are a total of 9,870 positive samples (meaning that the sentence provides a good definition for the corresponding concept) and 33,498 negative samples. The average length of sentences in this dataset is 24.5 words. In addition to this dataset, participants are encouraged to use existing datasets extracted from other resources such as the WCL dataset [20] to train the definition generation model. Participants are also encourage to use gazetteers, wikification resources as well as resources for abbreviation deciphering.

Evaluation. As in 2022, we will evaluate complex concept spotting in terms of their complexity and the detected concept spans [11]. For the explanations of difficult terms, the evaluation set will contain 1,000 concepts and their definitions extracted by subject matter experts. We will automatically evaluate provided explanations by comparing them to references (e.g. ROUGE, cosine similarity etc.). We will provide manual evaluation the provided explanations in terms of their usefulness with regard to a query as well as their complexity for a general audience. Note that the provided explanations can have different forms, e.g. definition, abbreviation deciphering, examples, use cases etc.

3.3 Task 3: Text Simplification - Rewriting Scientific Text

The goal of this task is to provide a simplified version of sentences extracted from scientific abstracts. Participants will be provided with the popular science articles and queries and matching abstracts of scientific papers, split into individual sentences.

Data. Task 3 uses the same corpus based on the sentences in high-ranked abstracts to the requests of Task 1, supplemented with additional training data from the health domain. Our training data is a truly parallel corpus of directly simplified sentences (648 sentences for now) coming from scientific abstracts from the DBLP Citation Network Dataset for *Computer Science* and Google Scholar and PubMed articles on *Health and Medicine* [7,8,10,11]. These text passages were simplified either by master students in Technical Writing and Translation or by a domain expert (a computer scientist) and a professional translator (English native speaker) working together [8,10,11].

All the existing large corpora used post-hoc aligned sentences [2,3,30,32,33]. The SimpleText corpus [11] contains directly simplified sentences, and is not much smaller than existing high-quality corpora like NEWSELA [30] (2,259 sentences).

---

[5] https://www.sciencedirect.com/.

Our track is the first to focus on scientific text simplification rather than news articles. In 2023, we will expand the training and evaluation data.

Evaluation. In 2023, we will emphasize large-scale automatic evaluation measures (SARI, ROUGE, compression, readability) that provide a reusable test collection.

These will be supplemented with small-scale detailed human evaluation of other aspects, essential for deeper analysis. As in 2022, we evaluate the complexity of the provided simplifications in terms of vocabulary and syntax as well as the errors (Incorrect syntax; Unresolved anaphora due to simplification; Unnecessary repetition/iteration; Spelling, typographic or punctuation errors) [11]. Rather than focus only on this evaluation which is similar to easy-to-read guidelines suggested in previous research [34], we prefer to assess the results according to information distortion which can be bring during simplification process. We distinguish the following types of information distortion with corresponding severity level: Style (1); Insertion of unnecessary details with regard to a query (1); Redundancy (without lexical overlap) (2); Insertion of false or unsupported information (3); Omission of essential details with regard to a query (4); Overgeneralization (5); Oversimplification(5); Topic shift (5); Contra sense / contradiction (6); Ambiguity (6); Nonsense (7).

4   Conclusions

This paper described the setup of the CLEF 2023 SimpleText track, which contains three interconnected tasks on scientific text simplification. Within the SimpleText track, we have already released extensive corpora and manually labeled data:

- a large corpus of over 4 million scientific abstracts that can be used for popular science;
- scientific terms from sentences coming from scientific abstracts with manually attributed difficulty scores;
- a parallel corpus of manually simplified sentences from scientific literature;
- a parallel corpus of sentences with different types of information distortion and simplification level.

Please visit the SimpleText website (http://simpletext-project.com) for more details on the track.

Acknowledgment. This track would not have been possible without the great support of numerous individuals. We want thank in particular Silvia Araujo, Patrice Bellot, Julien Boccou, Pierre De Loor, Radia Hannachi, Helen McCombie, Diana Nurbakova, Irina Ovchinnikov, and Léa Talec; the students of the Université de Bretagne Occidentale; and all the 2022 track participants for their great help in discussing and shaping the track, and in creating all the evaluation data and training data for 2023. We also thank the MaDICS (https://www.madics.fr/ateliers/simpletext/) research group and the French National Research Agency (project *ANR-22-CE23-0019-01*).

References

1. August, T., Reinecke, K., Smith, N.A.: Generating scientific definitions with controllable complexity. In: Proceedings of the 60th Annual Meeting of the Association for Computational Linguistics (Volume 1: Long Papers), pp. 8298–8317 (2022)

2. Bott, S., Saggion, H.: An unsupervised alignment algorithm for text simplification corpus construction. In: Proceedings of the Workshop on Monolingual Text-To-Text Generation, pp. 20–26 (2011)

3. Cardon, R., Grabar, N.: French biomedical text simplification: when small and precise helps. In: Proceedings of the 28th International Conference on Computational Linguistics, Barcelona, Spain, pp. 710–716. International Committee on Computational Linguistics (2020). https://www.aclweb.org/anthology/2020.coling-main.62

4. Chandrasekaran, M.K., et al.: Overview of the first workshop on scholarly document processing (SDP). In: Proceedings of the First Workshop on Scholarly Document Processing, pp. 1–6. Association for Computational Linguistics (2020). https://doi.org/10.18653/v1/2020.sdp-1.1. https://aclanthology.org/2020.sdp-1.1/

5. Chen, P., Rochford, J., Kennedy, D.N., Djamasbi, S., Fay, P., Scott, W.: Automatic text simplification for people with intellectual disabilities. In: Artificial Intelligence Science and Technology, pp. 725–731. World Scientific (2016). https://www.worldscientific.com/doi/abs/10.1142/9789813206823_0091

6. Cruz, F., Coustaty, M., Augereau, O., Kise, K., Journet, N.: An interactive recommendation system for 2nd language vocabulary learning-vocabulometer 2.0. In: 2019 International Conference on Document Analysis and Recognition Workshops (ICDARW), vol. 3, pp. 28–32. IEEE (2019)

7. Ermakova, L., et al.: Overview of SimpleText 2021 - CLEF workshop on text simplification for scientific information access. In: Candan, K.S., et al. (eds.) CLEF 2021. LNCS, vol. 12880, pp. 432–449. Springer, Cham (2021). https://doi.org/10.1007/978-3-030-85251-1_27

8. Ermakova, L., et al.: Automatic simplification of scientific texts: SimpleText lab at CLEF-2022. In: Hagen, M., et al. (eds.) ECIR 2022. LNCS, vol. 13186, pp. 364–373. Springer, Cham (2022). https://doi.org/10.1007/978-3-030-99739-7_46

9. Ermakova, L., Ovchinnikova, I., Kamps, J., Nurbakova, D., Araújo, S., Hannachi, R.: Overview of the CLEF 2022 SimpleText task 2: complexity spotting in scientific abstracts. In: Faggioli et al. [12]

10. Ermakova, L., Ovchinnikova, I., Kamps, J., Nurbakova, D., Araújo, S., Hannachi, R.: Overview of the CLEF 2022 SimpleText task 3: query biased simplification of scientific texts. In: Faggioli et al. [12]

11. Ermakova, L., et al.: Overview of the CLEF 2022 SimpleText lab: automatic simplification of scientific texts. In: Barrón-Cedeño, A., et al. (eds.) CLEF 2022. LNCS, vol. 13390, pp. 470–494. Springer, Cham (2022). https://doi.org/10.1007/978-3-031-13643-6_28

12. Faggioli, G., Ferro, N., Hanbury, A., Potthast, M. (eds.): Proceedings of the Working Notes of CLEF 2022: Conference and Labs of the Evaluation Forum. CEUR Workshop Proceedings (2022)

13. Gala, N., Tack, A., Javourey-Drevet, L., François, T., Ziegler, J.C.: Alector: a parallel corpus of simplified French texts with alignments of misreadings by poor and dyslexic readers. In: Language Resources and Evaluation for Language Technologies (LREC) (2020)

14. Grabar, N., Saggion, H.: Evaluation of automatic text simplification: where are we now, where should we go from here. In: Actes de la 29e Conférence sur le Traitement Automatique des Langues Naturelles. Volume 1: conférence principale, pp. 453–463 (2022)

15. Inui, K., Fujita, A., Takahashi, T., Iida, R., Iwakura, T.: Text simplification for reading assistance: a project note. In: Proceedings of the Second International Workshop on Paraphrasing - Volume 16, PARAPHRASE 2003, pp. 9–16. ACL, USA (2003). https://doi.org/10.3115/1118984.1118986

16. Kochmar, E., Gooding, S., Shardlow, M.: Detecting multiword expression type helps lexical complexity assessment. In: LREC 2020: Proceedings of the 12th Conference on Language Resources and Evaluation (2020)

17. Monteiro, J., Aguiar, M., Araújo, S.: Using a pre-trained SimpleT5 model for text simplification in a limited corpus. In: Proceedings of the Working Notes of CLEF 2022 - Conference and Labs of the Evaluation Forum, Bologna, Italy, 5–8 September 2022, Bologna, Italy. CEUR Workshop Proceedings, CEUR-WS.org (2022)

18. Mostert, F., Sampatsing, A., Spronk, M., Kamps, J.: University of Amsterdam at the CLEF 2022 SimpleText track. In: Proceedings of the Working Notes of CLEF 2022 - Conference and Labs of the Evaluation Forum, Bologna, Italy, 5–8 September 2022, Bologna, Italy. CEUR Workshop Proceedings, CEUR-WS.org (2022)

19. Nakatani, M., Jatowt, A., Tanaka, K.: Easiest-first search: towards comprehension-based web search. In: Proceedings of the 18th ACM Conference on Information and Knowledge Management, pp. 2057–2060 (2009)

20. Navigli, R., Velardi, P.: Learning word-class lattices for definition and hypernym extraction. In: ACL, pp. 1318–1327 (2010)

21. Ravana, S.D., Moffat, A.: Score aggregation techniques in retrieval experimentation. In: Proceedings of the Twentieth Australasian Conference on Australasian Database, vol. 92, pp. 57–66 (2009)

22. Rello, L., Baeza-Yates, R., Bott, S., Saggion, H.: Simplify or help? Text simplification strategies for people with dyslexia. In: Proceedings of the 10th International Cross-Disciplinary Conference on Web Accessibility, pp. 1–10 (2013)

23. Rigouts Terryn, A., Hoste, V., Drouin, P., Lefever, E.: Termeval 2020: shared task on automatic term extraction using the annotated corpora for term extraction research (ACTER) dataset. In: 6th International Workshop on Computational Terminology (COMPUTERM 2020), pp. 85–94. European Language Resources Association (ELRA) (2020)

24. Robertson, S., Zaragoza, H., et al.: The probabilistic relevance framework: BM25 and beyond. Found. Trends® Inf. Retrieval 3(4), 333–389 (2009)

25. Rubio, A., Martínez, P.: HULAT-UC3M at SimpleText@CLEF-2022: scientific text simplification using BART. In: Proceedings of the Working Notes of CLEF 2022 - Conference and Labs of the Evaluation Forum, Bologna, Italy, 5–8 September 2022. CEUR Workshop Proceedings, CEUR-WS.org (2022)

26. SanJuan, E., Huet, S., Kamps, J., Ermakova, L.: Overview of the CLEF 2022 SimpleText task 1: passage selection for a simplified summary. In: Faggioli et al. [12]

27. Sheang, K.C., Saggion, H.: Controllable sentence simplification with a unified text-to-text transfer transformer. In: Proceedings of the 14th International Conference on Natural Language Generation, pp. 341–352 (2021)

28. Siddharthan, A.: An architecture for a text simplification system (2002). https://citeseerx.ist.psu.edu/viewdoc/summary?doi=10.1.1.1.9968&rank=1

29. Tang, J., Zhang, J., Yao, L., Li, J., Zhang, L., Su, Z.: ArnetMiner: extraction and mining of academic social networks. In: KDD 2008, pp. 990–998 (2008)
30. Xu, W., Callison-Burch, C., Napoles, C.: Problems in current text simplification research: new data can help. Trans. ACL 3, 283–297 (2015). https://www.mitpressjournals.org/doi/abs/10.1162/tacl_a_00139
31. Yimam, S.M., et al.: A report on the complex word identification shared task 2018. In: The 13th Workshop on Innovative Use of NLP for Building Educational Applications (NAACL2018 Workshops) (2018)
32. Zhang, X., Lapata, M.: Sentence simplification with deep reinforcement learning. In: EMNLP 2017: Conference on Empirical Methods in Natural Language Processing, pp. 584–594. Association for Computational Linguistics (2017)
33. Zhu, Z., Bernhard, D., Gurevych, I.: A monolingual tree-based translation model for sentence simplification. In: Proceedings of the 23rd International Conference on Computational Linguistics (Coling 2010), Beijing, China, pp. 1353–1361. Coling 2010 Organizing Committee (2010). https://www.aclweb.org/anthology/C10-1152
34. Štajner, S., Sheang, K.C., Saggion, H.: Sentence Simplification Capabilities of Transfer-Based Models (2022)

Science for Fun: The CLEF 2023 JOKER Track on Automatic Wordplay Analysis

Liana Ermakova[1]([✉])[iD], Tristan Miller[2][iD], Anne-Gwenn Bosser[3],
Victor Manuel Palma Preciado[4], Grigori Sidorov[4][iD], and Adam Jatowt[5][iD]

[1] Université de Bretagne Occidentale, HCTI, Brest, France
`liana.ermakova@univ-brest.fr`
[2] Austrian Research Institute for Artificial Intelligence (OFAI), Vienna, Austria
[3] École Nationale d'Ingénieurs de Brest, Lab-STICC CNRS UMR 6285,
Plouzané, France
[4] Instituto Politécnico Nacional (IPN), Centro de Investigación
en Computación (CIC), Mexico City, Mexico
[5] University of Innsbruck, Innsbruck, Austria

Abstract. Understanding and translating humorous wordplay often
requires recognition of implicit cultural references, knowledge of word
formation processes, and discernment of double meanings – issues which
pose challenges for humans and computers alike. This paper introduces
the CLEF 2023 JOKER track, which takes an interdisciplinary approach
to the creation of reusable test collections, evaluation metrics, and methods for the automatic processing of wordplay. We describe the track's
interconnected shared tasks for the detection, location, interpretation,
and translation of puns. We also describe associated data sets and evaluation methodologies, and invite contributions making further use of our
data.

Keywords: Wordplay · Puns · Humour · Wordplay interpretation ·
Wordplay detection · Wordplay generation · Machine translation

1 Introduction

Humour remains one of the most thorny aspects of intercultural communication.
Understanding humour often requires recognition of implicit cultural references
or, especially in the case of wordplay, knowledge of word formation processes and
discernment of double meanings. These issues raise the question not only of how
to translate humour across cultures and languages, but also how to even recognise
it in the first place. Such tasks are challenging for humans and computers alike.

The goal of the JOKER track series at the Conference and Labs of the Evaluation Forum (CLEF) is to bring together linguists, translators, and computer
scientists in order to create reusable test collections for benchmarking and to
explore new methods and evaluation metrics for the automatic processing of
wordplay. In the 2022 edition of JOKER (see Sect. 2), we introduced pilot shared

© The Author(s), under exclusive license to Springer Nature Switzerland AG 2023
J. Kamps et al. (Eds.): ECIR 2023, LNCS 13982, pp. 546–556, 2023.
https://doi.org/10.1007/978-3-031-28241-6_63

tasks for the classification, interpretation, and translation of wordplay in English and French, and made our data available for an unshared task [9].[1] For JOKER-2023, we intend to expand the set of languages in our tasks to include Spanish. We also somewhat simplify and streamline the slate of shared tasks, more closely patterning them after the high-level process used by human translators and focusing them on one type of wordplay – puns.

We choose to focus on puns because, despite recent improvements in the quality of machine translation based on machine learning, puns are often held to be untranslatable by statistical or neural approaches [1, 26, 33]. Punning jokes are a common source of data in computational humour research, in part because of their widespread availability and in part because the underlying linguistic mechanisms are well understood. However, past pun detection data sets [29, 39] are problematic because they draw their positive and negative examples from texts in different domains. In JOKER-2023 we attempt to avoid this problem by generating our negative examples by using naïve literal translations, or by slightly editing our positive examples, a technique pioneered by *Unfun.me* [38].

The three shared tasks of JOKER-2023 can be summarised as follows:

1. **Detection and location of puns** in English, French, and Spanish;
2. **Interpretation of puns** in English, French, and Spanish; and
3. **Translation of puns** from English to French and Spanish.

The unshared task of JOKER-2022 saw its data used for a pun generation task potentially aimed at improving interlocutor engagement in dialogue systems. JOKER-2023 will likewise have an unshared task that aims at attracting runs with other, possibly novel, use cases, such as pun generation or humorousness evaluation.

While JOKER-2022 proved to be challenging (with only 13% of evaluated translations being judged successful), this round's larger data set and more constrained, interconnected tasks may present opportunities for better performance.

2   JOKER-2022: Results and Lessons Learnt

Forty-nine teams registered for JOKER-2022, 42 downloaded the data and seven submitted official runs for its shared tasks: nine for Task 1 on classification and interpretation of wordplay [10], four for Task 2 on wordplay translation in named entities [8], and six for Task 3 on pun translation [11]. One additional run was submitted for Task 1 after the deadline. Two runs were submitted for the unshared task, and new classifications were proposed by participants.

---

[1] In a shared task, the organisers define the evaluation criteria for an open problem in AI and produce a human-annotated data set for training and testing purposes; task participants then use the publically released training data to develop systems for solving the problem, which the organisers evaluate on the unpublished test data. In an unshared task, the organisers provide annotated data without a particular problem in mind, and participants are invited to use this data to propose and solve novel problems.

Participants' scores on the wordplay classification part of Task 1 were uniformly high, which we attribute to the insufficient expressiveness of our typology and the class imbalance of our data. Due to the expense involved in revising the typology and applying it to new data, we have decided to drop wordplay classification from JOKER-2023. However, the interpretation part of the task – which required participants to determine both the location and (double) meaning of the wordplay instances – proved to be more challenging, and provoked great interest from the participants. Besides this, we note that providing the location and interpretation of a play on words may be more relevant for downstream processing tasks such as translation [24, p. 86]. For this reason, this part of the task will be repeated in JOKER-2023, albeit with new data.

JOKER-2022's Task 2, on named entity translation, did not see much variety in the participants' approaches, and their low success rates may be due to a lack of context in the data that would be too expensive for us to source. For these reasons, we have opted to discontinue this task for JOKER-2023.

Like Task 1, Task 3 of JOKER-2022 proved to be both popular and challenging, and so we are rerunning it in JOKER-2023 with new data. Task 3 moreover had the side-effect of producing a French-language corpus with positive and negative examples of wordplay, which some participants endeavoured to use for wordplay generation in French (following methods developed for English). The corpus was also reused by the French Association for Artificial Intelligence to organise a jam on wordplay generation in French during a week-long conference [3]. Of particular interest is how humans perceive the generated wordplay. Participants in the jam, for example, raised questions about how to evaluate the humorousness of the system output. Furthermore, a curated selection of sentences generated using our corpus with a large language model[2] was used by some of the present authors during an outreach event, where a public audience was asked to guess if a given humorous sentence was created by an AI or a human. In JOKER-2023, we thus encourage unshared task submissions describing the use of our data for user perception studies and wordplay generation.

3   Shared Tasks

3.1   Task 1: Pun Detection and Location

Description. A *pun* is a form of wordplay in which a word or phrase evokes the meaning of another word or phrase with a similar or identical pronunciation [19]. *Pun detection* is a binary classification task where the goal is to distinguish between texts containing a pun and texts not containing a pun. *Pun location* is a finer-grained task, where the goal is to identify which words carry the double meaning in a text known *a priori* to contain a pun.

---

[2] The corpus provides numerous instances of particular wellerisms, and thus lends itself well to prompt engineering using large language models. (Wellerisms are a type of humour in which a proverb, idiom, or other well-known saying is subverted, for example by resegmenting it or by reinterpreting it literally.).

For example, the first of the following sentences contains a pun where the word *propane* evokes the similar-sounding word *profane*, and the second sentence contains a pun exploiting two distinct meanings of the word *interest*:

(1) When the church bought gas for their annual barbecue, proceeds went from the sacred to the propane.

(2) I used to be a banker but I lost interest.

For the pun detection task, the correct answer for these two instances would be "true", and for the pun location task, the correct answers are respectively "propane" and "interest".

Data. The positive examples for Task 1, which will be used for both the detection and location subtasks, consist of short jokes (one-liners), each containing a single pun. These positive examples will be drawn from previously constructed corpora as well as collections that may not have been used in previous shared tasks.[3] In contrast to previously published punning data sets, our negative examples will be generated by the data augmentation technique of manually or semi-automatically editing positive examples in such a way that the wordplay is lost but most of the rest of the meaning remains.[4] In this way, we hope to better minimise the differences in length, vocabulary, style, etc. that were seen in previous pun detection data sets and that could be picked up on by today's neural approaches. Negative examples will be used only for the pun detection subtask.

As usual with shared tasks, data for all tasks will be split into training and test sets, with the training set (including gold-standard labels) published as soon as available, and the test data withheld until evaluation phase.

English. Our training data will include positive examples from the corpora of SemEval-2017 Task 7 [29], SemEval-2021 Task 12 [35], and various other collections. Positive examples in the test data will be drawn, to the extent possible, from jokes not present in past data sets. As mentioned above, negative examples in both the training and test data will be produced by slightly perturbing the positive examples via data augmentation.

French. In 2022, we created a corpus for wordplay detection in French [9,11] based on the translation of the corpus of English puns introduced at SemEval-2017 Task 7 [29]. Some of the translations were machine translations, and others

---

[3] Admittedly, it may be impossible for us to source positive examples that are not discoverable online, unless we pay experienced comedians to produce a large collection of completely novel jokes, which is costly. We will have to rely on participants' good faith that their systems will not detect punning jokes by matching them against a database of web-scraped examples.

[4] The data augmentation technique will be fully described in the task overview paper in the CLEF 2023 proceedings; we hold off on presenting the details here to discourage participants from reverse-engineering it in their classifiers.

were human translations sourced from a contest or from native francophone students translators. The majority of human translations (90%) preserved wordplay in some form, while only 13% of the machine translations did so. The resulting corpus is homogeneous, across positive and negative examples, in terms of vocabulary and text length, and it maintains the class balance of the original. However, there was an imbalance across the training and test sets with respect to machine vs. human translations, with more machine translations in the test set. This corpus will be improved and extended for use with JOKER-2023. In particular, we will correct the machine vs. human translation imbalance by sourcing additional, manually verified machine translations for the training set. We will also source new positive examples for our test set, and will apply the same data augmentation technique used for our English data.

Spanish. Our Spanish data set is collected from various web sources (blogs, joke compilations, humour forums, etc.) to which we apply the same data augmentation techniques as for the English data.

Evaluation. We follow (and thereby facilitate comparison with) SemEval-2017 Task 7 [29] by evaluating pun detection using the precision, recall, accuracy, and F-score measures as used in information retrieval (IR) [25, Sect. 8.3], and pun location using the corresponding variants of precision, recall, and F-score from word sense disambiguation (WSD) [31].[5]

3.2   Task 2: Pun Interpretation

Description. In *pun interpretation*, systems must indicate the two meanings of the pun. The pun interpretation task at SemEval-2017 required systems to annotate the pun with senses from WordNet, and JOKER-2022 expected annotations according to a relatively complex, structured notation scheme. In JOKER-2023, semantic annotations will be in the form of a pair of lemmatised word sets. Following the practice used in lexical substitution data sets [27], these word sets will contain the synonyms (or absent any, the hypernyms) of the two words involved in the pun, excepting any synonyms/hypernyms that happen to share a spelling with the pun as written.[6] This annotation scheme removes the need for participating systems to directly rely on a particular sense inventory or notation scheme.

For example, for the punning joke introduced in Example 1 above, the word sets are {*gas, fuel*} and {*profane*}, and for Example 2, the word sets are {*involvement*} and {*fixed charge, fixed cost, fixed costs*}.

---

[5] The difference between IR-style and WSD-style metrics is that the former require the system to make a prediction for every instance in the data set, whereas the latter do not. IR-style accuracy is equivalent to WSD-style recall.

[6] Synonyms and hypernyms will be sourced preferentially from WordNet (or similar resources for data sets in other languages), and via manual annotation for those words not present in WordNet.

Data. The data will be drawn from the positive examples of Task 1, with the pun word annotated with two sets of words, one for each sense of the pun. Each set of words will contain synonyms or hypernyms of the sense or (in the case of heterographic puns) the latent target word.

Evaluation. Task 2 will be evaluated with the precision, recall, and F-score metrics as used in word sense disambiguation [31], except that each instance will be scored as the average score for each of its senses. Systems need guess only one word for each sense of the pun; a guess will be considered correct if it matches any of the words in the gold-standard set. For example, a system guessing {*fuel*}, {*profane*} would receive a score of 1 for Example 1, and a system guessing {*fuel*}, {*prophet*} would receive a score of 1/2.

3.3   Task 3: Pun Translation

Description. The goal of this task is to translate English punning jokes into French and Spanish. The translations should aim to preserve, to the extent possible, both the form and meaning of the original wordplay – that is, to implement the PUN→PUN strategy described in Delabastita's typology of pun translation strategies [5,6]. For example, Example 2 might be rendered into French as *J'ai été banquier mais j'en ai perdu tout l'intérêt*. This fairly straightforward translation happens to preserve the pun, since *interest* and *intérêt* share the same ambiguity. Needless to say, this is coincidence does not hold for the majority of punning jokes in our data set (or generally, for that matter).

Data. We will provide an updated training and test set of English-French translations of punning jokes, and new sets of English-Spanish ones, similar to English-French data sets we produced for JOKER-2022 [9,11].

Evaluation. As we have previously argued [9,11], vocabulary overlap metrics such as BLEU are unsuitable for evaluating wordplay translations. We will therefore continue JOKER-2022's practice of having trained experts manually evaluate system translations according to features such as lexical field preservation, sense preservation, wordplay form preservation, style shift, humorousness shift, etc. and the presence of errors in syntax, word choice, etc. The runs will be ranked according to the number of successful translations – i.e., translations preserving, to the extent possible, both the form and sense of the original wordplay. We will also experiment with other semi-automatic metrics.

4   State of the Art

Humour is part of social coexistence and therefore is part of interpersonal interactions. This places it in a complicated position, since the perception of humour can be somewhat ambiguous and depends on a number of subjective factors.

Thus, dealing with humour, even in its written form, becomes a rather complex undertaking, even for those (computational) tasks that at the first sight seem trivial. Various studies have addressed these tasks, including the detection, classification, and translation of humour, and also determining whether the intention or interpretability of the translated humour is maintained. Some of the present authors have even designed evaluation campaigns for some of these tasks (e.g., [8,10,11,29,35]), aiming not just to support traditional NLP applications, but also to gain a broader knowledge of the structure and nuances of verbal humour.

Nevertheless, relatively few studies have been carried out on the machine translation (MT) of wordplay. One of the earliest of these [12] proposed a pragmatic-based approach to MT, but no working system was implemented. An interactive method for the computer-assisted translation of puns was recently implemented [24], but it cannot be directly applied for MT. Four teams participated in the pun translation task of JOKER-2022 [7,14,16]; their approaches relied variously on applications of transformer-based models or on DeepL.

Automatic humour recognition has become an emerging trend with the rise of conversational agents and the need for social media analysis [13,15,18,22,23,30, 34]. While some systems have achieved decent performance on humour detection, location, and classification tasks [10,29], the lack of high-quality training data has been a limiting factor for further progress, and especially in case of languages other than English [9]. As with translation, many of the JOKER-2022 classification task participants [2,16] favoured applications of large language models such as Google T5 and Jurassic-1.

Other popular application areas in computational humour include humour generation and humorousness evaluation. Recent work in the former area includes template-based approaches for pun generation in English and French [17,20, 36], as well as injecting humour into existing non-humorous English texts [37]. Though these tasks have been studied in a monolingual setting, it may be possible to adapt them for a translation task. Work in humorousness evaluation covers methods that attempt to quantify the level of humour in a text, or to rank texts according to their level of humour [4,21,28,32,40]. Such methods also have possible applications in humour translation (e.g., by verifying that a translated joke preserves the level of humour of the original).

5   Conclusion

This paper has described the prospective setup of the CLEF 2023 JOKER track, which features shared tasks on pun detection, location, interpretation, and translation. We will also welcome submissions using our data for other tasks, such as pun generation, offensive joke detection, or humour perception. Please visit the JOKER website at http://joker-project.com for further details on the track.

Acknowledgment. This project has received a government grant managed by the National Research Agency under the program "Investissements d'avenir" integrated into France 2030, with the Reference ANR-19-GURE-0001. JOKER is supported by *La Maison des sciences de l'homme en Bretagne.*

References

1. Ardi, H., Al Hafizh, M., Rezqi, I., Tuzzikriah, R.: Can machine translations translate humorous texts? Humanus **21**(1), 99 (2022). https://doi.org/10.24036/humanus.v21i1.115698

2. Arroubat, H.: CLEF workshop: automatic pun and humour translation task. In: Proceedings of the Working Notes of CLEF 2022 - Conference and Labs of the Evaluation Forum, Bologna, Italy, 5–8 September 2022. CEUR Workshop Proceedings, CEUR-WS.org, Bologna, Italy (2022)

3. Bosser, A.G., et al.: Poetic or humorous text generation: jam event at PFIA2022. In: Faggioli, G., Ferro, N., Hanbury, A., Potthast, M. (eds.) 13th Conference and Labs of the Evaluation Forum (CLEF 2022), pp. 1719–1726. No. Working Notes: JokeR: Automatic Wordplay and Humour Translation in CEUR Workshop Proceedings, CEUR-WS.org, Bologna, Italy, September 2022. https://hal.archives-ouvertes.fr/hal-03795272

4. Castro, S., Chiruzzo, L., Rosá, A.: Overview of the HAHA task: humor analysis based on human annotation at IberEval 2018. In: Rosso, P., Gonzalo, J., Martinez, R., Montalvo, S., de Albornoz, J.C. (eds.) Proceedings of the Third Workshop on Evaluation of Human Language Technologies for Iberian Languages. CEUR Workshop Proceedings, vol. 2150, pp. 187–194. Spanish Society for Natural Language Processing, September 2018. https://ceur-ws.org/Vol-2150/overview-HAHA.pdf

5. Delabastita, D.: There's a Double Tongue: An Investigation into the Translation of Shakespeare's Wordplay, with Special Reference to Hamlet. Rodopi, Amsterdam (1993)

6. Delabastita, D.: Wordplay as a translation problem: a linguistic perspective. In: Ein internationales Handbuch zur Übersetzungsforschung, vol. 1, pp. 600–606. De Gruyter Mouton, July 2008. https://doi.org/10.1515/9783110137088.1.6.600

7. Dhanani, F., Rafi, M., Tahir, M.A.: FAST$_M$T participation for the JOKER CLEF-2022 automatic pun and human translation tasks. In: Proceedings of the Working Notes of CLEF 2022 - Conference and Labs of the Evaluation Forum, Bologna, Italy, 5–8 September 2022, p. 14. CEUR Workshop Proceedings, CEUR-WS.org, Bologna, Italy (2022)

8. Ermakova, L., Miller, T., Boccou, J., Digue, A., Damoy, A., Campen, P.: Overview of the CLEF 2022 JOKER Task 2: translate wordplay in named entities. In: Faggioli, G., Ferro, N., Hanbury, A., Potthast, M. (eds.) Proceedings of the Working Notes of CLEF 2022 - Conference and Labs of the Evaluation Forum, Bologna, Italy, 5–8 September 2022. CEUR Workshop Proceedings, vol. 3180, pp. 1666–1680, August 2022

9. Ermakova, L., et al.: Overview of JOKER@CLEF 2022: automatic wordplay and humour translation workshop. In: Barrón-Cedeño, A., et al. (eds.) Experimental IR Meets Multilinguality, Multimodality, and Interaction. Proceedings of the Thirteenth International Conference of the CLEF Association (CLEF 2022). LNCS, vol. 13390. Springer, Cham (2022). https://doi.org/10.1007/978-3-031-13643-6_27

10. Ermakova, L., et al.: Overview of the CLEF 2022 JOKER Task 1: classify and explain instances of wordplay. In: Faggioli, G., Ferro, N., Hanbury, A., Potthast, M. (eds.) Proceedings of the Working Notes of CLEF 2022: Conference and Labs of the Evaluation Forum. CEUR Workshop Proceedings (2022)

11. Ermakova, L., et al.: Overview of the CLEF 2022 JOKER Task 3: pun translation from English into French. In: Faggioli, G., Ferro, N., Hanbury, A., Potthast, M. (eds.) Proceedings of the Working Notes of CLEF 2022: Conference and Labs of the Evaluation Forum. CEUR Workshop Proceedings (2022)

12. Farwell, D., Helmreich, S.: Pragmatics-based MT and the translation of puns. In: Proceedings of the 11th Annual Conference of the European Association for Machine Translation, pp. 187–194, June 2006. https://www.mt-archive.info/EAMT-2006-Farwell.pdf

13. Francesconi, C., Bosco, C., Poletto, F., Sanguinetti, M.: Error analysis in a hate speech detection task: the case of HaSpeeDe-TW at EVALITA 2018. In: Bernardi, R., Navigli, R., Semeraro, G. (eds.) Proceedings of the 6th Italian Conference on Computational Linguistics, November 2018. https://ceur-ws.org/Vol-2481/paper32.pdf

14. Galeano, L.J.G.: LJGG @ CLEF JOKER Task 3: an improved solution joining with dataset from task. In: Proceedings of the Working Notes of CLEF 2022 - Conference and Labs of the Evaluation Forum, Bologna, Italy, 5–8 September 2022, p. 7. CEUR Workshop Proceedings, CEUR-WS.org, Bologna, Italy (2022)

15. Ghanem, B., Karoui, J., Benamara, F., Moriceau, V., Rosso, P.: IDAT@FIRE2019: Overview of the track on irony detection in Arabic tweets. In: Proceedings of the 11th Forum for Information Retrieval Evaluation, pp. 10–13. Association for Computing Machinery (2019). https://doi.org/10.1145/3368567.3368585

16. Glémarec, L.: Use of SimpleT5 for the CLEF workshop JokeR: automatic pun and humor translation. In: Proceedings of the Working Notes of CLEF 2022 - Conference and Labs of the Evaluation Forum, Bologna, Italy, 5–8 September 2022, p. 11. CEUR Workshop Proceedings, CEUR-WS.org, Bologna, Italy (2022)

17. Glémarec, L., Bosser, A.G., Ermakova, L.: Generating humourous puns in French. In: Proceedings of the Working Notes of CLEF 2022 - Conference and Labs of the Evaluation Forum, Bologna, Italy, 5–8 September 2022, p. 8. CEUR Workshop Proceedings, CEUR-WS.org, Bologna, Italy (2022)

18. Guibon, G., Ermakova, L., Seffih, H., Firsov, A., Le Noé-Bienvenu, G.: Multilingual fake news detection with satire. In: CICLing: International Conference on Computational Linguistics and Intelligent Text Processing, La Rochelle, France, April 2019. https://halshs.archives-ouvertes.fr/halshs-02391141

19. Hempelmann, C.F., Miller, T.: Puns: taxonomy and phonology. In: Attardo, S. (ed.) The Routledge Handbook of Language and Humor, pp. 95–108. Routledge Handbooks in Linguistics, Routledge, New York, NY, February 2017. https://doi.org/10.4324/9781315731162-8

20. Hong, B.A., Ong, E.: Automatically extracting word relationships as templates for pun generation. In: Proceedings of the Workshop on Computational Approaches to Linguistic Creativity, pp. 24–31. Association for Computational Linguistics, Boulder, Colorado, June 2009. https://aclanthology.org/W09-2004

21. Hossain, N., Krumm, J., Gamon, M., Kautz, H.: SemEval-2020 Task 7: assessing humor in edited news headlines. In: Proceedings of the Fourteenth Workshop on Semantic Evaluation, pp. 746–758. International Committee for Computational Linguistics, December 2020. https://doi.org/10.18653/v1/2020.semeval-1.98

22. Karoui, J., Benamara, F., Moriceau, V., Patti, V., Bosco, C., Aussenac-Gilles, N.: Exploring the impact of pragmatic phenomena on irony detection in tweets: a multilingual corpus study. In: 15th Conference of the European Chapter of the Association for Computational Linguistics, vol. 1, pp. 262–272. Association for Computational Linguistics, April 2017. https://aclanthology.org/E17-1025.pdf

23. Karoui, J., Farah, B., Moriceau, V., Aussenac-Gilles, N., Hadrich-Belguith, L.: Towards a contextual pragmatic model to detect irony in tweets. In: Proceedings of the 53rd Annual Meeting of the Association for Computational Linguistics and the 7th International Joint Conference on Natural Language Processing, vol. 2, pp. 644–650. Association for Computational Linguistics (2015). https://doi.org/10.3115/v1/P15-2106

24. Kolb, W., Miller, T.: Human-computer interaction in pun translation. In: Hadley, J.L., Taivalkoski-Shilov, K., Teixeira, C.S.C., Toral, A. (eds.) Using Technologies for Creative-Text Translation, pp. 66–88. Routledge (2022). https://doi.org/10.4324/9781003094159-4

25. Manning, C.D., Raghavan, P., Schütze, H.: Introduction to Information Retrieval. Cambridge University Press, Cambridge (2008)

26. Miller, T.: The Punster's Amanuensis: the proper place of humans and machines in the translation of wordplay. In: Proceedings of the Second Workshop on Human-Informed Translation and Interpreting Technology, pp. 57–64, September 2019. https://doi.org/10.26615/issn.2683-0078.2019_007

27. Miller, T., Benikova, D., Abualhaija, S.: GermEval 2015: LexSub - a shared task for German-language lexical substitution. In: Proceedings of GermEval 2015: LexSub, pp. 1–9, September 2015

28. Miller, T., Do Dinh, E.L., Simpson, E., Gurevych, I.: Predicting the humorousness of tweets using Gaussian process preference learning. Procesamiento del Lenguaje Natural **64**, 37–44 (2020). https://doi.org/10.26342/2020-64-4

29. Miller, T., Hempelmann, C.F., Gurevych, I.: SemEval-2017 Task 7: detection and interpretation of English puns. In: Proceedings of the 11th International Workshop on Semantic Evaluation, pp. 58–68, August 2017. https://doi.org/10.18653/v1/S17-2005

30. Nijholt, A., Niculescu, A., Valitutti, A., Banchs, R.E.: Humor in human-computer interaction: a short survey. In: Joshi, A., Balkrishan, D.K., Dalvi, G., Winckler, M. (eds.) Adjunct Proceedings: INTERACT 2017 Mumbai, pp. 199–220. Industrial Design Centre, Indian Institute of Technology Bombay (2017). https://www.interact2017.org/downloads/INTERACT_2017_Adjunct_v4_final_24jan.pdf

31. Palmer, M., Ng, H.T., Dang, H.T.: Evaluation of WSD systems. In: Agirre, E., Edmonds, P. (eds.) Word Sense Disambiguation: Algorithms and Applications, Chapter 4, vol. 33, pp. 75–106. Text, Speech, and Language Technology. Springer, Cham (2007). https://doi.org/10.1007/978-1-4020-4809-8_4

32. Potash, P., Romanov, A., Rumshisky, A.: SemEval-2017 Task 6: #HashtagWars: learning a sense of humor. In: Proceedings of the 11th International Workshop on Semantic Evaluation, pp. 49–57, August 2017. https://doi.org/10.18653/v1/S17-2004

33. Regattin, F.: Traduction automatique et jeux de mots : l'incursion (ludique) d'un inculte, March 2021. https://motsmachines.github.io/2021/en/submissions/Mots-Machines-2021_paper_5.pdf

34. Reyes, A., Rosso, P., Buscaldi, D.: From humor recognition to irony detection: the figurative language of social media. Data Knowl. Eng. **74**, 1–12 (2012). https://doi.org/10.1016/j.datak.2012.02.005

35. Uma, A., et al.: SemEval-2021 Task 12: learning with disagreements. In: Proceedings of the 15th International Workshop on Semantic Evaluation, pp. 338–347, August 2021. https://doi.org/10.18653/v1/2021.semeval-1.41

36. Valitutti, A., Toivonen, H., Doucet, A., Toivanen, J.M.: "Let everything turn well in your wife": generation of adult humor using lexical constraints. In: Proceedings of the 51st Annual Meeting of the Association for Computational Linguistics, vol. 2, pp. 243–248. Association for Computational Linguistics, August 2013. https://aclanthology.org/P13-2044

37. Weller, O., Fulda, N., Seppi, K.: Can humor prediction datasets be used for humor generation? Humorous headline generation via style transfer. In: Proceedings of the Second Workshop on Figurative Language Processing, pp. 186–191. Association for Computational Linguistics, Online, July 2020. https://doi.org/10.18653/v1/2020.figlang-1.25

38. West, R., Horvitz, E.: Reverse-engineering satire, or "Paper on computational humor accepted despite making serious advances". In: AAAI 2019/IAAI 2019/EAAI 2019: Proceedings of the Thirty-Third AAAI Conference on Artificial Intelligence and Thirty-First Innovative Applications of Artificial Intelligence Conference and Ninth AAAI Symposium on Educational Advances in Artificial Intelligence, pp. 7265–7272, January 2019. https://doi.org/10.1609/aaai.v33i01.33017265

39. Yang, D., Lavie, A., Dyer, C., Hovy, E.: Humor recognition and humor anchor extraction. In: Proceedings of the 2015 Conference on Empirical Methods in Natural Language Processing, pp. 2367–2376. Association for Computational Linguistics, September 2015. https://doi.org/10.18653/v1/D15-1284

40. Zhao, Z., Cattle, A., Papalexakis, E., Ma, X.: Embedding lexical features via tensor decomposition for small sample humor recognition. In: Proceedings of the 2019 Conference on Empirical Methods in Natural Language Processing and the 9th International Joint Conference on Natural Language Processing (EMNLP-IJCNLP), pp. 6376–6381, November 2019. https://doi.org/10.18653/v1/D19-1669

ImageCLEF 2023 Highlight: Multimedia Retrieval in Medical, Social Media and Content Recommendation Applications

Bogdan Ionescu[1], Henning Müller[2], Ana Maria Drăgulinescu[1(✉)],
Adrian Popescu[3], Ahmad Idrissi-Yaghir[4], Alba García Seco de Herrera[5],
Alexandra Andrei[1], Alexandru Stan[6], Andrea M. Storås[7], Asma Ben Abacha[8],
Christoph M. Friedrich[4], George Ioannidis[6], Griffin Adams[9],
Henning Schäfer[10], Hugo Manguinhas[11], Ihar Filipovich[12], Ioan Coman[1],
Jérôme Deshayes[3], Johanna Schöler[13], Johannes Rückert[4],
Liviu-Daniel Ştefan[1], Louise Bloch[4], Meliha Yetisgen[14], Michael A. Riegler[7],
Mihai Dogariu[1], Mihai Gabriel Constantin[1], Neal Snider[8,15],
Nikolaos Papachrysos[13], Pål Halvorsen[7], Raphael Brüngel[4], Serge Kozlovski[16],
Steven Hicks[7], Thomas de Lange[13], Vajira Thambawita[7], Vassili Kovalev[16],
and Wen-Wai Yim[8]

[1] Politehnica University of Bucharest, Bucharest, Romania
{bogdan.ionescu,ana.dragulinescu}@upb.ro
[2] University of Applied Sciences Western Switzerland (HES-SO),
Sierre, Switzerland
[3] CEA LIST, Gif-sur-Yvette, France
[4] University of Applied Sciences and Arts Dortmund, Dortmund, Germany
[5] University of Essex, Colchester, UK
[6] IN2 Digital Innovations, Lindau, Germany
[7] SimulaMet, Oslo, Norway
[8] Microsoft, Redmond, USA
[9] Columbia University, New York, USA
[10] University Hospital Essen, Essen, Germany
[11] Europeana Foundation, Hague, Netherlands
[12] Belarus State University, Minsk, Belarus
[13] Sahlgrenska University Hospital, Gothenburg, Sweden
[14] University of Washington, Seattle, USA
[15] Nuance, Burlington, USA
[16] Belarusian Academy of Sciences, Minsk, Belarus

Abstract. In this paper, we provide an overview of the upcoming ImageCLEF campaign. ImageCLEF is part of the CLEF Conference and Labs of the Evaluation Forum since 2003. ImageCLEF, the Multimedia Retrieval task in CLEF, is an ongoing evaluation initiative that promotes the evaluation of technologies for annotation, indexing, and retrieval of multimodal data with the aim of providing information access to large collections of data in various usage scenarios and domains. In its 21st edition, ImageCLEF 2023 will have four main tasks: (i) a *Medical* task addressing automatic image captioning, synthetic medical images

J. Kamps et al. (Eds.): ECIR 2023, LNCS 13982, pp. 557–567, 2023.
https://doi.org/10.1007/978-3-031-28241-6_64

created with GANs, Visual Question Answering for colonoscopy images, and medical dialogue summarization; (ii) an *Aware* task addressing the prediction of real-life consequences of online photo sharing; (iii) a *Fusion* task addressing late fusion techniques based on the expertise of a pool of classifiers; and (iv) a *Recommending* task addressing cultural heritage content-recommendation. In 2022, ImageCLEF received the participation of over 25 groups submitting more than 258 runs. These numbers show the impact of the campaign. With the COVID-19 pandemic now over, we expect that the interest in participating, especially at the physical CLEF sessions, will increase significantly in 2023.

Keywords: Information retrieval · Medical AI · Image captioning · GANs · Visual question answering · Dialogue summarization · Social media · User awareness · Late fusion · Cultural heritage · Content recommending · ImageCLEF benchmarking · Annotated data

1   Introduction

The ImageCLEF evaluation campaign has been organised each year since 2003 and continues to enable the benchmarking activities and research tasks on the cross-language annotation, indexing and retrieval of multimodal data. As part of the Conference and Labs of the Evaluation Forum (CLEF) [17,18], the 21st edition of ImageCLEF will be hosted by Centre for Research & Technology Hellas (CERTH) in Thessaloniki, Greece, in September 2023[1]. A set of benchmarking tasks was designed to test different aspects of mono- and cross-language information retrieval systems [14,17,18]. Target communities involve (but are not limited to): information retrieval (e.g., text, vision, audio, multimedia, social media, sensor data), machine learning, deep learning, data mining, natural language processing, image and video processing; with special emphasis on the challenges of multi-modality, multi-linguality, and interactive search. Both, Image-CLEF lab [29] and CLEF campaign, have important scholarly impact with 407 publications on Web of Science (WoS) mentioning ImageCLEF (with 2801 WoS citations) and 6730 results on Google Scholar.

The following sections introduce the four tasks that are planned for 2023, namely: ImageCLEFmedical, ImageCLEFaware, ImageCLEFfusion, and the new ImageCLEFrecommeding. Figure 1 captures representative images for the aforementioned proposed tasks.

2   ImageCLEFmedical

The ImageCLEFmedical task has been carried out every year since 2004 [18]. The 2023 edition will include the following tasks: (i) a sequel of the caption task with medical concept detection and caption prediction, (ii) a new task on synthetic medical images generated with GANs, (iii) a new Visual Question Answering and generation task for colonoscopy images, (iv) a new pilot task on medical doctor-patient conversation summarization for generation of clinical notes.

---

[1] https://clef2023.clef-initiative.eu/.

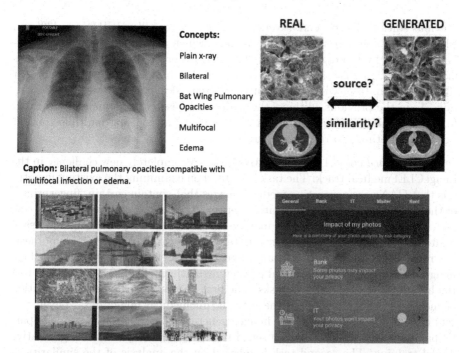

Fig. 1. Sample images from (left to right, top to bottom): ImageCLEFmedical-caption with an image with the corresponding CUIs and caption, ImageCLEFmedical-GAN with an example of real and generated images, ImageCLEFrecommending task with an example of editorial "European landscapes and landmarks" Gallery, and ImageCLEFaware with an example of user photos and predicted influence when searching for a bank loan.

ImageCLEFmedical-caption[2]. The topic of the *caption* task resides in the interpretation of the insights gained from radiology images. In the 7th edition of the task [8,10,21–23,25], there will be two subtasks: concept detection and caption prediction. The *concept detection* subtask aims to develop competent systems that are able to predict the Unified Medical Language System (UMLS®) Concept Unique Identifiers (CUIs) based on the visual image content. The F1-Score [11] will be used to evaluate the participating systems in this subtask. The *caption prediction* subtask focuses on implementing models to predict captions for given radiology images. After using the BLEU [20] score in previous editions, it was decided to change the primary scoring metric for the 2023 challenge because recent studies [3,15,31] that investigated the relationship between the BLEU score and human judgment found that there was only a moderate correlation. In reviewing BLEU as an appropriate metric, it was also found that semantically correct sentences tend to be disadvantaged when they differ from the reference in terms of morphology [32]. This is also consistent with recent feedback from the

---

[2] https://www.imageclef.org/2023/medical/caption.

previous edition. To this end, several different metrics alike BERTScore [32] and BLEURT [26] are currently being evaluated, which aim to capture the underlying semantics by leveraging state-of-the-art transformer-based language models like BERT [6]. In 2023, an updated version of the Radiology Objects in Context (ROCO) [24] dataset will be used, further extended in comparison to 2022's edition. As in the previous editions, the updated dataset will be manually curated (e.g., image modalities, anatomy in x-ray images) after using multiple concept extraction methods to retrieve accurate CUIs.

ImageCLEFmedical-GAN[3]. The *GANs* task is a completely new challenge in the ImageCLEFmedical track. The task is focused on examining the existing hypothesis that GANs are generating medical images that contain certain "fingerprints" of the real images used for generative network training. If the hypothesis is correct, artificial biomedical images may be subject to the same sharing and usage limitations as real sensitive medical data. On the other hand, if the hypothesis is wrong, GANs may be potentially used to create rich datasets of biomedical images that are free of ethical and privacy regulations. The participants will test the hypothesis by solving two tasks. The first task is dedicated to the detection of mentioned "fingerprints" in the artificial biomedical image data. Given two sets of real images and a set of images generated by some GAN models, participants will try to detect which set of real images was used for the generative model training. The second task is focused on the analysis of the similarity of output produced by generative models with different architectures and/or with different training strategies. In this task participants will be given a set of artificial images produced by a set of different generative models and participant will try to group images by their source model. Possible options for a data type in both tasks are histology, X-ray, CT scans.

ImageCLEFmedical-VQA[4]. The *VQA* task, also a new task in this format, combines the task of visual question answering and question generation with detecting diseases within the gastrointestinal (GI) tract. Medical doctors usually examine the GI tract using colonoscopy, gastroscopy or capsule endoscopy. For the VQA task, we combine images taken from the procedures with medically relevant questions and answers. In total, the task has three subtasks: (i) The visual question answering (VQA) subtask asks participants to generate text answers given a text question and image pair. For example, we provide an image containing a colon polyp with the following question: "Where in the image is the polyp located?". Here, the answer should be a textual description of where in the image the polyp is located, like the upper-left or in the center of the image. Example questions could be "How many findings are in the image?", "What are the colors of the findings?", etc. (ii) The visual question generation (VQG) subtask requires participants to generate text questions based on a given text answer and image pair. This task can be seen as the inverse of VQA, where instead of generating the answer, we are asking for the question. An example

---

[3] https://www.imageclef.org/2023/medical/gans.

[4] https://www.imageclef.org/2023/medical/vqa.

could be that given the answer "The image contains a polyp" and an image containing a polyp, the question should be "Does the image contain an abnormality?". (iii) The visual location question answering (VLQA) subtask where the participants get an image and a question and are required to answer it by providing a segmentation mask for the image. Example questions are: "Where in the image is the polyp?", "Where in the image is the normal and the diseased part?", "What part of the image shows normal mucosa?" The data is based on the HyperKvasir dataset [2] with additional question-and-answer ground truth verified by medical doctors. It includes images spanning the entire GI tract and will include abnormalities, surgical instruments, and normal findings from gastroscopy, colonoscopy and capsule endoscopy procedures. For VQA and VQG, at least 5,000 image samples, each with five question-and-answer pairs will be provided. For VLQA at least 1,000 images with question and segmentation mask pairs are given. Evaluation will be performed using well know metrics suitable for medical applications such as precision, recall, F1 score, and Matthew correlation coefficient [12].

ImageCLEFmedical-mediqa[5]. The *MEDIQA-Sum* task is a new pilot task that focuses on automatic note generation from patient-clinician conversations, a challenging task that encompasses spoken language understanding and clinical note generation. MEDIQA-Sum 2023 will include three subtasks: (i) the Dialogue2Topic Classification subtask focuses on identifying the topic associated with a conversation snippet between a doctor and patient, (ii) the Dialogue2Note Summarization subtask focuses on producing a clinical note section text summarizing a conversation snippet between a doctor and a patient, and (iii) the Full-Encounter Dialogue2Note Summarization subtask tackles the generation of a full clinical note summarizing a full doctor-patient encounter conversation. New datasets have been created for the MEDIQA subtasks. The Dialogue2Note dataset was created based on clinical notes and corresponding conversations written by domain experts. The Full-Encounter Dialogue2Note dataset consists of full doctor-patient encounters and corresponding notes written by medical scribes. We will measure topic prediction with standard classification metrics such as F1 and accuracy. The two Dialogue2Note subtasks will use SOTA language generation metrics including ROUGE [16], BERTScore [32], and BLEURT [26].

3   ImageCLEFaware

When users contribute to online platforms they share data in a given context, which is controlled and understood by them. These data are then stored and can later be reused in contexts which were not anticipated initially. Such reuse can be directed toward impactful situations, and can have serious consequences for the users' real lives. For instance, future employers often search online information about prospective candidates. This process can involve humans or be automated using Artificial Intelligence tools. Users should be aware that such inferences are possible, with potentially detrimental effects for them.

---

[5] https://www.imageclef.org/2023/medical/mediqa.

Photos represent a large part of the data shared online, and the ImageCLE-FAware[6] task focuses on their usage. Given a set of user photographic profiles, and four modeled situations (search for a bank loan, an accommodation, a waiter job, or an IT job), the objective consists in providing feedback to the users about how their profiles compare to those of a community of reference. Users' photos are manually labeled with an appeal score by several annotators, and an average score per situation is computed and serves as a ground truth. User profiles are created by automatically detecting visual objects in users' images, and the resulting profiles can be used to automatically rate and rank profiles. Correlation between automatic and manual profile rankings will be measured using a classical measure such as the Pearson correlation coefficient.

While the global objective remains unchanged since the first edition, the dataset will be enriched and updated for the third edition of the task. The main changes refer to: (1) a larger number of user profiles to make the dataset more robust, and (2) a new object detector based on EfficientDet, to provide better profiles. Among the resources associated to the proposed tasks that will be provided to the participants in different communities, we include: (i) visual object ratings per situation obtained through crowdsourcing; (ii) automatically extracted visual object detections for over 350 objects which have non-null rating in at least one situation, using a new object detector compared to 2022.

The dataset includes personal data, and strong anonymization is performed in order to comply with EU's General Data Protection Regulation. Participants receive only the object detections which compose the profile. Furthermore, the names of these objects are anonymized.

4   ImageCLEFfusion

Late fusion approaches represent one of the goto methods of improving machine learning performance in particular domains, where single-system performance may not be acceptable, or even in critical systems, where every improvement is vital. There are numerous examples in the literature where the top performers on specific datasets or even entire domains are represented by late fusion systems that use several models and fuse their predictions. Some of these examples would be the prediction of media memorability [1] and interestingness [30], the detection of violent video scenes [5], and human action recognition [28]. As the previous edition of the ImageCLEFfusion task shows [27], numerous types of approaches to fusion exist, ranging from simple statistical approaches to more complex systems that use traditional machine learning methods like k-Nearest Neighbours (kNN) or Support Vector Regression (SVR), or deep neural network-based learning, and even using more than one stage of fusion in order to create the final set of predictions.

In this context we propose the second edition of the ImageCLEFfusion[7] task, a follow-up to last year's edition [27]. In the first edition, two tasks are defined as

---

[6] https://www.imageclef.org/2023/aware.

[7] https://www.imageclef.org/2023/fusion.

two different machine learning task types, namely: (i) a regression scenario that uses data associated with the prediction of multimedia interestingness, extracted from the Interestingness10k dataset [4], and (ii) a retrieval scenario, using data that targets the retrieval of diverse social images extracted from the Div150 challenge [13]. Annotation data, metrics, inducers, and all other tools are developed and published during the respective benchmarking campaigns, and are provided to participants to the fusion task. For this edition, we propose integrating a third task, that targets another type of machine learning task, namely multi-label classification. Thus, we will integrate data associated with the Concept Detection task from the ImageCLEFmedical caption task [25].

5 ImageCLEFrecommending

ImageCLEFrecommending[8] is a new task which focuses on content-recommendation for cultural heritage content. Despite current advances in content-based recommendation systems, there is limited understanding how well these perform and how relevant they are for the final end-users. This task aims to fill this gap by developing ground truth data of recommendations and by allowing benchmarking different recommendation systems and methods.

The task targets a key infrastructure for researchers and heritage professionals, namely Europeana [9]. With over 53 million records, the single search bar that served as the main access point was identified as a bottleneck by many users. Thus, the strategy has gradually shifted towards exploration of the available collections based on themes. Now users can explore over 60 curated digital exhibitions, countless galleries and blog posts (to be refered to as *editorials*). The metadata of the content items available on Europeana is in most cases very rich and must follow a very well defined structure given by the Europeana Data Model [7]. The current Europeana Recommendation Engine [19] focuses only on providing recommended items based on the content of a gallery, which is a collection of items with a title and optional description. The recommendations are based on similarity of the most important metadata fields of the items (*dc:title, dc:creator, dc:subject, dc:date and dc:description*) into a sentence embedding. However, recommendations for editorials are done at the moment only manually. For instance when a new blog is created, the author would manually provide a list of related galleries, blogs or exhibitions that have been already published.

The task requires participants to devise recommendation methods and systems, apply them in the supplied data set gathered from Europeana and provide a series of recommendations for items and editorials. The task is thus divided into two sub-tasks: (i) given a list of items, provide a list of recommended items; (ii) given an editorial (Europeana blog or gallery), provide a list of recommended editorials. For the task a new dataset based on Europeana items and editorials will be provided to the participants. The individual items in the dataset will include a wealth of metadata based on the Europeana Data Model schema. Performance will be evaluated on the basis of the recommendations that are

---

[8] https://www.imageclef.org/2023/recommending.

provided computing Mean Average Precision at X (Map@X) compared to the ground truth. Moreover, because black-box systems make it difficult for users to assess why the recommendation should be trusted, the systems in this task that can provide an explanation for the results provided will be awarded additional points in terms of evaluation metrics.

6   Conclusions

The current paper provides an overview of the tasks proposed by the 2023 Image-CLEF evaluation campaign organised in the framework of the 14th CLEF Conference and Labs of the Evaluation Forum scheduled for September 2023, in Thessaloniki, Greece. Since 2003, the tasks proposed by the ImageCLEF lab gained popularity being held for the evaluation of technologies for *annotation*, *indexing, classification* and *retrieval* of multimodal data, with the objective of providing information access to large collections in various usage scenarios and domains. The 21st edition brings several new tasks, e.g., content recommending, generative adversarial networks, visual question answering and generation, medical dialogue summarization, while some old tasks were discontinued. All the tasks are solving challenging current issues and will provide a set of new test collections simulating real-world situations. Such collections are important to enable researchers to assess the performance of their systems and to compare their results with others following a common evaluation framework.

Acknowledgement. The lab is supported under the H2020 AI4Media "A European Excellence Centre for Media, Society and Democracy" project, contract #951911, as well as the ImageCLEFaware, ImageCLEFfusion and ImageCLEFrecommending tasks. The work of Louise Bloch and Raphael Brüngel was partially funded by a PhD grant from the University of Applied Sciences and Arts Dortmund (FH Dortmund), Germany. The work of Ahmad Idrissi-Yaghir and Henning Schäfer was funded by a PhD grant from the DFG Research Training Group 2535 Knowledge- and data-based personalisation of medicine at the point of care (WisPerMed).

References

1. Azcona, D., Moreu, E., Hu, F., Ward, T.E., Smeaton, A.F.: Predicting media memorability using ensemble models. In: Working Notes Proceedings of the MediaEval 2019 Workshop. CEUR Workshop Proceedings, vol. 2670. CEUR-WS.org (2019)
2. Borgli, H., et al.: Hyperkvasir, a comprehensive multi-class image and video dataset for gastrointestinal endoscopy. Sci. Data **7**(1), 1–14 (2020)
3. Cao, Y., Shui, R., Pan, L., Kan, M.Y., Liu, Z., Chua, T.S.: Expertise style transfer: a new task towards better communication between experts and laymen. In: Proceedings of the 58th Annual Meeting of the Association for Computational Linguistics, pp. 1061–1071. Association for Computational Linguistics (2020). https://doi.org/10.18653/v1/2020.acl-main.100. https://aclanthology.org/2020.acl-main.100
4. Constantin, M.G., Ştefan, L.D., Ionescu, B., Duong, N.Q., Demarty, C.H., Sjöberg, M.: Visual interestingness prediction: a benchmark framework and literature review. Int. J. Comput. Vis. 1–25 (2021)

5. Dai, Q., et al.: Fudan-Huawei at MediaEval 2015: detecting violent scenes and affective impact in movies with deep learning. In: Working Notes Proceedings of the MediaEval 2015 Workshop. CEUR Workshop Proceedings, vol. 1436. CEUR-WS.org (2015)

6. Devlin, J., Chang, M.W., Lee, K., Toutanova, K.: BERT: pre-training of deep bidirectional transformers for language understanding. In: Proceedings of the 2019 Conference of the North American Chapter of the Association for Computational Linguistics: Human Language Technologies, Volume 1 (Long and Short Papers), Minneapolis, Minnesota, pp. 4171–4186. Association for Computational Linguistics (2019). https://doi.org/10.18653/v1/N19-1423. https://aclanthology.org/N19-1423D

7. Doerr, M., Gradmann, S., Hennicke, S., Isaac, A., Meghini, C., Sompel, H.: The Europeana data model (EDM). In: World Library and Information Congress: 76th IFLA General Conference and Assembly, pp. 10–15 (2010)

8. Eickhoff, C., Schwall, I., García Seco de Herrera, A., Müller, H.: Overview of Image-CLEFcaption 2017 - the image caption prediction and concept extraction tasks to understand biomedical images. In: Working Notes of Conference and Labs of the Evaluation Forum (CLEF 2017). CEUR Workshop Proceedings, vol. 1866. CEUR-WS.org (2017)

9. Europeana Foundation: Europeana (2022). https://www.europeana.eu/

10. García Seco De Herrera, A., Eickhof, C., Andrearczyk, V., Müller, H.: Overview of the ImageCLEF 2018 caption prediction tasks. In: Working Notes of Conference and Labs of the Evaluation Forum (CLEF 2018). CEUR Workshop Proceedings, vol. 2125. CEUR-WS.org (2018)

11. Goutte, C., Gaussier, E.: A probabilistic interpretation of precision, recall and F-score, with implication for evaluation. In: Losada, D.E., Fernández-Luna, J.M. (eds.) ECIR 2005. LNCS, vol. 3408, pp. 345–359. Springer, Heidelberg (2005). https://doi.org/10.1007/978-3-540-31865-1_25

12. Hicks, S.A., et al.: On evaluation metrics for medical applications of artificial intelligence. Sci. Rep. **12**(1), 1–9 (2022)

13. Ionescu, B., Rohm, M., Boteanu, B., Gînscă, A.L., Lupu, M., Müller, H.: Benchmarking image retrieval diversification techniques for social media. IEEE Trans. Multimedia **23**, 677–691 (2020)

14. Kalpathy-Cramer, J., García Seco de Herrera, A., Demner-Fushman, D., Antani, S., Bedrick, S., Müller, H.: Evaluating performance of biomedical image retrieval systems: overview of the medical image retrieval task at ImageCLEF 2004–2014. Comput. Med. Imaging Graph. **39**, 55–61 (2015)

15. Li, J., Jia, R., He, H., Liang, P.: Delete, retrieve, generate: a simple approach to sentiment and style transfer. In: Proceedings of the 2018 Conference of the North American Chapter of the Association for Computational Linguistics: Human Language Technologies, Volume 1 (Long Papers), New Orleans, Louisiana, pp. 1865–1874. Association for Computational Linguistics (2018). https://doi.org/10.18653/v1/N18-1169. https://aclanthology.org/N18-1169

16. Lin, C.Y.: ROUGE: a package for automatic evaluation of summaries. In: Text Summarization Branches Out, Barcelona, Spain, pp. 74–81. Association for Computational Linguistics (2004). https://aclanthology.org/W04-1013

17. Müller, H., Clough, P., Deselaers, T., Caputo, B. (eds.): ImageCLEF - Experimental Evaluation in Visual Information Retrieval. The Information Retrieval Series, vol. 32. Springer, Heidelberg (2010). https://doi.org/10.1007/978-3-642-15181-1

18. Müller, H., Kalpathy-Cramer, J., García Seco de Herrera, A.: Experiences from the ImageCLEF medical retrieval and annotation tasks. In: Information Retrieval Evaluation in a Changing World. TIRS, vol. 41, pp. 231–250. Springer, Cham (2019). https://doi.org/10.1007/978-3-030-22948-1_10

19. Pangeanic, Anacode and Europeana Foundation: The recommendation system (2022). https://pro.europeana.eu/page/the-recommendation-system

20. Papineni, K., Roukos, S., Ward, T., Zhu, W.J.: BLEU: a method for automatic evaluation of machine translation. In: Proceedings of the 40th Annual Meeting of the Association for Computational Linguistics (ACL 2002), pp. 311–318 (2002)

21. Pelka, O., Abacha, A.B., García Seco de Herrera, A., Jacutprakart, J., Friedrich, C.M., Müller, H.: Overview of the ImageCLEFmed 2021 concept & caption prediction task. In: Working Notes of Conference and Labs of the Evaluation Forum (CLEF 2021). CEUR Workshop Proceedings, vol. 2936. CEUR-WS.org (2021)

22. Pelka, O., Friedrich, C.M., García Seco de Herrera, A., Müller, H.: Overview of the ImageCLEFmed 2019 concept detection task. In: Working Notes of Conference and Labs of the Evaluation Forum (CLEF 2019). CEUR Workshop Proceedings, vol. 2380. CEUR-WS.org (2019)

23. Pelka, O., Friedrich, C.M., García Seco de Herrera, A., Müller, H.: Overview of the ImageCLEFmed 2020 concept prediction task: medical image understanding. In: Working Notes of Conference and Labs of the Evaluation Forum (CLEF 2020). CEUR Workshop Proceedings, vol. 2696. CEUR-WS.org (2020)

24. Pelka, O., Koitka, S., Rückert, J., Nensa, F., Friedrich, C.M.: Radiology objects in COntext (ROCO): a multimodal image dataset. In: Stoyanov, D., et al. (eds.) LABELS/CVII/STENT -2018. LNCS, vol. 11043, pp. 180–189. Springer, Cham (2018). https://doi.org/10.1007/978-3-030-01364-6_20

25. Rückert, J., et al.: Overview of ImageCLEFmedical 2022 - caption prediction and concept detection. In: CLEF2022 Working Notes, Bologna, Italy, 5–8 September 2022. CEUR Workshop Proceedings, CEUR-WS.org (2022)

26. Sellam, T., Das, D., Parikh, A.: BLEURT: learning robust metrics for text generation. In: Proceedings of the 58th Annual Meeting of the Association for Computational Linguistics, pp. 7881–7892. Association for Computational Linguistics (2020). https://doi.org/10.18653/v1/2020.acl-main.704. https://aclanthology.org/2020.acl-main.704

27. Ştefan, L.D., Constantin, M.G., Dogariu, M., Ionescu, B.: Overview of imagecleffusion 2022 task-ensembling methods for media interestingness prediction and result diversification. In: Working Notes of Conference and Labs of the Evaluation Forum (CLEF 2022). CEUR Workshop Proceedings, CEUR-WS.org (2022)

28. Sudhakaran, S., Escalera, S., Lanz, O.: Gate-shift networks for video action recognition. In: Proceedings of the IEEE/CVF Conference on Computer Vision and Pattern Recognition (CVPR 2020), pp. 1102–1111 (2020)

29. Tsikrika, T., de Herrera, A.G.S., Müller, H.: Assessing the scholarly impact of ImageCLEF. In: Forner, P., Gonzalo, J., Kekäläinen, J., Lalmas, M., de Rijke, M. (eds.) CLEF 2011. LNCS, vol. 6941, pp. 95–106. Springer, Heidelberg (2011). https://doi.org/10.1007/978-3-642-23708-9_12

30. Wang, S., Chen, S., Zhao, J., Jin, Q.: Video interestingness prediction based on ranking model. In: Proceedings of the Joint Workshop of the 4th Workshop on Affective Social Multimedia Computing and First Multi-Modal Affective Computing of Large-Scale Multimedia Data (ASMMC-MMAC 2018), pp. 55–61. Association for Computing Machinery (ACM) (2018)

31. Xu, W., Saxon, M., Sra, M., Wang, W.Y.: Self-supervised knowledge assimilation for expert-layman text style transfer. In: Proceedings of the AAAI Conference on Artificial Intelligence, vol. 36, no. 10, pp. 11566–11574 (2022). https://doi.org/10.1609/aaai.v36i10.21410. https://ojs.aaai.org/index.php/AAAI/article/view/21410

32. Zhang, T., Kishore, V., Wu, F., Weinberger, K.Q., Artzi, Y.: Bertscore: evaluating text generation with BERT. In: 8th International Conference on Learning Representations, ICLR 2020, Addis Ababa, Ethiopia, 26–30 April 2020 (2020). https://openreview.net/forum?id=SkeHuCVFDr

LifeCLEF 2023 Teaser: Species Identification and Prediction Challenges

Alexis Joly[1,13(✉)], Hervé Goëau[2,13], Stefan Kahl[6,13], Lukáš Picek[9,13],
Christophe Botella[13,14], Diego Marcos[13,16], Milan Šulc[12,13],
Marek Hrúz[9,13], Titouan Lorieul[1,13], Sara Si Moussi[17],
Maximilien Servajean[7,13], Benjamin Kellenberger[13,15], Elijah Cole[8,13],
Andrew Durso[10,13], Hervé Glotin[3,13], Robert Planqué[4,13],
Willem-Pier Vellinga[4,13], Holger Klinck[6,13], Tom Denton[11,13], Ivan Eggel[5,13],
Pierre Bonnet[2,13], and Henning Müller[5,13]

[1] Inria, LIRMM, Univ Montpellier, CNRS, Montpellier, France
alexis.joly@inria.fr
[2] CIRAD, UMR AMAP, Montpellier, Occitanie, France
[3] Univ. Toulon, Aix Marseille Univ., CNRS, LIS, DYNI Team, Marseille, France
[4] Xeno-canto Foundation, The Hague, The Netherlands
[5] HES-SO, Sierre, Switzerland
[6] KLYCCB, Cornell Lab of Ornithology, Cornell University, Ithaca, USA
[7] LIRMM, AMI, Univ Paul Valéry Montpellier, Univ Montpellier, CNRS,
Montpellier, France
[8] Department of Computing and Mathematical Sciences, Caltech, Pasadena, USA
[9] Department of Cybernetics, FAV, University of West Bohemia, Pilsen, Czechia
[10] Department of Biological Sciences, Florida Gulf Coast University,
Fort Myers, USA
[11] Google LLC, San Francisco, USA
[12] Rossum.ai, Prague, Czechia
[13] Listening Observatory for Hawaiian Ecosystems, Univ. of Hawai'i at Hilo,
Hilo, USA
[14] Centre for Invasion Biology, Stellenbosch University, Stellenbosch, South Africa
[15] Department of Ecology and Evolutionary Biology, Yale University,
New Haven, USA
[16] Inria, TETIS, Univ Montpellier, Montpellier, France
[17] Mansfield, USA

Abstract. Building accurate knowledge of the identity, the geographic
distribution and the evolution of species is essential for the sustainable
development of humanity, as well as for biodiversity conservation. However, the difficulty of identifying plants, animals and fungi is hindering
the aggregation of new data and knowledge. Identifying and naming living organisms is almost impossible for the general public and is often
difficult, even for professionals and naturalists. Bridging this gap is a
key step towards enabling effective biodiversity monitoring systems. The
LifeCLEF campaign, presented in this paper, has been promoting and
evaluating advances in this domain since 2011. The 2023 edition proposes
five data-oriented challenges related to the identification and prediction of biodiversity: (i) PlantCLEF: very large-scale plant identification

from images, (ii) BirdCLEF: bird species recognition in audio sound-scapes, (iii) GeoLifeCLEF: remote sensing based prediction of species, (iv) SnakeCLEF: snake recognition in medically important scenarios, and (v) FungiCLEF: fungi recognition beyond 0–1 cost.

Keywords: Biodiversity · Machine learning · AI · Species identification · Species prediction · Plant identification · Bird identification · Species distribution model · Snake identification · Fungi identification

1 Introduction

Accurately identifying organisms observed in the wild is an essential step in ecological studies. Unfortunately, observing and identifying living organisms requires high levels of expertise. For instance, plants alone account for more than 400,000 different species and the distinctions between them can be quite subtle. Since the Rio Conference of 1992, this *taxonomic gap* has been recognized as one of the major obstacles to the global implementation of the Convention on Biological Diversity [1]. In 2004, Gaston and O'Neill [6] discussed the potential of automated approaches for species identification. They suggested that, if the scientific community were able to (i) produce large training datasets, (ii) precisely evaluate error rates, (iii) scale-up automated approaches, and (iv) detect novel species, then it would be possible to develop a generic automated species identification system that would open up new vistas for research in biology and related fields.

Since the publication of [6], automated species identification has been studied in many contexts [5,8,13,23,25,27,28,33]. This area continues to expand rapidly, particularly due to advances in deep learning [4,7,24,26,29–32]. In order to measure progress in a sustainable and repeatable way, the LifeCLEF [3] research platform was created in 2014 as a continuation and extension of the plant identification task [12] that had been run within the ImageCLEF lab [2] since 2011

Table 1. Overview of the data and tasks of the five LifeCLEF challenges

	Modality	#species	#items	Task	Metric
PlantCLEF	images	80,000	4.0M	Classification	Macro-Average MRR
BirdCLEF	audio	100–500	10K–50K	Multi-Label Classification	F1 score
GeoLifeCLEF	images time-series tabular	11,340	5.3M	Multi-Label Classification	Jaccard index
SnakeCLEF	images metadata	1,500	150–200K	Classification	ad-hoc metric
FungiCLEF	images metadata	1,600	300K	Classification	ad-hoc metric

[9–11]. Since 2014, LifeCLEF expanded the challenge by considering animals and fungi in addition to plants, and including audio and video content in addition to images [14–22]. LifeCLEF 2023 consists of five challenges (PlantCLEF, BirdCLEF, GeoLifeCLEF, SnakeCLEF, FungiCLEF), which we will now describe in turn. Table 1 provides an overview of the data and tasks of the five challenges.

2   PlantCLEF 2023 Challenge: Identify the World's Flora

Motivation: It is estimated that there are more than 300,000 species of vascular plants in the world. Automatic identification has made considerable progress in recent years as highlighted during previous editions of PlantCLEF. Deep learning techniques now seem mature enough to address the ultimate but realistic problem of global identification of plant biodiversity despite many problems that the data may present (a huge number of classes, very strongly unbalanced classes, partially erroneous identifications, duplicate pictures, variable visual quality, diversity of visual contents, e.g., photos, herbarium sheets, etc.).

Data Collection: The training dataset that will be used this year can be distinguished into two main categories: trusted and web (i.e., with or without a strong revision of species names by human experts). The trusted training subset will be based on a dataset of more than 2.8 million images, covering more than 80,000 plant species, shared and aggregated by the Global Biodiversity Information Facility (GBIF) platform. The web training dataset will be based on a web crawl with Google & Bing search engines. All datasets provided in previous editions of PlantCLEF can also be used; the use of external data will be possible. Finally, the test set will contain more than 60k pictures verified by world class experts related to various regions of the world and taxonomic groups.

Task Description: The task will be evaluated as a plant species retrieval task based on multi-image plant observations from the test set. The goal will be to retrieve the correct plant species among the top results of a ranked list of species returned by the evaluated system. Participants will initially have access to the training set, followed up a few months later by the whole test set. Self-supervised, semi-supervised or unsupervised approaches will be strongly encouraged and a starter package with pre-trained models will be provided.

3   BirdCLEF 2023 Challenge: Bird Species Identification in Soundscape Recordings

Motivation: Recognizing bird sounds in complex soundscapes is an important sampling tool that often helps reduce the limitations of point counts. In the future, archives of recorded soundscapes will become increasingly valuable as the habitats in which they were recorded will be lost. In the past few years, deep learning approaches have transformed the field of automated soundscape analysis. Yet, when training data is sparse, detection systems struggle with the recognition of rare species. The goal of this competition is to establish training and test datasets that can serve as real-world applicable evaluation scenarios for endangered habitats and help the scientific community to advance their conservation efforts through automated bird sound recognition.

Data Collection: We will build on the experience from previous editions and adjust the overall task to encourage participants to focus on few-shot learning and task-specific model designs. We will select training and test data to suit this demand. As in previous iterations, Xeno-canto will be the primary source for training data, expertly annotated soundscape recordings will be used for testing. We will focus on bird species for which there is limited training data, but we will also include common species so that participants can train good recognition systems. In search of suitable test data, we will consider different data sources with varying complexity (call density, chorus, signal-to-noise ratio, anthropophony ...), and quality (mono and stereo recordings). We also want to focus on very specific real-world use cases (e.g., conservation efforts in India) and frame the competition based on the demand of the particular use case. Additionally, we are considering including unlabeled data to encourage self-supervised learning regimes.

Task Description: The challenge will be held on Kaggle and the evaluation mode will resemble the 2022 test mode (i.e., hidden test data, code competition). We will use established metrics like F1 score and LwLRAP which reflect use cases for which precision is key and also allow organizers to assess system performance independent of fine-tuned confidence thresholds. Participants will be asked to return a list of species for short audio segments extracted from labeled soundscape data. In the past, we used 5-second segments, and we will consider increasing the duration of these context windows to better reflect the overall ground truth label distribution. However, the overall structure of the task will remain unchanged, as it provides a well-established base that has resulted in significant participation in past editions (e.g., 1,019 participants and 23,352 submissions in 2022). Again, we will strive to keep the dataset size reasonably small (<50 GB) and easy to process, and we will also provide introductory code repositories and write-ups to lower the entry level of the competition.

4   GeoLifeCLEF 2023 Challenge: Species Presence Prediction Based on Occurrences Data and High-Resolution Remote Sensing Images

Motivation: Predicting which species are present in a given area through species distribution modeling is a central problem in ecology and a crucial issue for biodiversity conservation. Such predictions are a fundamental element of many decision-making processes, whether for land use planning, the definition of protected areas or the implementation of more ecological agricultural practices. The models classically used in ecology are well-established but have the drawback of covering only a limited number of species at spatial resolutions often coarse in the order of kilometers or hundreds of meters at best. The objective of GeoLife-CLEF is to evaluate models with orders of magnitude hitherto unseen, whether in terms of the number of species covered (thousands), spatial resolution (on the order of 10 m), or the number of occurrences used as training data (several million). These models have the potential to greatly improve biodiversity management processes, especially at the local level (e.g. municipalities), where the need for spatial and taxonomic precision is greatest.

Data Collection: A brand new dataset will be built for the 2023 edition of GeoLifeCLEF in the framework of a large-scale European project on biodiversity monitoring (MAMBO, Horizon EU program). It will contain about 5 million species occurrences extracted from various selected datasets of the Global Biodiversity Information Facility (GBIF) and covering the whole EU territory (38 countries including E.U. members). For the explanatory variables, we will provide both high resolution remote sensing data (i.e., Sentinel-2 RGB, Near-IR, Red-Edge and SWIR, along with altitude) and coarser resolution environmental raster data (e.g., Chelsa climate, SoilGrids, land use, etc.). An important change this year will be the evaluation and test set composition. We will evaluate model ability to predict the whole set of species present in local sites using presence-absence data.

Task Description: Given a test set of locations (i.e., geo-coordinates) and corresponding high-resolution remote sensing images and environmental covariates, the goal of the task will be to return for each location the set of species that are were inventoried at that location. The test set will include only locations for which an exhaustive plant species inventory is available (i.e., in the form or presence/absence data).

5   SnakeCLEF 2023 Challenge: Snake Identification in Medically Important Scenarios

Motivation: Developing a robust system for identifying species of snakes from photographs is an important goal in biodiversity but also for human health. With over half a million victims of death & disability from venomous snakebite annually, understanding the global distribution of the $>$4,000 species of snakes and differentiating species from images (particularly images of low quality) will significantly improve epidemiology data and treatment outcomes. We have learned from previous editions that "machines" can accurately recognize ($F_1^C \approx 90\%$ and Top1 Accuracy $\approx 90\%$) even in scenarios with long-tailed distributions and $\approx 1,600$ species. Thus, testing over real Medically Important Scenarios and specific countries (primarily tropical and subtropical) and integrating the medical importance of species is the next step that should provide a more reliable machine prediction.

Data Collection: The dataset of the previous year will be extended up to $\approx 1,800\%$ snake species from around the world (minimum 10 images per species). The images will be divided into observations that depict the same snake individual. Additionally, medical importance (i.e. how venomous the species is) and country-species relevance will be provided for each species. The evaluation will be done on various subsets of a newly created "secret" test set with around 50k images.

Task Description: Given the set of authentic snake species observations and corresponding locations, the goal of the task is to create a classification model that, for each observation, returns a ranked list of predicted species. The classification model will have to fit limits for memory footprint (ONNX model with max size of 1 GB) and prediction time limit (will be announced later) measured on the submission server. The model should have to consider and minimize the danger to human life and the waste of antivenom if a bite from the snake in the image were treated as coming from the top-ranked prediction.

6 FungiCLEF 2023 Challenge: Fungi Recognition Beyond 0–1 Cost

Motivation: Automatic recognition of species at scale, such as in popular citizen-science projects, requires efficient prediction on limited resources. In practice, species identification typically depends not solely on the visual observation of the specimen but also on other information available to the observer, e.g., habitat, substrate, location and time. Thanks to rich metadata, precise annotations, and baselines available to all competitors, the challenge aims at providing a major benchmark for combining visual observations with other observed information. Additionally, since mushrooms are often picked for consumption, misclassification of edible and poisonous mushrooms is an important aspect for the evaluation of the practical prediction loss.

Data Collection: The dataset comes from a citizen science project, the Atlas of Danish Fungi, where all samples went through an expert validation process, guaranteeing a high quality of labels. Rich metadata (Habitat, Substrate, Timestamp, GPS, EXIF etc.) are provided for most samples. The training set will be the union of the training and test set (without out-of-scope samples) from the 2022 challenge (i.e. 295,938 training images belonging to 1,604 species observed mostly in Denmark). The test set will consist of new fungi observations from the citizen science project.

Task Description: Given the set of real fungi species observations and corresponding metadata, the goal of the task is to create a classification model that, for each observation, returns a ranked list of predicted species. The classification model will have to fit limits for memory footprint (ONNX model with max size of 1 GB) and prediction time limit (will be announced later) measured on the submission server. The model should have to consider and minimize the danger to human life, i.e., the confusion between poisonous and edible species. Baseline procedures of how metadata can help the classification and pre-trained baseline classifiers will be provided as part of the task description to all participants.

7   Timeline and Registration Instructions

All information about the timeline and participation in the challenges is provided on the LifeCLEF 2023 web page [3].

8   Discussion and Conclusion

To fully reach its objective, an evaluation campaign such as LifeCLEF requires a long-term research effort so as to (i) encourage non-incremental contributions, (ii) measure consistent performance gaps, (iii) progressively scale-up the problem and (iv) enable the emergence of a strong community. The 2023 edition of the lab supports this vision and also includes the following innovations:

– The GeoLifeCLEF challenge will be entirely revisited towards running at the scale of the whole Europe thanks to a newly created dataset with millions of occurrences paired with high-resolution remote sensing data.
– The BirdCLEF challenge will include new data with a focus on Central African species.
– The world's coverage of PlantCLEF test set will be improved with a focus on tropical regions and biodiversity hotspots.
– The inclusion of time- and memory-limits within several challenges to encourage the use of frugal methods rather than ensembles of tens of models.

Acknowledgements. This work has received funding from the European Union's Horizon research and innovation program under grant agreement No 101060639 (MAMBO project).

References

1. Convention on Biodiversity. www.cbd.int/
2. ImageCLEF. www.imageclef.org/
3. LifeCLEF. www.lifeclef.org/
4. Bonnet, P., et al.: Plant identification: Experts vs. Machines in the era of deep learning. In: Joly, A., Vrochidis, S., Karatzas, K., Karppinen, A., Bonnet, P. (eds.) Multimedia Tools and Applications for Environmental & Biodiversity Informatics. MSA, pp. 131–149. Springer, Cham (2018). https://doi.org/10.1007/978-3-319-76445-0_8

5. Cai, J., Ee, D., Pham, B., Roe, P., Zhang, J.: Sensor network for the monitoring of ecosystem: bird species recognition. In: 3rd International Conference on Intelligent Sensors, Sensor Networks and Information, ISSNIP 2007 (2007). https://doi.org/10.1109/ISSNIP.2007.4496859

6. Gaston, K.J., O'Neill, M.A.: Automated species identification: why not? Philos. Trans. R. Soc. London B: Biol. Sci. **359**(1444), 655–667 (2004)

7. Ghazi, M.M., Yanikoglu, B., Aptoula, E.: Plant identification using deep neural networks via optimization of transfer learning parameters. Neurocomputing **235**, 228–235 (2017)

8. Glotin, H., Clark, C., LeCun, Y., Dugan, P., Halkias, X., Sueur, J.: Proceedings of 1st Workshop on Machine Learning for Bioacoustics - ICML4B. ICML, Atlanta, USA (2013). sabiod.org/ICML4B2013_book.pdf

9. Goëau, H., et al.: The imageCLEF 2013 plant identification task. In: CLEF Task Overview 2013, CLEF: Conference and Labs of the Evaluation Forum, September 2013, Valencia, Spain (2013)

10. Goëau, H., et al.: The imageCLEF 2011 plant images classification task. In: CLEF Task Overview 2011, CLEF: Conference and Labs of the Evaluation Forum, September 2011, Amsterdam, Netherlands (2011)

11. Goëau, H., et al.: ImageCLEF 2012 plant images identification task. In: CLEF Task Overview 2012, CLEF: Conference and Labs of the Evaluation Forum, September 2012, Rome, Italy (2012)

12. Goëau, H., et al.: The imageCLEF plant identification task 2013. In: Proceedings of the 2nd ACM International Workshop on Multimedia Analysis for Ecological Data, pp. 23–28. ACM (2013)

13. Joly, A., et al.: Interactive plant identification based on social image data. Ecol. Informatics **23**, 22–34 (2014)

14. Joly, A., et al.: Overview of LifeCLEF 2018: a large-scale evaluation of species identification and recommendation algorithms in the era of AI. In: Jones, G.J., et al. (eds.) CLEF: Cross-Language Evaluation Forum for European Languages. Experimental IR Meets Multilinguality, Multimodality, and Interaction. LNCS, vol. 11018. Springer, Cham (2018). https://doi.org/10.1007/978-3-319-98932-7_24

15. Joly, A., et al.: Overview of LifeCLEF 2019: identification of Amazonian plants, South & North American birds, and Niche prediction. In: Crestani, F., et al. (eds.) CLEF 2019 - Conference and Labs of the Evaluation Forum. Experimental IR Meets Multilinguality, Multimodality, and Interaction, Lugano, Switzerland. LNCS, vol. 11696, pp. 387–401. Springer, Cham (2019). https://doi.org/10.1007/978-3-030-28577-7_29, hal.umontpellier.fr/hal-02281455

16. Joly, A., et al.: LifeCLEF 2016: multimedia life species identification challenges. In: Fuhr, N., et al. (eds.) CLEF: Cross-Language Evaluation Forum. Experimental IR Meets Multilinguality, Multimodality, and Interaction, Évora, Portugal. LNCS, vol. 9822, pp. 286–310. Springer, Cham (2016). https://doi.org/10.1007/978-3-319-44564-9_26, hal.archives-ouvertes.fr/hal-01373781

17. Joly, A., et al.: LifeCLEF 2017 lab overview: multimedia species identification challenges. In: Jones, G.J., et al. (eds.) CLEF: Cross-Language Evaluation Forum. Experimental IR Meets Multilinguality, Multimodality, and Interaction, Dublin, Ireland. LNCS, vol. 10456, pp. 255–274. Springer, Cham (2017). https://doi.org/10.1007/978-3-319-65813-1_24, hal.archives-ouvertes.fr/hal-01629191

18. Joly, A., et al.: LifeCLEF 2014: multimedia life species identification challenges. In: CLEF: Cross-Language Evaluation Forum. Information Access Evaluation. Multilinguality, Multimodality, and Interaction, Sheffield, United Kingdom. LNCS, vol. 8685, pp. 229–249. Springer, Cham (2014). https://doi.org/10.1007/978-3-319-11382-1_20, hal.inria.fr/hal-01075770

19. Joly, A., et al.: LifeCLEF 2015: multimedia life species identification challenges. In: Mothe, J., et al. (eds.) CLEF 2015. LNCS, vol. 9283, pp. 462–483. Springer, Cham (2015). https://doi.org/10.1007/978-3-319-24027-5_46

20. Joly, A., et al.: Overview of LifeCLEF 2020: a system-oriented evaluation of automated species identification and species distribution prediction. In: Arampatzis, A., et al. (eds.) CLEF 2020. LNCS, vol. 12260, pp. 342–363. Springer, Cham (2020). https://doi.org/10.1007/978-3-030-58219-7_23

21. Joly, A., et al.: Overview of LifeCLEF 2022: an evaluation of machine-learning based species identification and species distribution prediction. In: International Conference of the Cross-Language Evaluation Forum for European Languages. LNCS, vol. 13390, pp. 257–285. Springer, Cham (2022). https://doi.org/10.1007/978-3-031-13643-6_19

22. Joly, A., et al.: Overview of LifeCLEF 2021: an evaluation of machine-learning based species identification and species distribution prediction. In: Candan, K.S., et al. (eds.) CLEF 2021. LNCS, vol. 12880, pp. 371–393. Springer, Cham (2021). https://doi.org/10.1007/978-3-030-85251-1_24

23. Lee, D.J., Schoenberger, R.B., Shiozawa, D., Xu, X., Zhan, P.: Contour matching for a fish recognition and migration-monitoring system. In: Optics East, pp. 37–48. International Society for Optics and Photonics (2004)

24. Lee, S.H., Chan, C.S., Remagnino, P.: Multi-organ plant classification based on convolutional and recurrent neural networks. IEEE Trans. Image Process. **27**(9), 4287–4301 (2018)

25. NIPS International Conference: Proceedings of Neural Information Processing Scaled for Bioacoustics, from Neurons to Big Data (2013). sabiod.org/nips4b

26. Norouzzadeh, M.S., Morris, D., Beery, S., Joshi, N., Jojic, N., Clune, J.: A deep active learning system for species identification and counting in camera trap images. Methods Ecol. Evol. **12**(1), 150–161 (2021)

27. Towsey, M., Planitz, B., Nantes, A., Wimmer, J., Roe, P.: A toolbox for animal call recognition. Bioacoustics **21**(2), 107–125 (2012)

28. Trifa, V.M., Kirschel, A.N., Taylor, C.E., Vallejo, E.E.: Automated species recognition of antbirds in a Mexican rainforest using hidden Markov models. J. Acoust. Soc. Am. **123**, 2424 (2008)

29. Van Horn, G., et al.: The iNaturalist species classification and detection dataset. CVPR (2018)

30. Villon, S., Mouillot, D., Chaumont, M., Subsol, G., Claverie, T., Villéger, S.: A new method to control error rates in automated species identification with deep learning algorithms. Sci. Rep. **10**(1), 1–13 (2020)

31. Wäldchen, J., Mäder, P.: Machine learning for image based species identification. Methods Ecol. Evol. **9**(11), 2216–2225 (2018)

32. Wäldchen, J., Rzanny, M., Seeland, M., Mäder, P.: Automated plant species identification-trends and future directions. PLoS Comput. Biol. **14**(4), e1005993 (2018)

33. Yu, X., Wang, J., Kays, R., Jansen, P.A., Wang, T., Huang, T.: Automated identification of animal species in camera trap images. EURASIP J. Image Video Process. **2013**, 52 (2013)

BioASQ at CLEF2023: The Eleventh Edition of the Large-Scale Biomedical Semantic Indexing and Question Answering Challenge

Anastasios Nentidis[1,2(✉)], Anastasia Krithara[1], Georgios Paliouras[1], Eulalia Farre-Maduell[3], Salvador Lima-Lopez[3], and Martin Krallinger[3]

[1] National Center for Scientific Research "Demokritos", Athens, Greece
{tasosnent,akrithara,paliourg}@iit.demokritos.gr
[2] Aristotle University of Thessaloniki, Thessaloniki, Greece
nentidis@csd.auth.gr
[3] Barcelona Supercomputing Center, Barcelona, Spain
{eulalia.farre,salvador.limalopez,martin.krallinger}@bsc.es

Abstract. The large-scale biomedical semantic indexing and question-answering challenge (BioASQ) aims at the continuous advancement of methods and tools to meet the need of biomedical researchers and practitioners for efficient and precise access to the ever-increasing resources of their domain. With this purpose, during the last ten years a series of annual challenges have been organized with specific shared tasks on large-scale biomedical semantic indexing and question answering. Benchmark datasets have been concomitantly provided in alignment with the real needs of biomedical experts. BioASQ provides a unique common testbed where different teams around the world can investigate and compare new approaches for identifying and accessing biomedical knowledge. The eleventh version of the BioASQ Challenge will be held as an evaluation Lab within CLEF2023. In this version, three shared tasks will be presented: (i) the automated retrieval of relevant material for biomedical questions, and the generation of comprehensible answers. (ii) the synergistic retrieval of relevant material and generation of answers for open biomedical questions about developing topics, in collaboration with the experts posing the questions. (iii) the automated indexing of unlabelled clinical procedures-specific medical documents, primarily clinical case reports written in Spanish, with biomedical concepts and the extraction of human-interpretable evidence. As BioASQ rewards the methods that outperform the state of the art in these shared tasks, it pushes the research frontier towards approaches that accelerate access to biomedical knowledge.

Keywords: Biomedical information · Semantic indexing · Question answering

© The Author(s), under exclusive license to Springer Nature Switzerland AG 2023
J. Kamps et al. (Eds.): ECIR 2023, LNCS 13982, pp. 577–584, 2023.
https://doi.org/10.1007/978-3-031-28241-6_66

1   Introduction

BioASQ[1] [20] is a series of international challenges and workshops on biomedical semantic indexing and question answering. Each edition of BioASQ is structured into distinct but complementary tasks and sub-tasks relevant to biomedical information access. As a result, the participating teams can focus on particular tasks of interest to their specific area of expertise, including but not limited to hierarchical text classification, machine learning, information retrieval, and multi-document query-focused summarization. The BioASQ challenge has been running annually since 2012, with the participation of more than 100 teams from 28 countries. The BioASQ workshop has been taking place at the CLEF conference till 2015. In 2016 and 2017 it took place in ACL, in conjunction with the BioNLP workshop [1]. In 2018 and 2019, it took place respectively in EMNLP and ECML as an independent workshop. Since 2020 the BioASQ workshop become again part of CLEF [5,7,15].

BioASQ allows multiple teams that work on biomedical information access systems around the world, to compete in the same realistic benchmark datasets and share, evaluate, and compare their ideas and approaches. Therefore, a key contribution of BioASQ are the benchmark datasets developed for its tasks, as well as the corresponding open-source infrastructure developed for running the challenges. In particular, as BioASQ consistently rewards the most successful approaches in each task and sub-task, it eventually pushes toward systems that outperform previous approaches. Such successful approaches for semantic indexing and question answering can eventually lead to the development of tools to support more precise access to valuable biomedical knowledge and to further improve health services.

Notably, the performance of MTI, the system developed by the National Library of Medicine (NLM) for assisting the manual semantic indexing of MEDLINE, has improved by almost 10% during the last 10 years, largely due to the adoption of ideas from the systems that compete in the large-scale biomedical semantic indexing task (*Task a*) of the BioASQ challenge [13,22]. The high point for this task was the recent adoption of fully automated indexing by NLM in mid-2022[2]. In short, ten years after its initial introduction, this task fulfilled its goal of facilitating the advancement of biomedical semantic indexing research. However, major advancement is still needed regarding biomedical question-answering, as well as the semantic indexing of other types of documents, such as clinical case reports and documents in languages beyond English.

2   BioASQ Evaluation Lab 2023

The eleventh BioASQ challenge (BioASQ11) will consist of three tasks that are central to biomedical knowledge access and the question-answering process: (i)

---

[1] http://www.bioasq.org.

[2] https://www.nlm.nih.gov/pubs/techbull/nd21/nd21_medline_2022.html.

Task b[3] on the processing of biomedical questions, the generation of answers, and the retrieval of supporting material, (ii) *Task Synergy* on biomedical question answering for developing problems under a scenario that promotes collaboration between biomedical experts and question-answering systems, and (iii) *Task MedProcNER* on text mining and semantic indexing of clinical procedures in medical documents in Spanish, including the annotation of concepts in unlabeled documents and the subsequent normalization of these concept annotations. *Task MedProcNER* can be considered as a follow-up task of the previous DisTEMIST task on disease mentions [10]. As all three tasks have also been organized in the context of previous editions of the BioASQ challenge, we respectively refer to their current version, in the context of BioASQ11, as *task 11b*, *task Synergy 11* and *task MedProcNER*.

2.1 Task 11b: Biomedical Question Answering

BioASQ *task 11b* takes place in two phases. In the first phase (Phase A), the participants are given questions in English formulated by biomedical experts. For each question, the participating systems have to retrieve relevant documents (from PubMed) and relevant snippets (passages) of the documents. Subsequently, in the second phase (Phase B) of *task 11b*, the participants are given some relevant documents and snippets that the experts themselves have identified (using tools developed in BioASQ [16]). In this phase, they are required to return 'exact' answers, such as names of particular diseases or genes, depending on the type of the question, and 'ideal' answers, which are paragraph-sized summaries of the most important information of the first phase for each question, regardless of its type. A training dataset of 4,721 biomedical questions will be available for participants of *task 11b* to train their systems and about 500 new biomedical questions, with corresponding golden annotations and answers, will be developed for testing the participating systems.

The evaluation of system responses is done both automatically and manually by the experts employing a variety of evaluation measures [8]. In phase A, the official evaluation for document retrieval is based on the Mean Average Precision (MAP) and for snippet retrieval with the F-measure. In phase B, for the exact answers, the official evaluation measure depends on the type of question. For yes/no questions the official measure is the macro-averaged F-Measure on questions with answers *yes* and *no*. The Mean Reciprocal Rank (MRR) is used for factoid questions, where the participants are allowed to return up to five candidate answers. For List questions, the official measure is the mean F-Measure. Finally, for ideal answers, even though automatic evaluation measures are provided and semi-automatic measures [19] are also considered, the official evaluation is still based on manual scores assigned by experts estimating the readability, recall, precision, and repetition of each response.

---

[3] Since the introduction of BioASQ, the task on large-scale biomedical semantic indexing is called *Task a*, and the task on biomedical question answering is called *Task b*, for brevity. Despite the completion of *Task a* last year, we keep this naming convention for *Task b*, for the sake of uniformity with previous versions.

2.2   Task Synergy 11: Question Answering for Developing Topics

The original BioASQ *task b* is structured in a sequence of phases where the experts and the participating systems have minimal interaction. This is acceptable for research questions that have a clear, undisputed answer. However, for questions on developing topics, such as the COVID-19 pandemic, that may remain open for some time and where new information and evidence appear every day, a more interactive model is needed, aiming at a synergy between the automated question-answering systems and the biomedical experts.

In this direction, since 2020 we introduced the BioASQ *task Synergy* which is designed as a continuous dialog, that allows biomedical experts to pose unanswered questions for developing problems and receive the system responses to these questions, including relevant material (documents and snippets) and potential answers [5]. Next, the experts assess these responses, and provide feedback to the systems, in order to improve their responses. This process repeats iteratively with new feedback and new system responses for the same questions, as well as with new questions that may have arisen. In each round of this task, new material is also considered based on the current version of the resources. Initially, the task was focused on COVID-19 considering documents from the COVID-19 Open Research Dataset (CORD-19) [21]. This year, the topic of the questions in the *task Synergy 11* will be open to any developing problem considering documents from the current version of PubMed that will be designated for each round. As in previous versions of the task, the questions are not required to have definite answers and the answers to the questions can be more volatile.

The same evaluation measures used in *task 11b* are also employed in *task Synergy 11* for comparison. However, in order to capture the iterative nature of the task, only new material is considered for the evaluation of a question in each round, an approach known as *residual collection evaluation* [18]. In parallel, additional evaluation metrics are also examined in this direction. Through this task, we aim to facilitate the incremental understanding of new developing public health topics, such as COVID-19, and contribute to the discovery of new solutions.

2.3   Task MedProcNER: Medical Procedure Text Mining
and Indexing Shared Task

Despite the importance of medical procedures for a diversity of topics such as health data mining, analytics and research, limited efforts have been made so far to automatically extract, index, or identify medical procedure mentions from clinical documents. Clinical procedures are a critical concept type for clinical coding systems and can be considered one of the most significant medical entity types to characterize medical tests and therapeutic or surgical aspects associated to patient care. Moreover, medical procedures are of uttermost importance for determining, measuring, or diagnosing a patient's condition and characterize clinical aspects of relevance for medical and surgical treatments of patients. Medical procedures also have a direct practical relevance regarding the use and safety of medical implants and devices. They are undoubtedly transversal medical entity types, of relevance for all medical specialties, including cardiology, oncology, psychiatry and surgery-related clinical specialties such as gynecology and urology.

Correct detection and normalization of medical procedure terms is critical for clinical coding and medical information retrieval systems. The novel *task MedProcNER* will focus on the recognition and indexing of medical procedures in clinical documents in Spanish, by posing subtasks on (1) indexing medical documents with controlled terminologies (2) automatic detection (indexing) of textual evidence, i.e. mentions of medical procedure entities, in text and (3) normalization of these medical procedure mentions to terminologies.

The BioASQ *task MedProcNER* will rely primarily on 1,000 clinical case report publications in Spanish (SciELO [17] full text articles) for indexing diseases with concept identifiers from SNOMED-CT [2], MeSH and ICD10[4]. A large silver standard collection of additional case reports and medical abstracts will also be provided [9]. A silver standard can be described as a set of annotations provided automatically by state-of-the-art algorithms as opposed to manually annotated by experts. The evaluation of systems for this task will use flat evaluation measures following the *task a* [4] track (mainly micro-averaged F-measure, MiF).

2.4 BioASQ Datasets and Tools

During the ten years of BioASQ, hundreds of systems from research teams around the world have been evaluated on the indexing, retrieval, and analysis of hundreds of thousands of biomedical publications and on answering thousands of biomedical questions. In this direction, BioASQ has developed a lively ecosystem of tools that facilitate research, such as the BioASQ Annotation Tool [16] for question-answering dataset development and a range of evaluation measures for automated assessment of system performance in all tasks. All BioASQ software[5] and datasets[6] are publicly available.

In particular, for *task b* on biomedical question answering, BioASQ employs a team of trained biomedical experts who provide a set of about 500 questions on their specialized field of expertise annually for evaluating the performance of participating systems. For *task 11b*, a set of 4,721 realistic questions accompanied by answers, and supporting evidence (documents and snippets) is already available as a unique resource for the development of question-answering systems [6]. In addition, from previous versions of *task Synergy*, which took place in twelve rounds over the last two years, a dataset of 258 questions on COVID-19 is already available. These questions are incrementally annotated with different versions of exact and ideal answers, as well as documents and snippets assessed by the experts as relevant or irrelevant. During the *task Synergy 11* this set

---

[4] https://www.cdc.gov/nchs/icd/icd10cm.htm.

[5] https://github.com/bioasq.

[6] http://participants-area.bioasq.org/datasets.

will be extended with more than fifty new open questions on COVID-19 and other developing health topics. Meanwhile, any existing questions that remain relevant may be enriched with more updated answers and more recent evidence (documents and snippets) [14].

In addition, for biomedical semantic indexing, a training dataset of more than 16.2 million articles and fifteen weekly test sets of around 6,000 articles each are available from the tenth edition of *task a (task 10a)* [14]. Even though the *task a* completed its life cycle as a BioASQ task in 2022, the corresponding resources are still useful for related tasks such as semantic indexing in other languages, other types of biomedical documents, or specific sets of labels. Similarly, the datasets from the previous versions of the *task MESINEP*, on medical semantic indexing in Spanish, are also available [3].

For the *task MedProcNER*, a new dataset of semantically annotated medical documents in Spanish labeled with text-bound evidence mentions of medical procedures together with concept identifiers for entity linking and semantic indexing will be released. This year, the dataset will additionally focus on high-impact clinical specialties, such as cardiology. In addition, a dataset of 1,000 clinical cases in Spanish is already available from the previous edition of *task DisTEMIST*, together with corresponding concept recognition and linking annotations, as well as a set of disease-relevant mentions from over 200,000 biomedical articles in Spanish [10]. This will allow the exploration of disease-medical procedure relations.

3   Conclusions

BioASQ facilitates the exchange and fusion of ideas, providing unique realistic datasets and evaluation services for research teams that work on biomedical semantic indexing and question answering. Therefore, it eventually accelerates progress in the field, as indicated by the gradual improvement of the scores achieved by the participating systems [13]. An illustrative example is the Medical Text Indexer (MTI) [12], which achieved significant improvements [13] largely due to the adoption of ideas from the systems that compete in the BioASQ challenge [11], eventually reaching a performance level that allows the adoption of fully automated indexing in NLM[7].

Similarly, we expect that the new version of BioASQ will allow the participating teams to bring further improvement to the open tasks of biomedical question answering (*task 11b*), answering open questions for developing topics (*task Synergy 11*), and clinical procedure text mining and semantic indexing of medical documents in Spanish (*task MedProcNER*). In conclusion, BioASQ aims to assist participating teams in their approach to the challenge's tasks, which represent key information needs in the biomedical domain.

Acknowledgments. Google was a proud sponsor of the BioASQ Challenge in 2022. The eleventh edition of BioASQ is also sponsored by Atypon Systems inc. The *task*

---

[7] https://www.nlm.nih.gov/pubs/techbull/nd21/nd21_medline_2022.html.

MedProcNER is supported by the Spanish Plan for the Advancement of Language Technologies (Plan TL), the 2020 Proyectos de I+D+i-RTI Tipo A (Descifrando El Papel De Las Profesiones En La Salud De Los Pacientes A Traves De La Mineria De Textos, PID2020-119266RA-I00). This project has received funding from the European Union Horizon Europe Coordination and Support Action under Grant Agreement No 101058779 (BIOMATDB) and DataTools4Heart - DT4H, Grant agreement No 101057849.

References

1. Cohen, K.B., Demner-Fushman, D., Ananiadou, S., Tsujii, J. (eds.): BioNLP 2017. Association for Computational Linguistics, Vancouver, Canada, August 2017. https://doi.org/10.18653/v1/W17-23, aclanthology.org/W17-2300
2. Donnelly, K., et al.: SNOMED-CT: the advanced terminology and coding system for eHealth. Stud. Health Technol. Inform. **121**, 279 (2006)
3. Gasco, L., et al.: Overview of BioASQ 2021-MESINESP track. Evaluation of advance hierarchical classification techniques for scientific literature, patents and clinical trials. In: CEUR Workshop Proceedings (2021)
4. Kosmopoulos, A., Partalas, I., Gaussier, E., Paliouras, G., Androutsopoulos, I.: Evaluation measures for hierarchical classification: a unified view and novel approaches. Data Min. Knowl. Disc. **29**(3), 820–865 (2015)
5. Krallinger, M., Krithara, A., Nentidis, A., Paliouras, G., Villegas, M.: BioASQ at CLEF2020: large-scale biomedical semantic indexing and question answering. In: Jose, J.M., et al. (eds.) ECIR 2020. LNCS, vol. 12036, pp. 550–556. Springer, Cham (2020). https://doi.org/10.1007/978-3-030-45442-5_71
6. Krithara, A., Nentidis, A., Bougiatiotis, K., Paliouras, G.: BioASQ-QA: a manually curated corpus for biomedical question answering. bioRxiv (2022)
7. Krithara, A., Nentidis, A., Paliouras, G., Krallinger, M., Miranda, A.: BioASQ at CLEF2021: large-scale biomedical semantic indexing and question answering. In: Hiemstra, D., Moens, M.-F., Mothe, J., Perego, R., Potthast, M., Sebastiani, F. (eds.) ECIR 2021. LNCS, vol. 12657, pp. 624–630. Springer, Cham (2021). https://doi.org/10.1007/978-3-030-72240-1_73
8. Malakasiotis, P., Pavlopoulos, I., Androutsopoulos, I., Nentidis, A.: Evaluation measures for task b. Technical report, BioASQ (2018). participants-area.bioasq.org/Tasks/b/eval_meas_2018
9. Ménard, P.A., Mougeot, A.: Turning silver into gold: error-focused corpus reannotation with active learning. In: Proceedings of the International Conference on Recent Advances in Natural Language Processing (RANLP 2019), pp. 758–767 (2019)
10. Miranda-Escalada, A., et al.: Overview of DisTEMIST at BioASQ: automatic detection and normalization of diseases from clinical texts: results, methods, evaluation and multilingual resources. In: Working Notes of Conference and Labs of the Evaluation (CLEF) Forum. CEUR Workshop Proceedings (2022)
11. Mork, J., Aronson, A., Demner-Fushman, D.: 12 years on-Is the NLM medical text indexer still useful and relevant? J. Biomed. Semant. **8**(1), 8 (2017)
12. Mork, J., Jimeno-Yepes, A., Aronson, A.: The NLM medical text indexer system for indexing biomedical literature (2013)

13. Nentidis, A., et al.: Overview of BioASQ 2022: the tenth BioASQ challenge on large-scale biomedical semantic indexing and question answering. In: Experimental IR Meets Multilinguality, Multimodality, and Interaction (Including Subseries Lecture Notes in Artificial Intelligence and Lecture Notes in Bioinformatics). LNCS, vol. 13390, pp. 337–361. Springer, Cham (2022). https://doi.org/10.1007/978-3-031-13643-6_22

14. Nentidis, A., Katsimpras, G., Vandorou, E., Krithara, A., Paliouras, G.: Overview of BioASQ tasks 10a, 10b and Synergy10 in CLEF2022. In: CEUR Workshop Proceedings, vol. 3180, pp. 171–178 (2022)

15. Nentidis, A., Krithara, A., Paliouras, G., Gasco, L., Krallinger, M.: BioASQ at CLEF2022: the tenth edition of the large-scale biomedical semantic indexing and question answering challenge. In: Hagen, M., et al. (eds.) ECIR 2022. LNCS, vol. 13186, pp. 429–435. Springer, Cham (2022). https://doi.org/10.1007/978-3-030-99739-7_53

16. Ngomo, A.C.N., Heino, N., Speck, R., Ermilov, T., Tsatsaronis, G.: Annotation tool. Project deliverable D3.3, February 2013. www.bioasq.org/sites/default/files/PublicDocuments/2013-D3.3-AnnotationTool.pdf

17. Packer, A.L., et al.: SciELO: uma metodologia para publicação eletrônica. Ciência da informação 27, nd-nd (1998)

18. Salton, G., Buckley, C.: Improving retrieval performance by relevance feedback. J. Am. Soc. Inf. Sci. 41(4), 288–297 (1990). https://doi.org/10.1002/(SICI)1097-4571(199006)41:4⟨288::AID-ASI8⟩3.0.CO;2-H

19. ShafieiBavani, E., Ebrahimi, M., Wong, R., Chen, F.: Summarization evaluation in the absence of human model summaries using the compositionality of word embeddings. In: Proceedings of the 27th International Conference on Computational Linguistics. pp. 905–914. Association for Computational Linguistics, Santa Fe, New Mexico, USA, August 2018. www.aclweb.org/anthology/C18-1077

20. Tsatsaronis, G., et al.: An overview of the BioASQ large-scale biomedical semantic indexing and question answering competition. BMC Bioinform. 16, 138 (2015). https://doi.org/10.1186/s12859-015-0564-6

21. Wang, L.L., et al.: CORD-19: the COVID-19 open research dataset. ArXiv (2020). arxiv.org/abs/2004.10706v2

22. Zavorin, I., Mork, J.G., Demner-Fushman, D.: Using learning-to-rank to enhance NLM medical text indexer results. ACL 2016, 8 (2016)

eRisk 2023: Depression, Pathological Gambling, and Eating Disorder Challenges

Javier Parapar[1]([✉]) [iD], Patricia Martín-Rodilla[1] [iD], David E. Losada[2] [iD], and Fabio Crestani[3] [iD]

[1] Information Retrieval Lab,
Centro de Investigación en Tecnoloxías da Información e as Comunicacións (CITIC), Universidade da Coruña, A Coruña, Spain
{javierparapar,patricia.martin.rodilla}@udc.es
[2] Centro Singular de Investigación en Tecnoloxías Intelixentes (CiTIUS), Universidade de Santiago de Compostela, Santiago, Spain
david.losada@usc.es
[3] Faculty of Informatics, Università della Svizzera italiana (USI), Lugano, Switzerland
fabio.crestani@usi.ch

Abstract. In 2017, we launched eRisk as a CLEF Lab to encourage research on early risk detection on the Internet. Since then, thanks to the participants' work, we have developed detection models and datasets for depression, anorexia, pathological gambling and self-harm. In 2023, it will be the seventh edition of the lab, where we will present a new type of task on sentence ranking for depression symptoms. This paper outlines the work that we have done to date, discusses key lessons learned in previous editions, and presents our plans for eRisk 2023.

1 Introduction

The eRisk Lab[1] is a forum for exploring evaluation methodologies and effectiveness metrics related to early risk detection on the Internet (with past challenges particularly focused on health and safety). Since the pilot edition in 2017 in Dublin, we have been part of CLEF, the Conference and Labs of the Evaluation Forum. Along the different editions [7–10,12,13] many collections and models have been presented under the eRisk banner. Our dataset construction approach and evaluation strategies are broad, meaning they might be applied to various application domains.

Our interdisciplinary Lab addresses tasks touching areas such as information retrieval, computational linguistics, machine learning, and psychology. Participants with heterogeneous expertise collaborate to design monitoring solutions for critical social worrying problems. Ideally, the developed models may be used in systems that, for instance, will alert when someone shows suicidal ideas or on social media. Previous eRisk editions included joint work on depression, eating disorders, gambling, and self-harm detection.

---

[1] https://erisk.irlab.org.

J. Kamps et al. (Eds.): ECIR 2023, LNCS 13982, pp. 585–592, 2023.
https://doi.org/10.1007/978-3-031-28241-6_67

So far, eRisk eRisk has proposed early alert and severity estimation tasks. Early risk tasks (Sect. 2.1) involve evidence-building risk prediction. For that, participant systems must automatically analyse a temporal text stream from a source (e.g., social media posts) while accumulating evidence to decide about a specific risk (e.g. developing depression). On the other hand, in the severity estimate challenges (Sect. 2.2), participants use all user writings for computing a fine-grained estimate of the symptoms of a specific risk. With that, models must fill out a standard questionnaire as real users would.

2   A Brief History of eRisk

Since the Lab's inception, we have created numerous reference collections in the field of risk prediction in depression, anorexia, self-harm, and pathological gambling disorders.

In the first eRisk [7], early risk of depression was the only pilot task. This edition released temporal data chunks in sequential order (one chunk per week). Following each release, participants submitted their predictions. This demanding procedure resulted in only eight (up to 5 systems each) of the thirty participating groups completing the tasks by the deadline. The evaluation methodology and metrics were those defined in [6]. In 2018 [8], we maintained the same setup for a continuation of the early detection of depression task and a new one on the early detection of signs of anorexia. Participants submitted 45 systems for Task 1 (depression) and 35 for Task 2 (anorexia).

It was in 2019 [9] when we moved from the weekly chunk release of users' writings to a fine-grained release of user posts using a server. We used that approach for task 1 on early risk detection of anorexia and task 2, a new task on self-harm. Another important change was the introduction of a new task on severity estimation using clinically validated questionnaires. We presented this new kind of task for the case of depression. In this new challenge, the participants received the whole writing history of the users. We received 54, 33, and 33 for tasks 1,2 and 3. In 2020 [10], we continued the tasks of early detection of self-harm and estimating the severity of depression symptoms. Participants submitted 46 system variants for the early risk task and 17 different runs for the severity estimation one.

We proposed three tasks in 2021 [12]. Following our three-year-per-task cycle, we closed the tasks on the early detection of signs of self-harm challenge and the estimation of the severity of the symptoms of depression. We also presented a new domain for early detection, in this case, pathological gambling. We received 115 runs from 18 teams out of 75 registered. In 2022, we continued the task of early risk detection of pathological gambling. We closed the cycle for early risk detection of depression (the first edition under the new fine-grained setup). Additionally, we presented a new severity estimation task using, in this case, a standard questionnaire on eating disorders. The proposed tasks received 117 runs from 18 teams in total.

Over these five years, eRisk has received a steady number of active participants, slowly placing the Lab as a reference forum for early risk research. We
summarised the eRisk experience and the best models presented so far in our
recent book [4].

2.1 Early Risk Prediction Tasks

The primary goal of eRisk was to develop practical algorithms and models for
tracking social network activity. Most of the presented challenges were early
predicting risk in various domains (depression, anorexia, self-harm). They were
all organised the same way: the teams had to analyse social media writings (posts
or comments) sequentially (in chronological order) to spot signs of risk as early
and feasible as possible.

All shared tasks in the different editions were sourced from the social media
platform Reddit. It is critical to highlight that data extraction for research purposes is permitted under Reddit's terms of service. Reddit does not permit unau-
thorised commercial use or redistribution of its content except as authorised by
the concept of fair use. eRisk's research activities are an example of fair usage.

Redditors tend to be prolific in writing, being common to have many posts
published over several years. There are many communities (subreddits) dedicated
to mental health disorders such as depression, anorexia, self-harm, or pathological gambling. We leverage that for obtaining the writing history of redditors
(posts or comments) for building the eRisk datasets [6]. All our datasets, redditors are divided into positive class (e.g., depressed) and negative class (control
group). To obtain them, we followed Coppersmith and colleagues [3] methodology. For instance, when looking for positive redditors for depression, we searched
for in the writings for explicit strings (e.g. "Today, I was diagnosed with depression") about the redditors being diagnosed with depression. For example, "*I am
anorexic*", "*I have anorexia*", or "*I believe I have anorexia*" were not deemed
explicit affirmations of a diagnosis. We followed this semi-supervised method for
extracting information about patients diagnosed with different conditions. Since
2020, we have used Beaver [11], a new tool for labelling positive and negative
instances, for aiding us in this task.

In terms of evaluation methodology, in the first edition of eRisk, we presented
a new measure called ERDE (Early Risk Detection Error) for measuring early
detection [6]. Contrary to standard classification metrics that ignore prediction
latency, ERDE considers both the correctness of the (binary) decision and the
latency. In the original ERDE, the latency corresponds with the counting of posts
(k) processed before reaching the decision. In 2019, we also adopted $F_{latency}$,
an alternative assessment metric for early risk prediction proposed by
Sadeque et al. [14].

With the introduction of the writing-level release of user texts in 2019, we
could produce user rankings by the participants-provided estimated degree of

risk. Since then, we have evaluated those ranks using common information retrieval metrics (for example, P@10 or nDCG) [9].

2.2 Severity Level Estimation Task

In 2019 we introduced a new task on estimating the severity level of depression that we continued in 2020 and 2021. In the last edition, we introduced a new task in the eating disorder domain where participants had to fill out the EDE-Q questionnaire automatically. Those tasks investigate the feasibility and potential methods for automatically measuring the occurrence and severity of various well-known symptoms for the mentioned disorders. In this task, participants had access to the history of writings of some redditors who have volunteered to fill out the standard questionnaire. Participants had to produce models that answered each of the questions of the corresponding standard based on the evidence found in the provided writings.

In the case of depression, we used the Beck's Depression Inventory (BDI) [1]. It presents 21 questions regarding the severity of depression signs and symptoms (with four alternative responses corresponding to different severity levels) (e.g., loss of energy, sadness, and sleeping problems). For eating disorders, we used questions 1–12 and 19–28 from the Eating Disorder Examination Questionnaire (EDE-Q) [2].

To produce the ground truth, we compiled a series of surveys by social media users with their writing history. Because of the unique nature of the task, we presented new evaluation measures for evaluating the participants' estimations. We defined four metrics in the depression scenario: Average Closeness Rate (ACR), Average Hit Rate (AHR), Average DODL (ADODL), and Depression Category Hit Rate (DCHR), details can be found in [9]. Last year, for the eating disorder results, we also adopted new metrics: Mean Zero-One Error (MZOE), Mean Absolute Error (MAE), Macroaveraged Mean Absolute Error (MAE$_{macro}$), Global ED (GED), and the corresponding Root Mean Square Error (RMSE) for the four sub-scales: Restraint, Eating Concern, Shape Concern, Weight Concern [13].

2.3 Results

According to the CLEF tradition, Labs' Overview and Extended Overview papers compile the summaries and critical analysis of the participants' systems results [7–10,12].

So far, ten editions of early detection tasks on four mental health disorders have been celebrated. We have received a diverse range of models and methods. Many of them rely on traditional classification techniques. That is, most participants focused on improving classification accuracy on training data. As this task tries to promote fast-responding models to signs of the disorder, we missed more systems concerned with the accuracy-delay trade-off in general. In any case, we have observed a non-homogeneous system performance along the different disorders over the years. For instance, anorexia and pathological gambling seem to be more manageable tasks than depression detection. These discrepancies could be attributed to the amount and quality of released training data and the illness itself. We hypothesise that, depending on the illness, patients are more or less likely to leave traces of their social media language. The results reveal a pattern in how participants improved detection accuracy from edition to edition. This pattern encourages us to continue funding research on text-based early risk detection in social media. Furthermore, based on the performance of some participants, automatic or semi-automatic screening systems that predict the onset of specific hazards appears to be within reach.

The results also demonstrate that automatic analysis of the whole user's writing history could be a complementary technique for extracting some indicators or symptoms connected with the disorder when determining disease severity. In the case of depression, for example, where participants had access to training data, some systems had a 40% hit rate (the systems answered 40% of the BDI questions with the exact same response as the real user). Although there is still much room for improvement, this demonstrates that the participants were able to extract some signals from the jumbled social media data. For the eating disorder questionnaire, the results for the first edition are still very modest, considering that participants of the first edition had no access to training data.

The difficulties in locating and adapting measures for these novel challenges have prompted us to develop new metrics for eRisk. Some eRisk participants [14,15], were also engaged in proposing novel modes of evaluation, which is yet another beneficial outcome of the Lab. As commented, we incorporated error metrics in the new severity estimation task last year. Both MAE and RMSE are two widely used metrics in rating prediction for users in recommendation systems [5].

3   The Tasks of eRisk 2023

The outcomes of previous editions have encouraged us to continue the Lab in 2023 and examine the interaction between text-based screening from social media and risk prediction and estimation. The following is the task breakdown for eRisk 2023:

3.1   Task 1: Search for Symptoms of Depression

This is a new type of challenge. The task consists of ranking sentences from a collection of user writings according to their relevance to a depression symptom. The participants will have to provide rankings for the 21 symptoms of depression from the BDI questionnaire [1]. A sentence will be deemed relevant to a BDI symptom when it conveys information about the user's state concerning the symptom. That is, it may be relevant even when it indicates that the user is ok with the symptom.

We would release a sentence-tagged dataset (based on eRisk past data) together with the BDI questionnaire. Participants would be free to decide on the best strategy to derive queries from describing the BDI symptoms in the questionnaire. After receiving the runs from the participating teams, we would create the relevance judgements with the help of human assessors using pooling. We will use the resulting *qrels* to evaluate the systems with classical ranking metrics (e.g. MAP, nDCG, etc.). This new corpus with annotated sentences would be a valuable resource with multiple applications beyond eRisk.

3.2   Task 2: Early Detection of Pathological Gambling

In 2023 it will be the third edition of the task. It follows the early detection challenge. It consists of sequentially processing pieces of evidence and detecting early traces of pathological gambling as soon as possible. Participants must process Social Media texts in the order the users wrote them. In this way, systems that effectively perform this task could be applied to sequentially monitor user interactions in blogs, social networks, or other types of online media. We will provide the data from 2021 and 2022 as training data. The test stage will consist of a period where the participants have to connect to our server[2]. and iteratively get user writings and send decisions.

3.3   Task 3: Measuring the Severity of the Signs of Eating Disorders

The task consists of estimating the severity level of the eating disorder given a user history or written submissions. For that, we provide participants with the postings history, and the participants will have to fill out a standard eating disorder questionnaire (based on the evidence found in texts). The EDE-Q assesses the range and severity of features associated with the diagnosis of eating disorders. It is a 28-item questionnaire with four subscales (restrain, eating, concern, shape concern, and weight concern) and a global score [2]. The questionnaires filled by the users (ground truth) will be used to assess the quality of the responses provided by the participating systems. Participants will have training data from last year.

4   Conclusions

The results obtained so far under eRisk and the research community's participation drive us to continue proposing new challenges related to risk identification in Social Media. We sincerely thank all participants for their contributions to eRisk's success. We want to encourage the research teams to keep improving and developing new models for future tasks and dangers. Even while creating new resources is time-consuming, we believe that the societal benefits outweigh the costs.

---

[2] http://early.irlab.org/server.html.

Acknowledgements. The first and second authors thank the financial support supplied by the Consellería de Cultura, Educación, Formación Profesional e Universidades (accreditation 2019–2022 ED431G/01, ED431B 2022/33) and the European Regional Development Fund, which acknowledges the CITIC Research Center in ICT of the University of A Coruña as a Research Center of the Galician University System. The third author thanks the financial support supplied by the Consellería de Cultura, Educación, Formación Profesional e Universidades (accreditation 2019–2022 ED431G-2019/04, ED431C 2022/19) and the European Regional Development Fund, which acknowledges the CiTIUS-Research Center in Intelligent Technologies of the University of Santiago de Compostela as a Research Center of the Galician University System. The first, second, and third author also thank the funding of project PLEC2021-007662 (MCIN/AEI/10.13039/501100011033, Ministerio de Ciencia e Innovación, Agencia Estatal de Investigación, Plan de Recuperación, Transformación y Resiliencia, Unión Europea-Next Generation EU).

References

1. Beck, A.T., Ward, C.H., Mendelson, M., Mock, J., Erbaugh, J.: An inventory for measuring depression. JAMA Psychiat. **4**(6), 561–571 (1961)
2. Carey, M., Kupeli, N., Knight, R., Troop, N.A., Jenkinson, P.M., Preston, C.: Eating disorder examination questionnaire (EDE-Q): norms and psychometric properties in UK females and males. Psychol. Assess. **31**(7), 839 (2019)
3. Coppersmith, G., Dredze, M., Harman, C.: Quantifying mental health signals in Twitter. In: ACL Workshop on Computational Linguistics and Clinical Psychology (2014)
4. Crestani, F., Losada, D.E., Parapar, J. (eds.): Early Detection of Mental Health Disorders by Social Media Monitoring. Springer, Heidelberg (2022). https://doi.org/10.1007/978-3-031-04431-1
5. Herlocker, J.L., Konstan, J.A., Terveen, L.G., Riedl, J.T.: Evaluating collaborative filtering recommender systems. ACM Trans. Inf. Syst. **22**(1), 5–53 (2004)
6. Losada, D.E., Crestani, F.: A test collection for research on depression and language use. In: Proceedings Conference and Labs of the Evaluation Forum CLEF 2016, Evora, Portugal (2016)
7. Losada, D.E., Crestani, F., Parapar, J.: eRISK 2017: CLEF lab on early risk prediction on the internet: experimental foundations. In: Jones, G.J.F., et al. (eds.) CLEF 2017. LNCS, vol. 10456, pp. 346–360. Springer, Cham (2017). https://doi.org/10.1007/978-3-319-65813-1_30
8. Losada, D.E., Crestani, F., Parapar, J.: Overview of eRisk: early risk prediction on the internet. In: Bellot, P., et al. (eds.) CLEF 2018. LNCS, vol. 11018, pp. 343–361. Springer, Cham (2018). https://doi.org/10.1007/978-3-319-98932-7_30
9. Losada, D.E., Crestani, F., Parapar, J.: Overview of eRisk 2019 early risk prediction on the internet. In: Crestani, F., et al. (eds.) CLEF 2019. LNCS, vol. 11696, pp. 340–357. Springer, Cham (2019). https://doi.org/10.1007/978-3-030-28577-7_27
10. Losada, D.E., Crestani, F., Parapar, J.: Overview of eRisk 2020: early risk prediction on the internet. In: Arampatzis, A., et al. (eds.) CLEF 2020. LNCS, vol. 12260, pp. 272–287. Springer, Cham (2020). https://doi.org/10.1007/978-3-030-58219-7_20

11. Otero, D., Parapar, J., Barreiro, Á.: Beaver: efficiently building test collections for novel tasks. In: Proceedings of the Joint Conference of the Information Retrieval Communities in Europe (CIRCLE 2020), Samatan, Gers, France, 6–9 July 2020 (2020). https://ceur-ws.org/Vol-2621/CIRCLE20_23.pdf

12. Parapar, J., Martín-Rodilla, P., Losada, D.E., Crestani, F.: Overview of eRisk 2021: early risk prediction on the internet. In: Candan, K.S., et al. (eds.) CLEF 2021. LNCS, vol. 12880, pp. 324–344. Springer, Cham (2021). https://doi.org/10.1007/978-3-030-85251-1_22

13. Parapar, J., Martín-Rodilla, P., Losada, D.E., Crestani, F.: Overview of eRisk 2022: early risk prediction on the internet. In: Experimental IR Meets Multilinguality, Multimodality, and Interaction - 13th International Conference of the CLEF Association, CLEF 2022, Bologna, Italy, 5–8 September 2022, Proceedings, pp. 233–256. Springer, Heidelberg (2022). https://doi.org/10.1007/978-3-031-13643-6_18

14. Sadeque, F., Xu, D., Bethard, S.: Measuring the latency of depression detection in social media. In: Proceedings of the Eleventh ACM International Conference on Web Search and Data Mining, WSDM 2018, pp. 495–503. ACM, New York (2018)

15. Trotzek, M., Koitka, S., Friedrich, C.: Utilizing neural networks and linguistic metadata for early detection of depression indications in text sequences. IEEE Trans. Knowl. Data Eng. **32**, 588–601 (2018)

Overview of EXIST 2023: sEXism Identification in Social NeTworks

Laura Plaza[1,2]([⊠]), Jorge Carrillo-de-Albornoz[1,2], Roser Morante[1],
Enrique Amigó[1], Julio Gonzalo[1], Damiano Spina[2], and Paolo Rosso[3]

[1] Universidad Nacional de Educación a Distancia (UNED), 28040 Madrid, Spain
{lplaza,jcalbornoz,rmorant,enrique}@lsi.uned.es
[2] RMIT University, 3000 Melbourne, Australia
damiano.spina@rmit.edu.au
[3] Universidad Politécnica de Valencia (UPV), 46022 Valencia, Spain
prosso@dsic.upv.es

Abstract. The paper describes the lab on Sexism identification in social
networks (EXIST 2023) that will be hosted as a lab at the CLEF 2023
conference. The lab consists of three tasks, two of which are continuation of EXIST 2022 (*sexism detection* and *sexism categorization*) and
a third and novel one on *source intention identification*. For this edition new test and training data will be provided and some novelties
are introduced in order to tackle two central problems of Natural Language Processing (NLP): bias and fairness. Firstly, the sampling and
data gathering process will take into account different sources of bias in
data: seed, temporal and user bias. During the annotation process we will
also consider some sources of "label bias" that come from the social and
demographic characteristics of the annotators. Secondly, we will adopt
the "learning with disagreements" paradigm by providing datasets containing also pre-aggregated annotations, so that systems can make use of
this information to learn from different perspectives. The general goal of
the EXIST shared tasks is to advance the state of the art in online sexism detection and categorization, as well as investigating to what extent
bias can be characterized in data and whether systems may take fairness
decisions when learning from multiple annotations.

Keywords: Sexism detection · Data bias · Learning with
disagreements

1 Introduction

Sexism[1] is pervasive, even in modern developed societies and among young generations who have been born in democratic societies. Recently, a video went viral
where male students of a prestigious hall of residence in Spain were addressing

---

[1] The Oxford English Dictionary defines sexism as "prejudice, stereotyping or discrimination, typically against women, on the basis of sex".

J. Kamps et al. (Eds.): ECIR 2023, LNCS 13982, pp. 593–599, 2023.
https://doi.org/10.1007/978-3-031-28241-6_68

extremely sexist and insulting messages to the girls of the neighbouring residence as a way of welcoming them for the new academic year.[2] Inequality and discrimination against women that remain embedded in society are increasingly being replicated online [1]. Internet perpetuates and even naturalizes gender differences and sexist attitudes [2]. Moreover, given that an important percentage of Internet users (especially social networks users) are teenagers, the increasing sexism on the Internet requires urgent study and social debate that leads to actions, especially from an educational point of view.

Social networks are the main platforms for social complaint, activism, etc., and movements like #MeTwoo, #8M or #Time'sUp have spread rapidly. Under this umbrella, many women all around the world have reported abuses, discrimination and sexist experiences suffered in real life. Social networks are also contributing to the transmission of sexism and other disrespectful and hateful behaviours. Even though social platforms such as Twitter are continuously creating new ways to identify and eradicate hateful content, they are facing many difficulties when dealing with the huge amount of data generated by users [3]. In this context, automatic tools not only may help to detect and alert against sexism behaviours and discourses, but also to estimate how often sexist and abusive situations are found in social media platforms, what forms of sexism are more frequent and how sexism is expressed in these media.

Up to date, most work dealing with sexism in online media is focused on detecting misogyny or hatred towards women [4–6]. However, sexism does not always imply misogyny. Sexism may sound "friendly" as in (1), which may seem positive, but is actually transmitting that women are weaker than men. Sexism may sound "funny", as reflected on sexist jokes or humour (2). Or sexism may sound "offensive" and "hateful", as in (3). However, even the most subtle forms of sexism can be as pernicious as the most violent ones and affect women in many facets of their lives, including domestic and parenting roles, career opportunities, sexual image and life expectations, to name a few.

(1) *Women must be loved and respected, always treat them like a fragile glass.*
(2) *You have to love women... just that... You will never understand them.*
(3) *Humiliate, expose and degrade yourself as the fucking bitch you are if you want a real man to give you attention.*

Until recently, no NLP shared tasks had addressed the detection of sexism in social media. To fill this research gap, in 2021 and 2022 the sEXism Identification in Social neTworks (EXIST) shared tasks were proposed at the IberLEF forum [7,8]. The EXIST challenge was the first shared task on sexism detection in social networks whose aim was to identify and classify sexism in a broad sense, from explicit and/or hostile to other subtle or even benevolent expressions that involve implicit sexist behaviours. In the framework of these competitions, the first dataset of broad sexism was released. The 2021 and 2022 EXIST editions welcomed more than 50 teams from research institutions and companies from all around the world, demonstrating the importance of the problem, as well as

---

[2] https://www.theguardian.com/world/2022/oct/06/spanish-pm-leads-outcry-over-students-filmed-chanting-abuse-at-womens-halls. Accessed 14 October 2022.

the great interest of the research community around it. Given the success of the tasks, in 2023 a new edition of EXIST will take place as a lab at CLEF, which will continue with the tasks addressed in previous years, while facing yet a new challenge: the identification of the intention of the author of the sexist message. However, the main novelty will be the adoption of the "learning with disagreements" paradigm [9] for the development of the dataset and, optionally, for the evaluation of the systems. The adoption of this paradigm along with our effort to control bias in the annotations (see Sect. 3) will allow us to evaluate whether including the different views and sensibilities of the annotators contributes to the development of more accurate and fairer NLP systems.

In what follows we present the EXIST task for the 2023 edition.

2   EXIST Tasks

The two first editions of EXIST focused on detecting sexist messages in two social networks, Twitter and Gab[3] as well as on categorizing these messages according to the type of sexist behaviour they enclose. For the 2023 edition, we will focus on Twitter and we will address an additional task, namely "source intention classification". Therefore, three tasks are proposed which are described below.

2.1   Sexism Detection

The first task is a binary classification task where systems must decide whether or not a given tweet is sexist. The following statements show examples of sexist and not sexist messages, respectively.

(4) **Sexist:** *Woman driving, be careful!.*
(5) **Not Sexist:** *Just saw a woman wearing a mask outside spank her very tightly leashed dog and I gotta say I love learning absolutely everything about a stranger in a single instant.*

2.2   Source Intention Classification

This task aims to categorize the message according to the intention of the author, which provides insights in the role played by social networks on the emission and dissemination of sexist messages. In this task, we propose a ternary classification task: (i) direct sexist message, (ii) reported sexist message and (iii) judgemental message. This distinction will allow us to differentiate sexism that is actually taking place in online platforms from sexism which is being suffered by women in other situations but that is being reported in social networks with the aim of complaining and fighting against sexism. The following categories are defined:

– **Direct** sexist message: The intention was to write a message that is sexist by itself or incites to be sexist, as in:

---

[3] https://gab.com/. Accessed 14 Oct 2022.

(6) *A woman needs love, to fill the fridge, if a man can give this to her in return for her services (housework, cooking, etc.), I don't see what else she needs.*

(7) *Women shouldn't code...perhaps be influencer/creator instead...it's their natural strength.*

- **Reported** sexist message: The intention is to report and share a sexist situation suffered by a woman or women in first or third person, as in:

(8) *I doze in the subway, I open my eyes feeling something weird: the hand of the man sat next to me on my leg #SquealOnYourPig.*

(9) *Today, one of my year 1 class pupils could not believe he'd lost a race against a girl.*

- **Judgemental** message: The intention was judgmental, since the tweet describes sexist situations or behaviours with the aim of condemning them.

(10) *As usual, the woman was the one quitting her job for the family's welfare...*

(11) *21st century and we are still earning 25% less than men #Idonotrenounce.*

2.3 Sexism Categorization

Many facets of a woman's life may be the focus of sexist attitudes including domestic and parenting roles, career opportunities, sexual image, and life expectations, to name a few. Automatically detecting which of these facets of women are being more frequently attacked in social networks will facilitate the development of policies to fight against sexism. According to this, each sexist tweet must be categorized in one or more of the following categories:

- **Ideological and Inequality**: This category includes all tweets that discredit the feminist movement in order to devalue, belittle and defame the struggle of women in any aspect of their lives. It also includes messages that reject inequality between men and women, or present men as victims of gender-based oppression. Some examples of this type of sexism discourse can be found in the following tweets:

(12) *#Feminism is a war on men, but it's also a war on women. It's a war on female nature, a war on femininity.*

(13) *Think the whole equality thing is getting out of hand. We are different, that's how were made!*

- **Role Stereotyping and Dominance**: This category includes messages that express false ideas about women that suggest they are more suitable or inappropriate for certain tasks. It also includes any claim that implies that men are somehow superior to women.

(14) *Most women no longer have the desire or the knowledge to develop a high quality character, even if they wanted to.*

(15) *I feel like everytime I flirt with a girl they start to imagine all the ways they can utilize me.*

- **Objectification**: Objectification and physical stereotyping includes messages where women are presented as objects apart from their dignity and personal aspects. We also include messages that assume or describe certain physical qualities that women must have in order to fulfill traditional gender roles, for example, ideas that suggest that women should maintain a standard and ideal of beauty or attacks on a woman's physique.

(16) *No offense but I've never seen an attractive african american hooker. Not a single one.*

(17) *Most of the times I just want women for sex just because everytime I try to make a relationship out of it they always break my heart.*

– **Sexual Violence**: This category includes messages where sexual suggestions, requests or harassment of a sexual nature (rape or sexual assault) are made:

(18) *I wanna touch your tits..you can't imagine what I can do on your body.*

(19) *Fuck that cunt, I would with my fist.*

– **Misogyny and Non Sexual Violence**: This category includes expressions of hatred and violence towards women.

(20) *Domestic abuse is never okay.... Unless your wife is a bitch.*

(21) *Some woman are so toxic they don't even know they are draining everyone around them in poison.*

3   The EXIST 2023 Dataset: Controlling Bias in Data and Annotation

An important problem in NLP that has gained attention in the recent years in parallel to the growing protagonism of big language models is bias, both in the data that are used to train and test systems, and in the way algorithms learn, mainly due to the bias in the data [14]. In EXIST 2023 we tackle one aspect of this problem, i.e., the data bias that may be introduced both during the data selection and during the labeling process.

Firstly, the sampling and data gathering process will take into account different sources of bias: seed, temporal and user bias. To mitigate the seed bias, we will use as seeds a wide range of terms that are employed in both sexist and non-sexist contexts. Temporal bias between training, validation and test data will be mitigated by selecting texts from different time spans, with a temporal gap between the sets. We will also check the temporal gap between tweets for each seed to ensure that data are spread all over a certain period. Finally, we will check the author of the messages to ensure an appropriate balance in the contribution of the different types of users. We will also avoid having tweets by the same user in the different subsets.

Secondly, during the annotation process we will consider some sources of "label bias" [10]. Label bias may be introduced by socio-demographic differences of the persons that participate in the annotation process, but also when more than one possible correct label exists or when the decision on the label depends on subjectivity. During the development of the datasets, we will devote special attention to avoid annotation bias. Some of the following socio-demographic sources of bias will be considered: gender, ethnicity, country, education and age. In order to avoid the bias derived from the individuality of the annotators, we will adopt the "learning with disagreements" paradigm (see Sect. 4).

The labelling process will be carried out by crowd-workers, selected according to their different social and demographic parameters in order to avoid the label bias. Each tweet will be annotated by enough crowd-workers to ensure diversity. At least 5,000 tweets will be crawled from Twitter in two different languages: English and Spanish. A train, a validation and a test set will be provided. To retrieve the tweets, more than 200 potentially sexist phrases will be used as seeds. These phrases have been extracted from different sources: (a) previous works in the area; (b) Twitter accounts (journalist, teenagers, etc.) or hashtags used to report sexist situations; (c) expressions extracted from the EveryDay-Sexism project[4]; d) a compendium of feminist dictionaries. We will also include other common hashtags and expressions less frequently used in sexist contexts to ensure a balanced distribution between sexist/not sexist expressions.

4   Learning with Disagreements

As stated in [9], the assumption that natural language expressions have a single and clearly identifiable interpretation in a given context is a convenient idealization, but far from reality. To deal with this, Uma et al. [9] have proposed the learning with disagreements paradigm (LeWiDi) , which consists mainly in letting systems learn from datasets where no gold annotations are provided but information about the annotations from all annotators, in an attempt to gather the diversity of views. In the case of sexism identification, this is particularly relevant, since the perception of a situation as sexist or not can be subjective and may depend on gender, age and cultural background of the person who is judging it. Following methods proposed for training directly from the data with disagreements, instead of using an aggregated label [11–13], we will provide multiple annotations per example, as was done in the SemEval-2021 shared task [9] on learning with disagreements that provided a unified testing framework for methods for learning from data containing multiple and possibly contradictory annotations. The LeWiDi paradigm is a first step towards more equitative and fairer algorithms that reflect multiple points of view.

The selection of annotators for the development of the EXIST 2023 dataset will take into account the heterogeneity necessary to avoid bias. Rather than eliminating disagreements by selecting the majority vote (as done in EXIST 2021 and 2022), EXIST 2023 will preserve the multiple labels assigned by an heterogeneous and representative group of annotators, so that participants may exploit disagreement in their systems.

5   Evaluation Methodology and Metrics

As in SemEval 2021, we will carry out a "hard evaluation" and a "soft evaluation". The hard evaluation will assume that a single label is provided by the systems for every example in the dataset. The soft evaluation is intended to measure the ability of the model to capture disagreements, by considering the distribution of labels in the output as a soft label and comparing it with the distribution of the annotations. Different metrics for both types of evaluations will be employed in order to facilitate comparison among participants and with previous works.

---

[4] Everyday sexism project: https://everydaysexism.com/. Accessed 14 October 2022.

Acknowledgments. This work is supported by the Spanish Ministry of Science and Innovation (project FairTransNLP (PID2021-124361OB-C31 and PID2021-124361OB-C32)) and the Spanish Ministry of Economy and Competitiveness (project pace for Observation of AI in Spanish (UNED and RED.ES, M.P., ref. C039/21- OT)).

References

1. Social Media and the Silencing Effect: Why Misogyny Online is a Human Rights Issue. NewStatesman. http://bit.ly/3n3ox68. Accessed 25 Sept 2022
2. Burgos, A., et al.: Violencias de Género 2.0, pp. 13–27 (2014)
3. Twitter's Famous Racist Problem. The Atlantic. https://bit.ly/38EnFPw. Accessed 25 Sept 2022
4. Anzovino, M., Fersini, E., Rosso, P.: Automatic identification and classification of misogynistic language on twitter. In Proceedings of the International Conference on Applications of Natural Language to Information Systems (NLDB 2018), pp. 57–64 (2018)
5. Pamungkas, E., Basile, V., Patti, V.: Misogyny detection in twitter: a multilingual and cross-domain study. Inf. Process. Manag. **57**(6), 102360 (2020)
6. Guest, E., Vidgen, B., Mittos, A., Sastry, N., Tyson, G., Margetts, H.: An expert annotated dataset for the detection of online misogyny. In: Proceedings of the 16th Conference of the European Chapter of the Association for Computational Linguistics: Main Volume (EACL 2021), pp. 1336–1350 (2021)
7. Rodríguez-Sánchez, F., et al.: Overview of EXIST 2021: sexism identification in social networks. Procesamiento del Lenguaje Natural **67**, 195–207 (2021)
8. Rodríguez-Sánchez, F., et al.: Overview of EXIST 2022: sEXism identification in social neTworks. Procesamiento del Lenguaje Natural **69**, 229–240 (2022)
9. Uma, A., et al.: SemEval-2021 Task 12: learning with disagreement. In: Proceedings of the 15th International Workshop on Semantic Evaluation (SemEval-2021), pp. 338–347 (2021)
10. Hovy, D., Prabhumoye, S.: Five sources of bias in natural language processing. Lang. Linguist. Compass **15**(8), e12432 (2021)
11. Sheng, V.S., Provost, F., Ipeirotis, P. G.: get another label? improving data quality and data mining using multiple, noisy labelers. In: Proceedings of the 14th ACM SIGKDD International Conference on Knowledge Discovery and Data Mining (KDD 2008), pp. 614–622 (2008)
12. Rodrigues, F., Pereira. F. C.: Deep learning from crowds. In: Proceedings of the 32nd AAAI Conference on Artificial Intelligence, pp. 1611–1618 (2018)
13. Peterson, J. C., Battleday, R.M., Griffiths, T.L., Russakovsky, O.: Human uncertainty makes classification more robust. In: Proceedings of the 2019 IEEE/CVF International Conference on Computer Vision, pp. 9616–9625 (2019)
14. Roselli, D., Matthews, J., Talagala, N.: Managing bias in AI. In: Companion Proceedings of The 2019 World Wide Web Conference (WWW 2019), pp. 539–544 (2019)

DocILE 2023 Teaser: Document Information Localization and Extraction

Štěpán Šimsa[1]([✉]) [iD], Milan Šulc[1] [iD], Matyáš Skalický[1] [iD], Yash Patel[2] [iD], and Ahmed Hamdi[3] [iD]

[1] Rossum.ai, Prague, Czech Republic
{stepan.simsa,milan.sulc,matyas.skalicky}@rossum.ai
[2] Visual Recognition Group, Czech Technical University in Prague, Prague, Czech Republic
patelyas@fel.cvut.cz
[3] University of La Rochelle, La Rochelle, France
ahmed.hamdi@univ-lr.fr
http://www.rossum.ai/

Abstract. The lack of data for information extraction (IE) from semi-structured business documents is a real problem for the IE community. Publications relying on large-scale datasets use only proprietary, unpublished data due to the sensitive nature of such documents. Publicly available datasets are mostly small and domain-specific. The absence of a large-scale public dataset or benchmark hinders the reproducibility and cross-evaluation of published methods. The DocILE 2023 competition, hosted as a lab at the CLEF 2023 conference and as an ICDAR 2023 competition, will run the first major benchmark for the tasks of *Key Information Localization and Extraction* (KILE) and *Line Item Recognition* (LIR) from business documents. With thousands of annotated real documents from open sources, a hundred thousand of generated synthetic documents, and nearly a million unlabeled documents, the DocILE lab comes with the largest publicly available dataset for KILE and LIR. We are looking forward to contributions from the Computer Vision, Natural Language Processing, Information Retrieval, and other communities. The data, baselines, code and up-to-date information about the lab and competition are available at https://docile.rossum.ai/.

Keywords: Information extraction · Dataset · Benchmark · KILE · LIR · Business documents · Natural language processing · Computer vision

1 Introduction

The majority of business-to-business (B2B) communication takes place through the exchange of semi-structured documents such as invoices, purchase orders, and delivery notes. Information from the documents is typically extracted by humans and entered into information systems. This process is time-consuming, expensive, and repetitive. Automating the information extraction process has the potential to considerably reduce manual human labor, allowing people to focus on more creative and strategic tasks.

J. Kamps et al. (Eds.): ECIR 2023, LNCS 13982, pp. 600–608, 2023.
https://doi.org/10.1007/978-3-031-28241-6_69

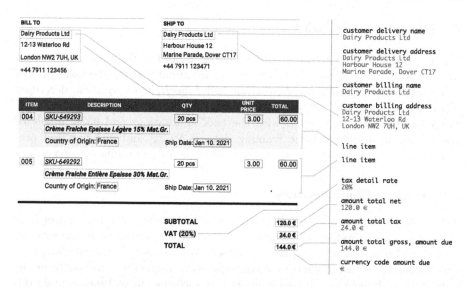

Fig. 1. Example invoice with annotations of fields and line items (LIs). Categories of fields within the LI are depicted by their color. Values of fields in the LI are not visualized in the Figure, but they are annotated in the dataset.

Automating document information extraction is challenging because semantic and syntactic understanding is required. These documents are designed to be interpretable by humans, not machines. An example with semantic information is shown in Fig. 1. Information extraction approaches must handle varying layouts, semantic fields and multiple input modalities at the intersection of computer vision, natural language processing and information retrieval. While there has been progress on the task [4,7,14,15,18,19,25,34], there is no publicly available large-scale benchmark to train and compare these approaches, an issue that has been noted by several authors [5,16,24,26,29]. Existing approaches are trained on privately collected datasets, hindering their reproducibility, fair comparisons and tracking field progression [11,23,24].

To mitigate the aforementioned issues, the DocILE lab provides a public research benchmark on a large-scale dataset. This benchmark was built by knowing the domain- and task-specific aspects of business document information localization and extraction. The DocILE benchmark will allow cross-evaluation and enable the reproducibility of experiments in business document information extraction. The dataset is the largest public source of densely annotated business documents. It consists of 8715 annotated pages of 6680 real business documents along with 100,000 synthetic documents and 3.4 million unlabeled pages of nearly a million real business documents. To mimic the real world use case, the dataset emphasizes layout diversity and contains over a thousand unique layouts. With the large amount of diverse documents and high-quality annotations, the dataset will allow researchers to investigate different aspects of document information extraction, including supervised, semi-supervised and unsupervised learning and domain adaptation.

2   Dataset and Tasks

The DocILE benchmark comes with a *labeled* dataset of 6680 documents from publicly available sources which were manually annotated for the tasks of *Key Information Localization and Extraction* and *Line Item Recognition*, described below in Sects. 2.2 and 2.3 respectively. Additionally, we provide a set of 100K *synthetic* documents generated with the task annotations and 932K *unlabeled* documents, as both synthetic training data [6,10,22] and unsupervised pre-training [33] have demonstrated to aid machine learning in different domains.

2.1   Dataset Characteristics

Table 1 shows the size of the challenge dataset. All documents in the dataset were classified[1] as invoice-like documents (i.e., tax invoice, order, proforma invoice) by a model pre-trained on a private dataset. Additionally, in the labelling process, documents misclassified as invoice-like were manually removed from the dataset (e.g., budgets or financial reports, as such document types contain different information than standard invoice-like documents).

To ensure a high variance of document layouts in the dataset, *unlabeled* documents were clustered into layouts[2]. Only a limited number of documents per layout were selected for annotation. The clustering is based on the location of field detections[3] predicted by a proprietary model for KILE pre-trained on a private dataset. Furthermore, to encourage solutions that generalize well to previously unseen layouts, the train./val./test split is done such that the validation and test sets contain layouts unseen in the training set (to measure the model's generalization) as well as some seen layouts (in practice, it is common to observe known layouts and important to read them out perfectly). Meta-information describing the layouts is included in the dataset annotations. The *synthetic* documents were generated using an unpublished rule-based document synthesizer based on layout annotations of 100 documents from the *labeled* set.

The dataset will be shared in the form of pre-processed[4] document PDFs with task annotations in JSON. As an additional resource, we will also provide predictions of text tokens (using OCR) including the location and text of the detected tokens.

---

[1] Using a proprietary document type classifier from Rossum.ai.

[2] We loosely define layout as the positioning of fields of each type in a document. Rather than requiring absolute positions to match perfectly, we allow transformations caused by different length of values, translations of whole sections (e.g. vertical shift caused by different lengths of tables) and translation, rotation and scaling of the whole document.

[3] The distance used for clustering relates to the difference in the relative x-translations between pairs of fields within a document. Vertical shifts are not penalized, since they commonly appear among documents of the same layout.

[4] Pre-processing consists of correcting page orientation, fixing or discarding broken pdfs and of de-skewing scanned documents and normalizing them to 150 DPI.

Table 1. Overview of the three parts of the challenge dataset.

	Labeled	Synthetic	Unlabeled
Documents	6680	100 000	932 467
Pages	8715	100 000	3.4M
Layout clusters	1152	100	*Unknown*
Pages per doc.	1–3	1	1–884

The data was sourced from two public data sources: UCSF Industry Documents Library [30] and Public Inspection Files (PIF) [32]. The UCSF Industry Documents Library contains documents from industries that influence public health, such as tobacco companies. The majority of the documents are from the 20th century. This source was previously used to create document datasets: RVL-CDIP [12] (subset of IIT-CDIP [17] and superset of FUNSD [9]) and DocVQA [21]. Filters in the UCSF public API [31] were used to retrieve only publicly available invoice-like documents with at most 3 pages, no redacted information and a threshold on document date[5]. PIF contains documents (invoices, orders, "contracts") from TV and radio stations for political campaign ads. This source was previously used to create Deepform [28].

2.2 Track 1: Key Information Localization and Extraction

The goal of the first track is to localize key information of pre-defined categories (field types) in the document. It is derived from the task of *Key Information Localization and Extraction* (KILE), as defined in [26].

KILE extends the common definition of Key Information Extraction (KIE) by additionally requiring the location of the extracted information within the document. Such annotation is missing even in the KIE datasets [3,27]. While localization is typically not needed at the end of document processing, it plays a vital role in applications that require human validation, and it is a valuable form of supervision for vision-based methods. Compared to *Semantic Entity Recognition*, as defined by [33], bounding boxes in KILE are not limited to individual words (tokens).

We focus the challenge on detecting semantically important values corresponding to tens of different field types rather than fine-tuning the underlying text recognition. Towards this focus, we provide word-level text detections for each document, we choose an evaluation metric (below) that doesn't pay attention to the text recognition part, and we simplify the task in the challenge by only requiring correct localization of the values in the documents in the primary metric. Text extractions are checked besides the locations and field types in a separate evaluation (the leaderboard ranking does not depend on it) and any post-processing of values (deduplication, converting dates to a standardized

---

[5] Old documents from this source are not included, since e.g. typewriter documents differ from today's document distribution.

Fig. 2. Each word is split uniformly into pseudo-character boxes based on the number of characters. Pseudo-Character Centers are the centers of these boxes.

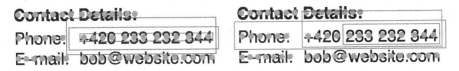

(a) Correct extraction examples.   (b) Incorrect extraction examples.

Fig. 3. Visualization of correct and incorrect bounding box predictions to capture the email address. Bounding box must include exactly the Pseudo-Character Centers that lie within the ground truth annotation. Note: In 3a, only one of the predictions would be considered correct if all three boxes were predicted.

format etc.) that is otherwise needed in practice is omitted. With the simplifications, the main task can also be viewed as a detection problem.

Challenge Evaluation Metric: Since the task is framed as a detection problem, the standard *Average Precision* metric will be used as the main evaluation metric. Unlike the common practice in object detection, where true positives are determined by thresholding the Intersection-over-Union, we use a different criterion tailored to better evaluate the usefulness of detections for text read-out. Inspired by the CLEval metric [1] used in text detection, we measure whether the predicted area contains all related character centers (and none others). Since the character-level annotations are hard to obtain, we use CLEval's definition of Pseudo-Character Center (PCC) (see Fig. 2). See Fig. 3 for examples of correct and incorrect detections.

Beyond the challenge leaderboard based on the metric described above, we set up a secondary benchmark for end-to-end KILE, where a correctly recognized field also needs to exactly read out the text. We invite all participants to provide the text value predictions, but it is not required for challenge submissions.

2.3   Track 2: Line Item Recognition

The goal of the second track is to localize key information of pre-defined categories (field types) and group it into line items [2,4,13,20,24]. A *Line Item* (LI) is a tuple of fields (i.e., *description, quantity,* and *price*) describing a single object instance to be extracted, e.g., a row in a table, as visualized in Fig. 1.

This track is derived from the task of *Line Item Recognition* (LIR) [26] and is related to *Table Understanding* [13] and *Table Extraction* [8,35]—problems where the tabular structure is also crucial for IE. Unlike these tasks, LIR does not explicitly rely on the structure but rather reflects the information to be extracted and stored.

Challenge Evaluation Metric: The main evaluation metric is the micro F1 score over all line item fields. A predicted line item field is correct if it fulfills the requirements from Track 1 (on field type and location) and if it is assigned to the correct line item. Since the matching of ground truth (GT) and predicted line items may not be straightforward due to errors in prediction, our evaluation metric chooses the best matching in two steps:

1. For each pair of predicted and GT line items, the predicted fields are evaluated as in Track 1.
2. Find the maximum matching between predicted and GT line items, maximizing the overall recall.

Similarly to the previous track, an out-of-competition end-to-end benchmark will assess the correctness of the extracted text values.

3   Conclusions

The first edition of the DocILE lab at CLEF 2023 and the ICDAR 2023 Competition on Document Information Localization and Extraction will present the largest benchmark for information extraction from semi-structured business documents, and will consist of two tasks: *Key Information Localization and Extraction (KILE)* and *Line Item Recognition (LIR)*. Participants will be given a collection of thousands of labeled documents, together with a hundred thousand of synthetic documents and nearly a million unlabeled real documents that can be used for unsupervised pre-training.

This Teaser paper summarizes the motivation and the main characteristics of the tasks. Given the input documents are practically a combination of visual- and text- inputs, we are looking forward to the contributions of several communities, including *Information Retrieval, Natural Language Processing*, and *Computer Vision*.

To access the data, the repository, baseline implementations, and updates regarding the challenge, please refer to https://docile.rossum.ai/.

References

1. Baek, Y., et al.: CLEVAL: character-level evaluation for text detection and recognition tasks. In: Proceedings of the IEEE/CVF Conference on Computer Vision and Pattern Recognition Workshops, pp. 564–565 (2020)
2. Bensch, O., Popa, M., Spille, C.: Key information extraction from documents: evaluation and generator. In: Abbès, S.B., et al. (eds.) Proceedings of DeepOntoNLP and X-SENTIMENT. CEUR Workshop Proceedings, vol. 2918, pp. 47–53. CEUR-WS.org (2021)
3. Borchmann, Ł., et al.: DUE: end-to-end document understanding benchmark. In: Proceeedings of NeurIPS (2021)
4. Denk, T.I., Reisswig, C.: Bertgrid: contextualized embedding for 2D document representation and understanding. arXiv preprint arXiv:1909.04948 (2019)
5. Dhakal, P., Munikar, M., Dahal, B.: One-shot template matching for automatic document data capture. In: Proceedings of Artificial Intelligence for Transforming Business and Society (AITB), vol. 1, pp. 1–6. IEEE (2019)
6. Dosovitskiy, A., et al.: Flownet: learning optical flow with convolutional networks. In: Proceedings of the IEEE International Conference on Computer Vision, pp. 2758–2766 (2015)
7. Garncarek, Ł, et al.: LAMBERT: layout-aware language modeling for information extraction. In: Lladós, J., Lopresti, D., Uchida, S. (eds.) ICDAR 2021. LNCS, vol. 12821, pp. 532–547. Springer, Cham (2021). https://doi.org/10.1007/978-3-030-86549-8_34
8. Göbel, M.C., Hassan, T., Oro, E., Orsi, G.: ICDAR 2013 table competition. In: Proceedings of ICDAR, pp. 1449–1453. IEEE Computer Society (2013). https://doi.org/10.1109/ICDAR.2013.292
9. Jaume, G., Ekenel, H.K., Thiran, J.P.: Funsd: a dataset for form understanding in noisy scanned documents. In: Accepted to ICDAR-OST (2019)
10. Gupta, A., Vedaldi, A., Zisserman, A.: Synthetic data for text localisation in natural images. In: Proceedings of the IEEE Conference on Computer Vision and Pattern Recognition, pp. 2315–2324 (2016)
11. Hamdi, A., Carel, E., Joseph, A., Coustaty, M., Doucet, A.: Information extraction from invoices. In: Lladós, J., Lopresti, D., Uchida, S. (eds.) ICDAR 2021. LNCS, vol. 12822, pp. 699–714. Springer, Cham (2021). https://doi.org/10.1007/978-3-030-86331-9_45
12. Harley, A.W., Ufkes, A., Derpanis, K.G.: Evaluation of deep convolutional nets for document image classification and retrieval. In: International Conference on Document Analysis and Recognition (ICDAR) (2015)
13. Holeček, M., Hoskovec, A., Baudiš, P., Klinger, P.: Table understanding in structured documents. In: 2019 International Conference on Document Analysis and Recognition Workshops (ICDARW), vol. 5, pp. 158–164. IEEE (2019)
14. Holt, X., Chisholm, A.: Extracting structured data from invoices. In: Proceedings of the Australasian Language Technology Association Workshop 2018, pp. 53–59 (2018)
15. Katti, A.R., et al.: Chargrid: towards understanding 2D documents. In: Riloff, E., Chiang, D., Hockenmaier, J., Tsujii, J. (eds.) Proceedings of the 2018 Conference on Empirical Methods in Natural Language Processing, Brussels, Belgium, 31 October–4 November 2018, pp. 4459–4469. Association for Computational Linguistics (2018). https://aclanthology.org/D18-1476/

16. Krieger, F., Drews, P., Funk, B., Wobbe, T.: Information extraction from invoices: a graph neural network approach for datasets with high layout variety. In: Ahlemann, F., Schütte, R., Stieglitz, S. (eds.) WI 2021. LNISO, vol. 47, pp. 5–20. Springer, Cham (2021). https://doi.org/10.1007/978-3-030-86797-3_1

17. Lewis, D., Agam, G., Argamon, S., Frieder, O., Grossman, D., Heard, J.: Building a test collection for complex document information processing. In: Proceedings of the 29th Annual International ACM SIGIR Conference on Research and Development in Information Retrieval, pp. 665–666 (2006)

18. Lin, W., Gao, Q., Sun, L., Zhong, Z., Hu, K., Ren, Q., Huo, Q.: ViBERTgrid: a jointly trained multi-modal 2D document representation for key information extraction from documents. In: Lladós, J., Lopresti, D., Uchida, S. (eds.) ICDAR 2021. LNCS, vol. 12821, pp. 548–563. Springer, Cham (2021). https://doi.org/10.1007/978-3-030-86549-8_35

19. Liu, W., Zhang, Y., Wan, B.: Unstructured document recognition on business invoice. Mach. Learn., Stanford iTunes Univ., Stanford, CA, USA, Technical Report (2016)

20. Majumder, B.P., Potti, N., Tata, S., Wendt, J.B., Zhao, Q., Najork, M.: Representation learning for information extraction from form-like documents. In: Jurafsky, D., Chai, J., Schluter, N., Tetreault, J.R. (eds.) Proceedings of the 58th Annual Meeting of the Association for Computational Linguistics, ACL, pp. 6495–6504 (2020). https://doi.org/10.18653/v1/2020.acl-main.580

21. Mathew, M., Karatzas, D., Jawahar, C.V.: Docvqa: a dataset for VQA on document images. In: Proceedings of WACV, pp. 2199–2208. IEEE (2021). https://doi.org/10.1109/WACV48630.2021.00225

22. Nassar, A., Livathinos, N., Lysak, M., Staar, P.W.J.: Tableformer: table structure understanding with transformers. CoRR abs/2203.01017 (2022). https://doi.org/10.48550/arXiv.2203.01017

23. Palm, R.B., Laws, F., Winther, O.: Attend, copy, parse end-to-end information extraction from documents. In: 2019 International Conference on Document Analysis and Recognition (ICDAR), pp. 329–336. IEEE (2019)

24. Palm, R.B., Winther, O., Laws, F.: Cloudscan - a configuration-free invoice analysis system using recurrent neural networks. In: Proceedings of ICDAR, pp. 406–413. IEEE (2017). https://doi.org/10.1109/ICDAR.2017.74

25. Schuster, D., et al.: Intellix-end-user trained information extraction for document archiving. In: 2013 12th International Conference on Document Analysis and Recognition, pp. 101–105. IEEE (2013)

26. Skalický, M., Šimsa, Š., Uřičář, M., Šulc, M.: Business document information extraction: towards practical benchmarks (2022). https://doi.org/10.48550/ARXIV.2206.11229, https://arxiv.org/abs/2206.11229

27. Stanisławek, T., et al.: Kleister: key information extraction datasets involving long documents with complex layouts. In: Lladós, J., Lopresti, D., Uchida, S. (eds.) ICDAR 2021. LNCS, vol. 12821, pp. 564–579. Springer, Cham (2021). https://doi.org/10.1007/978-3-030-86549-8_36

28. Stray, J., Svetlichnaya, S.: Deepform: Extract information from documents (2020). https://wandb.ai/deepform/political-ad-extraction, benchmark

29. Sunder, V., Srinivasan, A., Vig, L., Shroff, G., Rahul, R.: One-shot information extraction from document images using neuro-deductive program synthesis. arXiv preprint arXiv:1906.02427 (2019)

30. Web: Industry Documents Library. https://www.industrydocuments.ucsf.edu/. Accessed 20 Oct 2022

31. Web: Industry Documents Library API. https://www.industrydocuments.ucsf.edu/research-tools/api/, Accessed 20 Oct 2022
32. Web: Public Inspection Files. https://publicfiles.fcc.gov/. Accessed 20 Oct 2022
33. Xu, Y., et al.: LayoutXLM: multimodal pre-training for multilingual visually-rich document understanding. CoRR (2021)
34. Zhao, X., Wu, Z., Wang, X.: CUTIE: learning to understand documents with convolutional universal text information extractor. CoRR abs/1903.12363 (2019). https://arxiv.org/abs/1903.12363
35. Zheng, X., Burdick, D., Popa, L., Zhong, X., Wang, N.X.R.: Global table extractor (GTE): a framework for joint table identification and cell structure recognition using visual context. In: Proceedings of WACV, pp. 697–706. IEEE (2021). https://doi.org/10.1109/WACV48630.2021.00074

Author Index

© The Editor(s) (if applicable) and The Author(s), under exclusive license
to Springer Nature Switzerland AG 2023
J. Kamps et al. (Eds.): ECIR 2023, LNCS 13982, pp. 609–616, 2023.
https://doi.org/10.1007/978-3-031-28241-6

Printed in the United States
by Baker & Taylor Publisher Services